Electrophoresis '81
Advanced Methods
Biochemical and Clinical Applications

# Electrophoresis '81

Advanced Methods
Biochemical and Clinical Applications

Proceedings of the Third International
Conference on Electrophoresis
Charleston, SC, April 7–10, 1981

Editors
R. C. Allen · P. Arnaud

Walter de Gruyter · Berlin · New York 1981

*Editors:*

Robert C. Allen, Ph. D.
Professor of Biochemistry
Department of Laboratory Animal Medicine and Pathology
Medical University of South Carolina
171 Ashley Avenue
Charleston, South Carolina 29403, USA

Philippe Arnaud, Ph. D.; M. D.
Professor of Biochemistry
Department of Basic and Clinical Immunology and Microbiology
Medical University of South Carolina
171 Ashley Avenue
Charleston, South Carolina 29403, USA

*CIP-Kurztitelaufnahme der Deutschen Bibliothek*

**Electrophoresis** ... : advanced methods, biochem. and clin.
applications ; proceedings of the Internat. Conference on Electrophoresis. – Berlin ; New York : de Gruyter
3. 1981. Charleston, SC, April 7–10, 1981. – 1981.
   ISBN 3-11-008155-5
NE: International Conference on Electrophoresis

*Library of Congress Cataloging in Publication Data*

International Conference on Electrophoresis (3rd : 1981 :
   Charleston, S.C.)
   Electrophoresis '81.
   Includes bibliographical references and indexes.
   1. Electrophoresis–Congresses. 2. Biological chemistry–
Technique–Congresses. 3. Chemistry, Clinical–Technique–Con-
gresses. I. Allen, R. C. (Robert Chadbourne), 1924–   . II.
Arnaud, P. (Philippe) III. Title.
QP519.9.E43I57 1981     574.19'283     81-5456
ISBN 3-11-008155-5               AACR2

PREFACE

The International Congress Electrophoresis `81 was held at the
Sheraton Charleston Hotel in Charleston, South Carolina from
April 6-10, 1981. It was the third of a projected series of
international meetings organized with the objective of stimulating information exchange and advancement of knowledge and
techniques in all areas of electrophoresis.

This Congress was held in conjunction with the first annual
meeting of the Electrophoresis Society and in future, these
Congresses will be held as part of the annual meeting of the
Electrophoresis Society.

This Congress was attended by over 300 participants which represented almost 75 per cent of the society membership. Also,
due in large measure to the generous support of NASA, a complete
session devoted to free flow electrophoresis was able to be
included.

The format of the meeting was in part experimental and based on
suggestions of many of the members. Thus, a greater emphasis
was placed on posters and following round tables discussions.
This format, with all of the attendees staying in a single location, proved to have the desired effect of producing a lively
exchange of ideas and views among the participants, well on
into the evening. The added atmosphere of a society meeting
certainly aided in the success of the participation in the Congress.

This volume contains over 100 of the presentations made either
as oral reports or posters at "Electrophoresis `81". The manu-
scripts have been compiled into four sections entitled (I)
Theory and Methods, (II) High Resolution Two-Dimensional Electrophoresis, (III) Biomedical and Biological Applications and
(IV) Isotachophoresis and Free Flow Electrophoresis. Not every
manuscript fits precisely in a given section, but the most

appropriate has been attempted in order to provide continuity of subject material.

Production of this proceedings has been made possible only by cooperation of the authors who turned their manuscripts in on time and of which only a minority required serious editing. We wish to thank all of the contributors for their splendid efforts to which the many users of this technique are indebted. We also appreciate, in no small way, the efforts of the staff of Walter de Gruyter, Berlin, which led to the rapid publication of this volume.

Charleston, South Carolina

Robert C. Allen

Philippe Arnaud

We would like to acknowledge the outstanding help from Mrs. Susan Haskill, Mr. Michael Lack, Mr. Peter Brady of the Department of Pathology and our wives, Mrs. Carol Allen and Mrs. Marie-Laure Arnaud, in the organization and running of this meeting and Congress. We also wish to acknowledge the valiant secretarial efforts of Ms. Brenda Altman for preparation of the abstracts and proceedings.

Financial aid for the Congress was provided by NASA, Beckman Instruments, Beta Analytical, Biorad, Bio Products, Brinkman Instruments-Desaga, DAKO, de Gruyter Publishers, E C Apparatus Corporation, Gelman Sciences, Helena Laboratories, Isolab, LKB Instruments, Marine Colloid FMC Corporation, Pharmacia Fine Chemicals, Serva Feinbiochemicals, Shandon Company, Upjohn Diagnostics and Verlag-Chemie Publishers.

CONTENTS


SECTION I, THEORY AND METHODS

XVIII

# SECTION I

Theory and Methods

ON THE PORE SIZE AND SHAPE OF HYDROPHILIC GELS FOR ELECTROPHORETIC ANALYSIS

Pier Giorgio Righetti

Department of Biochemistry, University of Milano, Via Celoria 2, Milano
20133, Italy and

NASA, Marshall Space Flight Center, Separation Processes Branch,
Huntsville, Alabama 35812, USA


Introduction

Since the introduction of starch gels (1), polyacrylamide and agarose (2, 3)
matrices for electrophoretic separations, and of cross-linked dextrans (4)
for gel filtration, considerable interest has focused on the structure of
these hydrophilic support media. Fundamental equations have been described
linking the partition coefficient ($K_{av}$ or $\sigma$) in the latter or the mobility
(m) in the former technique to the molecular weight of the fractionated macromolecule. Indeed, a unified theory for gel filtration and gel electrophoresis has been proposed (5), which provides equations for inter-relationships
between mobility, partition coefficient, gel concentration and molecular radius. Extensive reviews have been published by Ackers (6, 7), Rodbard (8),
Kremmer and Boross (9), Fischer (10) and Determann (11). However, considerable disagreement still exists as to the actual pore geometry and to the maximum pore size which can be obtained with hydrophilic gels. In the latter
case, the research groups who have attempted to measure pore sizes have given all possible ranges of values, from extremely small (12) to extremely
large (13, 14) with variations of as much as 60-70 fold  for a given gel concentration. I shall attempt here to review this field, also at the light of
some recent data obtained in collaboration between my laboratory and NASA,
as the necessary ground work for future separations in microgravity (15).


Results

1) Pore shape

The series of models proposed to describe pores in gels stem from early observations by Flodin (16), who has considered the partition of a macromolecule between the gel and the liquid to be entirely governed by steric factors. The gel matrix chains form a network of varying density. Large molecu-

4

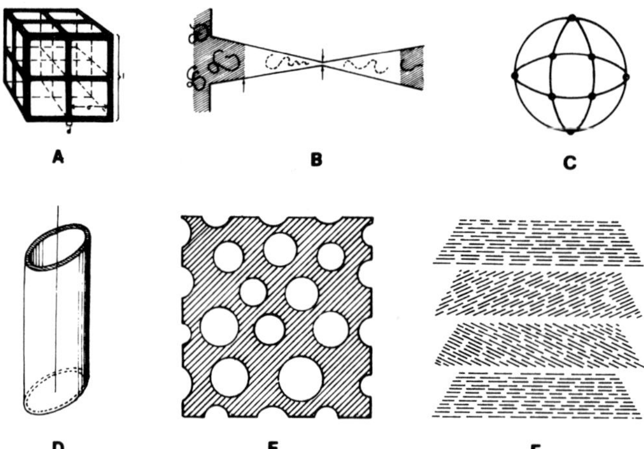

Fig. 1. Geometric pore shapes in gels as suggested by Ornstein (A)(21), Porath (B)(17), Casassa (C, D, F)(20), Ackers (D, E)(14) and Squire (B, D, F) (19).

les can only penetrate into regions where the meshes in the net are large. On the other hand, small molecules find their way into more tightly knit regions of the network, closer to the cross links. The partition coefficient, thus, corresponds to the part of the whole space that is "permitted" to a given macromolecule. The later theories have been largely concerned with different models of the gel structure which could explain the mesh width distribution giving the proportions of permitted volume found empirically. Three groups of models can be described: a) geometric; b) statistical and c) thermodynamic models.

*a) Geometric models.* Fig. 1 groups the most common pore geometries described in the literature. The earliest geometrical model for calibration of gels visualized the penetrable voids within the matrix as a collection of conical pores (17)(Fig. 1B). According to this geometry, a plot of $\sigma^{1/3}$ against $\overline{M}_r^{-\frac{1}{2}}$ should result in a straight line. This seems to hold for randomly coiled macromolecules (17) and even for compact globular proteins (18). Squire (19) has used a broader collection of constraining shapes, by assuming that the pore size distribution corresponds to that produced by equal amounts of cones, hollow cylinders (Fig. 1D) and crevices. This last pore model could be seen as "cracks on the wall" and could be exemplified as parallel planes (Fig. 1F). Cylindrical pores have also been assumed by Ackers (14), who has

also regarded them as circular and lying in a plane perpendicular to the direction of movement of the molecules (Fig. 1E). The distribution probability of flexible macromolecules within voids represented as spherical cavities (Fig. 1C), cylindrical pores and slab-shaped cavities has been calculated by Casassa (20) using random-flight statistics. Ornstein (21) has assumed that the chains in  a polyacrylamide gel follow the edges of cubes in a cubical matrix (Fig. 1A). Actually, one of the oldest models is the "random meshwork of fibers" of Ogston (22), who has calculated the fraction of space available to a sphere with radius $r_s$ if straight cylindrical rods with radius $r_r$ are distributed in the space in a random fashion with an average density of L length units of rods per volume unit of space. Laurent and Killander (23) have extended this model by assuming that the molecule is a sphere of a given radius, while the network is described as infinitely long straight rods that are randomly located in space. Among the "random" models, Giddings *et al.* (24) have also hypothesized a gel structure consisting of an isotropic network of random planes in which all plane orientations are equally repre-
·sented.

We will discuss further on critically the various models, but at the moment we can ask ourselves how would a macromolecule fare within the different geometries. We can here distinguish between two limiting particle shapes, spherical molecules and thin rods. Fig. 2 shows how $K_{av}$ (and therefore m in electrophoresis) varies as a function of particle size in the two cases. Spheres move with extreme difficulties in parallel planes, and most easily within random planes. Thin rods don't wander well through spherical structures but, above a critical length, move with the same difficulty within several geometries (parallel and random planes and rectangular shapes). Interestingly, as the sphere diameter $(L_o)$ or the cylinder length $(L_1)$ become small, all the curves tend to converge and approach unity along a common curve. In other words, as the macromolecule becomes sufficiently small, as compared to the gel pore size (less than 10% of this value), wall curvature and corners boceme unimportant: the molecule is so small that nearly all elements of surface appear to it as plane areas.

*b) Statistical models.* In order to circumvent the necessity of postulating specific geometric shapes for the pores within gel partitioning systems, statistical calibrating functions, which relate the molecular size of the ma-

6

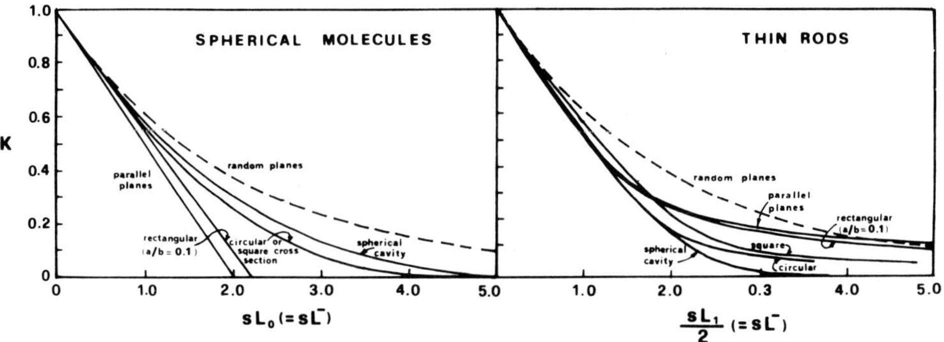

Fig. 2. Partition coefficient (K) of spheres and thin rods, as a function
of diameter ($L_0$) or length ($L_1$), respectively, within pores of different geometries (From Giddings *et al.*) (24).

cromolecule to the volume available to it, have been proposed. Hohn and Pol-
lman (25) assumed that $K_{av}$ is related to molecular weight by a Boltzmann distribution, and found this relationship to fit well for oligonucleotides. A-
ckers (26) assumed that the volume of the domains in the gel, where molecules of a certain radius could just reach, could be described by a normal
(gaussian) distribution. Acker's "gaussian pore distribution" model has proven to be quite satisfactory for many purposes but, as pointed out by Rodbard (8) it has one minor technical flaw: the gaussian distribution ranges
from minus infinity to plus infinity, which means that a small fraction of
the pores would be negative. In order to avoid this paradox, Rodbard (8) has
suggested that the pore sizes follow a "log-normal" distribution, i.e. it is
the logaritmm of the pore size that obeys a gaussian curve. In this way, no
negative pore sizes can occur; moreover, the "log-normal" statistics provide a more satisfactory fit for the data of Fawcett and Morris (27) on pore
radius distribution in polyacrylamide gels. As a further refinement of this
model, Rodbard (8) has proposed a "logistic" distribution, since it provides
simple equations which can be fitted more easily to small desk-top calcula-
tors, or drawn directly on logit-log paper.

*c) Thermodynamic models.* Another way to avoid specific pore geometries, is
to apply classical thermodynamic theories to the partitioning of a solute
between two phases. Thus, Albertsson (28) has proposed a thermodynamic relationship of the Brönsted-type to the partitioning of cells between two non
miscible liquid phases. This approach has been extended by Fischer (29) who

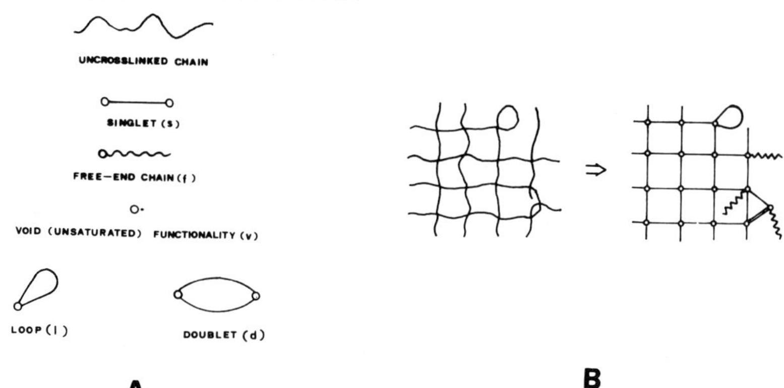



Fig. 3. A: structural elements in a gel with a bifunctional cross-link (e.g. Bis). B: gel formation by intersection of long chains with a bifunctional cross-link. Notice how singlets largely predominate, forming a regular network of chains (from Ziabicki)(32).

has calculated thermodynamic parameters for solute transfer between bulk liquid and gel phase. In another line of thinking, Bode (20) has explained molecular sieving in polyacrylamide gels as stemming from the properties of a hypothetical "viscosity-emulsion" composed of two interlacing fluid compartments endowed with different frictional coefficients. In this model, the gel is imagined as a visco-elastic matrix composed of layers of parallel sheets each of which comprises fluctuating polymer chains inserted into an unspecified backbone. Fluctuations due to thermal agitation are centred symmetrically around the median plane of each sheet. The motions of the polymer chains give raise to elastic forces directed towards any compact object which tends to invade the volume otherwise available to the polymers for molecular reorientation.

## 2) Pore shape

What is the biggest pore size that can be obtained with hydrophilic gels at 1 g gravity? I will narrow my discussion to two types of gels, which have proven over the years to be the most popular: polyacrylamide and agarose gels.

*a) Polyacrylamide gels.* So much has been written about them that we can confidently state that we know very little on their properties. Before discus-

8

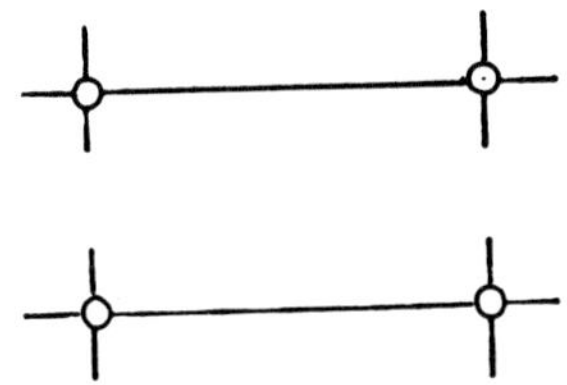 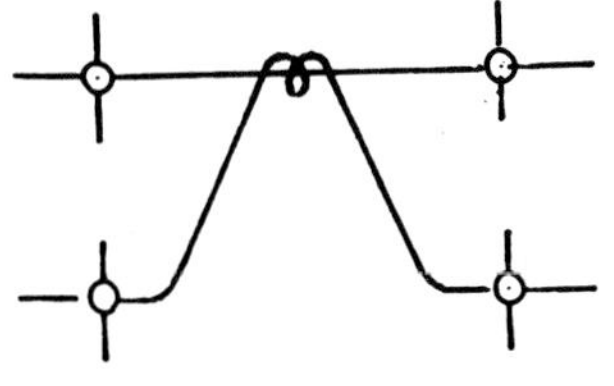



Fig. 4. Formation of entangled chains during polymerization of polyacrylamide gels (from Ziabicki) (32).

sing polyacrylamide pore size, a few words should be spent on their topological structure. When a gel is formed with a bifunctional cross-link (e.g.
bisacrylamide, Bis) we can distinguish several structural elements attached
to it (Fig. 3A): uncrosslinked chains (strictly speaking, though, they do
not form part of the cross-linked system); singlets (s), i.e. chains connected with their two ends to two different cross-links; free-end chains (f),
i.e. chains connected with one end to one junction only; void (unsaturated)
functionalities (v); doublets (d), i.e. chains connecting the same pair of
jnctions, and loops (1), i.e. chains connected with both ends to the same
cross-link (31, 32). Fig. 3B shows how a gel is formed by intersection of
long chains with bifunctional bridges. It is assumed here that at the inst-
ant of introducing a cross-link, macromolecules exhibit equilibrium confor-
mations, so that the resulting distribution of cross-links corresponds to
the equilibrium distribution of temporary contacts between macromolecules.
The situation is further complicated by the fact that, in flexible-chain gels
, entanglements in the fibers can result, which are "frozen" in the gel structure by the cross-links (Fig. 4). It has been calculated that the total sum
of structural elements in a gel could add up to a fantastic number: for a
macroscopic sample of 1 $cm^3$ volume, this number could be as high as $10^{35}$-
$10^{43}$. A town like Milano (2 million inhabitants) can probably be described
by no more than $10^5$ topological features (including small one-way and dead-
end streets). So, a tiny 1 ml gel volume is just as complicated as our entire galaxy.

Notwithstanding the "random" models proposed in the past (22-24) it is temp-
ting, however, to represent a gel structure as a rather "regular array" of

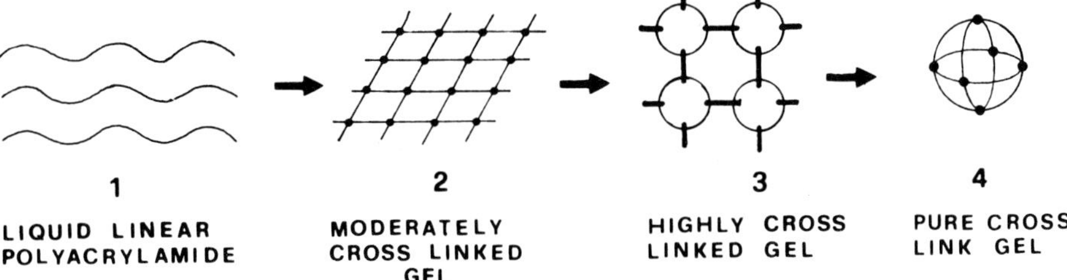



Fig. 5. Proposed variations of gel structure from an un-cross-linked gel to a pure cross-link gel. Structure 1 is not a gel, of course, but a highly viscous solution. Notice the shortening and thickening of gel fibers and the growth of beads in going from structures 1 to 4.

chains (Fig. 3B) which, indeed, has a much reduced complexity since, by far, singlets predominate and "irregularities" in the gel structure (d, l and f topologies) are rather scarce. A regular lattice of this kind, however, cannot be made highly-porous (by that I mean any gel structure with an average pore diameter, $\bar{p}$, greater than 100 nm) also because the matrix chains are not rigid and tend to fluctuate and collapse. One way to increase pore size in polyacrylamide gels is to progressively decrease the total content of solids (%T), the lowest concentration being 2%T, with a total amount of cross-linker (%C) of 2.2% (33). However, even in these highly diluted gels, it is doubtful that any pore diameter greater than 50 nm can be reached, and indeed much lower values have been given (27). Another way to increase pore size, though, is to progressively increase the percent of cross-linker, at fixed amounts of solids (%T) (21, 34). We have in fact recently demonstrated that, in highly cross-linked gels (3%T, 50%$C_{Bis}$) the pore diameter is as high as 500 nm, much higher, thus, than the highly diluted, low %C, gel series. How can we increase so considerably the pore size in high %C gels? I am proposing here the model depicted in Fig. 5. In going from a pure monofunctional monomer gel (un-cross-linked polyacrylamide) to a pure bifunctional monomer gel (pure cross-link gel) the gel matrix keeps changing its structure. Indeed, at one extreme, no gel is formed, but a highly viscous solution of liquid linear polyacrylamide (a visco-elastic continuum, a la Bode, but with no gel formation!). As cross-links are introduced, a regular array of chains, tied up by junctions (knots or cross-links), is formed which, at 5%C, has maximum sieving properties, i.e. minimum pore size, within a family of gels of equal %T. Now, as %C is increased above 10%, we a-

$$\overline{p} = Kd/\sqrt{C} \qquad \text{( RAYMOND and NAKAMICHI, 1962 )}$$

$$\overline{p} = \frac{K'd}{C} + K'' \qquad \text{( TOMBS, 1965 )}$$

$$\overline{p} \propto 1/\sqrt[3]{C} \qquad \text{( RODBARD and CHRAMBACH, 1970 )}$$

$$\overline{p} = 140.7 \times C^{-0.7} \qquad \text{( RIGHETTI, BROST \& SNYDER, 1981 )}$$

Fig. 6. Proposed equations linking the mean pore size ($\overline{p}$) to the gel concentration (C). The first three equations have been derived for polyacrylamide gels, the fourth for agarose matrices (see refs. 40, 39, 5 and 15, respectively).

re diminishing the population of singlets (s) in favor of the "irregularities" in the gel, namely doublets (d) and loops (1). As d and 1 increase, two phenomena occur simultaneously: the linear chains grow shorter and thicker, and the knots grow larger. This automatically drives the system towards bigger pore sizes (Fig. 5, 3). As the system approaches a composition of pure cross-link, the chains tend to disappear and the predominant topological elements left are loops, or better highly-concatenated loops which grow into beads or spheres. There is ample evidence in the literature for the existence of concatenated chains in living and synthetic polymers (35) and there is now also morphological evidence to the existence of such beads or spheres in pure cross-link gels (26, 37).

This explains the enormous porosity of very high %C gels. In regularly cross-linked gels, the macromolecules have to move *through* the gel network, through pores within the chain framework; in highly or pure cross-link gels, the macromolecule will move *around* the gel grains (or beads), in the void volume in between the gel spheres. Thus, paradoxically, highly-cross-linked gels resemble gel filtration media. This, in a way, could have been predicted: as shown in Fig. 6, among the four current equations linking the mean pore diameter ($\overline{p}$) to gel concentration (C), two of them suggest that this can simply be achieved by increasing the diameter (d) of the gel fibers. It is just too bad that highly cross-linked gels do not work properly for electrophoretic separations (38).

*b) Agarose gels.* With the new, almost charge-free agaroses today available on the market, this polymer has become again very popular for electrophoretic separations, including isoelectric focusing. We have recently "titrated"

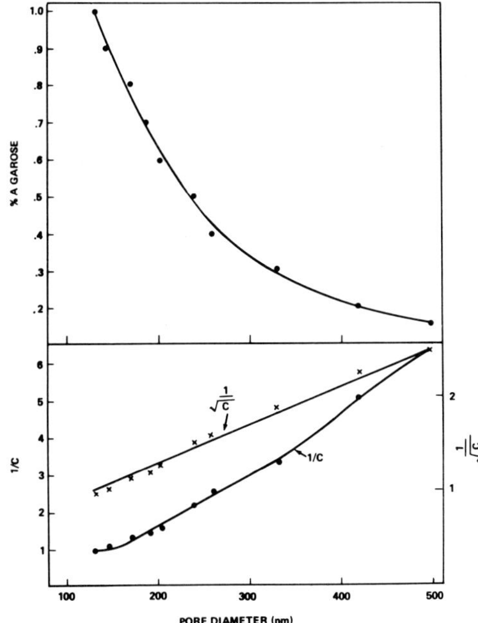



Fig. 7. Plot of limiting pore size *vs.* % agarose. The experimental points
have been obtained by moving electrophoretically latex particles through a-
garose gels ranging from 0.16% up to 1%. In the lower box, the experimental
points have been re-plotted according to p $\propto$ C$^{-1}$ or to p $\propto$ C$^{-0.5}$ (from Righetti *et al.*) (15).

the limiting pore size of this matrix, as a function of % solids in the gel,

by moving electrophoretically polystyrene particles through it. The data a-

re summarized in Fig. 7. It can be seen that, at the lowest % agarose (0.16%)

compatible with gel formation, a maximum pore diameter of 500 nm is reached.

We have tried to fit our data to $\bar{p} \propto C^{-1}$ (39) or to $\bar{p} \propto C^{-0.5}$ (40) (Fig. 7,

lower case) but with limited success. Our values seem to fit best the propor-

tionality $\bar{p} \propto C^{-0.7}$ (15). This maximum pore size obtained (500 nm) in agaro-

se is quite remarkable, since no other hydrophilic gel can reach such a po-

rosity (indeed, high %C polyacrylamides reach this size, too, but they are

useless for any practical purpose). How can agarose present such an open po-

re structure? It has been demonstrated (41) that this polysaccharide in so-

lution exists as a double helix (Fig. 8, left) and therefore it is conside-

rably more rigid than a polyacrylamide strand. Moreover, 7 to 11 such heli-

ces form bundles which extend as long rods (Fig. 8, right) thus further

strengthening the architectural framework of the gel. I believe it will be

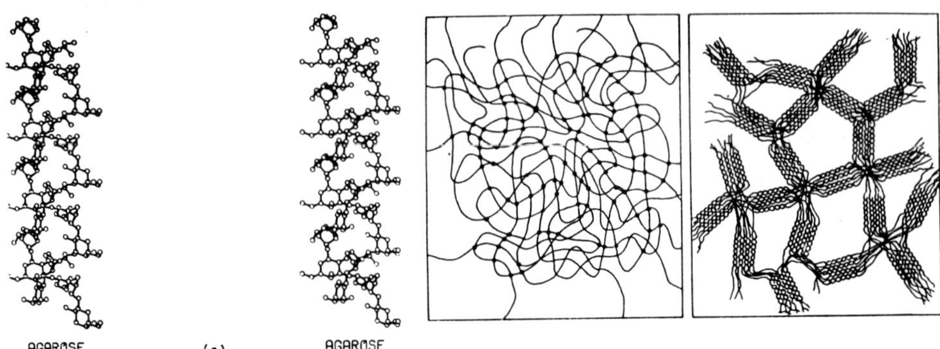



Fig. 8. Left: the agarose double helix viewed perpendicular to the helix a-
xis. The hydroxymethyl groups are located along the helix perimeter. Right:
a schematic representation of the agarose gel network (far right), in comparison with a network such as Sephadex, formed from free chains at similar
polymer concentration. Note the lateral aggregation of double helices in a-
garose gels, which strengthens the supporting gel structures (from Arnott *et
al.*) (41).

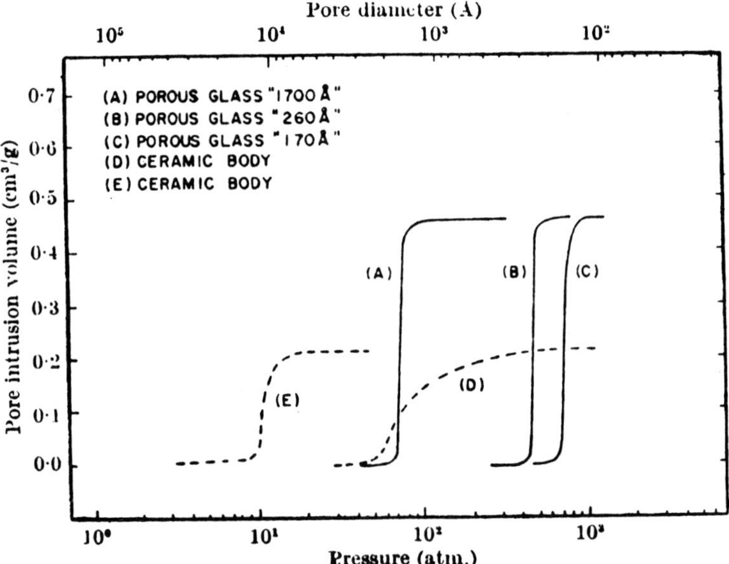



Fig. 9. Pore diameters in three types of porous glasses (A–C) and two different ceramic bodies (D, E). Notice that a pore size of ca. 2 μm is only obtained in ceramic body E. The pore size is obtained by mercury intrusion
(from Haller) (43).

quite difficult to be able to device new hydrophilic gels with a porosity

much greater than the limiting value of 500 nm we have found. Even if it we-

re possible, the structure will collapse, crashed by the 1 g gravity on earth.

Whether gels with pore sizes of 1 μm or more can be cast in microgravity, in the space shuttle, remains to be seen. At the moment, here on earth, the only way to make porous bodies ($\bar{p}$ > 1 μm) appears to be linked to ceramics (Fig. 9) and this belongs more to the art of pottery-making than to the art of biochemistry.

## Discussion

*1) Gel structure.* Whether the gel pores are seen as cones, or cubes, or spheres, or circular holes, or according to any other geometry, at least one model, the "random meshwork of fibers" of Ogston (22) seems to break down. Even though the Gidding's theory (Fig. 2) seems to support this model (whether they are spheres or thin rods, macromolecules seem to wander about most freely within random planes and to experience strong hindrance within an ordered structure of parallel planes or other geometrical shapes) I have an opposite point of view. I think this theory can be developed by people who live in non-seismic areas, but they could hardly survive in the earthquake-prone mediterranean basin. As people who have been buried in a collapsed building during an earthquake well know, it is extremely difficult to move around , even to crawl, within this "random meshwork of rubble". The pore sizes we have measured in polyacrylamide and agarose gels, and the mobilities exhibited by proteins in these matrices are not consistent with a "random meshwork of fibers" model, but rather with a sort of regular structural lattice. Evidence to this fine structure and gel organization has been given by electron microscopy (36, 37).

*2) Pore size.* Our data show that the upper pore limit which can be obtained with any hydrophilic gel is around 0.5 μm. The only way to achieve that is to strengthen the beams supporting the gel cavities, and this is nicely accomplished in agarose since the pillar (a double helix) is already rigid and is further hardened by lateral aggregation of 7 to 11 helices. Assuming a diameter of 1 nm for a double helix, then the supporting column should have a thickness of at least 10 nm, which would give a ratio beam:hole diameter of 1:50. Architecturally, this is quite an achievement, and should be confronted with examples of some of the finest gothic architecture (Fig. 10): in the nave and in the transept, very rarely the ratio column:surrounding cavity exceeds 1:10. In regularly cross-linked acrylamides (5%C) much smaller pore diameters are obtained, because the fiber is very long and thin

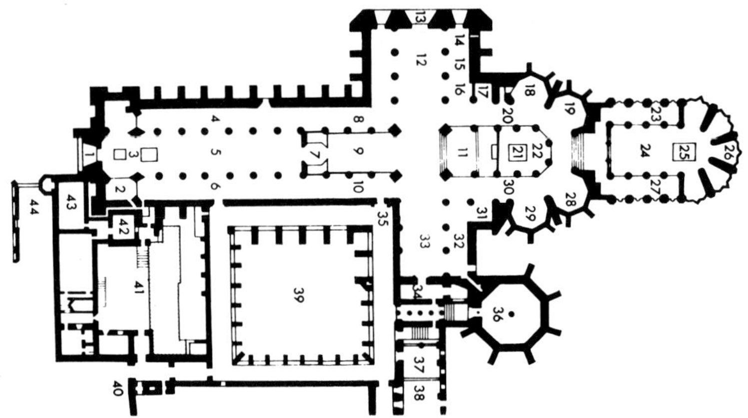

Fig. 10. Plan of Westminster Abbey (London, G.B.). Excerpta: 4: north aisle; 5: nave; 6: south aisle; 7: organ loft; 9: choir; 12: north transept; 33: south transept; 36: chapter house; 39: cloisters; 42: Jericho Parlour; 43: Jerusalem Chamber (sorry, 42 & 43 not open to public) (courtesy of C.A. Fox, sub-dean of Westminster and of Pitkin Pictorials Ptd.). Notice how the ratio pillar:surrounding cavity does not exceed 1:10.

(ca. 0.5 nm) is not rigid and therefore keeps fluctuating and vibrating in the space between the cross-links. But as the %C is increased, the fiber grows thicker and shorter, probably also more rigid, and the pores progressively open up. The limit to this structure is the fact that the fibers grow into big boulders, which weaken the gel consistency. The matrix becomes hydrophobic, and exudes water, probably not only due to the increased hydrophobicity of Bis, but also due to the fact that the boulders are tightly annealed into loops, into which the water "icebergs" cannot any longer penetrate (42).

## Acknowledgements

Supported in part by grants from Consiglio Nazionale delle Ricerche (CNR), Ministero della Pubblica Istruzione (MPI, Roma) and NASA (Huntsville, Ala.). I thank J. Biochem. Biophys. Methods for allowing reproduction of Fig. 7 before publication.

## References

1.   Smithies, O.: Biochem. J. 61, 629-641 (1955)
2.   Raymond, S., Weintraub, L.: Science 130, 711-713 (1959)

3.  Hjertén, S., Jersted, S., Tiselius, A.: Anal. Biochem. 27, 108-115 (19
    (1969)

4.  Porath, J.K., Flodin, P.: Nature 183, 1657-1658 (1959)

5.  Rodbard, D., Chrambach, A.: Proc. Natl. Acad. Sci. 65, 970-977 (1970)

6.  Ackers, G.K.: in Anfinsen, C.B., Edsall, J.T., Richards, F.M. (eds.)
    Advances in Protein Chemistry, vol. 24, pp. 343-446, Academic Press,
    New York, 1970

7.  Ackers, G.K.: in Neurath, H., Hill, R.L. (eds.) The Proteins, pp. 1-
    94, Academic Press, New York, 1975

8.  Rodbard, D.: in Catsimpoolas, N. (ed.) Methods of Protein Separation,
    vol. 2, pp. 145-179, Plenum Press, New York, 1976

9.  Kremmer, T., Boross, L.: Gel Chromatography, John Wiley & sons, Chiche-
    ster, 1979

10. Fischer, L.: An Introduction to Gel Chromatography, North Holland Publ.
    Co., Amsterdam, 1969

11. Determann, H.: Gel Chromatography, Springer Verlag, Berlin, 1967

12. White, M.L.: J. Phys. Chem. 64, 1563-1565 (1960)

13. Ackers, G.K., Steere, R.L.: Biochim. Biophys. Acta 59, 137-149 (1962)

14. Ackers, G.K.: Biochemistry 3, 723-730 (1964)

15. Righetti, P.G., Brost, B.C.W., Snyder, R.S.: J. Biochim. Biophys. Me-
    thods (1981) in press

16. Flodin, P.: J. Chromatogr. 5, 103-111 (1961)

17. Porath, J.: J. Applied. Chem. 6, 233-236 (1963)

18. Andrews, P.: Biochem. J. 91, 222-230 (1964)

19. Squire, P.G.: Arch. Biochem. Biophys. 107, 471-480 (1964)

20. Casassa, E.F.: J. Polymer Sci. B5, 773-780 (1967)

21. Ornstein, L.: Ann. N. Y. Acad. Sci. 121, 321-349 (1964)

22. Ogston, A.G.: Trans. Faraday Soc. 54, 1754-1756 (1958)

23. Laurent, T.C., Killander, J.: J. Chromatogr. 14, 317-325 (1964)

24. Giddings, J.C., Kucera, E., Russell, C.P., Myers, M.N.: J. Phys. Chem.
    72, 4397-4408 (1968)

25. Hohn, T., Pollmann, W.: Z. Naturforsch. 18B, 919-925 (1963)

26. Ackers, G.K.: J. Biol. Chem. 242, 3237-3241 (1967)

27. Fawcett, J.C., Morris, C.J.O.R.: Separ. Sci. 1, 9-20 (1966)

28. Albertsson, P.A.: The Partitioning of Cell Particles and Macromolecu-
    les, Academic Press, New York, 1960

29. Fischer, L.: in ref. 10, pp. 338-351

30. Bode, H.J.: in Radola, B.J. (ed.) Electrophoresis '79, pp. 39-52, de
    Gruyter, Berlin, 1980

31. Ziabicki, A., Walasek, J.: Macromolecules 11, 471-476 (1978)

32. Ziabicki, A.: Polymer 20, 1373-1381 (1979)

33. Shaaya, E.: Anal. Biochem. 75, 325-328 (1976)

34. Rodbard, D., Levitov, C., Chrambach, A.: Separ. Sci. 7, 705-723 (1972)

35. Lehninger, A.L.: Biochemistry, pp. 859-890, Worth Publ., New York, 1975

36. Rüchel, R., Brager, M.D.: Anal. Biochem. 68, 415-425 (1975)

37. Rüchel, R.; Steere, R.L., Erbe, E.F.: J. Chromatogr. 166, 563-575 (1978)

38. Bianchi Bosisio, A., Loeherlein, C., Snyder, R.S., Righetti, P.G.: J. Chromatogr. 189, 317-330 (1980)

39. Tombs, M.P.: Anal. Biochem. 13, 121-132 (1965)

40. Raymonds, S., Nakamichi, M.: Anal. Biochem. 3, 23-32 (1962)

41. Arnott, S., Fulmer, A., Scott, W.E., Dea, I.C.M., Moorhouse, R., Rees, D.A.: J. Mol. Biol. 90, 269-284 (1974)

42. Porath, J.: Lab. Pract. 16, 838-841 (1967)

43. Haller, W.: Nature 206, 693-696 (1965)

GENERALIZATION OF THE DEBYE-HÜCKEL THEORY TO MACROION
SOLUTIONS OF FINITE SIZE AND CONCENTRATION AS APPLIED TO
THEIR ELECTROPHORETIC MOBILITY AND TRANSLATIONAL DIFFUSION
COEFFICIENT


Eugene N. Serrallach
Naval Blood Research Laboratory, Boston University School of
Medicine, 615 Albany St., Boston, Mass.  02118, U.S.A.

Robert Schor
Physics Department and Institute of Materials Science,
University of Connecticut, Storrs, Connecticut 06268, U.S.A.



Introduction

The quantitative treatment of the Coulombic interactions among
salt ions and macroions in aqueous solutions has been a fundamental field of concern for many decades in physico-chemistry
(1,2).  Several reports (3-13) have summarized and discussed
the large amount of experimental and theoretical work done in
this field particularly with reference to colloidal particles
and biological macromolecules.  In addition to the interactions which arise from the molecular dimensions of the
macroions (excluded volume), the inter- and intramolecular
interactions arising from the electrical charge on the surface
of the macroions, also contribute to the free energy of the
solution and thus alter the chemical potential (10) given by

$$\mu_1 - \mu_1^\circ = RT \cdot \ln a = -RTV_1^\circ c_2 \left( \frac{1}{M_w} + Bc_2 + Cc_2^2 + \cdots \right) \tag{1}$$

where $\mu_1$ is the chemical potential of the macromolecular solution, $\mu_1^\circ$ the standard chemical potential, R the universal gas
constant, T the Kelvin temperature, a the activity coefficient
of the macroions, $V_1^\circ$ the molar volume of the water, $c_2$ the
concentration of the macroions and $M_w$ their molecular

weight. B and C represent the second and third virial coefficients respectively, which are expressible in terms of the
intermolecular potential. All the physico-chemical properties
of the macroions in solution such as electrophoretic mobility
$\mu$, electric conductance $\lambda$, translational diffusion coefficient
$D_T(\vec{K})$ ($\vec{K}$ being the scattering vector), structure factor $S(\vec{K})$,
activity coefficient a, osmotic pressure $\pi$, transmembrane
potential $E_M$, Donnan effect, solubility S, electroviscosity
$\eta_{el}$ , critical temperature $T_C$ of the sol - gel transition and
titration curves will depend upon the steric and electrostatic
interactions.

The Debye-Hückel (DH) theory (1) was developed in 1923 for
strong electrolytes such as $Na^+$ and $Cl^-$ and is ordinarily applied to solutions of ionized salts, where the central ion is
one of many similar ions in dimension and charge. Subsequently, the identical formalism was applied (3,6) to the field of
colloidal particles and proteins in salt solutions. **The theory
was used under the implicit assumption of infinite macroion
dilution, or equivalently, the absence of any macroion-**
macroion interactions. Since then, the DH-theory has become
one of the key elements in the computation and analysis of the
physico-chemical quantities mentioned above.

Many experiments, however, have been conducted in solutions of
finite macroion concentration, where macroion interactions
were definitely present. Under these conditions, the agreement between the experimental and the predicted values is in
general semi-quantitative. For example, the ratio between the
electrokinetic and the titration charge was reported (4,8,10,
13) to be in the range of $\approx0.6$ to $\approx0.8$ for a variety of different globular proteins such as Ovalbumin (14), Albumin (15),
β-Lactoglobulin (16), Trypsin (17), Aldolose (18), and Lysozyme (19), over the entire range of pH's on both sides of the
isoelectric point. Also, the electric conductance measured on
salt free albuminates (20) yielded, when compared to the theoretical values, a ratio between 0.8 and 0.9 for a wide range

of protein charge and concentration. In addition, the coefficient w, which accounts for the electrostatic intraparticle interactions of the charged groups in the Linderstrøm-Lang theory, is about 0.8 of the calculated value for several proteins such as Ovalbumin and Ribonuclease (9).

We (21) have previously treated the problem of predicting the translational diffusion coefficient $D_T$ of uniformly charged spherical macroions in solution, and discussed the values of the calculations in the case of bovine serum albumin (BSA) in aqueous solution with added NaOH - minimum salt point - as well as at higher values of the ionic strength. By introducing the hard sphere approximation for the radial distribution function $G(r)$ of the macroions, which includes only the excluded volume due to the dimension of the particle, into the Phillies theory (22), we were able to reproduce the predictions of the Doherty-Benedek-Stephen theory (23,24). The ratio between the experimental and the calculated values of $D_T^{BSA}$ is between 0.13 and 0.27 for 18, 10 and 7 elementary charges at the minimum salt point. In order to take into account correlations in the positions of the macroions, due to the electrostatic interactions, we then introduced the more realistic dilute gas approximation for $G(r)$ into the Phillies theory. Significantly better agreement with the experimental data was obtained. The predicted values of $D_T$, however, were systematically higher than the experimental values, the ratio being about 0.34.

In our opinion, an important reason for the discrepancy between the theoretical and experimental values of the different physico-chemical coefficients is that the DH-theory does not include the contribution of the charged macroions to the charge density $\rho(r)$ and therefore to the screening length $1/K$ of the solution. In fact, to the best of our knowledge no one has yet adequately accounted for this effect.

Theory

In order to take into account the effects of the finite size
and concentration of the charged macroions, we have generalized the usual III region Debye-Hückel ($DH^{III}$) model (Figure 1)
to include IV regions ($DH^{IV}$) (Figure 2). Region I represents
the central macroion of radius R, region II the water layer of
thickness $\delta$, region III from r=a to r=2a contains coions and
counterions in solution but does not include the macroions.
This region results from the excluded volume of two hard
spheres. The new feature of the model is the introduction
of region IV from r=2a to r=∞ which encompasses the macroions
in addition to the coions, counterions and water. In order to
obtain the electrical potential $\Psi(r)$ around the uniformly
charged central macroion, Laplace's equation must be solved
in regions I and II since no ions can penetrate into these
regions. The Poisson-Boltzmann equation holds in regions III
and IV. The usual boundary conditions for the continuity of
the potential and the discontinuity of the normal derivative
of the potential must be applied at r=R. In addition, the potential and its normal derivative must be continuous at r=a and
r=2a. Also $\Psi(r)$ must be finite everywhere and in particular at
r=0 and r=∞. As has already been mentioned, the charge density
$\rho(r)$ is zero in regions I and II and Laplace's equation is
satisfied:

$$\nabla^2 \Psi(r) = 0 \qquad (0 \leq r < R) \text{ and } (R < r \leq a) \qquad (2)$$

The assumption that the surface of the macroions are spherical
and uniformly charged implies a spherically symmetric potential function $\Psi(r)$. The appropriate spherically symmetric
solutions to Laplace's equation are:

$$\Psi_0 = A_1 \qquad (0 \leq r \leq R) \quad \text{and} \quad \Psi(r) = A_2 + A_3/r \qquad (R \leq r \leq a) \qquad (3)$$

$A_1$ and $A_2$ are as yet undetermined constants. Since $\Psi(0)$ must
be finite, the potential is constant in region I. In regions
III and IV the situation is more complex due to the presence

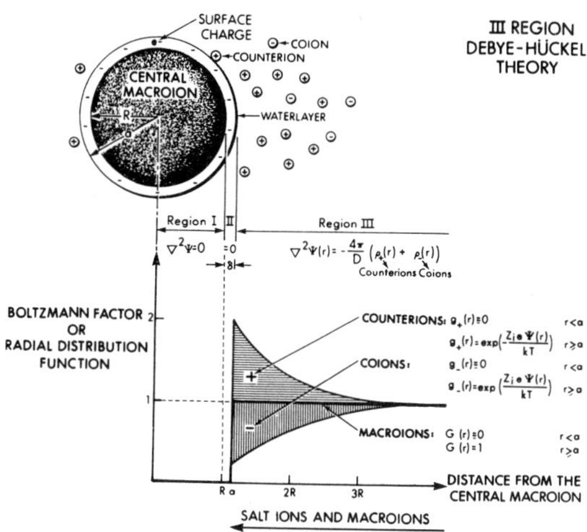


Figure 1: Ordinary III Region Debye-Hückel (DH$^{III}$) model and theory.

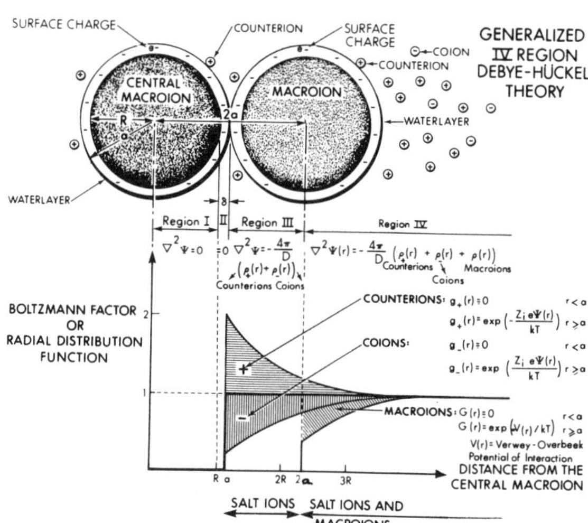


Figure 2: Generalized IV Region Debye-Hückel (DH$^{IV}$) model and theory.

of the ionic atmosphere. Let N represent the number density
of negatively charged macroions in the bulk solution. We
assume that each macroion carries a net charge $-Ze$ where Z is
the number of excess electrons and e is the electronic charge.
We consider the situation at the minimum salt point. The
generalization to the case of added salt is straightforward
and it shall not be done here. Let us assume a negatively
charged central macroion surrounded by positive univalent
counterions, e.g., in the case of BSA, the mobile counterions
are $Na^+$.. In order to ensure electrical neutrality, the number density of univalent mobile counterions must be NZ. According to the Boltzmann distribution law we have:

$$\rho_+^{III}(r) = NZe\, g_+(r) = NZe\, \exp\left(-e\, \psi(r)/kT\right) \qquad \text{(Region III)}$$

$$\rho_-^{III}(r) = -NZe\, g_-(r) = 0 \qquad \text{(no added salt)} \quad \text{(Region III)}$$

$$\rho_+^{IV}(r) = NZe\, g_+(r) = NZe\, \exp\left(-e\, \psi(r)/kT\right) \qquad \text{(Region IV)} \qquad (4)$$

$$\rho_-^{IV}(r) = -NZe\, G(r) = -NZe\, \exp\left(-V(r)/kT\right) \qquad \text{(Region IV)}$$

$$\rho(r) = \rho_+(r) + \rho_-(r) = NZe\left(g_+(r) - g_-(r) - G(r)\right) \qquad \text{(Region III and IV)}$$

In equation (4), $\rho_+(r)$ represents the charge density of the
positive salt ions and $\rho_-(r)$ represents the sum of the charge
densities of the negative salt ions and the macroions. $\rho(r)$ is taken as zero in region III because the centers of the macroions cannot penetrate into this region due to their excluded volume. Actually, some of the negative charge on the surface of the macroion can penetrate into region III.
$G(r)$, $g_+(r)$ and $g_-(r)$ represent the radial distribution
functions for the macroions , the mobile counterions and the
coions respectively, which we have replaced in equation (4) by
the Boltzmann factors using the dilute gas approximation. T
is the Kelvin temperature and k the Boltzmann constant. $V(r)$
is the potential energy of interaction of two macroions of
finite size with their corresponding double layers. Verwey
and Overbeek (25) derived an expression for $V(r)$ using the
DH-theory for the double layer which is valid for very dilute
macroion concentrations. The expression reads

$$V(r) = \psi_0^2 D a^2 \exp(-K(r-2a))/r \qquad (r > 2a)$$

$$= Z e/(1+Ka) \cdot \psi_{DH}^{III}(r) \cdot \exp(Ka) \qquad (5)$$

Here $\psi_0$ represents the surface potential, a the hydrodynamic radius, $1/K$ the DH screening length and r the distance between the macroions. The term $Ze/(1+Ka)$ corresponds to the effective charge as used in the electrophoretic calculations. The term $\psi(r)$ represents the potential at the distance r, and $\exp(Ka)$ is a "size factor" which accounts for the dimension of the macroion. The numerical calculations show that the simplified potential of interaction $V(r)=Ze\psi(r)$, which consideres the macroion as a point charge with unscreened charge Z, yields comparable values for $\psi_0$ as does the more physical Verwey-Overbeek interaction potential. The Poisson-Boltzmann equation is satisfied in regions III and IV:

$$\nabla^2 \psi(r) = -4\pi \rho(r)/D = -4\pi(\rho_+(r) + \rho_-(r))/D$$

$$\nabla^2 \psi(r) = -4\pi NZe \exp(-e\psi(r)/kT)/D \qquad \text{(Region III)} \qquad (6)$$

$$\nabla^2 \psi(r) = -4\pi NZe \left\{ \exp(-e\psi(r)/kT) - \exp(Ze\psi(r)/kT) \right\}/D \quad \text{(Region IV)}$$

Expanding the exponentials to first order in Equation (6) leads to the linearized Poisson-Boltzmann equation which is valid if $Ze\psi(r)/kT \ll 1$

$$\nabla^2 \psi(r) - K_3^2 \psi(r) = -4\pi NZe/D$$

$$\nabla^2 \psi(r) - K_4^2 \psi(r) = 0 \qquad (7)$$

where $K_3^2 = 4\pi NZe^2/DkT$ and $K_4^2 = K_3^2 (Z+1)$.

The solutions for $\psi(r)$ in regions III and IV are essentially screened Coulomb potentials.

$$\psi(r) = A_4 \exp(-K_3 r)/r + A_5 \exp(K_3 r)/r + 4\pi NZe/DK_3^2 \qquad \text{(Region III)}$$

$$\psi(r) = A_6 \exp(-K_4 r)/r \qquad \text{(Region IV)} \qquad (8)$$

In region III the term $4\pi NZe/DK_3^2 = kT/e$ represents a positive constant background potential due to the unbalanced counterions. A term of the form $A_7 \cdot \exp(K_4 r)/r$ in region IV is precluded by the condition that $\psi_{IV}(\infty)$ is finite. $A_4, A_5,$

and $A_6$ are as yet undetermined constants. The set of coefficients $\{A_1, A_2, \ldots A_6\}$ are determined by imposing the boundary conditions for the electrical potential $\psi(r)$ and for the discontinuity in its normal derivatives $\partial \psi / \partial n$ at $r=R$, $r=a$, and $r=2a$.

$$\psi_\alpha = \psi_\beta \quad \text{and} \quad \partial \psi_\beta / \partial n - \partial \psi_\alpha / \partial n = -4\pi\sigma/D \qquad (r=R)$$
$$\psi_\alpha = \psi_\beta \quad \text{and} \quad \partial \psi_\beta / \partial n - \partial \psi_\alpha / \partial n = 0 \qquad (r=a \text{ and } r=2a)$$

(9)

In Equation (9) $\vec{n}$ refers to the unit outward normal to the spherical surface. The subscripts $\alpha$ and $\beta$ refer to approaching the spherical surface from the inside and outside respectively. $\sigma$ represents the surface charge density on the protein ion at $r=R$. From Equations (3, 8, and 9) we have:

$$A_1 = A_2 + A_3/R \quad ; \quad A_3 = -Ze/D \quad ; \quad \tau_1 = K_3 a \quad ; \quad \tau_2 = K_4 a$$
$$a A_2 + A_3 = A_4 \exp(-\tau_1) + A_5 \exp(\tau_1) + 4\pi NZea/DK_3^2$$
$$A_3 = A_4 (1+\tau_1) \exp(-\tau_1) + A_5 (1-\tau_1) \exp(\tau_1)$$
$$A_4 \exp(-2\tau_1) + A_5 \exp(2\tau_1) + 8\pi NZea/DK_3^2 = A_6 \exp(-2\tau_2)$$
$$A_4 (1+2\tau_1) \exp(-2\tau_1) + A_5 (1-2\tau_1) \exp(2\tau_1) = A_6 (1+2\tau_2) \exp(-2\tau_2)$$

(10)

The solution to these 6 simultaneous algebraic equations is straightforward and yields $\{A_1, A_2, \ldots A_6\}$ and finally the electrical potential $\psi(r)$. The function $\psi(r)$ is thus determined as is usual by the parameters $N, Z, R, \delta, T$ and $D$. $Z$ is determined from the titration curve.


Results


In this section, we examine the numerical consequences of the introduction of the IV region model. In particular, we consider the effect on the surface potential $\psi_0$ and more generally on the dependence of the electrical potential $\psi(r)$. The calculations are based upon the following values of the parameters: $M_w^{BSA} = 69'000$ Daltons, $R = 34.5 \text{ Å}$, $\delta = 3 \text{ Å}$, $N = 4.37 \times 10^7$ particles/cm$^3$, $T = 25°C$, $D_{H_2O} = 78.54$ and no added salt, i.e., minimum salt point.

Figure 3 shows the results for the generalized Debye-Hückel IV region ($DH^{IV}$) model, which were derived on the basis of the theory presented in the preceding section. The dimensionless ratio $e\psi/kT$ is plotted as a function of the reduced distance $r/R$ for the charge states $z=1,4,7,10$ and $18$. In order to exhibit the differences between the results of the $DH^{IV}$- and the $DH^{III}$-model, we have plotted in Figure 4 the dimensionless ratio of both electrical potentials $\psi^{IV}(r)/\psi^{III}(r)$ versus the reduced distance $r/R$.

The most important points shown in Figures 3 and 4 are the following: 1) The surface potential $\psi_o^{IV}$ predicted by the $DH^{IV}$-model is in all cases about 30% lower than the $\psi_o^{III}$ predicted by the usual $DH^{III}$-model. The precise amount of the reduction decreases monatonically as a function of the charge state Z going from $\approx 0.65$ at $Z=1$ to $\approx 0.75$ at $Z=18$, 2) $\psi^{IV}(r)$ decays much more rapidly than $\psi^{III}(r)$ and, 3) the linearization procedure is crude at $r=2a$ when the macroions are in contact. In the worst case, the parameter $Ze\psi(2a)/kT$ is about 1.2 for $Z=18$ and the validity of the approximation rapidly improves with increasing r.

It is of interest to compare the calculated ratio of $e\psi_o/kT$ on the basis of four different approaches of accounting for the effect of the protein concentration on the ionic strength of the solution: 1) treat the proteins as point ions of charge Z and use the usual $DH^{III}$-model, 2) the present $DH^{IV}$-model, 3) treat the proteins as Z independent univalent point ions and use the usual $DH^{III}$-model, 4) complete neglect of the protein concentration. Table I shows the results as applied to a 5 g% solution of BSA at the minimum salt point. An examination of Table I shows that for the lowest charge states $Z=1$ and $Z=4$ the $DH^{IV}$-model predicts the lowest value of $\psi_o$. For the charge states $z=7, 10$ and $18$, however, the values of $\psi_o$ obtained on the basis of the $DH^{IV}$-model lie between the values obtained by treating the proteins as point ions of charge Z and those obtained by treating the proteins as Z independent univalent ions.

| Ze | 1) $e\psi_0/kT$ | 2) $e\psi_0/kT$ | 3) $e\psi_0/kT$ | 4) $e\psi_0/kT$ | $DH^{IV}/DH^{III}$ |
|---|---|---|---|---|---|
| Protein | $DH^{III}$ | $DH^{IV}$ | $DH^{III}$ | $DH^{III}$ | in % |
| Charge | Charge=Ze | Charge=Ze | Z Charges=e | Charge=0 | 2 /4 |
|  | Radius=0 | Radius=R | Radius=0 |  |  |
| 18 | .98 | 1.52 | 1.72 | 2.02 | 75% |
| 10 | .75 | .90 | 1.09 | 1.27 | 71% |
| 7 | .62 | .66 | .82 | .93 | 71% |
| 4 | .45 | .39 | .52 | .59 | 66% |
| 1 | .16 | .11 | .16 | .17 | 65% |

Table I: Calculated values of $|e\psi_0/kT|$ for different approximations of the charge contribution of the macroions to the screening length $1/K$ in the double layer. N is $4.39 \times 10^{17}$ particles/cm$^3$ (5 g% BSA) and the minimum salt point is considered. The electrophoretic mobility $\mu$ is proportional to the surface potential $\psi_0$ and and the electrokinetic charge of the macroion. $DH^{III}$ represents the III region Debye-Hückel model and $DH^{IV}$ the generalized IV region Debye-Hückel model.

| Ze | 1) $\alpha^{III}$ | 2) $\alpha^{IV}$ | $\alpha^{III}$ | 3) $\alpha^{III}$ | 4) $\alpha^{III}=\alpha^{th}_{D3}$ | $\alpha^{IV}/\alpha^{exp}_{D3}$ | $\alpha^{exp}_{D3}$ |
|---|---|---|---|---|---|---|---|
| Protein | $DH^{III}$ | $DH^{IV}$ | $DH^{III}$ | $DH^{III}$ | $DH^{III}$ |  |  |
|  | G(r)=1 | G(r)= $*$ | G(r)= $*$ | G(r)= $*$ | G(r)=1 | in % |  |
| Charge | Charge=Ze | Charge=Ze | Charge=Ze | Z Charges=e | Charge=0 | 2 / $\alpha^{exp}_{D3}$ |  |
|  | Radius=0 | Radius=R | Radius=R | Radius=0 |  |  |  |
| 18 | .77 | 1.39 | 7.2 | 4.1 | 18 | 60% | 2.3 |
| 10 | .81 | 1.22 | 6.4 | 3.5 | 10 | 55% | 2.2 |
| 7 | .32 | .97 | 5.4 | 3.0 | 7 | 53% | 1.8 |
| 4 | .78 | .62 | 3.5 | 1.9 | 4 | 56% | 1.1 |

$\alpha^{III}$ = The value of $\alpha$ calculated on the basis of the $DH^{III}$-theory
$\alpha^{IV}$ = The value of $\alpha$ calculated on the basis of the Generalized $DH^{IV}$-theory
using equations 4 and 5 where $\psi_0$ represents the value of the IV region model and $K=K_4$.

$\alpha^{exp}_{D3}$ = experimental values ( 23 ); $q^{th}_{D3}$ = theoretical values ( 23 ); $* = \exp(-V(r)/kT)$

Table II: Calculated values of $\alpha(N,K,Z)$ for different approximations of the charge contribution of the macroions to the screening length $1/K$ in the double layer. G(r) represents the radial distribution function of the macroions. $\alpha$ is a function of the radial dependence of the electrical potential $\psi(r)$. $DH^{III}$ represents the III region Debye-Hückel model and $DH^{IV}$ the generalized IV region Debye-Hückel model.

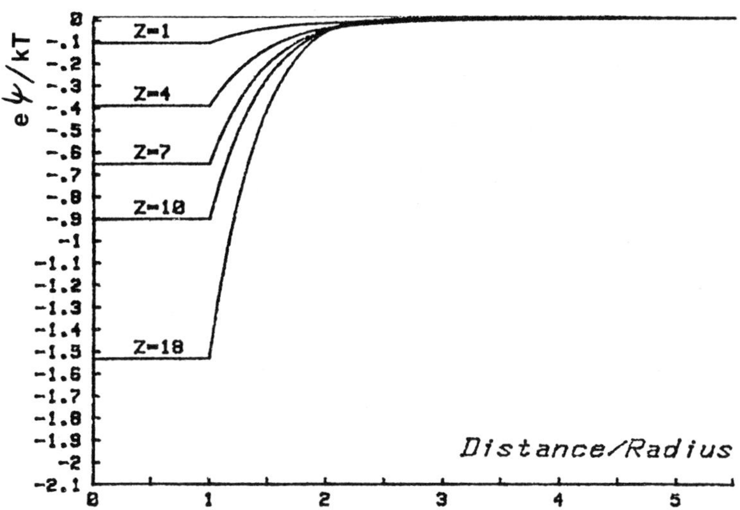



Figure 3:   Electrical potential $\Psi$(r) of the generalized DH$^{IV}$-model versus the distance from the central macroion using the linearized Poisson-Boltzmann equation for charge states of the macroion **Z**=18,10,7,4, and 1 elementary charges at the minimum salt point.   In addition, N is 4.37x10$^7$ particles/cm$^3$, R=34.5Å $\delta$=3Å, T=25°C and D=78.54.

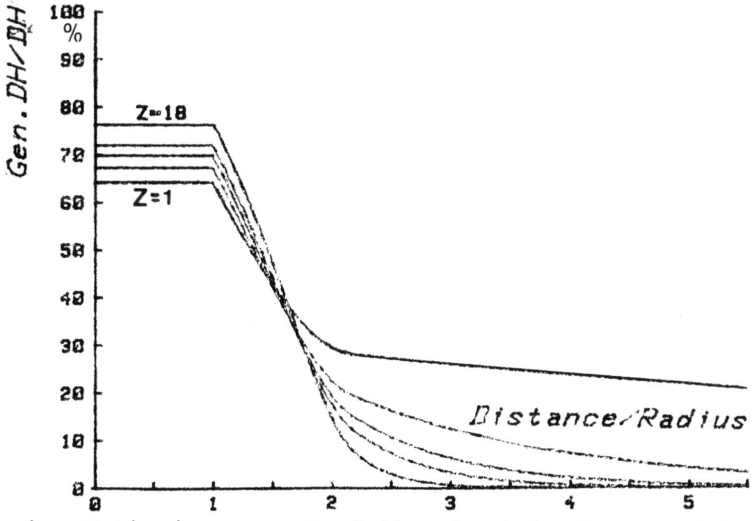



Figure 4:   Ratio in percent of the electrical potentials of the generalized DH$^{IV}$-model and the DH$^{III}$-model   $\Psi(r)^{IV}/\Psi(r)^{III}$ versus the distance from the central macroion using the linearized Poisson-Boltzmann equation for charge states of the macroion **Z** =18,10,7,4 and 1 elementary charges at the minimum salt point.

We have also calculated the quantity $\alpha(N,K,Z) = (D - D_0)/D_0$ at the minimum salt point on the basis of the $DH^{IV}$ model. $\alpha(N,K,Z)$ is a measure of the fractional increase of $D_T$ due to macroion interactions. Here D represents the modified diffusion coefficient and $D_0$ the zero charge or high ionic strength diffusion coefficient. We (21) have introduced equation (5) in equations (4,5) of our previous work for this calculation. The only modification was the replacement of $K_3^2 = 4\pi Ne^2 Z/DkT$ by $K_4^2 = K_3^2(Z+1)$. The results are given in Table II for the charge states Z=4, 7, 10 and 18. The $DH^{IV}$-theory reduces analytically and numerically to the ordinary $DH^{III}$-theory in the limit of zero macroion concentration, i.e., $N \rightarrow 0$. In addition, it reduces to the results that some reseachers have obtained in interpreting the DH-theory for charged protein solutions as well when region III collapses and G(r) of the macroions is set equal to 1 for $r>a$. They used $K^2 = 4\pi Ne^2 Z/DkT$ instead of $K^2 = 8\pi Ne^2 Z/DkT$. This approach completely neglects the contribution of the macroions to the ionic strength of the solution. Introducing G(r) overcomes most of the intrinsic physical difficulties of the application of the $DH^{III}$-theory to macroion solutions, namely the violation of charge neutrality. This is most clearly seen in the $DH^{III}$-theory at the minimum salt point since $\rho_+(r)$ of the positive salt ions contains a constant term which does not cancel out in the expression for $\rho(r)$ between r=a and $r=\infty$. Therefore, the integral $\int_a^\infty \rho(r)\, 4\pi r^2\, dr$ diverges instead of yielding the value of the macroion surface charge Ze.

Discussion

While it is premature to make a detailed assessment of the $DH^{IV}$-model, it is encouraging to note that the introduction of the contribution of the macroions of radius R and charge Z into the screening length 1/K of the <u>new model</u> yields results smaller in value, but of the same order of magnitude, as the ones predicted by the $DH^{III}$-theory, which completely neglects the macroion-macroion interactions. It is also of interest to examine the possible experimental implications of the model with respect to predicting $\mu$ and $D_T$ of uniformly charged spherical macroions.   $\mu$ depends upon $\psi_0$ and $D_T$ upon $\psi(r)$.  Let us qualitatively explore this question.  Although the experimental situation is complicated in detail, it is known that for a number of important globular proteins such as Albumin, Ovalbumin, Lysozyme, and $\beta$-Lactoglobulin the electrokinetic charge is significantly lower than the titration charge.  This effect is not understood at the present time and is often attributed to ion binding (4,10,13).  In the case of egg albumin (4), for example, the electrokinetic charge is approximately 60% of the titration charge for all values of the pH between 3 and approximately 11.7.  The titration charge varies continuously from about +23 elementary charges at a pH of 3 to about -33 elementary charges at a pH of approximately 11.7.  The usual interpretation attributes the discrepancy mentioned above to ion binding. The present approach offers a complementary interpretation and more strongly, an alternative one in the cases where ion binding is not found.

We now briefly consider $D_T$.  $\psi^{IV}(r)$ is considerably lower than $\psi^{III}(r)$ which is calculated on the basis of the complete neglect of the protein concentration at the minimum salt point. One would expect, therefore, a marked reduction in the predicted values of $\alpha$.  Inspection of Table II shows, however, that the  experimental values of $D_T$ are bracketed by the predictions of our previous work (21) and the present approach.  Although the calculations were based on a crude assumption for the interaction potential between two macroions V(r), our results suggest that the actual screening of the central macroion lies somewhere between that used in the two approaches.

In conclusion, while the comparison between the experimental and the theoretical results of the present work suggest a better agreement on the basis of the IV region model, they must be regarded as tentative.  The underestimation of the electrical screening involved in linearizing the Poisson-Boltzmann equation is difficult to assess. It should be mentioned that this approach is also used in the $DH^{III}$-theory. From the results of the work of

Hoskins (26) on the solution of the Poisson-Boltzmann equation for the potential distribution in the double layer of a single spherical colloidal particle, it appears unlikely to us that the errors would be "very large".  Nevertheless, it seems useful to attempt the difficult problem of solving the unlinearized Poisson-Boltzmann equation subject to the usual boundary conditions in the IV region model.  In addition, it is important to introduce the fact that somewhat less than half of the charge of the macroions, which are in contact with the central macroion, penetrates into region III.  If these and other refinements can be successfully carried out, it will be possible to make a more quantitative assessment of the extent to which the predictions based on the IV region model agree with the experimental data.  In this connection, it would be highly desirable to have dynamic laser light scattering data of both $\mu$ and $D_T$ as well as low angle neutron or X-ray scattering measurements of G(r) from solutions of charged globular macromolecules such as BSA, Ovalbumin, and Lysozyme for wide ranges of protein concentrations, pH and ionic strength.

Acknowledgements:  Robert Schor acknowledges financial support from the University of Connecticut Research Foundation during the early stages of this work and Eugene Serrallach wishes to thank Prof. Ch. P. Emerson and Prof. C. R. Valeri for much generous support and encouragement during this work. This work was supported by the U.S. Navy (Naval Medical Research and Development Command and Office of Naval Research Contract No.  N00014-79-C-1068).

References

1. Debye, P., Hückel, E.: Physik. Zeitschrift 24, 185-206 and 305-325 (1923).

2. Pusey, P.N.: J. Phys. A., Math. Gen., 11, No. 1, 119-135 (1978).

3. Abramson, H.A., Moyer, L.S., Gorin, M.H.: Electrophoresis of Proteins, Hafner Publ. Inc., New York, 1964.

4. Overbeek, J.Th.G.: Quantitative Interpretation of the Electrophoretic Velocity of Colloids in Advances in Colloid Science Vol. III, 97-135, Mark, H. et al. Ed., Interscience Publ. Inc., New York, 1950.

5. Doty, P., Edsall, J.T.: Light Scattering in Protein Solutions in Advances in Protein Chemistry, Vol. VI, 35-121, Anson, M.L. et. al. Ed., Academic Press Inc., New York, 1951.

6. Overbeek, J.Th.G.: Colloid Science, Kruyt, J.R. Ed., Elsevier Publ., Amsterdam 1952.

7. Overbeek, J.Th.G., Lijklema, J.: Electric Potential in Colloid Systems in Electrophoresis, Vol. I, 1-33, Bier, M. Ed., Academic Press Inc., New York, 1959.

8. Brown, R.A., Timasheff, S.N.: Applications of Moving Boundary Electrophoresis to Protein Systems in Electrophoresis, Vol. I, 317-365, Bier, M. Ed., Academic Press Inc., New York, 1959.

9. Linderstrøm-Lang, K., Nielsen, S.O.: Acid-Base Equilibria of Proteins in Electrophoresis, Vol. I, 35-89, Bier, M. Ed., Academic Press Inc., New York, 1959.

10. Tanford, Ch.: Physical Chemistry of Macromolecules, John Wiley & Sons Inc., New York, 1961.

11. Martin, R.B.: Introd. to Biophysical Chemistry, McGraw-Hill Inc., New York, 1964.

12. Overbeek, J.Th.G., Wiersema, P.H.: The Interpretation of Electrophoretic Mobilities in Electrophoresis, Vol. II, 1-52, Bier, M. Ed., Academic Press Inc., New York, 1967.

13. Van Holde, K.E.: Physical Biochemistry, Prentice-Hall Inc., Englewood Cliffs, New Jersey, 1971.

14. Longsworth, L.G.: Ann. N.Y. Acad. Sci. 41, 267 (1941).

15. Gorin, M.H., Moyer, L.S.: Electrophoresis of Proteins p. 158, Hafner Publ. Inc., New York, 1964.

16. Cannan, R.K., Palmer, A. H., Kibrick, A.C.: J. Biol. Chem. 142, 803 (1942).

17. Duke, J.A., Bier, M., Ford, F.F.: Arch. Biochem. Biophys. 40, 424 (1952).

18. Velich, S.F.: J. Phys. & Colloid Chem. <u>53</u>, 135 (1949).

19. Beychok, S., Warner, R.C.: J. Am. Chem. Soc. <u>81</u>, 1892-1897 (1959).

20. Möller, W.J.H.M., Van Os, G.A.J., Overbeek, J.Th.G.: Trans. Faraday Soc. <u>57</u>, 312, 325 (1961).

21. Schor, R., Serrallach, E.N.: J. Chem. Phys. <u>70</u>, 3012-15 (1979).

22. Phillies, G.D.J.: J. Chem. Phys. <u>60</u>, 976 (1974).

23. Doherty, P.M., Benedek, G.B.: J. Chem. Phys. <u>61</u>, 5426 (1974).

24. Stephen, M.J.: J. Chem. Phys. <u>66</u>, 1837 (1977).

25. Verwey , E.J.W., Overbeek, J.Th.G.: Theory of the Stability of Lyophobic Colloids, Elsevier Publ., Amsterdam, 1948.

26. Hoskin, N.E.: Trans. Faraday Soc., <u>49</u>, 1471, (1953).

CALCULATION OF THE THERMODYNAMIC CONSTANTS OF CONCANAVALIN
A - CARBOHYDRATE INTERACTIONS BY MEANS OF AFFINITY ELECTRO-
PHORESIS.


Kazusuke Takeo, Masanori Fujimoto, Akira Kuwahara, Ryosuke
Suzuno  and Kazuyuki Nakamura
Department of Biochemistry, Yamaguchi University School of
Medicine, Kogushi-1144. Une. 755-Japan


Introduction

Thermodynamic parameters of biochemical interactions such as
lectin-carbohydrate and antigen antibody reactions are
determined by calorimetric treatment, or from van't Hoff
plots, in which dissociation constants and association constants are obtained by equilibrium dialysis or photometric
titrations such as difference spectrum and fluorescence
quenching.

In 1972 we developed a new electrophoresis technique(1),
which is called affinity electrophoresis.  It combines the
principles of electrophoresis with affinity chromatography.
When electrophoresis was carried out in a gel containing
glycogen, the mobility of phosphorylase was retarded.  This
retardation was reversed by addition  of a competitive
inhibitor such as maltotriose or cyclodextrin to the gel.
From the variation in mobility as a function of the substrate or inhibitor concentration, dissociation  constants
were calculated(1, 2, 3).

The technique of affinity electrophoresis is very simple.
It requires  only a small amount of protein, and it is not
necessary to purify the protein.  When a thermostatic electrophoresis  apparatus is available, thermodynamic para-

34

meters can be determined from the van't Hoff plot. This is
the first report on determination of thermodynamic parameters of lectin-carbohydrate interaction by electrophoresis.   The procedure may be useful for determination of
thermodynamic parameters for antigen-antibody interactions.

Method

1.  The thermostatic electrophoresis apparatus.

In Fig. 1, a diagram of a simple thermostatic polyacrylamide
gel disc electrophoresis apparatus is presented.  All electrophoresis tubes are immersed in the buffer solution in the lower elec-
trode vessel.  The temperature of the buffer solution is regulated by circulating water at a constant temperature through a coiled condensor and by simulta-
neous stirring with a mag-
netic stirror.  When 3 mA per tube or more electricity is applied, the temperature of the gel rises and in these conditions the temperature difference between the gel and the buffer solution exceeds 0.3°C.  At

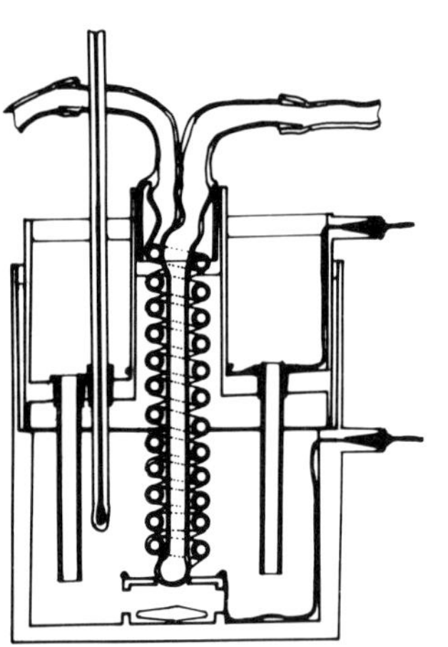

Fig. 1.  *Diagram of thermostatic electrophoresis apparatus*

electricity under 2.5 mA per tube, the temperature difference is maintained under 0.2°C.  In this study, 2.0 mA per
tube was applied throughout the experiment.

2.  Affinity electrophoresis.

Polyacrylamide gel disc electrophoresis was carried out by a
modified procedure(4) of Reisfeld et al(5).  The separting

gel, 5 cm in height, was prepared as a 5.0% gel, pH 4.3 and
the spacer gel, 1 cm in height, was 2.5% acrylamide, pH 6.7.
Dextran-T-2000 (Pharmacia pure Chem.) was added to the
separating gel.  To ensure a uniform concentration of dextran throughout the separating gel, it was prepared by overlaying a dextran solution of the same concentration as that
of the gel instead of water.  The oligosaccharides or glycosides were added to both the separating and spacer gels to
minimize dilution during electrophoresis.

Con A (Sigma Chem. Co.) and egg white lysozyme (Sigma Chem.
Co.) as a reference were dissolved in the buffer solution at
the same concentration as the spacer gel, containing 10%
sucrose and 0.001% methylene blue as the stacking dye.  0.1
ml of this solution, containing about 1 μg Con A and 1 μg
lysozyme, was applied on the spacer gel.  Electrophoresis
was run in β-alanine-acetic acid buffer at pH 4.5.  Electrophoresis gels were stained overnight in 0.05% Coomassie
BBR-250 in 7% acetic acid.

3.  Calculation of the dissocia tion constants for dextran
(Kd).

One set of electrophoresis was run with 12 tubes, duplicated
gels of 6 different concentrations of dextran (0, 0.1, 0.2,
0.3, 0.4, and 0.5%) in the separating gels.  Kd was calculated from equation 1(1),

$$1/Rm_i = 1/Rm_o ( 1 + c/Kd ) \dots\dots\dots\dots\dots (1),$$

where $Rm_o$ and $Rm_i$ are the relative migration distances of
Con A in the absence or the presence of dextran.  The relative migration distance is the ratio of the distance of
migration of Con A to that of lysozyme.  C was the concentration of dextran in the separating gel.  When the
reciprocal values of $Rm_i$ are plotted against c, a straight
line is obtained.  Its intercept on the c-axis gives -Kd.

4.  Calculation of the dissociation constant for oligosaccharides or glycosides (Ki).

36

Electrophoresis was run with 12 tubes; 2 contained neither
dextran nor oligosaccharide nor glycoside, the other 10
contained 0.5% dextran and different concentrations of
oligosaccharide or glycoside.  The Ki value was calculated
from equation 2(2, 3),

$$Rm_i/(Rm_o - Rm_i) = (Kd/c)(1 + i/Ki) \dots\dots\dots (2),$$

where i is the concentration of oligosaccharide or glycoside
in the gel.  When $Rm_i/(Rm_o - Rm_i)$ is plotted against i, a
straight line is obtained.  Its intercept on the i-axis
gives -Ki.

Results and Discussion

In Fig. 2, the affinity electrophoresis patterns of Con A -
dextran interaction at varying temperatures are presented.

*Fig. 2.  Affinity electrophoresis patterns of Con A-dextran interaction*

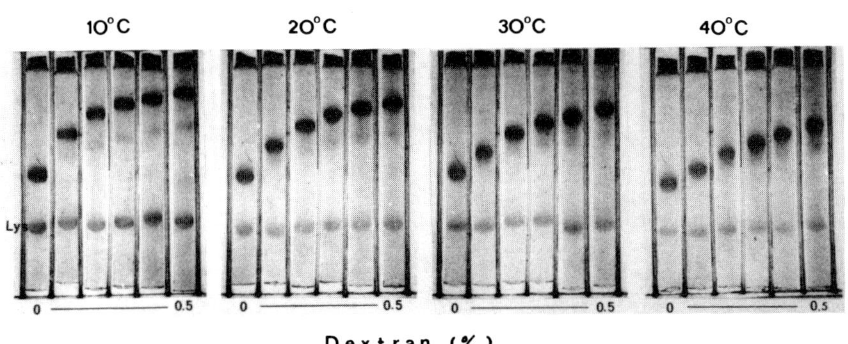


Mobility of Con A is decreased in the gel containing dextran, while that of lysozyme(Lys) does not change.  Hence, it
is clear that the decrease of Con A mobility is not based on
a change of the physico-chemical properties of the gels,
such as increase of viscosity, but based on its specific
interaction to dextran.  The rate of the mobility decrease
became less when the electrophoresis was carried out at
higher temperatures.  Thus, affinity of Con A to dextran
diminished progressively at higher temperatures.

In Fig 3A a van't Hoff plot of Con A - dextran interaction

is shown.  In the figure, Kd and temperature are presented.
Kd values at 11 and 59°C were 0.39 and 3.83 mM, respecti-

*Fig. 3.  Van't Hoff plots for Con A-carbohydrate interactions*

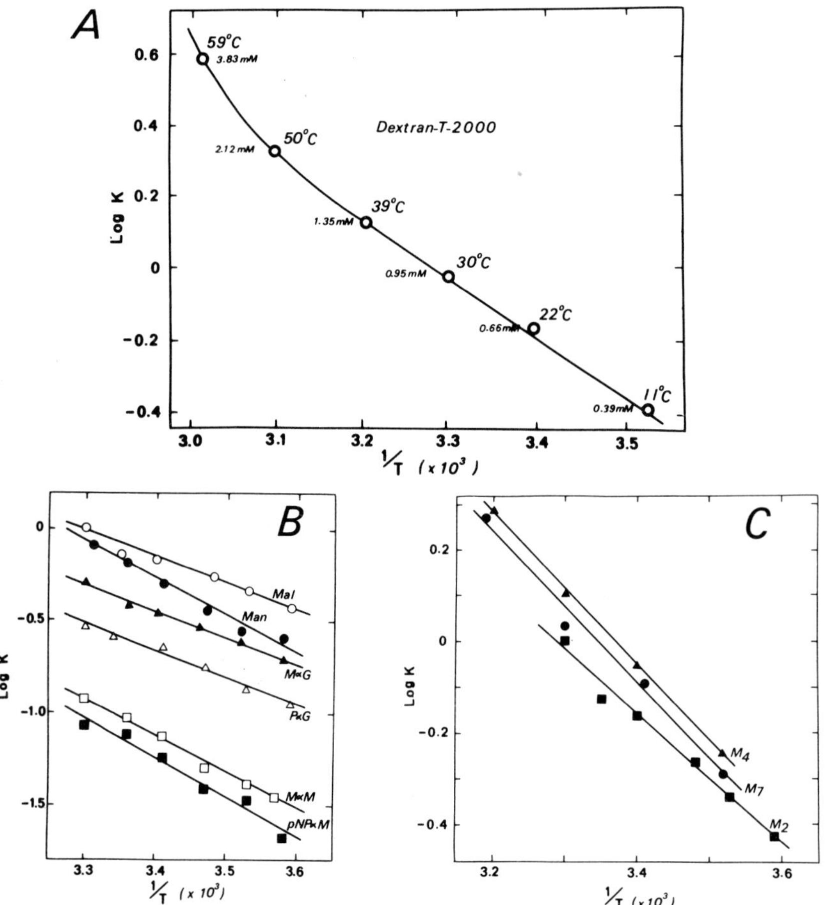


vely.  Thus, Kd value increases about 10 fold.  In other
words, the affinity of Con A to dextran is decreased about
to one tenth when the temperature increases from 11 to 59°C.
The plotted line is straight in the temperature range between 10 and 50°C. From its slope, enthalpy change, $\Delta H°$, is

calculated.  Fig. 3B shows van't Hoff plots for small molecular sugars such as maltose, mannose, $\alpha$-D-glucopyranosides, and $\alpha$-D-mannopyranosides.  The plots give fairly good straight lines.  Fig. 3C shows van't Hoff plots for maltose oligosaccharides.  In these cases, the plots also give fairly good straight lines.  From the slopes of the plots, $\Delta H^\circ$ are calculated.  The standard free energy change, $\Delta G^\circ$, and entropy change, $\Delta S^\circ$, at 20°C were calculated from Kd values at 20°C and $\Delta H^\circ$.

In Table 1, dissociation constants and thermodynamic parameters for Con A - carbohydrate interactions are summarized.

*Table 1.*

Thermodynamic Data for Con A - Carbohydrate Interaction

| Carbohydrates | Kd or Ki at 20°C (mM) | $-\Delta H^\circ$ (kcal/mol) | $-\Delta G^\circ$ (kcal/mol) | $-\Delta S^\circ$ (EU) |
|---|---|---|---|---|
| Dextran T-2000 | 0.55 | 6.2 | 4.3 | 6.3 |
| Glycogen, s.f. | 0.81 | 11.0 | 4.1 | 23.4 |
| Maltose | 0.69 | 6.1 | 4.2 | 6.4 |
|  | 0.75*[1] | – | – | – |
| Maltotriose | 0.77 | 6.8 | 4.2 | 8.9 |
| Maltotetraose | 0.89 | 7.5 | 4.1 | 11.4 |
| Maltopentaose | 0.94 | 7.2 | 4.1 | 10.8 |
| Maltoheptaose | 0.83 | 7.5 | 4.1 | 11.3 |
| Me-$\alpha$-D-Glc | 0.32 | 6.2 | 4.7 | 5.2 |
|  | 0.202*[2] | – | – | – |
| Ph-$\alpha$-D-Glc | 0.23 | 6.7 | 4.9 | 6.3 |
| Mannose | 0.34 | 8.4 | 4.7 | 12.3 |
| Me-$\alpha$-D-Man | 0.074 | 9.4 | 5.5 | 13.4 |
|  | 0.156*[3] | 9.07*[3] | 5.18*[3] | 13.37*[3] |
| $p$-NP-$\alpha$-D-Man | 0.057 | 8.9 | 5.7 | 10.9 |
|  | 0.067*[4] | – | – | – |

*1 pH 7.0 at 27°C.  Bessler et al. (1974) (6).
*2 pH 6.2 at 2°C, So and Goldstein (1968) (7).
*3 pH 5.5 at 25.3°C, Landschoot et al. (1980) (8).
*4 pH 5.4 at 25°C. Loontiens et al. (1973) (9).

Kd and $\Delta G^\circ$ values for maltose oligosaccharides and for glycogen were nearly the same regardless of the number of glucosyl residues.  On the other hand, $\Delta H°$ values for maltose oligosaccharides negatively increased from -6.1 to - 7.5 kcal per mol, when the number of glucosyl residues increased from 2(maltose) to 7(maltoheptaose). For glycogen it amounted to -11.0 kcal per mol.

Entropy changes for all Con A - carbohydrate interactions are negative.  For the maltose oligosaccharides $-\Delta S°$ values are also increased, when the numbers of glucosyl residues are increased.  Thus, both parameters, $\Delta H°$ and $\Delta S°$, compensate each other, making $\Delta G°$ and Kd values remain constant.

In regard to the glycosides, thermodynamic parameters of methyl-$\alpha$-D-mannopyranoside coincided fairly well with those obtained by Landschoot et al.(8) by use of substitution titration of fluorescence quenching by methyl-umberyfellyl-$\alpha$-D-mannopyranoside.  $\Delta G°$ values for methyl-and phenyl glucopyranosides are larger than those for maltose oligosaccharides, whereas their $\Delta H°$ and $\Delta S°$ values are pratically equal to that for maltose.  The same results are observed with mannose glycosides.  Introduction of these hydrophobic aglycons to glycosides decreased Kd values, while it seemed to have no effect on $\Delta H°$ values.

Entropy change seems to correspond to conformational change of the protein.  When carbohydrate combines with Con A, its conformation may change into a  more ordered form, making the $\Delta S°$ value negative.  On the other hand, in hydrophobic interactions, the more ordered form of water on the surface of the protein becomes less ordered.  Thus, entropy changes should be positive.  In phenyl-$\alpha$-D-glucopyranoside and *p*-nitro-phenyl-$\alpha$-D-mannopyranoside, $-\Delta H°$ are not so different from maltoriose, nevertheless their affinity is much stronger than the latter, because of the smaller change of entrophy values.  A similar tendency is seen with methyl and *p*-nitro-phenyl mannosides.

References


1.  Takeo, K. and Nakamura, S.: Arch. Biochem. Biophys.
    17 (1972).

2.  Takeo, K. and Nakamura, S.: in O. Hoffmann-Ostenhof, M.
    Breitenbach, F. Koller, D. Kraft and O. Scheiner (Edi-
    tors), Profeedings International Symposium on Affinity
    Chromatography (Vienna, September 20-24, 1977), Perga-
    mon Press, Oxford, New York, 1978. p. 67-70.

3.  Horejsi, H., Ticha, M. and Kocourek, J.: Biochim. Bio-
    phys. Acta. 499, 290-300 (1977).

4.  Takeo, K., Fujimoto, M., Suzuno, R., and Kuwahara, A.:
    The Physico-Chemical Biol. 22, 139-144 (1978).

5.  Reisfeld, R. A., Lewis, U. J., and Williams, D. E.:
    Nature 201, 281 (1962).

6.  Bessler, W., Shafer, J. A., and Goldstein, I. J.:
    J. Biol. Chem. 249, 2819-2822(1974).

7.  So, L. L., and Goldstein, I. J.: Biochim. Biophys. Acta
    165, 398-404 (1968).

8.  Van Landschoot, A., Loontiens, F. G., and De Bruyne,
    C. K.: Eur. J. Biochem. 103, 307-312 (1980).

9.  Loontiens, F. G., Van Wauwe, J. P., De Gussem, R., and
    De Bruyne, C. K.: Carbohyd. Res. 30, 51-62 (1973).

# HIGH RESOLUTION ELECTROPHORESIS OF PROTEINS IN SDS POLYACRYLAMIDE GELS


Dwight Anderson and Charlene Peterson
Departments of Microbiology and Dentistry
University of Minnesota
Minneapolis, Minnesota 55455


## Introduction

Polyacrylamide gel electrophoresis (PAGE) has become a powerful
routine tool for identification, quantification and isolation
of specific proteins in complex mixtures.  To identify the gene
products of the <u>Bacillus</u> <u>subtilis</u> bacteriophage Ø29 and to characterize their functions in studies of morphogenesis, we have
attempted to optimize the use of SDS-PAGE to separate viral and
host proteins in unfractionated cell extracts.  The procedure
described in this paper is based upon the method of Laemmli (1)
and includes conditions of gel composition, polymerization, and
electrophoresis that in concert provide the capability to
resolve more than 150 proteins in one dimension.

## Materials and Methods

<u>Chemicals</u>.  Electrophoresis grade acrylamide (3X recrystallized),
N,N'-methylenebisacrylamide, N,N,N',N'-tetramethylethylenedia-
mine (TEMED), and glycerol were obtained from Polysciences, Inc.
Tris (hydroxymethyl)aminomethane and glycine were obtained from
Fisher Scientific.  Sodium lauryl sulfate (SDS) was from Gallard
Schlesinger (BDH).  Ammonium persulfate was from Mallinckrodt.
Bromophenol blue and 2-mercaptoethanol were from Eastman Organic
Chemicals, and Coomassie brilliant blue was from Sigma Chemical
Co.

42

<u>Gel Composition</u>. The separating and stacking gels had the com-
position described by Laemmli (1) except for the incorporation
of glycerol into the separating gel (Susan Hawkes, personal
communication), the use of 0.03% by volume of ammonium persulfate in the polymerization of both gels, and the use of 0.1% by
volume of tetramethylethylenediamine in polymerization of the
stacking gel.  The separating gels were linear polyacrylamide
gradients (2); glycerol was incorporated at 30% by weight into
the Laemmli 30% acrylamide stock solution, so that gradients of
polyacrylamide contained parallel gradients of glycerol.

<u>Forming the Gel</u>.  A Model SE 520 slab gel unit (Hoefer Scien-
tific, San Francisco, CA) was used to form 14 cm X 28 cm gels
having a thickness of 0.75 mm or 0.37 mm.  The glass plates
were washed with Ivory Liquid (Procter & Gamble), rinsed with
distilled $H_2O$, immersed in 0.1% Photo-Flo 200 (Eastman Kodak
Co.) for 30 sec, air dried, and assembled on the gel device.
The assembled plates were cooled to $10^{\circ}C$ with a Lauda K2-R
refrigerated circulating water bath.

For linear gradient gels, the two acrylamide solutions were
mixed separately, deaerated on ice, and pipetted into a Buchler
density gradient maker, which was fixed in a lucite container
filled with ice.  A Gilson Model HP4 peristaltic pump delivered
the acrylamide at about 1 ml/min to the gel device through
tubing (I.D. of 0.76 mm) attached to a 22 gauge needle in the
support sealing bar of the gel device.  The 28 cm gel was overlaid  with water, and 1 hr and 20 min later, following polymerization at $10^{\circ}C$, the water was replaced with lower Tris buffer
(pH 8.8).  After an additional 10 hrs and 40 min, the buffer was
removed and the space above the gel rinsed quickly with water
and blotted dry.  The stacking gel mixture was added without
deaeration to give a gel about 1 cm high, and polymerization
was at $10^{\circ}C$ for 1 hr.

<u>Sample Preparation and Loading</u>.  Samples in 0.06 M Tris-HCl
(pH 6.8) were stored at 4$^O$C.  Just prior to electrophoresis,
two volumes of sample were combined with one volume of 3X
Laemmli sample buffer (1X with respect to Tris-HCl), and the
mixture was immersed in boiling water for 2 1/2 min.

The well-forming comb was removed from the stacking gel in the
presence of upper Tris buffer (pH 6.8).  The buffer was removed
from the wells by use of bibulous paper, and electrode buffer
(0.05 M Tris, pH 8.3 - 0.38 M glycine - 0.1% SDS) was added to
the upper chamber of the device to fill the sample wells.  For
the 0.75 mm gels, the samples were loaded through the electrode
buffer with a 1 ml syringe and a 22 gauge needle connected to
a length of intramedic polyethylene tubing (I.D. of 0.58 mm
and O.D. of 0.96  mm) that was stepped down at the tip to a
smaller piece of polyethylene tubing (I.D. of 0.28 mm and O.D.
of 0.61 mm).  The 0.37 mm gels were loaded using a mechanical
pipettor with a Pasteur pipette that was drawn out to a fine
diameter.  For 0.75 mm and 0.37 mm gels, sample volumes applied
to the gels were approximately 15 ul and 10 ul, respectively,
and samples generally contained less than 10 ug of protein.

<u>Electrophoresis</u>.  Proteins loaded onto the 0.75 mm gels were
stacked at 50-60 V, constant voltage, for about 3 hrs and separated at 10 W, constant power, for 9 hrs at 10$^O$C.  An initial
separating current of 25 mA was a requisite parameter for reproducible and optimal separation of proteins, and 13 mA/700-750 V
typified the ending current and voltage, respectively.  Proteins
loaded onto the 0.37 mm gels were stacked at a constant voltage
of 90 V for about 3 hrs and separated at a constant power of
6-10 W for 7-9 hrs.

<u>Gel Processing</u>.  The gels were fixed and stained for at least
60 min in 45% methanol - 10% acetic acid containing 0.25%
Coomassie blue and destained in 30% methanol - 10% acetic acid
before drying onto porous cellophane in a Hoefer Model SE 540

44

gel dryer.  Unstained gels could be dried for autoradiography
after fixing in 45% methanol - 10% acetic acid.  Autoradiographs of dried gels containing radiolabeled proteins were
prepared by use of Kodak XAR-5 X-Ray film.


## Results and Discussion

Autoradiographs of representative separations of $^{14}$C- or $^{35}$S-
labeled proteins from Bacillus subtilis or bacteriophage Ø29-
infected B. subtilis in 0.75 mm or 0.37 mm gels are illustrated
in Figs. 1-4.  To produce the autoradiographs of Ø29-specific
proteins, we have infected UV-irradiated B. subtilis SpoAl2 in
the presence of a mixture of $^{14}$C-labeled amino acids or $^{35}$S-
methionine.  The Ø29 gene products range in molecular weight
from approximately 88,000 to less than 5,000 (3,4).  The method
is applicable to unfractionated mixtures of proteins, and the
gels are capable of resolving more than 150 proteins.  With
B. subtilis, only lysozyme and DNaseI treatments precede heating in sample buffer for electrophoresis.

Fig. 2 illustrates the utility of the method in visualizing
subtle changes in the positions of Ø29 structural proteins that
occur as a result of point mutations, in this case in the gene
for the head fiber protein (gp8.5).  In Fig. 2a the fiber protein is shown at its normal position, in Fig. 2b the protein is
missing, and in Figs. 2c and 2e the proteins have a slightly
increased mobility.

Polyacrylamide concentrations in gradients have been varied to
focus on and improve the resolution of proteins of specific
molecular weight range.  As an example, an autoradiograph of
Ø29 structural proteins on a 16% to 20% linear gradient of
polyacrylamide is shown in Fig. 3.  In this case, 4 proteins
in the 35,000 to 36,000 molecular weight range are resolved,
whereas  only two bands in this area are observed in the 12%

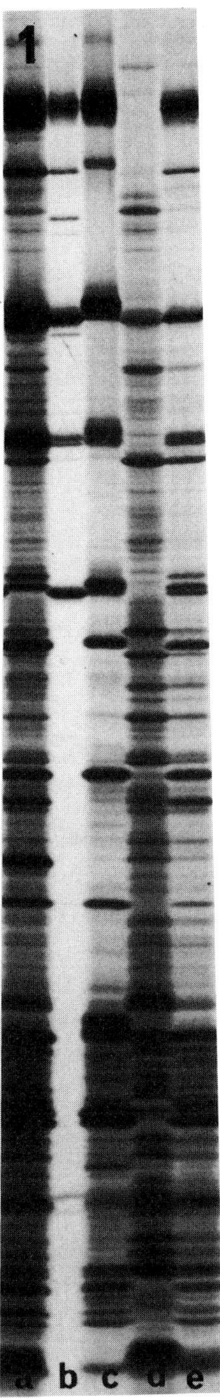

Fig. 1. Autoradiographs of $^{14}$C-labeled proteins produced in sus 8.5(900) and $\emptyset 29^+$ infections of UV-irradiated B. subtilis SpoA12. Cells were infected at an input multiplicity of 50 and labeled with a mixture of $^{14}$C-amino acids from 10 to 45 min after infection. Labeling was terminated as described (3), and the cell or supernatant fractions were analyzed by SDS-gel electrophoresis (12% to 19% linear polyacrylamide gradient, 0.75 mm thick) followed by autoradiography. Profiles are of lysates of cell pellets except for (c), which represents a supernatant. Profile (a) is of sus 8.5(900) infection. Profile (b) is of purified $\emptyset 29^+$ virions. Profiles (c) and (e) are of $\emptyset 29^+$ infection. Profile (d) represents an uninfected control culture. Viral-specific proteins can be identified by comparison of infected and uninfected lysates; $\emptyset 29$ gene products and their molecular weights have been described (3,4,6,7). The autoradiographs have been deliberately overexposed to reveal a larger number of protein bands. The tops and bottoms of the autoradiographs are not shown, but at least 120 bands can be observed.

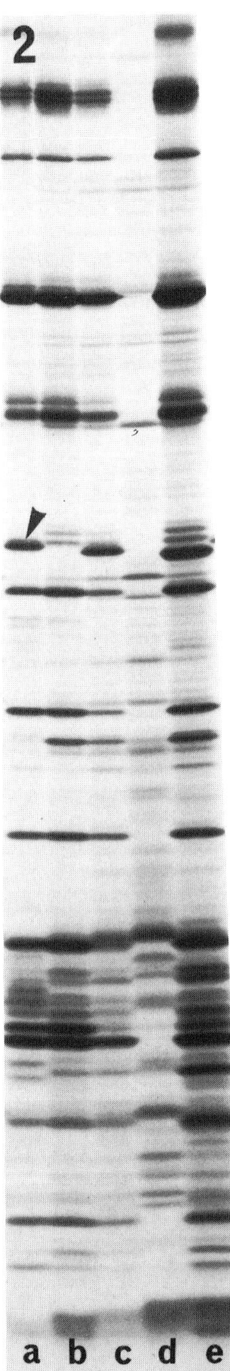


Fig. 2. Autoradiographs of [14]C-labeled proteins produced in Ø29 mutant infections of UV-irradiated B. subtilis SpoAl2. Cells were infected, labeled, processed and analyzed by electrophoresis as described in the Fig. 1 legend. Profiles are of lysates of cell pellets except for (a), which represents a supernatant. Profile (a) is of sus 2(628) infection. The arrow designates the normal position of the Ø29 head fiber protein, gp8.5. Profiles (b), (c), and (e) are of infections with "900 series" phages having point mutations in gene 8.5(5). Profile (d) represents an uninfected control culture.



Fig. 3. Autoradiographs of [14]C-labeled proteins produced in sus mutant infections of UV-irradiated B. subtilis SpoAl2. Cells were infected, labeled, processed and analyzed by electrophoresis as described in the Fig. 1 legend except that separation was achieved on a 16% to 20% linear polyacrylamide gradient. Profiles are of TCA precipitates of cell supernatants (3). Profiles (a), (b), (c), and (d) are of infections with the sus mutants 6(626), 6(626), 6(748), and 6(727), respectively. Profile (e) represents an uninfected control culture, and profile (f) is of purified Ø29[+] virions. The 16% to 20% gradient resolves 4 proteins in the region defined by the arrows that are not well resolved in the gels shown in Figs. 1 and 2. These are, in order of decreasing molecular weight, the Ø29 proteins gp10 and gp16, a host protein, and Ø29 gp11 (7).

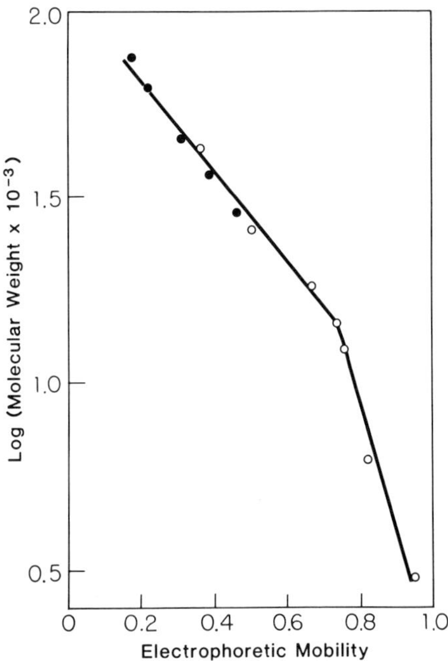

Fig. 4. Autoradiographs of $^{35}$S-labeled proteins produced in uninfected B. subtilis SpoA12. Cells were labeled with $^{35}$S-methionine (7) and processed as described (3), and proteins in the cell pellet were analyzed by electrophoresis on a 12% to 19% 0.37 mm linear polyacrylamide gradient. The profiles represent two concentrations of the same sample. The tops and bottoms of the autoradiographs are not shown.



Fig. 5. Relationship between electrophoretic mobility and molecular weight for reference and Ø29 proteins in a 12% to 19% 0.37 mm linear polyacrylamide gradient. Points (o) are given for: ovalbumin (43,000), $\alpha$-chymotrypsinogen (25,700), B-lactoglobulin (18,400), lysozyme (14,300), cytochrome c (12,300), bovine trypsin inhibitor (6,200), and insulin (3,000). Molecular weights of Ø29 proteins (●), determined previously (3), include gp12 (75,300), gp9 (62,300), gp8 (45,000), gp11 (35,200), and gp8.5 (28,500).

48

to 19% gradients (see Fig. 1b).

A plot of electrophoretic mobility versus molecular weight for
7 reference proteins (Bethesda Research Laboratories, Inc.)
spanning the 3,000 to 43,000 molecular weight range and for ∅29
structural proteins of known molecular weight (3) is shown in
Fig. 5 for a 0.37 mm gel.  The curve is biphasic and breaks at
a molecular weight of about 14,500.


Acknowledgements

The use of glycerol and 10°C polymerization/separation was
described in a procedure provided by Susan Hawkes, and the composition of methanol-acetic acid mixtures for staining/destain-
ing were provided by Robert Trimble.  We also acknowledge
profitable discussions with Jeffrey Reinhart.


References

1.  Laemmli, U.K.:  Nature (London) 227, 680-685 (1970).

2.  Margolis, J., Kenrick, K.G.:  Anal. Biochem. 25, 347-362
    (1968).

3.  Hawley, L.A., Reilly, B.E., Hagen, E.W., Anderson, D.L.:
    J. Virol. 12, 1149-1159 (1973).

4.  Anderson, D.L., Reilly, B.E.:  Microbiology 1976, American
    Society of Microbiology, Washington D.C., 254-274.

5.  Reilly, B.E., Nelson, R.A., Anderson, D.L.:  J. Virol. 24,
    363-377 (1977).

6.  Anderson, D.L., Reilly, B.E.:  J. Virol. 13, 211-221 (1974).

7.  Reilly, B.E., Tosi, M.E., Anderson, D.L.:  J. Virol. 16,
    1010-1016 (1975).

8.  Nelson, R.A., Reilly, B.E., Anderson, D.L.:  J. Virol. 19,
    518-532 (1976).

CONVENIENT PROCEDURES FOR SDS AND CONVENTIONAL DISC ELECTROPHORESIS

Jerry L. Neff, Nehemias Muniz, Joseph L. Colbourn, Aurora F. de Castro
Miles Laboratories, Inc., Research Products Division
Elkhart, IN 46515

Introduction

The biochemical isolation and separation of molecules uses several methods.
Gel filtration, adsorption, affinity and ion-exchange chromatography are
among them.  Electrophoresis is another fractionation methodology which
has been greatly developed.

Electrophoretic separations have been done on a variety of media like paper,
agar and polyacrylamide gels.  Paper electrophoresis has been used for separating and characterizing molecules, including metabolites of drugs.
Polyacrylamide electrophoresis has been widely used for separation of
peptides, proteins and DNA sequences.  The theory of using discontinuous
buffers for electrophoretic separations (disc electrophoresis) was first
introduced by Ornstein and Davis (1,2), was further defined physico-
chemically by Jovin (3) and was objectively defined as a fractionating
route by Chrambach (4).  The present presentation uses polyacrylamide disc
electrophoresis in a convenient way to achieve macromolecular separations.
This is done with preformulated reagents included in the Canalco® SAGE[TM]
Kit.[*]  Both conventional and sodium dodecyl sulfate (SDS) electrophoresis
can be performed.

The electrophoresis of SDS proteins in polyacrylamide gels has grown to
be the most widely used method for the determination of molecular weight
of proteins.  This method was first introduced by Shapiro, et al (5)
and further developed by others (6, 7, 8, 10).

[*]Canalco® SAGE[TM] Kit is a product of the Research Products Division,
Miles Laboratories, Inc., Elkhart, IN 46515

Materials and Methods

The materials and methods are given in detail in the instructions to the Canalco® SAGE[TM] Kit for SDS and conventional disc electrophoresis.  The following reagents are used:

1.  Gel Buffer Solution - 1.5 M tris (hydroxymethyl) aminomethane (tris) chloride buffer containing 0.026 M TEMED (N,N,N',N'tetramethylethylene-diamine), pH 8.9.

2.  Separating Gel Monomer Solution - Aqueous solution of acrylamide (mono-mer) and N,N'methylenebisacrylamide (bis) (co-monomer) 40% T* and 2.75% C**.

3.  Catalyst - 140 mg of ammonium persulfate crystals in 100 ml of distilled water.

4.  Stacking Gel Monomer  Solution - Aqueous solution of acrylamide and bis, 25% T and 20% C.

5.  Sample Buffer Tracking Dye Solution - 0.064 M tris chloride buffer containing 0.2 mM thymol blue.

6.  Sodium Dodecyl Sulfate (SDS) Solution - Aqueous solution of 4% SDS.

7.  Glycerol

8.  Electrolyte Buffer - Dry blend of 17.2% tris and 82.8% glycine. Dissolve 17.4 gm in 2000 ml distilled water.

9.  Stain Concentrate - Aqueous solution of 0.4% coomassie blue R250, 32% methanol, and 28% acetic acid.  A working solution is made by mixing one part of the stain concentrate with three parts of distilled water.

10.  Dithiothreitol (DTT) Solution - 0.364 M solution of DTT.

11.  Fixative Solution - 12% trichloroacetic acid (TCA).

12.  Destaining Solution - Mixture of acetic acid, methanol and water(7:5 85).

*%T - The amount of monomer plus co-monomer in grams per deciliter of final gel solution.

**%C - The amount of co-monomer as a percent of T.

The separating gels were prepared according to Table 1.  This table
indicates the units of reagents Solution 1 (buffer), Solution 2 (monomers), Solution 3 (catalyst) and water required to prepare gels from
3% to 25% T.

Table 1.

## Gel Preparation Table

| Gel %T | Parts of Solution (total 8) | | | |
| --- | --- | --- | --- | --- |
| | Buffer | Monomer | Catalyst | $H_2O$ |
| 3 | 2.0 | 0.6 | 2.7 | 2.7 |
| 4 | 2.0 | 0.8 | 2.6 | 2.6 |
| 5 | 2.0 | 1.0 | 2.5 | 2.5 |
| 6 | 2.0 | 1.2 | 2.4 | 2.4 |
| 7 | 2.0 | 1.4 | 2.3 | 2.3 |
| 8 | 2.0 | 1.6 | 2.2 | 2.2 |
| 9 | 2.0 | 1.8 | 2.1 | 2.1 |
| 10 | 2.0 | 2.0 | 2.0 | 2.0 |
| 11 | 2.0 | 2.2 | 1.9 | 1.9 |
| 12 | 2.0 | 2.4 | 1.8 | 1.8 |
| 13 | 2.0 | 2.6 | 1.7 | 1.7 |
| 14 | 2.0 | 2.8 | 1.6 | 1.6 |
| 15 | 2.0 | 3.0 | 1.5 | 1.5 |
| 16 | 2.0 | 3.2 | 1.4 | 1.4 |
| 17 | 2.0 | 3.4 | 1.3 | 1.3 |
| 18 | 2.0 | 3.6 | 1.2 | 1.2 |
| 19 | 2.0 | 3.8 | 1.1 | 1.1 |
| 20 | 2.0 | 4.0 | 1.0 | 1.0 |
| 21 * | 2.0 | 4.2 | 1.0 | 0.8 |
| 22 * | 2.0 | 4.4 | 1.0 | 0.6 |
| 23 * | 2.0 | 4.6 | 1.0 | 0.4 |
| 24 * | 2.0 | 4.8 | 1.0 | 0.2 |
| 25 * | 2.0 | 5.0 | 1.0 | 0.0 |

Thymol blue does not migrate at the leading
front at these concentrations.  Instead, add
1 µl of 1% phenol red per sample.

52

The stacking gels were prepared by combining Solutions 1, 3 and 4 in a
1:2:1 ratio. The sample preparation was done by combining the sample
with the tracking dye and glycerol, in the case of conventional electrophoresis, and in addition with sodium dodecyl sulfate (SDS) and a reducing
agent (dithiothreitol or 2-mercaptoethanol), in the case of SDS electrophoresis.

Electrophoresis was performed in the tris-glycine electrolyte buffer,
fix, stained with coomassie blue and destained with a mixture of acetic
acid, methanol and water.


Principles

Discontinuous (disc) electrophoresis is a very effective method for the
separation of charged components. As the name indicates, disc electrophoresis employs discontinuous multiphasic buffers varying in composition
and physical properties:
1. Disc electrophoresis utilizes polyacrylamide gel as support medium.
   Polyacrylamide is optically clear, thermostable, non-ionic, chemically
   inert and easy to prepare and handle.
2. Separating gel (lower gel) is where the actual separation of molecular
   species takes place. The separating buffered gel has a higher concentration of acrylamide monomers and, consequently, a smaller pore
   size. The restriction created by the small pores of the gel is what
   endows polyacrylamide gel electrophoresis (PAGE) with its high resolving power.
3. Stacking gel (upper gel) or second gel is where the sample concentrates itself into tightly packed starting zones. The concentration
   effect of the stacking gel is what provides disc electrophoresis
   with the capability of handling large sample volumes. The narrow
   starting zone obtained by stacking further improves the resolution
   obtainable by PAGE.
4. The gels are placed in contact with an electrolyte buffer with a
   certain pH, ionic species and ionic strength.
5. The sample, with or without SDS treatment or reducing agent, is

over the stacking gel and below the electrolyte buffer.

6. Upon application of an electrical potential, the sample goes through the stacking, unstacking and resolving stages.

7. After electrophoresis, the gels are stained for visualization.

8. A tracking dye is used to follow the progress of the electrophoresis. It also serves as a reference point for the measurement of the relative mobility of the bands ($R_f$).

9. Resolution between proteins should exploit their maximal charge differences. Conventional separation based on net charge between proteins is frequently maximal at pH's close to their isoelectric points (pI's).

10. In resolution of proteins in sodium dodecyl sulfate (SDS) containing buffers, pH of fractionation is not very important. Here, separation based on net charge of individual proteins becomes negligible since SDS imparts a large negative charge to all proteins. Proteins are then mainly separated by size differences.

It is the systematic use of buffer and pH discontinuities that characterize the technique of disc electrophoresis. This technique is needed when the species to be separated are available at a relatively low concentration, as is usually the case with components of biological systems. If concentration is not needed, a continuous zone electrophoresis can be carried out in a single buffer system (identical) in both buffer reservoirs and in the gel) at any pH, with characterization of bands in terms of absolute mobility.

Results

Figure 1 shows the separation of proteins when conventional electrophoresis was performed in human serum, human spinal fluid and human saliva with this system.

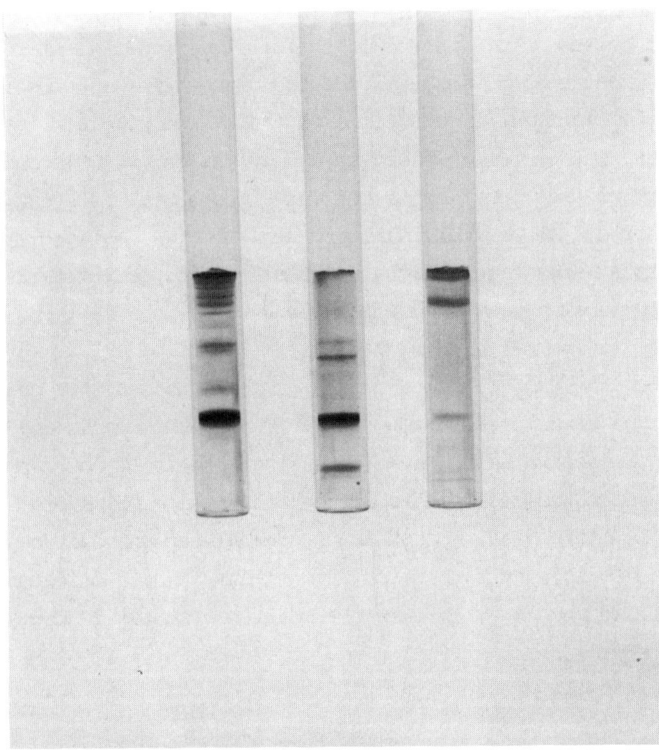

Fig. 1. Conventional disc electrophoresis of human serum proteins (1/4 diluted), human spinal fluid proteins and proteins in human saliva. Separations obtained using 7% gels.

In addition, conventional electrophoresis was also performed at different purification steps of an enzyme such as polynucleotide phosphorylase (PNPase). Figure 2 shows PNPase, a polynucleotide synthesizing enzyme, after DEAE cellulose treatment and after Sephadex G-100 column treatment. This allows an estimation of the degree of purity of PNPase during the purification process.

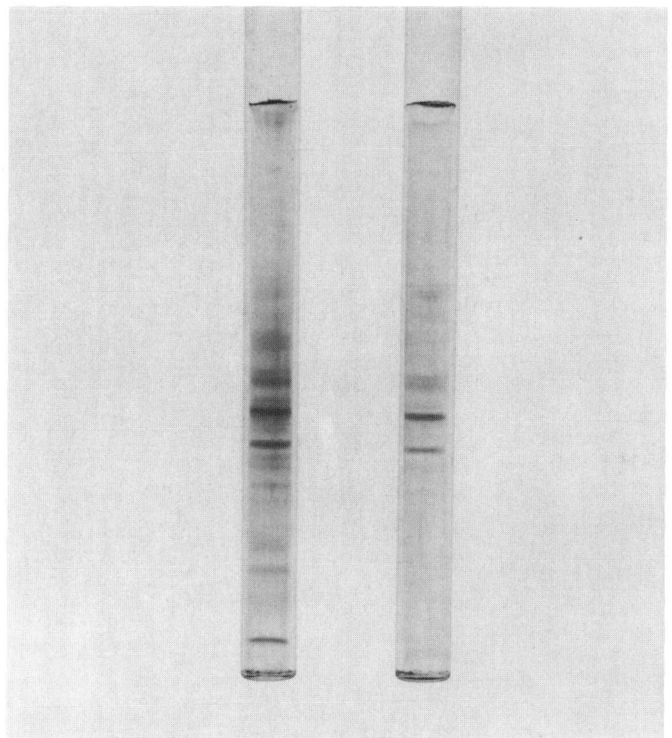

Fig. 2. Two separate stages of purification of polynucleotide phosphorylase (PNPase) shown on 5% T gels at 2.75% C.
Left. PNPase after DEAE cellulose column treatment.
Right. PNPase after Sephadex G-100 column treatment.

When samples are treated with SDS in the manner detailed in the SAGE$^{TM}$ Kit, there is a breakdown of polymeric molecular forms.  As shown in Figure 3, when RNA polymerase is treated with SDS, the beta, alpha and sigma sub-units appear.

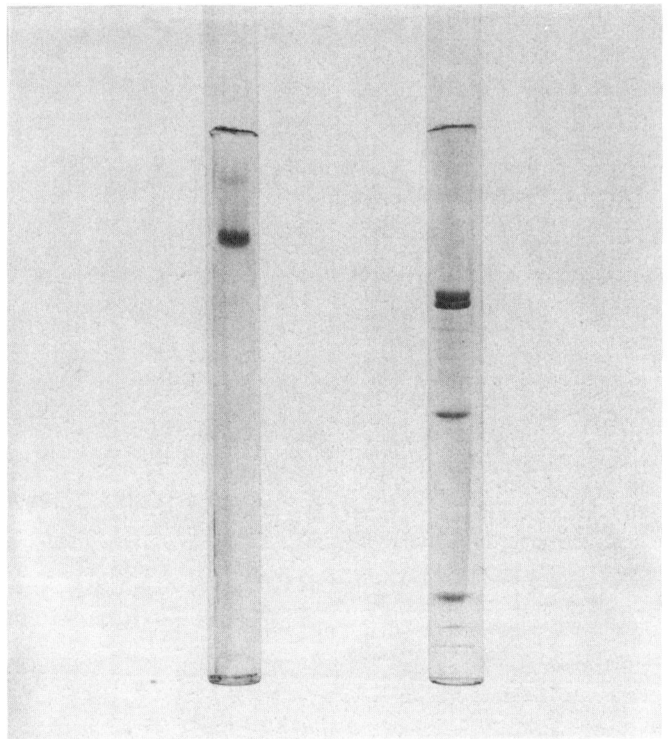

Fig. 3.  Separation of RNA polymerase by discontinuous PAGE and discontinuous SDS-PAGE methods shown in 5% T gels at 2.75% C.  Left. RNA polymerase without SDS.  Right.  RNA polymerase with SDS.

SDS disc electrophoresis is also used for molecular weight determinations. Figure 4 shows twelve different proteins used as molecular weight markers, separated in different percentage gels.  The $R_f$ values of these proteins are shown in Table 2.

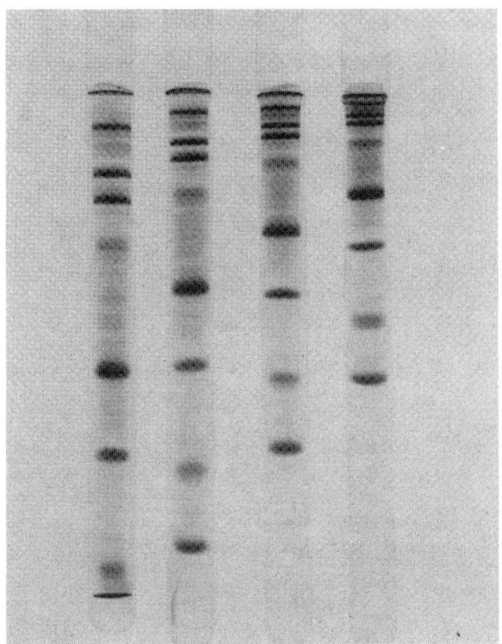

Fig. 4.  Protein markers after SDS-PAGE on 9, 11, 13 and 15% T gels at 2.75% C, (left to right). Tracking dye was run the same distance in all gels. The markers shown (top to bottom) are: 1) myosin 194,000; 2) β-galactosidase 116,000 (9); 3) phosphorylase B 94,000; 4) bovine albumin 68,000; 5) ovalbumin 43,000; 6) carbonic anhydrase 30,000; 7) soybean trypsin inhibitor 21,000; 8) lysozyme 14,500.

Table 2.

## $R_f$ VALUES OF TWELVE DIFFERENT PROTEINS ELECTROPHORESED IN SDS GELS 5%T TO 14%T AND 2.75%C

| Protein | Molecular Weight (Log MW) Daltons | + Mean $R_f$ at Various % T | | | | | | | | | |
|---|---|---|---|---|---|---|---|---|---|---|---|
| | | 5%T | 6%T | 7%T | 8%T | 9%T | 10%T | 11%T | 12%T | 13%T | 14%T |
| Myosin | 194.000 (5.238) | 0.235 | 0.170 | 0.112 | 0.082 | 0.068 | 0.055 | 0.041 | 0.032 | 0.028 | 0.028 |
| RNA Polymerase (β+β subunits) | 160,000 (5.204) | 0.348 | 0.263 | 0.177 | 0.136 | 0.113 | 0.032 | 0.062 | 0.054 | 0.038 | 0.034 |
| β-Galactosidase | 116.000 (5.064) | 0.474 | 0.358 | 0.254 | 0.183 | 0.151 | 0.120 | 0.092 | 0.076 | 0.058 | 0.050 |
| Phophorylase B | 94,000 (4.973) | 0.594 | 0.453 | 0.324 | 0.242 | 0.200 | 0.160 | 0.124 | 0.100 | 0.080 | 0.068 |
| RNA Polymerase (ϭ subunit) | 95,000 (4.978) | 0.590 | 0.458 | 0.321 | 0.248 | 0.200 | 0.158 | 0.118 | 0.100 | 0.076 | 0.060 |
| Bovine Serum Albumin | 68,000 (4.832) | 0.728 | 0.586 | 0.440 | 0.336 | 0.288 | 0.230 | 0.183 | 0.157 | 0.122 | 0.105 |
| Ovalbumin | 43,000 (4.633) | ★ | 0.901 | 0.730 | 0.606 | 0.514 | 0.414 | 0.352 | 0.296 | 0.248 | 0.210 |
| RNA Polymerase (α subunit) | 38.400 (4.584) | ★ | 0.913 | 0.742 | 0.625 | 0.518 | 0.416 | 0.357 | 0.310 | 0.253 | 0.218 |
| Carbonic Anhydrase | 30,000 (4.477) | ★ | ★ | 0.898 | 0.758 | 0.665 | 0.564 | 0.478 | 0.411 | 0.346 | 0.302 |
| Trypsinogen | 24,500 (4.389) | ★ | ★ | ★ | 0.908 | 0.806 | 0.699 | 0.602 | 0.532 | 0.460 | 0.402 |
| β-Lactoglobin | 17.500 (4.243) | ★ | ★ | ★ | ★ | 0.952 | 0.846 | 0.744 | 0.652 | 0.578 | 0.514 |
| Lysozyme | 14,500 (4.161) | ★ | ★ | ★ | ★ | ★ | 0.929 | 0.818 | 0.733 | 0.651 | 0.580 |

★ did not unstack
+ Mean of at least three independent determinations

Ferguson Plots are employed to detect systematic error of the charge density of proteins (10). Figure 5 shows the Ferguson Plots of twelve proteins used as molecular weight markers. A common point of intersection at or near the y-axis indicates the absence of systematic error.

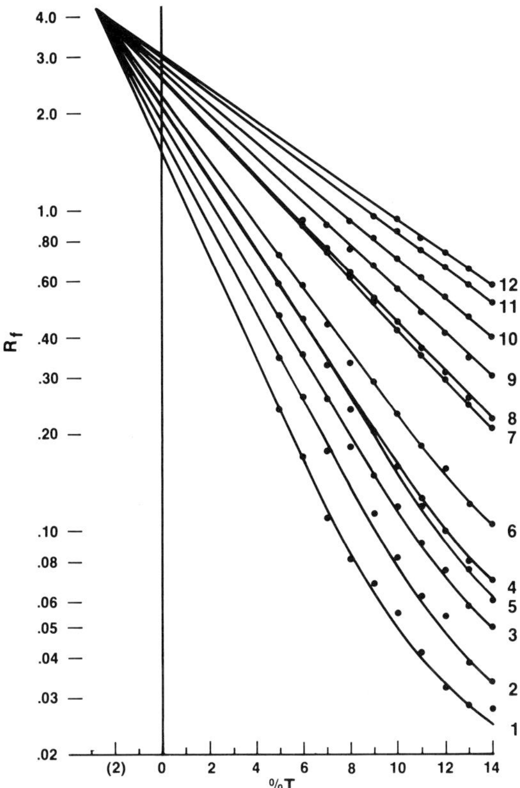



Fig. 5. Ferguson Plots of the following twelve proteins: 1) myosin, 2) RNA polymerase ($\beta_1\ \beta_2$), 3) β-galactosidase, 4) phosphorylase B, 5) RNA polymerase (α subunit), 6) bovine serum albumin, 7) ovalbumin, 8) RNA polymerase (σ subunit), 9) carbonic anhydrase, 10) trysinogen, 11) β-lactoglobin, 12) lysozyme.

Figures 6 and 7 show two methods of molecular weight calculation (11) from the $R_f$ values.  The twelve proteins shown in these figures are the ones mentioned in Table 2.

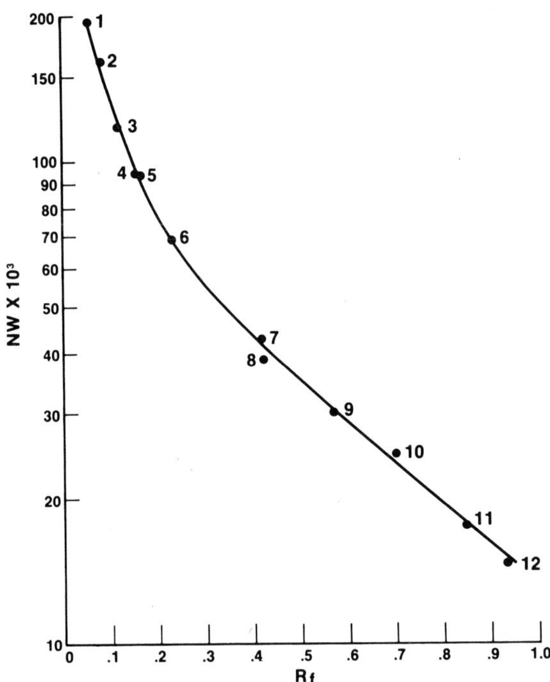


Fig. 6.    (Method 1) $R_f$ vs. log M.W. (linear or parabolic
$R_f = a + b$ log M.W.
or:
log M.W. $= a + bR_f + c(R_f)^2$

This method is most widely used for protein molecular weight estimation by SDS.  The fact that these plots are linear only over short range of M.W. is in agreement with theory.

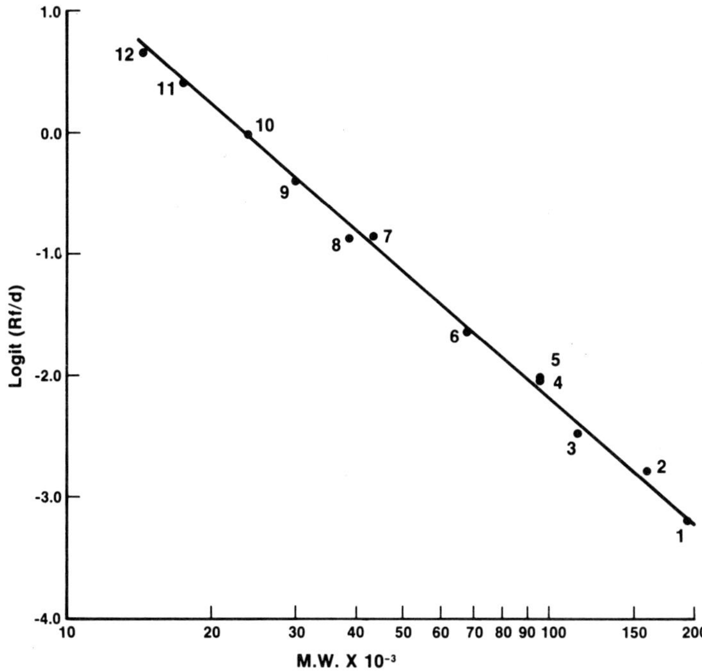


Fig. 7.    (Method 2) $R_f$ vs. log M.W. (sigmoidal)
logit $R_f/\alpha$= d + b log M.W.

This method provides optimal curve fitting for data from
a single gel concentration.  This is theoretically justified if the size of the pores of the gel follow a log-
normal or a "logistic" distribution.  The value of $\alpha$ for
a set of data at any particular %T must be determined by
computer methods.  This may be done on a desk top pro-
grammable calculator.

Discussion

Data from SDS polyacrylamide gel electrophoresis for the determination of molecular weights of proteins is most commonly analyzed by comparison with a calibration curve. This curve is constructed from data representing the mobility of a number of marker proteins of known molecular weights under like conditions. The most common plot for this curve is the log M.W. versus protein (subunit) mobility measured by $R_f$. The $R_f = \frac{d}{d_f}$ where d is the distance that each band has traveled from the top of the separating gel to the center of the band and $d_f$ is the distance that the tracking dye has traveled from the top of the separating gel.

The molecular weight of the unknown protein is determined by measuring its mobility $R_f$ and reading its molecular weight from the calibration curve. In order to accurately calculate the M.W. of an unknown protein, one should be aware of several considerations. When the charge density of a conjugated protein deviates from the common value exhibited by the proteins used as standards, a systematic error is introduced. In this case, the molecular weight of the unknown cannot be derived from $R_f$ values at a single gel concentration. One must measure M.W. at several gel concentrations (%T) and show that there is negligible systematic error.

To determine if there is a systematic error, one must construct Ferguson Plots (10) (log $R_f$ = log $Y_o$-$K_r$T) for the standards and unknown proteins. If a common point of intersection at or very near the y-axis of the Ferguson Plots is found, with all the lines emanating like spokes of a wheel, then one can conclude that there is negligible systematic error. This allows one to derive molecular size data from mobilities at a single gel concentration and use calculation methods as in Figures 6 and 7 (11). If this is not the case, the molecular weight must be calculated from the standard curve for random-coiled molecules, $K_r$ vs. M.W. (12).

In the disc electrophoretic format of the present system, the flexibility afforded by making it easy to construct gels of different percentages gives simplicity to the construction of Ferguson Plots. In addition, the pre-formulated reagent system of the Canalco® SAGE[TM] Kit gives the flexibility of obtaining tube or slab formats.

Extreme simplicity of operational instructions are available with the system. The system also offers new and unique tracking dyes which provide added accuracy in the measurement of the true front. The tracking dyes commonly used in SDS-PAGE, bromophenol blue and pyronin-Y, migrate with the true front only over a limited range of gel concentrations (13). The dye used with this system, thymol blue (patent applied for), has the characteristic of migrating with the true front at gel concentrations of up to 20% T. This fact allows for more accurate measurement of the $R_f$'s over a broader gel concentration range and more linear Ferguson Plots.

References


1.  Ornstein, L.: Ann. N.Y. Acad. Sci., <u>121</u>, 321 (1964).

2.  Davis, B.J.: Ann. N.Y. Acad. Sci., <u>121</u>, 404 (1964).

3.  Jovin, T.M., Dante, M.L. and Chrambach, A.: Multiphasic Buffer Systems Output, National Technical Information Service, Springfield, Va.22151, PB No. 196085-196091, 203016.

4.  Chrambach, A., Jovin, T.M., Svendsen, P.J. and Rodbard, D.: Methods of Protein Separation, Vol.2, Edited by N. Catsimpoolas (1976).

5.  Shapiro, A.L., Vinuela, E. and Jaizel, J.V. Jr.: Biochem. Biophys. Res. Comm., <u>28</u>, 815 (1967).

6.  Weber, K. and Osborn, M.: J. Biol. Chem., <u>244</u>, 4406 (1969).

7.  Maizel, J.V.: Nature (London), <u>227</u>, 680 (1970).

8.  Neville, D.M. Jr.: J. Biol. Chem., <u>246</u>, 6328 (1971).

9.  King, J. and Laemmli, U.K.: J. Mol. Biol., <u>62</u>, 465 (1971).

10. Ferguson, J.A.: Metabolism, <u>13</u>, 985 (1964).

11. Rodbard, D.: Methods of Protein Separation II, Chapters 3 & 4, Edited by Nicholas Catsimpoolas, Plenum, New York (1976).

12. Rodbard, D. and Chambach, A.: Anal. Biochem., <u>40</u>, 95 (1971).

13. Wyckoff, M., Rodbard, D. and Chrambach, A.: Anal. Biochem., <u>78</u>, 459 (1977).

THE DESIGN AND APPLICATIONS OF A UNIVERSAL GEL ELECTROPHORESIS
APPARATUS


Gregory Bambeck and Jeffrey Black
Hematology Institute, Department of Biological Sciences,
Kent State University, Kent, Ohio U.S.A.


Introduction

The great versatility of polyacrylamide gel electrophoresis
(PAGE) permits highly reproducible and resolved separations of
biological molecules with a precision not yet achieved with
methodologies using any other matrix or substratum (1, 3, 5, 6).
Polyacrylamide is the only medium ideal for nucleotide sequencing, disc, urea, SDS, isoelectric focusing (IEF) and enzyme
kinetic electrophoresis techniques (2, 3, 4, 5, 6). However
this large variety of PAGE techniques is not reflected in the
existing apparatuses upon which PAGE is performed. What is
needed is a simple standard system that is versatile enough to
accept most, if not all, of the more common PAGE techniques
that can be applied to the variety of demands certain to occur
soon in the clinical, industrial and governmental sectors (7, 8).

At present there are two general gypes of gel electrophoresis machines in common use:  the horizontal type and the
vertical type[1*].  Each of these systems has its advantages and
disadvantages.  Although horizontal systems yield superior
resolution with agarose, and are commonly used for IEF-PAGE and
immunoelectrophoresis, vertical machines have greater versatility
in that nucleotide sequencing PAGE, PAGE in gels saturated in
urea, 2D-PAGE, disc PAGE, SDS-PAGE, gradient PAGE as well as
PAGE in tubes and thick slabs can be accommodated.  Due to the

---

[1*]Literature can be obtained from a number of manufacturers of
gel electrophoresis equipment and supplies.

immersion of gels in buffers and the superior electrical
conductivity of vertical systems, the need for cooling coils,
pumps, humidity control and expensive high voltage power supplies
is obviated.  Thus, the simplest and most versatile system
should be modeled after a vertical device similar to that
described by O'Farrell (3).


Methods and Materials


The apparatus.  With the aforementioned considerations in mind,
we have designed a two chamber gel electrophoresis apparatus
which can be used to perform a large variety of PAGE techniques
(Figure 1).  This system has a unique clamping arrangement

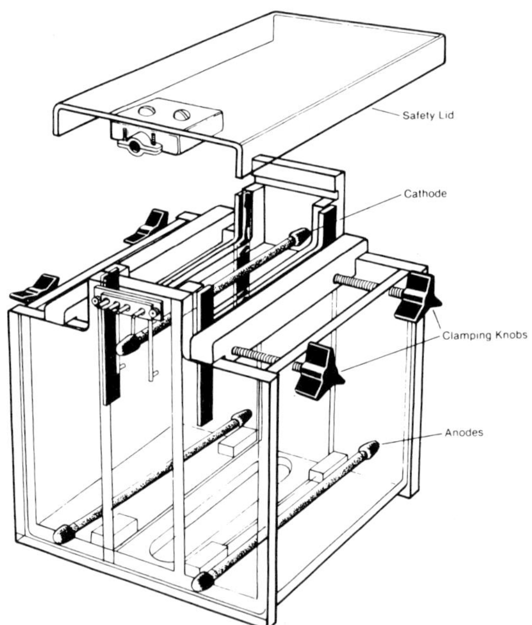


Figure 1.  Three dimensional drawing of the Isophore$^{TM}$ two
           chamber vertical gel electrophoresis apparatus.

capable of accepting a set of adaptors making the machine suitable for a variety of PAGE methodologies.  Thus, the machine is
of modular design, permitting the device to be disassembled and

reassembled into a "new" apparatus.  The device can accept slabs less than 13 cm wide, less than 15 mm thick and of any length (using the nucleotide sequencer adaptor).  Tubes less than 7 mm O.D. and less than 13 cm long can also be accommodated[2*].

Electrophoresis methods[2*].  All isoelectric focusing (IEF) was performed on 7% T and 3% C polyacrylamide gels containing 2-4% ampholytes w/v. Tubes were 3 mm I.D. and 8 cm long while slabs were 13 cm wide and 7 cm in length.  Gradient gel electrophoresis was performed on 8 cm by 8 cm gels of 3 mm thickness. SDS tube gels were 10 cm long and 5 mm I.D.

IEF-PAGE of lyophilized hemoglobin controls (Isolab, Inc.) was performed using pH 6-8 Servalytes (Serva, Inc.).  Each vial was reconstituted with 0.5 ml of a solution consisting of 1 M $\beta$-alanine, 25% glycerol and 0.05% KCN.  Cathode buffer was 0.1 M in ethanolamine and 0.02 M in $Ca(OH)_2$.  Anode buffer was 0.033 M in citric acid.  The gel was overlayed with 1 M $\beta$-alanine, 5 $\mu$l of sample was placed in each well (cathode side) and electrophoresis was performed at 4 W constant power for 70 minutes.

Oat endosperm protein was prepared by separately grinding 0.1 g of eight oat varieties in a pepper mill followed by homogenization with mortar and pestel in 100 $\mu$l of 2 M urea. The slurry was incubated at room temperature for 2 hours and then centrifuged at 10,000 g for 15 minutes.  The supernatant was saved and made 25% in glycerol and 5% in $\beta$-mercaptoethanol. Gels were overlayed with 2 M urea and 20 $\mu$l of sample was applied to each well (cathode side).  The anode buffer was 0.02 M in NaOH and 0.33% in pH 2-4 Servalytes.  Cathode buffer was 0.02 M in NaOH and 0.02 M in $Ca(OH)_2$.  Gels contained 2 M urea and pH 3-10 ampholytes synthesized in our laboratory (5). Electrophoresis was performed for 90 min. at 4 W constant power.  Seminal fluid and isoelectric focusing protein standards (Pharmacia, Inc.) were electrophoresed in the same manner except that the homogenization step was skipped for both, and $\beta$-mercaptoethanol was not added to the protein standards.

---

[2*]A detailed description of this apparatus (Isophore[TM]) in addition to specific methodologies for its operation can be obtained from Isolab, Inc., Drawer 4350, Akron, Ohio, U.S.A. 44321.

68

Oat endosperm proteins were also subjected to second
dimension SDS-PAGE on 10-20% linear gradient gels containing
0.375M Tris·HCl pH 8.8 and 0.1% SDS.  IEF-PAGE pH 3-10 tube
gels were electrophoresed in the first dimension and shaken
for one hour in a solution composed of 10% glycerol, 5%
β-mercaptoethanol, 2.3% SDS and 0.625 M Tris·HCl pH 6.8 (3).
Gels were then affixed to the top of the second dimension gel
with 1% agarose.  Both buffer tanks were filled with a solution
composed of 0.025 M Tris base, 0.192 M glycine and 0.1% SDS.
Gels were electrophoresed for 2.5 hours at 40 mA constant
current.

The polypeptides of murine lymphoblastic lymphoma mitochondria and molecular weight standards (Schwarz Mann, Inc.)
were electrophoresed by SDS-PAGE on 10% tube gels.  Mitochondria
were isolated by differential centrifugation according to the
method of Heidrich et al. (9).  Sample solubilization, gel
preparation and electrophoresis was performed according to
Keleti and Lederer (10).

Electrophoresis of serum and plasma oligomeric proteins
was performed on 2½-27% concave gradient gels.  Both the gels
and the buffer tank solution were 0.09 M in Tris base, 2.6 mM
in EDTA, 0.08 M in boric acid and 0.005% in Na·azide.  Samples
were made 3% in sucrose and 0.01% in bromophenol blue tracking
dye.  Electrophoresis was performed at 100 V until the tracking
dye migrated to within 1 cm of the bottom of the gel.

All gels were fixed in a solution containing 100 g trichloroacetic acid:  17.25 g Sulfosalicylic acid:  150 ml
methanol:  350 ml $H_2O$.  Destain solution was composed of 80 ml
acetic acid, 250 ml ethanol and distilled $H_2O$ to one liter.
Staining solution contained 0.11% Coomassie Brilliant Blue
R-250 in destain solution.


Results


The results of pH 6-8 IEF-PAGE of hemoglobin variants are shown
in Figure 2.  Densitometric scans of each sample demonstrated
that return to baseline was achieved in nearly every case, and
that the probability of confusing one hemoglobinopathy with

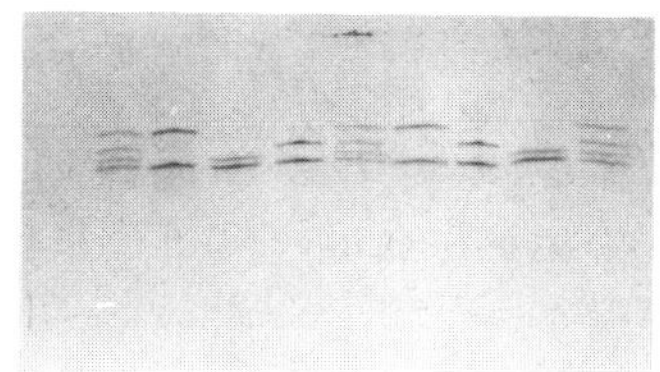

Figure 2.   pH 6-8 IEF-PAGE of hemoglobins AF, AS, AC, AFSC.

another was less than 0.01%.   The pH 3-10 urea IEF-PAGE gels
of seed proteins and seminal fluid samples are shown in Figures
3a and 3b.   In all samples, 35-45 separate bands can be

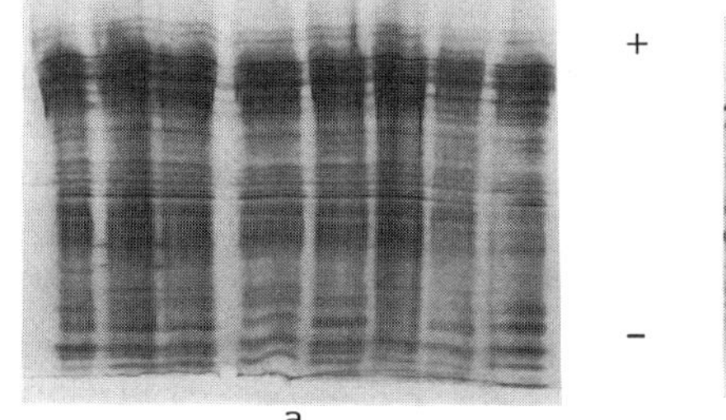

+

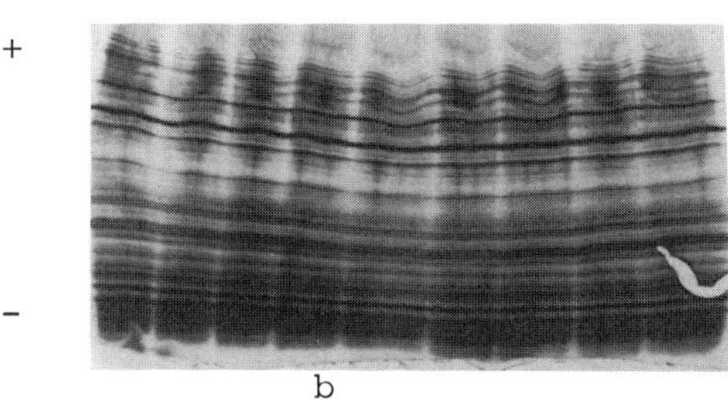

-

a                                           b

Figure 3.   pH 3-10 urea IEF-PAGE of a) eight oat varieties and
            b) seminal fluid.

visually detected.   All eight seed varieties can be distinguished.
Most distinct differences can be observed at the basic end of
the gel.   Sample to sample consistancy is observed.

     The 2D PAGE gel of an oat endosperm sample and the SDS-
PAGE tube gels are shown in Figures 4a and 4b.   We were able to

a

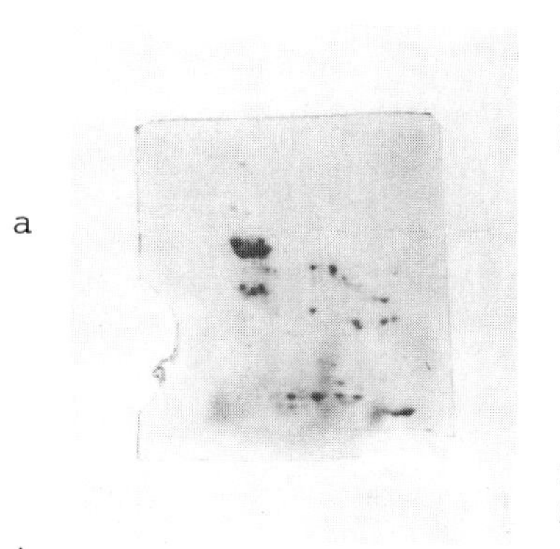

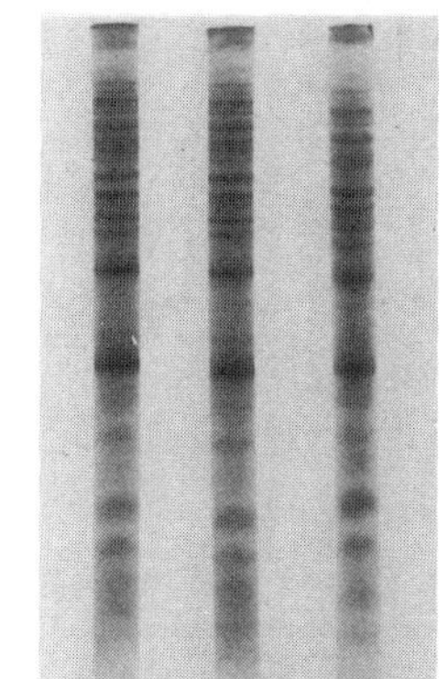

Low
M.W.

High
M.W.

b

+                    -

Figure 4.   a) Second dimension 10-20% linear gradient SDS gel
            of oat endosperm by 2D PAGE b) 10% acrylamide SDS-
            PAGE tubes of murine lymphoblastic lymphoma cell
            mitochondria.

discern over 130 polypeptide species in the 2D gel.  We were
easily able to distinguish numerous differences between the
eight oat varieties subjected to this technique.  The tube to
tube replicability varied by less than 2% for each band
detected.  This was verified by scanning densitometry.

Graphic display of the lineararity of both our pH 3-10
gradients and molecular weight separations by 10% SDS-PAGE are
displayed in Figure 5.  In both cases, average standard error

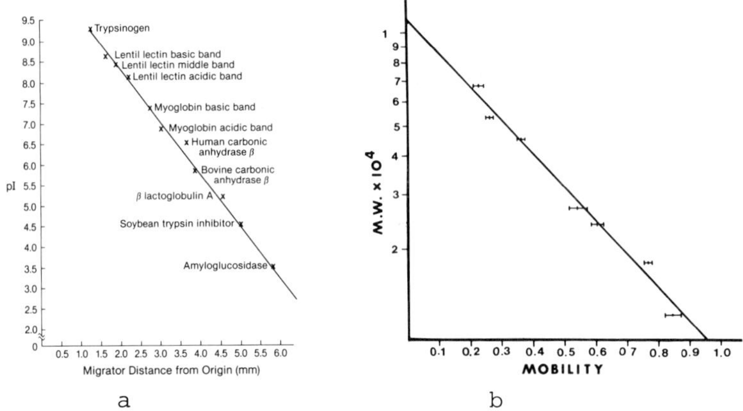


Figure 5.    a) pH as a function of distance migrated for IEF
protein standards.  Proteins and pI's were:
trypsinogen, 9.30; lentil lectins, 8.65, 8.45,
8.151; myoglobins, 7.35, 6.55; carbonic anhydrases
6.55, 5.85; β lactoglobulin A, 5.20; soybean
trypsin inhibitor, 4.55; and amyloglucosidase, 3.50.
b) molecular weight as a function of mobility.
Proteins and molecular weights were (in kilodaltons);
albumin, 67; γ globulin heavy chain, 53; ovalbumin,
45; γ globulin light chain, 27; apoferritin, 24,
myoglobin, 17.8; cytochrome c, 12.4.

of the mean deviation was less than ± 2%.  Both give clear
evidence that highly stable and reproducible linear separations
can be achieved with our system.

The 2½-27% PAGE of serum and plasma proteins is displayed
in Figure 6.  We were able to distinguish 24 bands on plasma
samples and 17 bands on serum.  Several proteins seem to be
altered in either charge or molecular weight during the
conversion of plasma to serum.

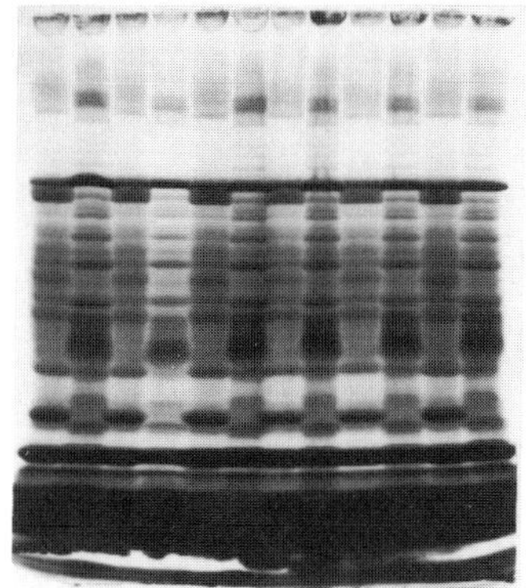

Figure 6.   Oligomeric serum and plasma proteins on a 2½-27%
concave gradient gel.


## Discussion

In addition to the techniques outlined here, we have successfully completed a number of other PAGE methodologies on our
apparatus.  Following the procedure of Ames and Nikaido (11),
we have performed 2D-PAGE on highly hydrophobic membrane bound
polypeptides.  We have also performed enzyme assays for
haptoglobin,lactate dehydrogenase and creatine phosphokinase.
We have identified various species of fish and plants other
than oats by pH 3-10 IEF-PAGE.  Chen and Roe (12) have used a
prototype of the Isophore$^{TM}$ to determine the base sequence of
tRNA.  Other techniques are forthcoming.
     To make our point clear, we performed a variety of high
resolution electrophoretic techniques that are potentially
related to some important social, medical and/or economic
problem.  We separated both hydrophilic and hydrophobic
oligomeric proteins and polypeptides in tubes and slabs on the
basis of both charge and molecular weight.  Samples were of
both plant and animal origin.  These techniques and others are
relevant to medicine, agriculture, food processing, criminology,
taxonomy, microbiology, genetic engineering, animal husbandry,
physiology, etc.  We believe that there is value in a highly
versatile gel electrophoresis module that is prepared to meet

the variety of needs in the many disciplines employing high resolution electrophoresis both now and in the future.


References

1.   Basset,P.,Beuzard, M.D., Rosa, L.: Blood $\underline{5}$, 971-973 (1978).

2.   Maxam, A., Gilbert, W.: Proc. Natl. Acad. Sci. U.S.A. $\underline{74}$, 560-564 (1977).

3.   O'Farrel, P.H.: J. Biol. Chem. $\underline{25}$, 4007-4021 (1975).

4.   Weber, K., Osborne, M.: J. Biol. Chem. $\underline{244}$, 4406-4412 (1969).

5.   Rhighetti, P.G., Drysdale, J.W.: Laboratory Techniques in Biochemistry and Molecular Biology:  Isoelectric Focusing, North-Holland Publishing Company, Amsterdam (1976).

6.   Maurer, H.R.: Disc Electrophoresis and Related Techniques of Polyacrylamide Gel Electrophoresis, First Edition, Walter de Gruyter, Berlin · New York (1974).

7.   Lundstrom, R.C., Roderick, S.A.: Science Tools $\underline{26}$, 38-43 (1979).

8.   Bishop, R.: Science Tools $\underline{26}$, 2-8 (1979).

9.   Heidrich, H., Stahn, R., Hannig, K.: Jour. of Cell Biol. $\underline{46}$, 137-150 (1970).

10.  Keleti, G., Lederer, W.: Handbook for Micromethods for the Biological Sciences, Van Nostrand Reinhold, 139-142 (1974).

11.  Ames, G.F., Nakaido, K.: Bioch. $\underline{15}$, 616-622 (1976).

12.  Chen, E., Roe, B.: Bioch. Biophys. Res. Comm. $\underline{32}$, 235-246 (1978).

A SIMPLE METHOD FOR CASTING POLYACRYLAMIDE GELS OF VARYING THICKNESS AND
SIZE FOR FLAT-BED ISOELECTRIC FOCUSING

Asim Esen

Department of Biology, Virginia Polytechnic Institute and State University
Blacksburg, Virginia  24061, USA

Analytical isoelectric focusing (IEF) is commonly performed in polyacrylamide gel slabs run horizontally on a flat-bed apparatus.  There are many
different flat-bed IEF systems using polyacrylamide or granulated gels as a
supporting matrix.  All but one system (Model 1415, BioRad Laboratories) require that polyacrylamide gels be cast in a vertical position.  There are
several problems associated with casting gels in "glass sandwich" systems
held vertically, i.e. fixed gel dimensions, long set-up time, accidental
toppling and breakage, and leakage of toxic acrylamide monomers due to im-
proper placement of the gasket or clamps.  Polyacrylamide gels of varying
size can be cast in the horizontal position by constructing a simple, inexpensive and versatile gel casting system which is made up of two glass
plates and polyvinyl chloride (PVC) spacer strips.


Materials and Methods

Glass plates (260 X 125 X 3.2 mm) were purchased from LKB-Produkter.  The
full-size (260 X 125 mm) (Fig. 1A) plates were used as the bottom plate of
the glass sandwich system while the top plates were either 240 mm (Fig. 1B)
or shorter.  In addition, larger size plates (200 X 200 X 3.2 mm) were cut
from window glass to form longer pH gradients.  Spacer strips that are 6 mm
wide and the same length as the bottom glass plate (Fig. 1A) were cut from
0.8, 1.6, 2.4 mm thick PVC sheets.  Irregularities on cut edges were removed
by fine grade Emery cloth and the edges were further smoothened by shaving
with a razor blade.  The "end strips" (113 X 6 mm) (Fig. 1D) were cut from
1.6 mm thick PVC sheets.


<u>Construction and Gel Casting</u>.  Construction involves glueing 0.8 mm thick

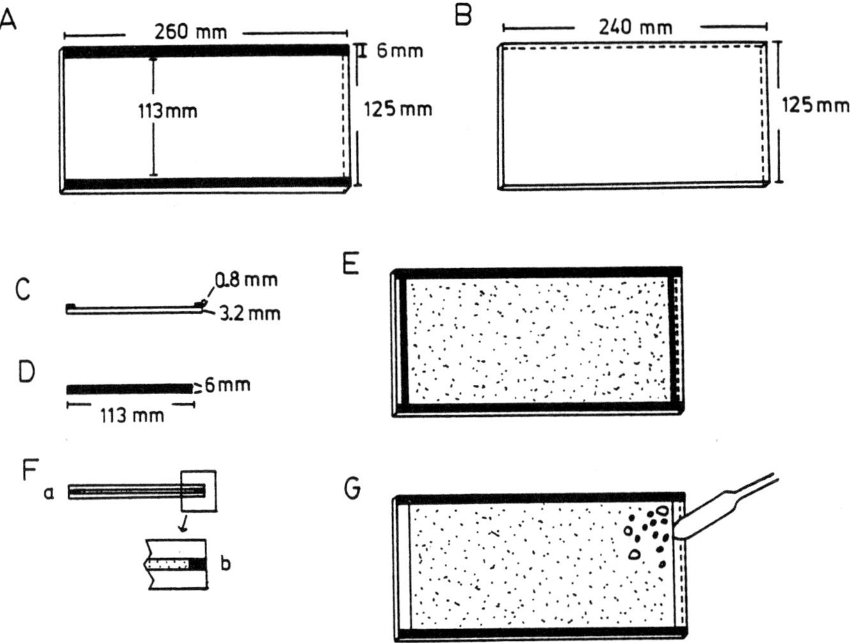



Fig. 1. Schematic representation of components of a horizontal gel casting system. (A) The bottom glass plate with PVC spacer strips (black) glued on the long side. (B) The top plate with the same dimension but 20 mm shorter than the bottom plate. (C) The end or cross sectional view of the bottom plate showing the alignment of the edge of the spacer strip and that of the glass plate with each other. (D) The top view of the PVC end strip. (E) The glass sandwich with the gel in between and the end strips in place. (F) The end or cross sectional view of E. (G) Insertion of spatula between the gel and bottom plate and the entry of air to remove the top plate with the gel adhered to it.

spacer strips to the bottom glass plate with Epoxy glue. Gel casting was performed as follows: 1. Place the bottom plate, spacer strip up, on a bench top. 2. Lay the top plate across the 0.8 mm thick spacer strip to form a glass sandwich and place a weight (e.g. another glass plate) on the top plate to hold it in place. 3. Pipet the required amount of gel mixture into the space between plates. For this, hold the pipet at an angle by resting the air-free tip on the bottom plate. Start the flow at one corner and move the tip along the edge of top plate to the other side while discharging the gel mixture.  After a front of solution across the plate has been formed, continue pipetting either at one side or at the midpoint of the plate until the gel mixture slightly protrudes beyond both open ends.  Then place the end strips at both open ends so that they will rest both on the bottom plate and against the edge of the top plate to insulate the gel mixture from the air (Fig. 1E).  4.  After the gel mixture has polymerized, insert a spatula between the bottom plate and the gel and slowly pry it against the top plate (Fig. 1G).  This operation lets the air enter between the gel and bottom plate which then enables one to remove the top plate with the gel adhered to it.  When gels thicker than 0.8 mm are needed, a small amount of vacuum grease was applied on the permanent spacer strip glued to the bottom plate and then another strip(s) of desired thickness was placed on it before the top plate is laid on.  Gels of 1.6, 2.4, 3.2, and 4.0 mm thickness were cast by placing 1 or 2 additional strips of the same or different thicknesses on the first spacer.  Isoelectric focusing was performed across the width on the Multiphor (LKB-Produkter), or on a home-made apparatus when gel plates were 200 mm wide.  Thin gels (0.8 and 1.6 mm) were used for analytical separations while thicker ones (2.4 to 4.0 mm) for preparative ones.


Results and Discussion

The glass sandwich system described permitted casting gels with the same width but varying lengths depending on the length of the top plate used. Only one bottom plate (Fig. 1A), which serves as a tray, and one top plate (Fig. 1B) are needed for casting gels of a fixed size.  If gels of different lengths with the same width are required, a separate top plate is needed for each length.  The reason for use of top plates at least 20 mm shorter than bottom ones is that the excess length of the bottom plate at both ends serves as a platform for resting the tip of the pipet during pipetting the gel mixture.

Among various materials (teflon, plexiglass, rubber, and PVC) tested as spacers, PVC was superior to others.  This was because PVC did not inhibit polymerization while other spacer materials resulted in gels 3-5 mm shorter in width than needed due to inhibition of polymerization along their surface contacting the gel mixture.

Having a spacer strip set (e.g. 0.8 mm) that is routinely used glued to the bottom plate not only minimizes the set-up time but it also serves as a fixed foundation for adding other strips on to cast thicker gels. It was essential that the glue would not protrude beyond the inner edge of the spacer. Otherwise, polymerization was inhibited and air bubbles trapped around such protrusions resulting in nicks on the edge of the gel. Occasionally bubbles were trapped in the gel solution during or after it has been pipetted. The major reason for bubbles was rapid pipetting. In this case the solution front moved faster in the middle than the sides. Improperly cleaned glass plates and spacers as well as an air bubble present at the tip of the pipet were also responsible for trapping bubbles. Bubbles are formed more often in larger size glass sandwiches (e.g. 200 X 200 mm) than smaller ones (e.g. 240 X 125 or 120 X 125 mm). This is because it is more difficult to form an even solution front and to maintain it over a surface with greater width and length than otherwise would be the case.

The end strips were required to cast gels of intended length. Otherwise, the exposure of the gel solution to the air at open ends inhibited polymerization there and more so at corners leading to gels 10-15 mm shorter than intended.

The thickest gels that could be cast in this system are 4 mm. Above 4 mm the gel solution moved beyond the open ends of the glass and thus could not be retained in the sandwich.

In summary, the gel casting system described is simple, inexpensive, versatile, free from accidental toppling, breakage or from leakage, and requires a short set-up time and yields gels with desired dimensions.

# ISOELECTRIC FOCUSING ON CELLULOSIC MEMBRANES

Borek Janik and Robert G. Dane
Gelman Sciences Inc.
Ann Arbor, MI  48106, U.S.A.

Introduction

Cellulose acetate usually contains residual and contaminant
charged groups which are responsible for a strong electro-
endomosis and interaction with ampholytes (1, 2) and render
it unsuitable for isoelectric focusing (IEF).  Inherently,
cellulosic membranes would have many advantages, e.g., they
are ready to use, nontoxic and nonsieving, would require
only minute amounts of ampholytes, and in overall are economical and convenient to use.  Because of their unques-
tionable appeal, we decided to continue in the efforts initiated by J. Ambler (2, 3).  Here, we wish to demonstrate
usefulness of modified cellulose acetate membranes for IEF
and point out some of their specific features.

Materials

Iso Sepraphore (Gelman Sciences) is a boron trifluoride
treated cellulose acetate membrane designed for IEF.  The
ampholytes tested were Servalyte (Serva), Ampholine (LKB)
and Pharmalyte, (Pharmacia).  The stains were Coomasie Blue
R-250 (Bio-Rad) and Violet 49 (Serva).  The pI indicator
mixture was from Serva and the individual components from
Sigma.

The Model 1405 Electrophoresis Cell and surface electrodes

were from Bio-Rad. The samples were applied with a four (Sepratek-4) or eight place (Sepratek-8) applicators (Gelman) which deliver a 0.50 $\mu$l and 0.25 $\mu$l sample volume, respectively. Sepraphore III and Tuffryn HT-200 (both Gelman) were tested as wicking material.


## Results and Discussion


### Equilibration of membranes with ampholytes

Iso Sepraphore membranes are packaged in 100% methanol which was removed by a light blotting immediately followed by a 10 min. immersion in distilled water agitating occasionally. Insufficient removal of methanol (e.g., blotting only) accelerated evaporation at the membrane edges and caused distortion of the pH gradient. The membranes were equilibrated with ampholyte mixture (<u>ca</u>. 1 ml per a 6 x 11 cm membrane) for at least one hour. The mixture for a pH 2-11 gradient contained: 1.25 ml ampholyte (A) pH 2-11, 0.25 ml A 4-6, 0.25 ml A 3-5, 2.0 ml 50% (v/v) glycerol, and distilled water up to 10 ml. Higher glycerol concentrations slowed down the focusing and made sample applications difficult. Other gradients were constructed similarly. In terms of the stability of pH gradient and insensitivity to bacterial contamination the ampholytes ranked as follows: Servalytes $\geq$ Ampholines $\gg$ Pharmalytes.


### Electrolyte solutions

A number of solutions were tested such as those recommended for $BF_3$ treated cellulose acetate membranes (2,3) and ultrathin polyacrylamide gels (4), and 0.1 M phosphoric acid with pH adjusted (NaOH) to the terminal values of the 2-11 gradient. A 0.1 M citric acid and a 0.25 M ethanolamine (<u>cf</u>. ref. 3) for the anolyte and catholyte, respectively, gave the best results with virtually no cathodic

drift and pH gradient distortion. The amino acid mixtures
(4) did not perform satisfactorily under our conditions.


Focusing with wicks
Sepraphore III strips could not withstand well the effects
associated with high field strengths. The wicks made from
Tuffryn HT-200 (a high temperature aromatic polymer) remained unaffected during IEF and had good mechanical properties. Their relatively low retention capacity for the
electrolyte solutions was overcome by sandwiching the air
suspended portion of the wick between two absorbent paper
strips.


The ampholyte soaked membrane was placed onto the cooling
plate, the excess of ampholyte and air bubbles trapped
underneath were rolled out with a glass rod and the membrane surface was blotted free from surface solution. The
wicks were placed to overlap 0.5 cm of the membrane ends.
The anticondensation lid was 2-4 mm from the membrane surface.


Prefocusing was carried out for about 1/3-1/2 of the focusing time at the initial focusing voltage of 300-500V. The
IEF of proteins in the pH 2-11 gradient required about 60
min. The current was kept at 2 mA/6 x 11 cm membrane; the
final voltage was 800V. At 4 mA/membrane, the focusing
time was somewhat shortened but the gradient at acid pH's
was prone to distortions. Temperature of the cooling water
was kept at about 5$^{\circ}$C for all gradients.


Focusing with surface electrodes
Under the above, otherwise identical conditions, the use of
surface electrodes resulted in gross distortions. Satisfactory results were achieved when (a) the anolyte was
0.25M rather than 0.1M citric acid (both the anolyte and
catholyte were soaked into 1.5 cm wide absorbent paper pads

placed underneath the electrodes and (b) the prefocusing
and the final focusing voltages were about 100-300V, and
800V, respectively, at 2mA/membrane.  Under the conditions,
for the effective membrane length of 6 cm and 10 cm, the
prefocusing was carried out for ca. 15 min. and 30 min.,
and the IEF was completed within ca. 15 min. and 40 min.,
respectively.

Positions of sample application
Fig. 1 shows focusing in progress from various positions of
sample application.  Since the sample contained hemoglobins, the progress could be observed visually.  The closer
to anode the sample was applied, the faster it focused due
to the pH gradient formed and expanding from the anode.
Eventually, all samples focused at about the same position.

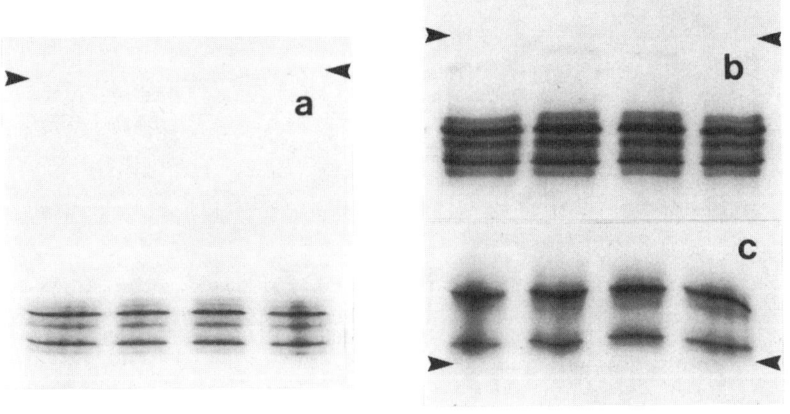

Fig. 1   IEF from various positions of sample applications.  Distances of the point of application (arrow) from
the anodic end of the membrane (picture top) and electrophoresis times: (a) 1 cm, 45 min.; (b) 5 cm, 75 min.; (c)
9 cm, 90 min.  Effective membrane length 10 cm.  Gradient
pH 2-11.  Wicks.  Hemoglobin A, C.

Staining and Storage

A 0.25% Coomasie Blue R-250 in MeOH:H$_2$O:HOAc (5:4:1) performed the best.  Violet 49 (4) destained too quickly causing loss of fine bands.

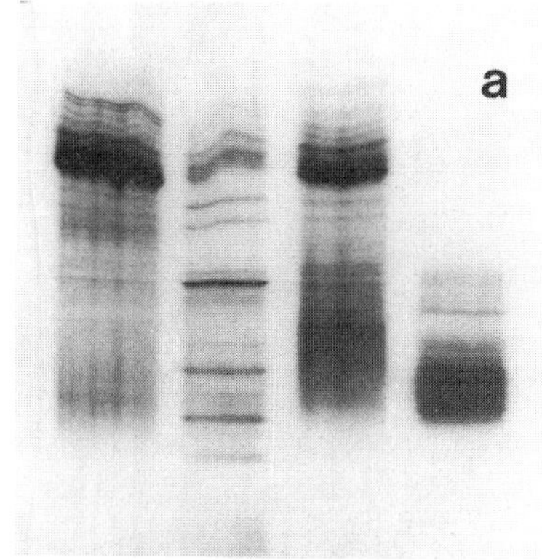
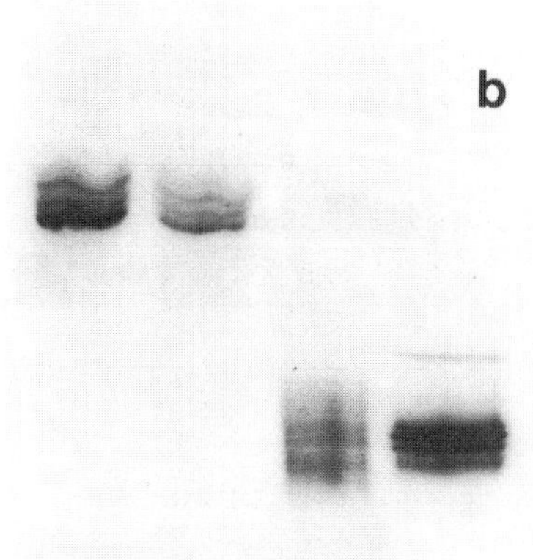
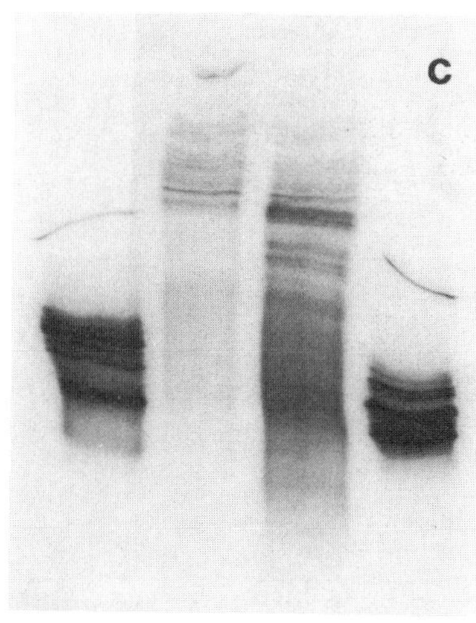

Fig. 2 IEF of proteins.  Left to right: (a) human serum; pI test protein mixture; goat serum; hemoglobin A, S, G Philly; (b) bovine serum albumin; cerebrospinal fluid (lyophylized control, Ortho); human serum Ig; blood hemolyzate, beta thalassemic patient; (c) hemoglobin A, C; rabbit serum Ig; goat IgG; hemoglobin S, C.  Gradient pH 2-11 (a, b) and 5-9 (c).  Anode is at the picture tops.  Effective membrane lengths 10 cm.  Wicks.

82

The membranes can be stored in 10% glycerol or, after a 1-2 min. immersion in 30% aq. diacetone alcohol, placed on a glass slide and dried at 80-90$^{\circ}$C to transparency.

Isoelectric focusing of proteins
Fig. 2 shows IEF patterns of proteins obtained in a setup with wicks. Similar results (not shown) were obtained with surface electrodes.

Linearity of the pH gradient was demonstrated by using a protein pI test mixture. Identity of individual bands was confirmed from separate runs of each protein component. The plot pH <u>vs</u>. d. (distance from the anode, cm) gave a line pH = 1.14 + 1.15d; r = 0.993.


Conclusions

The presented results demonstrate that modified cellulose acetate membranes can be sucessfully used for IEF at relatively low currents and voltages, and short focusing times. Their use is convenient and economical. Similar results were obtained with wicks and surface electrodes although the latter represent certain advantages.


References


1.  Harada, S.: Clin. Chim. Acta <u>63</u>, 275-283 (1975).
2.  Ambler, J.: Clin. Chim. Acta <u>85</u>, 183-191 (1978); <u>88</u>, 63-70 (1978).
3.  Ambler, J., Walker, G.: Clin. Chem. <u>25</u>, 1320-1322 (1979).
4.  Radola, B. J.: Electrophoresis <u>1</u>, 43-56 (1980).

A METHOD FOR PHENOTYPING OF ALPHA 1-ANTITRYPSIN VARIANTS USING
SEPARATOR ISOELECTRIC FOCUSING ON AGAROSE

Rauf A. Qureshi and Hope H. Punnett
Genetics Laboratory, St. Christopher's Hospital for Children
and Department of Pediatrics, Temple University School of
Medicine, Philadelphia, Pennsylvania


## Introduction


Alpha 1-antitrypsin (AAT) is the major protease inhibitor (Pi)
present in human plasma.  The association between AAT deficiency
and emphysema was first described by Laurell and Eriksson (1).
The genetic polymorphism recognized on acid starch gel electrophoresis by Fagerhol and Laurell (2) created great interest in
the separation of Pi variants.  Over the past two decades
several techniques have been developed (2-7) for the phenotyping of Pi variants.  Isoelectric focusing (IEF) on thin layer
polyacrylamide gel is probably the most widely used at present.
This technique requires 5-6 hours of electrophoresis in addition
to the time spent for the making of gels, staining, destaining
and preserving of gels.  Moreover acrylamide and bisacrylamide
are neurotoxic; thus their use constitutes a great health
hazard.

Use of low-endosmosis agarose as a support medium in isoelectric
focusing has been shown by Rosen, et. al. (8), Saravis and
Zamchek (9).  We describe here a procedure for Pi phenotyping
by isoelectric focusing on thin layer agarose gels.  The present
technique has been found to be suitable for separating Pi
variants.  The method is fast, convenient, reliable and reproducible.  Its advantages over polyacrylamide are the shortened
times for preparation, focusing, staining and destaining, and

84

the use of non-toxic substances.


## Materials and Methods

Ampholines and Agarose-EF were purchased from LKB Instruments,
Inc., Rockville, Maryland. Acrylamide/Bis (preweighed mixture,
ratio 29:1, 3.3%C) was purchased from BioRad Laboratories,
Rockville Centre, N. Y. Plastic plates (Gel Bond) were pur-
chased from FMC, Marine Colloids Division, Rockland, Maine.
Dithiothretol (DDT), iodoacetamide, N-(2-acetamido)-2-amino-
ethane sulfonic acid (ACES) were purchased from Polysciences
Inc. Warrington, Penna. Fast Green FCF was purchased from
Allied Chemical Corporation, New York.

All other reagents were of AR (ACS) grade.


## Preparation of Agarose Gel

1.65g Sorbitol (10%) and 0.124g ACES (0.75%) were dissolved in
10ml of deionized water in a 50ml Erlenmeyer flask. Then 0.132g
Agarose (0.8%) was sprinkled over solution and allowed to
sediment. The flask was sealed and the agarose was dissolved
by heating in a boiling water bath until bubbling ceased. The
clear agarose solution was allowed to cool (60-70°C). Carrier
ampholines (0.36ml pH3.5-5, 0.87%; 0.34 ml pH 4-6, 0.82%; and
.05ml pH 6-8, 0.12%) were then mixed with agarose solution and
the volume of this mixture was adjusted to 16.5ml with warm
deionized water. The agarose solution was poured onto a hydrophilic plastic sheet (11x12.5cm) and allowed to gel at room
temperature. The gel was aged for 1 hour at 4°C before use.
Alternatively, several gels were made and stored in a humidity
box at 4°C until used.

## Isoelectric Focusing

Isoelectric focusing was performed on LKB 2117 Multiphor System using an LKB 2103 Power Supply at 6.25W and a maximum of 2000 volts and a 20mA current with cooling at 4°C. The anode electrolyte was 1M phosphoric acid and cathode electrolyte was 1M glycine. The gels were prefocused for 30 minutes. The sera were reduced and alkylated (10). The sample pads (5x10mm, 3MM Whatman filter paper) were soaked in the sera and then applied to the cathodal end of the gel. After focusing for 30 minutes, the sample pads were removed and the gels were focused for another 30 minutes.

## Fixing and Staining

After focusing, the agarose plate was immediately soaked in fixative solution (12.5% TCA, 5% SSA) for 10 minutes. The plate was then washed in 95% ethanol (10 minutes). Filter paper (Whatman #1 or Schleicher and Schuell #557) soaked in 95% ethanol was placed on gel surface, followed by several layers of absorbent paper towel and a glass plate. A weight of 1-2kg was then placed on top of the glass plate and left for 15-20 minutes. Filter paper was removed and the agarose plate was dried with hot air using a hair dryer. The dried plates were stained with Fast Green (0.25% in 10% acetic acid) for 10-15 minutes at room temperature (11). The gels were destained (ethanol: water: acetic acid, 4:4:1, 2x5 minutes) and dried with hot air.

## Results

A typical isoelectric focusing pattern obtained for Pi phenotyping is shown in the figure below:

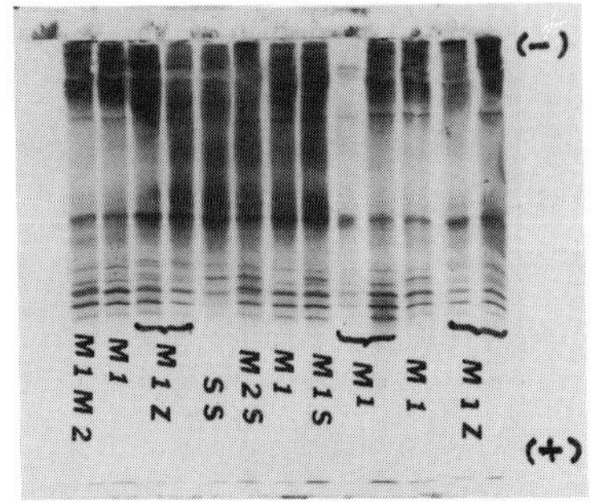



The identity of phenotypes MS, SZ, and Pi M subvariants was
confirmed by isoelectric focusing on acrylamide gel (10), and
the characteristic phenotypes could be easily recognize . Therefore, the method was found to be well suited for the screening
of Pi variants.


## Discussion

The basic banding pattern of Pi typing by isoelectric focusing
on agarose gel was found to be similar to that seen on polyacrylamide gel (5,6,7 and 10). Under the present conditions
used, the Pi M subvarants were well discriminated inspite of
the fact that the split in $M_4$ bands was not as prominent as in
$M_6$ bands. The overall resolution by the present technique resulted in distinct banding, providing unequivocal identification
of the Pi M subvariants in various heterozygous combinations.

Low-endosmosis agarose gel was found to be suitable as a support
medium for isoelectric focusing of AAT (Pi) variants. The use
of agarose over polyacrylamide as a support for thin layer
isoelectric focusing has the advantages of shorter times for
preparation of gels, focusing, staining and destaining procedures
It also avoids the use of toxic substances. The method is fast,
convenient, reliable and reproducible. Because of its convenienc
this method has great scope in children's hospitals where Pi

typing is sometimes urgently needed on short notice in emergency
cases, and in the research laboratories for mass screening.

This research was supported in part by U.S. Public Health
Service Grants CA 19834 and AM 26606 from the National Institutes
of Health and the National Cystic Fibrosis Foundation.

References

1.      Laurell, C.B. and Eriksson, S.: Scand. J. Clin. Lab.
        Invest. 15, 132-140 (1963).

2.      Fagerhol, M.K. and Laurell, C.B.: Clin. Chim. Acta. 16,
        199-203 (1967).

3.      Fagerhol, M.K.: Ser. Haematol. 1, 153-161 (1968).

4.      Fagerhol, M.K.: Scand. J. Clin. Lab. Invest. 23, 97-103
        (1969).

5.      Allen, R.C., Harley, R.A. and Talamo, R.C.: Am. J. Clin.
        Path. 62, 732-739 (1974).

6.      Kueppers, F.: J. Lab. Clin. Med. 88, 151-155 (1976).

7.      Arnaud, P., Chapuis-Cellier, C. and Creyssel, R.: Parotides
        of Biological Fluids. H. Peeters (Ed.)  Pregamon Press,
        Oxford. 22, pp. 515-520 (1975).

8.      Rosen, A., Ek, K. and Aman, P.: J. Immunol. Methods. 28,
        1-11 (1979).

9.      Saravis, C.A. and Zamchek, N.: J. Immunol. Methods. 29,
        91-96 (1979).

10.     Frants, R.R., Noordhoek, G.T. and Ericksson, A.W. : Scand.
        J. Clin. Lab. Invest. 38, 457-462 (1978).

11.     Allen, R.E., Masak, K.C. and McAllister, P.K.: Anal. Biochem.
        104, 494-498 (1980).

# QUANTIFICATION OF PROTEINS WITH ZONE IMMUNOELECTROPHORESIS ASSAY (ZIA)


Olof Vesterberg
Chemistry Division, Occupational Health Dept., The National Board of Occupational Safety and Health, S 171 84 Solna, Sweden


Only twenty years have passed since Ressler described how antigens (Ag) and antibodies (Ab) could form antigen- antibody complexes in a gel with the aid of an electric field (1). Laurell showed how this could be used for quantitative determination of proteins - a procedure now called rocket immunoelectrophoresis or electroimmuno assay (2). This technique is currently used in many laboratories. However, it has some limitations, e.g. nonlinear calibration curves, and a quite low sensitivity.

## PRINCIPLE OF ZONE IMMUNOELECTROPHORESIS

In order to improve quantitative determination of proteins the conditions have been considerably modified. An instrument was constructed with a set of 20 vertical glass tubes (inner $\emptyset$ 2 mm). The tubes were arranged close to each other and in parallel. The upper orifices of the tubes opened into a box - the upper electrode vessel - and the lower ends of the tubes were placed in another box - the lower electrode vessel (Fig. 1). The principle of the method is shown in Fig. 2.

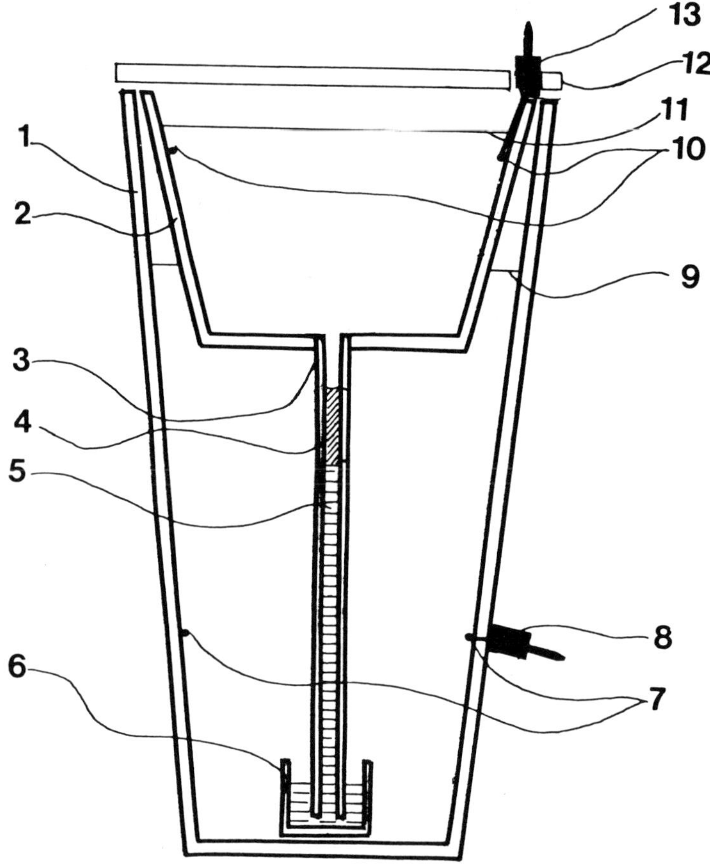

Fig. 1.

A cross sectional schematic view of an apparatus for zone immunoelectrophoresis, where 1 is the outer and 2 is the inner
electrode vessel. 3 is one of the many tubes, in which the
sample solution 4 is added on top of the gel 5, which is also
present in the tray 6. The upper surface of the buffer solutions is shown at 9 and 11. Electrodes of platinum wires 7 and
10, which run around the inside walls of each vessel, are
connected to the banana plugs 8 and 13. The lid 12 covers
the electrode vessels.

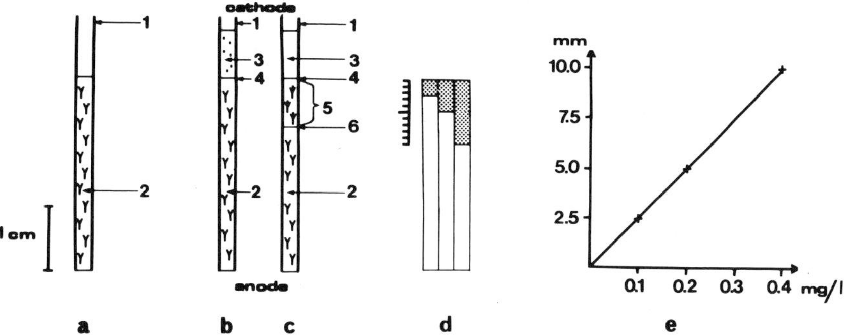


Fig. 2.

A schematic illustration of the steps at quantification of
proteins by zone immunoelectrophoresis.

a) Capillary tube 1, filled with antibody (Y) in agarose gel 2.

b) Same tube where the proteins to be quantified = antigens
(•) dissolved in a water solution 3, have been pipetted on
top of the agarose gel surface at 4. Electrophoresis is then
made during a few hours (or overnight) with the anode and
cathode positioned as indicated in the Figure.

c) The antigens (•) are at electrophoresis transported into
the agarose gel where immunoprecipitates (Y) are formed as a
zone 5, having a front at 6.

d) The distance from the gel surface 4, to the front of the
immunoprecipitate 6 is measured in each gel rod.

e) The distances mentioned are plotted in a graph (the concentration in the standards on x axis against the respective
distance in mm on the y axis).

The method has recently been described (3, 4). The buffer pH
was 8.6 and it consisted of Tris, N-tris (hydroxymethyl-amino-
methane) 0.07 M/L and Tricine, N-tris (hydroxymethyl)-methyl-
glycine 0.04 M/L in deionized water.

A 1% (w/v) of agarose (Litex, type HSA) was prepared in the
buffer by boiling. Polyethyleneglycol Mw 6000 or 20000 (Sigma)
was then added to obtain 4 or 2% (w/v), respectively. A sample
of this solution e.g. 10.0 ml was taken into a test tube. When
the temperature was 56° a suitable volume of antiserum was
added and the contents of the tube mixed well. The concentration of antiserum to be used depends on the titre of the antibodies and the desired sensitivity of the assay. The sensitivity seem to be inversely proportional to the concentration of
antiserum used in the gel.                              For many
commercial antisera, e.g. from DAKO, 2 $\mu$l/ml of gel solution
is appropriate. The agarose solution (5 ml) filled into a
syringe could be simultaneously pressed up into the twenty tubes by using a special device with a canal connecting the lower
orifices of the tubes. Alternatively the solution could be
simultaneously sucked up with the aid of an empty syringe into
the tubes by using another device connecting the upper orifices.
The lower orifices in the latter case were placed in a tray
with the agarose solution. The agarose solution was usually
taken up in the tubes to a level two centimetres below the
upper orifices.

Sample solutions could be applied on top of the gels in each
tube with a pipett having a capillary tip in two different modes. 1. Directly above the gel surface. Buffer was then carefully overlayered above the sample so as to fill the tube.
2. Under the buffer solution first applied above the gel. In
order to get the sample solution to stay just above the gel
surface its density was increased by adding e.g. 14 % sucrose
solution so as to obtain a final concentration of about 7 %.
When using alternative 1 the sucrose could be replaced by a

final concentration of 0.1 % agarose in the sample solution. A stain dissolved in the sample solutions or in the buffer used over the samples facilitates the pipetting. I recommend methylene blue 0.005 % (w/v) because upon electrophoresis it migrates up into the upper electrode vessel and does not pass the antibodies in the gel.

The electrophoresis is usually made with the anode in the lower vessel so that the proteins to be determined are transported down into the agarose gel because of their negative net charge in the buffer of pH 8.6. As in rocket electrophoresis the antibodies are expected to remain essentially immobile or have only a slow migration against the cathode due to endomosis. This can be effected by selecting an agarose with a proper endosmotic flow and/or by changing the pH of the buffer. The endosmosis increases also with the electric potential.

When the protein to be determined meet antibodies directed against it a zone of immunoprecipitates (ZIP) with a sharp front is formed. The more of the protein there is in the sample the longer will the ZIP be as can be seen in Fig. 2.

Because immunoprecipitation continues until there is no free antigen left the method is some sort of an end point procedure. Hence neither the voltage nor the time is very critical. This means that the electrophoresis may be run during a few hours or with a proportionally lower voltage during several hours,e.g. over night. Our experiments have indicated that in order to obtain straight calibration curves  at higher protein concentrations, a certain minimum value is needed for the product Voltage x hours. It seems logical to assume that the lower the electrophoretic mobility (i.e the higher the isoelectric point of the protein) the higher value of V x h is required. Suitable conditions are easily found after some experiments. The ZIP can often be seen already in the tubes directly in oblique light. Different antigen antibody systems show different direction li-

94

mits.

After the electrophoresis the gel rods are easily taken out and
placed in parallel on a glass or plastic plate (Gel Bond) by
using a special device.

In order to preserve the gels with the ZIP it is advantageous to
press and dry the gels with methods analogous to those currently used for flat bed agarose gels (6). We have compared some
different procedures for staining the ZIP and found one that
has a comparatively low detection limit. Staining is made at
$40^{\circ}$ C for 15 minutes.

Staining solution: Coomassie R 250, Serva Blau R (Serva, Heidelberg) 0.4 g, acetic acid 8 ml, ethanol 25 ml and water 67 ml.
Destaining is made at room temperature in the same solution
without stain.

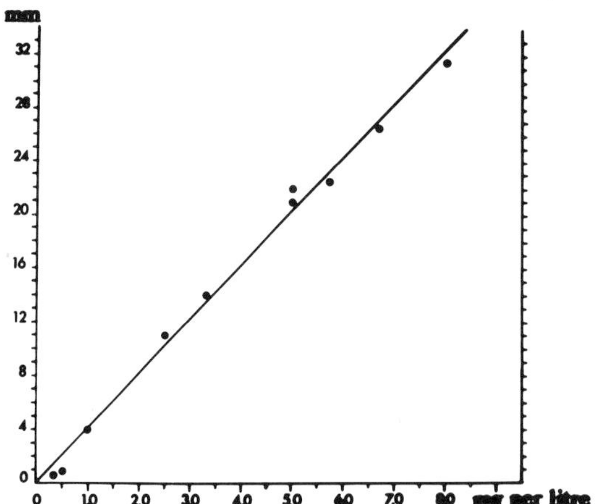


Fig. 3
Calibration curve for human serum albumin with ZI. Sample volume 50 µl. For other details see text. The equation of the line
was y = 4.02 x + 0.08 and the regression coefficient was 0.997.

Already in early experiments it was found that the distance
from the upper gel surface to the front of the ZIP was proportional to the amount of protein in a rather wide concentration
range as can be seen in Fig. 3. A wide applicability of the
method is indicated by the fact that successful results have
been obtained with several proteins, e.g. albumin (5) $\alpha_1$-anti-
trypsin, $\alpha$-fetoprotein, $\beta_2$-microglobulin, ferritin, HDL (high
density liprotein), the immunoglobulins IgE, and IgM and also
transferrin.

The buffer composition and ionic strength may be of interest to
discuss here. Laurell and several others have used 5,5-diethyl-
barbiturate (Veronal) buffers with a concentration of 0.07 mo-
les/l containing calciumlactate at 0.1 m mole/l (6). Many others
omitt this additive and use instead $Mg^{2+}$ (for studies in serum
of complement) while others use EDTA (ethylenediamine tetra-
acetic acid). Many Danish scientists use 0.02 mole/l of Veronal,
i.e. almost 1/4 of the ionic strength mentioned above (6). When
using large volumes of samples containing salt an unacceptable
low ionic strength in the buffer may result in undesirable pH
shifts and less suitable conditions at quantification (4). If
increasing the Veronal concentration above 0.04 moles/l one may
run into problems when drying the gels due to chrystal formation.
Veronal at pH 8.6 has only a small buffering capacity towards
the alcaline side. The situation is more favourable for Tricine
and Tris as can be seen from their pK values beeing at 25°C
8.10 and 8.15, respectively. In order to get a comparatively
high buffering capacity Tris - Tricine buffers were mostly used.
As no inorganic salt is used here and Tricine is an ampholyte
this buffer has a comparatively low ionic strength and thus the
electrical Joule heating is less, which abolishes the need for
complicated cooling devices. Futtermore, Veronal has the drawback of being a pharmaceutical drug substance often bound with
strict regulations at buying and use in the laboratory. When
using Tris it is valuable to know that some pH electrodes show
too low pH readings, which might give in fact too high pH at

preparation of the buffer (7,8).

I have also recently shown that the concentration of precipitating antibodies can with advantage also be determined by using
an analogous procedure called reversed zone immunoelectrophoresis assay (RZIA( (9). In this case a buffer is used with a pH
just above the pI of the protein to be determined. This protein
is incorporated in the gel which is made in the tubes of the
same instrument just described. Suitable dilutions of the antibodies are applied above each gel. The principle is illustrated
in Fig. 4. Electrophoresis is made with the cathode in the lower
electrode vessel. As the pH is lower than the pI of the antibodies these migrate down into the gel where ZIP are formed with
the almost stationary antigens. The ZIP are usually visualized
after pressing and drying by staining.

When using this procedure as well as ZIA it is possible to
shift the pI of antigens or antibodies, e.g. by carbamylation,
so as to obtain more suitable conditions for electrophoresis (5).

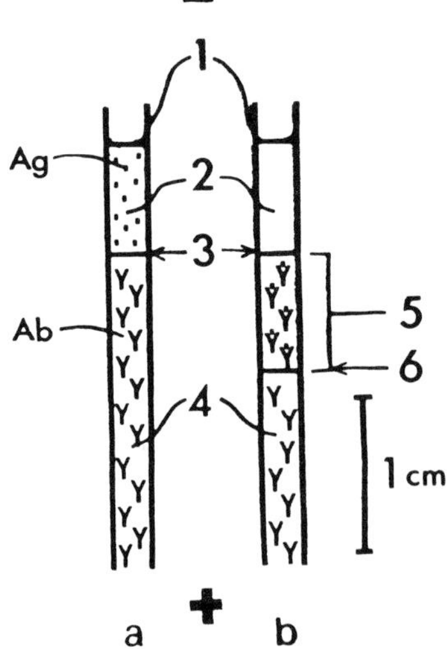


Fig. 4
Principle of quantification of
Ab=antibodies by reversed zone
immunoelectrophoresis (RZIA).
Tubes (1) are filled with agarose gel (4) containing antigen
to within a few centimetres of
their tops (3). On top of the
agarose sample solution (2) is
pipetted. The electrode vessels
(not shown) contain buffer solutions (also present in the gel
and on top of the sample solution) which make electric contact with the cathode (-) at the
lower end of the gel and with
the anode (+) through the sample
solution. a: before electrophoresis; b: after electrophoresis,
showing the zone of immunoprecipitate (5) which has a sharp
front (6). The distance 5, is
proportional to the quantity of
antibodies in the sample.

## Discussion and conclusion

In conventional rocket immunoelectrophoresis the immunoprecipitates first formed in the gel dissolve in antigen excess. This
process goes on until a final front of immunoprecipitate is obtained. However, there are often uneven current- and field
strength distributions in crossections perpendicular to the
current direction at and anodal to the sample application well.
Furthermore, sideway spreading of proteins occurs continuously.
These phenomena explain why the rocket technique is bound with
complications, e.g. curved calibration lines. The fact that
the field strength distribution is even through the whole cros-
section of a gel rod in ZIA and RZIA, and that no sideway spreading of salts or protein can occur, makes the conditions here
more favourable. Making the electrophoresis in narrow bore tubes also means less consumption of agarose and antibodies per
sample (see Table 1). Furthermore, this means that large sample
volumes can easily be applied (typically 100 $\mu$l) to be compared
with about 5 $\mu$l in the rocket method. These facts explain why
the detection limit is often one, or up to two orders of magnitude higher with ZIA than with the rocket method. The advantages
have been summarized in Table 2. In addition it can be mentioned
that the electrophoresis apparatus (Quantiphore) offers excellent cooling, is simple to operate and is easily adapted to
many different types of electrophoresis such as disc-, SDS-,
and isoelectric focusing. Many such applications have been
made and will be described in forthcoming publications.

TABLE 1

Consumption of antisera in different procedures for quantification of proteins

|  | $\mu$l of antiserum per sample |
|---|---|
| Nephelometry [1] | 10 - 100 |
| Turbidimetry [1] | 30 - 100 |
| Radial immunodiffusion | 5 - 10 |
| Rocket immunoelectrophoresis | 5 - 10 |
| Zone — " — | ~0.3 |

[1] That method requires firstly expensive instruments; secondly specially prepared or at least filtered solutions of antiserum; thirdly each sample must often be determined (read) at least twice, i.e. with and without antibody to be able to subtract the background of each sample.


TABLE 2

Advantages of Zone Immunoelectrophoresis (ZIA) over Rocket Electroimmuno Assay

1. Simpler working procedures, no making of sample wells and no electrode wicks.

2. Less consumption of agarose, buffer and antibodies per sample. The small need of antibodies (about 0.3 ul of antiserum per sample) with ZIA is not only economically attractive (especially for expensive highly specific antibodies) but it may also be of the utmost importance when the total available amount is limited.

3. The amount of antigen is directly related to a linear distance, i.e. a linear calibration curve over a wider concentration range than the other method. This means simple evaluation, no area measurements or calculations, and less rerunning of samples and thus also a better economy with respect to point 2.
   Furthermore, a lower frequency of samples where the result is delayed an additional day(s).

4. Large sample volumes 50-150 $\mu$l can easily be used for each test.

5. Much lower detection limit when calculated in $\mu$g of protein per ml of sample solution.

PRACTICAL HINTS

## Preventing losses of proteins at low proteinconcentrations

Due to the high sensitivity of the ZIA method it is possible to
assay proteins at low concentrations. When samples or standards
are diluted very much one may often observe in such solutions
lower than expected concentrations due to variable adsorption
of protein to the walls of the vessels or syringe tips. This
may show up as nonlinear standard curves with lower than expected readings for low concentrations of the assay protein. To
counteract this we have used tubes of polyethylene or polypro-
pylene (e.g. Eppendorf micro sample tubes), which have shown
much less adsorption than polystyrene. This was helpful when
using standards of liver ferritin and $\beta_2$-microglobulin in urine.
We have also tried the addition of ballast proteins such as
eggalbumin and bovine serum albumin at concentrations of 1-2 %
(w/v). Another hypothesis was that heavy metals or oxygen in the
diluting solutions might cause aggregation or other undesirable
effects. To counteract this we have used dithiotreitol at concentrations 1-10 mM. In our experience 1 mM/l of dithioteitol
was sufficient. It did also decrease unspecific precipitates,
which otherwise often are seen in the upper ends of the gel
rods, when using sera diluted < 1:8. In conclusion different
proteins may require different actions to prevent losses of the
protein to be assayed.

## Removal of disturbing proteins by washing the gels

At relatively high protein concentrations the immunoprecipitates my be read without staining. When low concentrations of
specific proteins are to be determined in samples with a high
total protein concentration, e.g. blood serum, (50 $\mu$l samples
diluted < 1:25), staining of unspecific (ballast-)proteins may
cause a high background. To reduce this twenty gel rods were
first taken out of the electrophoresis apparatus, and were

aligned in parallel on a glass or plastic plate (Gel Bond[R], Marine Colloids, Maine, USA). The gel rods were pressed for 10 minutes under a thick blotting paper using a weight of about 100 g. This step removed most of the water and much of the ballast proteins in the gels. In order to fix the gel rods in position during the subsequent washing they were covered with a net of soft plastic, which was kept in place by rubber bands (cf. Fig. 5). The assembly was put in a tray and covered with about 400 ml of saline. Washing was usually made at $40^{\circ}$ C for two hours with gentle stirring. The time required to wash at $20^{\circ}$ is about three times longer.

If it should be desirable to shorten the diffusion process even further one can proceed as follows. Press the gels for 5 minutes. (Not too long time and do not press too hard because this makes their swelling at washing slower.) Allow for diffusion and swelling during 20 min. at $40^{\circ}$ C in saline. Press for 5 min. and wash for another 20 min. at $40^{\circ}$ C. Finally, press for 10 min. and dry in an oven at $50^{\circ}$ C for 25 min. or under a hair dryer during 5-10 min.

If the precipitates are very well stained a shorter time for diffusion and/or staining than mentioned in ref. 4, or another less sensitive staining procedure may be tried. This in turn means, that less clearing of the background is required.

When the temperature of the washing solution is above $5^{\circ}$ C there excists risks for dessolution of the immunoprecipitates by hydrologic enzymes from microorganisms. To minimize the risk always use fresh solutions and cleaned equipment and do not wash longer time than necessary to remove the background of non precipitated proteins. Inhibitors of microbial growth and/or hydrolytic enzyme may be used.

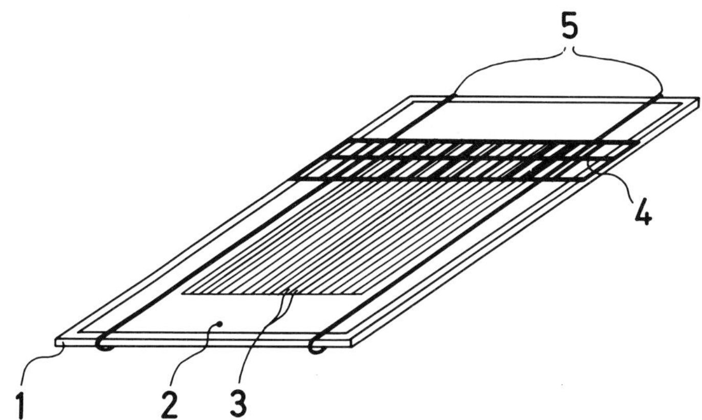

Fig. 5
Device for facilitating washing of agarose gel rods, where 1 is
a glass plate; 2 a thin plastic plate (Gel Bond[R]; 3 the agarose
gel rods; 4 a net of soft plastic which is kept in place by the
rubber bands 5.

REFERENCES


1.  Ressler, N.: Clin. Chim. Acta $\underline{5}$, 795 (1960)

2.  Laurell, C.-B.: Anal. Biochem. $\underline{10}$, 358 (1965)

3.  Vesterberg, O.: Fresenius Z. Anal. Chemie $\underline{301}$ 134 (1980)

4.  Vesterberg, O.: Hoppe-Seyler´s Z. Physiol. Chem. $\underline{361}$, 617
    (1980)

5.  Vesterberg, O.: Clin. Chim. Acta in press 1981

6.  Verbruggen, R.: Clin. Chem. $\underline{21}$, 8 (1975)

7.  Sigma Tech. Bulletin No 106 B (11-78)

8.  Ryan, M.F.: Science $\underline{165}$, 851 (1969)

9.  Vesterberg, O.: J. Immunol. Meth- $\underline{37}$, 311-314 (1980)

SPECIFIC AND EFFICIENT PROCEDURE FOR QUANTIFICATION OF PROTEINS
SEPARATED IN GEL BY ELECTROPHORETIC METHODS USING ZONE IMMUNO-
ELECTROPHORESIS ASSAY (ZIA)


O. Vesterberg and U. Breig
Chemistry Division, Occupational Health Dept., The National
Board of Occupational Safety and Health, S 171 84 Solna, Sweden


Introduction

Electrophoretic methods are increasingly used for separation
of proteins in gels. The separated proteins are almost always
visualized by staining. The results are usually only qualitively
evaluated by comparison of zone patterns and rough estimates
of relative stain intensity of certain zones of interest. Although qualified densitometers can supply figures of some significance, such instruments are expensive and other problems remain. Different protein molecules bind different number of stain
molecules. This makes it very difficult to determine the absolute amount of **protein** in each separated band. Almost one set
of standard would be required for each protein. The number of
bound molecules remaining after destaining of the background
may vary from one experiment to another, which means difficulties in standardization. There is a very serious problem if
a certain protein of interest occurs in a band which also
contains other proteins, because the protein stains do not
differentiate the proteins. Thus, for quantitative studies,
the generally used protein staining procedures have severe
limitations. There is thus a great need for a better quantitative determination of proteins after separation by electrophoretic methods.

A rapidly increasing number of reports describe proteins occurring in multiple molecular forms showing certain biological features, which may be very interesting and have biological significance (1-3). For example, it has been reported that isoelectric focusing (IF) in polyacrylamide gel followed by staining of the proteins has shown one form of transferrin in blood, which occurs in increasing concentration with increasing alcohol abuse (4). We have studied this in some detail and developed a new procedure that seems generally applicable for quantification of multiple molecular forms of proteins after separation. To indicate the position of a certain protein after IF is used coloured proteins as isoelectric point (pI) markers. Knowing the relative positions it is thereafter easy to locate and cut out a certain gel segment with the protein of interest, which can be quantitatively determined.

This is done in a subsequent step by using gel pieces containing the protein forms of interest as samples in zone-immunoelectrophoresis assay (ZIA). That method, as well as its advantages has recently been published (5, 6). In this paper we describe a new procedure for the quantitative determination of multiple molecular forms of transferrin. This procedure seems applicable for the quantification of several proteins separated in gels by electrophoretic procedures.

## Material and methods

Ampholine[R] carrier ampholytes, Ultrodex[R] and agarose for isoelectric focusing, type EF as well as Multiphore[R] and Power Supply, were form LKB Produkter (Bromma, Sweden). The apparatus for ZIA was a Quantiphor[R] from Desaga (Heidelberg, Germany). Agarose for ZIA was type HSA from Litex (Glostrup, Denmark). The agarose gels were pressed with a thick blotting paper, Munktell[R] (Grycksbo, Sweden). Samples were pipetted on the surface of gels for IEF with a Micopipett (Oxford Lab. Int. Corp., Ireland). Dilutions of standards and samples for ZIA were pipetted with Hamilton syringes (Boraduz, Switzerland).

The gel pieces were disintegrated with a special 1 ml injection syringe, Plastipak[R], Becton-Dickinson (Rutherford, N.J., USA).

Gel Bond[R], 0.2 mm thick was obtained from FMC, Marine Colloids Div. (Rockland, Maine, USA). Rabbit anti-human transferrin was from DAKO (Copenhagen, Denmark). Standard serum was Seronorm[R] Protein from Nyegaard (Oslo, Norway). Polyethylene glycol Mwt 6000 was purchased from Sigma (St. Louis, USA).

<u>Preparation of a coloured myoglobin marker</u>. The isoelectric point (pI) of myoglobin was shifted from above 7 to below pH 6 by carbamylation (7). Horse myoglobin was first prepared by ammonium sulphate precipitation (8). Dilution with water was made to obtain 5 mg/ml. Twenty ml of the myoglobin solution was mixed with forty ml of a borate buffer (0.1 mol/l) of pH 8.6 containing 1 mol/l of potassium isocyanate (KCNO). The mixture was kept at 45°C for 15 minutes, and then dialysed at +4°C overnight. It was then centrifuged to remove precipitates. The supernatant was submitted to isoelectric focusing in Ultrodex using Ampholine pH 5-7 as described in the LKB Application Note No. 198. A component with pI 5.65 was isolated and used as an isoelectric point (pI) marker. A solution of this was eluted with water, and then stored frozen in small aliquots until used.

<u>Isoelectric focusing in agarose gels and determination of a transferrin component</u>. Flat beds of agarose gel for isoelectric focusing were made on polyester film (Gel Bond[R]) cut to 125 - 130 mm, essentially as described earlier (9, 10).

Before isoelectric focusing serum samples were diluted 1 + 1 in water containing 1% (v/v) Ampholine (pH 6-8) and 4 mmoles/l of $FeCl_3$. Iron saturation ensured quantitative conversion of the transferrin to the diferric state ($Tf-Fe_2$). The monoferric transferrin molecules ($Tf-Fe_1$) focusing at pH 5.7 were thus eliminated.

Of each diluted serum sample 2.0 µl was pipetted on the agarose
gel as a 2.5 cm long streak starting 0.5 cm from the cathode
strip and perpendicular to it. The centre to centre distances
of the samples were 7.5 mm.

Of the pI marker solutions 10 µl samples were soaked into
paper pieces, which were applied on the gel 1 cm from each
short side (cf. Fig. 1). The electrode wicks were soaked in 1%
(v/v) Ampholine, pH 2.5 - 4 for the anode and pH 9 - 11 for the
cathode. Isoelectric focusing was made during 55 minutes in a
Multiphore essentially as described earlier (10). The final
voltage was about 1250 V. A gradual increase of the power was
made to allow better conditions for the proteins to go into the
gel. After half the focusing time the current was turned off
and excess liquid on the gel surface close to the cathode was
soaked up with soft paper e.g. Kleenex. This way only required
when the humidity in the air was high.

The pH gradient obtained was determined after some focusing
experiments. A long razorblade was used to cut out a 1 cm
wide strip extending from anode to cathode. Transverse cutting
at each 0.5 cm gave gel pieces, which were transferred to small
test tubes. To each of these was added 0.5 ml of 0.1 moles/l
of KCl. After incubation at 4° over night pH was determined
in the solutions.

In some experiments the transferrin components were visualized
by direct immunofixation in principle as described by Thymann
and Henningsen (11). The antiserum was diluted 4 times. It was
then soaked into a cellulose acetate membrane, "Sartorius
Membrane Filter type SM 11200 (Sartorius, Göttingen) to obtain
10-15 µl/cm$^2$. This was placed directly on the gel surface
3.5 - 6 cm from the cathode. Diffusion was allowed to continue
for 40 minutes in a humid chamber. The gel was then pressed
with blotting paper, washed in 0.9 % (w/v) NaCl at 40°C for
at least 3 hours, pressed again and stained. After having seen

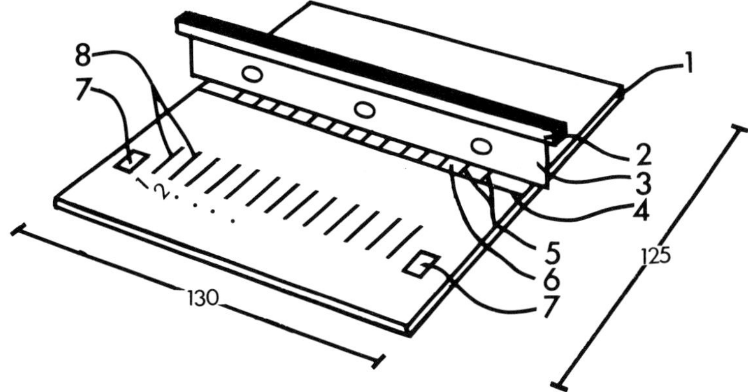

Fig. 1
A view of a "sandwich" indicated by 1, consisting of: the gel, its plastic support (Gel Bond$^R$), a glass plate and a template of plastic film on which the slicing pattern is drawn. A rigid plastic support, 2, is glued to the razorblade 3, which is used to cut the gel at its position and also along the line 4. The gel is also cut transversely between the lanes illustrated by the lines, 5, to obtain the gel pieces 6. Before focusing the marker myoglobin was applied in paper pieces at 7 and the serum samples as streaks (lanes) illustrated by the lines shown at 8. The cathode was at the gel edge parallel to the blad 3 and facing the viewer. Distances in mm.

the positions of the transferrin components it was easy to draw a template showing the pattern of the sample applications, the marker protein positions and a suitable slicing pattern. It was found that our marker myoglobin focused with its maximum very close to the transferrin of interest. With a long razor blade the gel was cut through the centre of the myoglobin zones and parallel to the electrode strips as shown in Fig. 1. Another cut was made 5 mm cathodal to the first one. Transverse slicing was made between each lane so as to obtain 7.5 mm times 5 mm gel segments. Each of these was lifted with a thin, 2 mm wide spatula and put into a Plastipak syringe and twisted. A 2 cm long polyethylene tubing with an outer diameter of 1.8 had been fitted onto the needle of the syringe in advance. A schematic drawing is shown in Fig. 2. From the electrophoresis buffer, 30 µl was applied as a droplet into the syringe. The plunger

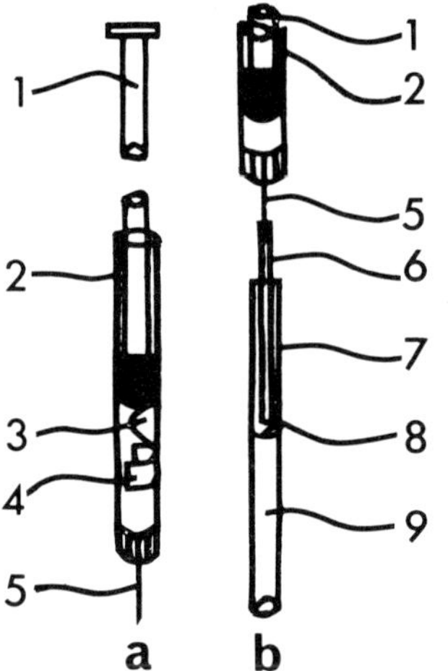

Fig. 2.
An illustration of the syringe
a; and b how it is used for
disintegration of a gel piece.
The plunger, goes into the
barrel of the syringe 2. Below
the piston there is a droplet
of buffer, 3, for embedment
of the gel piece 4. The syringe ends with the needle 5. On
this is fitted a plastic tubing, 6, which is shown inserted into one of the glass tubes,
7, of a Quantiphor. This tube
is partly filled with antibody
containing agarosgel, 9, with
an upper surface at 8.

was then put into the syringe. Holding the syringe with the
plunger below and slowly pushing it up all air was expelled.
Then the opening of the tubing was positioned just above the
gel for zone immunoelectrophoresis (ZIA) as shown in Fig 2b.
The gel piece was slowly forced out through the needle while
slowly retracting the syringe. The needle and tubing were cleared with 20 $\mu$l of buffer, which was also applied above the
gel for ZIA. The small dead volume of the syringe allowed the
gel piece to be efficiently pressed out. The space in the tubes
of the Quantiphore above the gels was subsequently filled with
electrophoresis buffer. To prevent the disintegrated gelpieces
from escaping upwards out of the tubes in the ZIA apparatus we
placed over the tube openings a piece of chromatography paper,
which in turn was held in place with a U-shaped glassrod. The
standard sample solutions were pipetted with a Hamilton syringe.
Then ZIA was made essentially as described earlier (6). The gel
rods were pressed and dried, and the immunoprecipitates stained
and read as described.

ZIA allowed a quantitative determination of the transferrin
with pI 5.7. The concentration of Tf 5.7 in a serum sample was

calculated from the dilution and application volumes used at isoelectric focusing. The concentration of total transferrin, determined by ZIA, and that of the transferrin component, as described above were used for calculation of the percentage of Tf 5.7 in each serum sample.

In order to check that the gel pieces had been cut correctly immunofixation, or Coomasie Blue protein staining (10) was often made just after cutting.

## Results

At the determination of total transferrin in serum with ZIA the relative standard deviation was about 4 %. The pH gradient obtained after focusing is shown in Fig. 3. At immunofixation after isoelectric focusing, it was found that the relative concentrations of transferrin and antiserum had to be balanced carefully to obtain sharp zones of immunoprecipitates. At high concentrations of transferrin a clearing up in the centre of the precipitate was obtained. Due to the fact that the difference in amount of the multiple forms of transferrin was large (e.g. Tf 5.7 was often only 2 % of the total) it was not possible to show all components without the major one showing a central clearing up. This did not matter because our main interest was to localize the minor form, Tf 5.7. For the isolation of this, the marker myoglobin was very valuable as immunofixation could not be applied before cutting the gel and removing the gel pieces containing the component Tf 5.7. A lot of experimentation with variable conditions at carbamylation had to be made to arrive at a good yield of the marker myoglobin with suitable pI. When the time at carbamylation was doubled most of the myoglobin obtained a pI < 5.6.

Isoelectric focusing in agarose gel was found to be a critical step. In order to be able to cut out the component Tf 5.7 this protein should preferably focus at the same distance from the

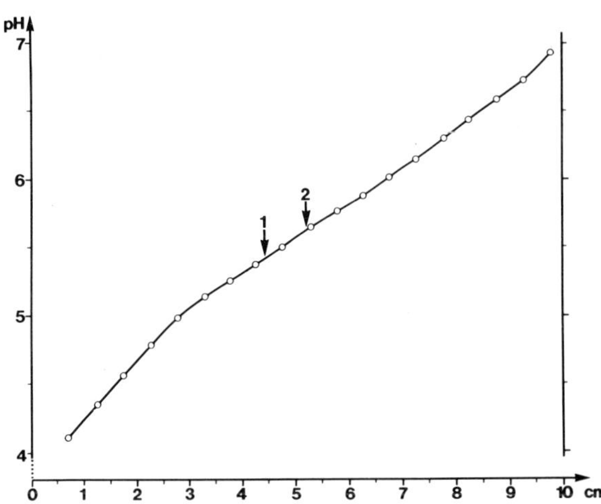

Fig. 3
A typical pH course in the gel obtained after isoelectric focusing. A cm scale is shown on the X-axis with 0 at the anode strip and a pH scale on the Y-axis. The arrows 1 and 2 indicate the positions of the major and minor (Tf 5.7) transferrin components, respectively.

the cathode in all lanes. To achieve this it seemed important: 1. to use a gel of even thickness; 2: to use a rather strictly controlled voltage increase; 3: not to use too high total electric load; 4: not to run more than ten minutes longer than necessary; 5: to soak up excess fluid at the cathode at half the focusing time. By carefully standardizing all steps of focusing and slicing we arrived at a good reproducibility. Typical results are presented in Fig. 4a and b. Immunofixation is shown in Fig. 5.

It is advantageous that the sensitivity of the ZIA can easily be selected. The distance from the gel surface to the front of an immunoprecipitate was found to be inversely proportional to the concentration of antiserum in the gel as shown in Fig. 6.

When using gel with a surface area of 12.5 x 24.5 $cm^2$ it was possible to focus 26 serum samples. One person could make two such separations per day, cut out the gel pieces, and apply these to ZIA. This makes a capacity of 52 samples per person and day.

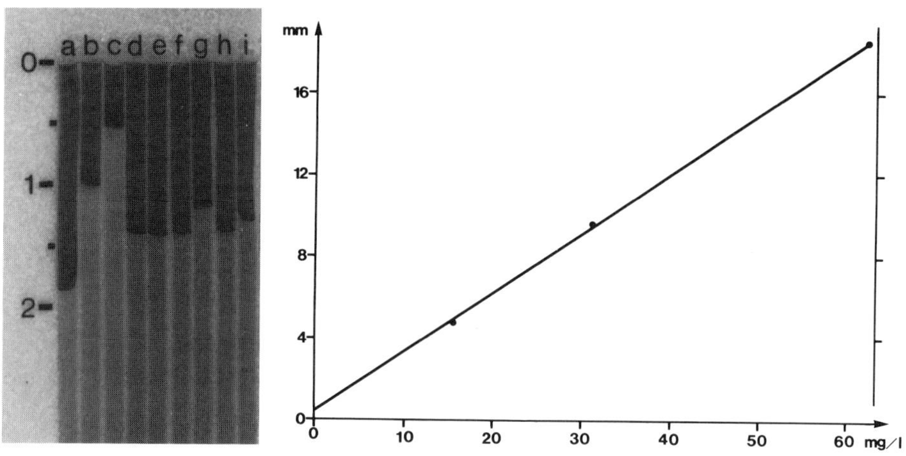

Fig. 4a.
To the left is shown a photograph of ZIA gels with stained immunoprecipitates, where a cm scale is shown to the left and the
gel rods are indicated with letters above (a-i). Standard serum
samples diluted 1:40, 1:80 and 1:160 are shown in a, b and c,
respectively. Different serum samples diluted 1:50 were applied
in d-i.

Fig. 4b.
A standard curve where the concentration of transferrin is shown
on the X-axis and the distance from the gel surface to the front
of the immunoprecipitate is shown on the Y-axis.

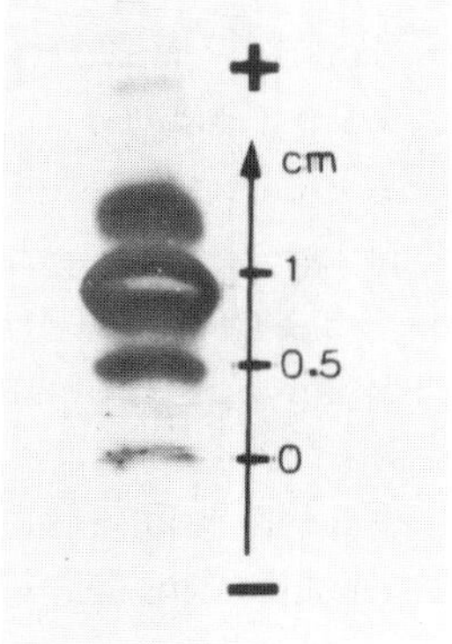

Fig. 5.

A photograph showing multiple molecular
forms of transferrin after isoelectric
focusing followed by immunofixation and
staining. A serum sample (2 $\mu$l) was used.
A central cleaning up of the major transferrin component is obtained because of
antigen excess. The anode was at the top.
A centimeter scale is shown to the right
with 0 at the minor component Tf 5.7 .

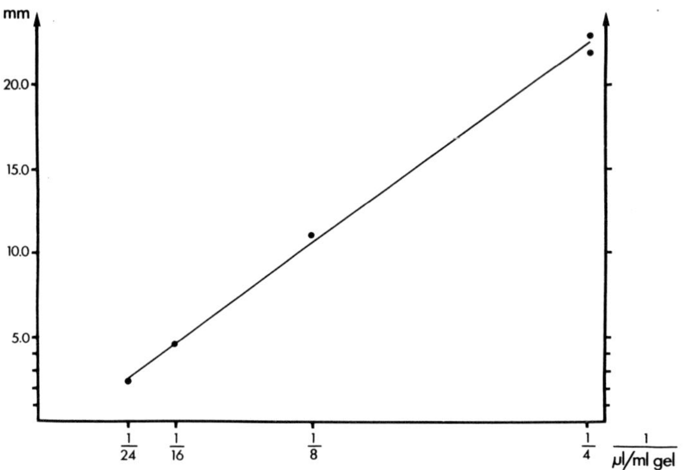


Fig. 6.
The inverse proportion between the concentration of antiserum
in µl/ml gel (on the X-axis) and the distance from the gel
surface to the front of the immunoprecipitates on the Y-axis.
The equation of the line was Y = 95.9 X - 1.5 and the regression coefficient = 0.999.


DISCUSSION


Multiple molecular forms of a protein may differ from each
other  for example in only one, or a few, of the hundreds of
aminoacids, or in one sialic acid. Separation may still be
possible with a modern method like isoelectric focusing (1). In
order to fully appreciate the high resolution obtained at separation in gels, it is desirable also to improve the quantitative
protein determination. Current evaluations of electrophoretic
protein separations are often insufficient or even misleading.

Till now most isoelectric focusing has been made in polyacrylamide gel. However, in cases where this matrix has been used it
is difficult to find reports on successful quantitative determination and/or recovery of separated components. The explana-

tion is probably that his type of gel consists of a rather tight matrix making elution difficult. Crossed immuno-isoelectric focusing, e.g. as described by Stibler (12), may be mentioned here because it gives the impression of providing quantitative results. However, practical experience has shown that this is technically difficult to achieve. Another limitation is that only a few samples can be evaluated per 24 hours. Approaches to quantitative evaluation requires determination of the area under the precipitation curves. Still the standard curves are often not linear and do not always pass close to origin. That technique has therefore been applied merely for recognition and identification. However, for these purposes immunofixation seems more rational although it also requires a long time for obtaining the final result. When using polyacrylamide gel two days are required to obtain the result (13). Densitometric evaluation is also bound with many problems and requires a very sophisticated and expensive instrument.

In order to circumvent the many problems with the use of polyacrylamide gels we prefer focusing in agarose gel. The polyacrylamide gels are more complicated to prepare and often contain substances that may react with the sample proteins at focusing and/or with the antibodies when immunotechniques are used. When using polyacrylamide gels long time periods are required for diffusion of the antibodies into the gel and excessive times (two days) are used for washing out unprecipitated proteins. A comparison is presented in Table 1. It can be added that the agarose matrix is used at less than 1 % (w/v) whereas acrylamide is often used at 5-10 % (w/v).

In addition we have direct access to the method of zone immunoelectrophoresis assay (5, 6) and relevant knowhow. This method has certain important advantages over radial immunodiffusion and Rocket electrophoresis. For example it often gives linear calibration curves and a much lower detection limit.

After cutting out the gel pieces containing the protein component of interest we first tried to elute them by incubation in
a buffer. When working with gel pieces i.e. 5 x 10 x 1.2 mm the
leaching required several hours at + 4°C to get a reasonable
yield. We have tried to shorten this time by using room temperature and above, as the diffusion rate is roughly increased
by a factor of more than two for every ten degrees´ increase
in temperature. However, this meant increased risk for microbial growth with production of hydrolytic enzymes, which might
cause degradation of the proteins. Therefore, it was tried to
facilitate diffusion by disintegrating the gel pieces with a
glas rod or a spatula. The recovery was variable partly because
disintegration was not made in a reproducible way. The shape of
the gel pieces influences to a high extent the mass of protein
that has diffused out of them during a certain time. Therefore,
in order to get reproducible recoveries a rather complete
equillibrium had to be obtained, which required at least 10 h.
Microbial growth and adsorption of proteins to the test tube
walls may then again cause problems.

Significant improvement was achieved after following the new
procedure, where each gel piece was first disintegrated by
pressing it out through the needle of a special syringe, and
depositing it just above a gel rod for ZIA. Here the proteins
were transported out of the disintegrated gel pieces with the
aid of the current. The advantages have been summarized up in
Table 2.

It is obvious that electrophoretic transport of the protein out
of a disintegrated gel is much quicker and more efficient than
diffusion. This may easily be hampered by even weak adsorption
of proteins to the gel matrix or other substances present therein. Since some years we have had experience reporting that electrophoresis is not only efficient for transport of protein
out of a gel but also for desorption when originally adsorbed
to a matrix by hydrophic interaction (14). The positive outcome

Table 1    Advantages of using agarose gel rather than polyacrylamide gel for the separation of proteins followed by immunofixation.

|  | Polyacrylamide gel (PAG) | Agarose gel |
|---|---|---|
| 1. Concentration of antibodies | high (often undiluted) | much lower (about 1/10 of that used for PAG) |
| 2. Hours for washing for the removal of proteins not precipitated | 24 – 48 | 1 – 3 |
| Hours for staining and destaining the gel | 8 – 18 | ~1 |
| 3. Hours required for obtaining a permanent storable gel | 4 – 8 | 0.5 |

Table 2    Advantages of using pieces obtained after isoelectric focusing as samples in ZIA are:

1. No leaching buffer and subsequent pipetting

2. No test tubes for leaching

3. No precision pipetting of samples for ZIA.

4. No critical parameters such as time, temperature, size of gel pieces at leaching, etc.

5. No extra time lost for manupulations and equillibrium at leaching.

6. When using this technique it is often of interest to quantify at least one out of some components, which means that relative amounts are of significance. This decreases the requirements for absolute determinations, which are not easy to make and often brings controversy at standardization within and between laboratories.

of the combined use of focusing in agarose followed by quantification with ZIA can furthermore be ascribed to the agarose gel allowing easy cutting, transfer and crushing. The introduction of a special syringe, with a small dead volume between the needle and the plunger facilitated disintegration of the gel pieces and their deposition above the gels of the ZIA method.

The method proposed here, due to its simplicity and advantages, seems generally applicable for proteins separated in gels by various electrophoretic methods. It may be used for many different applications, e.g. for studies of diseases, malfunction and in genetic investigations.


REFERENCES


1. Vesterberg, O.: Int. Lab., May/June, pp 61-68; also in Am. Lab. June. pp 13-24 (1978).

2. Basset, P., Beuzard, Y., Garel, M.C., Rosa, J.: Blood, 51, 971 (1978).

3. Jeppsson, J.-O., Franzén, B., Nilsson, K.O.: Science Tools, 25, 69 (1978).

4. Stibler, H., Allgulander, C., Borg, S., Kjellin, K.G.: Acta Med Scand 204, 49-56 (1978).

5. Vesterberg, O.: Fresenius Z. Anal. Chem. 301, 134-135 (1980a)

6. Vesterberg, O.: Hoppe-Seyler´s Z. Physiol. Chem., 361, 617-624 (1980b).

7. Bobb, D., Hofstee, B.H.J.: Anal. Biochem. 40, 209 (1971).

8. Vesterberg, O.: Acta Chem. Scand., 21, 206 (1967).

9. Rosén, A., Ek, K., Åman, P., Vesterberg, O. in: Protides Biol. Fluids, ed. H. Peeters (Pergamon Press, Oxford) p. 707 (1979)

10. Vesterberg, O.: Electrophoresis´ 79, Advanced Methods Biochem. and Clin. Applications, ed. B. Radola (Walter de Grueyter, Berlin) pp. 95-104 (1979).

11. Thymann, M., Henningsen, K.: Ärztl. Labor. 25, 47 (1979).

12. Stibler, H.: J. Neurol. Sci., 36, 273 (1978).

13. Stibler, H.: J. Neurol. Sci., 42, 275 (1979).

14. Vesterberg, O., Hansén, L.: Biochim. Biophys Acta, 534, 369, (1978).

POLYACRYLAMIDE DENSITY GRADIENT GEL ELECTROPHORESIS

A Simplified Polyacrylamide Density Gradient Gel Electrophoresis with a New Staining Method - Electrostaining

Tadao Hoshino
Pharmaceutical Institute, School of Medicine, Keio University
35, Shinanomachi, Shinjukuku,Tokyo, Japan 160

Setsuko Jitsukawa*, Masaru Tahara*, Gohki Yamada* and Kohichi
Sakakibara**
Department of Patho-biochemistry* and Department of Internal
Medicine**, Kumpuhkai Yamada Hospital
3-4-10, Minamichi, Tanashishi, Tokyo, Japan 188


Introduction

Polyacrylamide density gradient gel electrophoresis allows excellent separation of proteins with a wide range of molecular
weight on one gel column (1-3).  However, it is very trouble-
some to treat the gel columns for staining, and for comfirming
gel density at the positions of protein bands migrated.  The
density gradient gel column is too hard and fragile at the
higher density portion to strip out from the glass tube, and it
swells  disproportionally in the solution dipped for staining
and storage.  We made a device for polyacrylamide density
gradient gel electrophoresis which was designed to permit preparation of density gradient gel column, electrophresis and
staining, and designed to carry out these procedures one after
another without removing the gel column from the device (4).
With this device, we tried to stain the protein bands on the
gel column lying in glass tubes, namely electrostaining.  For
the feasibility of this new staining method, thirty six dye
stuffs were checked on human plasma proteins.

Methods

The device was made by hand of polymethacrylate. As the scheme (Fig-1) shown. The device consisted of one inlet, one anode chamber, one outlet, twenty holders of glass tubes and one cathode chamber. The glass tube holders were open to the both chambers, and glass tubes were fixed and sealed with rubber o-rings. The capacity of the anode chamber was 0.5ml. For the preparation of density gradient gel columns, a gradient maker was joined by a silicon tube to the inlet and the outlet was closed. And for the electrophoresis and the staining, the cathode chamber was connected to the inlet through a peristaltic pump.

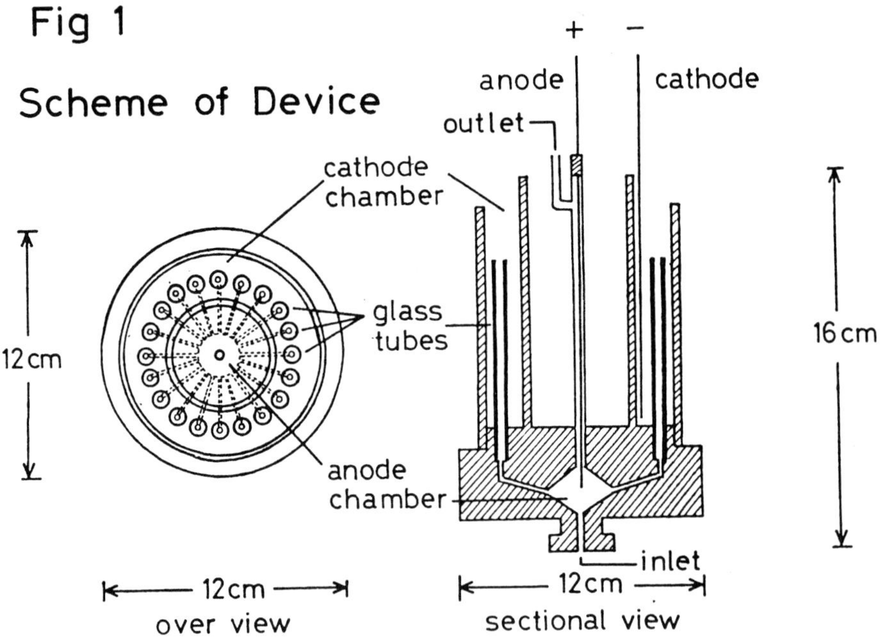


Fig-1. Scheme of the device: This device was designed to permit all the procedures of density gradient gel electrophoresis. See the text.

The buffer system was based on that of Ornstein (5) and Davis (6). Stock solutions were as follow:

A) Gradient gel buffer; 3M Tris-HCl containing 0.23%(v/v)

TEMED, pH 8.9.

B)    Stock gel solution; 60%(w/v) acrylamide monomer-3%(w/v) Bis in water.

C)    Electrode buffer; 0.05M Tris-0.38M Glycine, pH 8.3.

D)    Sucrose solution; 70%(w/v) sucrose in C-buffer.

E)    Ammonium persulfate(APS) solution; 200mg/dl APS in water.

F)    Methanol solution; 20%(v/v) methanol in water.

<u>Preparation of polyacrylamide density gradient gel column</u> was performed as follows. Twenty glass tubes(0.5 X 7cm) were fixed to the tube holders of the device, and the cathode chamber was filled with F-solution. It was necessary to make gel columns uniform. The higher and the lower density gel solutions were prepared by mixing A-buffer, B-solution and water at appropriate ratios according to the desired density gradient. The mixture of the higher and lower density gel solution from the gradient maker was mixed again with E-solution and led to the inlet of the device. When the mixture of gel solution was poured out into the device, D-solution was followed until the end of the mixture     reached to the bottom of the glass tubes.

<u>Electrophoresis of serum protein</u> was as follows:   After the gel formation, F-solution in the cathode chamber was discarded and replaced with C-buffer. Silicon tubes were set as mentioned above. 10µl of serum sample, made up to contain 20% of sucrose with D-solution, was applied to the gel top and electrophoresis was performed at 400V for 20hr. During the electrophoresis, D-solution was circurated between the two chambers to maintain the pH constant and remove the air bubbles in the anode chamber produced by the electrolysis at the electrode.

<u>The protein bands were developed</u> by the conventional dipping method and the new electrostaining method. The dipping method was as follows. The gel tube was removed from the device and the glass tube was cracked. The gel column was dipped in 0.025% (w/v) Coomassie Brilliant Blue R 250-50%(v/v) methanol-7%(v/v) acetic acid overnight for staining, and in 25%(v/v) methanol-7% (v/v) acetic acid also overnight for destaining. The electrostaining method was performed without removing the gel tube

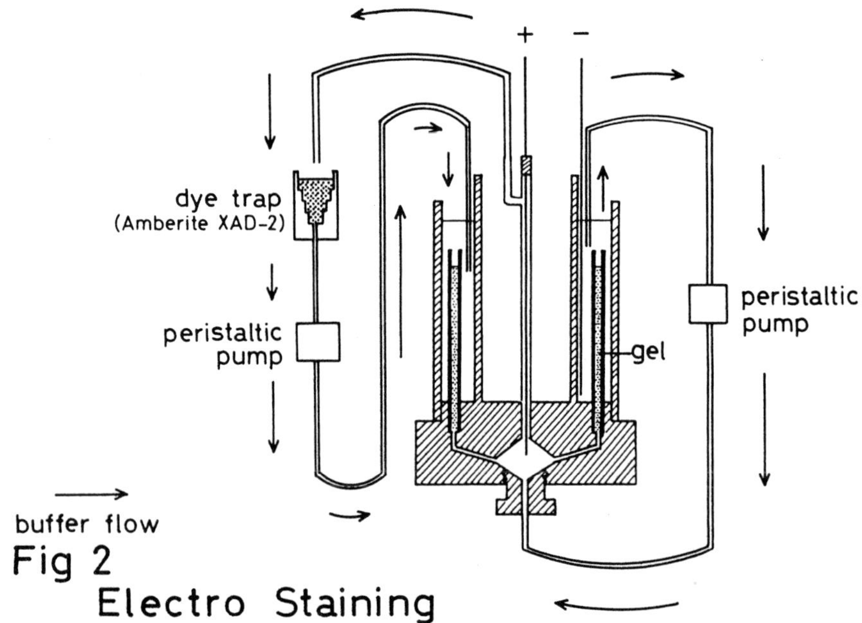



Fig-2. Electrostaining:  Flow diagram of electrode buffer was shown.  At the end of electrophoresis, a dye trap was set between the outlet and the cathode chamber.  Destaining was not nessesary, the excess dye stuff passed through the gel was adsorped by the trap.  See the text.

from the device as follows.  A dye trap which was easily made of a small glass column and a small amount of styrene resine (Amberite XAD-2) was set between the outlet and the cathode chamber (Fig-2).  20µl of dye solution containing 50-500µg of dye stuff and 20%(w/v) sucrose in electrode buffer was applied to the gel top just as the same way as the serum sample.  Then electric current was applied in the same direction as electrophoresis.  Proteins were stained and the stained bands appeared as the dye stuff moved to the anode.  Destaining was not necessary, the excess dye stuff passed through the gel was adsorped by the dye trap.

Gel density of the gel column was checked longitudinally.  As an indicator, Bromphenol Blue (BPB) was used.  Appropriate amount of BPB were added to the higher density gel solution previously, and after the gel formation the gel tubes were remov-

ed from the device and the gel density were checked at 570nm by
longitudinal scanning densitometry.


Results


The longitudinal scanning densitograms of the gel columns from
this device showed linear density gradient formation and good
uniformity, and permitted to estimate the gel density at any
point on the density gradient gel columns.  In the tested dye
stuffs, Coomassie Brilliant Blue R-250[R](R-250), Coomassie Brilliant Blue G-250[R](G-250), Remazol Turquoise Blue G[R](G), Direct
Deep Black E[R](E) and Lanasol Violet 3-B[R](3-B) were available
for the electrostaining method.

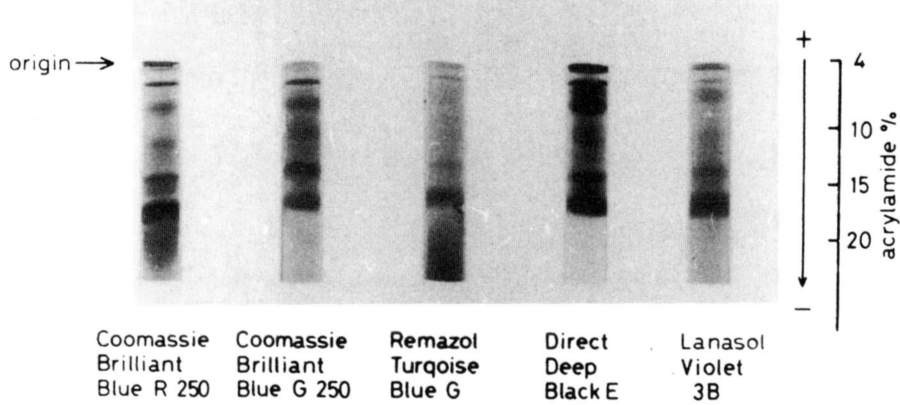


Fig-3.  Profiles of serum proteins stained by electrostaining:
Density gradient gel electrophoreses were performed on the
same human serum and the protein bands were developed by electrostaining method with R-250, G-250, G, E and 3-B.  See the
text.

The bands stained by this method were thinner than those stained by the conventional dipping method.  Lanasol Violet 3-B[R]
showed the clearest distribution pattern of bands particulary
in the high molecular weight region.  Direct Deep Black E[R]
showed the most intensive staining, but the back ground was
considerably high. When the 4-45% density gradient gel column
was used ten bands were developed by the conventional dipping

122

method, and eight to eleven bands by this electrostaining method; ten bands by R-250, ten bands by G250, eight bands by G,
ten bands by E and eleven bands by 3-B. Depending on the kinds
of dye stuffs different distribution profiles of stained bands
were appeared on the gel column with the same serum proteins.
Staining intensity against the same protein band were differ-
ernt among the dye stuffs (Fig-3). 100ng protein/gel or less
amounts of bovine serum albmin was detected by electrostaining
with these five dye stuffs. Staining sensitivity was somewhat
lower by this method than by the dipping method using the same
dye stuff. The positions of the protein bands did not changed
respectiv  when the time for electrophoresis increased from
20hr to 28hr. The migration distance of proteins depended on
their molecular weight and the gel density of the gel column.
As shown in Fig-4, log acrylamide density has a linear relationship to the log molecular weight.

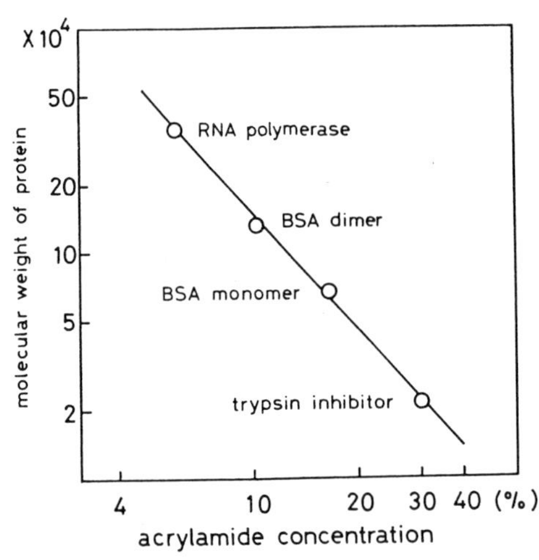


Fig-4. Relation between molecular weight and migration distance: Four proteins, shown in Fig., were charged on the separate gel column and stained by electrostaining for 20hr after 20hr of the electrophoresis. Migration distance was expressed in acrylamide percent at the point of protein band.

## Discussion

A number of the methods for the detection of proteins on polyacrylamide gel have been reported. These methods may be divided into two general groups; post-staining and pre-staining methods. The post-staining methods are available for almost all
kinds of polyacrylamide gels and used most widely. The gel is

dipped, for staining, in an acidic solution of dye stuff such as Amido Black (6,7), Fast Green (8,9) and Coomassie Blue (10-13), and destaining and pretreatments for dipping are necessary in these methods.  On the other hand, the pre-staining methods are suitable for SDS-polyacrylamide gel electrophoresis.  Proteins are treated with dye such as Remazol Brilliant Blue (14), cationic dyes (15), DANSYL-chloride (16-18), fluoresamine(19) and orthophthalaldehyde (20) before the electrophoresis. Our new staining method for proteins on polyacrylamide density gradient gel , namely electrostaining, is one    post-staining method.  This method is characterized by its simplicity. Pretreatment (i.e., prestaining of proteins and stripping out of the gel column from gel-holder) and destaining are not necessary.  Proteins on the polyacrylamide density gradient gel were stained under alkaline conditions.  In an acidic medium, cationic dye reacts with $NH_3^+$ and non-polar groups of protein (21). Under the alkaline conditions, positive charges are not expected and staining intensity must be low.  In fact, as mentioned above, electrostaining with R-250, G-250, G, E and 3-B showed lower staining intensity compaired  with the dipping method with the same dye stuffs.  The detection limit of the electrostaining method was 100ng-5µg of protein and differ with respective dye stuffs. This value is lower than that by the conventional dipping method, however, higher than that of the prestaining method with non-fluorescent dye (14).  Polyacrylamide density gradient gel electrophoresis is able to estimate molecular weight of proteins (3).  By a pore limit effect of the acrylamide-polymer, the velocities of the proteins in the electric field are reduced and proteins almost stopped at respective gel density on the gradient gel column according to their molecular sizes.  In our experiment, serum proteins almost stopped for 20-48hr with  a   potential gradient of 400V. They did not move practically during electrostaining.  The gel column did not swell, and the gel density at protein band was easily determined.  The relation of the gel density(%) at the band and the molecular weight was linear in the logarithm forms.

In the tested dye stuffs, R-250, G-250, G, E and 3-B are presently available for electrostaining. The protein-dye binding is affected by the velocity of the reaction between dye and protein, and also by the velocity as the dye passes by protein. By selecting more suitable dye stuffs, it  may be possible to shorten the staining time and to make the staining itself more sensitive.  The electrostaining method is very simple to carry out and easy to perform densitometry and storage of gel column.  The electrostaining method may become useful in density gradient gel electrophoresis.

References


1.    Lorentz, K.: Anal. Biochem. 76, 214 (1976)

2.    Reichel, W., Wolfran,D.E., Klein,R., Scheler, F.:Klin. Wschr. 54, 19 (1976)

3.    Slater, G.G.: Anal. Chem. 41, 1039 (1969)

4.    Hoshino, T., Kan, K.: Seibutsu Butsuri Kagaku. 22, 200 (1979)

5.    Ornstein,L.: Ann. N.Y. Acad. Sci. 121, 321 (1964)

6.    Davis, B.J.: Ann. N.Y. Acad. Sci. 121, 404 (1964)

7.    Kruski, A.V., Narayan, K.A.: Anal. Biochem. 60, 431 (1974)

8.    Gorovsky, M.A., Carlson, K., Rosenbaum, J.L.: Anal. Biochem. 35, 359 (1970)

9.    Bertolini, M.J., Tankersley, D.L., Schroeder, D.D.: Anal. Biochem. 71, 6 (1976)

10.   Diezel, W., Kopperschlaeger, G., Hoffman, E.:Anal. Biochem. 48, 617 (1972)

11.   Mayer, T.S. Lamberts, B.L.: Biochem. Biophys. Acta.: 107, 144 (1965)

12.   Weber, K., Osborn, M.: J. Biol. Chem.:244, 4406 (1969)

13.   Fairbanks, G., Steck, T.L., Wallach, D.F.H.: Biochemistry 10, 2606 (1971)

14.   Griffith, I.P.: Anal. Biochem. 46, 402 (1972)

15.   Datyner, A., Finnimore, E.D.: Anal. Biochem. 55, 479 (1973)

16.   Talbat, D.N., Yphantis, D.A.: Anal. Biochem. 44, 246 (1971)

17.  Stephens, R.E.: Anal. Biochem. 65, 369 (1975)

18.  Kato, T., Sasaki, M.: Anal. Biochem. 66, 515 (1975)

19.  Eng, P.R., Parkes, C.O.: Anal. Biochem. 59, 323 (1974)

20.  Weidekamm, E., Wallach, D.F.H., Flückiger, R.:Anal.
     Biochem. 54, 102 (1973)

21.  Maurer, H.R.: Disc Electrophoresis and Related Techniques
     of Polyacrylamide Gel Electrophoresis, pp72 Walter de
     Gruyter

22.  Lambin, P., Rochu, D., Fine, J.M.:Anal. Biochem. 74, 567
     (1976)

# ENHANCEMENT METHODS IN THE LOCALIZATION OF PROTEINS FOLLOWING ELECTROPHORESIS OR ISOELECTRIC FOCUSING

Andrew Myron Johnson

Department of Pediatrics, University of North Carolina School of Medicine
Chapel Hill, North Carolina  27514, USA

Introduction

Recent advances in the technology of fixation and staining have resulted
in markedly increased sensitivity in the detection of proteins in
separatory gels.  In particular, the development of rapid and inexpensive
silver stains is worthy of note (1, 2).  These nonspecific methods are
mentioned elsewhere in these Proceedings and will be discussed here only
in association with other methods.  (Diamine silver stains should not be
used with agar or agarose-containing gels, which are stained by the
silver.)  Techniques which have been utilized for the localization and
identification of specific proteins or groups of proteins include
immunological reactions, ligand binding, enzyme-substrate reactions, and
labels such as radioisotopes, fluorescent tags, and enzymes.

Immunological Reactions

Traditional methods have included immunoelectrophoresis (3) and crossed
immunoelectrophoresis (4).  The former of these suffers from a lack of
sensitivity and from the inability to detect minor differences in
electrophoretic ability.  The latter is, under proper conditions, a very
sensitive method which is particularly useful for detection of conversion
products, complexing, or other changes within a single sample.  However,
it does not permit accurate comparison of electrophoretic mobilities.

Alper and Johnson (6) modified the direct immunoelectrophoretic methods
of Wilson (7) and Afonso (8) for the identification of genetic variants

and conversion products of plasma proteins.  In their technique, called immunofixation electrophoresis, antigens are precipitated by the direct application of monospecific antiserum following high-resolution agarose gel electrophoresis.  This method has been utilized in the study of many proteins and is of particular usefulness in protein phenotyping and in classification of monotypic immunoglobulins, or M-components (9, 10). Modifications of the original technique include application of antiserum with cellulose acetate strips (11) or with plastic masks (10).  Either of these permits the use of antisera with several specificities in the same gel.  This is helpful especially in the classification of M-components.

Immunofixation may also be used following electrophoresis or isoelectric focusing in agarose-acrylamide (12) or acrylamide (13) gels.  Because of the smaller pore sizes of these gels, compared to agarose, more prolonged washing in saline is necessary to remove unwanted proteins and unprecitated antibodies.  As a result, Arnaud and coworkers proposed the use of a cellulose acetate overlay method with a resulting "immunofixation print" or "immuno-print" (14) of the isoelectric focusing patterns of $\alpha_1$-antitrypsin variants.  The same technique has been applied to group-specific component (15) and other proteins.  In this case, cellulose acetate strips (Sepraphore III, Gelman Instruments Co.; strips from other sources must be tested, since all do not work) are applied for only a few minutes, washed in saline, and stained.  In the Ritchie method (9, 11), antibodies are allowed to diffuse from the strip into the gel for a longer period of time, then the gel itself is washed in saline and stained.

Regardless of the primary method used, the sensitivity of immunofixation can be enhanced by using silver stains or enzyme-linked second antibodies (see below).  Because of the extensive washing required and the necessity of performing autoradiography, the author does not feel that the use of radioisotopically-labeled antisera is warranted.

The major problem likely to be encountered with immunofixation is antigen excess, particularly with isoelectric focusing.  In general, high-titered high-affinity antibodies should be used, and serial dilutions of antigen should be tested to establish optimal conditions.  Most antisera produced in rabbits or goats work well.  Prolonged washing in tap water following saline may result in loss of immunoprecipitate.


## Polymer Enhancement of Immune Complex Formation

The steric exclusion of antigen-antibody complexes by solutions of dextrans (16) or polyethylene glycol (A.M. Johnson, unpublished observations) may increase the sensitivity of immunoprecipitin techniques. Polyethylene glycol (2 to 4 percent final concentration, PEG 4000) may be added to the antibody-containing gels used in electroimmuno assay or crossed immunoelectrophoresis; however, the background will become translucent. Either enhancing agent may be applied to the gel following electrophoresis or isoelectrofocusing with any of the precipitin techniques. The polymer must be removed completely from the acrylamide gel if a silver stain is used; as noted previously, silver stains should not be used with agar or agarose.


## Ligand Binding

Ligands may be added to samples before electrophoresis or isoelectric focusing (17) or applied following the separatory run (18). In either case, if the ligand does not have an inmate characteristic (such as fluorescence) that permits its detection, a label must be used. In most reported cases, these have been radioisotopes. The addition of ligands before separation may affect the electrophoretic mobility, isoelectric point, effective molecular size, or combinations of these; as a result, samples should be saturated with ligand to avoid artifacts.


## Substrate/Product Reactions

The use of colorimetric or fluorometric substrates for the localization of isomers of enzymes is well-established in clinical chemistry as well as in research. Substrate may be applied directly to the separatory gel or by overlaying a second gel containing substrate, buffer, and necessary

Anode

Origin

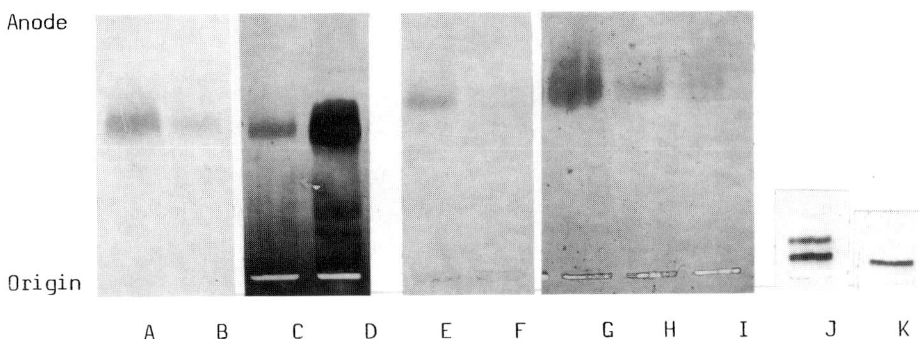

A    B    C    D    E    F    G    H    I    J    K

Electrophoresis of amniotic fluid from a pregnancy with an anencephalic fetus. A,B: Undiluted and 1:4, Coomassie R250 stain. C,D: 1:16 and 1:4, silver stain (ref. 1). E,F: Undiluted and 1:4, immunofixed with rabbit anti-human alpha fetoprotein (AFP), Coomassie R250 stain. G,H,I: 1:4, 1:16, and 1:64, immunofixed for AFP, silver stain. J,K: Cholinesterase (acetyl thiocholine substrate, dithiooxamide stain), with (K) and without (J) specific acetyl cholinesterase inhibitor. Note 5- to 10-fold increase in sensitivity with silver stain. Alpha-fetoprotein is not visible in the anodal portion of albumin without immunofixation. Pseudocholinesterase (cathodal band in J and K) and acetyl cholinesterase (anodal band in J) are visible only with substrate/product stain.

cofactors.

An interesting recent adaptation of the gel overlay technique has been the utilization of gels containing sensitized sheep erythrocytes and serum deficient in a single complement component for the detection of complement factor polymorphisms (19, 20). In the case of C4, this method has shown that the products of the C4A locus are less stable than those of the C4B locus. Detection of variants produced by the former is best performed using immunofixation (20).

The cleavage of acetyl thiocholine by nonspecific pseudocholinesterase or acetyl cholinesterase results in an insoluble product of thiocholine (21). Although this precipitate can be seen in separatory acrylamide gels, visualization is enhanced by staining with dithiooxamide (22), as shown in the Figure.

Labels

Immunologic precipitates and the products of enzymic cleavage may be
detected by the use of radioisotopic or fluorescent markers.  More
recently, enzyme-coupled antisera have been used for detection of
immunoprecipitates (23), whether in immunofixation, immunoelectrophoresis,
electroimmunoassay, or other techniques.  Best results are obtained with
horseradish peroxidase-coupled second antisera, which are applied
following the initial precipitation and washing in saline (A. Ingild,
personal communication; A. M. Johnson, unpublished observation).
Peroxidase-antiperoxidase complexes do not improve detection.
Diaminobenzidine is an excellent substrate for detection of the peroxidase
(24, 25), although it is carcinogenic and must be handled with care.

132

References


1.   Merril, C. R., Goldman, D., Sedman, S. A., Ebert, M. H.:  Science 211, 1437-1438 (1981).

2.   Switzer, R. C., Merril, C. R., Shifrin, S.:  Anal. Biochem. 98, 231-237 (1979).

3.   Scheidegger, J. J.:  Intern. Arch. Allergy Appl. Immunol. 7, 103-110 (1955).

4.   Laurell, C.-B.:  Anal. Biochem. 10, 358-361 (1965).

5.   Laurell, C.-B., Persson, U.:  Biochim. Biophys. Acta 310, 500-507 (1973).

6.   Alper, C. A., Johnson, A. M.:  Vox Sang. 17, 445-452 (1969).

7.   Wilson, A. T.:  J. Immunol. 92, 431-434 (1964).

8.   Afonso, E.:  Clin. Chim. Acta 10, 114-122 (1964).

9.   Ritchie, R. F., Smith, R.:  Clin. Chem. 22, 1982-1985 (1976).

10.  Sun, T., Lien, Y. Y., Degnan, T.:  Amer. J. Clin. Pathol. 72, 5-11 (1979).

11.  Ritchie, R. F., Smith, R.:  Clin. Chem. 22, 497-499 (1976).

12.  Johnson, A. M.:  J. Lab. Clin. Med. 87, 152-163 (1976).

13.  Johnson, A. M.:  Ann. Clin. Lab. Science 8, 195-200 (1978).

14.  Arnaud, P., Wilson, G. B., Koistinen, J., Fudenberg, H. H.:  J. Immunol. Meth. 16, 221-231 (1977).

15.  Constans, J., Viau, M., Cleve, H., Jaeger, G., Quilici, J. C., Palisson, M. J.:  Hum. Genet. 41, 53-60 (1978).

16.  Hyslop, N. St. G., Cochrane, D. G.:  J. Immunol. Meth. 6, 99-107 (1974).

17.  Daiger, S. P., Schanfield, M. S., Cavalli-Sforza, L. L.:  Proc. Nat. Acad. Sci. 72, 2076-2080 (1975).

18.  Allen, R. C.:  Presentation at Electrofocus/78, Atlanta, 1978.

19.  Raum, D., Glass, D., Carpenter, C. B., Schur, P. H., Alper, C. A.:  Amer. J. Hum. Genet. 31, 35-41 (1979).

20.  Awdeh, Z. L., Alper, C. A.:  Proc. Nat. Acad. Sci. 77, 3576-3580 (1980).

21.  Smith, A. D., Wald, N. J., Cuckle, H. S., Stirrat, G. M., Bobrow, M., Langercrantz, H.:  Lancet 1, 685-688 (1979).

22.  Chubb, I. W., Smith, A. D.:  Proc. R. Soc. Lond. B. 191, 245-261, (1975).

23.  Pinon, J. M., Dropsy, G.:  J. Immunol. Meth. 16, 15-22 (1977).

24.  Avrameas, S.:  Histochem. J. 4, 321-330 (1972).

25.  Malmgren, L., Olsson, Y.:  J. Histochem. Cytochem. 25, 1280-1283 (1977).

ONE AND TWO-DIMENSIONAL MICROELECTROPHORESIS AND

"STAINING" OF PROTEINS WITH A SILVER METHOD


Hans Michael Poehling and Volker Neuhoff
Max-Planck-Institut für experimentelle Medizin
Forschungsstelle Neurochemie
D-3400 Göttingen, Germany


Introduction

All widely used systems for PAGE, SDS-PAGE, PAGIF or 2-D electrophoresis can be converted to so called micro versions, simply
by reducing the gel dimensions. Proteins are concentrated in
bands or spots as small and distinct as possible in order to
achieve a maximal local extinction of light after staining procedures. A reduction of the gel length results in sharper bands
in PAGE or PAGIF, but in some instances, especially when complex
protein mixtures have to be fractionated, in incomplete resolution. In general the gel length should be held as short as
possible, but even when micro systems are used it must be varied
in order to achieve optimal resolution. The main reason for the
increased sensitivity of micro-systems is the small diameter of
the separation lane. This is determined in capillary electrophoresis by the inner diameter of the tube and in slab gel electrophoresis by the width of the sample application point, normally the width of the sample well. Ideal conditions for highly
sensitive micro-systems are provided by capillary electrophoresis (1,2,3,4,5,6,7) with the possibility to reduce the gel diameter to as little as 100 μm or even to 50 μm, with the additional advantage of easy sample application. In addition large
volumes of diluted samples can be applied due to the concentration effect of the multiphasic buffer systems.

134

When the direct comparison of different samples is necessary
fractionation in slab gels is superior to capillary gels. Micro
versions of slab gel electrophoresis have been recently described (8,9,10,11,12,13). It is nearly impossible to produce sample
wells in 3 or 6 % stacking gels with the usual comb technique
in vertical PAGE systems less than 500 μm in width. Therefore
the sensitivity of the micro slab gel system decribed cannot be
compared to micro electrophoresis in capillaries. However, it
should be pointed out that debating the term "micro" is not the
aim of this paper.

Next to the enhanced sensitivity, other important features of
the "micro" slab-system are: Short separation, staining and destaining times; easy adaptability for various separation problems, easy handling, low cost for chemicals and also technical
equipment which may be constructed in any small workshop.

Sensitivity of protein detection in micro slab gels can be enhanced by radioactive labelling or silver staining (14,15). In
particular silver staining by a modification of the original
Switzer method (16) (Oakley et al. (17) and Allen (18)) seems
to be suitable for routine analysis. When applying the Oakley
method we encountered problems with the reproducibility and
background staining. Therefore we attempted to find conditions
for a reproducible and fast silver method (19). Methods and the
problems associated with micro gel electrophoresis and silver
staining will be described and discussed in this paper.


Methods and Results

Details of the micro-slab gel system have been described (13).
Here we shortly repeat the basic principles and present examples
of application.

Gel chambers (Fig. 1-1) were constructed from microscope slides cut to a suitable size. Small strips of plastic sheets of various thickness (0.2 to 1 mm) were used as spacers. Special clamps (Fig. 1-3), melted dental wax, or electric tape were used to seal the chambers laterally. Sealing of the bases of the chambers was not necessary when spacers of less than 0.25 mm thickness were used, as then the capillary forces were sufficient to maintain the gel solution between the plates. Thicker chambers were sealed by pressing them into a cushion of plasticine coated with parafilm (Fig. 1-3), or electric tape. The upper reservoir

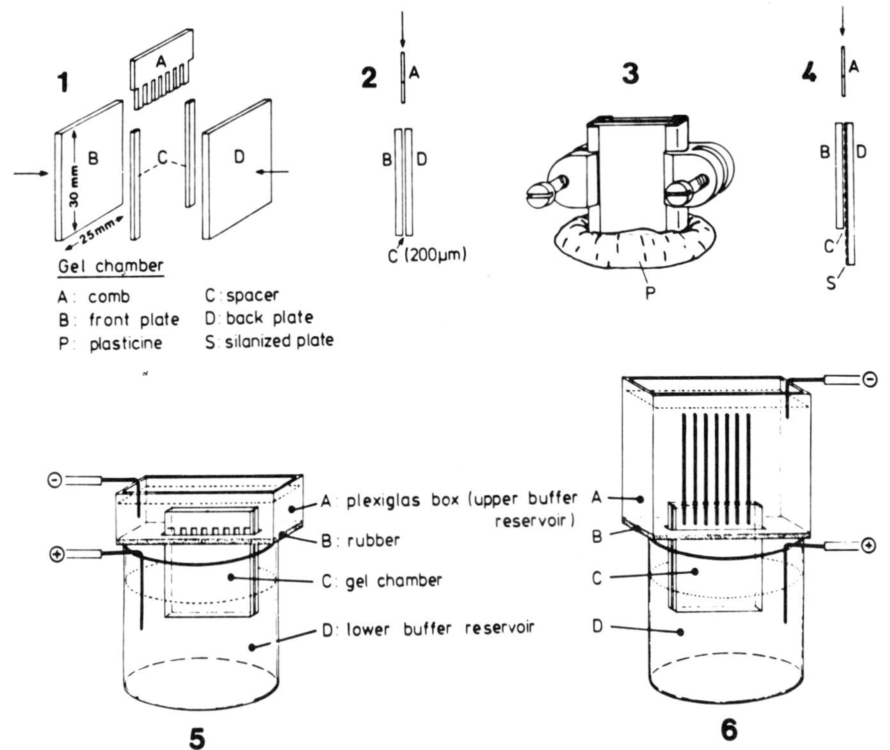


Fig. 1: Schematic drawing of the technical equipment and electrophoresis stand used for micro-slab gels. 1 and 2: Micro-gel chamber, 3: Teflon clamps for lateral sealing of the gel chamber, 4: Chamber for preparation of gradient gels on silanized glass plates, 5: Complete electrophoretic stand, 6: Electrophoretic stand for combined slab gel technique with sample application with μl capillaries.

was a simple plexiglass box with a bottom of soft rubber, into
which a slit was made to hold the gel chamber in position. The
lower buffer reservoir was a beaker of suitable size.

We often used silanized plates according to Radola (20) to fix
the gel to one plate of the gel sandwich. This prevented shrin-
king or swelling of the gel and provided ideal support after
drying of the gel for densitometric or fluorographic evaluation.

Homogenous gels: The buffer system of Laemmli (21) was used for
all slab gels with the exception that the stacking gel for 2-D
separations was buffered with tris/phosphate pH 6.9 according
to Klose (8). Homogenous slab gels were prepared by filling the
gel chambers with separation gel mixture of 10% or 15% acrylamide by either capillary attraction, or using fine pipettes to
approximately 2/3 of their height. After polymerisation a 3%
stacking gel solution was applied and teflon combs, forming
sample wells 1 mm wide and 3 mm high, were introduced. An example of the resolution which can be achieved in 10% homogenous
micro-slab gels (30x35x0.3 mm) is represented in Fig. 2 which
shows SDS-PAGE of proteins extracted from small plant-parasitic
nematodes, Xiphinema index feeding on root tips of fig seed-

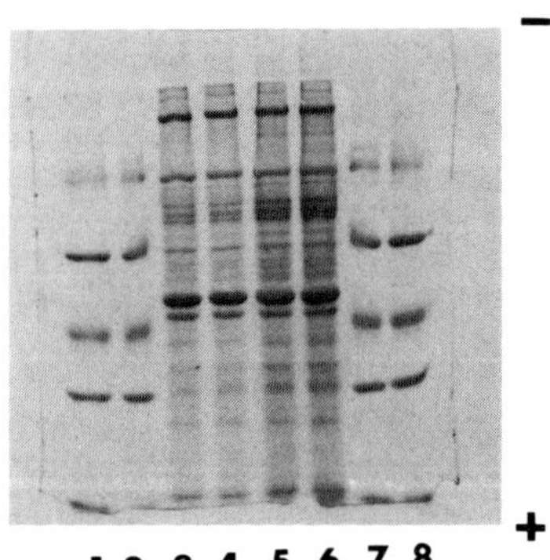


Fig. 2: One-dimensional separation of proteins extracted from
fed and starved X. index females. SDS-PAGE in a 10% micro-slab
gel (30x35x0.3 mm). 1,2,7, and 8 marker proteins; 3,4 starved
nematodes; 5,6 fed nematodes.

lings. About 0.5 µg of total nematode protein was loaded per
sample well. The separation time was 30 min. at room temperature
and the gel was stained with Coomassie Blue for 15 min. and destained for 1 hour after which it was ready for densitometric,
or photographic evaluation. Proteins with molecular weights of
more than 30,000 are usually very well separated by this system.

In Fig. 3 the influence of both decreasing sample well sizes
and increasing gel length on the sharpness of protein bands is
demonstrated. Equal amounts of marker proteins were fractionated
in every lane. With small sample wells, protein concentration
and local color development after Coomassie Blue staining increased. In 1 mm wells the limit of protein detection was about
five times more sensitive than in 5 mm wells. A reduction of the
gel length resulted in a slightly less pronounced effect. Even
smaller protein bands and conditions approaching those of micro-
electrophoresis in capillaries could be achieved when µl capillaries (Drummond) were used for sample application (13), see
Fig. 1-6.

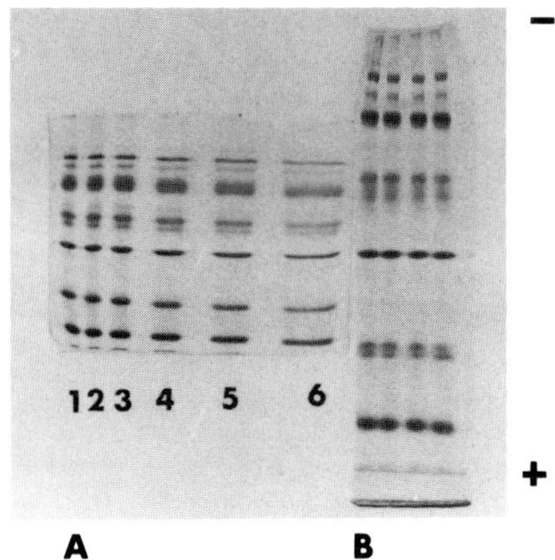


Fig. 3: Influence of sample well size on sensitivity of protein
detection. Each sample well was loaded with an equal amount of
marker proteins (low molecular weight protein calibration kit
from Pharmacia). A: Micro-slab gel (35x25x0.5 mm). Width of sample well: 1: 1 mm, 2: 1.5 mm, 3: 2 mm, 4: 3 mm, 5: 4 mm and 6:
5 mm. B: Strip of a normal sized slab gel (80x80x0.5 mm). Each
sample well 1 mm wide.

138

<u>Gradient gels:</u> For the fractionation of protein mixtures of wi-
dely differing molecular weights gradient gels are most suitable.
Gradient formation for batches of micro gels was performed with
a two chamber gradient mixer and a simple casting chamber.
Several gel chambers were assembled in a plexiglass box with a
removable front plate. The chamber was first filled completely
with water or 0.1% SDS to completely eliminate capillary forces,
afterwards a preformed gradient was slowly pumped into the chamber through a hole in the center of the bottom. The whole polymerized block was then removed from the chamber, stored in the
cold and slabs were cut out when required.

Fig. 4 shows the fractionation of total brain proteins from control rats of different ages and from rats with experimentally
induced phenylketonuria (22) in a 6 - 30% micro-gradient gel.
The variation of protein bands due to age and disease are clearly visible.

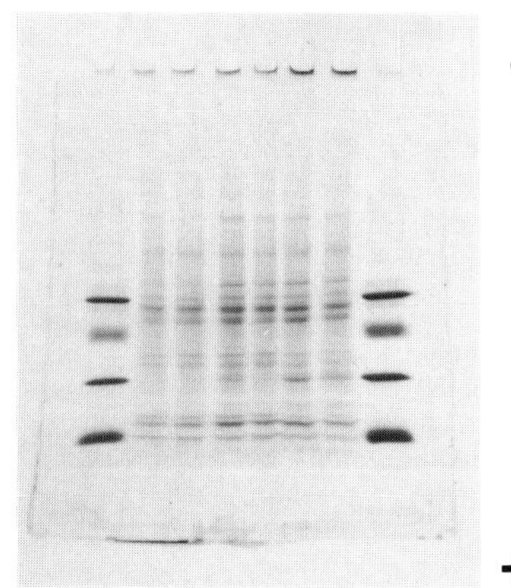


Fig. 4: SDS-PAGE in a 6 - 30% gradient micro-slab gel (30x35x
0.5 mm) of total brain proteins from rats with and without experimental phenylketonuria at different stages of development.
Lanes 1 and 8: Marker proteins (0.15 µg of each). Lanes 2,4 and
6: Control rats at 10, 15 and 20 days of age, respectively.
Lanes 3, 5 and 7: Experimental rats at 10, 15 and 20 days of age,
respectively. 0.5µg total brain protein was run in each case.

<u>2-D electrophoresis</u>: For 2-D electrophoresis, PAGIF in μl capillaries in the first dimension was combined with SDS-PAGE in homogenous, or gradient micro-slab gels in the second dimension, using the gel systems of Klose (8) and O'Farrel (14).

First dimension PAGIF was performed in μl capillaries with an inner diameter of 0.6 mm. Usually gels containing 4% acrylamid, 9M urea and 2% carrier ampholytes (Servalyte AG, various pH ranges) with or without 2% NP-40 were used. Proteins to be separated were dissolved in 9M urea, or 9M urea + 2% NP-40 with 2 - 5% mercaptoethanol at a concentration of 10 mg/ml. 1 μl of sample solution was applied to the acid end of the gel. If larger sample volumes have to be applied, an anodic shift of the protein pattern, due to pH gradient formation in the sample column, has to be corrected by adding more acid carrier ampholytes to the gel mixture. 0.5% $H_3PO_4$ and 0.5 N NaOH were used as anolyt and catholyte respectively.

PAGIF gels were transferred to the second dimension SDS-PAGE gel with fine forceps and held in place on the top of the stacking gel with 0.5% agarose in stacking gel buffer. The equilibration of the PAGIF gel in SDS solution is only recommended when proteins are fractionated, which are difficult to load with SDS. In our experiments only some membrane proteins from myelin preparations were not completely eluated from the PAGIF gel without intensive SDS loading.

For the second dimension homogenous or gradient gels of 0.7 mm thickness were used in most experiments. With smaller chambers the upper inner sides of the two plates had to be ground to form a V-shaped groove to accomodate the PAGIF gel. With these gels no increase in sensitivity could be achieved, however, but they are very suitable for autoradiographic protein detection. Fig. 5 and 6 show examples of 2-D micro-separations of proteins from rat brain and rat spinal cord. The micro-system has a similar resolving power as macro systems, but is about 10 times faster.

140

The gels shown in Fig. 6 were fixed to glass plates using silane
(see above).

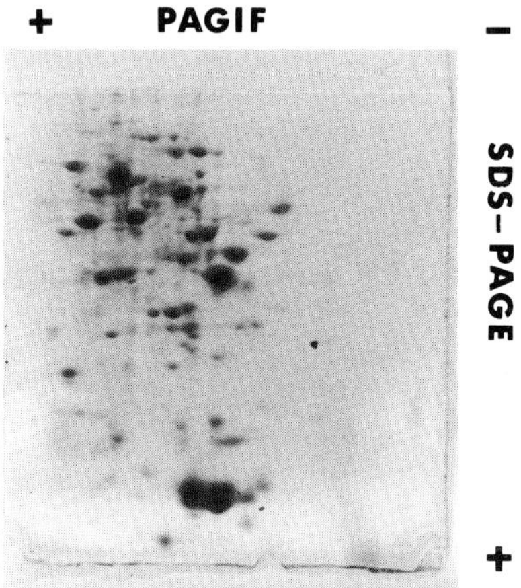


Fig. 5: 2-D electrophoresis of soluble rat brain proteins. First
dimension: PAGIF in 10 µl capillaries. Gel system: 4% acrylamide,
9M urea, 2% carrier ampholytes (Servalyte AG pH 2-4, 4-6, 6-8,
7-9 and 9-11 in equal parts). 25µg protein were fractionated.
Second dimension: SDS-PAGE in a 15% acrylamide micro-slab gel
(25x40x0.7 mm).

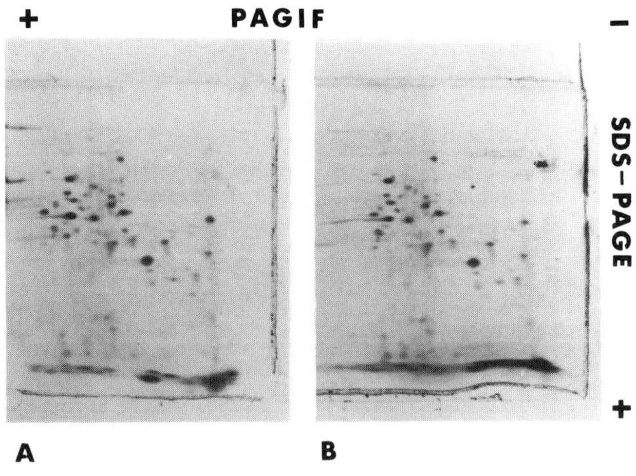


Fig. 6: 2-D electrophoresis of (A) total spinal cord protein
from 30 days old rats and (B) from rats of the same age but with
experimental phenylketonuria. 8 µg protein were used for PAGIF.
Gels as in Fig. 5 with the exception of slab size: 30x35x0.7 mm.

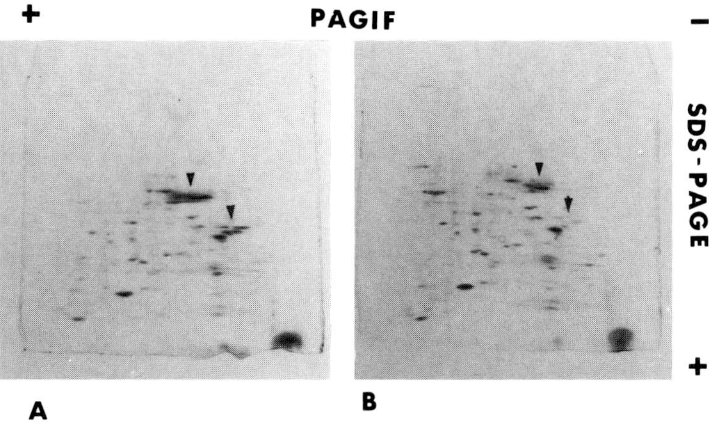



Fig. 7: 2-D electrophoresis of total proteins from X. index females. First dimension: PAGIF in 10 µl capillaries. Gel system:
4% acrylamide, 9M urea, 2% NP-40, 2% carrier ampholytes (Servalyte AG pH 2 - 11). Second dimension: SDS-PAGE in 6 - 25% micro-
slab gradient gels (30x35x0.7 mm). A: Fed nematodes, B: Starved
nematodes. Arrows: Selected proteins exhibiting distinct differences between A and B.

Fig. 7 shows proteins extracted from plant parasitic nematodes
(see above) fractionated by 2-D micro-electrophoresis. The long
time required to prepare the very small specimens could be greatly reduced with the micro-system. In contrast to 1-D electrophoresis in either capillaries (23), or slab gels (15; Fig. 2) in
which the differences in the protein patterns of fed and starved
nematodes could be only recognized in a few bands, striking differences were evident after 2-D electrophoresis. It could be
shown that specific proteins were accumulated in feeding nematodes which were continously metabolized during starvation or
egg development.

142

## Silver Staining

To further enhance the sensitivity of micro-electrophoresis we
examined the silver method of Oakley (17). The small micro-gels
should be very suitable for silver staining as only small amounts
of the expensive silver nitrate are required. The problems encountered with the 10% or 15% homogenous, or 6 - 30% gradient
micro-gels were severe background staining and bad reproducibility. The use of background reducers as proposed by Oakley and
Switzer is very questionable as the reduction in background
staining is linked with a reduction in the intensity of the protein stain. Furthermore this clearing depends on the acrylamide
concentration and thus gradient gels are not evenly destained.

After many trial and error experiments we were not able to find
a staining procedure which gave both an optimal silver deposition on the proteins and at the same time a clear background.
Factors which intensify the stain such as increased incubation
time in glutaraldehyde, concentration of the silver solution,
time of silver impregnation, the concentration of formaldehyde
in the developer and developing time also increase background
staining. Washing procedures used to eliminate interfering substances also reduce to some extent the staining intensity due
to loss of protein from the gel. Silver staining as presently
used is thus a very critical procedure as its reproducibility
depends on many factors. The procedure described here considers
these factors and requires a precise timing of all individual
steps and gives satisfactory results with good reproducibility.

<u>Staining</u>
1: Fix gels in 50% methanol/10% acetic acid for at least 30 min.
2: Wash in 15% methanol 3 x 10 min. (shaking).
3: Incubate in 10% glutardialdehyde for 15 min. at 40°C (shaking)
4: Wash in 15% methanol 3 x 15 min. (shaking).
5: Incubate in silver nitrate solution (40 ml 0.02 N NaOH + 1.5
   ml 25% $NH_4OH$ + 2 ml 20% $AgNO_3$ (stirring), add water to 50 ml)

for 12 min. at 20 - 22°C.

6: Wash in quartz destilled water for 5 min. (shaking).

7: Develop silver stain, shaking the gel for 3 - 4 min. in 100 ml 0.005% citric acid + 50 µl 37% formaldehyde.

8: Immediately after the bands became intensively stained with a clear background wash the gel with several changes in destilled water.

The procedure described here in contrast to Coomassie Blue staining does not stain proteins throughout the whole cross section of the gel. The silver deposition is more or less a reaction at the gel surface. The problem may be minimized when thin slab gels are used as described by Görg et al. (24) and Radola (20). However, initial experiments with such gels demonstrated that large amounts of protein were lost during the washing procedures. Furthermore, certain proteins were washed out to a far greater extend than others.

| Protein | Relative amounts in % of total protein | | |
|---|---|---|---|
| | A | B | C |
| Phos (Phosphorylase b) | 11.0 | 11.5 | 5.2 |
| BSA (Bovine Serum Albumin) | 14.3 | 15.8 | 17.4 |
| Ov (Ovalbumin) | 25.4 | 26.7 | 28.2 |
| CA (Carbonic Anhydrase) | 14.3 | 12.9 | 5.7 |
| ST (Trypsin Inhibitor) | 13.8 | 14.8 | 17.9 |
| Lac ($\alpha$-Lactalbumin) | 20.9 | 17.8 | 25.7 |

Tab. 1: Variation in protein stainability: Coomassie Blue versus the silver staining method. A: Percentage weight of proteins in the sample. B: Percentage of protein determined by evaluation of peak area after Coomassie Blue staining. C: Percentage of proteins determined by evaluation of peak area after silver staining.

The question arises whether or not the silver staining method could be considered suitable for comparative investigations quantifying the relative amounts of proteins in different samples. Tab. 1 gives the results of densitometric evaluation (25) of marker proteins fractionated in 15% homogenous gels by SDS-

PAGE, stained with either Coomassie Blue or the silver method. It becomes evident that the Coomassie Blue staining correlates well with the relative amounts (weight) of proteins, whereas silver staining shows remarkable variations between individual proteins. Without any doubt silver staining can considerably increase the detection limit for proteins, but not in a linear fashion as is shown in Tab. 2. The increase in sensitivity is strongly dependent on the protein concentration and is much higher for low concentrations.

| | Protein | Peak Area in Arbitrary Units | | |
| --- | --- | --- | --- | --- |
| | | Coomassie | silver | $\frac{silver}{Coomassie}$ |
| A | Phos | 669 | 4840 | 7.23 |
| | BSA | 917 | 16435 | 17.92 |
| | Ov | 1555 | 28421 | 18.29 |
| | CA | 753 | 7580 | 10.07 |
| | ST | 995 | 14411 | 14.48 |
| | Lac | 934 | 18245 | 19.55 |
| B | Phos | 324 | 2765 | 8.56 |
| | BSA | 374 | 12819 | 34.36 |
| | Ov | 446 | 20107 | 45.18 |
| | Ca | 357 | 4563 | 12.78 |
| | ST | 366 | 10782 | 29.45 |
| | Lac | 344 | 14172 | 41.19 |

Tab. 2: Comparison of the sensitivity of the Coomassie Blue and silver staining methods. A: Protein concentration approximately 25 ng/sample well; B: Protein concentration approximately 12.5 ng/sample well.

In contrast to silver staining, Coomassie Blue exhibits an almost linear correlation between the amount of protein (weight) and bound dye in the range of 10 to 200 ng loaded per 1 mm sample well. This correlation is shown in Fig. 8 for three different proteins. Densitometric evaluation (25) of silver stained proteins in various dilutions shows only limited linearity between 1 and 40 ng, the slope of the standard curve tends much more to a convex function and an S-shaped curve in the lower concentration range. This explains the high sensitivity of the silver stain for proteins at low concentrations. 0.1 to 0.2 ng of pro-

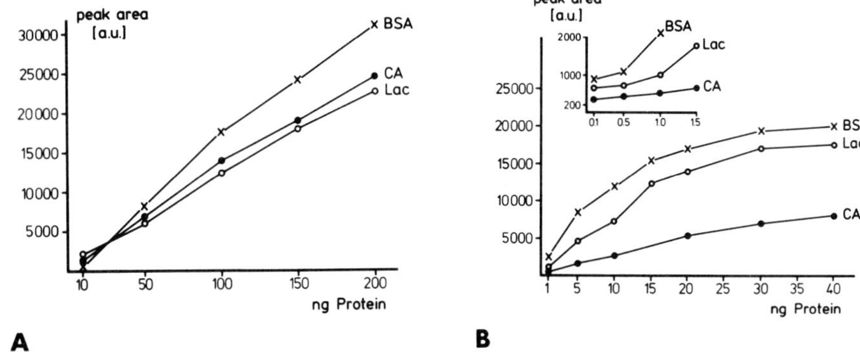

Fig. 8: Influence of protein concentration on the staining sensitivity in the Coomassie Blue (A) and silver (B) methods. Aliquots of serial dilutions of a standard protein solution were
electrophoresed on 15% acrylamide slab gels (80x80x0.7 mm),
peak area of individual proteins was plotted against the amount
of protein per sample well (1 mm wide). Small inset in B represents corresponding plot for the protein concentration between
0.1 and 1.5 ng. BSA: Bovine serum albumine; Lac: $\alpha$-Lactalbumine;
CA: Carbonic anhydrase.


tein per 1 mm sample well are detectable.


In contrast to conventional staining procedures the mechanism
of the silver stain is unknown and far from stoichiometry. The
term staining is rather misleading for the silver method, one
has to be very cautious using this silver stain for comparative
studies. The high sensitivity of the silver stain may be advantageous in connection with 2-D separations of high resolving
power, to detect as many proteins as possible, but the high sensitivity for lower protein concentrations creates problems.
Stained spots, especially in micro gels, are so close together
that evaluation and spot separation may be very difficult and
macro systems have to be used (Fig. 9). On the other hand streaking in either PAGIF or PAGE dimensions due to more or less incomplete stacking, reaggregation of proteins, unsuitable pore
sizes etc., which is hardly visible after Coomassie Blue staining, strongly disturb the spot pattern after silver staining.

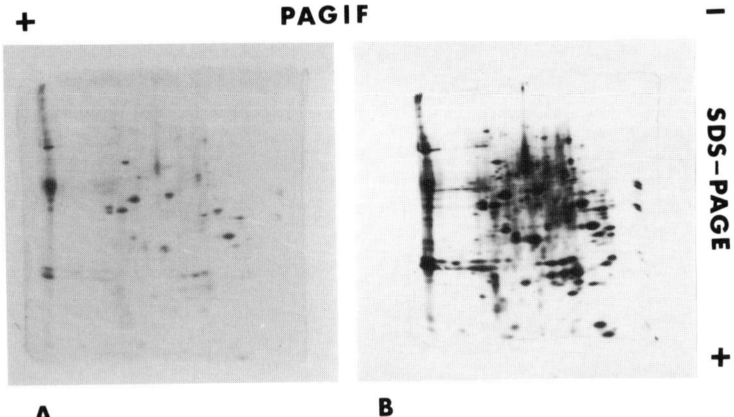



Fig. 9: 2-D electrophoresis of soluble rat brain proteins. 5 µg
of total protein were separated. Gel system as in Fig. 5. Slab
gel size: 30x30x0.7 mm. Gels were stained with Coomassie Blue
(A) or the silver method (B).


So the separation conditions have to be optimized or protein
concentrations should be reduced to achieve better separations.

In spite of the fact that the silver staining method poses many
problems as described, its high sensitivity may be of great advantage in many anlytical experiments which do not require quantification of the proteins.


References


1.   Neuhoff, V.: Micromethods in Molecular Biology, Springer
     Verlag, Berlin 1973
2.   Rüchel, R., Mesecke, S., Wolfrum, D.I., Neuhoff, V.: Hoppe-
     Seyler's Z. Physiol. Chem. 354, 1351-1368 (1973)
3.   Rüchel, R., Mesecke, S., Wolfrum, D.I., Neuhoff, V.: Hoppe-
     Seyler's Z. Physiol. Chem. 355, 997-1020 (1974)

4.  Grossbach, U., Kasch, K.: Techniques Biochem. Biophys. Morphol. $\underline{3}$, 81-161 (1977)

5.  Gainer, H.: Anal. Biochem. $\underline{44}$, 589-605 (1971)

6.  Condeelis, J.S.: Anal. Biochem. $\underline{77}$, 195-207 (1977)

7.  Bispink, G., Neuhoff, V.: Hoppe-Seyler's Z. Physiol. Chem. $\underline{357}$, 991-997 (1976)

8.  Klose, J., Blohm, J., Gerner, L.: In: Neuberg, D., Merker, H.J., Kwasigroch, T. (Eds.): Methods in Prenatal Toxicology, Georg Thieme, Stuttgart, 303-313 (1977)

9.  Been, A.C., Rasch, E.M.: J. Histochem. Cytochem. $\underline{20}$, 368-384 (1972)

10. Jones, M.G.K.: Physiol. Plant. Pathol. $\underline{16}$, 359-367 (1980)

11. Ogita, Z.I., Markert, C.L.: Anal. Biochem. $\underline{99}$, 233-241 (1979)

12. Matsudaira, P.T., Burgess, D.R.: Anal. Biochem. $\underline{87}$, 386-396 (1978)

13. Poehling, H.M., Neuhoff, V.: Electrophoresis $\underline{1}$, 90-102 (1980)

14. O'Farrel, P.H.: J. Biol. Chem. $\underline{250}$, 4007-4021 (1975)

15. Poehling, H.M., Wyss, U., Neuhoff, V.: Electrophoresis $\underline{1}$, 198-200 (1980)

16. Switzer, R.C., Merril, C.R., Shifrin, S.: Anal. Biochem. $\underline{98}$, 231-237 (1979)

17. Oakley, B.R., Kirsch, D.R., Morris, N.R.: Anal. Biochem. $\underline{105}$, 361-363 (1980)

18. Allen, R.C.: Electrophoresis $\underline{1}$, 32-37 (1980)

19. Poehling, H.M., Neuhoff, V.: Electrophoresis in prep.

20. Radola, B.J.: Electrophoresis $\underline{1}$, 43-56 (1980)

21. Laemmli, U.K.: Nature $\underline{227}$, 680-685 (1970)

22. Lane, J.D., Neuhoff, V.: Naturwissenschaften $\underline{67}$, 227-233 (1980)

23. Poehling, H.M., Wyss, U.: Nematologica $\underline{26}$, 230-242 (1980)

24. Görg, A., Postel, W., Westermeier, R.: Anal. Biochem. $\underline{89}$, 60-70 (1979)

25. Yakin, H.M., Zimmer, H.G., Neuhoff, V.: Electrophoresis, in prep.

# "COLD FOCUS" ISOELECTRIC FOCUSING OF SMALL SAMPLES OF CEREBROSPINAL FLUID


Linda L. Lorincz

Department of Pathology, Medical University of South Carolina,
Charleston, South Carolina 29425


## Introduction

Examination of cerebrospinal fluid (CSF) for protein abnormalities such as
oligoclonal immunoglobulins may provide information of diagnostic or
prognostic value in a variety of neurologic diseases.  However, the
techniques routinely used in clinical laboratories, cellulose acetate
electrophoresis and agarose gel electrophoresis, have limited resolving
power (5 to 10 protein bands) and require the use of concentrated CSF
samples.  Isoelectric focusing (IEF) offers greater resolution (1-4), but
IEF methods have not yet been widely applied for routine clinical analysis
of CSF proteins.  The recent developments of the "Cold Focus" apparatus,
which uses Peltier devices for cooling (5), a 3000 V power supply (6), and
a highly sensitive silver stain for polyacrylamide gels (7) have significantly improved IEF techniques, and in this report we describe the use of
this system for examination of CSF samples.

## Materials and Methods

Polyacrylamide gel slabs 250 $\mu$m thick (5% T, 3.5% C), prepared as described
previously by Allen et al. (8), were cast onto silanized glass lantern
slides, prepared according to the method of Radola (9).  The polymerization
solution (4 ml) was placed on a silanized slide whose edges were framed by
a three-layer parafilm gasket (0.5 cm wide).  A sandwich consisting of a
glass plate on the top, a piece of GelBond (Marine Colloids, Rockland,
Maine) on the bottom, and a few drops of ethanol to hold the GelBond flat
(hydrophobic side down) was carefully lowered, starting at one edge, onto
the glass.  A weight (500 g) was placed on the top, and the gel was allowed

150

to polymerize at room temperature.  Although polymerization was complete
within 1 hr, better results were obtained when the gel was allowed to set
for two days at room temperature in the gel casting unit.  The ampholine
composition of the gel was 3% Biolyte 4/9, 1% Biolyte 5/7, and 0.5% Biolyte
3/5.

Samples of CSF (2.0 μl) and Pharmacia pI standards (2.0 μl) were applied
directly on the gel surface with a Corning microsyringe.  The CSF samples
from patients with multiple sclerosis and other neurologic diseases were
obtained with informed consent by the Neurology Department at the Medical
University of South Carolina.  The pI standards were kindly supplied by
Pharmacia.  Bromphenol blue was added to each side of the gel to verify
even contact between the wicks and the electrodes.  Uneven contact was
corrected by placing a weight on the appropriate side of the electrode lid.
The cathode solution was 1.0 N NaOH and the anode solution was 1.0 M $H_3PO_4$.
The distance between the cathode and anode wick edges was 5.4 cm.

The gel was placed on the berrylium oxide component of a Cold Focus
apparatus (MRA Corporation, 1058 Cephas Road, Clearwater, Florida) using a
few drops of ethanol between the glass and the berrylium oxide plate to
ensure even heat conductance.  The electrodes in the lid of the Cold Focus
apparatus were 6 cm apart.  The power supply was a Phamacia ECPS 3000/150
with attached Volthour Integrator VH-1.

The number of volt-hours was read directly from the volt-hour integrator
(10).  The gels were run with maximal cooling at constant power, 8 W for
600 Vhr, 12 W for 250 Vhr, and 15 W for 250 Vhr.  At the end of the run,
the maximum voltage was 333 V/cm, and the average temperature was 20°C.
The holes in the electrode lid were covered with tape, except during brief
temperature measurements, to prevent spot drying of the gel.

Gels were fixed in 12.5% TCA, rinsed in deionized $H_2O$, and dried at 60°C.
They were then stained for 2 min in Coomassie Brillant Blue R-250, as
described previously (8), photographed, and destained.  Silver staining, as
described previously by Allen (11), was performed following destaining,
with the following modifications:  (a) gels were agitated for 2 min in a

10% ethanol solution at room temperature both before and after the silver diamine solution, and (b) gels were placed for 5 min each in paraformaldehyde solution, cupric nitrate/silver nitrate solution, and silver diamine solution. The intensity of the silver staining was greatly reduced unless the ammonium hydroxide used in the silver diamine solution was from a freshly opened bottle. Silver was recovered for reuse from all silver-containing waste solutions by standard chemical methods (12). Densitometry was performed with an ORTEC 4300 integrating microdensitometer using a 70-um spot with an image magnification of 35 diameters. Direct visible light, without filters, was used for silver stained gels.

Results and Discussion

A silver stained gel with CSF samples is shown in Figure 1. More than 50 protein bands were seen and were reproducible. Figure 2 shows a densitometric scan from a CSF sample (slot 4 of the gel shown in Figure 1). The sharpest protein bands were obtained by rapid focusing in a high field strength as previously described (10). The excellent resolution seen with this method, particularly at the cathodal region of the gel, appears to be related to more efficient cooling, due both to the "Cold Focus" apparatus and to the thinness of the gel, which allow application of such high field strength. Relatively high Ampholine concentrations were required to avoid areas of low conductance in the gel.

This method has several potential adavantages over previous methods for CSF protein studies. The entire procedure takes 4 hr and can be done with up to 20 samples per gel. In contrast, standard IEF of CSF proteins which requires a preliminary concentration step clearly takes much longer. Furthermore, six ultrathin gels can be made with the same amount of ampholine solution used in one normal thickness gel, thus saving considerable expense. The silver stain is up to 250 times more sensitive than Coomassie blue stains (7) and thus allows the use of microliter amounts of unconcentrated CSF. Handling of unwieldy gels is avoided, since gels are attached to a small glass plate and can be kept as permanent records once dried. The small size of the gels is also advantageous, as few micro-

152

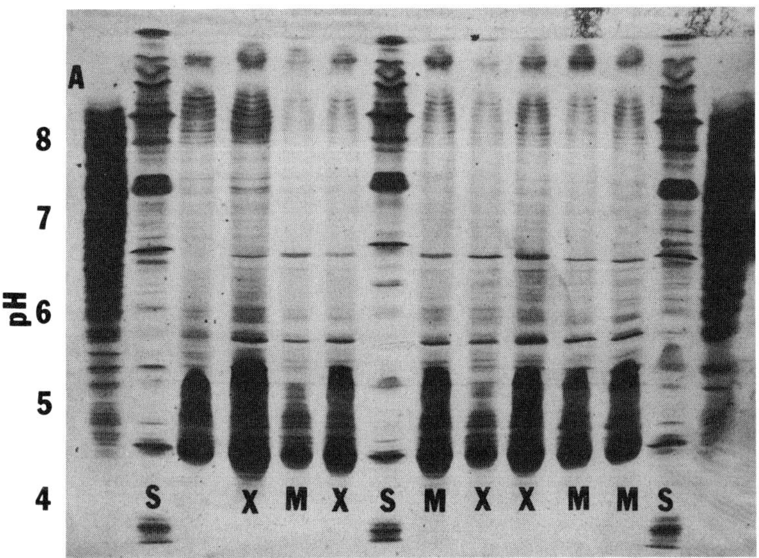

Figure 1. Isoelectric focusing of cerebrospinal fluids. 2.0-μl samples
were focused on ultrathin-layer gels for 1100 Vhr over a distance of 6 cm
with a Biolyte 3 to 9 gradient, and stained with silver. (S) Pharmacia pI
standard, (M) multiple sclerosis, (X) other neurologic disease, (A) point
of sample application.

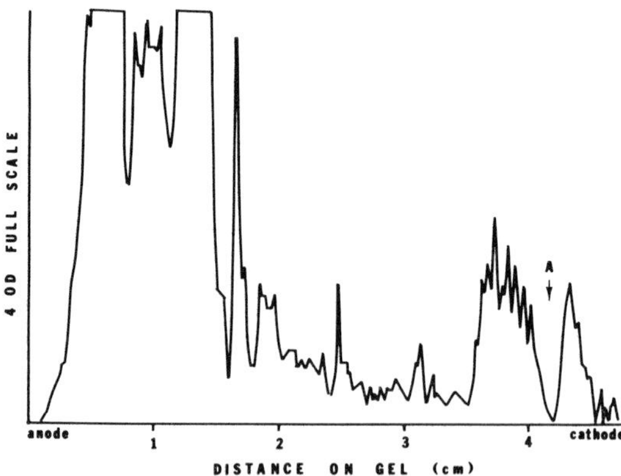


Figure 2. Densitometric scan of focused and silver stained cerebrospinal
fluid sample from a patient with neurologic disease. (A) Point of sample
application. The scan was made without filter, with a slit width of 70 μm
on an ORTEC 4300 integrating microdensitometer.

densitometers are designed to accommodate the large gels produced by other methods. In addition, detergents are not required and immunoenzyme methods may thus be used to identify proteins on the gel. The expense of the silver stain is greatly reduced by the silver recovery method described. The potential resolution of this method is more than 200 protein bands over a distance of 5.4 cm (11), far greater than possible with other one-dimensional electrophoretic methods.

Although even greater resolving power (300 polypeptides from 60 $\mu$l of CSF) is provided by the elegant and sensitive method of Merril, Switzer, and Van Keuren (7), which combines two-dimensional elelctrophoresis and a silver stain, the technical difficulties inherent in two-dimensional methods, the necessity for computerized image processing and data reduction, and the ability to process only one sample at a time limit the usefulness of the two-dimensional methods for clinical laboratories. The one-dimensional method described here avoids these difficulties. Further improvement of this method should be possible with the use of narrow-range pH gradients, which would allow better visualization of small regions of the gel, and selective absorption of albumin from the samples by affinity or dye-ligand chromatography, which would permit visualization of proteins of pI similar to that of albumin.

References


1. Fossard, C., Dale, G., Latner, A.L.: J. Clin. Pathol. 23, 586-589 (1970).

2. Stibler, H.: J. Neurol. Sci. 36, 273-288 (1978).

3. Delmotte, P., Gonsette, R.: J. Neurol. 215, 27-37 (1977).

4. Kjellin, K.G., Sidén, A.: Eur. Neurol. 16, 79-89 (1977).

5. Allen, R.C., Oulla, P.M., Arnaud, P., Baumstark, J.S. in: Radola, B.J., Graesslin, D. (Eds.), Electrofocusing and Isotachophoresis, Walter de Gruyter, Berlin, p. 256-264 (1977).

6. Pharmacia brochure. Electrophoresis constant power supply ECPS 3000/150 for isoelectric focusing and electrophoresis.

7. Merril, C.R., Switzer, R., Van Keuren, M.: Proc. Nat. Acad. Sci. U.S.A. 76, 4335-4339 (1979).

8.  Allen, R.C., Harley, R.A., Talamo, R.C.: Amer. J. Clin. Pathol. 62, 732-739 (1974).

9.  Radola, B.J.: Electrophoresis 1, 43-56 (1980).

10. Pharmácia Fine Chemicals.  Separation News 4 (1980).

11. Allen, R..: Electrophoresis 1, 32-37 (1980).

12. Tischer, T.N. in: Kirk-Othmer Encyclopedia of Chemical Technology, 2nd ed., Interscience Publishers, New York, p. 295-308 (1969).

A UNIQUE SILVER STAINING PROCEDURE FOR COLOR CHARACTERIZATION
OF POLYPEPTIDES

Lonnie D. Adams and David W. Sammons
Diabetes and Atherosclerosis Research, The Upjohn Company,
Kalamazoo, Michigan 49001, U.S.A.

Introduction

Under optimal conditions electrophoresis is perhaps the most
powerful separation technique ever devised. Certainly its
popularity and utility today has placed it into almost every
investigative subfield of biology and medicine. Full utilization of its potential has not yet been completely realized
because of inadequate equipment design, theoretical restraints,
and partly because successful utilization requires a differential stain that is specific for the polypeptides being separated.

Combining isoelectric focusing in one direction and sodium
dodecyl sulfate electrophoresis in the other (1) has stimulated a flurry of recent developments that include the two-
dimensional (2-D) gel (ISO-DALT) system (2-4). Advances in one
technology naturally lead to advances in another. Thus, as
2-D gel electrophoresis demonstrated high resolving power not
previously realized by prior techniques, the deficiencies of
staining methodologies became more apparent, thereby calling
for new ultrasensitive methods for staining polypeptides.
Monochromatic staining dyes such as Coomassie Blue are insensitive and stain only the most abundant polypeptides. Ultrasensitive polypeptide detection is possible with incorporation

of radiolabeled amino acids or other radiolabeled precursors
of proteins, however, the procedure requires active protein
synthesis.  Often the signal of the incorporated radiolabel
precursor is too weak and many of the cells' protein gene
products are not detected.  In addition, proteins present in
body fluids and in human biopsy samples cannot be practically
radiolabeled.  A series of procedures has evolved that
utilizes the binding of silver to protein reactive centers
(5-13).  The silver staining methods were borrowed from
histology and photography, are monochromatic, and are ultrasensitive if performed on gels less than 1 mm thick.  Serious
disadvantages peculiar to the methods included time-consuming
multistep processes, cost, special lighting or temperature
requirements, and unnecessary chemicals and steps that lead to
capricious results.  Goldman _et al_. (14) have described some
shades of color after silver staining spinal fluid and plasma
proteins, however, the method that was described suffers the
same disadvantages as the monochromatic silver stains.

This paper describes a new gel electrophoresis color development system for silver staining of polypeptides in one- and
two-dimensional polyacrylamide gels that is simple to use,
efficient (batch-wise staining of gels), reproducible, ultrasensitive, inexpensive, and is optimized for gels greater than
1.0 mm thick.  Most importantly, however, the polypeptides
complex with silver in a characteristic manner and the colored
polypeptide-silver complex is readily visualized while backlighting with white light.

The reproducible and characteristic color staining of a polypeptide adds a third dimension to the 2-D analysis and aids in
visual as well as computer-assisted characterization and identification of a particular polypeptide.  Polychromatic
staining of 2-D patterns from various tissues facilitates
tracking a protein gene product expression in different cell
types and makes possible the indexing of proteins heretofore

not readily discerned with prior monochromatic stains.


## Methods and Materials

Protein samples used were derived from human plasma, Chinese
hamster kidney and liver, rat heart, and human fibroblasts.
Human plasma from blood collected in sterile sodium citrate
(9 parts blood to 1 part citrate) was centrifuged for 3
minutes in a Beckman Microfuge Model B.  Two hundred milligrams of wet weight liver and kidney from the same animal
were minced with a razor and homogenized into 1 ml of ISO-
urea mixture (15).  After centrifugation at 200,000 x g, 15 µl
of sample were loaded into each ISO gel (LKB Ampholine, pH
3-10).  Rat heart standards were prepared as previously described (16) and diluted 1:25 with hot agarose without protein.
The warm rat heart protein-agarose solution was solidified by
cooling a filled tip of a Pasteur pipette in an ice bath.
After extrusion of the agarose rod from the pipette tip, a
15 mm piece was placed on top of an SDS polyacrylamide (10-20%
gradient) and fixed in place with a warm agarose solution.
Human fibroblasts, originally established from a skin biopsy,
were cultured to confluency in F-12-DMEM + 10% fetal calf
serum.  Immediately prior to L-[$^{35}$S]-methionine (1260 Ci/mmol)
addition the cells were rinsed several times with serum-free
culture medium (MEM without methionine).  Serum-free medium
which contained 10 µCi/ml of $^{35}$S-methionine was then added to
the cells and culture was continued for 20 hours.  The labeled
cells were scraped from the plastic culture flask (Falcon,
T-25) with a rubber policeman and pelleted in Hank's buffer
with a Beckman Microfuge.  The Hank's medium was aspirated
from the pellet and the cells were immediately vortexed in
100 µl of ISO-urea mix (15).  The sample was occasionally
mixed for 1 hour at room temperature, stored at -70°C, and
run on the ISO-DALT system (15).

Two-dimensional gel electrophoresis was performed with the
ISO-DALT system and solutions and procedures were prepared and
used according to directions (15).  DALT tank buffers were
changed immediately prior to electrophoresis and were not
reused.  All solutions except ammonium persulfate and TEMED
were filtered through 0.22 micron filters prior to use.  Electrophoresis quality reagents were used.  The sources of
reagents were SDS, Serva Chemicals; TEMED, BIORAD; 2-mercapto-
ethanol, Eastman Kodak; Tris base and glycine, Sigma Chemical.
DALT plates were washed in Micro detergent, rinsed with water,
finally rinsed with 95% ethanol, and air dried.  Cathode and
anode IEF solutions were sodium hydroxide and phosphoric acid,
respectively.  Ethylenediamine is not recommended as a cathode
solution since a high greenish interfering background will
result during staining with silver.  Similarly, it was observed that the use of ethylene dichloride solvent in repair of
the DALT tanks contaminated the buffer and subsequently the
DALT gels, thereby resulting in an interfering background.
For best results the IEF equilibration was done against fresh
equilibration buffer which contained 2% $\beta$-mercaptoethanol.
Bromophenol Blue dye was deleted from the equilibration buffer.
Equilibration was done exactly for 30 minutes and the IEF's
were frozen at -70°C until needed.  A minimum of agarose
(BIORAD) was used to bond the IEF to the second dimension.
The SDS second dimension consisted of a 10-20% gradient of
acrylamide.  The polyacrylamide gel is 1.5 mm thick with a
total gel volume of 54 ml.


Results and Discussion


After electrophoresis the polypeptides must be fixed in the
polyacrylamide gel with ethanol-acetic acid.  Groups of 5 gels
per tray (Pyrex, 12" x 17" x 3") were processed through the
fixation and washing procedure.  Fairly extensive washing is
recommended to completely remove the SDS.  Generally the gels

are left overnight in the first wash solution whereupon the
washing steps are completed the next day.  The gels may remain
in the last step of the wash indefinitely until staining is
desired.  Table I illustrates washing procedure.  The optimal
ratio of gel volume to solution volume is 1:5.5.

TABLE I.  Staining Procedure

**STAINING PROCEDURE**

| Steps | Solutions | Duration of Agitation |
|---|---|---|
| Fix | 50% ETOH 10% HAC | 2 hr or more |
| ↓ | ↓ | |
| Wash | 50% ETOH 10% HAC | 2 hr |
| | ↓ | |
| | 25% ETOH 10% HAC | 1 hr  2X |
| ↓ | ↓ | |
| | 10% ETOH 0.5% HAC | 1 hr  2X |
| | ↓ | |
| Equilibrate gel | $AgNO_3$ (1.9 g/l) | 2 hr or more |
| ↓ | ↓ | |
| Rinse | $H_2O$ | 10-20 sec |
| ↓ | ↓ | |
| Reduce silver | $NaBH_4$ (87.5 mg/l) | 10 min |
| | HCHO (7.5 ml/l) | |
| ↓ | in 0.75N NaOH | |
| | ↓ | |
| Enhance color | $Na_2CO_3$ (7.5 g/l) | 1 hr |
| | ↓ | |
| | $Na_2CO_3$ (7.5 g/l) | |

The first step of the staining procedure is equilibration in
silver nitrate.  A balance must be struck between concentration
of silver, gel thickness, and protein concentration.  We have
found that for gels that are 1.5 mm thick 1.9 g/l is ideal.
For 1.5 mm thick gels they should equilibrate for at least 2
hours.  The ratio of gel volume to solution volume is 1:3.
Prior to reduction the gels are briefly rinsed in $H_2O$ to
remove surface silver.

Reduction of silver is done in a basic environment (.75N NaOH)
and by formaldehyde (7.5 ml/l).  Addition of $NaBH_4$ (87.5 mg/l)
enhances the color slightly but is not absolutely necessary

for the staining. $NaBH_4$ alone will result in partial reduction of the silver. The spots will begin to appear within 5-6 minutes during the reducing step, however, the spots are only redish-brown on a rust colored background. For best results the gels should be removed from the reducing solution within 10 minutes and placed into the color enhancing solution.

The color develops as the sodium carbonate diffuses into the gel and continues for the next several hours. The two changes at 1 hour intervals are necessary to remove the excessive NaOH. After one hour of agitation of the gel in the last $Na_2CO_3$ step, the gel may be set aside for at least 6 hours. Generally, the colors are optimal and are best viewed with backlighting on a standard fluorescent viewer, 5000°K, daylight. The colors are stable for several days; in fact, the background tends to lighten as time passes without significant fading of colors.

The silver stain procedure is designed to optimize several advantageous features such as ease of use, time efficiency, reproducibility, and inexpensiveness.

Figure 1 illustrates the sensitivity of the silver stain process relative to Coomassie Blue and [35]S-autoradiography. Equal amounts of fibroblast protein were loaded onto the silver stained 2-D gel as the Coomassie Blue stained gel. It is obvious that the silver stain is far superior to Coomassie Blue staining. The autoradiograph was obtained by developing X-Omat film that was exposed to the Coomassie Blue 2-D gel for 9 days. Careful comparison of the silver stained gel to the autoradiograph shows comparable levels of sensitivity. There are some spots visible on the silver stained gel (designated A) that are not observed in the autoradiograph. Presumably these spots did not incorporate [35]S-methionine into the primary sequence of the polypeptide. In contrast, some spots clearly visible in the autoradiograph (designated B) do not appear or are barely visible in the silver stained gel.

Apparently, these proteins incorporate a disproportionately larger amount of methionine than other proteins.  These proteins are dim on the silver stain because they are less abundant or bind less silver than others.  In order to more easily visualize the protein a more concentrated protein sample should be added to the IEF gel.

A range of colors is seen in Figure 1 that is in stark contrast to either of the monochrome methods of protein detection.  The colors assist in the discrimination of overlapping spots, thereby resulting in resolution heretofore impossible.  This feature of color coding proteins is of significant assistance in the computer analysis and quantitation of overlapping spots (16).

Figure 2 shows the applicability of the silver stain to one-dimensional gels.  The rat heart standard was prepared and run exactly as described by Giometti et al. (14) except fiftyfold less protein was applied to the SDS gel.  The same protein bands were observed as previously reported by Coomassie Blue staining.  The colors of each protein may serve as internal color wedge as well as molecular weight markers.  In this way corrections can be made for slight variations in color Perhaps most importantly, color adds a third dimension and helps to characterize the human plasma proteins shown in Figure 3.  The proteins are tentatively named from the litera-ture (17).  Not all proteins were identified, however, several well known examples are shown to indicate the range of colors present.  The colors range from gray-blue for LDL, HDL, and one charge train of G4.  Another charge train of G4 appears orange in color and has a slightly lower molecular weight. This is a good example of how a monochromatic stain (Coomassie Blue or silver) could not distinguish the two species of G4 protein.  In fact, there appears to be a charge train of G4 which has even a lower molecular weight and also a slightly different hue of blue color.

Haptoglobin charge train is interesting since it appears to change colors within the train of spots. It begins as a run of 6-7 orange spots and trails off into 5 grayish-blue spots that have progressively higher molecular weights. Immediately below the bright orange spots are three spots that appear to be another class of haptoglobin spots that are a shade of yellow just different from the background.

Thus, it appears from the haptoglobin that proteins may be different colors depending upon their post-translation modification. The assignment of various sugars or lipids to a particular color will have important implications in predicting protein characteristics from 2-D gels. Alpha-1-antitrypsin has interesting staining characteristics since it appears to have two colors. There is a red center surrounded by a blue halo. Two spots of the alpha-1-antitrypsin dimer are blue with no red center. The blue form of this protein seems to be present in lower concentrations than the other spots with red centers. Fibrinogen alpha and beta chains also are similar to alpha-1-antitrypsin in that they have two colors. They are basically yellow proteins with a brown center. Similar to the alpha-1-antitrypsin dimer, some fibrinogen spots do not have the differently colored center. Whether the change in hue is a function of primary sequence, side chain modification, or is an artifact of the protein concentration remains to be determined.

Transferrin, C-3 activator, and alpha-2-macroglobin have brown centers and red halos. Albumin has primarily a "dirty" brown color at lower concentrations (data not shown) and has a "sheen" colored center surrounded by a brown halo when it is present at high concentrations.

Thus, change in hue may be a function of the type of moiety which has been covalently linked to the protein. It is also possible that other protein side chain modifications might

contribute to the color of the polypeptide-silver complex.  It
is clear, however, that the protein color code is a unique
property of a protein just as the molecular weight and isoelectric point is a distinguishing characteristic.

In Figure 4 a comparison of 2-D gels derived from kidney and
liver proteins is made.  One immediately sees that color is
essential if one wishes to find whether a protein gene product
occupying an identical or nearly identical X-Y coordinate position is the same protein or not.  For example, protein A is red
in the kidney and shares an X-Y coordinate position with a
yellow protein.  The yellow protein is clearly seen in the
liver, however, the kidney-specific red protein is absent from
this X-Y coordinate position.  One would mistakenly believe, if
one used a monochromatic stain, that the kidney merely has more
of the yellow protein than the liver.  Proteins labeled "B" do
not share an exact X-Y coordinate position, however, since the
spots overlap, it would appear that a single protein was
present if a monochromatic stain was utilized.  The area
designated "C" is two spots (one orange and the other blue)
that have overlapping X coordinate positions and identical Y
coordinates.  In a monochromatic stain it would appear that a
single protein is streaked in the IEF direction.  Thus,
Figure 4 convincingly demonstrates the importance of a third
dimension of color in mapping and identifying proteins in
various tissues.  The color parameter can also act as an
internal standard when comparing constellations of spots
between gels and between tissues, thereby assisting both
manual and computer-assisted comparisons of two different
2-D patterns.

164

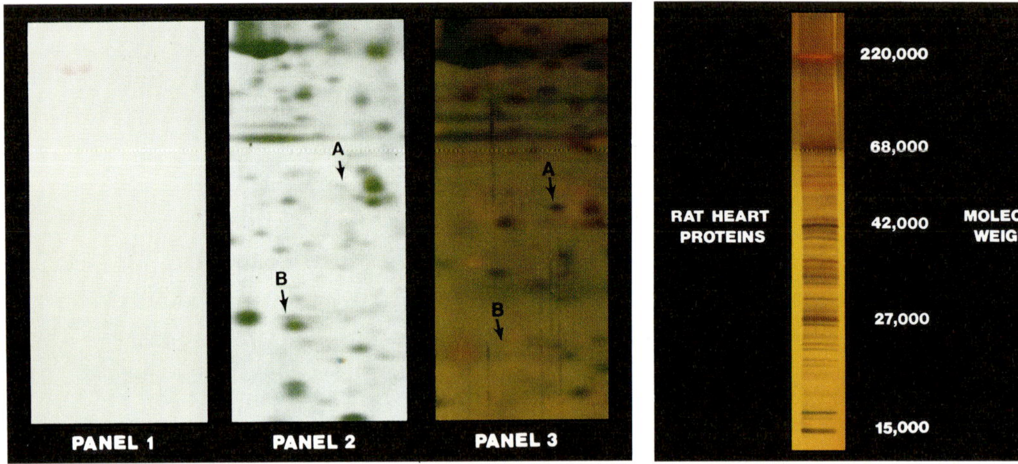



Fig. 1

Fig. 2

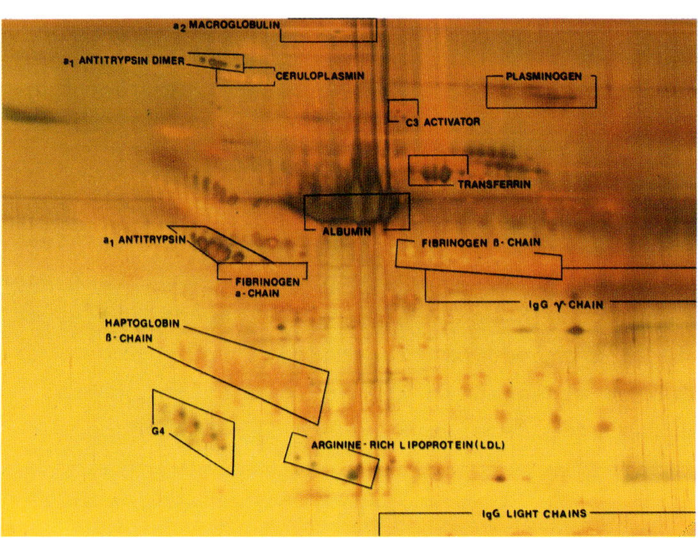



Fig. 3

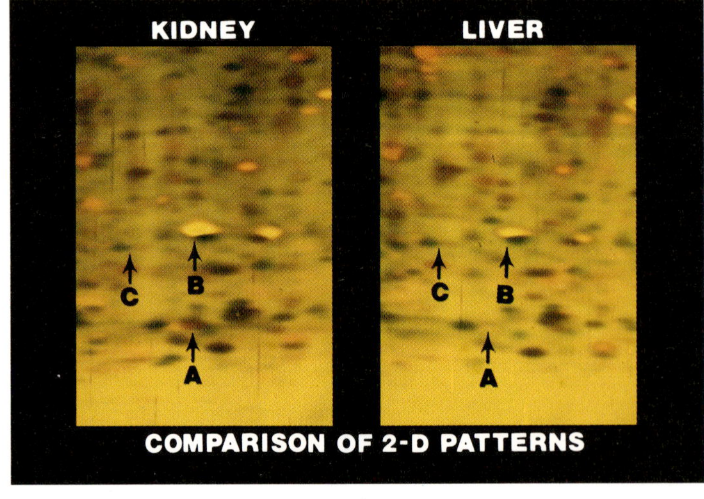



Fig. 4

## References


1. O'Farrell, P.H.:  J. Biol. Chem. $\underline{250}$, 4007-4021 (1975).

2. Anderson, N.G., Anderson, N.L.:  Anal. Biochem. $\underline{85}$, 331-341 (1978).

3. Anderson, N.L., Anderson, N.G.:  Anal. Biochem. $\underline{85}$, 341-354 (1978).

4. Anderson, N.G., Anderson, N.L.:  Behring Inst. Mitt. $\underline{63}$, 169-210 (1979).

5. Kerenyi, L., Gallyas, F.:  Clin. Chim. Acta $\underline{38}$, 465-467 (1972).

6. Verheecke, P.:  J. Neurol. $\underline{209}$, 59-63 (1975).

7. Switzer, R.C., Merril, C.R., Shifrin, S.:  Anal. Biochem. $\underline{98}$, 231-237 (1979).

8. Merril, C.R., Switzer, R.C., VanKeuren, M.L.:  Proc. Nat. Acad. Sci. USA $\underline{76}$, 4335-4339 (1979).

9. Karcher, D., Lowenthal, A., VanSoom, G.:  Acta Neurol. Belgium $\underline{79}$, 335-337 (1979).

10. Allen, R.C.:  Electrophoresis $\underline{1}$, 32-37 (1980).

11. Oakley, B.A., Kirsch, D.R., Morris, N.R.:  Anal. Biochem. $\underline{105}$, 361-363 (1980).

12. Merril, C.R., Dunau, M.L., Goldman, D.:  Anal. Biochem. $\underline{110}$, 201-207 (1980).

13. Merril, C.R., Goldman, D., Sedman, S.A., Ebert, M.H.:  Science $\underline{211}$, 1437-1438 (1981).

14. Goldman, D., Merril, C.R., Ebert, M.H.:  Clin. Chem. $\underline{26}$, 1317-1322 (1980).

15. Anderson, N.G., Anderson, N.L., Tollaksen, S.L.:  ANL-BIM-79-2, Argonne, Illinois (1979).

16. Giometti, C.S., Anderson, N.G., Tollaksen, S.L., Edwards, J.J., Anderson, N.L.:  Anal. Biochem. $\underline{102}$, 47-58 (1980).

17. Vincent, R.K., Hartman, J., Barrett, A.S., Sammons, D.W.:  Electrophoresis 1981 Symposium, Charleston, South Carolina, ed. Arnaud, Walter DeGruyter, Berlin · New York.

18. Anderson, N.L., Anderson, N.G.:  Proc. Nat. Acad. Sci. $\underline{74}$, 5421-5425 (1977).

AFFINITY-IMMUNODELETION (AID) ISOELECTRIC FOCUSING ON ULTRATHIN
GELS APPLIED TO THE IDENTIFICATION OF SWEAT SALIVA AND BLOOD
PROTEINS USING SILVER DIAMINE STAINING


R. C. Allen, P. Arnaud and S. S. Spicer
Department of Laboratory Animal Medicine
Pathology and Basic and Clinical Immunology and Microbiology
Medical University of South Carolina
Charleston, South Carolina  29425



INTRODUCTION

Separation of macromolecules by electrophoretic methods has
progressed in the last four decades from an initial method
capable of resolving some five serum proteins (1,2) to one
capable of resolving over 1,000 serum proteins using two-
dimensional PAGIF followed by SDS-PAGE (3).  Not only has resolution been greatly advanced, but also limits of detection
now have reached the sub-nanogram range with such staining procedures as enhanced immunostaining with Avidin-Biotin (4) and
with silver diamine staining of proteins (5,6).  Thus, dilute
body fluids such as saliva, CSF, sweat, tears, and joint fluids
are now amenable to direct separation and study in macromolecular detail without preliminary concentration procedures and
their attendant denaturation potential.

The recognition of the ever increasing role of molecular genetics in clinical and forensic medicine brings about the need to
be able to classify a greatly increased number of genetically
controlled protein and enzyme microheterogeneities.  The human
genome is presently thought to be able to code for some 30 to
50,000 proteins or protein sub-units and about ten per cent of
these may appear in any one cell type.  The Iso-Dalt system
described by Anderson has a resolution potential of some 10,000
proteins (7).  However, presently, its application to routine

clinical and forensic medicine for the average routine clinical laboratory is not practical.

Ultrathin layer isoelectric focusing in polyacrylamide gel offers a higher resolution capability due to the ability to use high voltage gradients, with its attendant increase in resolution. The addition of a more efficient heat dissipation using a Beryllium oxide heat exchange plate in conjunction with a Peltier cooling device (8), "Cold Focus", (MRA Corporation, Clearwater, Florida) allows the potential for even higher voltage gradients to be employed. According to the fundamental theory of isoelectric focusing (9), the width of a focused protein zone is inversely proportional to the square root of the field strength; thus, a fourfold increase in field strength will produce a protein zone half its original width, thus doubling the space in which to resolve an additional protein or proteins. The present availabliity of ampholytes with improved conductance profiles or the use of tailored ampholytes at higher ionic strengths, with the additional heat dissipation capability of the "Cold Focus" apparatus, allows the employment of voltage gradients three to five or more times higher than those commonly used in polyacrylamide gel isoelectric focusing (PAGIF).

As a clinical tool, this presents both the added advantages of reduced separation time and an increase in the information obtainable. In conjunction with the silver-diamine stain, it provides the means for detection at subnanogram levels of protein concentration.

The separation of up to 150 protein zones over a 5.4 cm distance and a theoretical resolution potential of 250-300 proteins is possible. However, it does not solve the problem of two or more proteins having a similar isoelectric point and, thus, occupying the same position on the gel.

Toward this aim, a method of selective removal by affinity chromatography (10) in combination with immunoprecipitation and deletion was developed, in order to identify proteins both by class and individually. This report concerns preliminary studies of an Affinity-Immunodeletion (AID) process coupled with the highly sensitive silver-diamine stain to extend the capability of PAGIF in clinical and forensic medicine, not only with side by side comparison of controls and specifically deleted samples, but also, where in certain affinity procedures, the deleted macromolecules can be eluted and classified in an additional track on the gel slab.

Methods and Materials

A 0.25 ml packed gel volume of Blue-Sepharose CL6B was equilibrated with phosphate buffered saline at pH 7.2. The gels then were centrifuged at 13,000 G for one minute and the excess PBS removed by a Pasteur pipette. Then 250 ul of serum, diluted 1:32 for silver staining, was added to the swollen gel still contained in a 1.5 ml microfuge tube. The samples were thoroughly mixed with the affinity gel by stirring, allowed to incubate at room temperature for 30 minutes and then centrifuged again. The supernatant consisted of sample less the bound albumin. The supernatants were then divided into 50 ul aliquots and mixed with equal volumes of specific anti-alpha-2-macroglobulin, Anti Gc, Anti-C3 and Anti-transferrin antibody (Dako IgG antibody in excess). They were allowed to incubate 30 minutes at room temperature and were then centrifuged at 13,000 G for one minute in a "Microfuge." The supernatants from the saline and specific antibody incubations were then compared by isoelectric focusing on various pH gradients. Alternatively, the sample or centrifuged antibody-sample mixture, or untreated sample was reacted similarly for 20 minutes with the concanavalin-A covalently bound to sepharose and again centrifuged and the supernatant collected. The gel with bound glycoprotein

was then washed three times with saline and the bound material
eluted with 10% ⍺-D methyl glucoside.  Then both bound and released material was separated side-by-side for comparison with
the original sample.  The latter step also allows sample concentration when the elution is carried out with a lesser volume
of ⍺-D methyl-glucoside.


Isoelectric Focusing


Isoelectric focusing was performed on 250u thick gels backed
on either "Gel Bond"(Marine Colloids) or silianized on glass
plates (Polyfix 1000). Separations were performed on a "Cold
Focus" (Apparatus kindly supplied by MRA Corporation) over a
5.4 cm separation distance with peak voltages ranging from
375-600 volts per centimeter depending on the ampholyte employed.  Final peak Voltage gradients of 375 V/cm with 4% Ampholine
(LKB) and 600 V/cm  with 4% Pharmalyte (Pharmacia) were used.
Ampholine containing gels at a pH range of 3.5-10 were rein-
forced with 2% additional pH 5-7 ampholine to prevent hot spots
while Pharmalyte pH 3-10 was used at a 4% concentration.  Run
times were approximately one hour in all cases with the maximum
voltages stated above used toward the end of the run to sharpen
the bands as previously reported (6).  A Pharmacia model ELPS
3000/150 power supply, kindly supplied by Pharmacia Fine Chemicals was used as was a Model 2103 LKB 2000 Volt supply.

Sample application was made on 4MM wide filter paper tabs 3-8
mm  in height, or directly on the gel depending on the sample
concentration required.  One ul of diluted serum, 6 ul of
diluted saliva and diluted sweat were used in these studies.


Staining


Staining of undiluted serum was performed using  Coomassie Blue

R250 at 55$^{\circ}$C for one minute after fixation by 12.5 TCA.  Staining for dilute body fluids was performed by the modified silver-diamine stain previously described (6) with the following modification:  After reaction at 55$^{\circ}$C for 5 minutes in the silver-copper nitrate solution, the gels were washed one minute at 55$^{\circ}$C with gels focused with ampholine or room temperature with Pharmalyte prior to a 2-3 minute reaction time in Ammonial silver nitrate.  They were again washed in 10% alcohol at the appropriate temperature and then placed in reducing solution number one 30 seconds with agitation, then changed to fresh reducing solution.  Four to five minutes in reducing solution one was normally sufficient for optimal development of the protein bnads.  Gels were then washed in distilled water and then into Kodak acid fixer for five minutes followed by thorough washing in distilled water.  Gels were then dried at 60$^{\circ}$C for subsequent densitometry and photography.


Densitometry


Densitometry was performed on an ORTEC Model 4300 integrating microdensitometer.


Results


Removal of albumin by reaction of serum with Cibacron Blue-Sepharose is shown in figures 1 and 2.  In Figure 1-B, a two-dimensional PAGIF-SDS separation is shown with the albumin present while in 1-A albumin was removed by Blue Sepharose treatment indicating the masking of Gc and AT III by albumin. In Figure 2, densitometric analysis of the albumin region of normal serum is indicated in panel A stained with silver-diamine.  This region is partially resolved into its components by reducing the sodium ion concentration   providing a greater

intensity of the silver stain for glycoproteins with reduced
staining of the albumin.  In panel C, the region is unmasked
completely by removal of the albumin with Blue-Sepharose.

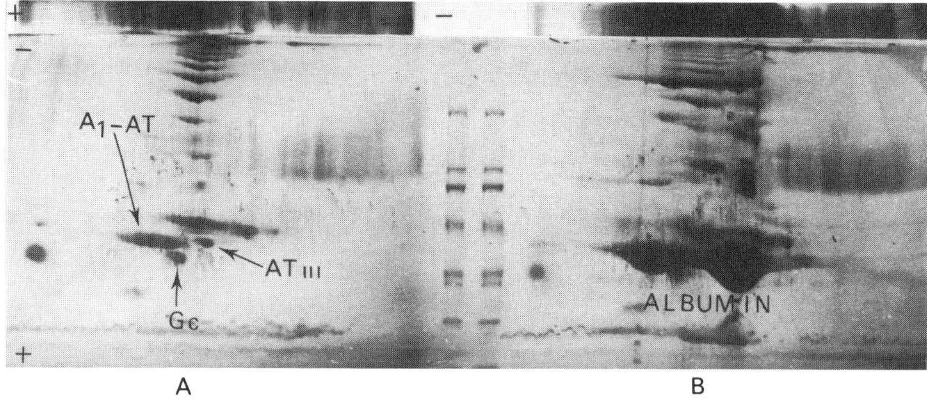


Figure 1.  Two-dimensional separation of human serum protein
carried out with flat slabs horizontally.  A is Blue-Sepharose
treated and B normal.

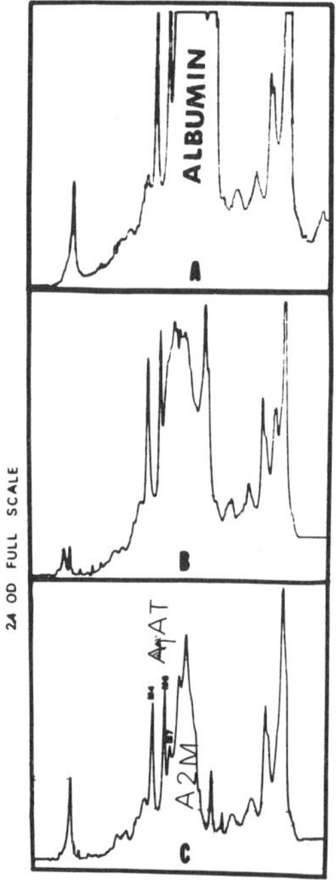


Figure 2.  Densitometric trace of
human serum separated on a pH
3.0-10 Pharmalyte gradient and stained with silver diamine.  A is normal
serum albumin region; B is lowered
Sodium concentration in the stain to
reduce albumin staining; C is Blue
Sepharose treated serum showing same
region.

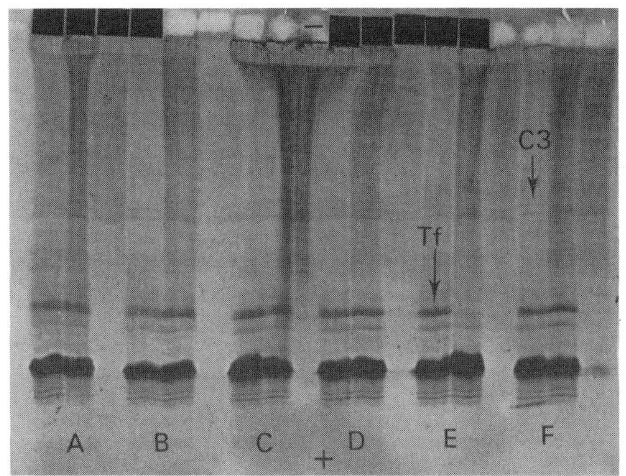

Figure 3. Immunodeletion of various serum proteins. A human serum + saline and serum + anti-A-1AT, B Serum + saline and serum + anti-alpha 2 macroglobulin, C serum and saline and serum + anti-haptoglobin, D serum + saline and serum + anti-Gc, E serum + saline and serum + anti-transferrin and F Serum + saline and serum + antibodies used.

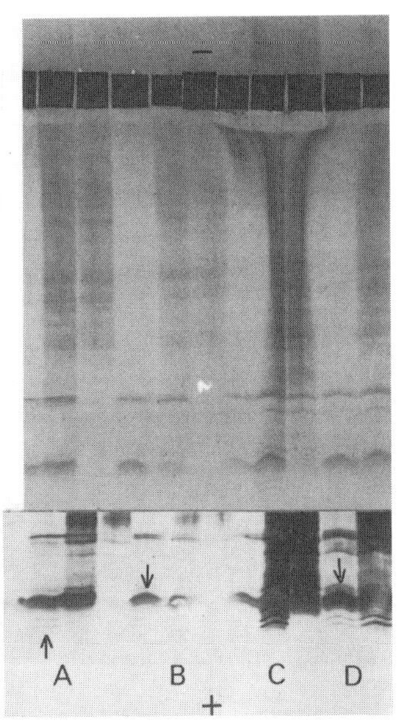

Figure 4. Similar to Figure 3 but reacted first with Blue-Sepharose to remove albumin. Insert at botton is the lower portion of the same gel counter stained with silver diamine.

A comparison of the combined affinity and immunodeletion technique is illustrated in Figures 3 and 4 employing antibodies against alpha-1-antitrypsin, alpha-2-macroglobulin, Gc globulin transferrin and the third component of complement C3. Proteins with isoelectric points outside of the range of albumin may be readily identified by deletion with specific antibody but the

174

alpha-2-macroglobulin and Gc are masked by the albumin.  With
affinity deletion as a first step, the alpha-2 diminution by
specific antibody is clearly evident in Figure 4 stained
with Coomassie Blue R250. The Gc, A-1AT are not in sufficient
concentration to stain with Coomassie Blue, and are shown in
the insert following counter staining with silver diamine.

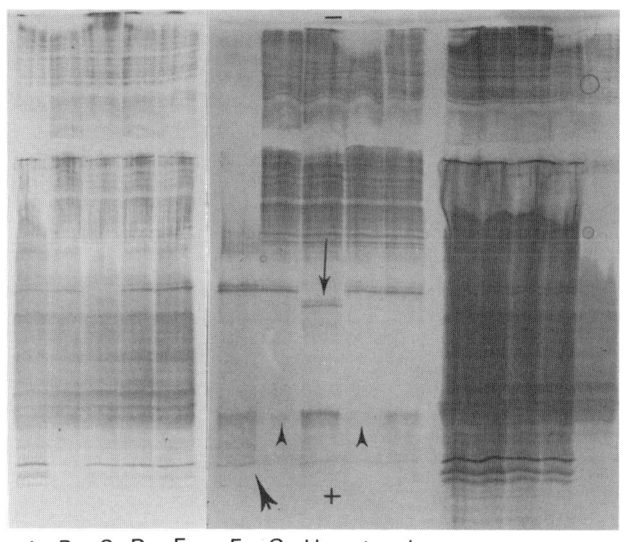



Figure 5.  A serum diluted 1:2 + Blue-Sepharose, B same as A
+ Anti-A-1AT, C plus anti-transferrin, D + anti-Gc, E + anti
C3, F Glycoproteins released from serum + concanavalin A-Seph-
arose, G serum + anti-A-1AT reacted with concanavalin A-Sepha-
rose and then released, H similar anti-transferrin, I anti-Gc,
J anti-C3, K serum, L serum + saliva, M,N serum + Matrix Gel A,
O serum treated with Matrix Gel A and concanavalin A-Sepharose.
The gradient was pH 3.5-8 and the gel stained with Coomassie
Blue R250.


In Figure 5, an additional affinity deletion step was performed following albumin subtraction and immunodeletion.  Here the
immunodeletion sample following centrifugation was reacted with
Concanavalin A covalently linked to Sepharose, the unbound material was discarded and the gel washed then three times with
PBS.  The bound glycoproteins were eluted then for separation.

Removal of the glycoproteins including the antigen-antibody
complex with concanavalin A-Sepharose and then eluting it for
PAGIF provided a much clearer background in which to observe
the deletion of the bound transferrin.  A comparison of track
C and H clearly shows two new bands with lower Isoelectric points
than transferrin present in the antisera  in H which is masked
with background protein stain in C.  Again, the anti-A-1AT
antigen  and Gc antigen appear to partially cross react while
C3 antibody produced no apparent deletion.

The lectins with their high binding affinity and ready release
of bound glycoproteins with the addition of appropriate sugars
offer the additional advantage of a concentration method of
dilute glycoprotein solutions in that in the elution step
smaller volumes of the eluting carbohydrate may be used, i.e.,
1 ml of sample originally bound may be eluted as here with 0.1
ml solution.

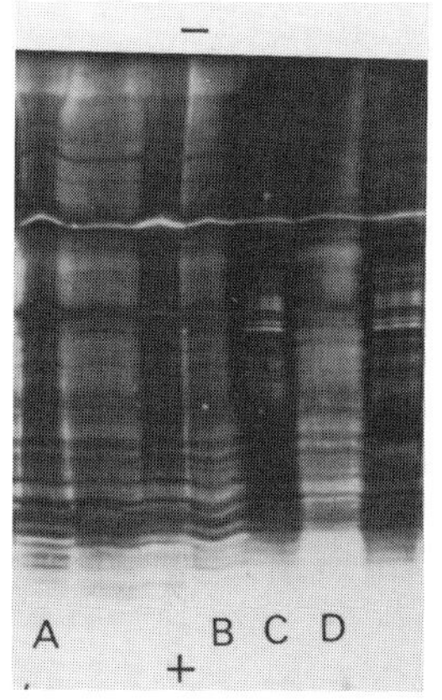


Figure 6.  Fresh saliva separated
on a pH 3.5-10 ampholine gradient.
A is normal saliva, B is saliva
from the same individual + saline,
C + anti-A-1AT and D trypsin.

Saliva separations showing the pH region of 3.5 to 7 are shown
in Figure 6.  Reaction with the IgG fraction of rabbit-anti-
alpha-1-antitrypsin and with trypsin directly indicates that
the proteins with isoelectric points similar to the Pi region
in serum are indeed A-1AT.

The use of both trypsin and the specific antibody cross confirm the specificity of the reaction.

Sweat was collected from the three authors after moderate
exercise and it was observed that those who perspired extensively had a much lower protein concentration and that various
locations on the face and neck tended to give slightly altered
patterns.  Separation of sweat obtained from the chin and forehead is shown in Figure 7.  The region marked by black arrows
was diminished by both Anti-A-1AT and Gc antisera but was not
deleted by trypsin, thus suggesting that this region is Gc
and that the specificity of these antisera may be suspect.
The region marked by the small white arrows shows the presence
of transferrin in the IgG fraction of the antisera similar to
to that found in serum as illustrated in Figure 5.

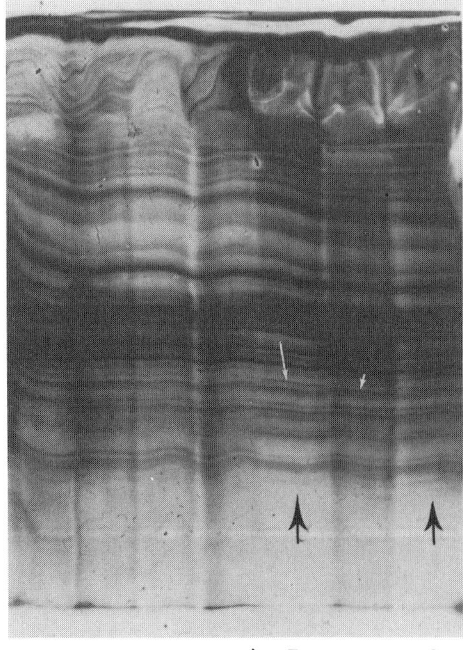

Figure 7.  Separation of sweat
on a pH 3.5-8 gradient with
proteins stained with silver
diamine.

A.    Sweat from chin
B.    Sweat from forehead
C.    Forehead sweat + saline
D.    Forehead sweat + trypsin
E.    Forehead sweat + anti-A-
      1AT antibody
F.    Forehead sweat + anti-
      transferrin
G.    Forehead sweat + anti-Gc
      antibody

## Discussion

These preliminary studies and studies by Altland (10) show
that a number of proteins can be directly identified in serum
and other body fluids by simply removing them following reaction with specific antibody or by a component to which they
complex strongly as trypsin (8).

To make this a more generally applicable diagnostic tool, the
combination of direct immuno deletion step with a preceding
step to remove selectively any interfering proteins enhances
the practicability of this procedure.  While  the preliminary
work described herein was limited to the removal of interfering albumin with Blue Sepharose, Matrix Gel A, concanavalin A
linked to Sepharose and trypsin, other ligands covalently linked to a gel matrix may be employed initially or at any stage
for the purpose of removing specific individual or classes of
proteins thus selectively clearing regions, particularly on
narrow range pH gels, to observe all the proteins which may
actually be present by simply using charge and specific binding affinity differences rather than more time consuming and
difficult charge size distribution such as is accomplished
with two-dimensional techniques.  In the AID technique, procedures may be selected also that avoid denaturation of the
proteins so that enzyme and other studies requiring proteins
in the native state may be accomplished.

This procedure would appear to be improved by the use of monoclonal antibodies where deletion of the complex and or identification of the soluble complexes should provide an additional
important tool to the immunologist.

These procedures in combination with high voltage isoelectric
focusing with its significantly higher resolution potential
offer a rapid method for genetic and forensic comparisons

where each sample serves as its own control and multiple treatments for identification of protein or complexes side by side on a single dimensional gel. The problem of establishing confidence limits between runs for exact comparative purposes or denaturation is thereby eliminated. It has the added advantage, in its simpler forms, of requiring only micro amounts of dilute samples due to the high sensitivity of the silver stain.

Although this approach was developed to extend uni-dimensional PAGIF resolution and protein separation, it is obvious that an extension of this approach to the second dimension in either SDS or gradient gels may offer an additional method with which to classify many if not all of the spots present in two-dimensional electrophoresis.

Acknowledgements

The authors wish to acknowledge the expert technical assistance of Ms. Margaret Ann Simmons and Mrs. Joyce Christopher and the secretarial assistance of Mrs. Brenda Altman.

References

1.  Tiselius, A.: Trans, Faraday Soc. $\underline{33}$, 524 (1937).

2.  Konig, P.:  Acts and Words of the Third Congress of South American Chemistry $\underline{2}$, 334 (1937).

3.  O'Farrell, P. H.:  J. Biol. Chem. $\underline{250}$, 4007 (1975).

4.  Guesdan, J. L., Ternynck, T. and Avrameas, S.:  J. Histochem. Cytochem. $\underline{27}$, 1131-1139 (1979).

5.  Switzer, R. C., Merrill, C. R. and Shifrin, S. A.: Anal. Biochem. $\underline{98}$, 231-237 (1979).

6.  Allen, R. C.:  Electrophoresis $\underline{1}$, 32-37 (1980).

7.  Anderson, N.G., Anderson, N. L. and Tollaksen, S. L.: Clin. Chem. $\underline{25}$, 1199 (1979).

8.  Allen, R.C., Oulla, P.M., Arnaud, P. and Baumstark, J.S.
    in: Radola, B.J. and Graesslin, D. (Eds.)Electrofocusing
    and Isotachophoresis, Walter de Gruyter, Berlin, pp. 256 -
    264 (1977).

9.  Swensson, H.: Acta Chem. Scand. $\underline{15}$, 325 - 341 (1961).

10. Arnaud, P., Galbraith, R.M., Chapuis-Cellier, C.,
    Galbraith, G.M.P., Fudenberg, H.H.: in Peeters, H. (Ed.),
    Protides of the Biological Fluids, Proc. 26th Colloquium
    Pergamon Press, New York, pp. 649 - 652 (1979).

ULTRATHIN-LAYER ISOELECTRIC FOCUSING IN 20 - 50 µm POLYACRYL-
AMIDE GELS: COMPARISON OF CONVENTIONAL AND MINIATURE SYSTEMS



Bertold J. Radola, Angelika Kinzkofer and Manuela Frey
Institut für Lebensmitteltechnologie und Analytische Chemie,
Technische Universität München, D-8050 Freising-Weihenstephan



Introduction

In the two most popular techniques of isoelectric focusing,
namely density gradient columns and gel-stabilized layers,
long separation times to date have been dictated by inherent
limitations in conductivity and resultant heat dissipation.
In thin-layer isoelectric focusing movable electrodes offer
high flexibility with respect to the choice of separation
distance and time. Surprisingly, most of the work with this
technique was limited to a separation distance of 10 - 12 cm.
Shorter separation distances were used only occasionally (1),
and aimed more at a reduction of costs rather than shortened
separation time and improved resolution. The traditional
1 - 2 mm gels thick appear unsuitable for rapid focusing
because high field strengths, imperative for the shorter
separation distances,would produce excessive heat. Although field
strengths as high as 150 - 300 V/cm, with power outputs of
0.3 - 0.5 W/cm$^2$ may be applied in the traditional gels (2-4)
they have not been widely used in practice. Ultrathin-layer
isoelectric focusing reduces the power output by a factor of
20-40 (5). Therefore much higher field strengths may be used
in standard equipment for thin-layer isoelectric focusing.
The approach chosen in this report was to reduce the separation distance and simultaneously to increase the field
strength to limits dictated by the conductivity of the carrier
ampholytes and heat dissipation of the system. To avoid ex-

cessive heating ultrathin-layer isoelectric focusing is carried out on 20-50 μm gels on 100 μm polyester films for better heat dissipation, instead of the previously employed 180 μm films (5). In a preliminary report it was shown that on a 3 cm gel isoelectric focusing of pH marker proteins and enzymes can be achieved in only 10 min (6,7).


Steady state ultrathin-layer isoelectric focusing

Isoelectric focusing on 10-12 cm gels requires 1500-3000 Vh to attain the steady state, judged by coalescence of proteins migrating from different application sites. Although in high porosity polyacrylamide gels (3%T, 20%C) coalescence of high molecular weigth marker proteins (e.g. ferritin, $M_r \geqslant$ 465000 and thyroglobulin, $M_r$ 790000) is achieved after isoelectric focusing for only 1300-1500 Vh (5), the usually employed gels (5%T, 3%C) require 2000-3000 Vh. With the equipment available to date the latter Vh products can be scarcely achieved in separation times shorter than 1-2 h, depending on pH range and power input. Experiments with 50 μm gels (5%T, 3%C) have shown that in the final stage much higher field strengths (300 V/cm) are possible than previously described (5). The main difficulty in shortening the separation time is that distorted patterns are observed when either prefocusing or focusing are performed at high initial field strengths. While 20-50 μm gels on the conventional 10-12 cm separation distance may be run at much higher field strength than usually employed the potential use of such gels has so far not been fully exploited. Power supplies yielding 5000-10000 V are available only for high voltage paper electrophoresis and the safety features of standard equipment for thin-layer isoelectric focusing is not compatible with voltages higher than 3000 V. In an attempt to shorten focusing time and fully exploit the capabilities of available equipment miniature

ultrathin-layer isoelectric focusing on 1-3 cm gels was
developed. Preliminary experiments have shown that samples
applied after prefocusing at relatively low field strength
gave optimal focusing patterns, and field strengths of 130 -
170 V/cm were adopted because of the consistency of results.
After 5 min prefocusing a linear gradient is observed in both
20 and 50 μm gels containing pH 4-9 Servalyt T carrier ampholytes. Fig. 1 shows the effect of time on focusing of pH marker
proteins (8) applied at different positions on the prefocused
gel. Samples applied anodically and in the middle coalesce
after 2 min focusing whereas the proteins starting from the
cathode are markedly retarded (Fig. 1B). All marker proteins
irrespective of application site, coalesced after 4 min.

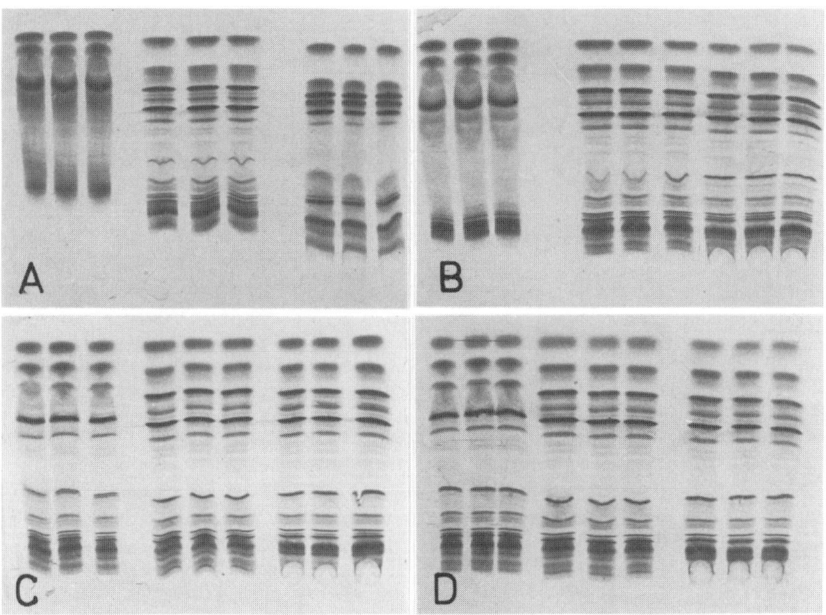

Fig. 1. The effect of time on miniature ultrathin-layer isoelectric focusing of pH marker proteins applied 0.7 cm from
the cathode, in the middle and 0.7 cm from the anode (from
left to right). Polyacrylamide gel - 5%T, 3%C, 20 μm layer.
Separation distance - 3 cm. Prefocusing - 130 V/cm, 5 min.
Focusing - 400 V/cm (A) 1 min, (B) 2 min, (C) 4 min, (D) 5 min.
Staining with Serva Violet 49.

These experiments demonstrate that on 3 cm gels prefocused for
5 min a steady state is reached after 4 - 5 min focusing time
corresponding to 110 - 130 Vh. An important characteristic of
the miniature system is that the pH gradient extends further
into the alkaline range. The pH gradient is probably stabilized by the high concentration of arginine and lysine of the
catholyte (5) and interference from $CO_2$ (9) is less owing to
short focusing time.


Isoelectric focusing on 1 and 2 cm gels


Coalescence of pH marker proteins on 1 and 2 cm gels in a
pH 4-9 Servalyt T gradient was achieved in 2-4 min total focusing time; 15 Vh (1 cm) or 40 Vh (2 cm) were adequate. The

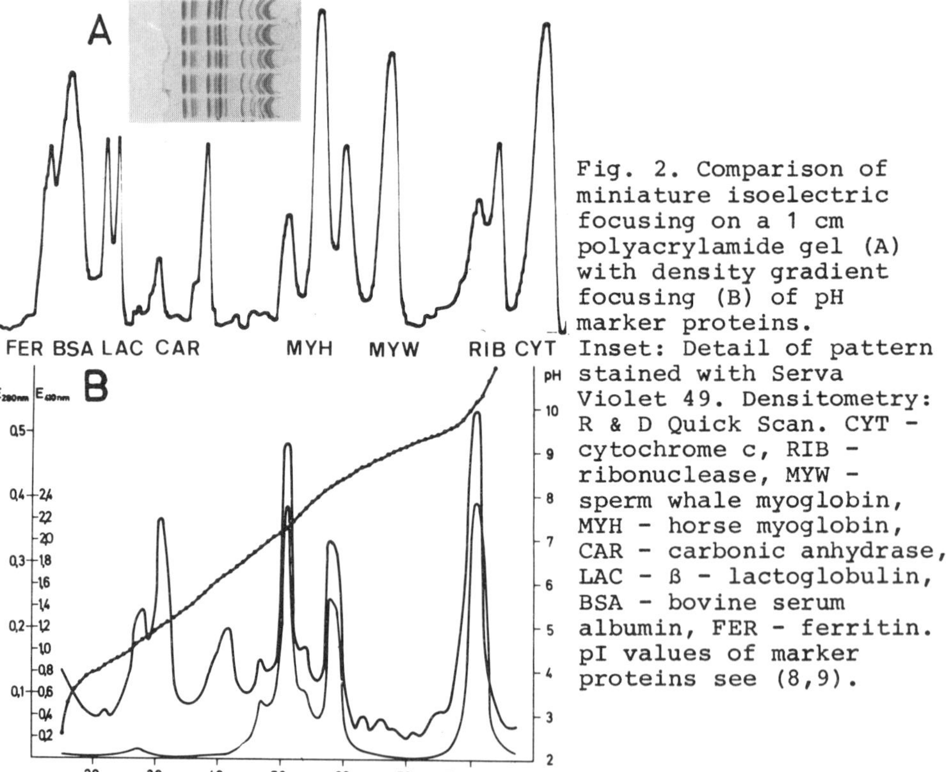


Fig. 2. Comparison of
miniature isoelectric
focusing on a 1 cm
polyacrylamide gel (A)
with density gradient
focusing (B) of pH
marker proteins.
Inset: Detail of pattern
stained with Serva
Violet 49. Densitometry:
R & D Quick Scan. CYT -
cytochrome c, RIB -
ribonuclease, MYW -
sperm whale myoglobin,
MYH - horse myoglobin,
CAR - carbonic anhydrase,
LAC - ß - lactoglobulin,
BSA - bovine serum
albumin, FER - ferritin.
pI values of marker
proteins see (8,9).

20 µm gels were prefocused at 150 V/cm for 1 min (1 cm gels)
or 2 min (2 cm gels). The prefocused 1 cm gels were focused
for an additional minute at 800-1000 V/cm. Densitometry of
the marker proteins reveals good resolution even on the 1 cm
gels, but a comparison of patterns over different separation
distances indicates that the number of components decreases
on reducing the separation distance. Even on 1 cm gels the
resolution of a mixture of marker proteins is better than in
a conventional 110 ml column operated under optimal conditions (Fig. 2). Thus isoelectric focusing can be a very rapid
method too, with fast generation of isoelectric spectra known
so far only from buffer pH gradients (10).


## Multiple samples

Miniature ultrathin-layer isoelectric focusing may be easily
adapted to an analysis of multiple samples (Fig. 3). Crude
fungal enzymes and marker proteins were focused on 3 x 10 cm
gels with excellent resolution and reproducibility. The samples may be applied with the aid of the multiple syringe as
originally suggested for multiple transfer of samples from
microtiter plates in thin-layer isoelectric focusing (1). When
the volume is limited to 0.2 - 0.4 µl up to four samples can
be applied directly on the gel surface. The densitometric
tracings of crude fungal enzymes closely resemble, with respect to quantitative distribution, the patterns over longer
separation distances (Fig. 5 in Ref. 5).


## Rapid staining

Staining of proteins in ultrathin gels offers considerable
advantages when compared with the traditional 1-2 mm gels (5).
With suitable carrier ampholytes and dyes, the total time

required for fixation, staining and complete destaining of gel
background is about 10 - 15 min. This is an impressive time
saving over traditional staining in which complete destaining
is rarely achieved in less than 20 h. In view of the potential
for routine application of ultrathin-layer isoelectric focusing, in particular the miniature system combining high resolution and short focusing time, a rapid staining method was
developed in which protein location is completed in only
3 - 5 min. Rapid staining of proteins in 20-50 µm gels involves

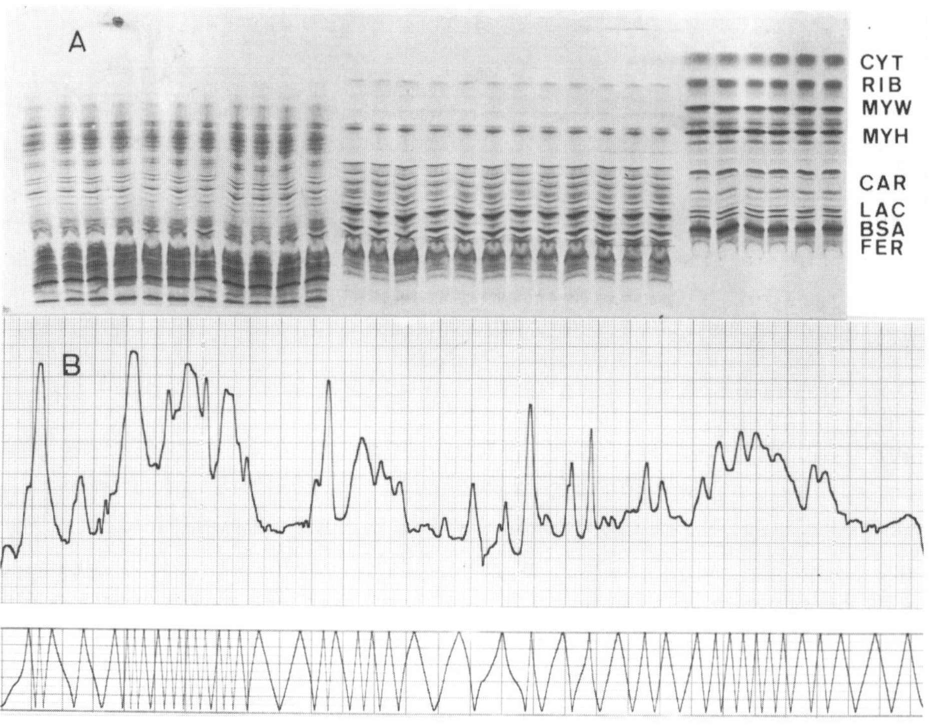



Fig. 3. Miniature ultrathin-layer isoelectric focusing of
multiple samples. Detail of a 3 x 10 cm gel with 36 samples
of crude pentosanase, Rohament P and pH marker proteins (from
left to right). Polyacrylamide gel - 5%T, 3%C, 50 µm layer.
Total focusing time - 10 min. (A) Proteins stained with
Serva Violet 49, (B) Scan of crude pentosanase, track indicated (●). R & D Quick Scan, scale expansion 1 : 20.
Abbreviations see Fig. 2.

the following steps: (i) Fixation of the gel in 20 % trichloroacetic acid, 30 sec, (ii) rinsing with destaining solution,
5 sec, (iii) drying on a HP Thermoplate (Desaga) at 80 $^{o}$C,
60 sec, a step previously described by (11), (iv) staining in
0.5 % Serva Violet 17 or Serva Violet 49, 20 sec, (v) destaining in methanol-water-acetic acid 25 : 65 : 10 v/v, 1 - 2 min.
With serial dilutions of a mixture of pH marker proteins and
a crude fungal enzyme (Rohament P) the method was found to be
highly sensitive. In miniature systems 10-20 ng protein were
detected on a 3 cm separation distance, and 100-200 ng on
10-12 cm gels. Fixation with acids and/or aldehydes (e.g.
glutaraldehyde, paraformaldehyde), diffusional losses of
proteins from the gel  in different solvents, without and
after fixation and methylacetate in the dye and/or destaining
solution were found to effect staining sensitivity (Frey, M.
and Radola, B.J., in preparation). In addition to speed and
high sensitivity, rapid staining may prove attractive for
those applications in which proteins soluble in dilute alcohol
are washed out from the gel when traditional staining procedures are employed.


Outlook


The results presented in this report demonstrate that optimal
resolution is contingent with conductance characteristics
and the voltage gradients applied to the system. The added
dimension of more effective cooling and heat dissipation
provides an important parameter to further improve the
efficiency of isoelectric focusing on ultrathin gels. Perhaps
the most important side effect of this approach is the fact
that sample size is also reduced by a factor of 10-50 times
thus allowing studies to be carried out with microquantities
of materials where staining techniques such as silver diamine
(11) can detect subnanogram amounts of protein. A major

advantage is the considerably smaller gel volume per analyzed
sample. On miniature gels 3 - 4 samples per cm are applied as
droplets resulting in 1.5 - 5 µl gel volumes per sample over
the 3 cm separation distance, and 0.5 - 1.5 µl gel volumes per
sample over a 1 cm separation distance in 20 - 50 µm gel
layers. This means considerable savings of carrier ampholytes
and reagents for gel preparation. Volumes in the range of
0.5 - 5 µl have been to date known only from microelectrophoresis in capillaries (12).

For the first time, miniature ultrathin-layer isoelectric
focusing combines high resolution with speed and operational
simplicity. Instead of hours of focusing time, miniature
systems require only minutes. This should prove an attractive
method for routine application, which in the area of electrophoresis has so far been dominated by continuous buffer zone
electrophoretic methods offering only low resolution. Of the
latter cellulose acetate membrane electrophoresis has remained
popular, at least to some extent due to such attributes as
short separation time, rapid location of separated substances
and partial automation. Miniature ultrathin-layer isoelectric
focusing can readily replace cellulose acetate membrane electrophoresis affording better resolution and a higher predictive
value at a comparable level of experimental effort. In addition to clinical laboratories the method may prove useful in
industrial laboratories for production and quality control.
Miniature ultrathin-layer isoelectric focusing may be an
attractive tool for research laboratories in a variety of
applications where either small amounts of material are
available or rapid monitoring of experiments is required. The
method is particularly suitable for studies of genetic polymorphisms of proteins and enzymes.

# References


1. Altland, K., in: Radola, B.J., and Graesslin, D. (Eds.), Electrofocusing and Isotachophoresis, Walter de Gruyter, Berlin, pp. 295-301 (1977).

2. Righetti, P.G., Righetti, A.B.B., in: Arbuthnott, J.P. and Beeley, J.A. (Eds.), Isoelectric Focusing, Butterworth, London, pp. 114-131 (1975).

3. Söderholm, J., Allestam, P., Wadström, T., FEBS Lett. 24, 89-92 (1972).

4. Allen, R.C., Oulla, P.M., Arnaud, P., Baumstark, J.S., in: Radola, B.J. and Graesslin, D. (Eds.), Electrofocusing and Isotachophoresis, Walter de Gruyter, Berlin, pp. 255-264 (1977).

5. Radola, B.J., Electrophoresis 1, 43-56 (1980)

6. Radola, B.J., in: Radola, B.J. (Ed.), Electrophoresis '79, Walter de Gruyter, Berlin, pp. 79-94 (1980).

7. Radola, B.J., in: Radola, B.J. (Ed.), Elektrophorese Forum '80, Technische Universität München, pp. 43-66 (1980).

8. Radola, B.J., Biochim. Biophys. Acta 295, 412-428 (1973).

9. Delincée, H., Radola, B.J., Anal. Biochem., 90, 609-623 (1978).

10. Kolin, A., Proc. Natl. Acad. Sci. USA, 41, 101-110 (1955).

11. Allen, R.C., Electrophoresis 1, 32-37 (1980).

12. Neuhoff, V., in: Radola, B.J. (Ed.), Electrophoresis '79, Walter de Gruyter, pp. 203-218 (1980).

# ULTRATHIN LAYER PAGIF: A COST BENEFIT ANALYSIS OF THE FOCUSING DISTANCE


Torgny Låås, Ingmar Olsson
Pharmacia Fine Chemicals AB, Electrophoresis group, Box 175,
751 04 Uppsala, Sweden.


## INTRODUCTION

Ever since the advent of isoelectric focusing it seems that the
majority of analytical experiments, whether performed in rods
or thin layer slabs, have been run over approximately 10 cm.
For preparative purposes, on the other hand, generally about
twice that distance have been used both in columns and Sephadex
layers.

However, there are by no means any _a priori_ reasons why 10 cm
should always be optimal for analytical separations. The effect
of the separation distance on the band width and resolution was
theoretically calculated by Svensson already twenty years ago
(1, 2). According to these theoretical calculations there is a
square root dependence between resolution and focusing distance
when all other properties are kept constant.

In practice, however, the incresed resolution by using longer
separation distances is only attained at the cost of, for
example, higher consumption of both carrier ampholytes and
sample and longer running times.

Moreover, with the presently available equipment, it is
difficult, if not impossible, to use the same field strength
over a long distance as over a short one. In such cases it is
not even theoretically possible to take full advantage of the
longer distance.

192

Consequently, the use of very long gels have never been very
popular except in conjunction with SDS-electrophoresis in the
O'Farrell (3) or Iso-Dalt technique where normally 18-20 cm
gels are used for the IEF separation.

Neither has the use of miniature gels as sometimes advocated
been generally accepted (4).

One reason for the conservatism may be the lack of direct and
objective comparisons between different focusing distances.

The aim with this work was therefore to give a direct, and as
far as possible, objective comparison between focusing of the
same samples over different distances. Parameters that were
compared include: time and number of volt-hours to achieve the
same degree of focusing, amount of sample, pH gradients
obtained, degree of resolution and carrier ampholyte
consumption.


METHODS

Gel casting

Ultrathin (0.2 mm) polyacrylamide gels were cast on silanized
(5) glass plates using a modification of the flap technique
described by Radola(6).

Gel composition was T5 C3  containing Pharmalyte 3-10 diluted
1:15 and 10% glycerol. The acrylamide and bisacrylamide were
from BioRad. The polymerization was initiated by adding
ammonium persulphate to 0,7 mM.

Gel lengths

Focusing experiments were performed over the following
interelectrode distances: 23, 47, 95 and 190 mm. This required

an apparatus with adjustable electrodes such as the Pharmacia
Flat Bed Aparatus FBE 3000.

Samples

The following samples were used: 1) elk (Swedish moose) muscle
extract, 2) the Pharmacia pI calibration kit and 3) rainbow
trout muscle extract. The sample amounts needed to get the same
band intensity in all experiments were experimentally found to
be:

Table 1
Amount of sample (µl) used for different gel lengths

| Sample | Gel length (mm) | | | |
|---|---|---|---|---|
| | 23 | 47 | 95 | 190 |
| 1. Elk | 0.15 | 0.5 | 1.5 | 6 |
| 2. pI kit | 0.25 | 0.7 | 2.5 | 10 |
| | (0.13) | (0.35) | (1.25) | (5) |
| 3. Rainbow trout | 0.15 | 0.5 | 1.5 | 6 |
| Relative amounts-found | 1 | 3 | 10 | 40 |
| Relative amounts-theor. | 1 | 3 | 8 | 23 |

Within parenthesis are given the amounts in µg/protein of the
calibration standards used.

Theoretically it was expected that sample amount would be
proportional to gel length x $\sqrt{1/dpH/dx}$.

The samples were applied with the aid of sample applicators
made of plastic film with cut-out holes. The width of the
holes were adjusted in proportion to the focusing distance so
that all separations would show the same physical
proportions.

RUNNING CONDITIONS

Duration of the experiments

The relevant parameter for controlling electrophoresis
experiments, including IEF, is volt-hours. However, when
comparing focusing lengths different volt-hours will be needed
for the same degree of focusing. These can only be found by
using some internal standard. We used the coalescence of two
haemoglobin samples focusing from the anodic and cathodic side
resp. as the internal standard. Since hemoglobin focuses faster
than most other proteins, we arbitrarily took twice the number
of volt-hours for haemoglobin samples to coalescence as the end
stop marker for each IEF distance (see Fig. 1).

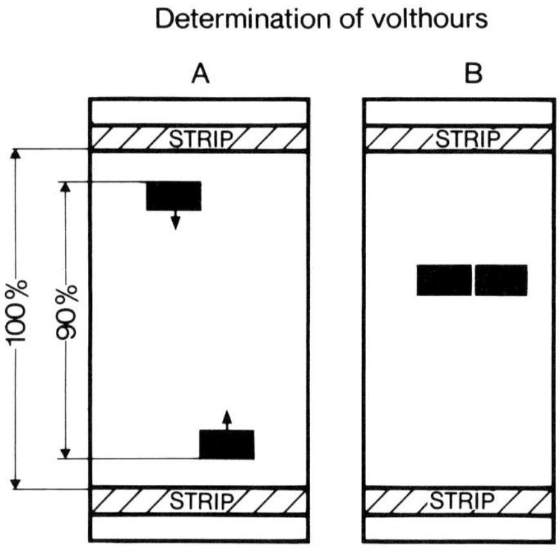

Fig. 1. Schematic illustration of determination of volt-hours.

The volt-hours were conveniently and accurately recorded by the
Pharmacia Volt-Hour Integrator VH-1 connected to the Pharmacia
Power Supply ECPS 3000/150, except when running the smallest
gels. With the extremely low power loading used in those cases
the Pharmacia ECPS 2000/300 was used instead and volt-hours
calculated manually.

## Electrical load

In the first set of experiments, the same number of watts/mm$^3$ (0.016) as recommended for PAGIF with Pharmalyte in 1 mm thick gels were used throughout. This makes evaluation of the effect of gel length possible since all other conditions (esp. field strength) were the same.

For the practical worker, however, it is perhaps more interesting to explore the optimal performance of each distance. So, in another set of experiments the effect of increased voltage was investigated.

## Staining

After focusing the gels were stained with Coomassie brilliant blue G 250 according to conventional procedures (5).

## Photographing

The gels were photographed so as to give separation patterns of the same size in order for relevant and objective comparisons to be done.

## RESULTS

That the experimental conditions used gave comparable results is indicated by Figure 2 which shows the pH gradients obtained with the longest and shortest gels, and by Figure 3 which shows the pH gradients obtained with the 2.3 cm gel at low and high field strength. The pH is plotted as a function of the relative electrode distance. pH was determined with the pI calibration standards  This is the only way for accurate pH gradient determinations with the smallest gels.

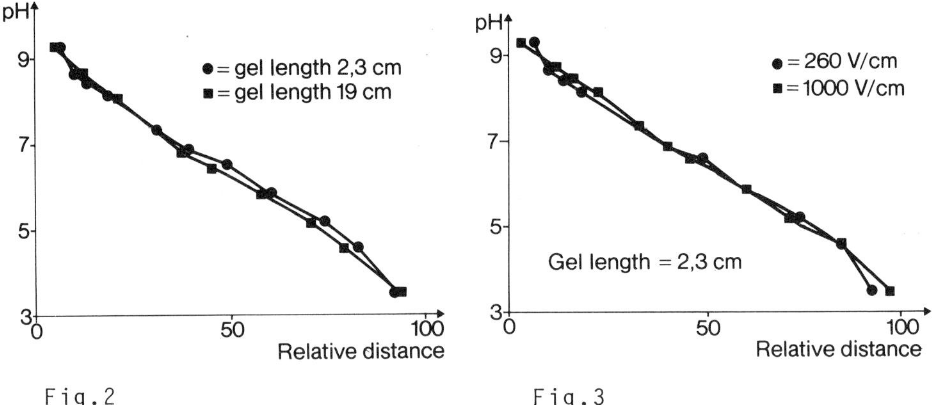

Fig.2

Fig.3

As can be seen, the pH gradients are practically identical. The gradients of gels of intermediate lengths and field strengths were also identical but excluded from the Figure for the sake of clarity.

Length of focusing (time, volt-hours)

The relation between focusing distance and the volt-hours and time is seen in Figure 4.

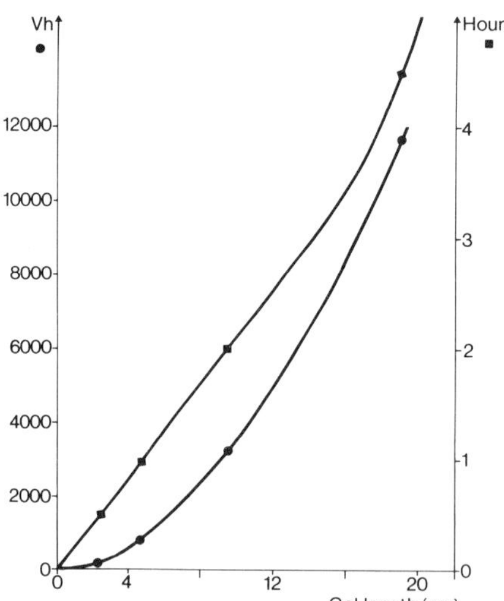

Fig.4

The results plotted here relate to the first set of experiments where the gels were run at an average of 0.016 watts/mm$^3$ in all cases.

It is seen that under these conditions there is a linear dependence between time and distance whilst the relationship between volt-hours and distance follows a completely different path. The time deviation from the straight line for the 19 cm gels is because in this case  the whole experiment could not be performed at constant power. That would have needed a power supply able of giving about 4500 Volts by the end of the experiment.

## EVALUATION OF RESULTS

### Constant field strength

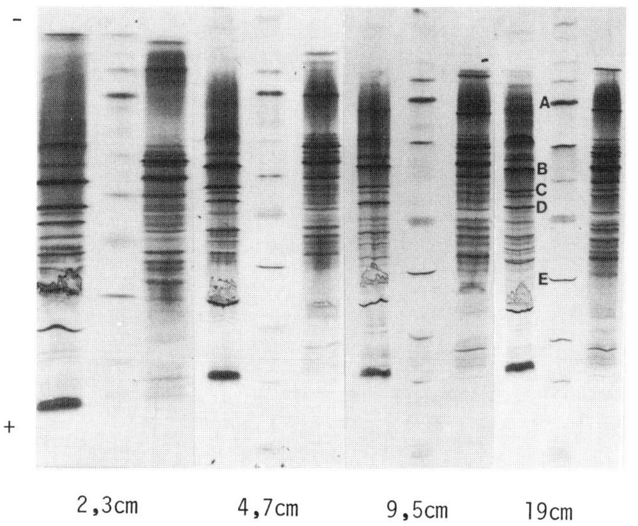


Fig. 5. Gels from the experiments described above were photographed so that the pictures could be corrected to the same size with a photographic enlarger in order to make comparisons easier

The band widths for five easily recognizable bands, marked as
A. B. C. D and E in Figure 5, were measured on enlarged
photographs. The relative band width (as % of focusing
distance) is plotted as a function of distance in Figure 6.

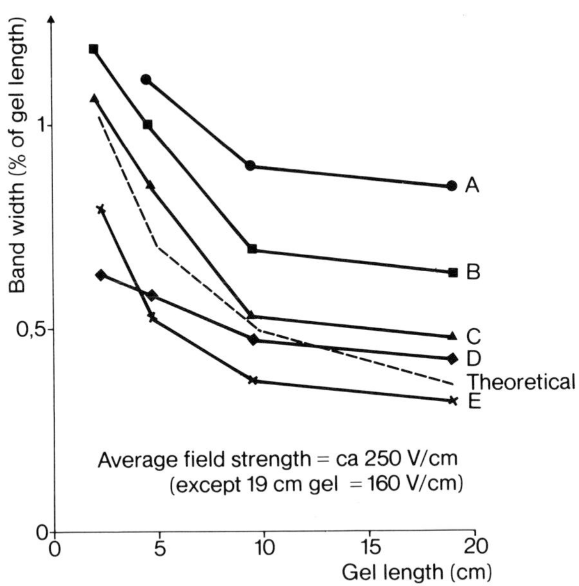


Fig. 6. Band width as a function of gel length at constant
field strength.

In the figure is also plotted the theoretical curve for a
hypothetical protein with a 0.5% band width when focused over
10 cm. The correlation between theoretical and observed band
widths is very good. Only for the 190 mm gels are the band
widths too broad. This is again depending on the fact that the
190 mm gels could not be run under power limited conditions
during the whole experiment because the available voltage was
limited to 3000 volts.

Maximal field strength

In the next set of experiments the 47 and 23 mm gels were run
at different powers.

The voltage change with time for the different gel lengths and
powers used is illustrated in Figure 7d. It is seen for
example, that with the maximal power used  the 2.3 cm gel is
ready after 15 minutes' focusing.

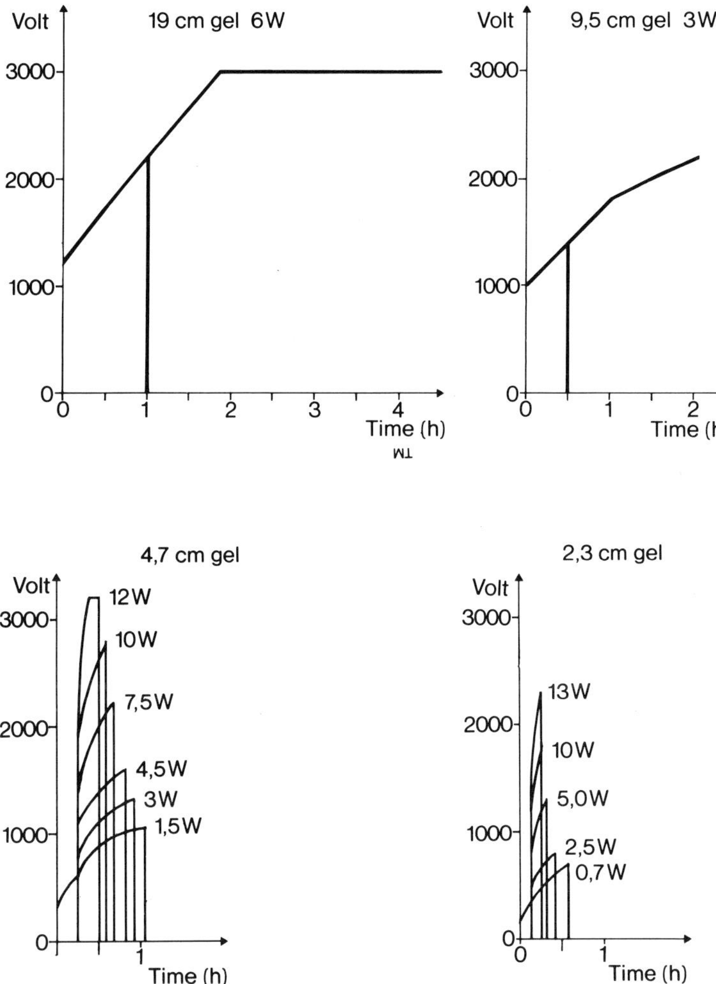


Fig. 7. Voltage as a function of time for different gel lengths
when run at different powers. The gels were always prefocused
for 15% of the total number of volt-hours before sample
application and power increase.

200

A few examples to illustrate typical protein patterns in a 2.3
cm long gel are given below.

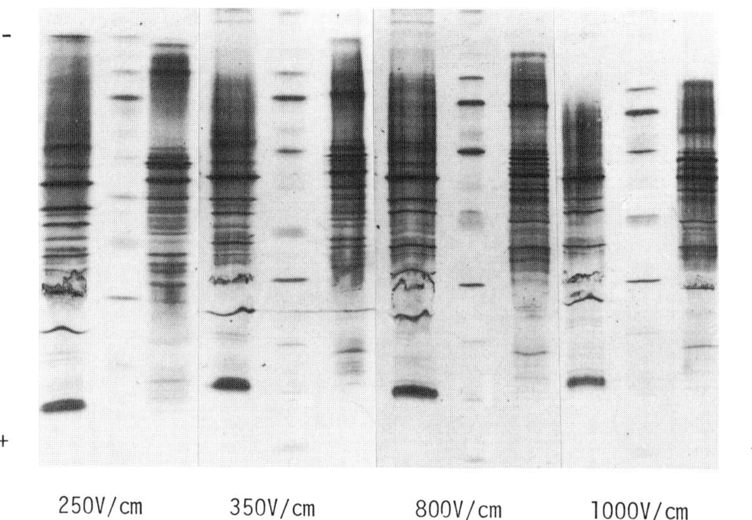



Fig. 8. Comparison of the results obtained with the 2.3 cm gel
when run under normal and high field strength.

A first glance of the patterns obtained reveals no big
differences. To more accurately evaluate the difference that
actually exist, we measured the band widths on magnified
pictures of the gels.

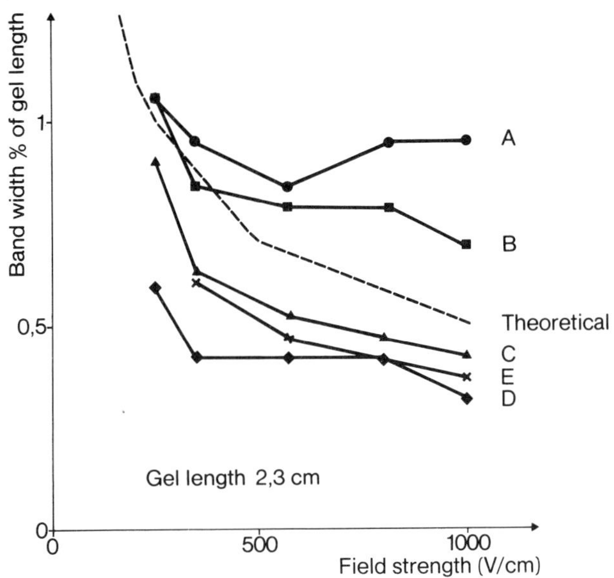



Fig. 9. The band widths of the same five proteins as previously
studied as a function of field strength. The theoretical band
width for a hypothetical protein with 0.5% bandwidth at 1000
Volts is also depicted. It is clear that the observed field
strength dependence on voltage agrees very well with the
theoretical.

As an illustration to the results obtained with maximal field
strengths, the rainbow trout extract patterns obtained with the
2 3  and 19 cm gel are shown side by side in Figure 10.

Fig. 10. Rainbow trout extract run on a 2.3 cm gel at a maximal
field strength of 1000 V/cm (top lane) and on a 19 cm gel at
the maximal field strength of 158 V/cm (bottom lane). Cathode
to the right. Protein patterns photographically enlarged to the
same size.

It is clear that a better separation is obtained with the longer gel but to the price of longer running time and higher chemical consumption.

The results are summarized in Table 2.

Table 2

| Quality of results | Focusing distance (mm) | | | | | |
|---|---|---|---|---|---|---|
| (Band width %) | 23 | | 47 | | 95 | 190 |
| Protein A* | 1.0 | 0.9 | 1.0 | 0 6 | 0.9 | 0.8 |
| B | 1.1 | 0.7 | 1.0 | 0.5 | 0.7 | 0.6 |
| C | 1.0 | 0.4 | 0.8 | 0.4 | 0.5 | 0.5 |
| D | 0.6 | 0.3 | 0.6 | 0.3 | 0.5 | 0.4 |
| E | 0.8 | 0.4 | 0.5 | 0 5 | 0.4 | 0.3 |
| | | | | | | |
| Economy | | | | | | |
| Carrier ampholyte (µl/sample) | 1.6 | | 5.5 | | 21 | 76 |
| Rel. sample amount | 1 | | 2.7 | | 9.6 | 38 |
| | | | | | | |
| Conditions | | | | | | |
| Volt-hours | 220 | 220 | 870 | 870 | 3270 | 11400 |
| Final av. field strength (V/cm*) | 260 | 1000 | 213 | 638 | 232 | 158 |
| Time min * | 33 | 15 | 63 | 31 | 122 | 270 |

* Left column refers to runs performed at the same power/mm$^3$. Right column to the highest used voltage.

CONCLUSIONS

Very long gels can at present hardly be recommended. The theoretically obtainable increase in resolution is practically impossible to achieve due to lack of power supplies giving sufficiently high voltages (4-10 kV) - only the drawbacks remain.

When extremely high resolution is needed in a specific pH
region this is more efficiently accomplished by expansion of
that particular pH region. Besides the use of narrow range pH
intervals, there now exist at least three principally different
methods for this (7).

However, we believe that our results show that smaller gels
than the standard 100 mm ones may be used advantageously for
many applications. A small sacrifice in resolution is
compensated for by increased economy in terms of shorter
running time, much smaller carrier ampholyte consumption and
higher sensitivity. The only significant problem being that of
detection when it comes to the very smallest miniature gels.


References

1. Svensson, H.: Acta Chem. Scand. 15, 325-341 (1961).

2. Svensson, H.: Acta Chem. Scand. 16, 456-466 (1962).

3  O'Farrell, P.H.: J  Biol. Chem. 250, 4007-4021 (1975)

4. Neuhoff V.: in Electrophoresis '79, Walter de Gruyter,
   Berlin-New York, 1980, ed  B.J. Radola, 203-318.

5. "Pharmalyte$^{TM}$ Carrier Ampholytes for Isoelectric
   Focusing. Instruction for Use", Pharmacia Fine Chemicals AB.

6. Radola, B.J.: Electrophoresis 1, 43-56 (1980).

7  Låås, T., Olsson, I.: Anal. Biochem. In press.

# ISOELECTRIC FOCUSING IN THIN-LAYER AGAROSE GELS

D. L. Harper

BioProducts Department, Marine Colloids Division, FMC Corporation, Rockland, ME, 04841, USA


## Introduction

Due to properties of chemical inertness and good mechanical
strength at low concentration, highly purified agarose in thin
layer gel slabs has been found to be a useful medium for the
separation of complex protein mixtures by the electrofocusing
technique (1). At the concentrations usually employed ($\leq 1\%$),
amphoteric species of widely varying molecular weight range are
readily taken up by and transported through the relatively macroporous gel matrix. We examine here separations in agarose gels
of a mixture of well-characterized, relatively homogeneous proteins, a number of which have been described previously as useful pH gradient markers (2, 3, 4, 5). The apparent isoelectric
point (pI) values of the proteins focused in agarose media are
compared with literature values derived from work employing
polyacrylamide, dextran, and density gradient systems at various
temperatures. We also note any apparent effect upon pI through
inclusion of the nonionic additives glycerol, sorbitol, and sucrose in the agarose media.


## Materials and Methods

1)  IsoGel$^{TM}$ agarose, IsoGel Ampholytes pH 3.5-9.5 for agarose,
    GelBond$^{TM}$ gel supporting film, an electrofocusing accessory
    kit consisting of instructions, plastic sample application
    mask and paper pieces, micropipet syringe, blotting papers,
    electrode strip wicks, lined grid for gradient measurement,

and spacers suitable for casting up to either 125 x 140 x 0.8
or 1.6 mm gels are all available from Marine Colloids, Rockland, ME. Ampholine$^{TM}$ pH 9-11 was obtained from LKB, Rockville, MD. Purified proteins examined are available from
various distributors of biochemicals. Wide-range protein
pI marker kits were obtained from Serva, Heidelberg, FRG,
and Pharmacia, Piscataway, NJ, and were prepared and used
according to recommended instruction. The catholyte used
was 1.0 M sodium hydroxide and the anolyte 0.5 M acetic acid.
Focusing was performed in a Bio-Rad, Richmond, CA, flat bed
coolant chamber model 1415, with adjustable ribbon electrode
assembly, in conjunction with an ISCO, Lincoln, NE, model
494 regulated power supply and a temperature-controlled circulating water bath. pH readings were obtained with an
Ingold, Lexington, MA, model 6122 surface micro-electrode
to the nearest one hundredth pH unit.

2) 0.8 mm thick IsoGel slabs at 1% concentration in distilled,
deionized water, with or without the neutral additive substances, and adhered to GelBond backing were formed according to the Marine Colloids recommended instructions. The
pH 3.5-9.5 ampholyte was used at 2.5% concentration and
supplemented at the alkaline end with an additional 0.12%
of pH 9-11 Ampholine (both expressed as dry weight basis of
respective ampholyte).

Aqueous solutions of varying dilution were prepared of the
following proteins, singly and in combination: Cytochrome C
(horse heart), myoglobin (whale skeletal muscle), myoglobin
(horse skeletal muscle), carbonic anhydrase (bovine erythrocytes)
β-lactoglobulin (bovine milk), ovalbumin (chicken egg), glucose
oxidase (<u>A.niger</u>), and amyloglucosidase (<u>A.niger</u>). Additionally,
0.1% aqueous solutions were made of the acidic dyes amaranth
red, bromophenol blue, and methyl red.

The ruled grid was pressed to the cooling platform of the
focusing chamber on a thin water film, to ensure good heat

transfer, and the coolant begun circulating at a thermostatted $10^{\circ}C$. The gels on GelBond backing were, in turn, secured upon the surface of the grid with an even water film layer. Samples were, in most instances, applied to the gels at approximately one third distance from the cathode by means of the slotted application mask at 2 μl load levels each.

Electrode paper strips 1 mm thick and cut to a length slightly less than the width of the gel were soaked briefly in the electrolyte solutions, blotted to remove excess fluid, and placed in position at the ends of the gel, using the lined grid as a guide. To ensure perfectly even contact between the ribbon electrodes and the strips, a small metal bar weight was laid across the electrode assembly for 1 minute, then removed.

The power supply was set to 1000 V and 1 W limiting, current unlimited, and power applied for 10 minutes, during which time the samples are taken into the gel. At the end of this period, the application mask was removed, wattage setting increased to 25 W upper limiting, and focusing conducted for an additional 35 minutes. During the course of the run, temperature levels in various regions of the gel were monitored with a surface probe of our manufacture. Upon completion of focusing, the temperature of the reservoir coolant was rapidly adjusted to that observed in the gel at termination. With the lined grid as guide, pH measurements were taken at progressive 1 cm distances from cathode to anode with the surface pH electrode, which was calibrated successively against a range of standard buffers chilled to the temperature of the gel. After pH measurement, the gel was refocused for 5 minutes.

The proteins were fixed by placing the gels into a mixture of sulfosalicylic acid (3%), trichloroacetic acid (10%), methanol (25%), and water (62%) for 10-15 minutes. Most of the fixative and ampholyte were then removed by inverting the gels upon blotting paper and absorbent toweling and pressing with a flat weight of $\leq 1$ kg for 20 minutes. The papers were then removed and residual ampholyte and fixative diffused out by placing the gels in distilled water

or a freshly-prepared mix of methanol (80%), glacial acetic acid
(10%), water (10%) for 5 minutes.  After rinsing, the gels were
dried at 60°C, then stained for 10-15 minutes in either Crowle's
double stain (Crocein Scarlet-0.25%, Coomassie Brilliant Blue
R-250-0.015%, glacial acetic acid-5%, trichloroacetic acid-3%,
water-91.8%) or Coomassie Brilliant Blue R-250 0.1% in glacial
acetic acid (10%), methanol (30%), water (60%).  Any background
stain uptake may be removed by a rinse in 0.3% aqueous solution
of acetic acid for the former or methanol for the latter.


Results and Discussion


1)  It was found that the mixture of visible proteins cytochrome
    C, whale and horse myoglobins, and the dye substances, which
    focus as a series of fine lines near the anode, were useful
    in estimating appropriate termination points of electrofocus-
    ing runs.  The remaining proteins, which are visualized after
    staining and whose pI's have been previously reported in the
    literature, may also serve as useful pH gradient reference
    points in estimating pI's of sample materials of known or un-
    known composition (Fig. 1).  A typical measured pH gradient i
    an additive-free gel is also shown (Fig. 2).

2)  Due to drop in conductivity through the gels, as the ampholyt
    focused at their respective pI's, terminal wattages were ob-
    served to fall within the range 4.5-6 W.  Under these final
    power conditions, average temperature along the lengths of th
    gels was determined to be 15±2°C.  This was then utilized as
    the maintenance and calibration temperature for determining
    the pH gradient profiles.  Apparent pI values of the proteins
    and dye substances were derived by interpolating the values o
    pH measured in bounding regions, using the dried, stained gel
    patterns.  Thin-layer agarose films on the GelBond backing dr
    down uniformly without shrinking.  This precludes any problem
    of linear distortion of the original points of measurement
    relative to banding placement.

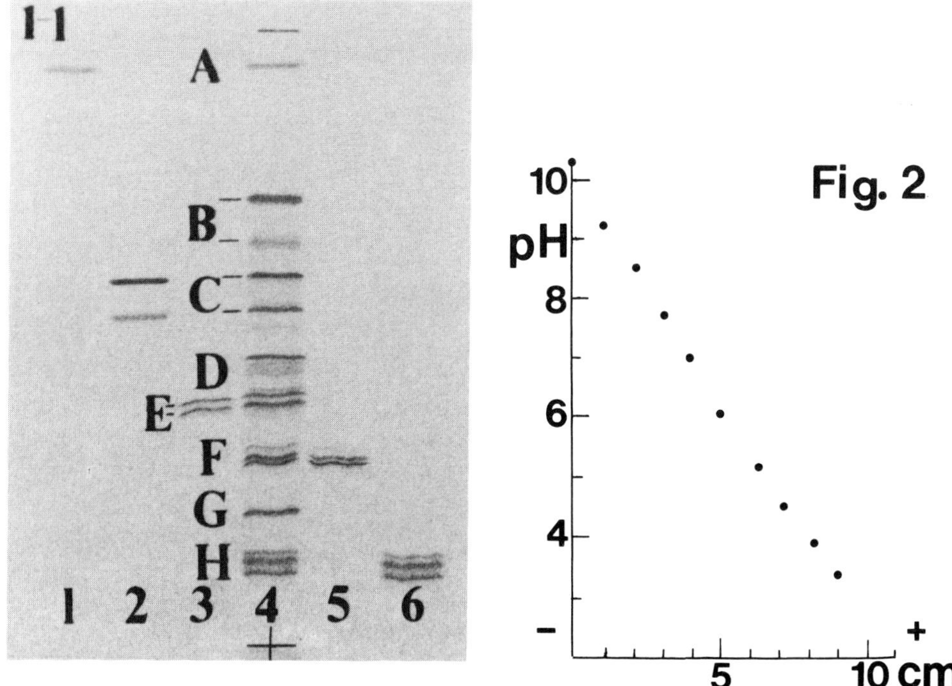



Figure 1 (1-1).  Focusing performed as described in text, no neutral additives.  Samples:  Lane (1) cytochrome C; (2) horse myoglobin; (3) β-lactoglobulin; (4) composite blend of (A) cytochrome C, (B) whale myoglobin, (C) horse myoglobin, (D) carbonic anhydrase, (E) β-lactoglobin, (F) ovalbumin, (G) glucose oxidase, (H) amyloglucosidase; (5) ovalbumin; (6) amyloglucosidase.  Proteins visualized in Coomassie stain.  Figure 1 (1-2).  Visualized in Crowle's stain.  Figure 1 (1-3).  Lane (1) Pharmacia wide-range test mixture, (2) Serva wide-range test mixture, (3) mixture A-H applied near cathode, (4) mixture A-H applied near anode.

Figure 2.  Observed pH at end of focusing period versus distance from cathode at 15°C, as described in text.

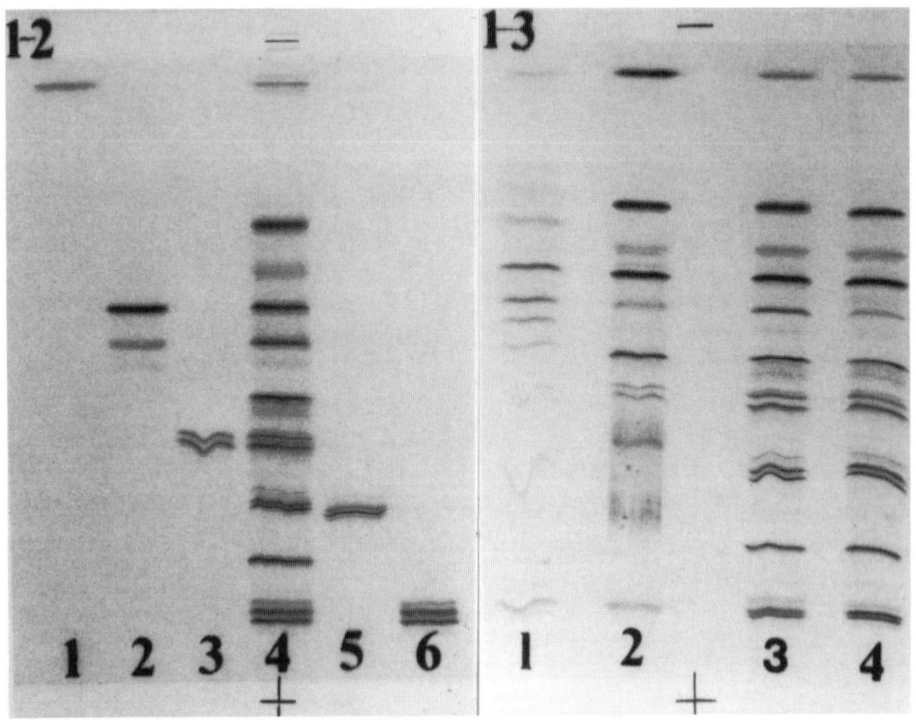


1-2
1 2 3 4 5 6
1-3
1 2 3 4

TABLE 1

| Protein or Dye | Molecular Weight | Literature pI Values | Apparent pI Values, Agarose Gels, 15°C | | | |
| --- | --- | --- | --- | --- | --- | --- |
| | | | No Additive | 10% Sorbitol | 10% Glycerol | 10% Sucrose |
| *Cytochrome C | 13,000 | 9.0-10.16 (2,3,4) | 10.2 | 10.2 | 10.2 | 10.2 |
| *Myoglobin (sperm whale) | | | | | | |
| Major band | 17,000 | 8.1-8.3 (2,3,4) | 8.2 | 8.1 | 8.3 | 8.2 |
| Minor band | | 7.66 (2) | 7.7 | 7.7 | 8.0 | 7.7 |
| *Myoglobin (horse) | | | | | | |
| Major band | 17,500 | 7.31-7.76 (2,3,4,5) | 7.4 | 7.3 | 7.5 | 7.2 |
| Minor band | | 6.47-7.30 (2,4,5) | 7.0 | 7.0 | 7.0 | 6.9 |
| Carbonic anhydrase | 31,000 | 5.9-6.21 (3,5) | 6.1 | 6.1 | 6.0 | 5.9 |
| β-Lactoglobulin (A,B) | 35,000 | 5.1-5.48 (2,3,4,5) | 5.4-5.5 | 5.4-5.5 | 5.3-5.4 | 5.1-5.2 |
| Ovalbumin | 45,000 | 4.70 (4) | 4.8 | 4.8 | 4.6 | 4.5 |
| Glucose oxidase | 186,000 | 4.2 (6) | 4.2 | 4.2 | 4.2 | 4.1 |
| Amyloglucosidase | 97,000 | 3.5 (7) | 3.6 | 3.6 | 3.6 | 3.6 |
| *Methyl red dye | | | 3.7 | 3.7 | 3.7 | 3.7 |
| *Amaranth red dye | | | 3.2 | 3.2 | 3.2 | 3.2 |
| *Bromophenol blue dye | | | 3.1 | 3.1 | 3.1 | 3.1 |

*Colored markers. Apparent "pI's" of the acidic dyes are believed to result from complex formation with certain carrier ampholyte species in this pH region (8).

212

Including glycerol or sorbitol at levels of 10% in the gels
seemed to have little effect on apparent pI's of the proteins,
compared with those gels without additive.  10% sucrose gels
did, on average, appear to lower observed pI's somewhat in the
pH range 4-6.  The apparent pI values derived from the fore-
going experiments, rounded to the nearest one tenth pH unit,
are compared with those from available literature (Table 1).
The literature values span working temperatures of 4-25$^{\circ}$C, a
variable which is known to influence observed protein pI (higher
values at lower temperatures).  Referenced protein pI values
are those where source tissue is known and the same as described
in materials and methods.


References


1.  Saravis, C.A. and Zamcheck, N., J. Immunol. Methods 29, 91-96
    (1979).

2.  Radola, B.J., Biochim. Biophys. Acta 295, 412-428 (1973).

3.  Radola, B.J., Electrophoresis 1, 43-56 (1980).

4.  Righetti, P.G. and Caravaggio, T., J. Chromatogr. 127, 1-28
    (1976).

5.  Bours, J., Sci. Tools 20, No. 2-3, 29-34 (1973).

6.  Bently, R., in: Boyer, P.D., Lardy, H.A. and Myrback, H., The
    Enzymes 7, Academic Press, New York-London, p. 567 (1963).

7.  Pharmacia and Serva descriptive literature, source work other-
    wise uncited.

8.  Gianazza, E. and Righetti, P.G., in: Radola, B.J. (Ed.),
    Electrophoresis '79, Walter de Gryter, Berlin-New York, 129-1
    (1980).

# SEAPREP[TM] 15/45: A NEW AGAROSE WITH LOW GELLING AND REMELTING PROPERTIES FOR PREPARATIVE ELECTROPHORESIS

Samuel Nochumson
Marine Colloids Division, FMC Corporation, Rockland, ME, 04841.

## Introduction

SeaPrep[TM] 15/45 is a unique hydroxyethylated-agarose derivative which, at a one percent concentration, forms a gel at $15^{\circ}$C and remelts at $45^{\circ}$C. The relatively low temperature required for thermoreversibility from the gel to the sol state is potentially useful for recovering biological macromolecules from the gel subsequent to electrophoretic separation. In order to demonstrate the usefulness of SeaPrep[TM] 15/45 as a preparative electrophoresis medium, several hemoglobin variants and alkaline phosphatase were electrophoretically resolved in the gel, followed by an extraction procedure which included remelting of the gel by a mild heat treatment.

## Materials and Methods

Horizontal gels containing 10 ml of 2% SeaPrep[TM] 15/45 in 0.08M Tris, 1.8M EDTA, 0.05M borate buffer (pH 8.6) were open cast on GelBond[TM] sheets (110 x 125 mm) and allowed to gel for one hour in the refrigerator. Red cell hemolysates were prepared as described previously (1) and human placental alkaline phosphatase was obtained from Sigma Chemical Company (St. Louis, MO). Alkaline phosphatase activity was determined by a reagent kit from Dow Diagnostics (Indianapolis, IN) and enzyme activity was located in the gel by a previously described method (1).

Sample application on to the gel was accomplished by using a plastic mask (Marine Colloids) holding up to ten microliters of either hemoglobin (10 mg/ml) or alkaline phosphatase (10 mg/ml). Electrophoresis was performed in a water-cooled horizontal chamber (Miles, Elkhart, IN) at 400 volts in Tris, EDTA, borate buffer (pH 8.6). Contact between the gel and buffer reservoirs was made with paper wicks and the cooling temperature was maintained at $10^{\circ}$C. After one hour of electrophoresis, the separated hemoglobin zones (visually determined) and the alkaline phosphatase zone (determined by staining for activity) were removed as a gel slice and transferred to small glass test tubes containing 0.5 ml of 0.05M Tris-HCl (pH 8.3). The test tube was then heated in a $45^{\circ}$C water bath for 3-5 minutes with occasional agitation with a glass stirring rod. After this time, the gel to sol transition had occurred and the solution was applied to 1 ml DEAE-Sephadex column packed in a 5 ml plastic syringe having a glass wool plug. The resin had previously been equilibrated with 0.05M Tris-HCl buffer (pH 8.3). Following sample application, the column resin wash washed with 50 ml of the Tris-buffer to remove the nonadsorbing SeaPrep$^{TM}$ 15/45. The protein fraction was removed from the DEAE-Sephadex resin by elution with 1M NaCl and concentrated by adding a saturated solution of ammonium sulfate to a final concentration of 70%. The resulting protein precipitation was centrifuged and redissolved in a minimum volume of distilled water and dialyzed against 1 liter of water for 3 hours.

The recovered proteins were analyzed for purity by isoelectric focusing in 1% IsoGel$^{TM}$ containing 2.5% IsoGel Ampholytes-pH 3.5-9.5 (Marine Colloids). All gels were cast on GelBond$^{TM}$ plastic at a thickness of 0.8 mm and focused as described previously (2), except power conditions were at 25 watts, 1000 volts for 30 minutes. Following focusing, the gel was fixed in a solution containing 30% methanol, 5% trichloroacetic acid, and 3.5% sulphosalicylic acid for 10 minutes. A sheet of filter paper (Schleicher and Schuell #577) was placed on the gels surface and pressed against several layers of adsorbent paper towels

with a 1 kg weight on top.  After 10 minutes of blot drying, the
film was placed in 1 liter of water for 15 minutes to remove any
remaining acid and ampholytes.  The gel was oven dried and stained
in 0.05% Coomassie Brilliant Blue R-250 dye in 25% isopropanol
and 10% acetic acid.  Destaining was performed in methanol/water/
acetic acid (5:5:1).  The film was air dried to yield a permanent
record of the pattern.


Results and Discussion

The gelling and remelting temperatures of SeaPrep[TM] 15/45 is
concentration dependent (Fig. 1) as is the case with any agarose.

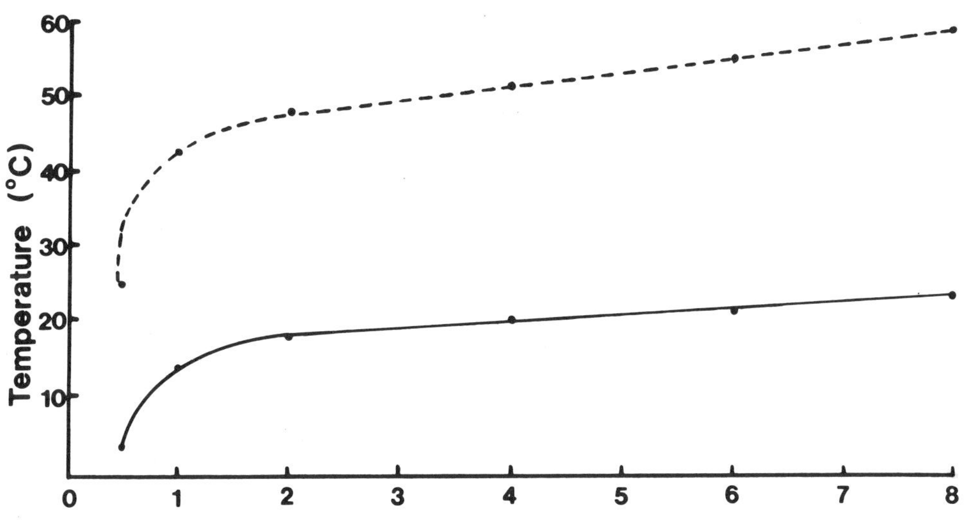


Fig. 1.  Gelling and remelting temperatures as a function of
SeaPrep[TM] 15/45 concentration.  Gelling ⚫—⚫ Remelting --⚫--⚫--

It can be seen that a one or two percent SeaPrep$^{TM}$ 15/45 gel is thermoreversible to the sol state within a 42$^{O}$C-48$^{O}$C temperature range.  This remelting temperature is considerably lower than other agaroses and would allow for the extraction of a number of proteins or nucleic acids which are stable under these heating conditions.  Once the gel to sol transition has occurred, then the protein may be separated from SeaPrep$^{TM}$ 15/45 by adsorbtion to an ion exchange resin.  Thus, hemoglobin variants and a commercial preparation of human placental alkaline phosphatase have been resolved in 2% SeaPrep$^{TM}$ 15/45 gels (Fig. 2) by electrophoresis and then recovered from the gel by remelting, followed by adsorption and elution on a DEAE-Sephadex column.

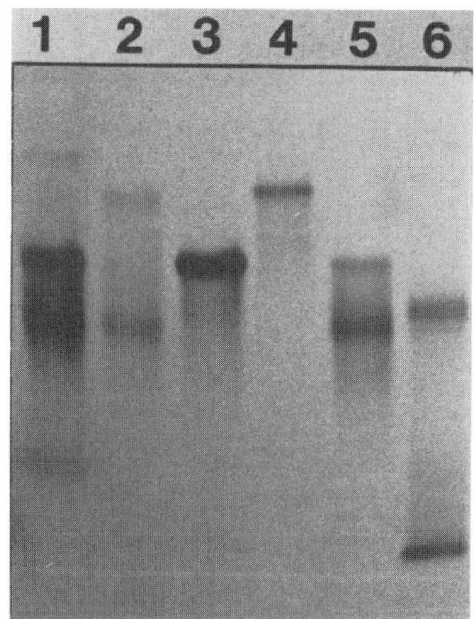


Fig. 2.  Electrophoretic separation of various proteins on a 2% SeaPrep$^{TM}$ 15/45 gel.  1) Hb A/S, 2) Hb A/A$_2$, 3) HbS, 4) Hb A$_2$, 5) Hb A/S, 6) alkaline phosphatase.

During this procedure, SeaPrep$^{TM}$ 15/45 does not adsorb to the
column resin and overall recoveries can be greater than 80%.

Increased purity of the recovered material was assessed by isoelectric focusing in IsoGel$^{TM}$ (Fig. 3).

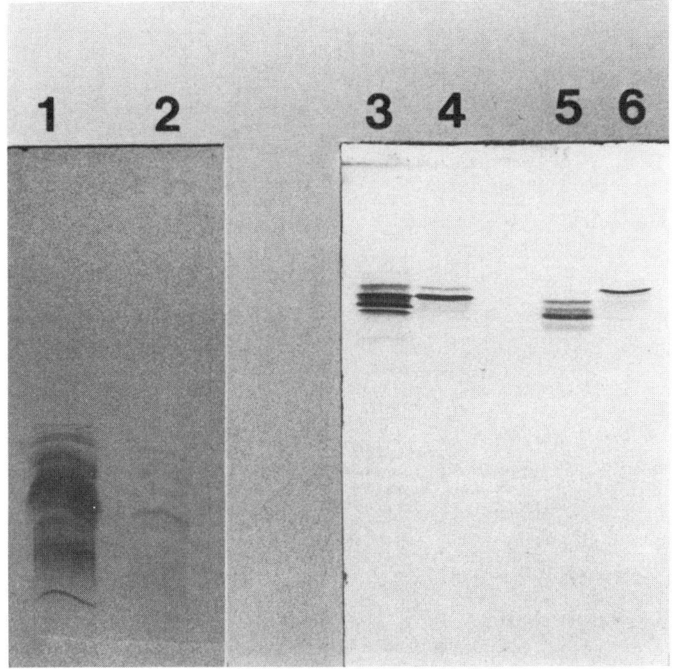


Fig. 3.  Isoelectric focusing of protein fractions in IsoGel$^{TM}$
1) crude alkaline phosphatase, 2) recovered alkaline phosphatase,
3) sickle cell trait red cell hemolysate, 4) recovered HbS,
5) recovered HbA, 6) recovered HbA$_2$.

In each instance, it was observed that the recovered fractions
were substantially more homogeneous than the starting material.
Since alkaline phosphatase is a glycoprotein, heterogeniety in
isoelectric focusing is commonly observed (3).  The hemoglobin
A fraction does contain hemoglobin F and probably some glycosylated forms.

In addition to its reduced remelting temperature, SeaPrep$^{TM}$ 15/45
also has an extremely low gelling temperature (Fig. 1).  This

property makes it useful for incorporating thermolabile material
into the gel matrix (i.e. antibodies, cells).  It could also be
particularly useful for performing DNA ligation reactions directly in the remelted gel, since the reaction is frequently run
below ambient temperature (4).


References

1.  Cawley, L.P.:  Electrophoresis and Immunoelectrophoresis,
    Little, Brown and Company, Boston, pp. 259 and 274 (1969).

2.  Saravis, C.A., Cunningham, C.G., Marasco, P.V., Cook, R.B.,
    and Zamcheck, N.:  Electrophoresis '79, Walter de Gruyter
    and Company, Berlin, New York, pp. 118-122 (1979).

3.  Williamson, A.R., Salaman, M.R., Kreth, H.R., Ann. N.Y. Acad
    Sci. 209, 210-224 (1973).

4.  Shimotohno, K., Mizutani, S., Temin, H.M., Nature 285,
    550-554 (1980).

AGAROSE GEL ELECTROENDOSMOSIS:  ITS ENHANCEMENT, REDUCTION AND
PRACTICAL RAMIFICATIONS


Richard B. Cook
BioProducts Department, Marine Colloids Division, FMC Corporation, Rockland, Maine


Introduction

Theoretical models for electroendosmosis (EEO) in glass capillaries are well known (1) but a comprehensive application and
expansion of this theory to agarose gels has not yet been
forthcoming.  This may result, in part, from the complexity
of the agarose gel system by comparison to the glass capillary.
Certainly it has not resulted from a lack of study since the
EEO of agarose gels has regularly been examined and elucidated
over the years (2, 3, 4, 5).  Some areas of controversy and
uncertainty still remain, however, and this paper will attempt
to address them.

The importance of EEO cannot be underestimated in the practical
application of agarose gels.  It is the basis on which some
commercial agaroses are named and continues to be one of the
most critical parameters to the success of many agarose applications.  Its importance in zone electrophoresis (6), counter-
electrophoresis (5), DNA electrophoresis (7), and, more recently, isoelectric focusing (8, 9) is well established.

This paper considers the fact that EEO can be both dramatically
increased and decreased by mechanisms which are consistent
with the predictions of the Helmholtz-Smoluchowski equation
(1).  In addition, data has been adduced to show that % $SO_4^=$
alone <u>cannot</u> reliably predict the EEO of representative

agarose gels since many agaroses contain at least as much pyruvate as they do organic ester sulfate ($ROSO_3^-$). <u>A</u> <u>priori</u>, one would expect both moieties to be associated with mobile cations which could influence EEO. It is not surprising, then, that if both % $SO_4^=$ and % pyruvate are considered by linear least squares, the resultant correlation coefficient is found to be significantly higher than with % $SO_4^=$ alone. Nevertheless, the combined % $SO_4^=$ and % pyruvate correlation coefficient is still far too low ($r = 0.60$) to reliably predict the EEO of agarose gels. These results are in sharp contrast with some that have appeared in the literature (5) that suggest that a strong ($r=0.927$) and relatively straightforward correlation exists between % $SO_4^=$ and EEO over a wide range of agars and agaroses. In this study we have examined this relationship for a much larger population of agaroses than reported to date and found that % $SO_4^=$ alone cannot be used to reliably predict the EEO of agarose gels.

In contrast, a high linear correlation coefficient ($r=0.92$) has been found between EEO and the total milliequivalents of $Na^+$ and $Ca^{+2}$ in a very diverse but limited set of commercial agaroses from different suppliers. This high correlation coefficient suggests that these cations are associated with some immobilized·anionic moieties, perhaps other than sulfate or pyruvate, that would account for their improved correlation with EEO.

In order to investigate EEO and accurately measure the EEO of certain new or experimental types of agarose-containing gels, modifications of conventional electroendosmosis test procedures have had to be made. These are briefly discussed in this study.

Materials and Methods

1) Sulfate analysis: Total sulfate (i.e. ester and inorganic) was assayed by N. Stanley's modification of the standard method for inorganic sulfate. This method has been described in detail in the literature (5). It entails the complete digestion of a relatively large (2g.) agarose sample (due to its low % $SO_4^=$ content) with $HNO_3$, reduction of the excess $HNO_3$ with HCHO, followed by the addition of $BaCl_2$ to produce a $BaSO_4$ precipitate from which the % $SO_4^=$ can be determined gravimetrically. Reproducibility of the method is ± 0.02%.

2) Pyruvate analysis: Pyruvic acid and its various salts can be present as the 4, 6 acetal of the D-galactose units in agarose. The pyruvic acid content of the agaroses in this study was assayed by a modification of the enzymatic procedure described by M. Duckworth and W. Yaphe (10). The procedure involves hydrolysis of the ketal with 0.02 M oxalic acid to produce free pyruvic acid which can then be assayed spectrophotometrically (340 nm) as a function of the amount of $\beta$-NADH (Sigma) consumed in the presence of lactic dehydrogenase (type II, Sigma). Reproducibility of the method is ± 0.02%.

3) Standard EEO analysis: The method used was a modification of Wieme's procedure (11) which employs the use of Dextran T 500 (Pharmacia Fine Chemicals) and bovine serum albumin (fraction V, Sigma). The EEO is determined in a 1% agarose gel containing barbital buffer (pH 8.2/0.05 M) as a ratio of the cathodal dextran movement to the total distance between the dextran and albumin spots at the end of the electrophoretic run (5V/cm; 90 min.).

4) Cation analysis: The sodium, calcium, potassium and magnesium analyses of agarose were done using a Perkin Elmer model 560 atomic absorption spectrophotometer set at the appropriate wavelength: sodium (589 nm; 0.7 nm slit) potassium (766.5 nm;

222

2.0 nm slit), calcium (422.7 nm; 0.7 nm slit), and magnesium
(285 nm; 0.7 nm). All agarose samples were dissolved in concentrated $HNO_3$ before analysis and assayed in an oxidizing air-
acetylene flame. Reproducibility of the method is $\pm 0.01\%$

5) The linear least squares correlation was done by means of
a Hewlett Packard model 9820 A calculator and a standard
Hewlett Packard model 20 Math-Pac program.

6) Agarose samples: Were obtained from the various commercial suppliers indicated and are representative of their respective product properties during the time period from 1974-
1976.
Note: Detailed copies of all of the above methods of analysis
(1-4) can be obtained from Marine Colloids Division, FMC Corporation, Rockland, Maine 04841.


Results and Discussion

In view of the very large number (138) of agarose samples analyzed for $\% \ SO_4^=$, $\%$ pyruvate and EEO in this study, only a representative selection of the data is shown in Table I due to limitations of space. By inspection, it can be seen that the $\%$
pyruvate is often comparable to the $\% \ SO_4^=$ for agaroses having
a gelling temperature $36 \pm 1^{\circ}C$. In contrast, agaroses having
a $41 \pm 1^{\circ}C$ gelling temperature typically exhibit a very low or
negligible level of pyruvate. Table II illustrates that the
very high correlation coefficient for $\% \ SO_4^=$ with EEO reported
by Hierholzer is lowered somewhat by elimination of the agars
from his samples population. If a very much larger group (138)
of representative agaroses containing significant pyruvate is
considered, it is clear that the correlation of $\% \ SO_4^=$ with EEO
is not strong ($<0.5$). If the combined $\%$ pyruvate and $\% \ SO_4^=$ of
these agaroses is considered, the correlation with EEO is sig-

TABLE I

| Agarose | | | Properties | | | |
|---|---|---|---|---|---|---|
| Supplier | Type | Lot | % $SO_4$ | % Pyruvate | EEO | Gel Temp |
| Marine Colloids | SeaKem (HE) | 146896-1 | 0.32 | 0.25 | 0.26 | 36°C |
| | SeaKem (ME) | 151759-4 | 0.29 | 0.26 | 0.17 | 36°C |
| | SeaKem (ME) | 153045 | 0.30 | 0.21 | 0.16 | 36°C |
| | SeaKem (ME) | 153036 | 0.26 | 0.12 | 0.15 | 36°C |
| | SeaKem (LE) | 153049 | 0.31 | 0.19 | 0.13 | 36°C |
| | SeaKem (ME) | 156138 | 0.25 | 0.25 | 0.16 | 36°C |
| | SeaKem (LE) | 153048 | 0.27 | 0.21 | 0.14 | 36°C |
| | SeaKem (HEEO) | 153050 | 0.14 | 0.22 | 0.44 | 36°C |
| | SeaKem (ME) | 153052 | 0.21 | 0.21 | 0.18 | 36°C |
| | SeaKem (LE) | 156118 | 0.25 | 0.26 | 0.14 | 36°C |
| | SeaKem (ME) | 156129 | 0.30 | 0.25 | 0.16 | 36°C |
| | SeaKem (HGT) | 153013 | 0.34 | <0.03 | 0.08 | 41°C |
| | SeaKem (HGT) | 153019 | 0.30 | <0.03 | 0.10 | 41°C |
| | SeaKem (HGT) | 153088 | 0.23 | <0.03 | 0.08 | 41°C |
| | SeaKem (HGT) | 153014 | 0.20 | <0.03 | 0.08 | 41°C |
| | SeaKem HGT-P | 60266 | 0.19 | <0.03 | 0.07 | 41°C |
| | SeaKem HGT-P | 60826 | 0.11 | <0.03 | 0.07 | 41°C |
| | SeaPlaque | 156127 | 0.04 | 0.21 | 0.07 | 28°C |
| CalBiochem | | | 0.28 | 0.03 | 0.15 | 42°C |
| IBF | A 45 | FF 8634 | 0.28 | 0.02 | 0.09 | 40°C |
| | A 45 | FF 6482 | 0.25 | 0.04 | 0.10 | 41°C |
| | A 37 | FF 2608 | 0.38 | 0.13 | 0.15 | 35°C |
| | A 37 | FF 5388 | 0.53 | 0.15 | 0.19 | 35°C |
| Litex | HSA | AGS 222 | 0.32 | 0.07 | 0.14 | 41°C |
| | HSB | AGS 219 | 0.24 | 0.05 | 0.12 | 41°C |
| | HSC | AGS 197 | 0.16 | 0.03 | 0.04 | 41°C |
| | LSA | AGS 234 | 0.75 | 0.12 | 0.29 | 36°C |
| | LSB | AGS 243X | 0.43 | 0.06 | 0.06 | 34°C |
| Miles | | | 0.79 | 0.14 | 0.19 | 35°C |

nificantly improved. For "high" gelling temperature ($41^{\circ}$C) agaroses, which contain virtually no pyruvate, the correlation of % $SO_4^=$ with EEO is approximately the same as the correlation of EEO with the sum of % $SO_4^=$ and % pyruvate for agaroses having appreciable pyruvate.

EEO($-M_R$) CORRELATION
WITH % $SO_4$ = AND % PYRUVATE (PYR.)

### Table II

| SAMPLES | N | RANGE | | CORRELATION (R) | | REF. |
| | | % $SO_4$ | % PYR. | EEO WITH % $SO_4$ | EEO WITH % $SO_4$ & % PYR | |
|---|---|---|---|---|---|---|
| VARIOUS COMMERCIAL AGARS AND AGAROSES | 23 | 0.06 - 2.69 | N.A. | 0.927 | N.A. | 1 |
| VARIOUS COMMERCIAL AGAROSES | 17 | 0.06 - 0.45 | N.A. | 0.790 | N.A. | 1 |
| MARINE COLLOIDS AGAROSES WITH $36^{\circ}$C GT | 72 | 0.10 - 0.37 | N.A. | 0.477 | 0.597 | 2 |
| SEAKEM HGT & HGT(P) AGAROSES; $41^{\circ}$C GT | 45 | 0.09 - 0.34 | $\leq$0.03 | 0.609 | 0.609 | 2 |
| VARIOUS COMMERCIAL AGAROSES (1974-1976) | 21 | 0.07 - 0.75 | 0.02-0.20 | 0.496 | 0.796 | 2 |

1. J. C. HIERHOLZER, J. IMM. METHODS 11 (1976) 63-76.
2. R. B. COOK, ELECTROPHORESIS '81 CONFERENCE, CHARLESTON, S.C.

Theoretically, one might expect a higher correlation of EEO with the cations associated with the $SO_4^=$ and pyruvate moieties than with the moieties themselves. Such an expectation would be based on the known differences in both hydration radii and binding coefficients for different cations. Indeed, it is the hydrated cations that actually move cathodally causing EEO to occur. Table III illustrates that for this limited but diverse sample population, the highest correlation exists between EEO and the cumulative $Na^+$ and $Ca^{+2}$ concentrations. For maximum predictive value, such a treatment would have to take into account the cation contribution of residual inorganic salts. This could be done by analyzing for free $Cl^-$, for example, and back-calculating the cation contribution. In general, however, the residual salts are likely to represent only a small percentage of the total cations associated with commercial agaroses.

Sulfate-containing or other inorganic salts would be expected
to be substantially less.

Table III          CATION CONTENT OF COMMERCIAL AGAROSES

| AGAROSE TYPE | $\%SO_4$ | $\%PYR$ | $\%Na^+$ | $\%Ca^{+2}$ | $EEO(-m_r)$ |
|---|---|---|---|---|---|
| SEAKEM (HE)  #1530100 | 0.26 | 0.24 | 0.19 | 0.08 | 0.20 |
| SEAKEM (HEEO) #153050 | 0.19 | 0.22 | 0.42 | 0.06 | 0.44 |
| SEAKEM (LE)   #153046 | 0.24 | 0.21 | 0.12 | 0.05 | 0.13 |
| SEAKEM (HGT)  #153089 | 0.22 | 0.03 | 0.07 | 0.02 | 0.07 |
| IBFA45; FF8634 | 0.28 | 0.02 | 0.01 | 0.00 | 0.09 |
| LITEX LSA #234 | 0.75 | 0.12 | 0.25 | 0.01 | 0.29 |

CORRELATION:
  EEO WITH % $SO_4$: r=0.204
  EEO WITH (% $SO_4$ + % PYR): r=0.454
  EEO $\bar{c}$ TOTAL MEQ ($Na^+$ + $Ca^{+2}$): r=0.922

The effect of gel concentration on the EEO of agarose gels has
not received much attention in the literature.  Graph I illustrates the dramatic and somewhat analogous effect that concentration plays on the EEO of agarose gels.  From the graph, it
is evident that 3% agarose gels, having more total sulfate(and
pyruvate) per unit gel volume than 1% gels, actually exhibit a
lower EEO.  This is clearly in contrast to the effect of gel
concentration on agar gels as reported by Wieme (3).  For a
suggested explanation of this phenomenon, one must return to
work of G. Kortum  (12) who indicates that for calculation of
EEO in "capillaries" having a radius (r) less than 50 nm, the
following equation must be used:

$$\frac{dV}{dT} = D = \frac{IF\ C}{8\ \eta_O K} \cdot r^2$$

where D = EEO, I is the electrical current, F is the product of
Avogadro's number and the elementary charge,  C  is the concentration of counterions, $\eta_O$ is the viscosity and K  is the
conductivity of the fluid in the capillaries.  For agarose gels
Rees, et al. (13) has described the formation of junction zones
and association of adjacent agarose helices.  As gel concentra-

tion is increased, the amount of junction zone formation would
be expected to increase.  As a result, many matrix anion-asso-
ciated cations might end up in very narrow"capillaries"
of the junction zone system and hence severely restrict their
mobility.

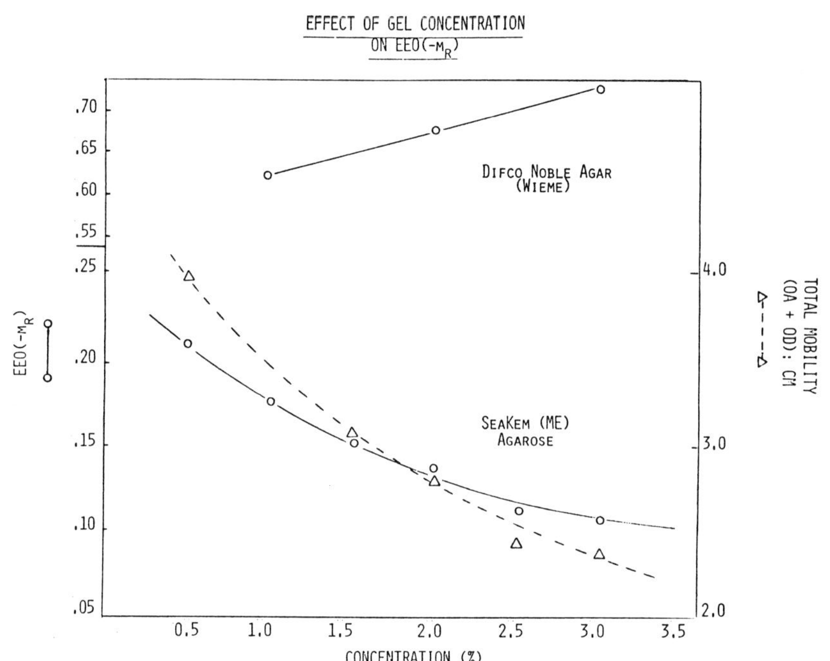



The fact that EEO is inversely proportional to the viscosity
of the fluid in the double layer is consistent with the reduction of agarose gel EEO by the corporation of such neutral gums
as clarified locust bean gum (CLBG) or clarified guar gum (CGG)
as shown in Table IV for SeaKem HGT agarose (14).  The
interpretation is further confirmed by the observation of Rees,
et al. that there is an association between such viscous gums
and the anionic sulfate groups on the sulfated polysaccharide
matrix (13).  In contrast, the incorporation of highly anionic
polymers into agarose gels can increase the EEO values beyond

anything obtainable in an agar.  (see Table IV)

ALTERATION OF AGAROSE EEO

BY POLYMER ADDITIVES

Table IV

| No. | AGAROSE | | POLYMER ADDITIVE | | EEO ($-M_R$) |
|---|---|---|---|---|---|
| # | TYPE | CONC. | TYPE | CONC. | |
| 1) | SEAKEM HGT | 1.0 | NONE | -- | 0.06 |
| | " | 0.5 | NONE | -- | 0.12 |
| 2) | " | 0.5 | CLBG; MCD | 0.5 | 0 |
| 3) | " | 0.5 | CGG; MCD | 0.5 | 0 |
| 4) | " | 0.5 | PEG, 20M | 0.5 | 0.10 |
| 5) | " | 0.5 | DOW MGL POLYACRYLAMIDE | 0.5 | 0.13 |
| 6) | " | 0.7 | K-CARRAGEENAN | 0.3 | 0.81* |
| 7) | " | 0.7 | $\lambda$-CARRAGEENAN | 0.3 | 1.20* |
| 8) | SEAKEM (ME) | 1.0 | NONE | -- | 0.17 |
| 9) | " | 0.7 | DOW NP-10 POLYACRYLAMIDE | 0.3 | 0.20* |
| 10) | " | 0.7 | ALGINATE | 0.3 | 0.74* |
| 11) | " | 0.7 | $\lambda$-CARRAGEENAN | 0.3 | 1.48* |

* MEASURED USING GELMAN RBY DYE MIXTURE

It should be recognized, however, that accurate measurement of
either very low or very high EEO's by the standard Wieme procedure runs the risk that either dextran or albumin may occasionally remain in the sample well and be washed out during
fixation and/or staining thereby precluding an unequivocal EEO
determination.  Instead, better results are obtained by either
a circular sample mask procedure which does not disrupt the
gel surface or an application of the EEO sample mixture by
means of a dried spot on GelBond film as described elsewhere
(14).


Acknowledgement

The analytical assistance of Mr. Cliff Harper and Mrs. Jean
Faustini was invaluable in the completion of this work.

References

1.  Abramson, H. A., Moyer, L. S., Gorin, M. H.:  Electro-
    phoresis of Proteins and the Chemistry of Cell Surfaces,
    Hafner, New York, 1964, Ch. 1 and 5.

2.  Kunkel, H. G., Trautman, R., Bier, M.:  Electrophoresis,
    Theory, Methods and Applications, Academic Press,
    New York, 1st. Ed., p 226 (1959).

3.  Wieme, R. J.:  Agar Gel Electrophoresis, Elsevier Pub. Co.,
    Amsterdam, PP 109, 111, 129, (1965).

4.  Quast, R.:  J. Chromatog. 54, 405-412 (1971).

5.  Hierholzer, J. C.:  J. Imm. Methods 11, 63-76 (1976).

6.  Jeppsson, J. O., Laurell, C. B., Franzen, B.:  Clinical
    Chemistry, 25 (No. 4) 629-638 (1979).

7.  Johnson, P. H., Miller, M. J., Grossman, L. I.:  Anal.
    Biochem. 102, 159-162 (1980).

8.  Saravis, C. A., Zamcheck, N.:  J. Imm. Methods, 29, 91
    (1979).

9.  Ebers, G. C., Rice, G. P., Armstrong, H.:  J. Imm. Methods,
    37, 315-323 (1980).

10. Duckworth, M., Yaphe, W.:  Chem. Ind. 747-748 (1980).

11. Wieme, R. J.:  Agar Gel Electrophoresis, Elsevier Pub. Co.,
    New York, P 100 (1965).

12. Kortum, G., Lehrbuch:  Der Elecktrochemie Verlag Chemie,
    Weinheim/Bergster, PP 171, 406, 413 (1966).

13. Rees, D. A., Arnott, S., Fulmer, A., et al.:  J. Mol.
    Biol., 90, 269 (1974).

14. Cook, R. B., Witt, H. J.,:  U. S. Patent Serial Number
    010,033 (filed 1979).

15. Renn, D. W., U. S. Patent #3,975,162.

ENHANCEMENT OF COMPLEX PATTERNS IN ISOELECTRIC FOCUSING WITH
IMPROVED ISOGEL PERFORMANCE



Susan E. Coulson
Marine Colloids Division - FMC Corporation
Rockland, Maine 04841 U.S.A.




Summary

The macroporous structure of an agarose gel allows for extremely rapid transport of macromolecules ($\leq 10^6$ molecular weight
through the medium and thus permits very short isoelectric
focusing times.  This paper examines an improved methodology
for analytical separations in ultrathin layers of IsoGel
(0.35-0.8 mm thickness).
This method for enhancing resolution applies to the low (3-5)
and mid (5-8) ranges of the pH gradient, and consists of
supplementing the agarose gel with 5% glycerol and 10% (w/v)
Sorbitol.  The influence which these additives exert upon the
sharpness and degree of separation of complex protein samples
is discussed; enhanced patterns were obtained with heterogeneous immunoglobulins, enzymes, and sera.



Introduction

As a high resolution electrophoretic technique, isoelectric
focusing (IEF) is one of the most sensitive means of differen-
tiating components of protein mixtures.  In recent years, the
emphasis upon optimization of resolving power has paralleled
the need to evaluate complex samples with greater accuracy and
specificity.
Many samples are handled as crude extracts, partially purified

mixtures, or as whole biological fluids, and the majority of
proteins currently studied focus on the mid and acid intervals
of the pH gradient. A method for improving resolution in these
regions of the gradient has been developed, and its application
to a broad range of subjects is promising.


Materials and Methods

Samples:

Sera were obtained from patients on a random basis from two
coastal Maine medical centers: IgA, IgM, and IgE (Atlantic Antibodies); antihuman whole serum (Dako). The remaining samples
were purchased from Sigma Chemical Company. Dried samples were
reconstituted in distilled water to a 5-10% (w/v) solution.

Isoelectric Focusing

IEF was performed in ultrathin ($\leq$0.8 mm) agarose slabs made of
0.1 g IsoGel (Marine Colloids) and 0.62 ml of ampholyte (40%
(w/v) commercial preparation in a total volume of 10 ml of distilled water. Supplemented gels were prepared by boiling 0.1 g
IsoGel and 1 g Sorbitol in a 10 ml volume of 5% glycerol. Sufficient evaporative loss was allowed to occur during heating
and verified by reweighing the entire solution before adding
the ampholytes. A final total volume of 10 ml and 2.5% (w/v)
ampholyte concentration was used for all gels. Ampholytes used
were IsoGel Ampholytes pH 3.5-9.5, Servalyt pH 3-10, and Sebia-
lyte pH 3.5-10.
Gels measuring 11 x 11 cm were cast on GelBond[TM] (110 x 125 mm,
7 mil thickness) according to the Marine Colloids published
procedure (1).
Gels were run on either the BioRad Electrophoresis Cell No. 1415
with the Isco power supply No. 494 or the LKB Multiphor No. 2117

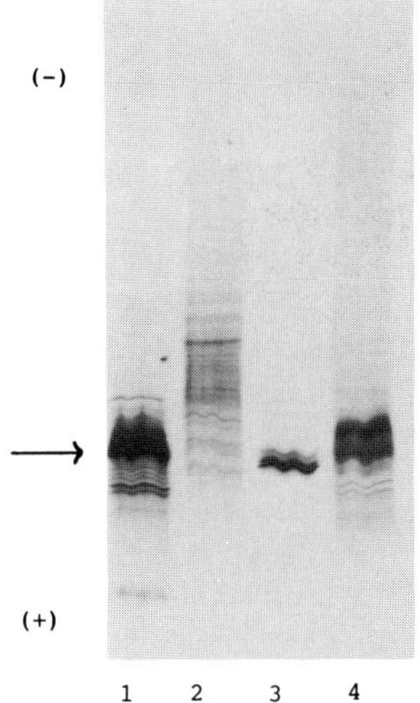



FIGURE 5.
1% IsoGel, 2.5% IsoGel Ampholyte pH 3.5-9.5
5% Glycerol, 10% Sorbitol.  Samples: 1. $\alpha$ -
Antitrypsin.  2. $\beta$-Glucuronidase.  3. Ovalbumin.
4. Human Serum.  Arrow indicates albumin.  Gel
was focused at 8w, 1500v for 47 min.

Alpha$_1$-antitrypsin is another sample which contains a number
of variants having their pIs within a narrow pH range.  The
improvement of the $\alpha_1$-antitrypsin pattern by gel supplementation was dramatic in this instance (Fig. 5).  The bands were
thin and sharply focused as well as distinctly separated on
the gel.  The unsupplemented $\alpha_1$-antitrypsin pattern (Fig. 1)
was overwhelmed by the heavily-stained albumin component, and
the narrower more acidic bands were likewise diffuse zones and
tended to run together.

## Discussion

The gradient profiles suggest that it is possible not only to
attain a stable pH gradient more slowly, but also to maintain
the linearity of the gradient over a more extended period of
time in gels containing both glycerol and Sorbitol.  Normally,

it is desirable to minimize the focusing time in agarose gels since this should also reduce acidification of the pH gradient by atmospheric $CO_2$, gel dehydration, and sample degradation due to overfocusing (3). However, it is known that proteins display somewhat different rates of migration; it is, therefore, reasonable to suppose that optimal length of a focusing run of complex samples is one where the fast focusing species do not reach their pI's long before the slower focusing materials since this could cause sample denaturation and subsequent band broadening.

Gel supplementation as described here provides a wider margin of time where all sample components are migrating in a linear gradient. The gel patterns presented in this paper reinforce the conclusion drawn from the pH profiles of unsupplemented and glycerol/sorbitol-containing gels. The gel patterns in the additive gels reveal less drift and improved band definition and sharpness.

The potential applications of this method are numerous, with the most practical one being the detection of minor components in heterogeneous samples focusing in the acid and neutral pH intervals. In some cases, implementation of this method provides a preliminary assessment of complex samples before resorting to a narrow pH interval gel, or a monospecific technique such as affinity chromatography.


References


1.  How to Cast Professional Agarose Plates:  Marine Colloids Division - FMC Corporation.

2.  IsoGel: How-to-Pamphlet:  Marine Colloids Division - FMC Corporation.

3.  Nguyen, N. Y., Chrambach, A.:  Journal of Electrophoresis 1980, Vol. 1, pp. 14-22.

SIEVING OF SPHERICAL VIRUSES AND RELATED PARTICLES DURING ELECTROPHORESIS
IN GELS OF AGAROSE


Philip Serwer and Shirley J. Hayes
Department of Biochemistry, The University of Texas Health Science Center
at San Antonio, San Antonio,Texas  78284


Introduction

Viruses and related particles have been fractionated by electrophoresis in
gels of agarose (1-3).  Theories for predicting the dependence of the sieving of gels on the size of particles subjected to electrophoresis (4) have
been applied to the study of the dimensions of virus-sized (but nonviral)
supramolecular complexes, using gels of agarose (5); the lowest concentration of agarose used was 0.60%.  More recently, agarose gel electrophoresis
of viruses and related particles has been conducted in gels with concentrations as low as 0.04% and improved procedures for minimizing experimental
errors in measuring electrophoretic mobility ($\mu$) as a function of agarose
concentration ($\underline{A}$) have been developed (6,7).  In the present communication,
these procedures have been used to quantitate the sieving of such particles
using spherical particles with radii of 13.3-41.9 nm.


Methods

1)  Bacteriophages.  The following roughly spherical particles, with radius
    (R) determined by low-angle x-ray scattering, have been used as sam-
    ples:  bacteriophage R17, a gift of Dr. J.A. Steitz, R = 13.3 nm (8);
    bacteriophage T7 missing its tail fibres (tail fibres are the product
    of T7 gene 17 [9]; therefore, T7 missing fibres are referred to as
    $17^-$T7), puriified as previously described (7), R = 30.1 nm (assumed to
    be the same as the R of wild-type T7[10]); capsid I, a DNA-free cap-
    side of bacteriophage T7 previously described (2,9), purified as pre-

viously described (2) without sedimentation in a sucrose gradient, R of glutaraldehyde-fixed capsid I (the sample used for electrophoresis was treated with glutaraldehyde as described in reference 10 and was then dialyzed against .20 M NaCl, 0.01 M sodium phosphate, pH 7.4, 0.001 M $MgCl_2$) = 26.1 nm (10). In addition, bacteriophage T5 missing its tail (to be referred to as T5 full head) was used. T5 full heads are particles that are roughly spherical in electron micrographs (11); they were purified as previously described (12). The radius of the T5 full head was estimated to be 41.9 nm by assuming: (a) that the packing density of the DNA, molecular weight = 77 x $10^6$ (13), is the same as for several other bacteriophages with duplex DNA (14), and (b) the thickness of the envelope of the T5 capsid is 2.5 ± .5 nm, observed in electron micrographs of negatively stained specimens (P. Serwer, unpublished observation).

2)  Apparatus. All gels were cast and electrophoresis was performed in a horizontal electrophoresis apparatus previously described (6). To control the temperature of gels, the electrophoresis apparatus was placed in a water bath with temperature controlled to 25.0 ± 0.3°C. Details of the design and construction of the water bath will be the subject of a future communication.

3)  Gels. A horizontal slab consisting of several agarose gels (running gels), each of a different concentration of agarose, embedded within a frame of agarose (frame gel), were prepared as previously described (7); the slab consisting of the running gels and the frame gel are referred to as a multigel. All gels were cast in an electrophoresis buffer and using a procedure previously described (6). Agarose was dissolved by boiling in a microwave oven; evaporative losses of water during boiling were monitored by weighing the solution of agarose, and water thus lost was replaced. After formation of multigels, these gels were covered with electrophoresis buffer and allowed to stand for 18-20 hr in the temperature-controlled water bath at 25.0°C to help avoid variability in the effects of syneresis (15).

4)  Electrophoresis. Samples in 0.20 M NaCl, 0.01 M Tris-Cl, pH 7.4, 0.001 M $MgCl_2$ were diluted with a 2x volume of a buffer containing: 0.005 M sodium phosphate, pH 7.4, 0.001 M $MgCl_2$, 4% sucrose, 400 $\mu g/cm^3$ bromophenol blue. Fifty μl of this mixture was layered in wells beneath 7-8

mm of electrophoresis buffer and electrophoresis was started 1.5 hr
later (the delay reduced discontinuities in the buffer near the sample).
Electrophoresis was performed at 0.71 ± .04 volts/cm for  20.0
hr.      To prevent changes in pH within the gel, circulation of the
buffer over the surface of the gel at 50 cm$^3$/min was started at 1.25 hr
after the start of electrophoresis; a metering pump (Cole-Parmer) was
used.  After electrophoresis, bacteriophage R17 and T5 full heads were
detected by staining with ethidium bromide (7); glutaraldehyde-fixed T7
capsid I and 17$^-$ bacteriophage T7 were detected by staining with Coomassie Blue (2,3).

5)  Agarose.  HGT[P] agarose manufactured by Marine Colloids, Rockland,
Maine (Lot 60689) was used for all gels, coefficient of electro-osmosis
$(E_r)$ (16) = 0.07 (determined by the manufacturer).

6)  Theory.  If a particle subjected to electrophoresis in gels does not
bind to the gel and if the particle moves in response only to the electrical field (i.e., electro-osmosis is neglected), the electrophoretic
mobility, $\mu$, of the particle in a gel as a function of $\underline{A}$ is, in theory,
described by the following equation (4):

$$\log \mu = \log \mu_o - K_R \cdot \underline{A} \qquad (1)$$

$\mu_o$ is the electrophoretic mobility in the absence of agarose and $K_R$ is
a constant whose value depends on the properties of the gel.  The $\mu$ of
all samples used here was independent of the concentration of sample
for concentrations 3-fold higher and lower than those used for determining $K_R$, suggesting that particle particle or particle-gel reversible
binding does not affect values of $K_R$.  By electron microscopy (2,6),
irreversible particle-particle aggregation also does not occur.
In theory, $K_R$ is related to the radius of a spherical particle subjected to electrophoresis by (4):

$$(K_R)^{1/n} = c \ (R + r) \qquad (2)$$

c and n are constants that depend on the model used to represent the
gel; r is the radius of a fibre of the gel.  For those models explored,
the value of n is either 2 or 3 (4).

7)  Effects of electro-osmosis.  It has been shown experimentally that $E_r$
in gels of agarose is independent of $\underline{A}$ for $\underline{A} \leq 2.0\%$ and an ionic
strength of 0.05 (17).  In theory, the dependence of $E_r$ on $\underline{A}$ decreases

as the ionic strength increases and as $\underline{A}$ decreases (18). Thus, because the electrophoresis buffer used here has an ionic strength of 0.12 and the $\underline{A}$'s of running gels used are $\leq$ 1.8%, the above data indicate that in the present study $E_r$ is independent of the A of the running gel. Therefore, measurements of $K_R$ will not depend on $E_r$. However, the $\mu_o$ measured by extrapolation to an $\underline{A}$ of 0% ($\mu_o'$) will depend on $E_r$ and therefore $\mu_o'$ is not the $\mu_o$ that would be measured in the absence of a gel (17).


Results


Semilogarithmic plots of $\mu/\mu_o'$ as a function of $\underline{A}$ for the above particles are in Figure 1. These plots are linear for $\underline{A} \leq$ 0.90% and then decrease more rapidly at higher $\underline{A}$'s. The extent of this nonlinearity increased as the size of the particle increased. In other experiments, the plots of Figure 1 remained linear to an $\underline{A}$ of 0.075% for bacteriophage R17, $17^-$ T7 and T5 full heads (not shown). Although sieving by the gel must occur for $\underline{A}$ < 0.075, the effects on $\mu$ are too small to be detected for $17^-$ bacteriophage T7 (7) and, therefore, data were not taken for $\underline{A}$ < 0.075.

Qualitatively, the $K_R$ determined from the linear region of the plots of Figure 1 increased as the R of the particle increased. Quantitatively, a plot of $(K_R)^{1/2}$ as a function of R is linear (Figure 2) with c = 7.97 x $10^{-3}$nm$^{-1}$, r = 23.8 nm; a plot of $(K_R)^{1/3}$ as a function of R is also linear (not shown) with c = 7.21 x $10^{-3}$nm$^{-1}$, r = 44.6 nm. Thus, values of c are comparatively insensitive to the model used to describe the gel, but in the absence of additional constraints, values of r calculated from the data in Figure 2 might have a model-derived uncertainty of a factor of 1.9. Electron micrographs of thin sections of agarose gels reveal fibres 1.0 - 15.0 nm in radius (19), suggesting that the model for which n = 2 is more accurate than the model for which n = 3, at least for $\underline{A}$'s in the linear region of the plots of Figure 1.

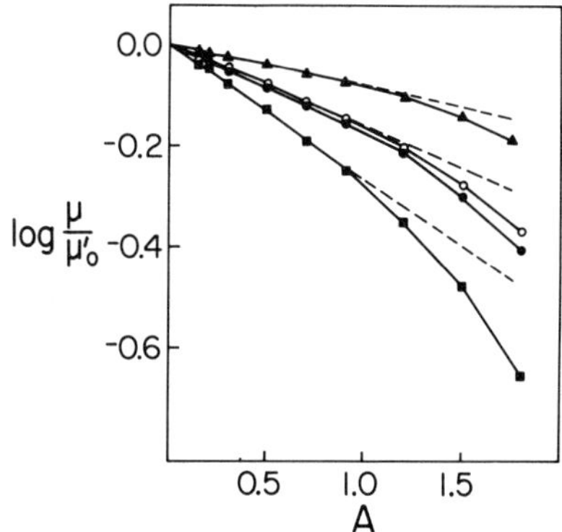

Figure 1.  Log $\mu/\mu_o'$ as a function of A.  Values of $\mu$ as a function of A were determined by measurement of the positions of bands after electrophoresis in multigels; electrophoresis was performed as described in the Materials and Methods.  Values of $\mu_o'$ were obtained by extrapolation and log $\mu/\mu_o'$ was plotted as a function of A.  -▲-, bacteriophage R17 ($\mu_o' = 1.7$ x $10^{-4}$ cm$^2$/v·sec.); -O-, T7 glutaraldehyde-treated capsid I ($\mu_o' = 1.8$ x $10^{-4}$ cm$^2$/v·sec.); -●-,17$^-$ bacteriophage T7 ($\mu_o' = 4.2$ x $10^{-5}$ cm /v·sec.); -■-, T5 full heads ($\mu_o' = 9.2$ x$10^{-5}$ cm$^2$/v·sec.).

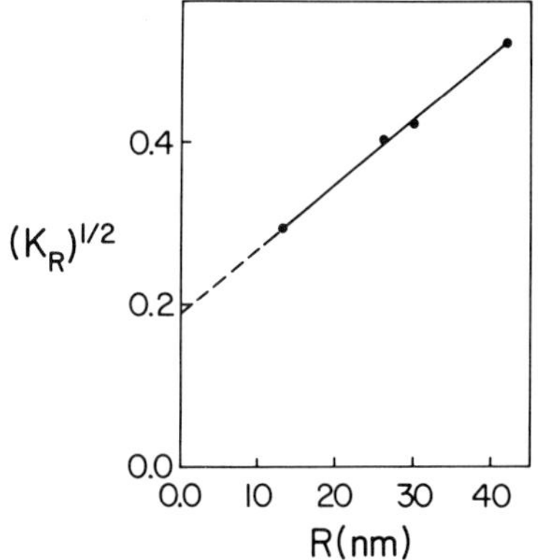

Figure 2.  $(K_R)^{1/2}$ as a function of R.  The $K_R$'s were determined from the linear region of the plots in Figure 1 and $(K_R)^{1/2}$ was plotted as a function of R.

242

## Discussion

The plot in Figure 2 is a calibration curve for determining the R of a particle from $K_R$ if it is known that the particle is spherical (from electron microscopy, for example). Values of R determined from $K_R$ will probably be more accurate than values of R determined from electron microscopy because of shrinkage and flattening of particles during preparation for microscopy; 24% error in radius has been observed after preparation of bacteriophage T7 for electron microscopy by negative staining, thin sectioning and freeze-drying (20). No more than a 12% error is expected if R is determined from $K_R$ measured in multigel. However, if the $K_R$ of a particle of unknown R is determined from the ratios of $\mu$'s for the unknown particle to $\mu$'s for a particle that is present in the same multigel and that has a known R, the maximum error expected is 8%.

The non-linearity in the plots of Figure 1 has previously been observed using T7 capsids as samples (6), but was not mentioned in another study (5)(although examination of Figure 1 of reference 5 does reveal some curvature). In the latter study, the largest particle for which data was plotted had an R of 25.0 nm; no data was taken below 0.60% agarose and only three values of $\underline{A}$ were used. The curvature detected in the Results would also probably not have not been noticed if the present experiments had been thus constrained. There are at least two possible reasons for the curvature of the plots of Figure 1: (a) values of r for agarose decrease as $\underline{A}$ increases above 0.90%; (b) the model used to represent the gel in the derivation of equation 1 (n = 2) becomes less realistic (or perhaps n changes) as $\underline{A}$ is increased.


## Acknowledgements

For typing this manuscript, we thank Lavonne Banse. Support was received from the National Institutes of Health (Grants GM 24365 and AI 16117) and from the Robert A. Welch Foundation.

References

1. Polson, A., Russell, B.: Methods in Virology, Vol 2 (Maramorosch, K. and Koprowski, H., eds.), pp. 391-426, Academic Press, New York (1967).

2. Serwer, P., Pichler, M.E.: J. Virol. **28**, 917-928 (1978).

3. Serwer, P.: J. Mol. Biol. **138**, 65-91 (1980).

4. Rodbard, D., Chrambach, A.: Proc. Nat. Acad. Sci. U.S.A. **65**, 970-977 (1970).

5. Ghosh, S., Basu, M.K., Schweppe, J.S.: Anal. Biochem. **50**, 592-601 (1972).

6. Serwer, P.: Anal. Biochem. **101**, 154-159 (1980).

7. Serwer, P.: Anal. Biochem., in press (1981).

8. Fishbach, F.A., Harrison, P.M., Anderegg, J.W.: J. Mol. Biol. **13**, 638-645 (1965).

9. Serwer, P.: J. Mol. Biol. **107**, 271-291 (1976).

10. Stroud, R.M., Serwer, P., Ross, M.J., Ruark, J.E.: Biophys. J., in press (1981).

11. Serwer, P.: Proc. Electron Microscopy Soc. Amer. **37**, 354-355 (1979).

12. Serwer, P., Graef, P.R., Garrison, P.N.: Biochem. **17**, 1166-1170 (1978).

13. Shaw, A.R., Lang, D., McCorquodale, D.J.: J. Virol. **29**, 220-231 (1979).

14. Earnshaw, W.C., Casjens, S.R.: Cell **21**, 319-331 (1980).

15. Guiseley, K.B., Renn, D.W.: Agarose - Purification, Properties, and Biomedical Applications, Marine Colloids, Inc., Rockland, Maine, p. 7 (1975).

16. Wieme, R.J.: Agar Gel Electrophoresis, Elsevier, Amsterdam, p. 111 (1965).

17. Ghosh, S., Moss, D.B.: Anal. Biochem. **62**, 365-370 (1974).

18. Quast, R.: J. Chromatog. **54**, 405-412 (1970).

19. Amsterdam, A., Er-el, Z., Shaltiel, S.: Arch. Biochem. Biophys. **171**, 673-677 (1975).

20. Serwer, P.: J. Ultrastruct. Res. **58**, 235-243 (1977).

# NEW ELECTROPHORETIC TECHNIQUES FOR THE SEPARATION AND CHARAC-
TERIZATION OF POLIOVIRUS PROTEINS


Rudolf Dernick[*], Klaus-Jochen Wiegers, Jochen Heukeshoven
Heinrich-Pette-Institut, Martinistrasse 52, D-2000 Hamburg 20,
Germany


## Introduction

Poliovirus is the infectious agent responsible for the cripp-
ling disease poliomyelitis, which caused paralysis and death
of thousands of children before the era of vaccination against
this virus began about twenty years ago. Today this virus is
an excellent model for studying the structure, infectivity and
antigenicity of an animal virus and the processing of virus
proteins within infected cells. The poliovirus particle is
one of the smallest (30 nm in diameter) and most stable animal virions. It contains one molecule of RNA per virus particle and 240 polypeptides of four different types, called
virion proteins, VP1 (34000), VP2 (28000), VP3 (24000) and
VP4 (7500). Their molecular weights are given in parenthesis
as determined by SDS-PAGE for the Mahoney strain, belonging
to the serological type 1.

## One-Dimensional Techniques

The command of one-dimensional separation techniques is a
necessary step for their combination into two-dimensional
methods.

---

[*] R. Drzeniek changed his family name to Dernick in
October 1980.

Polyacrylamide gel electrophoresis in the presence of sodium
dodecyl sulfate (SDS-PAGE)

The four polypeptides VP1-VP4 obtained by dissociation of
poliovirus particles with sodium dodecyl sulfate (SDS) and
electrophoresis in SDS-containing polyacrylamide gels (SDS-
PAGE) were the first viral proteins separated by this technique (1). SDS-PAGE of poliovirus-infected HeLa cells (2)
and the determination of the molecular weight of poliovirus
polypeptides by SDS-PAGE (3) were milestones for the use of
this technique in virology and biochemistry. We use SDS-PAGE
as reference for new techniques elaborated for the separation
of poliovirus proteins.


Isoelectric focusing

Isoelectric focusing of poliovirus proteins in polyacrylamide
gels (PAGIEF) was started in 1975 after a detailed analysis
of the dissociation of poliovirus particles by urea (4). The
four polypeptides of poliovirus particles, type 1, strain
Mahoney, separated by PAGIEF in 9 M urea, were identified by
subsequent SDS-PAGE in a 2D-analysis; their isoelectric points
were also determined (5). The multiple bands observed (up to
13) were analyzed and explained in a detailed methodological
study (6), in which the influence of ribonuclease, urea concentration, neutral detergents, SDS, riboflavine, iodoacetamide
etc. was described. It was made clear that fresh poliovirus
preparations, which were dissociated and focused under correct
conditions gave four bands only in PAGIEF. Different virus
strains revealed differences in the charge of their individual
polypeptides (7). Their apparent isoelectric points ($pI_{app}$),
which were measured in 9 M urea, always decreased in the order
VP1, VP4, VP2, VP3 for all poliovirus strains tested. VP1 was
the most basic, VP3 the most acidic protein (7, 8). Isoelectric focusing of poliovirus infected HeLa cells was already
presented in 1978 (9), a detailed publication including the
$pI_{app}$-values for 22 out of 24 identified proteins appeared

three months ago (10). We anticipate that PAGIEF of viral proteins will be applied in the future for a number of purposes. So far we have used its results for the isolation of pure poliovirus polypeptides VP1, VP2, VP3 and VP4 (never obtained before) by isoelectric focusing in urea containing sucrose gradients (SUGRIEF) (11).


Electrophoresis in formic acid (Fo-PAGE)

In order to avoid detergents or ampholytes present in protein preparations separated by SDS-PAGE or by isoelectric focusing, respectively, we have tried to find a substance with the following properties: (i) it should be volatile, i.e. capable of being removed by freeze-drying, (ii) it should dissociate poliovirus particles into protein subunits, and (iii) it should allow the separation of poliovirus proteins by electrophoresis. Such a substance is formic acid, known as an excellent solvent for proteins, also for proteins of high hydrophobicity. Formic acid solutions are used in chemical reactions, e.g. for CNBr cleavage or performic acid oxidation. Therefore, proteins separated in this solvent could be used directly, i.e. without removal of solvent. Recently, Mokrasch (12) has described a polyacrylamide gel electrophoretic separation of membrane proteins in 13 M formic acid.

We have elaborated the electrophoresis in formic acid in great detail, especially for the separation of the hydrophobic proteins of poliovirus particles (13). This method is completely different from techniques using PAGE in acidic aqueous medium. In our method the essential solvent is formic acid, not water, due to the high concentration (50-80%) of formic acid used in the experiments. The four structural proteins of poliovirus, type 1, strain Mahoney, were separated in polyacrylamide gel rods containing 6% polyacrylamide and 5% bisacrylamide (i.e. 6% T, 5% C) and 70% formic acid (Fo-PAGE). Interestingly, the order of the electrophoretic mobility in Fo-PAGE was VP2-VP1-VP3-VP4, instead of VP1-2-3-4 as found in SDS-PAGE (Fig. 1).

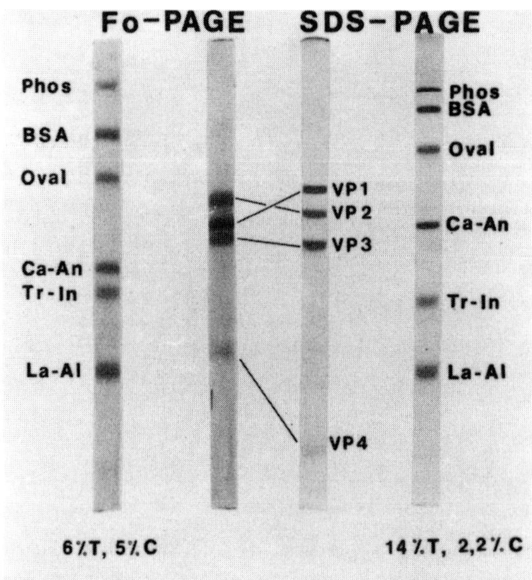



<u>Fig. 1</u>: Comparison of Fo-PAGE and SDS-PAGE. The two gels in the middle contain poliovirus, type 1, and the gels on both sides contain reduced molecular weight marker proteins. Conditions: Fo-PAGE 6% T, 5% C, 24 h, 50V, 70% formic acid. SDS-PAGE 14% T, 2.2% C, 20 h, 50V, Tris-buffer, pH 8.6; 0.1% SDS. Phos = phosphorylase b, BSA = bovine serum albumin, Oval = ovalbumin, Ca-an = carbonic anhydrase (bovine), Tr-in = soybean trypsin inhibitor, La-al = $\alpha$-lactalbumin.

In addition, the molecular weights of the four viral proteins were larger than estimated by SDS-PAGE using the usual molecular weight marker proteins as standard reference. A discrepancy was also found for bovine serum albumin (BSA). Under non-reducing conditions it migrated faster than ovalbumin, despite its higher molecular weight (66700 vs. 43000, not shown, see ref. 13). Treatment of proteins by reducing agents like mercaptoethanol or dithioerythritol, which split disulfide bridges, changed their electrophoretic mobility. Thus, BSA migrated slower than ovalbumin under reducing conditions, assuming its proper sequence (Fig. 1). Dramatic changes were also observed when Fo-PAGE was performed in the

presence of urea, whereas guanidine hydrochloride did not
change the migration patterns of marker and poliovirus proteins. The differences observed in the electrophoretic mobility
and the 'wrong' molecular weights, determined by Fo-PAGE, are
due to the different structures of the investigated proteins.
Therefore Fo-PAGE is not recommended for the determination of
molecular weights. Fo-PAGE, however, is very useful for the
separation of proteins having the same molecular weight, but
differing in structure. In the future it could be used to
study the conformation of proteins, even of those being highly hydrophobic and insoluble in aqueous solutions (13).

Comparison of one-dimensional techniques

The characteristics of these methods as revealed by their use
in the laboratory are summarized in Table 1. A compilation
of the conditions employed in one-dimensional techniques for
the separation of poliovirus proteins is given in Table 2. For
further details the reader is referred to the publications
cited.

Table 1: Characteristics of one-dimensional methods

|  | Electrophoresis | | Isoelectric Focusing | |
|---|---|---|---|---|
|  | SDS-PAGE | Fo-PAGE | PAGIEF | SUGRIEF |
| Applied to virion proteins | yes | yes | yes | yes |
| Applied to virus infected cells | yes | no | yes | no |
| Use of RNase | not necessary | | mandatory | |
| Tolerable salt conc. in the sample | 150 mM | 3M CsCl | 50 mM | 50 mM |
| Dissociating agent | SDS | formic acid | urea | urea |
| Incorporation of dissociating agent | before | after | before polymerization | into gradient |
| Additional dissociating agents | urea | urea guanidine·HCl | detergents should not be used | |

Table 2: Characterization of one-dimensional methods

| | Electrophoresis | | Isoelectric Focusing | |
| --- | --- | --- | --- | --- |
| | SDS-PAGE | Fo-PAGE | PAGIEF | SUGRIEF |
| Separation according to: | size | size/struct. | charge | charge |
| Determination of: | mol.weight | mol.weight structure | isoelec. point | isoelec. point |
| Conditions for the dissoc.of poliovirus | 1% SDS 2min,100°C | 70% formic acid at 4°C | 9M urea at 25°C | 9M urea at 25°C |
| Dissociating conditions during elec-trophoresis | 0.1% SDS at 20°C | 70% formic acid at 4°C | 9M urea at 20°C | 7M urea at 4°C |
| Electrophor.in | PAA* gel | PAA gel | PAA gel | sucrose gradient |
| Separation gel or gradient | 14%T;2.6%C 10-20%T, 2.6%C | 6%T,5%C | 5%C,2.6%T | 0-30% |
| Separation conditions | 0.1% SDS in TRIS-HCl buffer, pH 8.9 | 70% formic acid in $H_2O$ | 9M urea in 2% am-pholytes (pH 5-9 and 2-11) | 7M urea in 2% am-pholytes (pH 5-9 and 2-11) |
| Temperature | room | -6 to +4°C | room | 4°C |
| Standard conditions for electro-phoresis | 10 cm PAA gels | 10 cm PAA gels | 10 cm PAA gels | 15cm(7ml) sucrose gradients + 7M urea |
| sample applied at: | cathode | anode | anode | anode |
| anode solution | TRIS-glycine buffer pH 8.3 0.1% SDS | 70% formic acid | 0.1M $H_3PO_4$ | 0.1M $H_3PO_4$ |
| cathode solution | | 70% formic acid | 0.1M NaOH | 0.1M ethanol-amine |
| Voltage | 100V/250V | 50V-60V | 150V/500V | 150V-300V |
| Time | 1 h / 3 h | 16 - 24 h | 16 h/1 h | 20 h |

*PAA = polyacrylamide

Two-Dimensional Techniques

2D-analysis

Isoelectric focusing in polyacrylamide gels (PAGIEF) combined
with SDS-PAGE to a high resolving 2D-analysis was used for
the separation and characterization of viral proteins of poliovirus particles and infected HeLa cells. The first goal was
to correlate bands of poliovirus proteins obtained by PAGIEF
to patterns obtained by SDS-PAGE, because in the past poliovirus proteins were characterized exclusively by SDS-PAGE.
These experiments resulted in the identification of the four
proteins of poliovirus particles separated by PAGIEF and by
the determination of their apparent isoelectric points (5-8).
In addition, 24 virus-specific proteins of poliovirus infected
HeLa cells were characterized very recently by 2D-analysis
(10). The second goal was to use the high resolving power of
2D-analysis for the detection of poliovirus proteins not
separated or separated only partially by SDS-PAGE. This analysis showed about 15 new proteins, the viral origin of which
and their relationship to each other was demonstrated by peptide mapping as outlined in detail below.

2D-electrophoresis

The development of the electrophoresis of poliovirus proteins
in high concentrations (50-80%) of formic acid (Fo-PAGE)
forced us to combine this method with SDS-PAGE to a 2D-electro-
phoresis in order to identify the bands demonstrated by Fo-
PAGE (Fig. 2). We used the same approach as the one developed
for the 2D-analysis with some minor modifications specific
for the differences in the technique of the first dimension
(13).

252

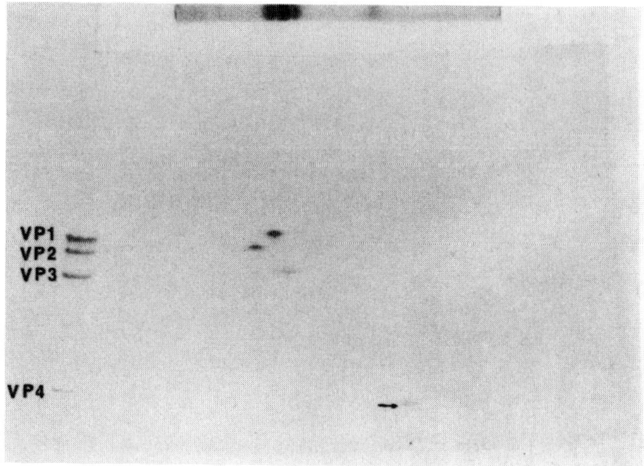


Fig. 2: Two-dimensional electrophoresis of poliovirus. First
        dimension (horizontal) Fo-PAGE (6%T, 5%C) in 70%
        formic acid; gel was cut longitudinally. One gel slice
        was stained and is shown on top as reference. Second
        dimension (vertical) SDS-PAGE (gradient gel 7.5-25%T).
        A reference sample of SDS-dissociated poliovirus is
        shown on the left.

Characterization of Proteins by Peptide Mapping

One of the most important goals for a critical evaluation and
broad exploitation of 2D-techniques is the problem of identifying the vast number of protein spots separated by these
powerful techniques. Different approaches are possible, ranging
from computer analysis of all detected protein spots to the
isolation of individual spots and their characterization. The
very low amount of protein present in spots of 2D-gels is a
major obstacle for the use of classical biochemical methods.
Therefore we have applied a direct electrophoretic separation
of peptides obtained by limited proteolysis to those minute
amounts of radioactive proteins separated by 2D-analysis. This
method is suitable for 2D-gels which were dried in order to
locate proteins by autoradiography (14). It was elaborated by
use and modification of existing methods (15,16) of limited
proteolysis in SDS, followed by electrophoresis in SDS-polyacrylamide gels. Its principle is shown in Fig. 3.

For the characterization of radioactively labeled proteins after 2D-analysis prolonged staining and destaining should be omitted to avoid possible acid hydrolysis. No hydrolysis was detected during fixation in 7% acetic acid for up to 90 min. Therefore 2D-gels containing $^{35}$S-methionine labeled proteins were not stained and destained, but fixed for 60-90 min in 7% acetic acid and dried. Dried gels were autoradiographed (6) and the obtained map of radioactive spots was used as template to locate proteins on the gel. Spots (marked by frames in Fig. 3) were excised from dried gels and rehydrated briefly (1-2 min) in sample buffer (1% SDS + 1 mM EDTA, TRIS-phosphate buffer, pH 6.8) until they regained their original shape. Rehydrated gel slices were inserted into sample wells of a stacking gel placed on a 10-20% polyacrylamide gradient slab gel. The slices were forced to the bottom of the wells and overlaid first with sample buffer containing 20% glycerol and then with the proteolytic enzyme (α-chymotrypsin or V8-protease from staphylococcus aureus) dissolved in sample buffer with 10% glycerol. After overlaying with electrode buffer (0.05 M TRIS base, 0.384 M glycine, 0.1% SDS, 0.001% bromophenol blue) electrophoresis was performed until the dye front moved to the lower end of the stacking gel and stopped for 30 min. This allowed the limited hydrolysis of the analyzed proteins concentrated at the end of the stacking gel. The peptides obtained were then separated by continuation of the electrophoresis for about 15 hrs at 300V. All steps were performed at room temperature. The separation gel was stained by Coomassie brilliant blue in order to visualize the proteolytic enzymes used. This permitted a quick judgment of the obtained gel before its preparation for fluorography. Fluorography was performed according to Laskey and Mills (17) after stepwise insertion of gels into dimethylsulfoxide (DMSO), into DMSO and PPO and then into $H_2O$. The dried gels were exposed to X-ray film (Kodak X-omat) and kept at $-70^{\circ}$C for 3 days to 4 weeks. This allowed the detection of a minimum amount of about 1000 cpm per protein spot of $^{35}$S-methionine labeled protein. We

254

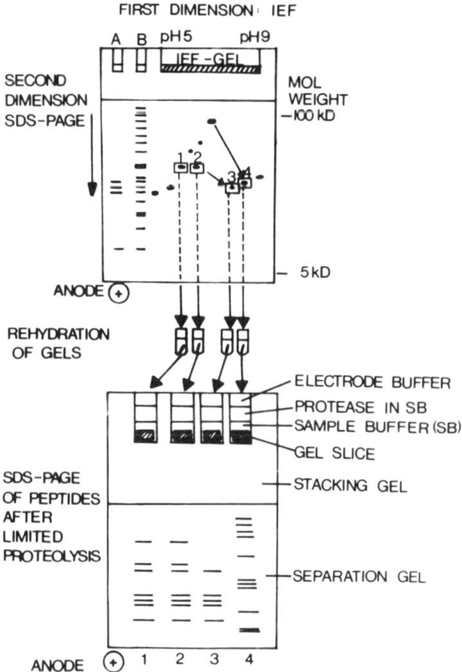



Fig. 3: Schematic representation of 2D-analysis followed by
limited proteolysis of selected proteins and SDS-PAGE
of the obtained peptides (peptide mapping).
Upper part: 2D-analysis, with two reference samples
(A = virion, B = infected cells; SDS-PAGE only).
Lower part: limited proteolysis followed by peptide
mapping by SDS-PAGE.
1,2,3,4: numbers of protein spots selected for peptide
mapping.

started the 2D-analysis with about $1 \times 10^6$ cpm in order to
characterize all spots, even very weak ones, present on the
2D-gel. Fig. 3 represents schematically an idealized pattern
of four proteins (marked by frames) separated by 2D-analysis
and submitted to limited proteolysis and subsequent SDS-PAGE.
From this scheme it is evident that proteins No. 1 and 2 give
identical peptide patterns. They represent charge modifications
of the same protein because they have identical molecular

weights but different charges. Proteins No. 2 and 3 are related
to each other, because they share a set of five identical peptides and differ in two peptides. Because protein No. 3 has a
lower molecular weight than protein No. 2, a precursor-product
relationship could be established, e.g. by a pulse-chase experiment. Therefore cleavage (processing) of protein No. 2 could
result in protein No. 3 and a small unidentified polypeptide.
Protein No. 4 is completely different from No. 3 because of
their different peptide patterns. The precursor for protein
No. 4 could be a protein not analyzed but marked by an arrow
to visualize the assumed precursor-product relationship.

An authentic figure of a 2D-gel and its analysis by limited
proteolysis is given in Fig. 4.

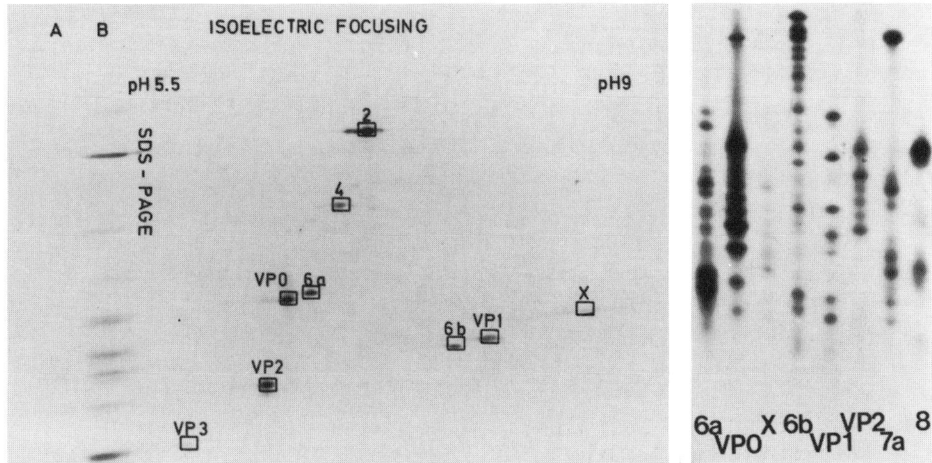


<u>Fig. 4</u>: 2D-analysis and peptide mapping of poliovirus proteins.
<u>Left side</u>: Photograph of a section of an authentic 2D-
gel of poliovirus proteins obtained from infected HeLa
cells, detected by autoradiography, marked by frames,
and numbered according to poliovirus nomenclature (14).
<u>Right side</u>: Photograph of a section of an authentic
peptide map obtained after treatment of selected
poliovirus proteins (numbers at the bottom) by V8-
protease and subsequent SDS-PAGE. Detection of radioactive spots by fluorography.

256

The relatedness and precursor-product relationship for all
poliovirus proteins detected by 2D-analysis can be established
as outlined above. The complete peptide mapping of all poliovirus proteins separated by 2D-analysis is in progress (Wiegers and Dernick, in preparation). The method of peptide
mapping and its application to a limited number of proteins
found on 2D-gels was already submitted for publication in
"Electrophoresis" (14).


References


1. Maizel, J.V.: Biochem. Biophys. Res. Commun. 13, 483-489
   (1963).

2. Summers, D.F., Maizel, J.V., Darnell, J.E.: Proc. Nat. Acad.
   Sci. 54, 505-513 (1965).

3. Shapiro, A.L., Vinuela, E., Maizel, J.V.: Biochem.Biophys.
   Res. Commun. 28, 815-820 (1967).

4. Drzeniek, R.: Z. Naturforsch. 30c, 523-531 (1975).

5. Hamann, A., Wiegers, K.J., Drzeniek, R.: Virology 78,
   359-362 (1977).

6. Hamann, A., Drzeniek, R.: J. Chromatogr. 147, 243-262
   (1978).

7. Hamann, A., Reichel, C., Wiegers, K.J., Drzeniek, R.:
   J. gen. Virol. 38, 567-570 (1978).

8. Drzeniek, R., Reichel, C., Wiegers, K.J., Hamann, A.,
   Hilbrig, M.: In: Electrophoresis '79, Ed. B.J. Radola,
   pp. 475-489, Walter de Gruyter, Berlin, New York (1980).

9. Wiegers, K.J., Drzeniek, R.: Abstracts of the Fourth
   International Congress for Virology. The Hague, Netherlands,
   W22A, p. 312, Wageningen; Centre for Agricultural Publi-
   shing and Documentation (1978).

10. Wiegers, K.J., Dernick, R.: J. gen. Virol. 52, 61-69 (1981).

11. Wiegers, K.J., Drzeniek, R.: J. gen. Virol. 47, 423-430
    (1980).

12. Mokrasch, L.C.: Anal. Biochem. 88, 263-270 (1978).

13. Heukeshoven, J., Dernick, R.: Electrophoresis (in press).

14. Wiegers, K.J., Dernick, R.: Electrophoresis (in press).

15. Cleveland, D.W., Fischer, S.G., Kirschner, M.W.,
    Laemmli, U.K.: J. Biol. Chem. 252, 1102-1106 (1977).

16. Burge, B.W., Huang, A.S.: Virology 95, 445-453 (1979).

17. Laskey, R.A., Mills, A.D.: Eur. J. Biochem. 56, 335-341
    (1975).

# SECTION II

High Resolution Two-Dimensional Electrophoresis

SDS-GEL GRADIENT ELECTROPHORESIS, ISOELECTRIC FOCUSING AND HIGH-RESOLUTION
TWO-DIMENSIONAL ELECTROPHORESIS IN HORIZONTAL, ULTRATHIN-LAYER POLYACRYLA-
MIDE GELS.

Angelika Görg, Wilhelm Postel, Rainer Westermeier
Lehrstuhl für Allgemeine Lebensmitteltechnologie der Technischen Universität, München, D-8050 Freising-Weihenstephan, G.F.R.

Elisabetta Gianazza and Pier Giorgio Righetti
Department of Biochemistry, University of Milano, Via Celoria 2, Milano
20133, Italy

Introduction

With the advent of ultrathin-layer isoelectric focusing (IEF) (1-4), in the
120-360 μm thickness range, the resolution of this technique has been considerably improved. Due to the extreme gel thinness, vertical thermal gradients within the gel layer (most gel slab apparatus for gel IEF provide
for cooling on one surface only) are practically abolished, so that the zones focus just perpendicular, and not slanted, to the applied voltage gradient. Higher field strengths are thus possible, with consequent sharpening
of the focused bands and shorter focusing times.This band sharpness can also be retained during subsequent fixing, staining and destaining, due to
the short time required by denaturing agents and stains to diffuse within
the gel layer. Most important, the resolution in the IEF dimension is also
maintained when developing activity stains, such as zymograms and immunoprints, due to the extremely short contact times (just a few minutes) needed
for their development.

We have recently extended this work (5) by running pore gradient gels in
ultrathin layers and by performing two-dimensional (2-D) techniques in horizontal, ultrathin gels. The novelty of the method consists not only in the
introduction of ultrathin matrices, but also in the possibility of performing 1-D and 2-D techniques horizontally, instead of vertically. It should
be appreciated that in the past, when the classical starch gel technique
was developed by Smithies (6), the method had to be modified from an horizontal to a vertical set-up (7), the reason being the electrodecantation of
the sample at the bottom of the pocket in the horizontal gel layer, result-

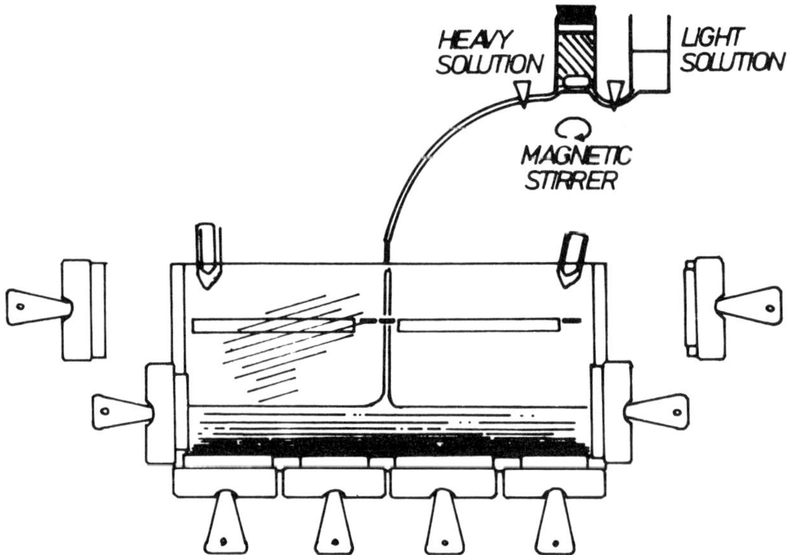



Fig. 1. Set-up for casting a concave exponential polyacrylamide gel gradient from 4% T to 22.5%T. The cell is from the LKB Multiphor apparatus and the U-gasket is cut out of 3 layers of parafilm (360 μm gel thickness). Mixing chamber: 5 ml of 22.5%T, 4%C, 55% glycerol; reservoir: 5 ml of 0% acrylami-de, 0% glycerol; both solutions buffered with 375 mM Tris-Gly, pH 8.8.

ing in an uneven distribution of the protein zones within the gel matrix. Since then, the vast majority of electrophoretic separations in hydrophilic gels have been run vertically, including disc electrophoresis and all 2-D techniques (9). Interestingly, we are now attempting a reversal of this trend. This is not just for an ebb- to flood-tide mechanism, nor just for the fun of creating a new fashion, but because horizontal, ultrathin systems have several inherent advantages, as we shall discuss further on. In the present report, we will describe new improvements of this methodology, also in connection with the silver stain (10), and its applications to different separation problems. Since the general technical aspects of ultrathin gels and 2-D separations have been already thoroughly described, we will report here only the latest innovations and refer the reader to previous publications (1-5) for the necessary background.


Results

1) Methodology

It might not appear an easy and routine task to cast an exponential gradient

gel of 240 or 360 μm thickness. In Fig. 1 we have drawn the set-up for casting a 4% to 22.5%T concave exponential gradient gel between glass slabs
at 360 μm apart. Indeed, while pouring the gradient, the two upper clamps are
removed and two paper clips inserted, so that the outflow needle can be inserted firmly at the chamber top. The gradient is calculated according to
the equations by Noll (11): the mixing chamber contains 5 ml of heavy solution (22.5%T, 4%C in 55% glycerol) while the reservoir is filled with 5
ml of light solution (0% acrylamide, 0% glycerol) (for a definition of T and
C see ref. 12). Both are buffered with 375 mM Tris-HCl, pH 8.8 containing,
when needed, 0.1% SDS. The mixing chamber is stoppered, so that the liquid
volume in it remains constant. The gradient is allowed to flow from the chamber top by gravity pull and capillary suction, at a rate of about 1 ml/min.
The actual gradient fills only 5/8 of the chamber volume, up to ca. 5 mm below the sample pockets, since in this region, and behind it, where the electrode connections will be placed, there is no need for a pore gradient. The
gel mould is then filled with the remaining solution in the mixing chamber.
Since this particular set-up will be used for a second dimension, SDS-electrophoresis, the cover glass plate is provided with two trench-formers (onto which will be loaded the first dimension, IEF gel strip) and three poc-
.ket-formers, which will accomodate liquid samples, such as the $M_r$ calibration mixture and/or the same sample fractionated in the first run, but omit-
ting the IEF step. The slot- or trench-formers are obtained by gluing 1 layer. of parafilm to the glass plate, so that, when opening the chamber, a
gel floor of 240 μm thickness is left in the sample-loading areas of the gel.
The facing glass slab is covered by a cellophane foil stretched taut over
it, or by a LKB foil, or by a Gel-Fix membrane (Serva), which will serve as
a mechanical support to the gel layer during subsequent staining and destaining steps.

In order to check the shape of the pore gradient, we have run an SDS-electrophoresis not along the porosity gradient, but perpendicular to it, as
suggested by Margolis and Kenrick (13). As shown in Fig. 2, the mobility
curve drawn in the gel slab by each protein marker depicts the shape of the
polyacrylamide gel gradient. In fact, all the curves have identical shapes
except that, as the molecular mass of each marker increases, they are progressively shifted toward the application trench (cathode). The importance
of a concave gradient for optimizing separation of small $M_r$ proteins can be

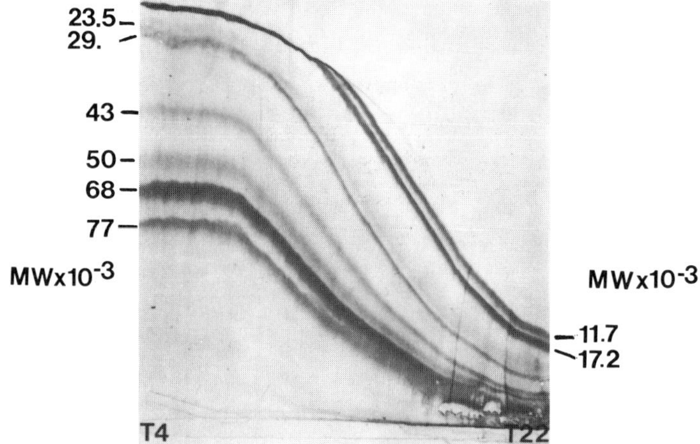



Fig. 2. Transverse SDS-gradient electrophoresis in a ultrathin-gel slab. The $M_r$ markers are: cytochrome C (11,700); horse myoglobin (17,200); γ-globulin (light chain, 23,500); carbonic anhydrase (29,000); ovalbumin (43,000); γ-globulin (heavy chain, 50,000); human albumin (68,000); human transferrin (77,000). The migration is toward the cathode for a total of 200 min at 500 V (constant).

appreciated from this figure. Cytochrome C ($\overline{M}_r$=11,700) and myoglobin ($\overline{M}_r$ = 17,200) travel as a single zone in the diluted portion of the gradient but, as they reach a critical pore size, the sieving becomes more pronounced and they fork into two lines which, up to the 22.5%T end of the gradient, stay parallel.

Other methodological aspects worth mentioning regard the handling and transfer of the 1-D gel strip onto the 2-D gel matrix. Previously, we used to cut away with a sharp razor blade the focusing gel strip, equilibrate it briefly in SDS-denaturing buffer (9 mM Tris-HCl, 2% SDS, 2% β-mercaptoethanol) and then load it onto the 2-D gel. However, after a report by Jäckle (14), we have found it convenient to fix and stain the entire IEF slab in Coomassie Brilliant Blue (CBB) G-250. This provides the experimenter with a reference pattern visible to the eye, so that the best sample track can be cut away to be loaded onto the      SDS-slab, discarding samples with wavy zones, or smears,or displaying poor resolution. As shown in Fig. 3, since the ultrathin gel layer is cast onto a plastic membrane, the strip to be further analyzed can simply be cut away with a pair of scissors. After equilibration in SDS-buffer, the same strip is loaded onto the SDS-

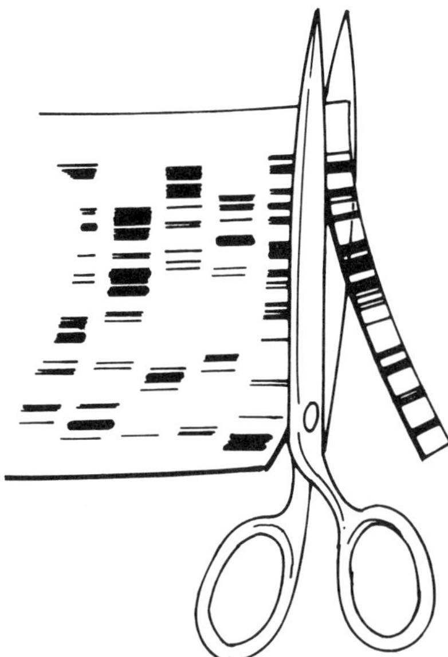



Fig. 3. Handling of the IEF gel slab. After IEF, the protein zones are fixed and stained in Coomassie Brilliant Blue G-250. For transfer to the 2-D
matrix, a gel strip as wide as the trench in the 2-D gel is cut away.

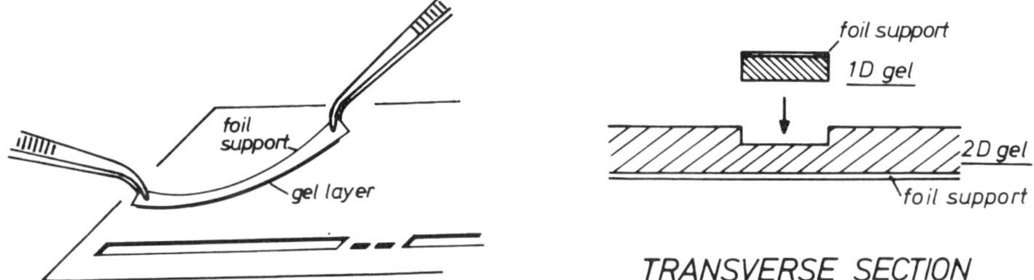



Transfer of the focusing gel strip to the electrophoresis gel

Fig. 4. Interfacing the 1-D with the 2-D gel. The stained IEF strip, rinsed
in water, equilibrated in SDS-buffer and blotted, is transferred, gel layer
facing down, to the trench of the 2-D matrix. The operation is greatly facilitated by the plastic backing of the gel. On the right, we can see a transverse section of the gels, indicating that there is a gel floor in the trench dug in the 2-D gel.

gel slab. We have shown in Fig. 4 how this is performed. The IEF gel strip

is lifted at the two extremes with tweezers which hold onto the two protru-

ding edges of the supporting plastic film, and lowered, *gel layer facing*

*down,* into the trench of the SDS-slab. On the right of Fig. 4, we have shown

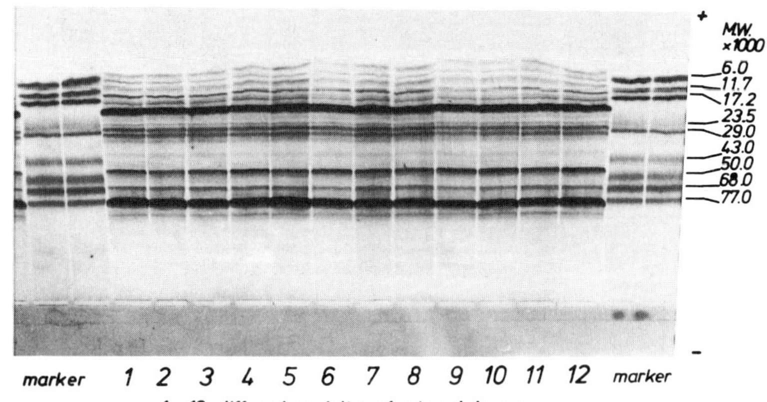



Fig. 5. SDS-pore gradient gel electrophoresis (4% to 22.5%T) of 12 varieties
of winged beans. All $M_r$ markers as in Fig. 2 plus an additional marker (apro-
tinin, $M_r$ = 6,000). Running conditions: 200 min at 500 V (constant). Gel
thickness: 240 µm.

also a transverse section of the gel; the 2-D matrix is 360 µm thick and contains a trench which is 120 µm deep, so that  a gel floor of 240 µm is left
at the bottom. The 1-D, IEF gel, is 240 µm thick, so that it fills completely the trench. For a proper electrical contact, and even gel conductivity,
it is important that the trench in the 2-D gel does not go entirely through
the gel thickness. The importance and advantages of such gel contact will be
discussed further on.

2) Applications

a) *1-D techniques*. Fig. 5 gives an example of an SDS-electrophoresis in a
ultrathin, concave exponential gradient gel. Here the matrix was 240 µm
and the sample was applied as a free liquid in tiny pockets, of 2 µl volume,
which can be seen ca. 8 mm **above** the cathodic symbol (-), where the zone
   of the cathodic contact ends. The extreme sharpness of the polypeptide
zones, the very high resolution and reproducibility of banding patterns
can    be appreciated. Even though the marker proteins have been loaded at
the two gel extremities, where, due to the meniscus-forming properties of
liquids, one might expect non-reproducible gel gradients, they indeed exhibit identical patterns and migration distances, suggesting highly-reprodu-

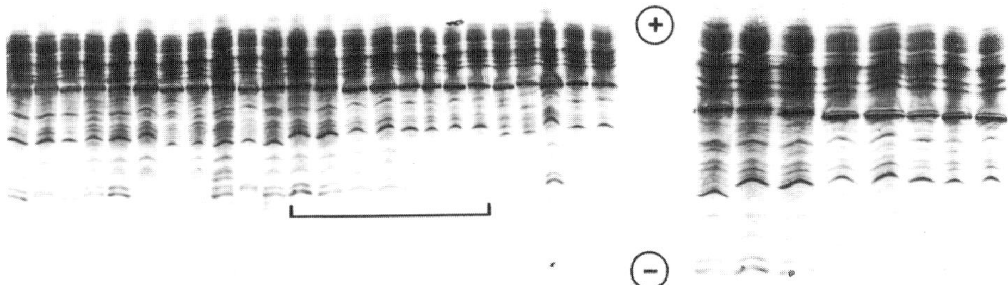

Fig. 6. IEF of 24 varieties of *pisum sativum* in an ultrathin gel (240 μm thick) (left side). On the right side, a blow-up section of 8 electrophoretic tracks (marked with a horizontal bar) is shown. The anode is uppermost. The sample has been loaded at the center of the gel onto rectangular pieces of Whatmann 3MM (10 μl volume, corresponding to ca. 20 μg protein). Ground peas were extracted with 25 mM Tris-Gly buffer, pH 8.6. The gel contained 2% Ampholine pH 3-10.

## Ultrathin-Layer Horizontal 2D Electrophoresis
### (Urea-IEF × SDS Pore Gradient PAGE)

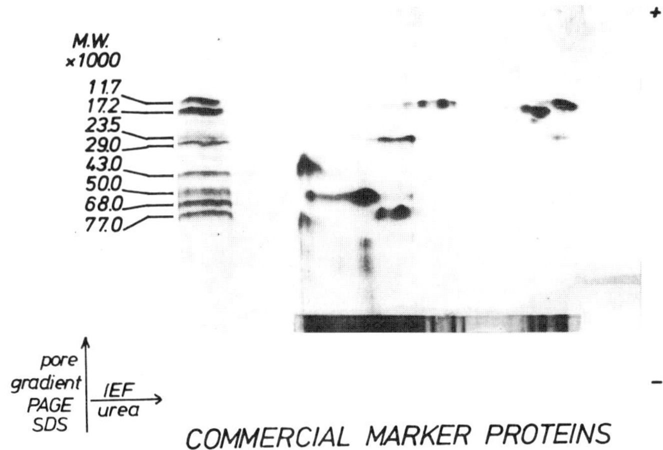



Fig.7. 2-D separation of commercial marker proteins. The markers have been listed in Fig.2. The first dimension was in 9M urea and 2% Ampholine pH 3-10. After fixing and staining, the IEF strip was rinsed 10 min in water, equilibrated 2 min in SDS-buffer, blotted and loaded on the 2-D gel. Running conditions: 10 min at 15mA (constant) and then 190 min at 500 V(constant).

cible pore gradients. Also the 12 varieties of winged beans show a remarkable constancy of SDS-pattern, indicating that their differences must be due to spot mutations in the genome, rather than to size differences in the constituent polypeptide chains.

Fig. 6 gives an analogous example of 1-D separation, but in the IEF dimen-

266

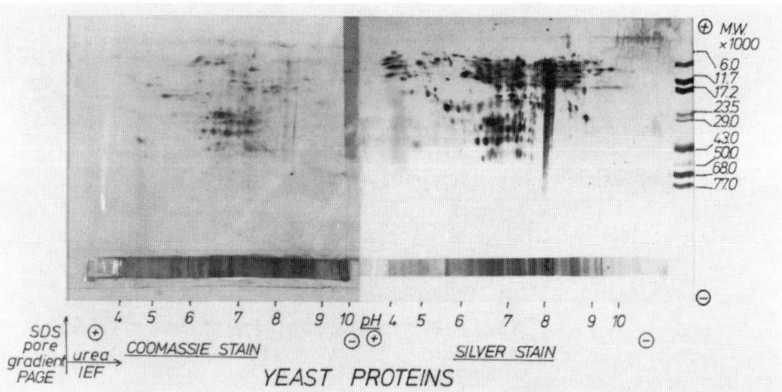



Fig. 8. 2-D separation of yeast proteins. Comparison of Coomassie Blue stain (left) and silver stain (right). Yeast cells were extracted in a sonifier in 25 mM Tris-Gly, pH 8.6, containing 9M urea. **The IEF gel was cast in 9 M urea.**

sion. 37 varieties of *Pisum sativum* soluble proteins were analysed by IEF in ultrathin gels. Here only 24 tracks are shown and, on the right side, a blow-up section of just 8 lines.The very high resolution is also coupled to a minimal sample amount required for analysis: 20 µg protein per each pea strain.

*b) 2-D techniques.* Fig. 7 displays the $pI/M_r$ map of a mixture of 8 marker proteins spanning the molecular mass range 11,700 to 77,000 daltons. The 1-D IEF gel was run in duplicate so that, before photography, the spare, stained strip could be fitted into the sample trench. In this way, the two-dimensional separation can be viewed together with the first dimension run, for easy comparison. Moreover, in the left side, the same mixture of marker proteins, not subjected to IEF fractionation, was loaded in a pocket and run in the SDS-dimension. In this way, calibration of both, the pH and molecular mass axes, is greatly facilitated. For best results, in the transfer of stained IEF strips to the 2-D run, we have found the following method to give highly reproducible data. The gel strip, removed from the destaining solution, is bathed for 10 min in distilled $H_2O$ (removal of ethanol and acetic acid seems to be a must), then incubated for 2 min in the denaturing SDS-buffer. If too wet, the strip should be gently blotted before transfer to the SDS-slab, otherwise diffuse spots will result in the second dimension. After 30 min of electrophoresis in the SDS-gel, the IEF gel strip should be remov-

## TABLE I

| Fields of Applications of Ultrathin-Techniques |
| --- |
| Ultrathin-Layer Isoelectric Focusing |
| Ultrathin-Layer Electrophoresis |
| Ultrathin-Layer SDS Electrophoresis |
| Ultrathin-Layer Gradient Gel Electrophoresis |
| Ultrathin-Layer SDS Gradient Gel Electrophoresis |
| Ultrathin-Layer 2D Electrophoresis |
| Ultrathin-Layer High Resolution 2D Electrophoresis |

ved and the trench filled with buffer solution.

In Fig. 8, we are comparing 2-D maps developed by either standard Coomassie stain, or by the recently developed silver stain (10). The sample used was an extract of yeast cytoplasmic proteins. The improvement in sensitivity is remarkable, approaching the detection limit of autoradiography. Its sensitivity is said to be 100 times higher than the conventional CBB stain. Thus ultrathin techniques, coupled to high sensitivity stains, appear to be very close to the limits of 2-D techniques in terms of the resolution and sensitivity achieved. In table I, we have summarized the various fields of application of ultrathin-techniques.

Discussion

We will comment here only upon 2-D techniques, since they are the most popular among biochemical separation methods and very nearly achieve the aim of spreading maximally the protein spots on the separation plane, since the fractionation parameters vary from net charge (or free mobility) along one axis, to size (or mass) along the other axis. Barrett and Gould (15) and McGilliwray and Rickwood (16) were among the first ones to introduce this method of charge/mass fractionation. However, it was only in 1975, when O'Farrell (17) described a simple method for casting SDS-pore gradient gels and assembling an electrophoretic cell for 2-D runs, and coupled this technique to autoradiography, thus revealing more than a thousand spots on the 2-D gel, that 2-D techniques became laboratory routine. When multiple sam-

268

ples have to be analyzed at once, Anderson and Anderson (18, 19) have devised a set-up for multiple, parallel IEF-SDS runs, which has been termed ISO-DALT system, to indicate that separation is based on ISOelectric focusing (charge) in the first dimension and on molecular mass (DALtons) in the second. Notwithstanding the incredible explosion of 2-D techniques (for a review see ref. 20) in the vast majority of cases they are performed in thick (1-3 mm) gels, in vertical systems, and utilizing gel rods for the IEF run and gel slabs for the SDS-dimension. This also means that the 1-D cylinder has to be loaded at one extreme of the 2-D slab since, given the geometry of these apparatus (17-19), no other arrangements are possible. This introduces some fundamental differences between the present and previous methodologies.

*a) 1-D to 2-D gel contact.* Due to the fact that our IEF dimension is run in ultrathin, flat slabs, this optimizes the contact between the first and second dimension. In fact, in our case, the interfacing is done on a flat to flat contact, while, in all other reports, only a round to flat contact is achieved in the transfer between the two dimensions. It is true that a gel cylinder, when used for the IEF run, could be sliced longitudinally before the transfer, but this operation is very difficult, the more so the thinner and softer the gel. In our case, the gel transfer is also greatly improved by the fact that it is bound to a rather rigid plastic support (see Fig. 4). Once the 1-D gel is lowered in the trench of the 2-D matrix, into which it penetrates with a rather snug fit, no further operation is necessary. In the case of cylinder/slab contacts, the gel continuity has to be ensured by hot agarose solution poured in the contact region and let to gel. Since in most cases the 2-D technique is performed in polyacrylamide gels, the introduction of a different matrix with different structure and adsorptive properties does not seem to be ideal. The situation was even worse in the past, in early 2-D separations of amino acids and peptides performed in big sheets of filter paper (the classical fingerprinting) (21). In this case, the contact between the 1-D, electrophoretic strip, and the transverse chromatography sheet, was ensured by physically sewing together the two foils.

*b) Protein transfer from first to second dimension.* In our system, due to the optimal flat to flat contact, there is a complete removal of proteins from the IEF gel strip into the SDS-gel. This was shown by staining again the first dimension strip upon completion  of the SDS-run: no Coomassie Blue positive material was revealed in it. On the contrary, when gel cylinders are tested by the same procedure, the core always stains faintly in blue, indicating uncomplete removal of protein zones. This could also be the cause

of sample **streaking** , often found in 2-D techniques utilizing a round to flat contact. In our case, the total sample mobilization from the 1-D matrix could be due also to the better surface/volume ratio of our ultrathin layers, allowing for complete SDS diffusion within the IEF strip during the equilibration step.

*c) General advantages.* As a result of shorter diffusion pathways, staining and destaining of ultrathin gel layers are completed in a fraction of the time necessary for conventional, 2 mm thick gel slabs. Also the drying process is considerably shortened, and this is very important when the spots are to be revealed by autoradiography. Moreover, the extreme gel thinness ensures uniform dye penetration within the gel matrix and an even dye uptake by the protein zone. We can recall here that at a meeting in Hamburg, in 1976, Dr. Vesterberg summarized the problems of staining gel rods by showing the audience a  hollow cylinder cut out of packing foam: only the outer skin was stained, but the core was unstained (22). Another advantage of the horizontal system, when running native proteins in the second dimension, is the possibility of separating and detecting also cationic proteins (at the pH of the electrophoretic separation) which would be lost in the cathodic compartment in conventional, vertical systems (5). In this connection, the fact that we don't need to seal the 1-D – 2-D contact with hot agarose is a safeguard against denaturation of heat-labile proteins.

*d) Gel prestaining.* It is quite convenient, at the end of the IEF run, to fix and stain the gel slab. Once the IEF matrix has been stained, it can be stored for several days in the destaining solution, without loss or diffusion of proteins. This allows ample time for the 2-D run, without resorting to freezing of the gel slab, a process known to alter the gel size and structure. Size constancy, in our ultrathin gels, is also guaranteed by the plastic foil support. We have found, however that, when prestaining the IEF gel strip, there is a difference between the CBB R-250 and CBB G-250 forms. CBB G-250 does not give **streaking**  in the SDS-run, forming a sharp zone which moves with the leading boundary, while CBB R-250 often gives smears

and a  curtain of dye molecules with wavy fronts, which severely hamper the
resolution and reproducibility of the 2-D run. We don't have a ready explanation of this, except to think that the R-250 form, due to the absence of
two $-CH_3$ groups in the chromophore structure, could spontaneously form aromatic stacking, leading to aggregation via $\pi-\pi$ interactions.

## References


1.  Görg, A., Postel, W., Westermeier, R.: Z. Lebensm. Unters. Forsch. 164,
    160-162 (1977)

2.  Görg, A., Postel, W., Westermeier, R.: Anal. Biochem. 89, 60-70 (1978)

3.  Görg, A., Postel, W., Westermeier, R.: GIT Labor Medizin 2, 32-40 (1979)

4.  Görg, A., Postel, W., Westermeier, R.: Z. Lebensm. Unters. Forsch. 168
    25-28 (1979)

5.  Görg, A., Postel, W., Westermeier, R., Gianazza, E., Righetti, P.G.:
    J. Biochem. Biophys. Methods 3, 273-284 (1980)

6.  Smithies, O.: Biochem. J. 61, 629-641 (1955)

7.  Smithies, O.: Biochem. J. 71, 585-587 (1959)

8.  Ornstein, L.: Ann. N. Y. Acad. Sci. 121, 321-349 (1964)

9.  Gianazza, E., Righetti, P.G.: in Righetti, P.G., Van Oss, C.J., Wander-
    hoff, J.W. (eds.) Electrokinetic Separation Methods, pp. 293-311, El-
    sevier/North Holland, Amsterdam (1979)

10: Switzer III, R.C., Merril, C.R., Shifrin, S.: Anal. Biochem. 98, 231-
    237 (1979)

11. Noll, H.: in Campbell, P.N., Sargent, J.R. (eds.) Techniques in Protein
    Biosynthesis, pp. 101-180, Academic Press, London (1969)

12. Hjertén, S.: Arch. Biochem. Biophys. Suppl. 1, 147-151 (1962)

13. Margolis, J., Kenrick, G.K.: Anal. Biochem. 25, 347-362 (1968)

14. Jäckle, H.: Anal. Biochem. 98, 81-84 (1979)

15. Barrett, T., Gould, H.J.: Biochim. Biophys. Acta 294, 165-170 (1973)

16. McGilliwray, A.J., Rickwood, D.: Eur. J. Biochem. 41, 181-190 (1974)

17. O'Farrell, P.H.: J  Biol. Chem. 250, 4007-4021 (1975)

18. Anderson , N.G., Anderson, N.L.: Anal. Biochem. 85, 331-340 (1978)

19. Anderson, N.G., Anderson, N.L.: Anal. Biochem. 85, 341-354 (1978)

20. Righetti, P.G., Gianazza, E., Ek, K.: J. Chromatogr. 184, 415-456
    (1980)

21. Wieland, T.: in Bier, M. (ed.) Electrophoresis, pp. 492-530, vol. I,
    Academic Press, New York (1959)

22. Vesterberg, O., Hansén, L.: in Radola, B.J., Graesslin, D. (eds.) Elec-
    trofocusing & Isotachophoresis,pp.123-133, deGruyter, Berlin (1977)

# TWO-DIMENSIONAL ELECTROPHORESIS ON CELLULOSE ACETATE MEMBRANES


Tosifusa Toda, Toshiko Fujita and Mochihiko Ohashi
Department of Biochemistry, Tokyo Metropolitan Institute of Gerontology
35-2 Sakae-cho, Itabashi-ku, Tokyo 173, Japan


## Abstract


We have developed a novel method of two-dimensional electrophoresis on cellulose acetate for analysis of protein. In the first-dimensional step, proteins in an applied specimen are concentrated to a narrow zone on a cellulose acetate strip prior to separation. Then, the stacked proteins are subjected to zone electrophoresis. We called this first-dimesional technique "concentrating electrophoresis". In the second-dimensional step, the strip from the first dimension is put on the sheet of cellulose acetate absorbing carrier-ampholytes, perpendicularly to the second-dimensional direction. Then, isoelectric focusing is carried out on the membrane sheet. The whole procedure of the two-dimensional electrophoresis requires only 5 h to be performed. More than 75 spots can be observed on the two-dimensional pattern of human serum proteins by staining with Coomassie Brilliant Blue G-250. This technique is suitable to the analysis of proteins at low concentrations.


## Introduction

Cellulose acetate electrophoresis (1-3) has been widely used in clinical laboratories because of its simplicity and high performance. Abnormalities, which occur along with many diseases can be screened out through the electrophoregrams of their serum proteins (4,5).     Isozyme patterns also offer many informations supporting the clinical diagnoses (6-10). Cellulose acetate electrophoresis, however, has two major inferiorities to other gel electrophoresis. First, it is not applicable to samples at low concentrations. We can not expect any good resolution, when we must apply a large volume of sample to the supporting medium. Second, resolution of the pattern obtained by cellulose acetate electrophoresis is inferior to those on gels. Micro-heterogenity of protein such as sub-fractions of transferrin (11) is hardly analyzed by the conventional cellulose acetate electrophoresis. We expect this novel two-dimensional electrophoresis to solve those problems. In this paper, apparatus, procedure, supporting membrane and the performance are investigated and discussed.


Materials


Specimens.  The tissues were obtained by autopsy and stored at -20°C until analysis.    Each tissue was cut into small pieces with scissors, and homogenized with a glass homogenizer by adding 4 vols of 5 mM Tris-HCl buffer (pH 8.5).  The homogenates were centrifuged at 25,000 x g for 20 min at 0-4°C.  The supernatants were applied to the two-dimensional electrophoresis.   Urine collections from healthy individuals were dialyzed against 5 mM Tris-HCl buffer (pH 8.5) overnight before analysis.   A kit of pI-marker proteins (acetylated horse cytochrome c, cytochrome c', *R. rubrum* cytochrome c , horse met-myoglobin, sperm whale met-myoglobin, and horse cytochrome c) was from Oriental Yeast Co., Ltd. (Tokyo, Japan).

Electrolytes and reagents.    Electrolytes for the experiment are listed on Table 1.  Precipitate of barium carbonate is formed in the terminating electrolyte, but it does not affect the experiment.   Reagents for protein staining are listed on Table 2.  These reagents should be prepared just before use.

**Table 1  Electrolytes for the two-dimensional electrophoresis**

**In concentrating electrophoresis**

| | |
|---|---|
| Leading electrolyte | 62 mM Tris-HCl, pH 6.7 |
| Terminating electrolyte | 50 mM arginine-Ba(OH)$_2$, pH 11.7 |
| Passing electrolyte | 100 mM Tris-H$_3$PO$_4$, pH 7.5 |

**In isoelectric focusing**

| | |
|---|---|
| carrier-ampholyte | 5%(w/v) Ampholine, pH 3.5-10 / 10%(w/v) sucrose |
| anode electrolyte | 1%(v/v) H$_3$PO$_4$ / 30%(v/v) sucrose |
| cathode electrolyte | 1%(v/v) ethylenediamine |

**Table 2  Reagents for protein staining**

| | |
|---|---|
| Protein fixer | 20%(w/v) sulfosalicylic acid |
| Dye solution | 0.06%(w/v) Coomassie Brilliant Blue G-250 / 5%(w/v) trichloroacetic acid / 10%(w/v) sulfosalicylic acid |
| Destainner | 5%(v/v) acetic acid / 10%(v/v) ethanol |

**Chemicals.**    Titan III cellulose acetate plates and Electra HR Buffer were purchased from kabushiki-kaisha Helena Kenkyujyo (Urawa-shi, Japan).   Separax and Separax EF cellulose acetate membranes were products of Fuji Photo Film Co., Ltd. (Tokyo, Japan).   Visking dialysis tubing was from Union Carbide Corp. (Chicago, U.S.A.).   Glass fiber filter GF/B was obtained from Whatman Ltd. (Maidstone, England).   Filterpaper No.51A was from Toyo Roshi Kaisha, Ltd. (Tokyo, Japan).   Coomassie Brilliant Blue G-250 was from Nakarai Kagaku Yakuhin, Ltd. (Kyoto, Japan).   Tris-(hydroxy-methyl) aminomethane  was from Merck (Darmstadt, F.R.G).   Ampholine carrier-ampholytes were from LKB Productor AB (Stockholm, Sweden).   Dow Corning High Vacuum Grease (silicon lubricant) was from Toray Silicon Co., Ltd. (Tokyo, Japan).   Merckogel PVA2000 (polyvinyl alcohol) and other chemicals of analytical reagent grade were purchased from Wako Pure Chem. Ind., Ltd.(Osaka, Japan).

**Equipment.**    Apparatus for the two-dimnsional electrophoresis were designed by us, and made by Jookoo Co., Ltd. (Tokyo, Japan).   The diagrams of the apparatus are shown in Fig. 1 and 2.   The details of the apparatus will be described in Method. Regulated  DC power supply  Model 2000-200 Auto  was  purchased

274

from Koike Precision Instruments (Kawasaki-shi, Japan).
Peristaltic pump Model P-3 was from Pharmacia Fine Chemicals
(Uppsala, Sweden). Temperature of water circulation was
regulated by Coolnics Model CTR-220 (Komatsu-Yamato, Tokyo,
Japan). Densitometry was carried out by TLC Scanner CS-910
(Shimazu Co., Ltd., Tokyo, Japan).


Method


First dimension

Isotachophoretic concentration. Isotachophoretic concentration
and the following zone electrophoretic separation were carried
out on a cellulose acetate strip (76 x 10 mm). The strip,
which was wetted with the leading electrolyte (table 1), was
put on the cooling bed of the apparatus A. The bed was cooled
by water circulation regulated at 4°C. The anode and the cathode vessels were filled with the leading and the terminating electrolytes, respectively (Table 1). The filterpaper, which was wetted with the terminating electrolyte, was set on the suspension bridge on the lid (see Fig. 1). Excess electrolyte on the strip was removed by blotting with a soft tissue paper. In this time, the strip was not in contact with the filter-

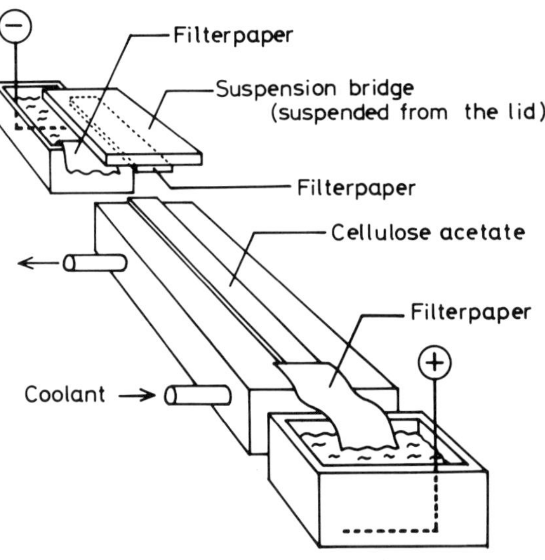


Fig. 1   Apparatus A for concentrating electrophoresis. Concentration and separation of protein are carried out on the cellulose acetate strip. The filterpaper on the suspension bridge touches the strip and the cathodic filterpaper by closing of the lid. The lid is not drawn.

paper on the suspension bridge. A constant voltage of 200 V was applied. Then, the lid was completely closed to obtain contact of the filterpaper with the strip and the cathodic filterpaper. Flow of electric current was initiated by the contact. Then, the strip was loaded with a specimen about 1-cm from the cathodic end, adjusted for each specimen. Serum was dropped on the strip using a micropipette. Tissue extract was preabsorbed onto a small piece of Separax EF. The piece was put on the strip. Dialyzed urine was applied to the strip by a drop-wise procedure using a peristaltic pump of controlled speed. After the final application, the electric current was continued for 30 min.

Zone-electrophoretic separation. After the concentration, the cathode vessel was replaced by another one containing the passing electrolyte as shown in Table 1. Both filterpapers on the suspension bridge and in the cathode vessel also replaced with others which were wetted with the passing electrolyte. Then, a constant current (0.4 mA) was applied for about 45 min.

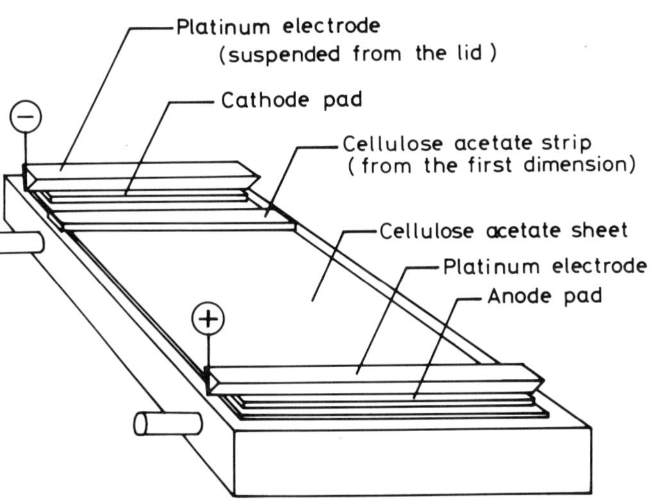


Fig.2 Apparatus B for isoelectric focusing. The strip from the first dimension is laid on the sheet. The lid is not drawn.

Second Dimension
Isoelectric focusing was carried out on the sheet of cellulose
acetate absorbing carrier-ampholyte (shown in Table 1). The
strip was laid on the sheet 1-cm from the cathodic end, perpendicularly to the second-dimensional direction. The pads of
glassfiber filter (59 x 4 mm), which were wetted with the anode electrolyte and with the cathode electrolyte were put on
anodic and cathodic ends of the sheet, respectively. By
closing the lid, the knife-edge-shaped platinum electrodes
gently touched the pads. Then, a constant current (1 mA) was
applied. Voltage rised gradually. When the voltage reached
800 V, the strip of cellolse acetate was removed and
isoelectric focusing was continued at the constant voltage (800
V) for 3 h in all.


Convensional cellulose acetate electrophoresis
It was carried out on Titan III strip (76 x 10 mm) equilibrated
with Helena Electra HR Buffer (Tris-Barbital-Sodium Barbital,
pH 8.8, ionic strength 0.075). Constant voltage of 100 V was
applied for 30 min. Proteins on the strip were stained with
0.5%(w/v) Ponceau S / 10%(w/v)sulfosalicylic acid. Excess dye
was removed by washing with 5%(v/v) acetic acid.


Staining of proteins on the two-dimensional map
After the two-dimensional electrophoresis, the membranes were
dipped into 20%(v/v) sulfosalicylic acid for 5 min in order to
fix proteins. Then the membranes were transfered into the Dye
solution (shown in Table 2). After 10-min dipping, excess dye
on the membrane was removed by washing with Destainner. The
membranes were then rinsed with water and dipped into drying
solution containing 1%(w/v) polyvinyl alcohol (Merckogel
PVA200) / 7.5%(v/v) glycerol / 5%(v/v) acetic acid for 5 min.
The membranes were sticked on a glass plate, and dried at room
temperature overnight.

Results and Discussion

1. Efficiency of the concentrating electrophoresis
The effiency of the electrophoresis was examined with diluted human serum and its original specimen. The electrophoregrams by the concentrating electrophoresis and by a conventional cellulose acetate electrophoresis were compared (Fig. 3). Diluted serum (*lefthand*) and the original specimen (*righthand*) were applied. By the concentrating electrophoresis, almost the same patterns were obtained from both the diluted and its original specimens. However, in the conventional method without concentrating process, the large volume of the diluted specimen impaired the resolution. This result shows that the concentrating electrophoresis is suitable to the analysis of proteins at low concentrations.

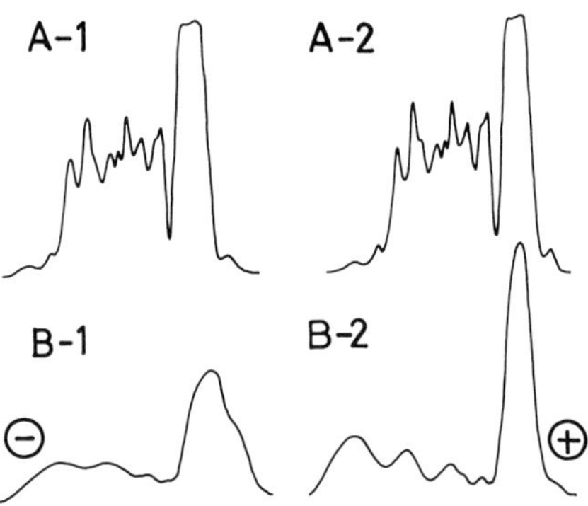


Fig.3 Electrophoregrams of human serum proteins. The upper two patterns are obtained by the concentrating electrophoresis, and the lower two are by a conventional cellulose acetate electrophoresis. In the experiments of A-1 and B-1, 10 μl aliquots of 20-fold diluted serum were applied, and in those of A-2 and B-2, 0.5 μl aliquots of the original specimen were applied by dropping with micropipette.

2. Stability of pH-gradient on the membrane
We examined the degree of cathodic drift of pH-gradient on several kinds of commercially available membranes. Isoelectric focusing was carried out as described in Method. Fig. 4 shows typically the pH-gradients on Separax EF membrane. The pH-gradient on Separax EF was more stable than that on Titan III (data not shown). On

278

Separax EF, the pH-gradient was formed within 1 h , and i¹
was kept stable for more than 3 h.  Therefore, in our standard method of the two-dimesional electrophoresis, Separax
EF was used as supporting medium for the second-dimensional
isoclcctric focusing.

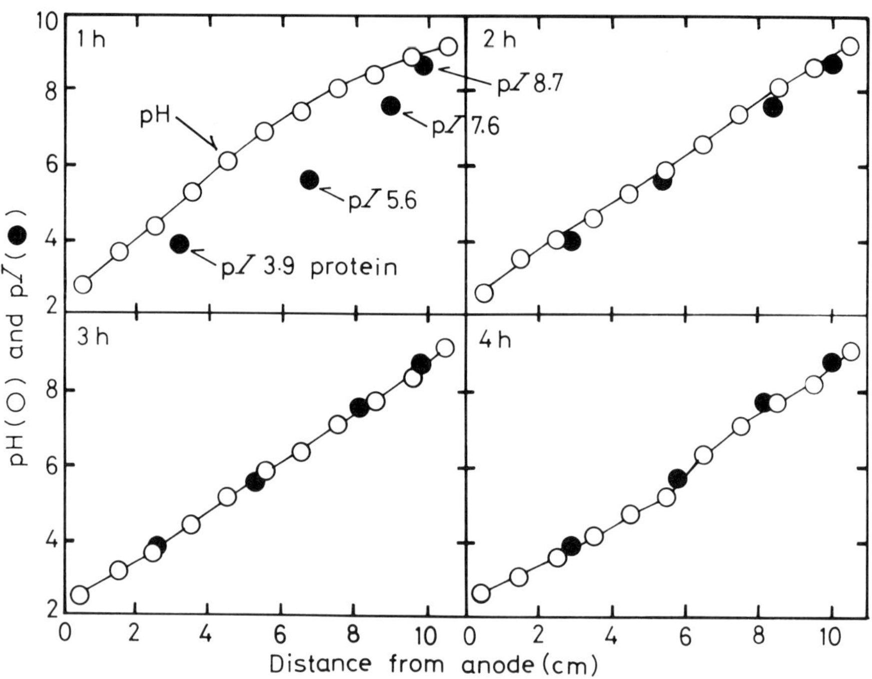


Fig. 4   The pH-gradients on Separax EF membrane, and focusing of
pI-marker proteins.   Isoelectric focusing was carried out as
described in Method.

3.  Two-dimensional  map  of  human  materials  at  various  con-
    centrations

    The two-dimensional mapping was applied to three kinds of
    human materials at wide variety of concentrations.   The
    resulting protein maps of serum, skeletal muscle extract
    and  urine  are  shown  in  Fig. 5.   By virtue of the con-
    centrating process at the first-dimensional step, well
    separated maps were obtained from any of those speci-
    mens.   More than 75 spots were observed on the map of human

serum protein. Almost all major spots of them have been assigned to some well-known serum proteins, using specific anti-sera. Further detailes of the result will be reported on a separate paper.

## Conclusion

Although the separation methods in either dimensions only depend on the electric property of protein, the electrophoregram by the concentrating electrophoresis was very different from that by iso-electric focusing. There were many components in the first-dimensional separation, besides they have the same isoelectric point. And, there were also many subfractions of different $pI$ values which can not be separated by the first-dimensional step. The novel technique,

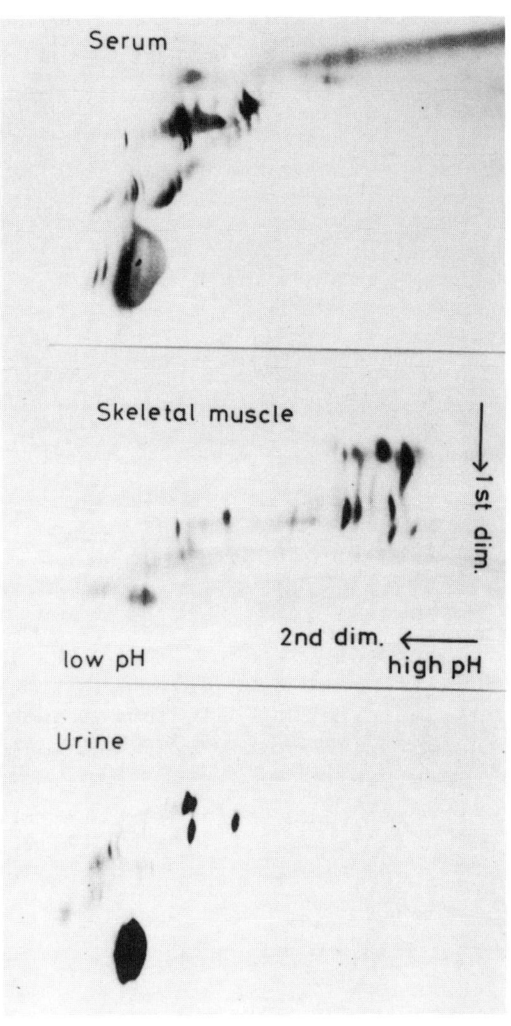



Fig. 5 Two-dimensional maps of human materials. Proteins were stained with Coomassie Brilliant Blue G-250.

280

studied here, has many remarkable properties. The high resolution, the high performance, and the wide applicability of the method are suitable for clinical surveys. By virtue of the concentrating process, urinary proteins at very low concentrations can be analyzed easily. In our preliminary study, the two-dimensional map of urine protein is likely to vary with progression of some marignant tumor and with aging process. We are now studing how the maps drift along with these processes.


References


1. Kohn, J.: Clin. Chem. Acta 2, 297-303 (1957).

2. Brackenridge, C. J.: Anal. Chem. 32, 1353-1356 (1960).

3. Mullan, F. A., Hancock, D. M., Neill, D. W.: Nature 194, 149-150 (1962).

4. Thomas, J. H., Powell, D. E. B.: Blood Disorders in the Elderly, pp. 189-209, John Wright & Sons Ltd., Bristol (1971).

5. Ritzmann, S. E., Daniels, J. C.: Serum Protein Abnormalities, Little Brawn and Co., Boston (1975).

6. Barnett, H.: J. Clin. Path. 17, 567-570 (1964).

7. Lainer, A. L., Skillen, A. W.: Isoenzymes in Biology and Medicine, pp. 146-169, Academic Press, New York (1968).

8. Vessell, E. S.: Isoenzymes (Markert, C. L., ed.), Vol. 2, pp. 1-28, Academic Press, New York (1957).

9. Posen, S., Neale, F. C., Birkett, D. J., Brudenellwoods, J.: Am. J. Clin. Path. 48, 81-86 (1967).

10. Smith, E. E., Rutenburg, A. M.; Nature, Lond. 197, 800-801 (1963).

11. Anderson, N. L., Anderson, N. G.: Biochem. Biophys. Res. Commun. 88, 258-265 (1979).

# ISOELECTRIC FOCUSING, SDS GEL GRADIENT ELECTROPHORESIS AND TWO-DIMENSIONAL ELECTROPHORESIS OF TROPICAL AND EUROPEAN LEGUME PROTEINS IN ULTRATHIN POLYACRYLAMIDE LAYERS.


Reiner Westermeier, Wilhelm Postel, Angelika Görg
Lehrstuhl für Allgemeine Lebensmitteltechnologie der Technischen Universität München, D - 8050 Freising-Weihenstephan, G.F.R.

Lehel Telek
Mayagüez Institute of Tropical Agriculture, P.O. Box 70,
Mayagüez, Puerto Rico 00708


## Abstract


Ultrathin-layer horizontal electrophoretic methods are applied on buffer extracted proteins of european and tropical legume seeds, which serve as protein sources in human nutrition. Isoelectric focusing of the seed proteins show great variations in protein patterns between the different species and also between the different varieties. Subsequent staining  specific for glycoproteins and trypsin inhibitors are usable methods for screening of antinutritional proteins. Ultrathin-layer sodium dodecyl-sulphate (SDS) gel gradient electrophoresis demonstrates the species specific molecular weight distributions of the legume seed proteins. Very high resolution of the proteins is obtained by the two-dimensional separations, with  isoelectric focusing in the first and SDS-gel gradient electrophoresis in the second dimension. The spots can be identified in an easy way, because the resulting pherograms are read as  coordinate systems with the proteins' isoelectric points in the x-axis and the molecular weights in the y-axis. In the case of two winged bean varieties the 2D-electrophoresis makes it clearly evident that  genetically  controlled  **proteins are only displaced by** different net charges within the same molecular sizes' files.

For the examples of pea and lentil proteins the 2D-electrophoresis patterns with IEF without urea and IEF under denaturing conditions in 9M urea (proteins extracted with 9M urea and 2 % $\beta$-mercaptoethanol) in the first dimension are compared. The different techniques show  very different spot patterns and    for these examples  the urea-technique is preferred.


Introduction


Many species of legumes (Leguminosae) are important protein sources in human nutrition in many parts of the world  and they are therefore subject of particular research interest (1, 2). Legumes are mostly consumed for their high protein content, but unfortunately  this  food component is  not easily digestible (3) and also contains toxic proteins, e.g. Lectins (4) and antinutrional proteins: (trypsin inhibitors)(5), which considerably reduce the protein efficiency ratio (PER). Different legume species and varieties have different protein patterns (6) and different toxic activities (7). Knowledge of the protein composition of different species and varieties can be useful for breeding legumes for high content and high quality proteins (8).

Isoelectric focusing of seed proteins and isoenzymes proved to be  a better   method for the differentiation of, for instance, soybean varieties (<u>Glycine</u> <u>max.</u> L.) (9) than disc electrophoresis (1o). Ultrathin-layer isoelectric focusing and electrophoresis in 24o µm thin polyacrylamide gels provide higher resolution of the protein patterns and markedly sharper bands than conventionally used methods in 1-3 mm thick gels (11, 12, 13). Further improvements in protein resolution were obtained by SDS gel gradient electrophoresis and two-dimensional electrophoresis in ultrathin polyacrylamide layers (14).

In the following short report we demonstrate the application

of ultrathin-layer electrophoresis methods on 8 different
european and tropical legume species and each 2 varieties of
2 species.


Materials and Methods


Plant material: White bean (Phaseolus vulgaris), runner bean
(Phaseolus coccineus), broad bean (Vicia faba), soya bean
(Glycine max.), pea (Pisum sativum), and lentil (Lens culinaris) from Europe, black bean (Phaseolus vulgaris) from Puerto
Rico, lablab (horse gram) (Dolichos lablab) from Israel, and
two varieties of winged bean (Psophocarpus tetragonolobus)
from Indonesia and New Guinea.


Equipment: Multiphor chamber LKB 2117, power supply LKB 21o3,
cooling apparatus K2R mgw Lauda.


Methods:
1. Sample solutions: o.1 g of ground seeds were extracted with
   o.5 ml 25 mM tris-gly buffer, pH = 8.6, (15) and 2o μl of
   each sample on Whatman paper 3 MM applied on isoelectric
   focusing. For SDS electrophoresis 2oo μl of SDS sample buffer (14) was added to each 5o μl of protein solution and heated with 1oo °C for 3 min.
2. Preparation of ultrathin-layer gels: (11, 13, 14, 15)
3. U.T.L. isoelectric focusing: 3o min 4oo V, 3o min 1ooo V,
   3o min 15oo V.
4. U.T.L. SDS gel gradient electrophoresis: 12o min max. 5o mA,
   max. 3o W, max. 5oo V.
5. 2D-electrophoresis: (14, 15)
6. Staining techniques:
   Proteins: Coomassie brilliant blue (11)
   Glycoproteins: (PAS) (17)
   Trypsin inhibitors: (18)

Results

The isoelectric focusing patterns of the seed proteins (fig. 1)
of 8 legume species and each 2 varieties of 2 species show a
great variation of their isoelectric points and, thus, provide
a good basis for the differentiation of species and also varieties.

Horizontal ultrathin-layer SDS gel gradient electrophoresis of
the same 1o samples (fig. 2) demonstrates the different molecular weight distributions of the different species and similar spectra of the same varieties of the legume seed proteins.

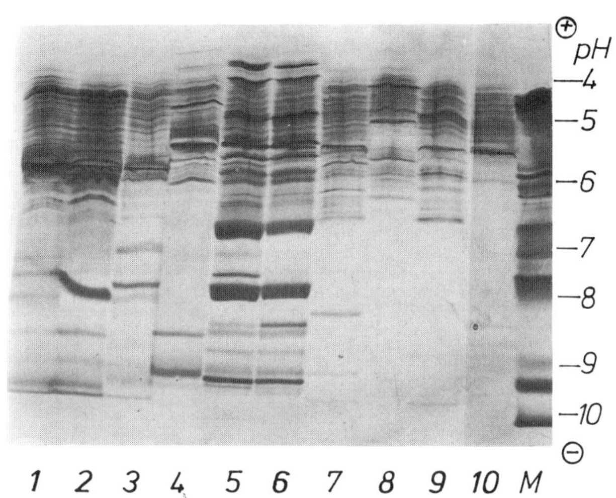


Fig. 1: U.T.L. isoelectric focusing of 1) white bean, 2) black
bean, 3) runner bean, 4) lablab, 5) winged bean from Indonesia, 6) winged bean from New Guinea, 7) broad bean, 8) soybean,
9) pea, 1o) lentil. PAA gel: 4 % T, 6 % C; 2 % LKB Ampholine
pH = 3.5 - 9.5; separation distance: 1o cm.

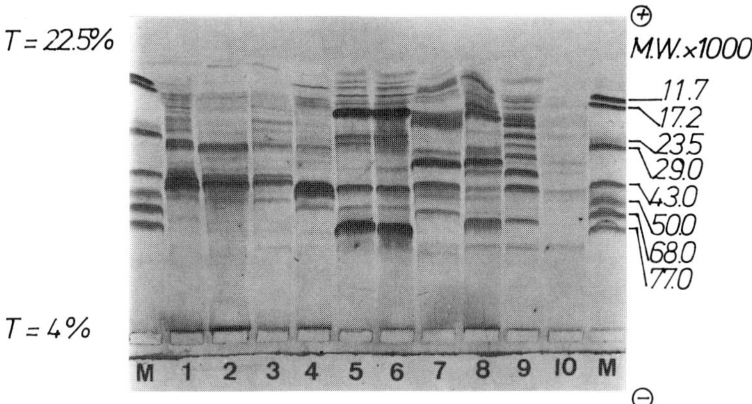



Fig. 2: Horizontal U.T.L. SDS gel gradient electrophoresis.
For samples see fig. 1.. PAG gradient 4-22.5 %, 2.5 % C.

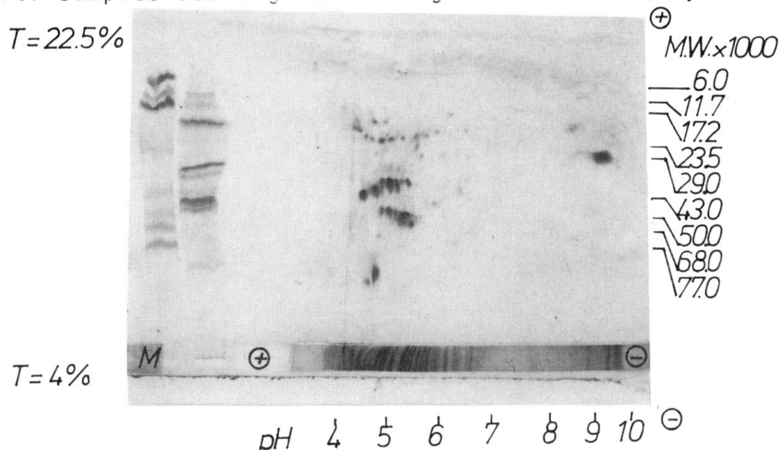



Fig. 3: Horizontal U.T.L. high-resolution 2D-electrophoresis
(urea IEF x SDS gel gradient PAGE) of lentil seed proteins
extracted with 9 M urea and 2 % β-mercaptoethanol.

The high resolution of a 2D-electrophoresis combining urea
isoelectric focusing and SDS gel gradient electrophoresis is
shown in figure 3 for the example of lentil seed proteins.

The <u>2D-separations</u> (fig.4 ) of two different winged beans
demonstrate that the genetic different seed proteins have the
same molecular sizes, but different isoelectric points.

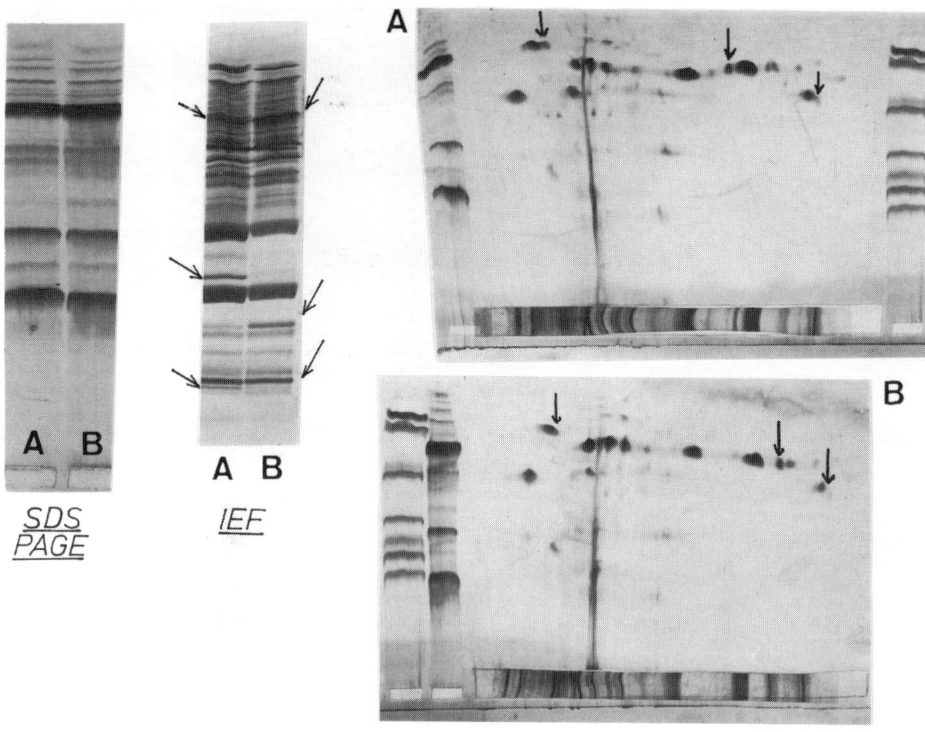


<u>Fig.4</u> : Horizontal U.T.L. 2D-electrophoresis (IEF x SDS gel
gradient PAGE) of 2 different varieties of winged beans.

# References


1. Aykroyd, W.R., Doughty, J.: Legumes in Human Nutrition, FAO Nutritional Studies 19, Rome (1974).

2. Patwardhan, V.N.: Amer. J. Clin. Nutr. 11, 12-3o (1962).

3. Bliss, F.A., Hall, T.C.: Cereal Foods World 22, 1o6-113 (1977).

4. Pusztai, A., Stewart, J.C.: Biochim. Biophys. Acta 536, 38 - 49 (1978).

5. Birk, Y.: Meth. Enzymol. 45, 697-722 (1976).

6. Fox, D.J., Thurman, D.A., Boulter, D.: Phytochemistry 3, 417-419 (1964)

7. Klozova, E., Turkova, E.: Biol. Plant 2o, 129-134 (1978).

8. Boulter, D.: Basic Life Sci. 8 (Genet. Diversity Plants), 387-396 (1977)

9. Postel, W., Westermeier, R., Görg, A.: Lebensm.-Wiss. u. -Technol. 11, 2o2-2o5 (1978).

1o. Larsen, A., Cadwell, B.E.: Crop. Science 9, 385-387 (1969).

11. Görg, A., Postel, W., Westermeier, R.: Anal. Biochem. 89, 6o-7o (1978).

12. Görg, A., Postel, W., Westermeier, R.: in Electrophoresis'79 B.J. Radola ed., Walter de Gruyter, Berlin, New York (198o).

13. Görg, A., Postel, W., Westermeier, R.: Science Tools 27, 14-18 (198o).

14. Görg, A., Postel, W., Westermeier, R., Gianazza, E., Righetti, P.G.: J. Biochem. Biophys. Methods 3, 273-284 (198o).

15. Görg, A., Postel, W., Westermeier, R., Gianazza, E., Righetti, P.G.: in Electrophoresis'81, R.C. Allen and P. Arnaud ed., Walter de Gruyter, Berlin, New York (1981).

16. Lowry, K.L., Caton, J.E., Foard, D.E.: J. Agr. Food Chem. 22, 1o43-1o45 (1974).

17. Kapitany, R.A., Zebrowski, E.J.: Anal. Biochem. 56, 361-369 (1973)

18. Uriel, J., Berges, J.: Nature (London) 218, 578-579 (1968)

# HIGH RESOLUTION POLYPEPTIDE MAPPING OF HUMAN HAIR ROOTS FROM HEALTHY INDIVIDUALS AND PATIENTS WITH GENETIC DEFECTS


Surjit Singh, Ingrid Willers, H. Werner Goedde
Institute of Human Genetics, University of Hamburg, Hamburg, F. R. G.


Joachim Klose
Institute of Toxicology and Embryonal Pharmacology, Department of Genetics, Free University of Berlin, Berlin, F. R. G.



## Introduction

Hair root cells constitute a readily available material for the detection of some genetic defects of metabolic origin and for studies of heterozygosity (1-6). The advantage of easy collection without requirement of sterile conditions for the biopsy and time consuming cell culture make this approach quite attractive. Uptil now most of the genetic studies with this material have dealt with the monitoring of individual enzyme proteins. In continuation of our studies on protein mapping of human cultured fibroblasts in health and genetic disease (7-11), we are now studying the polypeptide profiles of water soluble proteins from hair root cells. In this report we present the results pertaining to our collection of healthy controls and some patients with muscle dystrophy duchenne and hereditary ataxia. With our technique we have been able to monitor a large number of structural and functional proteins in a single polyacrylamide gel. As the methods of exact quantitative evaluation of the gels are being standardized, only the data processed for qualitative differences by visual and semiquantitative evaluation procedure (7) is being presented here.

Material and Methods

This examination comprises 12 healthy individuals, 4 patients
with duchenne muscular dystrophy and 3 patients with hereditary ataxia (2 of them belong to Friedreich's autosomal
recessive type and one patient with Marie's dominantly inherited type.

Preparation of hair root cell lysates. 100 hairs with intact
follicles were plucked from the head. The roots were cut
right behind the hair bulb, mixed with $H_2O$ and frozen and
thawed 3 times. The supernatant obtained after centrifugation
at 31000 g for 10 min. was used for protein mapping. The
protein content of this sample was about 100 to 120 $\mu$g.

Protein mapping. The sample was made upto 9 M urea, 5 %
mercaptoethanol and 2 % Ampholine pH 5 - 7 mixed 1 : 1 with
a Sephadex G 200 superfine solution swollen in 20 % sucrose
containing 9 M urea, 5 % mercaptoethanol and 1 % Ampholine.
60 $\mu$l sample/Sephadex solution were put on top of the focusing gel. Isoelectric focusing and subsequent SDS electrophoresis were carried out by the modification of the method of
Klose (12) exactly as described in detail before (7, 10).

Evaluation of the protein pattern. The absence or presence of
a new protein spot or change in the mobility of a spot
distinguished on a lightbox has been referred as qualitative
difference, while changes in the stain density has been
referred as quantitative variation.
The spots of some gels were semiquantitatively evaluated by
employing the photographic procedure explained previously
(7). For a precise quantitative analysis of the spots a
system composed of a video camera and a computer (13, 14),
which is currently being standardized, is proposed.

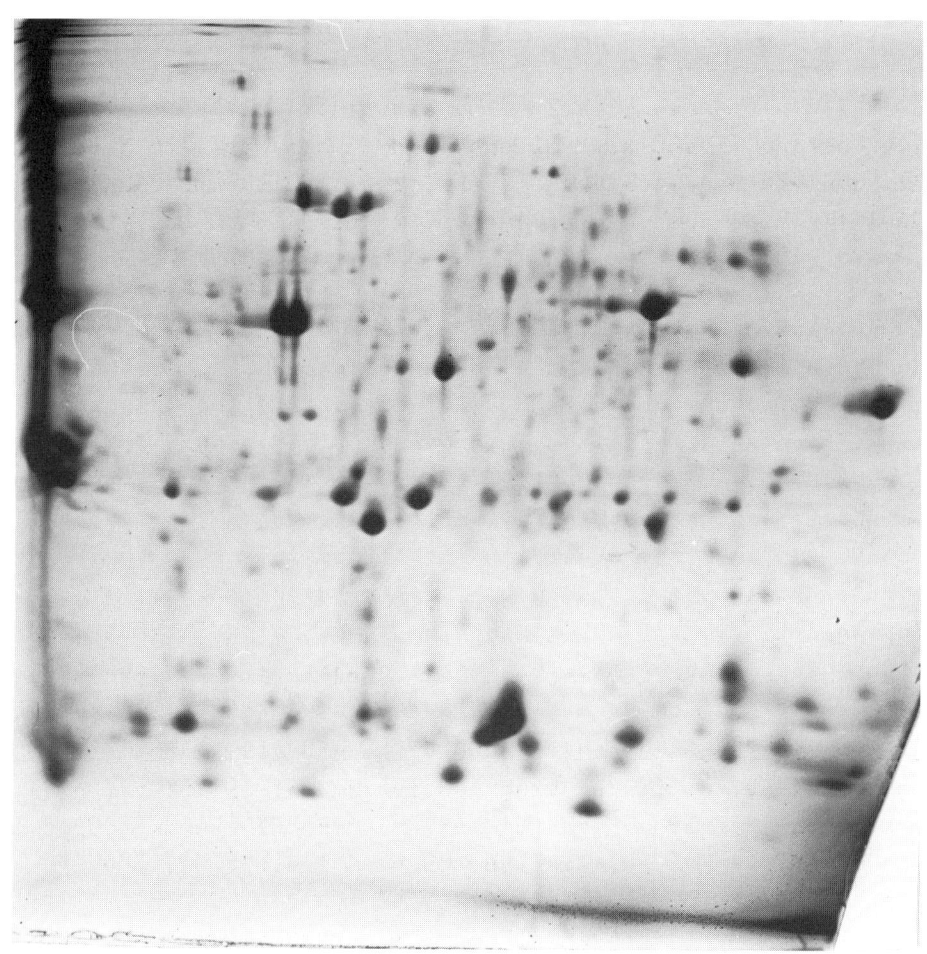

Results and Discussion

<u>Protein pattern of hair root lysates</u>. With this technique,
very well separated protein pattern have been obtained from
hair root lysates. The protein pattern of the same sample
running simultaneously in twin electrophoretic cells or on
different days were highly reproducible. About 250 protein
spots can easily be distinguished on the gel plate. With a
higher sample loading a more complex pattern can be obtained.
Fig. 1 above presents a typical pattern of this tissue. The

hair root proteins are distributed all over the gel surface
like the fibroblast proteins (7), yet they differ markedly
from these proteins in the placement and intensity of the
individual spots.
The interindividual variation with respect to qualitative
differences in the control group has been rather low (about
1 %) in this tissue as compared to the cultured fibroblasts
(about 4 % - see reference 9). No dramatic qualitative
differences have been observed in the protein pattern from
the patients with the genetic defects as compared to the
controls. The reasons for the low variability encountered in
this tissue seem to be the same as described before for the
fibroblasts: In this case too, the majority of the protein
spots represent structural proteins, which show less heterogeneity (15). On the other hand it is possible that these
genetic defects do not show qualitative differences in the
protein pattern, but may be showing specific quantitative
differences. This particular aspect has still to be studied
as the methods for the video aided computerized evaluation
of the fine quantitative differences in the protein spot
intensity are still under standardization by us. Nevertheless,
more patients with genetic defects of known etiology should
be studied for the qualitative and quantitative differences
to investigate the utility and limits of this approach. In
order to further sensitize this method we propose to modify
this method by marking the follicle proteins with $^{35}$S- Methionine   (16) and $^{14}$C-aminoacids and monitoring the protein
spots by subsequent autoradiography.

## Acknowledgements

We thank the Stiftung Volkswagenwerk for financial assistance and Dr. Doris Meier-Tackmann for preparation of the hair
root lysates, Mrs. Ilona Gerner, Bernadette Ressler and Gu-
drun Knaack for excellent technical assistance.

# References


1.  Gartler, S.M., Scott, R.C., Goldstein, J.L., Campbell,B., Sparkes, R.: Science $\underline{172}$, 572-574 (1971).

2.  Nwokoro, N., Neufeld, E.B.: Am. J. Hum. Genet. $\underline{31}$, 42-49 (1979).

3.  de Bruyn, C.H.M.M., Vermorken, A.J.M., Oei, T.L., Geerts, S.J.: Brit. J. Dermatol. $\underline{101}$, 111-113 (1979).

4.  Hösli, P., Schneck, L., Amsterdam, D., Volk, B.W.: Lancet $\underline{I}$, 285 (1977).

5.  Goedde, H.W., Agarwal, D.P., Harada, S.: Enzyme $\underline{25}$, 281-286 (1980).

6.  Grimm, I., Wienker, T.F., Ropers, H.H.: Hum. Genet. $\underline{32}$, 329-334 (1978).

7.  Singh, S., Klose, J., Willers, I., Goedde, H.W.: Electrophoresis '78, ed. Catsimpoolas, N. Elsevier North Holland Inc. 297-304 (1978).

8.  Singh, S., Willers, I., Klose, J., Goedde, H.W.: Fresenius Z. Anal. Chem. $\underline{301}$, 193-194 (1980).

9.  Singh, S., Willers, I., Goedde, H.W., Klose, J.: J. Inher. Metab. Dis. $\underline{4}$, in press (1981).

10. Singh, S., Willers, I., Klose, J., Goedde, H.W.: Biochem. Genet.,in press (1981).

11. Willers, I., Singh, S., Goedde, H.W., Klose, J.: Clin. Genet., in press (1981).

12. Klose, J.: Hum. Genet. $\underline{26}$, 231-243 (1975).

13. Garells, J.I.: J. Biol. Chem. $\underline{254}$, 7961-7979 (1979).

14. Klose, J., Nowak, J., Kade, W.: Radola, B.J. (ed.) Electrophoresis '79, Walter de Gruyter & Co., Berlin, New York (1980).

15. Walton, K.E., Steyer, D., Grunstein, E.I.: J. Biol. Chem. $\underline{254}$, 7951-7960 (1979).

16. Vermorken, A.J.M., Wertings, P.J.J.M., Bloemendal, H.: Molec. Biol. Rep. $\underline{4}$, 211-216 (1978).

ANALYSIS OF CULTURED SKIN FIBROBLASTS FROM PATIENTS WITH DUCHENNE MUSCULAR
DYSTROPHY USING ELECTROPHORETIC TECHNIQUES


A.H.M. Burghes, M.J. Dunn, H.E. Statham and V. Dubowitz
Jerry Lewis Muscle Research Centre, Department of Paediatrics and Neonatal
Medicine, Hammersmith Hospital, London W12 OHS, U.K.


Introduction

The molecular abnormality responsible for the progressive degeneration of
skeletal muscle characteristic of the x-linked disease Duchenne muscular
dystrophy (DMD) has not been identified.  The currently popular theory is
that the genetic defect in DMD affects an enzyme or structural protein
which results in an abnormal structure and function of the sarcolemma (for
reviews see 1,2).  As biochemical abnormalities of genetic disorders may be
manifest in cells other than the symptomatic tissue, considerable effort
has been devoted to investigations of cells other than muscle cells (3).

Skin fibroblasts grown _in vitro_ represent an attractive source of material
for biochemical investigations in DMD due to the facility with which skin
biopsy samples can be obtained and the ease of _in vitro_ culture of the cells.
In the present study we have employed gel electrophoresis in the presence
of sodium dodecyl sulphate (SDS), polyacrylamide gel isoelectric focusing
(IEF) and two-dimensional gel electrophoresis to compare the proteins of
skin fibroblasts in an attempt to identify differences between cultures
derived from normal and DMD individuals.  The principal finding from these
studies is that there is a difference in the protein of 230,000 molecular
weight which is associated with the outer surface of skin fibroblast cells
derived from patients with DMD.  We have tentatively identified this protein
as fibronectin.

## SDS polyacrylamide gel electrophoresis

In preliminary studies samples of fibroblasts were solubilised in SDS
sample buffer and electrophoresed in discontinuous 10% polyacrylamide gels
containing SDS by the method of Laemmali (4).  However, a single gel
concentration cannot optimally separate the complex mixture of macromolecules, with a wide range of molecular weights, present in skin fibroblasts.  In theory there is an optimal gel concentration which will provide
optimal separation of any two polypeptides.  Therefore improved resolution
of a complex protein mixture can be obtained using gels with a continuously
variable gel concentration (5,6).  Using SDS-containing slab gels (1.4mm
thick) with a linear 5 to 20% acrylamide gradient greatly improved
resolution was obtained.  Gels were stained with Coomassie brilliant blue
R250 and relative intensities of the bands were determined using a photometric gel scanner.  A typical stained slab gradient gel loaded with 100µg
samples of SDS-solubilised normal and DMD fibroblasts is shown in Figure 1.

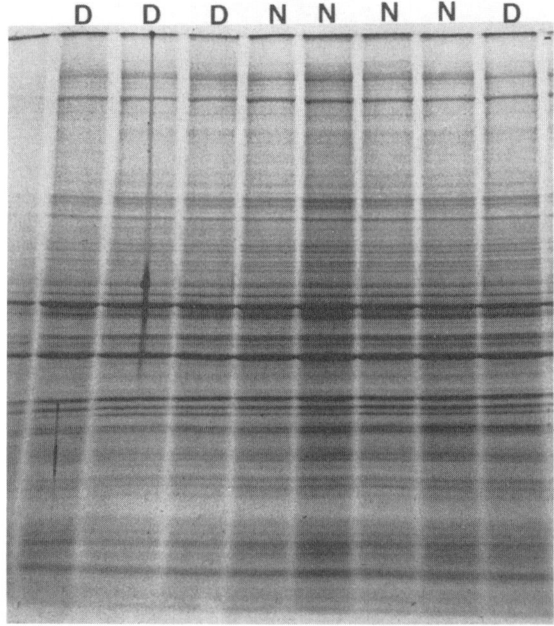


Figure 1 : 5 to 20% gradient polyacrylamide slab gel loaded with 100µg
samples of SDS-solubilised normal (N) and DMD (D) fibroblasts.

Both visual inspection and densitometric scanning of a series of such gels
has failed to reveal any qualitative or quantitative differences in the
protein profiles of normal and DMD fibroblasts.

Staining with Coomassie blue lacks the sensitivity to detect many proteins
that are present at low concentrations in fibroblast preparations. Recently
a highly sensitive silver staining method has been described (7,8), and we
are carrying out a study of skin fibroblast proteins from normal and DMD
individuals using this technique.  A 5 to 20% gradient gel slab (1.4mm
thick) loaded with **50**µg of SDS-solubilised proteins from normal and DMD
fibroblasts and stained by the silver procedure is shown in Figure 2.

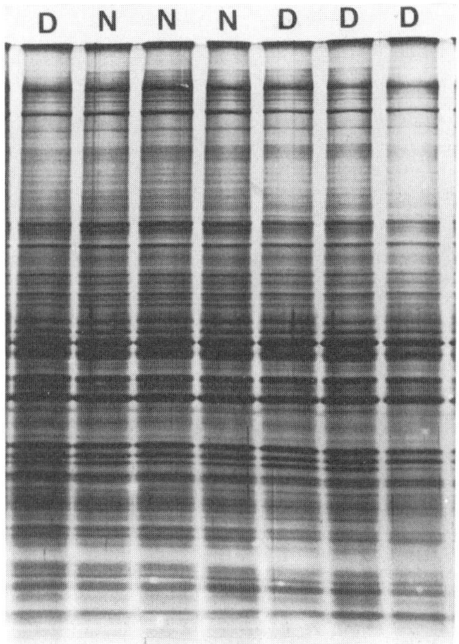


Figure 2 : 5 to 20% gradient gel slab stained by the silver procedure.
**50**µg samples of normal (N) and DMD (D) fibroblast proteins.  The gel was
fixed in 20% TCA, then washed 3x30min 10% ethanol and then put into the
paraformaldehyde step of the silver staining procedure.

Despite a report to the contrary (9), we have found that background staining
does not increase with increasing gel concentration.  However, development
of the stain was found to proceed more rapidly at low acrylamide
concentrations.  Standard photographic reducers can be used subsequent to

298

staining to control stain intensity.

In another attempt to improve detection sensitivity of the gels, radioactive
labelling methods were used (5).  Fibroblasts were grown in 15mm diameter
multi-well tissue culture plates and labelled for 24hr with 50μCi [35S]-
methionine in 0.5ml methionine-free MEM containing 10% newborn calf serum.
After labelling the cultures were washed with PBS, solubilised in SDS sample
buffer and samples containing $3 \times 10^5$ cpm electrophoresed in 5 to 20% gradient
gel slabs (1.4mm thick).  Molecular weight markers labelled with [14C]
(Radiochemical Centre) were electrophoresed on the same gels.  Subsequent
to electrophoresis the gels were dried under vacuum and the polypeptide
profiles revealed by autoradiography (10-day exposure).  Typical autoradiographic gel patterns of normal and DMD fibroblast proteins are shown in
Figure 3a and Figure 3b respectively.

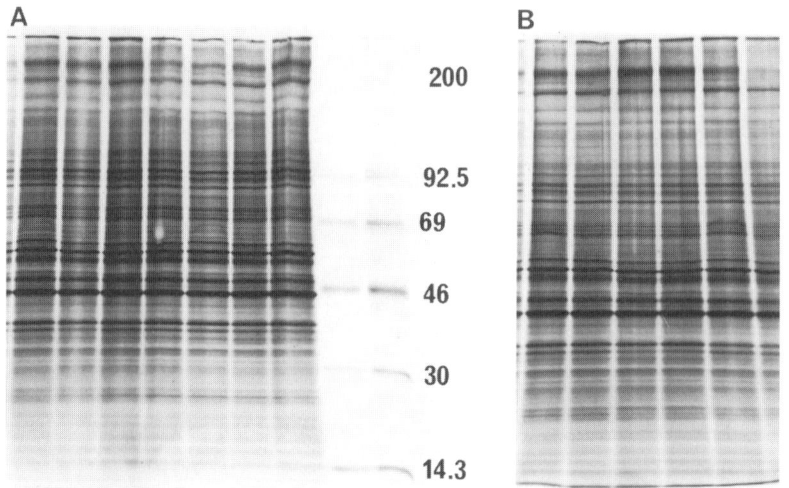


Figure 3 : Autoradiographs of gradient gel slabs of [35S]-methionine labelled
normal (a) and DMD (b) fibroblast proteins.  The molecular weights (x10⁻3)
of [14C]-labelled standards are indicated.

Although there appears to be some individual variation in the profiles,
careful comparison has failed to reveal any consistent differences related
to the disease state.

Samples labelled with [35S]-methionine were also electrophoresed on gradient

gel rods, (cast in siliconised tubes) which were subsequently sliced into
0.5mm segments.  The protein in each slice was eluted for 18hrs with 3ml
of scintillation cocktail (870ml Lipoluma (LKB), 100mls NCS tissue
solubiliser (Amersham Corporation), 20ml H₂0, 10mls hydroxide of Hyamine
10-X (Packard) ) and counted in a scintillation counter.  A typical plot
from such an experiment is shown in Figure 4.

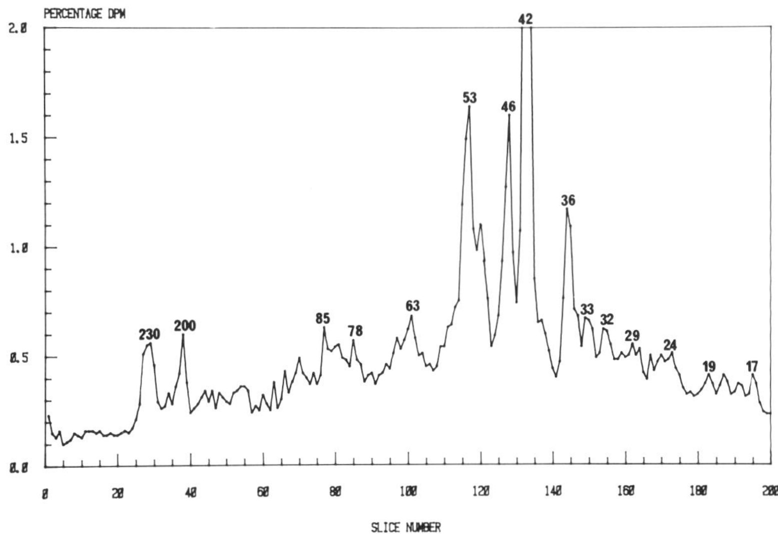


Figure 4 : Plot of [35S]-methionine labelled fibroblast proteins derived
from a gel slicing experiment.  The molecular weights (x10⁻3) of the major
peaks are indicated.


Analysis of such plots for a series of normal and DMD skin fibroblast samples
has failed to detect any significant difference between the normal and DMD
profiles.


In an attempt to improve the sensitivity of comparisons between gel
electrophoretic profiles of normal and DMD samples we have adopted a dual-
labelling method (10).  One type of culture (e.g."DMD") was labelled with
[3H]-leucine while the other (e.g."Normal") was labelled with [14C]-leucine.
After labelling the cultures were solubilised with SDS sample buffer,
pooled and subjected to electrophoresis on 5 to 20% gradient gels.  The
gels were cut into 0.5mm segments and dual-label counted in a beta-counter
(to <5% counting error).  Values for the dpm of [3H] and [14C] in each slice

were normalised and the $[^3H]/[^{14}C]$ ratio calculated.  A typical plot of normalised dpm values and $[^3H]/[^{14}C]$ ratios is shown in Figure 5a and 5b.

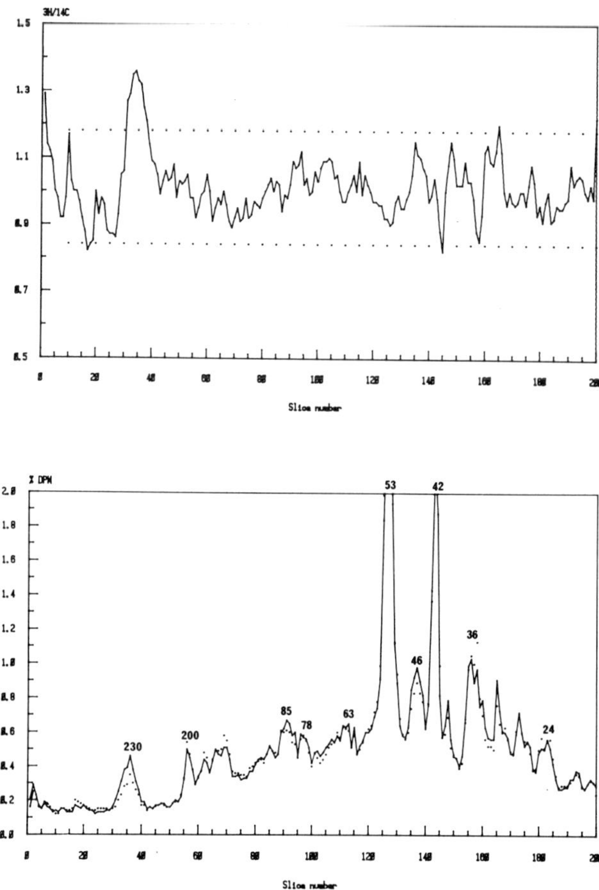

Figure 5 : Plots of normalised dpm values (a) and $[^3H]/[^{14}C]$ ratios (b) from a dual labelling experiment.  The normal culture was labelled with $[^{14}C]$-leucine and the DMD culture with $[^3H]$-leucine.  The molecular weights (x10$^{-3}$) of the major peaks are indicated.

The presence in a slice of a $[^3H]/[^{14}C]$ ratio deviating significantly from the mean value has been used as a criteria for assessing molecular abnormalities.  Although this study is still in progress, our initial results suggest that the major consistent difference in DMD cultures was an increase in the amount of labelling of a polypeptide of molecular weight

230,000.

The proteins exposed at the surfaces of skin fibroblasts from normal and
DMD individuals have been investigated by iodination of the proteins with
[125I]. The solid-phase reagent 1,3,4,6-tetrachloro-3α,6α-diphenylglycouril
(Iodogen) (11) was used to iodinate external proteins as we have found this
reagent to cause less cell damage and to produce a higher level of labelling
than did lactoperoxidase-catalysed iodination. Glass coverslips were coated
with 50μg Iodogen, and these were placed in contact with cell monolayers in
35mm petri dishes in the presence of 100μl PBS and 500μCi [125I] for 15mins.
Subsequent to labelling the cultures were solubilised and applied to gradient
SDS gels. After electrophoresis, slab gels (0.7mm thick) were dried and
exposed to x-ray film (5 days), whereas rod gels were sliced into 0.5mm
segments and counted in a gamma counter (to 2% counting error). Both plots
(Figure 6) of gel slicing experiments and autoradiographs (Figure 7) reveal
that a series of polypeptides were iodinated.

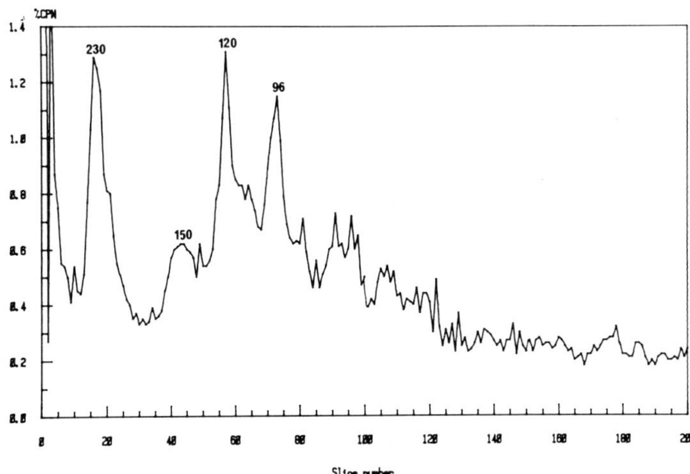


Figure 6 : Typical plot of [125I]-labelled cell surface proteins of skin
fibroblasts derived from a gel slicing experiment. The original sample
contained 9x10$^5$ cpm.

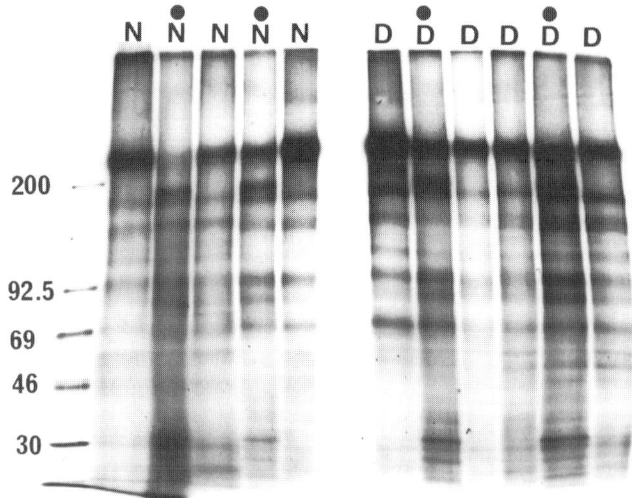



Figure 7 : Autoradiograph of a slab gel of [125I]-labelled cell surface
proteins of normal (N) and DMD (D) fibroblasts.  The molecular weights
(x10⁻³) of [14C]-labelled standard proteins are indicated.  Samples
contained 3x10⁵ cpm. [● + DNAase].

The proteins which were iodinated appear to be qualitatively identical in
normal and DMD samples.  However, our preliminary results suggest that a
polypeptide of molecular weight 230,000 is consistently more heavily
iodinated in cultures of skin fibroblasts from individuals with DMD.

Our studies using SDS gel electrophoresis suggest that there is a
difference in the protein of molecular weight 230,000 to 250,000 in skin
fibroblasts from individuals with DMD.  We are tentatively identifying
this protein as fibronectin.  Studies of total proteins indicate that there
is more of this protein associated with DMD cells.  The increased iodination
of this protein by surface-labelling procedures, however, could be due
either to an increase in the amount of this protein present at the cell
surface or to a structural reorganisation resulting in an altered
accessibility of reactive groups.

Polyacrylamide gel isoelectric focusing

As an alternative approach to the separation of proteins on the basis of molecular weight by SDS gel electrophoresis, we have utilized the high resolving power of IEF to separate skin fibroblast proteins on the basis of charge. The development of IEF gels containing urea and non-ionic detergents has offered the advantages of solubility and disaggregation of proteins to their constituent polypeptides (12), elimination of conformational varieties, and the possibility of resolving neutral mutations (13). In addition the use of ultra-thin gel slabs at high feild strengths (14) combines high resolution, speed, versatility and reagent economy with simplicity of operation. C4T4 gels (0.7mm thick) were prepared on glass plates pretreated with methacryloxypropyl-trimethoxy-silane (14). The gels contained 8M urea, 2% (w/v) N-P40 and 4% (w/v)Pharmalyte 3-10. The anolyte solution was 0.04M aspartic acid in 8M urea. The catholyte was 1M NaOH, 0.25M Ca(OH)$_2$, 0.025M lysine and 0.025M arginine (spun to remove ppt). The gels were placed on a flat-bed apparatus cooled to 15°C and prerun at 400V for 30mins. Fibroblasts labelled with [35S]-methionine were solubilised in 9.2M urea, 3% (w/v) N-P40, 5% (v/v) β-mercaptoethanol, 4% (w/v) Pharmalyte 3-10 and 0.1M arginine. The samples (3x10$^5$ cpm) were applied and the gel run at 15W (1400V limiting) for 5hr 40min at 10°C. The run was carried out in an atmosphere of CO$_2$-free nitrogen. Equilibrium was assessed by migration of samples from both anode and cathode. The gels were soaked for 2x30min in 45% methanol-10% acetic acid, then for 15min in 50% methanol, and dried in an oven at 60°C. The protein bands were revealed by autoradiography (1-day exposure).

In order to compare isoelectric points measured in 8M urea with those in the absence of urea, correction has to be made for the effect of urea on the measured pH because urea appreciably decreases the activity coefficient of hydrogen ions (15,16). This effect is well illustrated by the pH gradient profile, as measured using a surface electrode, shown in Figure 8. We have attempted to correct for this effect using a series of marker proteins of known pI (Pharmacia Ltd.) which indicate (Figure 8) that the gradient extended from pH 3.5 at the anode edge to above pH 9.5. The pH gradients which were obtained were found to be quite stable in agreement with other reports of increased gradient stability in the presence of 8M urea (17).

304

Autoradiographs of typical gels comparing the profiles of $[^{35}S]$-labelled
proteins from normal and DMD fibroblast cultures are shown in Figure 9.
These illustrate well the high resolving power of the technique.

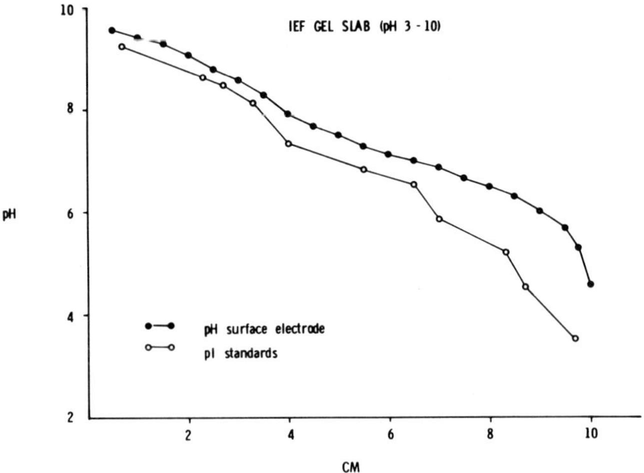


Figure 8 : pH gradient profiles of an IEF gel, containing 8M urea and 2%
N-P40, as measured with a pH surface electrode and by the use of pI marker
proteins (pH 3-10 kit, Pharmacia Ltd.).

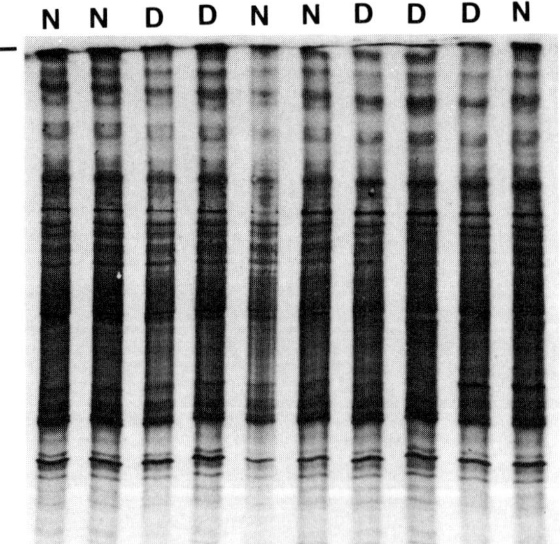


Figure 9 : Autoradiograph of IEF gel of normal (N) and DMD (D) skin fibroblast proteins labelled with 35S -methionine.

The majority of proteins were located in the pH 5-8 region of the gels, a distribution similar to that reported in a survey of pI values of a large number of proteins (18). Although some individual variation has been noted in the profiles we have so far been unable to detect any consistent differences between normal and DMD samples. In order to maximise the sensitivity of such comparisons it will be necessary to resort to dual-labelling techniques as described for SDS gel electrophoresis.

<u>Two-dimensional gel electrophoresis</u>

The resolution of complex protein mixtures in order to detect qualitative or quantitative changes can be greatly increased by the use of two-dimensional techniques. The optimum strategy for such preparations is to discriminate molecules on the basis of charge in the first dimension, followed by size fractionation in the second dimension. A combination of IEF and SDS gradient gels, as popularised by O'Farrell (19), has been found to be very useful in this respect. In the present study we have used this approach to compare [35S]-methionine labelled proteins of normal and DMD skin fibroblasts.

Glass tubes (3mm i.d.x20cm long) were treated to reduce electroendosmosis by a method (20) which results in a firmly bound layer of methyl cellulose. A 1cm plug of 18% acrylamide containing 0.04M aspartic acid and 8M urea was polymerised in the bottom of each tube. A C4T4 gel containing 8M urea, 2% (w/v) N-P40 and 4% (w/v) Pharmalyte pH 3-10 was then cast to a length of 16.5cm. The anolyte was 0.04M aspartic acid, 5mM $H_3PO_4$, 8M urea. The catholyte was 2.5% ethylenediamine, degassed and overlaid with Stoddard solvent. Gels were prerun at 200V for 1hr, the samples (solubilised as for IEF) ($1.5x10^6$ cpm) were loaded and run at 350V for 13hrs, 500V for 5hrs, and then 600V for a total of 10000Vhrs. The pH gradient profile for the run was established by slicing one gel into 0.5cm sections and eluting each slice in 1ml of degassed 8M urea at 10-15°C for 18hrs. A pH profile from a typical run (Figure 10) shows that the gradient, uncorrected for the presence of urea, extended from pH 4.5 to 10.

After the first dimension IEF run, gels were equilibrated for 40min in SDS

sample buffer (19) and applied to 7 to 20% gradient polyacrylamide gels
(1.4mm thick) containing SDS.  Subsequent to electrophoresis, the slabs were
dried and the polypeptide maps visualised by autoradiography (10-day exposure)

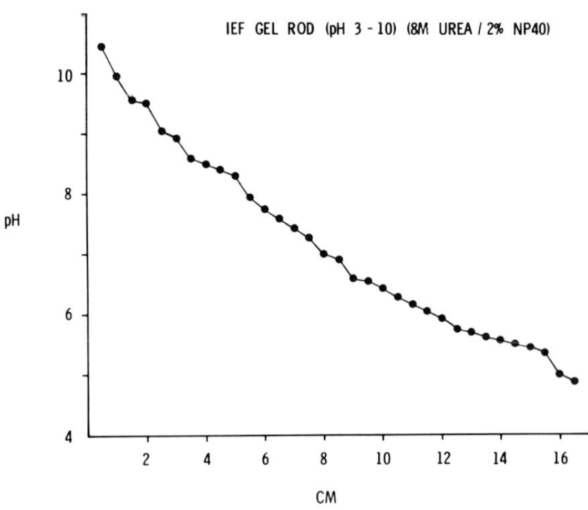


Figure 10 : Typical pH profile from an IEF first-dimension gel.


Typical two-dimensional electrophoretograms of the proteins of normal and
DMD fibroblasts are shown in Figure 11a and 11b respectively.  These clearly
demonstrate the high resolving power of the technique which makes it
potentially very powerful for the analysis of protein abnormalities in
genetic diseases.  However, the analysis of such complex maps and the
comparison of patterns from different samples to identify either
qualitative or quantitative differences between them presents a formidable
task.  A high resolution two-dimensional gel scanner linked to a
sophisticated image analysis computer facility appears to be the best method
to approach this problem.  Unfortunately such an apparatus is not available
to us, so we are at present pursuing the development of a suitable system.


<u>Note on drying gradient gels</u>


We have developed a special drying procedure in order to minimise the
hazards of shrinkage and cracking of regions of the gel with an acrylamide

concentration greater than 10%.  Gels were soaked 2x45min in 2% (v/v) DMSO containing 1% (w/v) glycerol, then placed on the drying apparatus with the concentrated acrylamide (>10%) region of the gel covered with a sheet of 4 mil Gel Bond film (FMC Corporation) and the whole gel covered with plastic wrapping film.  The gels were dried under vacuum.

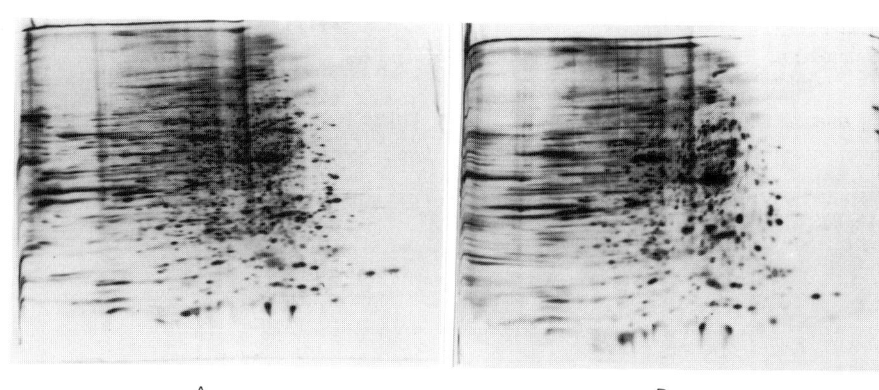

A                B

Figure 11 : Autoradiographs of two-dimensional electrophoretograms of [35S]-methionine labelled normal (a) and DMD (b) fibroblast proteins.

References


1.   Rowland, L.P. : Muscle and Nerve 3, 3-20 (1980).

2.   Lucy, J.A. : Brit.Med.Bull. 36, 187-192 (1980).

3.   Roses, A.D., Hartwig,G.B., Mabry, M., Nagano, Y. and Miller,S.E. : Muscle and Nerve 3, 36-54 (1980).

4.   Laemmli, U.K. : Nature (London) 227, 680-685 (1970).

5.   Kapadia, G., Chrambach, A. and Rodbard, D. : in Electrophoresis and Isoelectric Focusing in Polyacrylamide Gel (ed.R.C.Allen and H.R. Maurer), pp.115-144, de Gruyter, Berlin (1974).

6.   Poduslo, J.F. and Rodbard, D. : Analyt.Biochem. 101,394-406 (1980).

7.   Switzer, R.C., Merril,C.R. and Shifrin,S. : Analyt.Biochem.98, 231-237 (1979).

8.   Merril,C.R., Switzer, C. and Van Keuren, M.L. : Proc.Natl.Acad.Sci. U.S. 76, 4335-4339 (1978).

9.   Oakley,B.R., Kirsch,D.R. and Morris, N.R. : Analyt.Biochem. 105, 361-363 (1980).

10.  Pena,S.D.J. and Wrogemann, K. : Pediat.Res. 12, 887-893 (1978).

11.  Frakar, P.J. and Speck, J.C. : Biochem.Biophys.Res.Commun.80,849-857 (1978).

12.  Zechel, K. : Analyt.Biochem. 83, 240-251 (1977).

13.  Righetti,P.G. and Gianazza, E. : J.Chromatogr.184, 415-456 (1980).

14.  Radola,B.J. : Electrophoresis 1, 43-56 (1980).

15.  Ui, N. : Ann.N.Y.Acad.Sci. 209, 198-209 (1973).

16.  Gelsema,W.J., De Ligny,C.L. and Van der Venn, N.G. : J.Chromatogr. 171, 171-181 (1979).

17.  Gianazza, E., Astorri, C. and Righetti, P.G. : J.Chromatogr. 171, 161-169 (1979).

18.  Gianazza, E. and Righetti, P.G. : J. Chromatogr.193, 1-9 (1980).

19.  O'Farrell, P.H. : J.Biol.Chem. 250, 4007-4021 (1975).

20.  Vanderhoff, J.W., Micale, F.J. and Krumrine, P.H. : Sep.Pur.Meths.6, 61-87 (1977).

STUDIES OF GENE EXPRESSION IN HUMAN LYMPHOCYTES USING HIGH-
RESOLUTION TWO-DIMENSIONAL ELECTROPHORESIS


N. L. Anderson

Molecular Anatomy Program, Division of Biological and Medical
Research
Argonne National Laboratory
Argonne, Illinois 60439  U.S.A.



## Introduction

The introduction of high-resolution two-dimensional electrophoresis (1-4) and its subsequent development to allow
large-scale systematic studies (5,6) have made it possible to
consider cataloging a large proportion of the protein products
of the estimated 30-50,000 human structural genes (9).  Such a
catalog, or Human Protein Index (10), is a necessary prerequisite to the detailed understanding of cellular function and
malfunction.  It will perform the same basic service for
biology that the periodic table performs for chemistry or that
a complete list of parts serves in the maintenance of a 747.

Initially, two major problems arose concerning the feasibility
of constructing a Human Protein Index: (i) the two-dimensional
patterns might contain too much uninterpretable data and (ii)
there were no methodologies available for adequately characterizing the thousands of proteins resolved and  for gaining
useful insight into their biologically relevant properties.
Lately, it has become clear that computerized systems of
several types can quantitate and organize the data obtained
from 2-D gels, greatly facilitating the interpretation of
results (11-16).  The way is now open for the construction
of large and useful data bases containing a wide variety
of information about a large number of molecules.  It is also
apparent that a new spectrum of techniques can be devised

which yield specific information about many proteins simultaneously (i.e., without the need for the classical approach of biochemical isolation). The variety of such techniques currently in use is illustrated in Table I. Taken together, these developments suggest that detailed information on a large number of cellular proteins can be obtained and usefully managed. This type of data base, in association with the growing library of DNA sequence information, forms the core of a major expansion in our knowledge of human biology.


## Categorization of 2-D Gel Spots into Functional Sets

A major objective in the systematic analysis of cells is, of course, to know what each of the thousands of cellular proteins is doing and how its synthesis is controlled. As it is currently used, the concept of a protein's function is not very rigorous. Some proteins such as actin or calmodulin are known to have a variety of functions, and there seems to be no justification for assuming that evolution would not package several useful properties in each gene product. Even a single function is known for no more than 2 to 3% of the cellular proteins. Likewise, the current notions of gene regulation are rather vague, except in the cases of exceptional molecules like the globins, immunoglobulins, or ovalbumin (whose control may not be at all representative). It can therefore be concluded that very little is known or understood concerning the properties or control of the vast majority of cellular proteins. In this situation, the principal advantage of two-dimensional mapping as a means of exploring gene regulation and gene-product function is the ability of the technique to look in detail at a large part of the whole system. The major effects of manipulating various control systems can be determined by inspection.

Table I

Technique | Information Obtained
--- | ---

Physico-chemical

| | |
| --- | --- |
| 1. Various amino acid radiolabels | Partial amino acid composition (ratios) |
| 2. $^{32}PO_4$ labeling | Identification of phosphoproteins |
| 3. $^{125}I$ surface labeling | Identification of surface proteins |
| 4. $^{14}C$-iodoacetamide labeling | Identification of reactive <u>vs</u> total SH |
| 5. Thermal denaturation cofactors | Thermostability, interaction with cofactors |
| 6. Affinity chromatography | Identification of proteins binding to |
| | immobilized high or low MW molecules |
| 7. Immunoprecipitation | Identification of proteins reacting with |
| | a given antibody |

Cell Manipulation

| | |
| --- | --- |
| 8. Treatment with mitochondrial inhibitors | Mature mitochondrial proteins not produced |
| 9. Treatment with tunicamycin | Mature glycoproteins not produced |
| 10. Extraction with digitonin | Soluble proteins released |
| 11. Extraction with NP-40 | Cytoskeletal and nuclear proteins remain |
| 12. Pulse-chase labeling | Rate of synthesis and degradation |
| 13. Treatment with α-amantin, actinomycin D | Message lifetimes |
| 14. Cell fusion | Assignment of protein genes to chromosomes |
| 15. Gene amplification | Identification of amplified gene product and |
| | co-amplified genes |

Post-electrophoretic

| | |
| --- | --- |
| 16. Nitrocellulose or DBM transfers | Identification of proteins binding given |
| | antibodies, etc. |
| 17. Proteolytic digestion mapping of 2-D spots | Comparative positions of proteolytic cleavage |
| | sites (detection of protein homology) |
| 18. Microsequence analysis of individual | Partial amino-acid sequence. |
| spots | |

Use as Part of Assay System

| | |
| --- | --- |
| 19. Assay for monoclonal antibody specific | Identify antibody specific to desired |
| to given spots | protein(s) |
| 20. Assay for message (or gene) for specific | Identify nucleotide sequence associated with |
| spots after reticulocyte translation | desired protein |

311

312

As a first step toward characterization of some of the major regulation systems, a broad survey of possible regulatory effectors was undertaken. Three types of human cells have been used: fresh peripheral blood lymphocytes, a lymphoblastoid cell line (GM607), and a fibroblast line. Gene control in fresh lymphocytes, cultured _in_ _vitro_ for 18 hr or less, represents the closest approximation we now have to regulation under normal physiological conditions. Even so, there are some substantial differences in the patterns of gene expression in lymphocytes isolated in Ficol-Paque compared to lymphocytes labeled briefly in whole heparinized blood at 37°C. Lymphoblastoid cells represent a transformed (presumably by E-B virus) version of the lymphocyte, and fibroblasts a nontransformed, but also nonphysiological, cell for comparison. The survey of regulatory effectors has so far encompassed about 60 compounds or treatments. The majority do not appear to alter gene expression, even though at high enough levels most can kill the cells. Table II lists some of the treatments that I have found to alter gene expression in the lymphocyte.

The results of these exploratory experiments identify sets of proteins that are controlled in various ways. The different groups of spots are given six letter acronyms ("Group name" in Table II; 17) referring to the properties that define them. Interf:1 is, for example, the first protein of the set induced by interferon, while Calgon:3 is the third member of the set repressed by raised intracellular calcium. Each protein within a group has a unique number in that group. A particular protein may have several names in addition to its spot number and biochemical name. For example, Mitcon:1 is a protein that belongs to the set whose synthesis stops when cells are treated with mitochondrial poisons; its uncleaved precursor is Mitpro:1; its master spot number in the lymphocyte pattern is 10 and it is probably the $\alpha$-subunit of the $F_1$ATPase.

Table II.  <u>A List of Some Treatments Producing Changes in Gene Expression in Human Cells</u>

| Treatment | Group Name | No. of Proteins Affected |
|---|---|---|
| 1. Antimitochondrial agents (nonactin, oligomycin, DNP etc.) | Mitcon | 40 |
| 2. Chloramphenicol | Mitcod | 1 |
| 3. Heat shock | ShockH | 6 |
| 4. Sulfhydryl poisons (iodoacetamide, etc.) | ShockS | 8 |
| 5. Amino acid analogues | ShockA | 6 |
| 6. Cyclosporin A | CycspA | 2 |
| 7. Tunicamycin | Tglyco | 10 |
| 8. Colchicine | ColchA | 3 |
| 9. Dexamethasone | Dexmet | 1 |
| 10. Interferon | Interf | 7 |
| 11 Poly I:C | PolyIC | 7 |
| 12. Phorbol esters | Phorbl | 30 |
| 13. A23187 ($Ca^{++}$ ionophore) | Calgon | 10 |
| 14. Ouabain | Ouaban | 3 |
| 15. $CdCl_2$ | Hmetal | 5 |
| 16. Isolation (altered from rate of production in blood) | VitroA | 20 |
| 17. Urinary proteins | Urocon,Urocoff | 30 |
| 18. Variable (unknown control) | Varble | 3 |

Interesting relationships exist among some of these regulation sets.  The set of proteins induced in human lymphocytes by human interferon is substantially the same as the set induced by poly I:C.  However, human interferon does not produce the effect in cells of other species, while poly I:C still does. This is apparently an example of the control of a single set of genes at two levels, one level specific for cell type and species, the other nonspecific.  The relationship between

314

some peptide hormones and cAMP is formally similar. There are
overlapping sets, however, which share only one or a few
elements.  The Mitcon group, which appears to contain all the
mitochondrial matrix and inner membrane proteins (17),
shares only one major spot with the set of lymphocyte proteins
affected by treatment with the phorbol ester tumor promoters
(the Phorbl group).  This protein (Mitcon:5) is absent from
lymphoblastoid and fibroblast cells, and therefore is a
transformation-sensitive mitochondrial protein, possibly
lymphocyte specific.  Mitcon:5 is among the three or four most
abundant proteins of lymphocyte mitochondria, and its complete
repression by transformation must make a major difference to
cellular metabolism.  Further entirely unsuspected set relationships may be found.  An example is the effect of colchicine in shifting a particular group of fibroblast spots
(probably forms of a single protein) to a higher apparent
SDS-molecular weight.  This group happens to be one that is
shifted to lower SDS-MW by tunicamycin treatment (i.e., when
glycosylation is inhibited).  It seems likely therefore that
the effect of colchicine in this case is to cause hyperglyco-
sylation of the affected protein.  Since this protein is known
also to be a member of the class of spots that are relatively
invariant through the mammals, its function (and any disturbance of it) is likely to be important to the cell.


Discussion

The examples I have given here of results derived from protein
set comparisons are few, and as yet do not solve the major
problems of gene regulation.  However, they do begin to
demonstrate that the complex pattern of proteins produced by
living cells can be broken down into tractable, inter-related
subsets based on regulatory behavior.  Proteins at the intersection of several sets may possess sufficiently unique
properties as to suggest aspects of function in some cases.

Once sets are established in one cell type, interesting questions arise concerning the preservation of these sets unbroken in other cell types and other animal species. So far it appears that a number of sets (including Mitcon, ShockH, poly I:C) can be considered constant throughout the mammals and perhaps further. Such homologies, and the associated ability to identify analogous proteins in widely separated species, offer interesting opportunities for the comparative study of evolutionary variations on a constant functional theme.

A rudimentary, but self-consistent, picture of gene expression in some human cells is beginning to take shape. It provides sufficient insight into the "state of the cell" to determine objectively whether a number of major systems are functioning properly, and in some cases to differentiate various forms of cellular damage. Although the major goal is to understand the basic cellular systems almost in chemical engineering terms, the ultimate diagnostic usefulness of a Human Protein Index data base can not be ignored.

Acknowledgment

This work is supported by the U.S. Department of Energy under contract .No. W-31-109-ENG-38.

References.

1.  O'Farrell, P. H.: J. Biol. Chem. 250, 4007-4021 (1975).

2.  Klose, J.:  Humangenetik 26, 231-243 (1975).

3.  Scheele, G. A.: J. Biol. Chem. 250, 5375-5385 (1975).

4.  Iborra, G., Buhler, J.-M.: Anal. Biochem. 74, 503-511.

5.  Anderson, N. G., Anderson, N. L.:  Anal. Biochem. 85, 331-340 (1978).

6.  Anderson, N. L., Anderson, N. G.:  Anal. Biochem. <u>85</u>,
    341-354 (1978)

7.  Anderson, N. G., Anderson, N. L.: Behring Inst. Mitt. <u>63</u>,
    169-210, 1979.

8.  Anderson, N. L., Edwards, J. J., Giometti, C. S., Willard,
    K. E., Tollaksen, S. L., Nance, S. L., Hickman, B. J.,
    Taylor, J., Coulter, B., Scandora, A., Anderson N. G.:
    In Processing of Electrophoresis '79, B. Radola, ed.,
    W. deGruyter, pp. 313-328, 1980.

9.  Anderson, N. G., Anderson, N. L.: J. Automatic Chemistry <u>2</u>,
    177-178 (1980).

10. Report of the Human Protein Index: Anderson, N. G., chairma
    available from Dr. R. E. Stevenson, Director, American Type
    Cell Culture, Rockville, MD.

11. Lutin, W. A., Kyle, C. F., Freeman, J. A.:  In Proceedings
    of Electrophoresis '78, N. Catsimpodas, ed., Elsevier,
    North Holland, Amsterdam, pp 93-106, 1978.

12. Garrels, J. I.: J. Biol. Chem. <u>254</u>, 7961-7977 (1979).

13. Bossinger, J., Miller, M. J., Vo, K.-P., Geiduschek, E. P.,
    Xuong, N.-H.: J. Biol. Chem. <u>254</u>, 7986-7998.

14. Taylor, J., Anderson, N. L., Coulter, B. P., Scandora,
    A. E., Jr., Anderson, N. G.: In Proceedings of Electrophore
    '79, B. Radola, ed., W. deGruyter, 329-339, 1980.

15. Lester, E. P., Lemkin, P., Lipkin, L., Cooper, H. L.:
    Clin. Chem. <u>26</u>, 1392-1402 (1980).

16. Lipkin, L. E., Lemkin, P. F.: Clin. Chem. <u>26</u>: 1403-1412 (19

17. Anderson, N. L.: Proc. Nat. Acad. Sci., in press (1981).

ISOELECTRIC FOCUSING AND TWO-DIMENSIONAL MAPS OF NORMAL AND
CYSTIC FIBROSIS SALIVA

S. E. Bustos and L. Fung
Department of Oral Biology (Biochemistry)
Medical College of Georgia
Augusta, Georgia   30912


Introduction

Cystic Fibrosis (CF) is one of the most common inborn errors
of metabolism in persons of Caucasian ancestry.  Clinically
this disease is characterized by recurrent pulmonary complications, pancreatic deficiency with accompanying digestive and
nutritional disturbances, and a heightened concentration of
sweat electrolytes; all these manifestations being the result
of a generalized disorder of the exocrine glands (1,2).  Since
the first description of CF about four decades ago (3,4), it
has been recognized as the most lethal or semi-lethal genetic
disease affecting children in the U.S.  Since there is evidence that CF is inherited as an autosomal recessive trait (5,
2), it is generally agreed that the disease originates from a
single gene defect which in turn results in a modified gene
product or in a partial or total lack of the product (a single
enzyme or other protein).  In spite of about 40 years of research in CF and the findings of numerous biochemical abnormalities, the primary biochemical lesion has yet to be
ascertained.

This report deals with the search for genetic markers in
Cystic Fibrosis saliva.  The purpose of our investigation was
to determine whether abnormalities in the protein or glycoprotein make-up of the salivary secretions of CF individuals may
be detectable by two-dimensional electrophoresis in conjunc-

tion with lectin autoradiography. The specific aims of this research were: (a) to obtain a map of the proteins and glycoproteins present in the parotid saliva of CF and normal individuals and (b) to look for the presence of unique or abnormal protein species which may be present in the saliva of CF individuals and which may provide possible genetic markers for Cystic Fibrosis.


Methods


I. Collection of saliva

Parotid saliva from CF patients and control individuals was collected by means of Curby-modified Carlson-Crittenden cups placed over the Stenson duct. Saliva flowed from the cups through tubing connected to a test tube maintained in crushed ice. Salivary flow was stimulated by means of sour lemon drops. The saliva samples were extensively dialyzed against distilled water and lyophilized.

II. Analytical procedures

(a) <u>Protein determination</u>. The protein concentration in the saliva samples was estimated by the method of Lowry <u>et al</u> (6) using bovine serum albumin as a standard.

(b) <u>Two-dimensional electrophoresis</u>. Essentially the method of O'Farrell (7), with minor modifications to process salivary protein samples, was used.

Lyophilized parotid saliva samples were solubilized at room temperature in a solution containing SDS, urea, mercaptoethanol, Nonidet P-40 and Ampholines pH 3.5-10. Isoelectric focusing of these samples was carried out on gel rods (3 x 120 mm) made up of a 4% acrylamide, 9M urea, 2% NP-40 and 3.6% Ampholines (pH 3.5-10) solution. The anolyte was 0.01M $H_3PO_4$ and the catholyte was 0.02M NaOH. Runs were made with the anode or cathode at the top. Prior to sample loading, the gels were

electrophoresed for 1 hr. at a maximum of 400V; after loading
the samples, focusing was continued for 18 hrs. at 300V, plus
1 hr. at 800V. Subsequently, the gels were stained with
Coomassie Brilliant Blue R-250 by a modification of the procedure of Drysdale (8). Instead of 0.1% cupric sulfate in
acetic-acid-ethanol-water, we used TCA-sulfosalycilic acid-
methanol (Ampholine PAG plate instructions, LKB 1974). The
pH of serial 5mm gel sections was measured by means of a combination pH microelectrode, to determine the pH gradient
formed during isoelectrofocusing.

A concave exponential gradient of 10 to 16% acrylamide was
used to form a 1.5mm thick (12 x 14 cm) SDS slab gel in a
vertical electrophoretic apparatus. A 4.5% acrylamide stacking gel of pH 6.8 was then cast on top of the separating gel
and the first dimension isoelectrofocusing gel, equilibrated
with SDS buffer for 10-15 min., was then fused to the stacking
gel top surface by means of a 1% agarose solution. Electrophoresis in the second dimension was conducted in SDS running
buffer (0.192 M glycine - 0.25 M Tris - 0.1% SDS) at 25 mA/gel
for 3.5 hrs. at 4°C. The CF and control gels were stained in
Coomassie Brilliant Blue R-250 solution as before. Alternatively the gels were stained by a modification (9) of the
silver staining technique of Switzer et al (10).

To compare the spot distribution patterns, spots in the two-
dimensional patterns were classified according to migration,
density of staining and molecular weight coordinates.
(c) Incubation of gel slabs with $^{125}$I-lectins and autoradiography. The Coomassie-Blue-stained or unstained second dimension slab-gels were equilibrated with pH 7.4 Tris-HCl saline
(TBS) and subsequently incubated for 48 hrs. at room temperature with approximately 2.3 x 10$^8$ dpm of $^{125}$I-labeled
Concanavalin-A (Con-A) or Wheat Germ Agglutinin (WGA) (11)
(see below for details on radioiodination). This step was
followed by exhaustive washes with TBS until background

320

levels of $^{125}$I were attained.

The lectin-stained gels were soaked in 1% glycerol - 10%
acetic acid for 40 minutes, then transferred onto a piece of
wet Whatman 3 mm filter paper and dried under vacuum.  The
dried gels were then placed in contact with an X-ray film
(Kodak XPR-1) inside an X-ray film holder and left for 48 to
72 hour exposure or longer.

(d)  <u>Radioiodination of lectins</u>. A modification of the lacto-
peroxidase-catalyzed iodination method of Marchalanis (12) was
used to label lectins with $^{125}$I.

The iodination reaction was carried at 25°C in a microreaction
vial.  The reaction mixture contained 500 $\mu$Ci Na $^{125}$I(sp. act.
170 mCi/mM), 5 mg of Con-A, or WGA in 1.5 ml of 50 mM sodium
acetate (pH 6.8), 10 $\mu$l of 0.06% $H_2O_2$ and 500 $\mu$g of lacto-
peroxidase.  This reaction mixture was continuously stirred
while 0.06% $H_2O_2$ was added in 10 $\mu$l aliquots for 20 minutes.
The reaction was stopped by adding 2 mg of sodium azide.  The
iodination mixture obtained in this step was purified by molecular exclusion chromatography on G-25 Sephadex.  As explained
under (c), these radioiodinated lectins were used to stain resolved glycoproteins on gel slabs.

(e)  <u>Molecular weight estimation</u>.  A set of 7 molecular weight
protein markers was used to calibrate the SDS-polyacrylamide
gradient slab gel (13).  This set comprised Phosphorylase B
MW 94,000, Bovine serum albumin MW 67,000, Ovalbumin MW 43,000,
Aldolase MW 39,500, Carbonic Anhydrase MW 30,000, Soybean
Trypsin Inhibitor MW 20,100, and $\alpha$-Lactalbumin MW 14,400;
these values correspond to the subunit molecular weights.


Results and Discussion


Parotid saliva isofocusing patterns from normal and CF
patients are shown in Figure 1.  The gels are oriented
from left (basic end) to right (acidic end), with an average

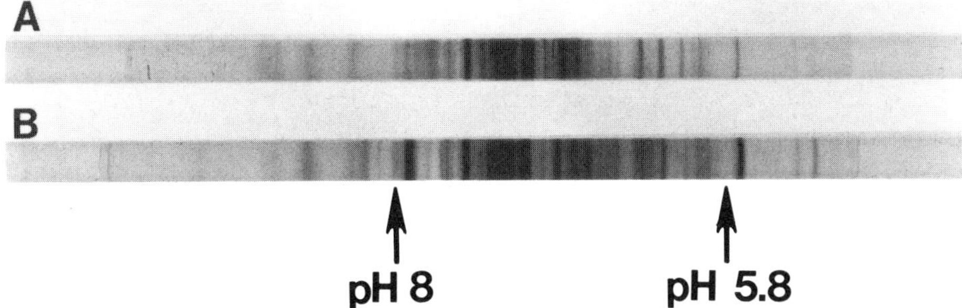


Figure 1.   Isoelectrofocusing patterns of parotid saliva proteins from normal and CF individuals.   The gels are oriented
with the basic end to the left and the acidic end to the right.
Arrows show the basic (pH 8) and acidic (pH 5.8) peptides.   (A)
Normal saliva.   (B)  Cystic fibrosis saliva.

pH gradient ranging from 9 to 4.   Actually the pH range of the

ampholytes used in the gel is 3.5 to 10 but the pH gradient

under our  experimental conditions is narrower (Figure 2).

Further resolution of the great majority of these bands was

accomplished using the SDS-polyacrylamide gradient gel.  The

two-dimensional electrophoretic parotid saliva maps from normal

and CF patients are shown in Figure 3.   The bands that

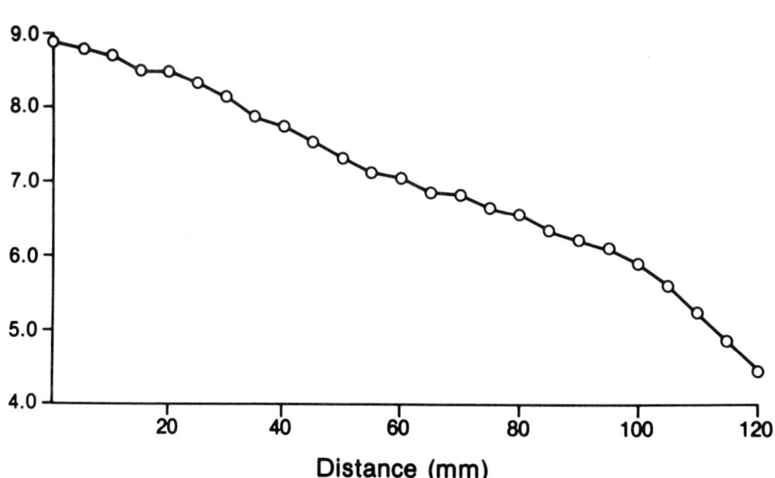


Figure 2.   Typical pH gradient formed after 6,800 volt-hours.
Ampholine pI range 3.5-10.   Catholyte:    0.02M NaOH:   Anolyte:
0.01M $H_3PO_4$.

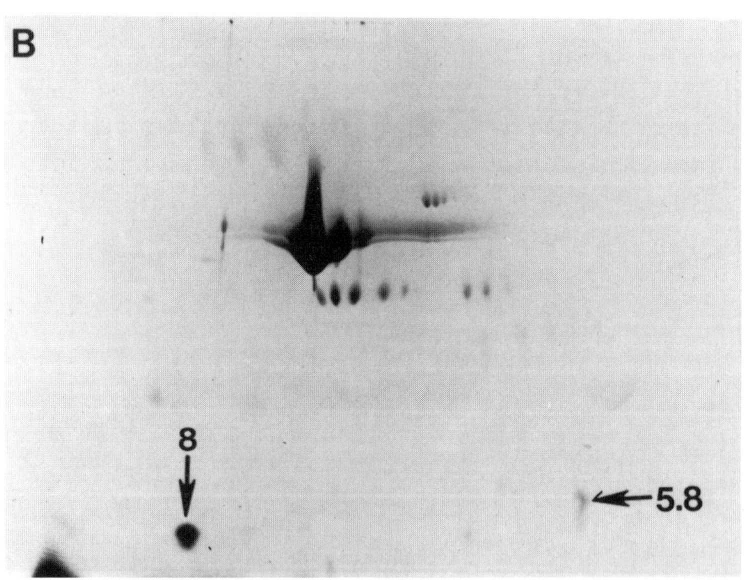



Figure 3. Two-dimensional patterns of parotid saliva proteins from normal and CF individuals. The gels are oriented with the basic end to the left and the acidic end to the right. Arrows indicate the position of the basic (pH 8) and acidic (pH 5.8) peptides. Stain: 0.05% Coomassie Brilliant Blue R-250. (A) Normal saliva. (B) Cystic fibrosis saliva.

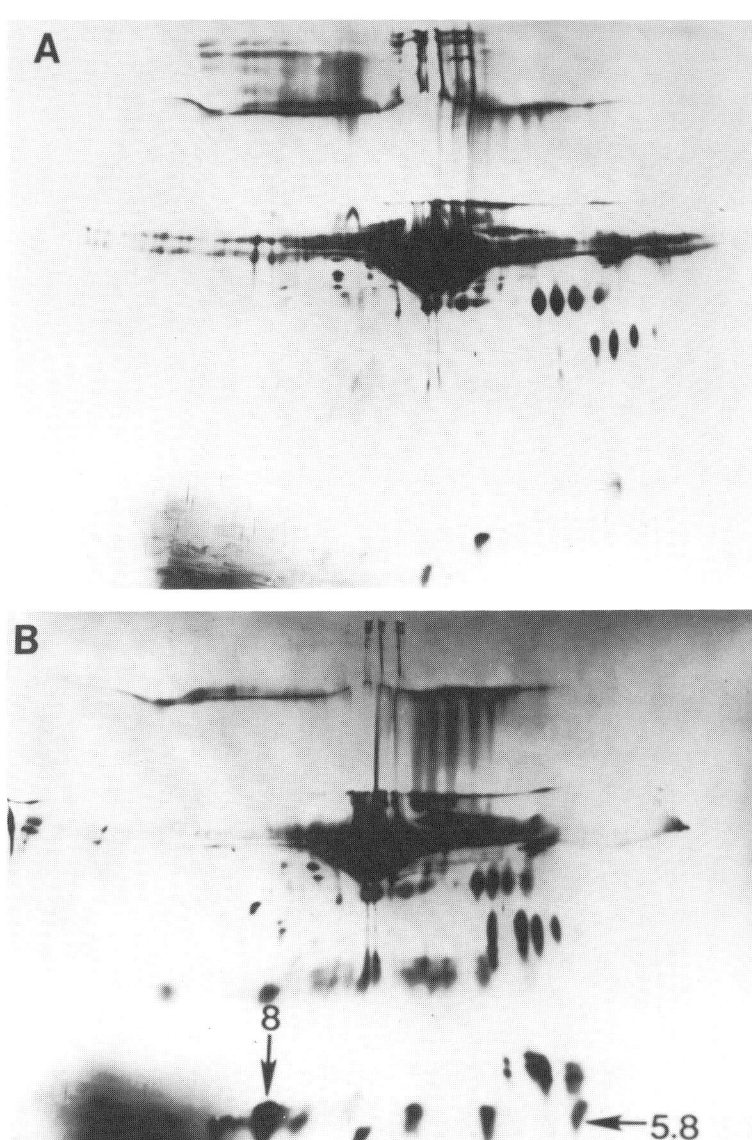



Figure 4.  Silver-stained two-dimensional patterns of parotid
saliva proteins from normal and cystic fibrosis patients.
The gels are oriented with the basic end to the left and the
acidic end to the right.  Arrows indicate the position of the
basic (pH 8) and acidic (pH 5.8) peptides.  (A)  Normal
saliva.  (B)  Cystic fibrosis saliva.

324

appeared as singlets in the IEF dimension were further resolved into 2 or more components by the SDS-PAGE dimension.

A semilogarithmic plot of MW of standard proteins against
distance of migration was used to estimate the molecular
weights of the parotid saliva proteins in the two-dimensional
maps.  The range of molecular weights of these proteins was
from 8,500 to 200,000 daltons.

The most prominent differences between the IEF patterns of
normal and CF saliva were two bands that focused at pH 8 and
5.8.  In the CBB-stained two-dimensional maps, the differences
were confined to two molecular species with pI's 8 and 5.8 and
approximate molecular weights of 10,000 daltons.  These differences, with the exception of the basic low MW peptide in the
2-D map, are quantitative.

The silver-stained control and CF gels are shown in Figure 4.
The increased sensitivity of this stain over Coomassie
Brilliant Blue is evidenced by the appearance of previously
undetected differences between the control and CF two-dimensional patterns.  Most were quantitative and corresponded to
spots in the MW range of 7,600 to 10,000 daltons and in the pI
range of 5.8 to 8.  The low molecular weight peptide that
focused at pI 8 in the Coomassie Blue stained gels appears
further resolved with the silver stain.  Due to the preliminary
nature of our data we cannot ascertain whether this basic
peptide represents a qualitative difference.

A typical chromatographic profile obtained during purification
of the radioiodinated lectins Con-A or WGA is shown in Figure
5.  The first peak to come off the column was the radioiodinated Con-A and the retarded peak corresponded to Na$^{125}$I.
Fractions 3 to 8 were pooled and dialyzed for 12 hours against
TBS at 4°C with frequent buffer changes to get rid of the free
$^{125}$I.  The ratio of bound/free $^{125}$I was found to be 95%.

The autoradiograms obtained after incubation of the normal and

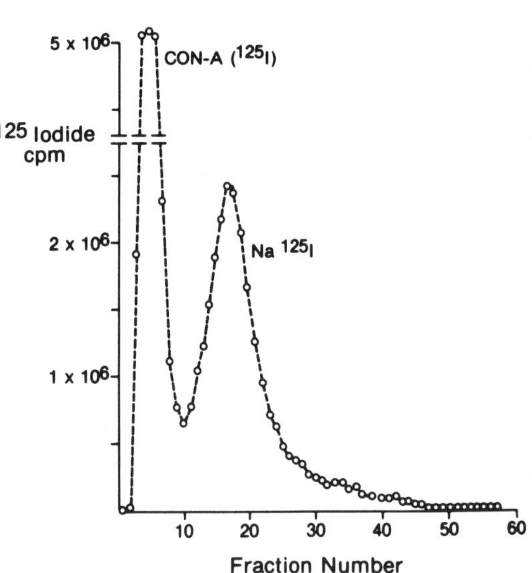

Figure 5. Chromatographic separation of radioiodinated Con-A on a G-25 Sephadex column (5x1.3cm). The elution was performed with TBS containing 0.2M α-methyl mannoside.

CF IEF-gels with $^{125}$I-Con-A are shown in Figure 6. The pI 8 band in control and CF saliva exhibited a slight affinity towards $^{125}$I-Con-A, whereas the pI 5.8 band did not bind this lectin. The majority of the control and CF bands focusing between pH 4.5 and 8 bound $^{125}$I-Con-A with different affinities;

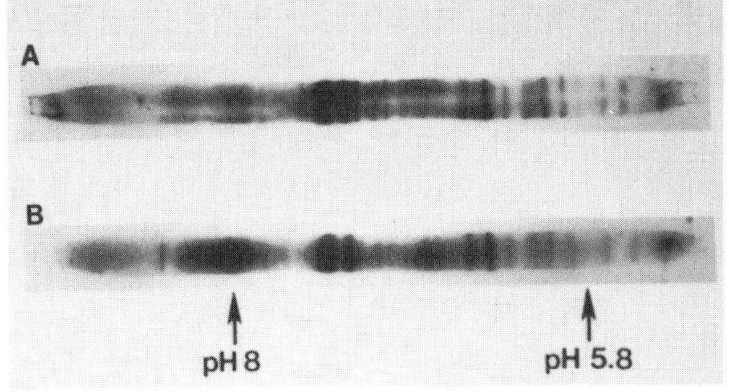


Figure 6. Autoradiograms from $^{125}$I-Con-A stained IEF patterns of parotid saliva proteins from normal and cystic fibrosis patients. The gels were oriented with the basic end to the left and the acidic end to the right. Arrows indicate the position of the basic (pH 8) and acidic (pH 5.8) peptides. (A) Normal saliva. (B) Cystic fibrosis saliva.

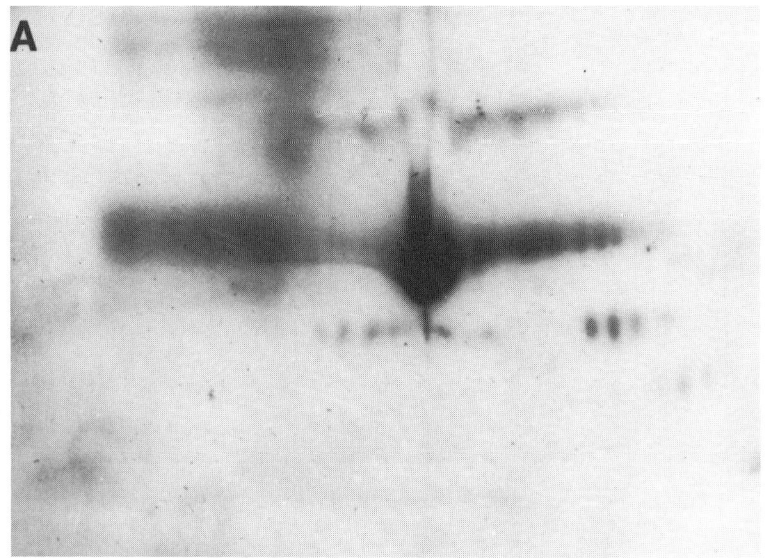

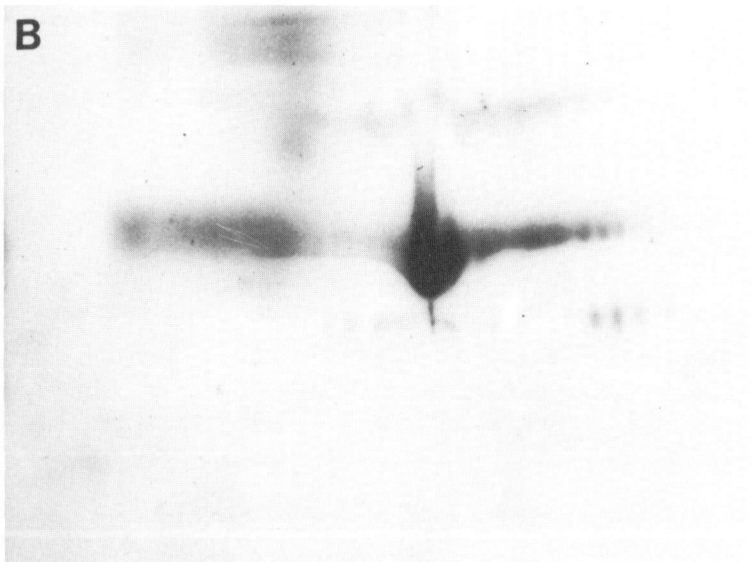

Figure 7. Autoradiograms from $^{125}$I-Con-A stained two-dimensional patterns of parotid saliva proteins from normal and cystic fibrosis patients. Gels are oriented as in previous figures. (A) Normal saliva. (B) Cystic fibrosis saliva.

their control and CF autoradiographic patterns, however, showed no quantitative or qualitative differences. Data from $^{125}$I-WGA binding experiments showed patterns closely resembling the ones obtained with $^{125}$I-Con-A. Binding of radioiodinated Con-A and WGA to protein bands indicated the presence of $\alpha$-D-glucose, $\alpha$-D-mannose and $\alpha$-N-acetyl glucosamine residues. Control experiments conducted in the presence of Con-A or WGA-inhibitory monosaccharides confirmed that the binding of these lectins to the proteins resolved in the IEF or SDS-PAGE dimension was indeed specific.

The autoradiograms of the $^{125}$I-Con-A stained two-dimensional maps from normal and CF parotid saliva are shown in Figure 7. The majority of the resolved Con-A-binding species had molecular weights from approximately 14,000 to 200,000 daltons. Both in the Coomassie-Blue stained two-dimensional maps and their autoradiograms, proteins with molecular weights of 175,000, 125,000, 100,000, 58,000 and 36,000 daltons were arranged in horizontal rows of the same or similar molecular weights. This may be attributed to charge heterogeneity, most likely due to post-translational modifications such as glycosylation or phosphorylation.

In our laboratory we have used two-dimensional electrophoresis and have clearly shown that the complexity of the salivary proteins is significantly greater than one-dimensional gel systems have been able to disclose (14,15); these results have been confirmed by Giometti and Anderson (16).

The availability of highly-sensitive and high-resolution two-dimensional electrophoretic methods for the analysis of proteins and glycoproteins has opened up new possibilities for defining genetic variation and for identifying molecular changes that may account for a defective function. The use of these techniques may reveal important though subtle, qualitative and quantitative molecular changes in the salivary proteins of CF individuals. The definition and identification of these markers is a goal of paramount importance and possibly a prerequisite for a better prognosis and treatment of the disease.


References


1. Di Sant'Agnese, P.A., David, P.B., N. Engl. J. Med., <u>295</u>, 481-485 (1976).

2. Nadler, H.L., G.H.S. Rao and Taussig, L.J. in The Metabolic Basis of Inherited Disease, 4th ed., ch. 73, ed. Stanbury, J.B., Wyngaarden, J.B. and Frederickson, D.S., McGraw-Hill, New York, 1978, pp. 1683-1710.

3. Blackfan, K.D. and May, C.D.  J. Pediat. <u>13</u>, 627-634 (1938).

4. Andersen, D.H., Amer. J. Dis. Child. <u>56</u>, 344-399 (1938).

5. Wright, S.W., Morton, N.E.  Amer. J. Hum. Genet. <u>20</u>, 157-169 (1968).

6. Lowry, C.H., Rosebrough, H.J., Farr, A.L. and Randall, R.J. J. Biol. Chem. <u>193</u>, 265-275 (1951).

7. O'Farrell, P.H.  J. Biol. Chem. <u>250</u>, 4007-4021 (1975).

8. Drysdale, J.W. in Methods of Protein Separation, Vol. 1, Ch. 4, ed. Catsimpoolas, N., Plenum Press, New York, 1975, pp. 83-126.

9. Switzer, R.C., Merril, C.R., Shifrin, S.  Anal. Biochem. <u>93</u>, 231-251 (1979).

10. Oakley, B.R., Kirsch, D.R.  Anal. Biochem. <u>105</u>, 361-363 (1980).

11. Horst, M.N., Branbach, G., Roberts, R.M., FEBS Letters <u>100</u>, 385-388 (1979).

12. Marchalonis, J.J.  Biochem. J. <u>133</u>, 299-305 (1969).

13. Weber, K. and Osborn, M.  J. Biol. Chem. <u>244</u>, 4406-4412 (1969).

14. Bustos, S.E. and Fung, L.  J. Dent. Res. <u>58A</u>, 391 (1979).

15. Bustos, S.E. and Fung, L.  J. Dent. Res. <u>59A</u>, 268 (1980).

16. Giometti, C.S. and Anderson, N.G. in Electrophoresis '79, ed. Radola, B.J., Walter de Gruyter, New York, 1980, pp. 395-404.

# ANALYSIS OF HUMAN AMNIOTIC FLUID PROTEINS BY TWO DIMENSIONAL ELECTROPHORESIS


Peter Burdett, Jorge Lizana
The Electrophoresis Group, Pharmacia Fine Chemicals AB,
Box 175, S-751 04 Uppsala, Sweden.

Peter Eneroth and Katarina Bremme
Department of Obstetrics and Gynecology, Karolinska Hospital,
S-104 01 Stockholm, Sweden.


## INTRODUCTION

The concentration of total proteins in amniotic fluid (AF) during normal pregnancy ranges between 2 to 8 g/l. Various techniques have been used in studying AF proteins including electrophoresis in agar, starch and polyacrylamide gels, and immunoelectrophoresis. These techniques have, in general, confirmed the apparent similarity between the AF proteins and a filtrate of maternal serum proteins (for review see ref. 1).

Two dimensional electrophoresis is a tool which permits extremely detailed studies of complex protein mixtures (2,3). With this method the proteins are separated by isoelectric focusing in the first dimension followed by electrophoresis in sodium dodecylsulphate (SDS) in the second dimension. This method does not necessarily demonstrate native proteins but rather constituent polypeptides.

In this communication we report studies of AF proteins patterns during pregnancy by use of gradient polyacrylamide gel electrophoresis, isoelectric focusing and two dimensional electrophoresis, in an attempt to gain further insight into the

composition and interrelation of the protein species present in
this fluid.


MATERIAL AND METHODS


<u>Samples</u>. Amniotic fluid (AF) was obtained by amniocentesis and
kept at -20°C until used. Except in the case of abnormal AF
samples, pools of AF were prepared from three to five donors.
These samples were taken as part of a screening program for
fetal malformation or fetal pulmonary maturation. The samples
were concentrated approximately ten fold at 4°C, using an
Amicon UM-2 membrane with a cut off at a molecular weight of
2,000.


<u>Reagents</u>. Pharmalyte  3-10, Silane A 174, High and Low
molecular weight calibration kits for electrophoresis, pI
calibration kit 3-10 and Polyacrylamide Gradient Gels PAA 4/30
(all from Pharmacia Fine Chemicals AB, Uppsala, Sweden).
Acrylamide (purum), N,N'-Methylene-bis-acrylamide (pract.) and
Coomassie Brilliant Blue R-250 (Fluka AG, Switzerland).
Ammonium persulphate p.a. and Glycerol analytical reagent (May
and Baker Ltd., England). All other chemicals used were of
analytical grade.


<u>Equipment</u>. Isoelectric focusing (IEF) was performed on a Flat
Bed Apparatus FBE 3000 using the Electrophoresis constant power
supply ECPS 3000/150 connected to a Volthour Integrator VH-1.
Gradient gel electrophoresis (PAGE) and 2-D electrophoresis
were performed in the Electrophoresis Apparatus GE 2/4 LS
together with an EPS 500/400 power supply. Gel Casting
Apparatus GSC-2 and other electrophoresis accessories were all
from Pharmacia. The pH gradient was measured with an Ingold
surface electrode type 10-403-3002 connected to a Beckman pH
meter Phasar 1.

<u>Gradient Gel Electrophoresis</u> on native proteins in one
dimension was performed in PAA 4/30 polyacrylamide gels as
described earlier (4). Gels were electrophoresed for 16 hours
at 125 volts and samples were run alongside the molecular
weight standards. Modified running conditions were used for the
analysis of low molecular weight proteins again using PAA 4/30
gels. Samples were applied with a tracking dye (1% bromophenol
blue) and were electrophoresed at 500 volts for 15 minutes,
followed by 125 volts until the marker reached the bottom of
the gel. The gels were stained in Coomassie Brilliant Blue
R 250. Once suitably destained they were preserved in 10%
acetic acid.

<u>Isoelectric focusing</u> of native proteins was performed as
previously described (13) using a 1 mm thick T5 C3 polyacrylamide (PAA) gel with Pharmalyte 3-10. Focusing was
carried out at constant power (30W, 150 mA, 2,000 V) for
3,000 volthours. Samples were applied 4 cm from the anode,
after the gel had been pre-focused for 500 volthours. Proteins
were visualised using Coomassie Blue.

<u>Two dimensional electrophoresis</u> (2-D) was performed according
to the method of Anderson and Anderson (3), as described in
detail elsewhere (4). Samples were pretreated with SDS -
dithiothreitol buffer (pH 9.5) at 95°C for 10 minutes. The
first dimension polyacrylamide gel rods (T5 C2.6) contained 8 M
urea, 2% Nonidet NP 40 and Pharmalyte 3-10. The gel rods were
run in the GE 2/4 LS apparatus, with the appropriate gasket.
Ethylenediamine 1 M was used as catholyte (upper buffer vessel)
and 0.01 M imidodiacetic acid as anolyte (lower buffer vessel).
Following IEF (10,000 volthours) the gel rods were removed and
stored frozen in SDS equilibration buffer until used in the
second dimension.

The second dimension gel (180x200x2.5 mm or 180x200x0.7 mm)
containing a polyacrylamide gradient from 10% to 20% were cast,
four at a time, in the GSC-2 gel casting apparatus. The gel
rods were applied on top of the second dimension slabs and were
cast in place using 1% agarose. The gels were run in the
GE 2/4 LS electrophoresis apparatus, fitted with the appropiate
gasket for running slabs, until the bromopherol blue marker dye
reached the base of the gel. Gels were stained, destained and
preserved as described previously (3,4).


RESULTS


Gradient gel electrophoresis. Figure 1 shows the separation of
proteins in both normal and abnormal amniotic fluids. In
general the pattern resembles that of maternal serum. In the
case of sample No 11, in which low loadings were applied, the
dominant proteins are albumin, transferrin and others in the
low molecular weight fraction. A high molecular weight fraction
(MW 669,000) is also present, which aligns with a band in
serum. In the abnormal AF sample, the proportion of proteins
other than albumin appear markedly increased and in particular
the IgG region. The presence of very high molecular weight
proteins was noted, and these are likely to include
lipoproteins and alpha$_2$-macroglobulin. The presence of low
molecular weight proteins is demonstrated in Fig. 2. The
pattern produced from normal AF (lane 2) resembles that of
human serum (lane 4), whilst that shown by abnormal AF (lanes 5
and 6) seems to include a variety of additional low molecular
weight proteins.

The subunit structure of human AF proteins is illustrated in
Fig. 3, using SDS-PAGE. Most of the subunits have a molecular
weight that lies within the range 67,000 to 14,000 but smaller
molecules were also present. When a sample is applied in large
quantities (lane 2) an extra band is seen near the origin.

Increased levels of low molecular weight material are readily apparent in the abnormal amniotic fluid. Many of these bands are confluent with those found in normal AF, whilst some new bands are also present (in particular at around 36,000 and 20,000 daltons).

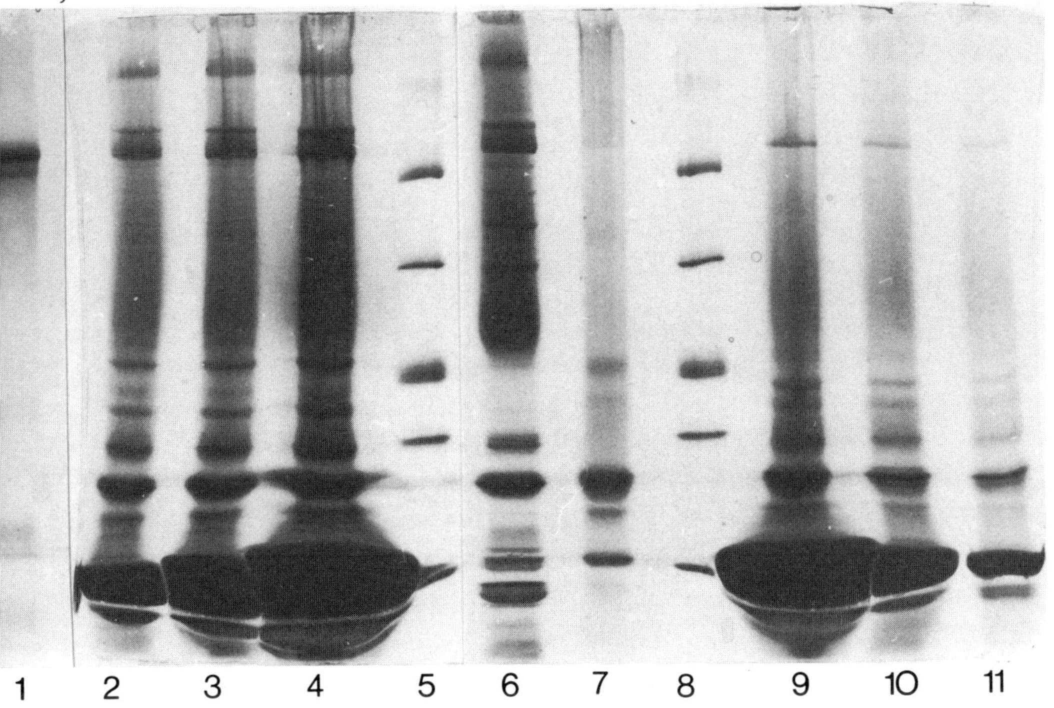

Fig.1. Separation of proteins in AF using PAA 4/30 gradient gels. See text for details. Samples: 1. human fibronectin, 15 µl (10 mg/ml); 2, 3 and 4. abnormal amniotic fluid (3,8 and 20 µl respectively); 5 and 8. HMW Calibration kit (5 µl); 6. Albumin-free serum (10 µl); 7. human transferrin fraction, 5 µl (10 mg/ml); 9, 10 and 11. normal amniotic fluid, third trimester of pregnancy (20, 2 and 1 µl respectively). The calibration proteins (samples 5 and 8) are from top to bottom thyroglobulin (669,000), ferritin (440,000), catalase 232,000, lactic dehydrogenase (140,000) and albumin (67,000).

334

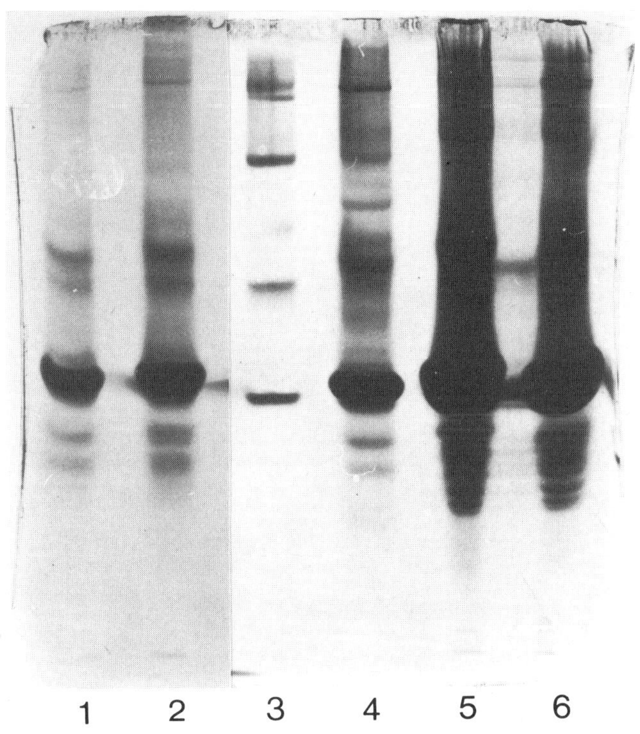



Fig. 2. Separation of amniotic fluid samples using 4/30 PAA
gradient gels specifically for low molecular weight proteins.
Samples: 1 and 2. normal amniotic fluid third trimester (5 and
20 µl); 3. HMW Calibration kit (5 µl); 4. normal human serum
(3 µl); 5 and 6. abnormal amniotic fluid (20 and 10 µl).
Calibration proteins as in Figure 1.

<u>Isoelectric focusing (IEF)</u>. IEF in Pharmalyte 3-10 of AF
proteins is illustrated in Fig. 4. By comparison with the
albumin free serum, purified proteins and pI markers, the
presence of $\alpha_1$-antitrypsin (pI 4.1-4.7), albumin (pI around
5.8), transferrin (pI 5.8) and IgG (pI 5.3-9) is seen,
particularly in lane 5. The majority of the AF proteins have
isoelectric points within the range 4 to 6.

these spots was made by running human serum (Fig. 10) as well
as purified proteins in the 2-D electrophoresis system and by
using the maps prepared by Anderson and Anderson (3), Latner et

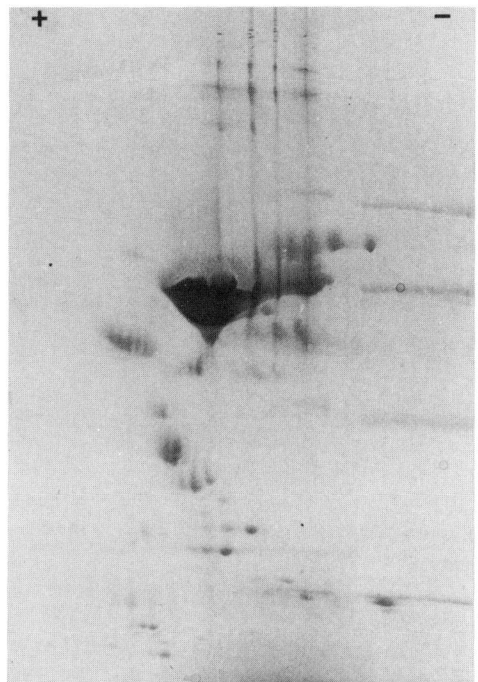


al (5) and Goldman et al. (6).
There remain however many spots
which cannot be identified with
any degree of certainty, in
particular those of low
molecular weight (around
14,000) and the prominant
compound(s) with an apparent
subunit molecular weight of
about 43,000.

Fig. 7. Two-dimensional
separation of amniotic fluid.
Normal pattern from the third
trimester of pregnancy.
Molecular weight calibration
proteins are visible on the
cathodic side of the gel.

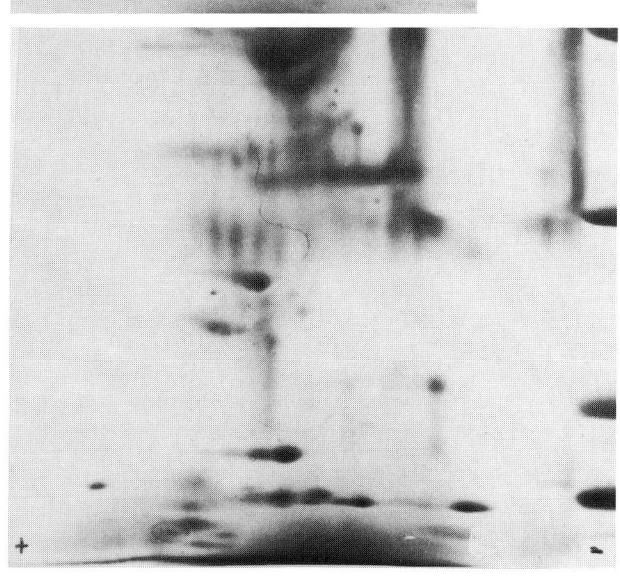


Fig. 8. Separation of
normal amniotic
fluid, showing the
lower part of the gel
only. The sample was
overloaded (400 µl)
to bring out the
weaker bands. A 5 mm
diameter gel rod was
used in this
instance. A map of
the complete gel is
shown in Fig. 9.

336

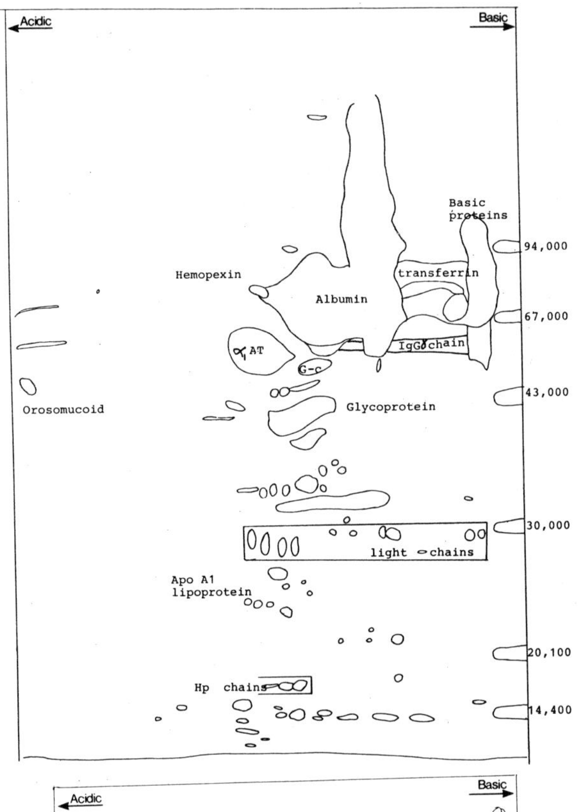



Fig. 9. Tentative 2-dimensional map of normal amniotic fluid. 400 µl of sample was applied to the gel rod. Molecular weight standards were incorporated into the gel (right side).

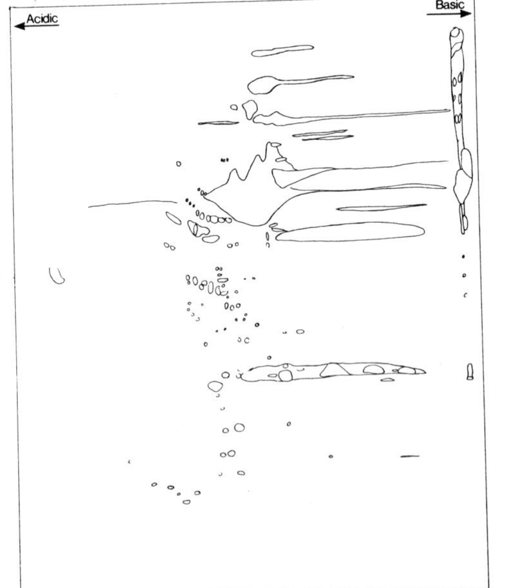



Fig. 10. 2-D map obtained from a fresh serum sample.

Amniotic fluid from two pregnancies with malformed fetuses was
also analysed using 2-D electrophoresis. The 2-D map shown in
Fig. 11 is taken from the same sample that was analysed by
SDS-PAGE (Fig. 3 lane 4). Notable features include raised
levels of all proteins relative to albumin, and the quantity of
the heavy and light chains of IgG is particularly noticeable.
AF protein patterns in the two abnormal cases were similar and
the presence of high levels of a band tentatively identified as
apo AI lipoprotein is striking. Low molecular weight polypeptides are also present. Furthermore, the two bands marked
with arrows were not found in amniotic fluid from normal
pregnancies.

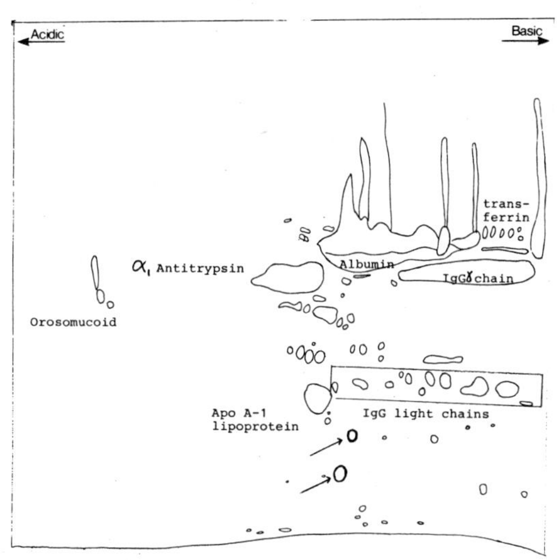


Fig. 11. 2-D map obtained from the analysis of abnormal
amniotic fluid, together with a tentative identification of
some proteins.


DISCUSSION

The amount of total proteins present in amniotic fluid during
normal pregnancy is considerable (2-8 g/l). Studies on the AF

proteins have revealed the presence of more than 30 species,
but the function and origin of many of these proteins remains
unknown (7-11). Our results with gradient polyacrylamide gel
electrophoresis (Figs. 1-3) confirmed the previous findings
that AF protein patterns resemble the maternal serum protein
pattern and that amniotic fluid contains an important number of
proteins with molecular weight below 15,000.

In this study we have also demonstrated the feasibility of
using two dimensional electrophoresis in the analysis of
polypeptides in AF during pregnancy. At least 200 peptide spots
have been seen in normal AF from the third trimester of
pregnancy. The positions of many of these spots correlates with
those seen in a 2-D map of serum and of purified serum
proteins, again demonstrating that AF resembles an ultrafiltrate of maternal serum proteins. In addition a number of
spots have been observed which appear to be "specific" to AF,
in particular those with a molecular weight of 15,000-25,000.
It is possible that this group contains for instance prolactin
and placental lactogen as well as beta$_2$-microglobulin which
is present in AF in appreciable quantities (45-190 mg/l) and
has a molecular weight of 11,800 and a pI of 5.6

The location of alpha-fetoprotein which is a single polypeptide
chain with a molecular weight of 74,000 and pI of 4.8 awaits
confirmation as proteins of this size and pI are masked by the
high amounts of albumin. Removal of albumin by use of for
example Blue-Sepharose 4B will allow a definitive
identification of proteins related to alpha-fetoprotein (12).
Furthermore the pH gradient will be less disrupted by albumin
and distortions are likely to be minimized. Such experiments
are currently under way in our laboratory.

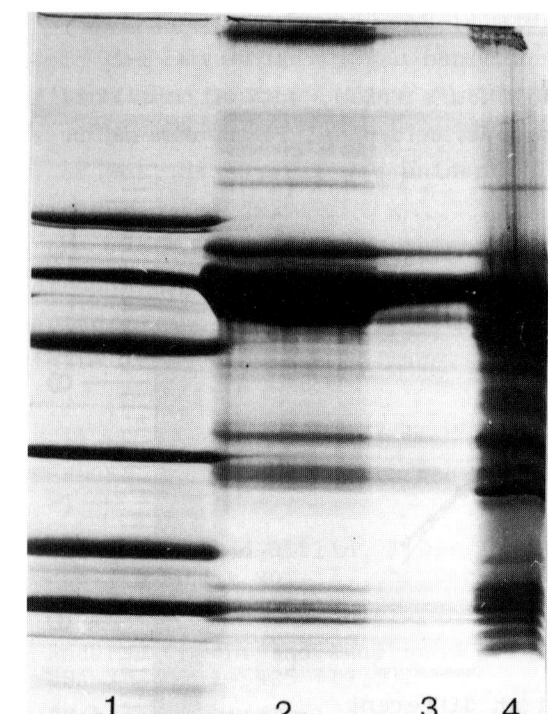

Fig. 3. SDS electrophoresis in a 10-20% PAA gradient gel (180-200 mm). Sample 1. Low molecular weight (LMW) Calibration kit, phosphorylase b, (94,000); albumin (67,000), ovalbumin (43,000), carbonic anhydrase (30,000), trypsin inhibitor (20,000) and $\alpha$-lactalbumin (14,400). Sample 2. Normal amniotic fluid, high loading sample; 3. Normal amniotic fluid, low loading sample; 4. Abnormal amniotic fluid.

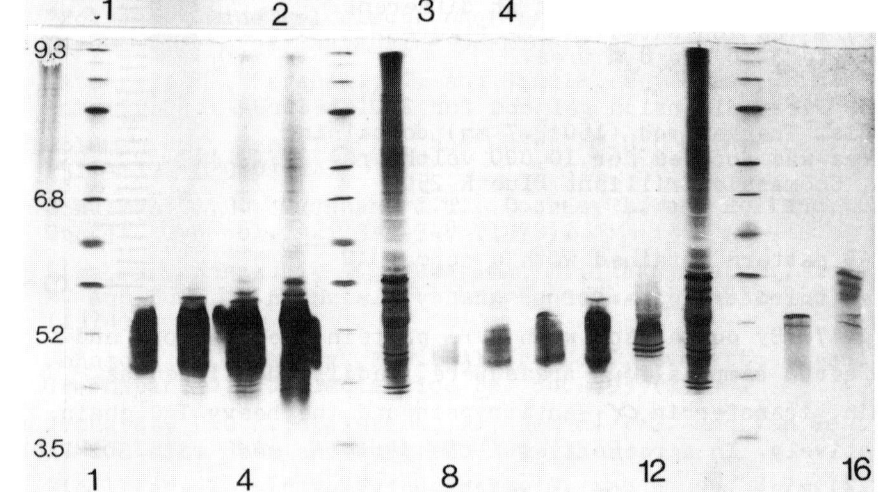

Fig. 4. IEF patterns of amniotic fluid samples using Pharmalyte 3-10, without denaturants. Samples were applied 35 mm from the anode. Samples: 1, 6 and 14. Broad pI marker kit containing 10 proteins covering the pH range 3.5 to 9.3 (15 µl); 2, 3, 4 and 5. normal AF sample (third trimester) 1,3,5 and 12 µl; 7 and 13. albumin-free human serum (10 µl); 8, 9, 10 and 11. normal AF (third trimester, but different to samples 2-5) aliquots of 1,3,6 and 12 µl respectively; 12. Haptoglobin rich serum fraction; 15. Ceruloplasmin (5 µl); 16. Transferrin (5 µl).

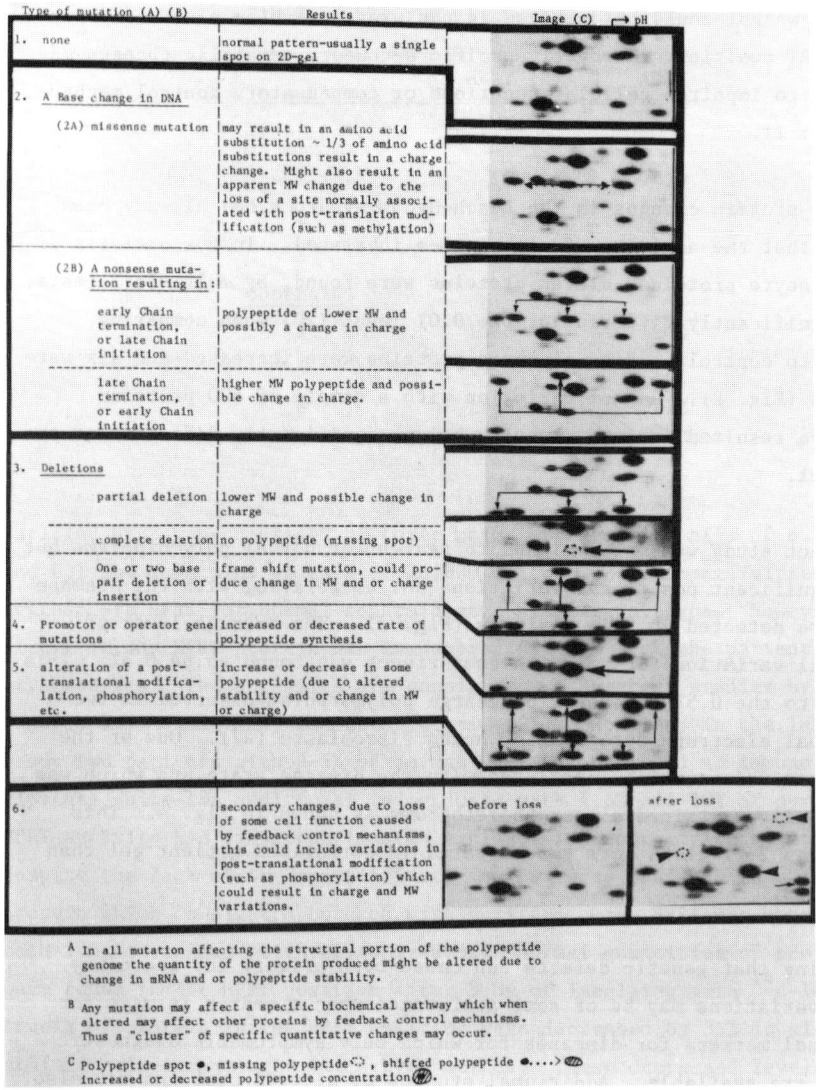

| Type of mutation (A) (B) | Results | Image (C) → pH |
|---|---|---|
| 1. none | normal pattern—usually a single spot on 2D-gel | |
| 2. A Base change in DNA | | |
|   (2A) missense mutation | may result in an amino acid substitution ~ 1/3 of amino acid substitutions result in a charge change. Might also result in an apparent MW change due to the loss of a site normally associated with post-translation modification (such as methylation) | |
|   (2B) A nonsense mutation resulting in: | | |
|     early Chain termination, or late Chain initiation | polypeptide of Lower MW and possibly a change in charge | |
|     late Chain termination, or early Chain initiation | higher MW polypeptide and possible change in charge. | |
| 3. Deletions | | |
|     partial deletion | lower MW and possible change in charge | |
|     complete deletion | no polypeptide (missing spot) | |
|     One or two base pair deletion or insertion | frame shift mutation, could produce change in MW and or charge | |
| 4. Promotor or operator gene mutations | increased or decreased rate of polypeptide synthesis | |
| 5. alteration of a post-translational modification, phosphorylation, etc. | increase or decrease in some polypeptide (due to altered stability and or change in MW or charge) | |
| 6. | secondary changes, due to loss of some cell function caused by feedback control mechanisms, this could include variations in post-translational modification (such as phosphorylation) which could result in charge and MW variations. | before loss / after loss |

A In all mutation affecting the structural portion of the polypeptide genome the quantity of the protein produced might be altered due to change in mRNA and or polypeptide stability.

B Any mutation may affect a specific biochemical pathway which when altered may affect other proteins by feedback control mechanisms. Thus a "cluster" of specific quantitative changes may occur.

C Polypeptide spot ●, missing polypeptide ⬭, shifted polypeptide ●....> ▰ increased or decreased polypeptide concentration ▰.

Fig. 1. Representation of possible effects of genetic disorders on polypeptide patterns in two-dimensional gels.

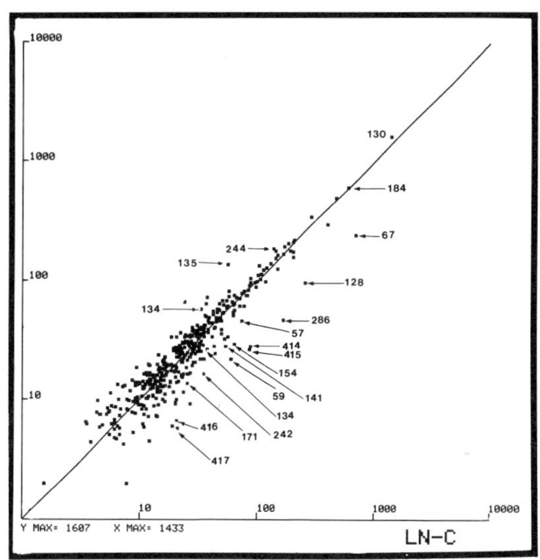



Fig. 2. Densitometric comparisons of normalized protein densities from three Lesch-Nyhan patients (ordinate) to three controls (abscissa). Background values were derived from the modal value of the local density histogram in the region adjacent to the measured protein. Identified proteins include:

|  |  | Density | | | |
|---|---|---|---|---|---|
| Protein | Protein index number | L-N | Control | T | 2P< |
| Actin | 130 | 1428 ± 184 | 1235 ± 6 | 0.977 | NS |
| β-Tubulin | 184 | 550.3 ± 263 | 533.9 ± 67 | 0.288 | NS |
| HPRT | 304 | 51 ± 5.8 | 43.1 ± 4.9 | 1.045 | NS |
| Intermediate filament protein | 209 | 156.3 ± 15.9 | 181.1 ± 5.7 | 1.475 | NS |

352

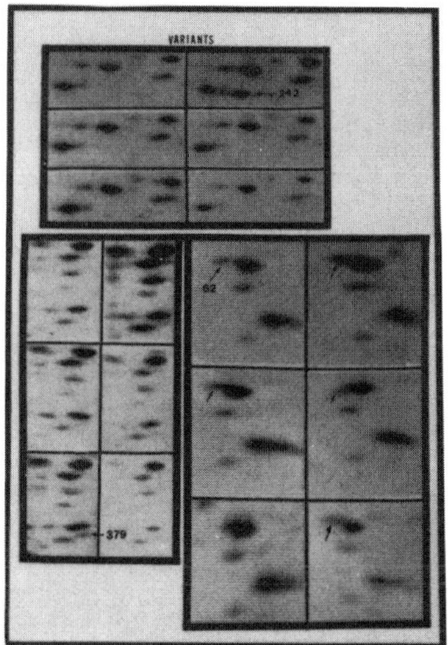

Fig. 3. Positional variations observed in this study which did not correlate with the disease state.

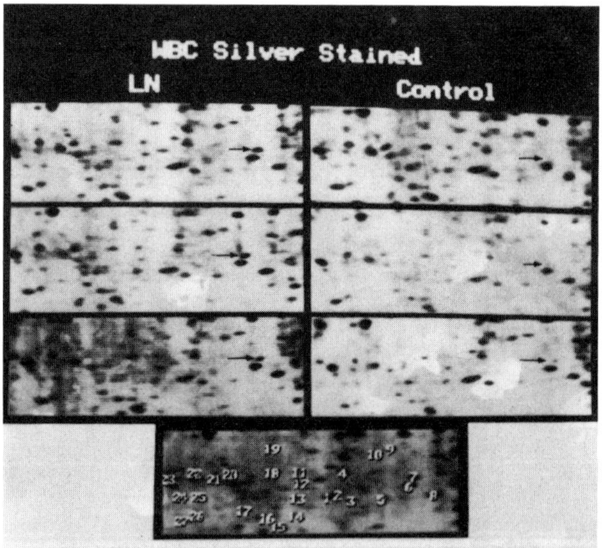


Fig. 4. Silver stained electrophoretograms demonstrating a protein variation which correlates with the disease state. This protein is indexed number seven and is indicated by arrows. It has an apparent molecular weight of approximately 30,000.

Acknowledgment

This work was supported in part by the National Institute on Alcohol Abuse and Alcoholism.

References

1.  Lesch, M., Nyhan, W. L.: Am. J. Med. 36, 561-570 (1964).

2.  Sorenson, L. B., Benke, P. H.: Nature 213, 1122-1123 (1967).

3.  Seegmiller, J. E., Rosenbloom, F. M., Kelley, W. N.: Science 155, 1682-1683 (1967).

4.  Kelley, W. N., Arnold, W. J.: Fed. Proc. 32, 1656-1659 (1973).

5.  Kelley, W. N., Green, M. L., Rosenbloom, F. M., Henderson, J. F., Seegmiller, J. E.: Ann. Intern. Med. 70, 155-206 (1969).

6.  Pehlke, D. M., McDonald, J. A., Holmes, E. W., Kelley, W. N.: J. Clin. Invest. 51, 1398-1404 (1972).

7.  Yip, L. C., Cancis, J., Balis, M. E.: Biochim. Biophys. Acta 293, 359-369 (1973).

8.  Greene, M. L., Boyle, J. A., Seegmiller, J. E.: Science 167, 887-889 (1970).

9.  Holmes, E. W., Pehlke, D. M., Kelley, W. N.: Biochim. Biophys. Acta 364, 209-217 (1974).

10. Becker, M. A., Argubright, K. F., Fox, R. M., Seegmiller, J. E.: Mol. Pharmacol. 10, 657-668 (1974).

11. Rosenbloom, F. M., Henderson, J. F., Caldwell, I. C., Kelley, W. N., Seegmiller, J. E.: J. Biol. Chem. 243, 1166-1173 (1968).

12. Reem, G. H.: Adv. Exp. Med. Biol. 41A, 245-253 (1974).

13. O'Farrell, P. H.: J. Biol. Chem. 250, 4007-4021 (1975).

14. Merril, C. R., Leavitt, J., Van Keuren, M. L., Ebert, M. H., Caine, E. D.: Neurology 29, 131-134 (1979).

15. Lowry, O. H., Rosebrough, N. G., Farr, A. L., Randall, R. J.: J. Biol. Chem. 193, 265-275 (1951).

16. Olsen, A. S., Milman, G.: Biochemistry 16, 2501-2505 (1977).

17. Ghangas, G. S., Milman, G.: Science 196, 1119-1120 (1977).

18. Merril, C. R., Switzer, R. C., Van Keuren, M. L.: Proc. Natl. Acad. Sci. U. S. A. 76, 4335-4339 (1979).

19. Switzer, R. C., Merril, C. R., Shifrin, S. A.: Anal. Biochem. 98, 231-237 (1979).

A QUANTITATIVE TWO-DIMENSIONAL ELECTROPHORETIC SURVEY OF PROTEINS
AFFECTED BY CHROMOSOME 21


Margaret L. Van Keuren, David Goldman*, and Carl R. Merril

Laboratory of General and Comparative Biochemistry, and *Laboratory of
Clinical Science, National Institute of Mental Health
Bethesda, Maryland  20205


## Introduction

The normal complement of human chromosomes consists of 22 pairs of
autosomal chromosomes and the sex chromosomes.  Normally there are two
copies of the autosomal chromosome 21.  An extra copy of chromosome 21,
a small acrocentric chromosome, or of a specific segment of it, is the cause
of a disorder known as Down's Syndrome.  The features of Down's Syndrome
include mental retardation, moderate growth retardation, epicanthal eye
folds, simian palmar creases, muscle hypotonia, an increased incidence
of congenital heart and intestinal tract malformations, an increased
incidence of leukemia, and premature aging. The pathogenesis of these
stigmata is unknown.  Complete monosomy 21, in which only one copy of
chromosome 21 is present, has been reported in only a few cases involving
liveborn infants.  The features of this disorder include growth
retardation, mental retardation, craniofacial anomalies, and failure
to thrive.

Genes which have been mapped to chromosome 21 include rRNA (1), the
cytoplasmic, or soluble form of superoxide dismutase ($SOD_s$) (2), the
interferon receptor (2), and two enzymes of the purine biosynthetic
pathway, phosphoriboslyglycineamide synthetase (3), and
phosphoribosylaminoimidazole synthetase (4).  Gene dosage effects have
been shown for some of these enzymes.  Erythrocytes, fibroblasts, lymphocytes
and granulocytes from individuals who are trisomic, disomic, and monosomic
(or who are monosomic for the region containing the $SOD_s$ gene) for
chromosome 21 display $SOD_s$ enzyme activity in the ratios 1.5:1.0:0.5,

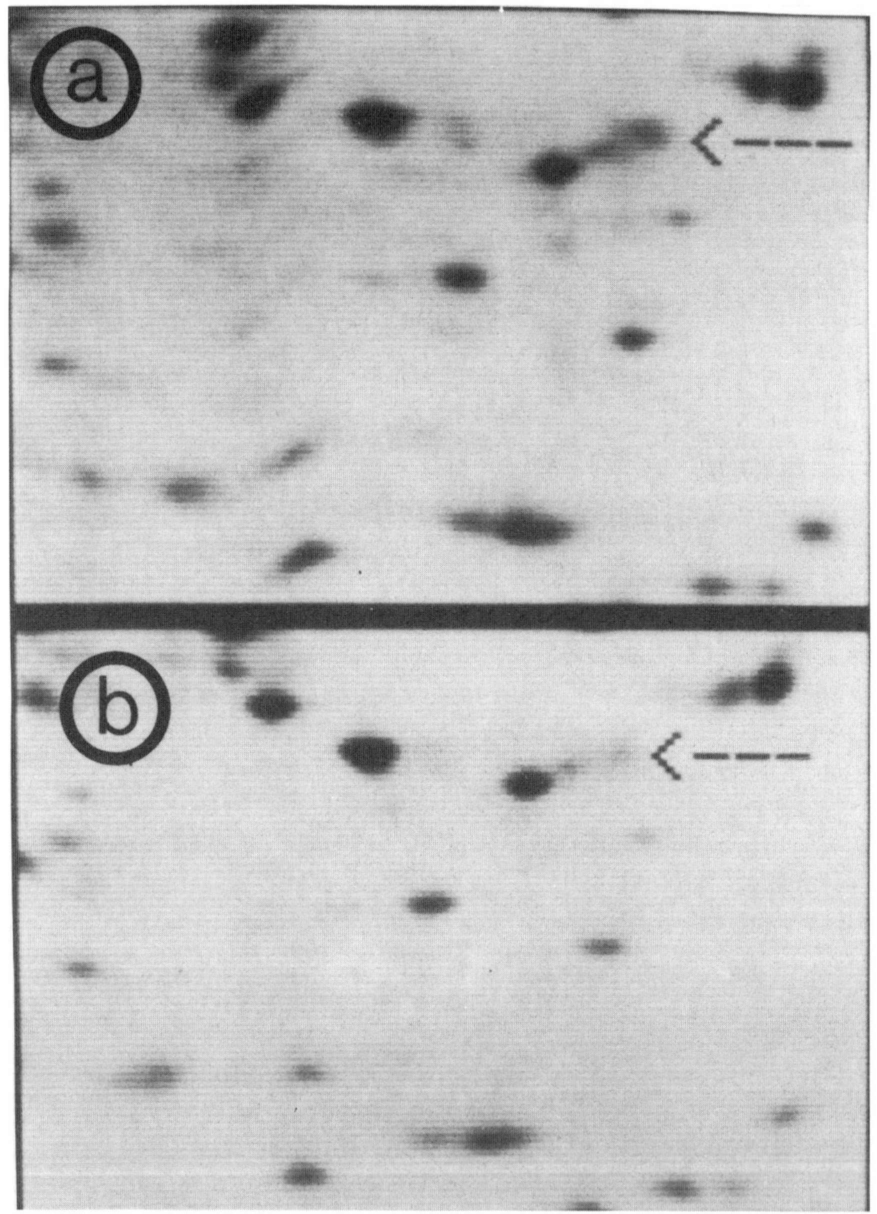

Figure 5. Autoradiograms showing protein density difference
between human-mouse hybrid cell and mouse parental cell lines.
a) proteins from the human-mouse hybrid cell line.
b) proteins from the mouse cell line. The molecular weight of
this protein (arrow) is approximately 47,000, and it is found
near the acid end of the gel.

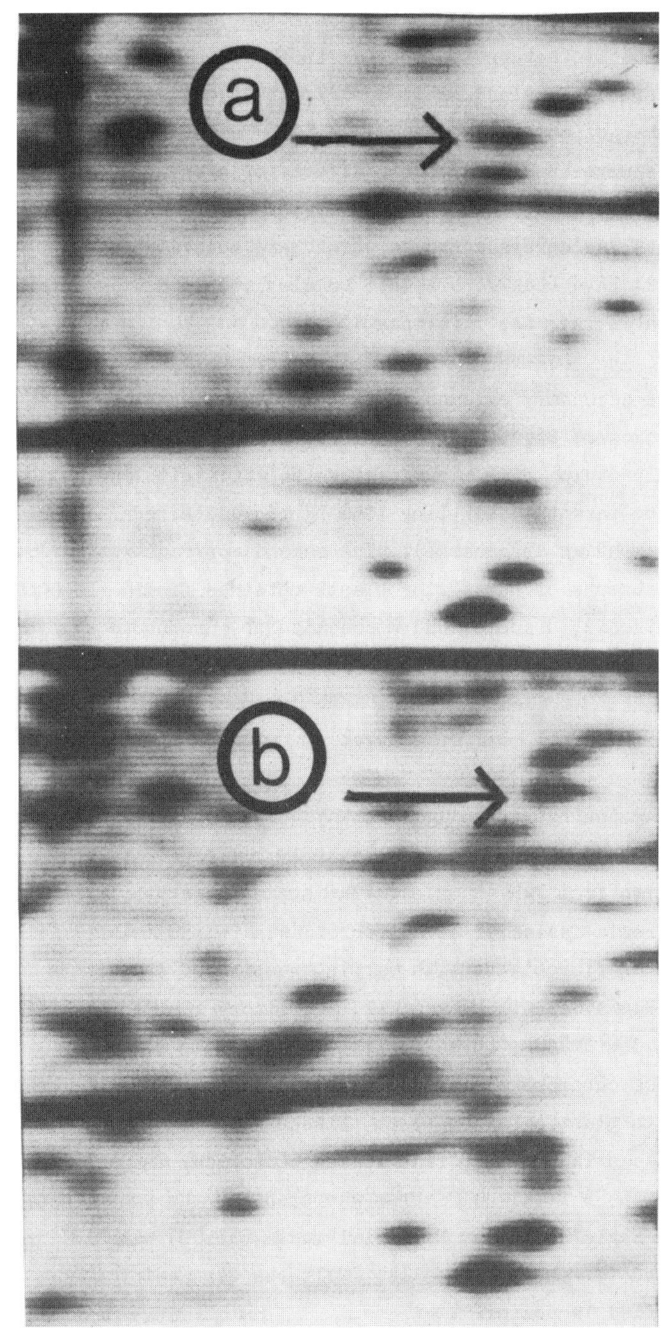

Figure 6. Silver stained gels of proteins from (a) human-mouse hybrid cells and (b)mouse parental cells. The arrow points to a protein with an apparent charge shift. The approximate molecular weight is 44,000 and approximate isoelectric point is 5.5. (It is more acidic in b.)

360

<u>Protein</u> <u>Density</u> <u>Measurement</u>. Autoradiograms and photographic negatives
of silver stained gels were scanned on an Optronics P-1000HS scanning
densitometer (Optronics) using the 0 to 3 optical density range at 100
microns resolution for photographic negatives and 300 microns for
autoradiograms. Digitalized image data was analyzed with a PDP 11/60
computer (Digital Equipment Corp.) equipped with a DeAnza IP5000 image
processor (DeAnza Systems) (16). The density standard was used to
normalize the computer images to correct for variations in photographic
and scanning techniques.

For the study involving the measurement of the human fibroblast $SOD_s$,
the densities of 30 proteins were measured in each silver stained gel or
autoradiogram. For the study of proteins in the hybrid and mouse
fibroblast lines, 500 protein densities were measured in each gel. To
correct for overall gel differences in staining and autoradiographic
intensity, densities were normalized by using the slope and y-intercept
derived from comparing non-saturating protein densities in each of the
gels followed by linear regression analysis.


Results


<u>Superoxide</u> <u>Dismutase</u>. Purified $SOD_s$ contained several protein species
as revealed by two-dimensional electrophoresis followed by silver staining.
The enzyme is known to be a dimer, containing two identical subunits
(with the exception of an electrophoretic variant found in Scandinavia (17))
of molecular weight approximately 16,000. The region of the gel
corresponding to this molecular weight is shown in figure 1. The arrow
points to the major, most densely stained, species. Upon co-electrophoresis
of purified $SOD_s$ with extract from hybrid cells, this major species
appeared in the same location as a protein consistently seen in silver
stained gels and autoradiograms of proteins from human-mouse hybrid
cells. There was only a very faint spot in this location on gels and
autoradiograms of proteins from the mouse parental line (fig. 2).

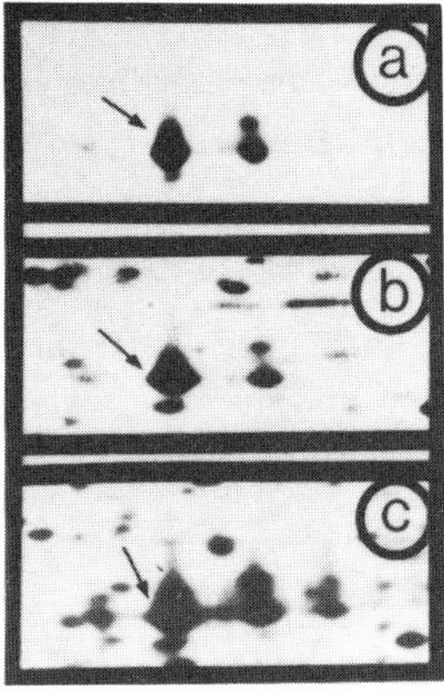

Figure 1. Localization of $SOD_S$ on two-dimensional gels.
a) 4 µg purified $SOD_S$ (silver stained).
b) 4 µg purified $SOD_S$ co-electrophoresed with proteins extracted from WAVR4dF9-4a hybrid cells.
c) 4 µg purified $SOD_S$ co-electrophoresed with proteins extracted from human fibroblasts (GM2504). (In all photographs the acid end is to the right.)

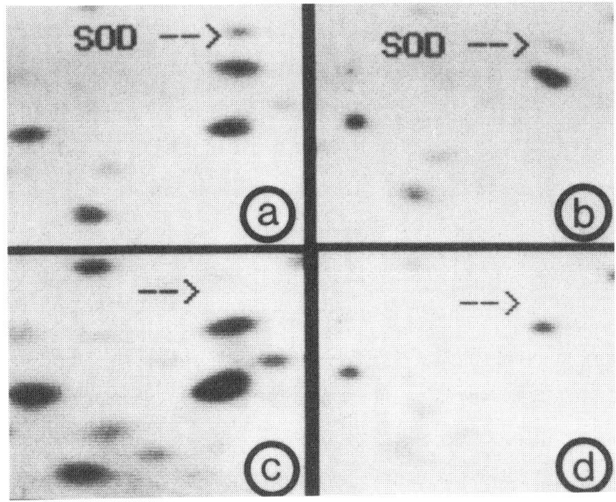


Figure 2. Comparison of the location of $SOD_S$ in the human-mouse hybrid cell line and in the mouse parental cell line. Proteins extracted from the hybrid line (a,b) and the mouse parental line (c,d). Silver stained gels: a) and c); Autoradiograms: b) and d). The arrow points to $SOD_S$. The normalized density ratios for the $SOD_S$ in the human-mouse hybrid cell to the mouse parental cell were 716:100 and 279:100 for the autoradiograms and silver stained gels, respectively. (Note the unusual protein which appears below $SOD_S$ in silver stained gels but is absent in autoradiograms.)

The density of $SOD_S$ was measured in silver stained gels and autoradiograms of human fibroblasts trisomic, disomic, and monosomic for chromosome 21. Proportionality between the number of gene copies and the density of $SOD_S$ was found, as shown in Table 1.

Table 1.  Percent Ratios of Densities of Cytoplasmic Superoxide Dismutase on Two-Dimensional Gels.

|                     | Silver Stained Gels | Autoradiograms |
|---------------------|:-------------------:|:--------------:|
| Trisomy 21          | 161.7               | 164.7          |
| Disomy 21 (Normal)  | 100.0               | 100.0          |
| Monosomy 21         | 68.2                | 33.0           |

(Proteins for silver stained gels were extracted with TX-100, and those for autoradiography by lysis in 10M urea.  The data were normalized by the slope and the y-intercept of the density graphs (not shown) of the proteins other than superoxide dismutase.)

Double Label Autoradiography.  Double label autoradiography was performed to demonstrate proteins expressed in human-mouse hybrid cells but not mouse parental cells.  Hybrid cell proteins labeled with $^3$H were mixed with $^{14}$C-labeled parental cell proteins as described previously (18). Using the computer, positive and negative images of fluorograms and autoradiograms were superimposed on one another at high magnification. The negative image of the fluorogram, which shows both $^3$H- and $^{14}$C-labeled proteins, and the autoradiogram, which shows only $^{14}$C-labeled proteins, are displayed in figure 3.  Proteins unique to the hybrid cell (and therefore labeled with $^3$H but not $^{14}$C) would be expected to appear as white spots unobscured by black spots.  Out of 400 proteins, only $SOD_S$ appeared as a unique protein in the human-mouse hybrid cell line.

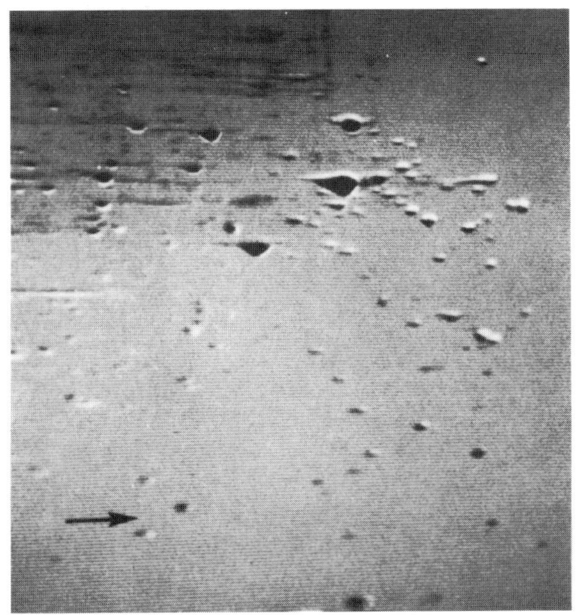

Figure 3. Double-label autoradiography of hybrid cells (labeled with $^3$H) and mouse parental cells (labeled with $^{14}$C). The negative image of the fluorogram (white spots) shows $^{14}$C- and $^3$-H labeled proteins, and the autoradiogram (black spots) shows $^{14}$C-labeled proteins only. $SOD_s$ (as indicated by the arrow) was seen in the fluorogram as a white spot which is too faint to be seen in this photograph.

<u>Qualitative</u> <u>and</u> <u>Quantitative</u> <u>Analysis</u> <u>of</u> <u>Hybrid</u> <u>and</u> <u>Mouse</u> <u>Parental</u> <u>Cell</u> <u>Lines</u>. Proteins were assigned an identification number and their densities in silver stained gel images and autoradiograms were measured. Some proteins which were resolved on silver stained gels were not resolved on autoradiograms, and the converse was also the case. Density versus density plots were constructed for proteins visualized either by silver staining or autoradiography and which were of non-saturating density (fig. 4). Protein densities were normalized using the slope and y-intercept obtained from linear regression analysis.

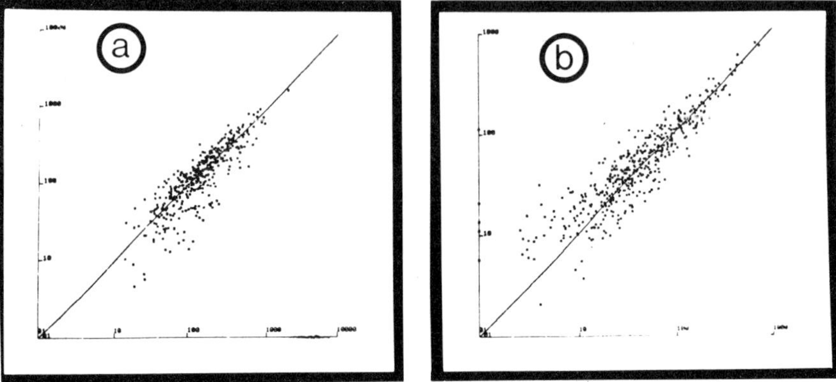

Figure 4.  Density versus density graphs of proteins from the human-mouse
hybrid cell (y-axis) and the mouse parental mouse cell (x-axis).
a) silver stained gel images.  b) autoradiograms.


Most of the normalized protein densities were approximately equal in the
two cell lines.  Examination of those normalized protein densities that
differed by more than 75% revealed that these were generally proteins
which were faint or in regions of exceptionally dark background or local
distortion.  One density difference was found which was clearly not due
to an image artifact (fig. 5). The normalized density of this protein
was twice as great in human-mouse hybrid cells than that found in mouse
cells.

A possible protein charge shift was found when comparing proteins from
the human-mouse hybrid cell and mouse parental cell lines as shown in
figure 6.  The proteins illustrated by the arrows may be the same protein.
If these are the same proteins, it is more acidic in the mouse parental
cell line.  Evidence for the equivalence of these proteins includes the
fact the their normalized densities are equal, they both appear only on
silver stained gels, not on autoradiograms, and they have the same molecular
weight.

A second apparent protein charge shift between the human-mouse hybrid
and mouse parental cell lines is shown in figure 7.  Two protein
species in the human-mouse hybrid cell electrophorese closely together
but apart in parental cells.  This interpretation is supported by
the fact that the sum of the densities of the two proteins in mouse
parental cells equals the density of the single protein in human-mouse
hybrid cells (fig. 7, panels b and d) in the autoradiograms.  However,
in silver stained gels, the summed density of the proteins in the mouse
parental cell was twice the density of the protein in the human-mouse
hybrid cell (fig. 7, panels a and c).

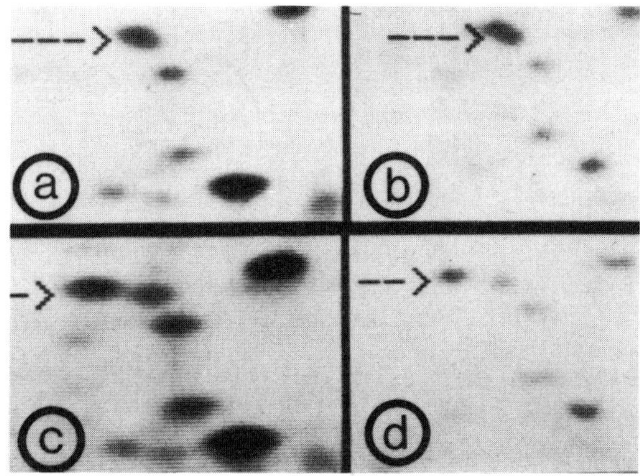

Figure 7.   Proteins from the human-mouse hybrid and mouse parental lines.
a) silver stained gel of human-mouse hybrid proteins.
b) autoradiogram of human-mouse hybrid proteins.
c) silver stained gel of mouse parental proteins.
d) autoradiogram of mouse parental proteins.
The illustrated proteins (arrow) are of molecular weight approximately
20,000 and apparent isoelectric point 5.0.

Discussion

Two dimensional electrophoresis, with the aid of a microdensitometer
and computer to measure protein spot densities, has been utilized to
search for proteins whose expression is affected by chromosome 21. The
protein spot corresponding to $SOD_S$, which is coded for by chromosome 21,
was located by co-electrophoresis of purified enzyme with cellular proteins.
The density of the $SOD_S$ was measured in fibroblast protein extracts from
cells which are trisomic, disomic, and monosomic for chromosome 21. The
densities on silver stained gels and autoradiograms, each prepared from
samples extracted by different methods, were measured. Normalized density
ratios (Table 1) closely approximated the previously reported enzyme
activity ratios of 1.5:1.0:0.5 for trisomic, disomic, and monosomic cells.

The gel pattern of purified human red blood cell $SOD_S$ revealed several
protein species of the same molecular weight but different charges.
The major species of purified $SOD_S$ was shown to co-electrophorese with
a specific protein on gels of protein extracts from the mouse-human
hybrid and human lines. However, multiple charge species of $SOD_S$ were
not observed in preparations of protein from these fibroblast lines.
Multiple protein species found in purified enzymes preparations may be
the result of modification, such as carbamylation, which occured during
purification. However, there may be a biological explaination for the
appearance of multiple spots in the purified $SOD_S$ preparation. An
example of a biological situation in which multiple charge species appear
was found in studies of hypoxanthine-guanine phosphoribosyl transferase
(HGPRT). Red blood cell HGPRT appears as three subunit species varying
only in isoelectric point, while only one appears in fibroblast
preparations (19). $SOD_S$ of human red blood cells has been reported to
have one major electrophoretic band and one or two minor ones (17).
Isoelectric focusing of $SOD_S$ from liver of cow, rat, mouse, and chicken
revealed 7,4,6, and 8 bands, respectively, with isoelectric points between
4.5 and 5.5 (20, 21). On our gels, the protein spot corresponding to
$SOD_S$ has an apparent isoelectric point of approximately 5.5. Gels of
mouse parental cells have a very faint spot at this location. It is

possible that one form of the mouse enzyme electrophoreses to the same location as the major form of the human enzyme.

The technique of double labeling was utilized to search for proteins coded for by human chromosome 21 or influenced by its presence in a human–mouse hybrid cell line.  No clear examples other than $SOD_s$ were found by double labeling in this study.  Our inability to detect such proteins by double labeling could be due insufficient production of human protein, production of human proteins which are not observed under the electrophoresis conditions employed in this study, or co-migration with mouse proteins.  It has been estimated that 50% of human and mouse proteins co-migrate on two-dimensional gels (18).  However, another double labeling study using a Chinese hamster–human cell hybrid line demonstrated proteins which were influenced by the presence of human chromosome 21 (22).  The Chinese hamster cell may be sufficiently different from the mouse cell to allow detection of differential protein expression with the double label technique.

Quantitative analysis of the proteins of human–mouse hybrid cells and mouse parental fibroblasts showed one protein spot to have twice the density in the human–mouse hybrid line.  This protein could be a gene product coded for by chromosome 21, and be of such composition that it co-migrates with an identical protein in the mouse.  Alternatively, this could be a mouse protein whose quantity is affected by the presence of human chromosome 21.

Two additional differences were found between the human–mouse hybrid cell and mouse cell proteins.  Both are possible charge shifts.  However, the second of these (fig. 7) may be a more complex alteration since the silver stained protein is also more dense in human–mouse hybrid cells. The protein which has an apparent charge shift as illustrated in figure 6 was observed only in silver stained gels.  In this case, double labeling was not capable of illustrating an effect of the presence of human chromosome 21.  In general, double labeling relies on qualitative differences, while methods employed in this study are capable of detecting

368

quantitative differences as well and may detect human proteins which
co-migrate with mouse proteins.

This study illustrates the need for multiple technologies to search for
proteins whose expression is influenced by the presence of human chromosomes
in somatic cell hybrids. Delineation of proteins affected by chromosome
21 may lead to a better understanding of the pathophysiology associated
with Down's Syndrome.


Acknowledgement

This work was supported in part by the National Institute on Alcohol
Abuse and Alcoholism.


References

1.  Schmickel, R.D., Knoller, M.; Pediat. Res. 11, 929-935 (1977).

2.  Tan, Y.H., Tischfield, J., Ruddle, F.H.: J. Exp. Med. 137, 317-330 (197

3.  Moore, E.E., Jones, C., Kao, F.-T., Oates, D.C.: Am. J. Hum. Genet.
    29, 389-396 (1977).

4.  Patterson, D., Graw, S., Jones, C.: Proc. Natl. Acad. Sci. 78, 405-409
    (1981).

5.  Schitiu, S., Sinet, P.M., Lejeune, J., Frezal, J.: Humangenetik 23,
    65-72 (1974).

6.  Feaster, W.W., Kwok, L.W., Epstein, C.J.: Am. J. Hum. Genet. 29, 563-570
    (1977).

7.  Scoggin, C.H., Beskan, J., Davidson, J.N., Patterson, D.: Clin. Res.
    28, 31A (1980).

8.  Bartley, J.A., Epstein, C.J.: Biochem. Biophys. Res. Comm. 93, 1286-
    1289 (1980).

9.  Tan, Y.H., Schneider, E.L., Tischfield, J., Epstein, C.J., Ruddle, F.H.:
    Science 186, 61-63 (1974).

10. Franke, U.: in Trisomy 21 (Down Syndrome) Research Perspectives. Ed.
    de la Cruz, F.F., Gerald, P.S.: University Park Press, Baltimore.
    237-251, (1981).

11. Milman, G., Lee, E., Ghangas, G.S., McLaughlin, J.R., George, M:
    Proc. Natl. Acad. Sci. $\underline{73}$, 4589-4593 (1976).

12. McConkey, E.H., Taylor, B.J., Phan, D.: Proc. Natl. Acad. Sci $\underline{76}$,
    6500-6504 (1979).

13. O'Farrell, P.H.: J. Biol. Chem. $\underline{250}$, 4007-4021 (1975).

14. Switzer, R.C., Merril, C.R., Shifrin, S.: Anal. Biochem. $\underline{98}$, 231-237
    (1979).

15. Merril, C.R., Switzer, R.C., Van Keuren, M.L.: Proc. Natl. Acad. Sci.
    $\underline{76}$, 4335-4339 (1979).

16. Goldman, D., Merril, C.R. (manuscript submitted).

17. Beckman, G., Pakarinen, A.: Hum. Hered. $\underline{23}$, 436-451 (1973).

18. McConkey, E.D.: Somat. Cell. Genet. $\underline{6}$, 139-147 (1980).

19. Zannis, V.I., Gudas, L.J., Martin, D.W.: Biochem. Genet. $\underline{18}$, 1-19 (1980).

20. Lönnerdal, B., Keen, C.L., Hurley, L.S.: Fed. Proc. $\underline{39}$, 836(abstract)(1980).

21. Lönnerdal, B., Keen, C.L., Hurley, L.S.: FEBS Letters $\underline{108}$, 51-55 (1979).

22. Scoggin, C.H., Davidson, J.N., Patterson, D.: Clin. Res, $\underline{28}$, 32A (1980).

# MULTISPECTRAL DIGITAL IMAGE ANALYSIS OF COLOR TWO-DIMENSIONAL ELECTROPHORETOGRAMS


Robert K. Vincent, John Hartman
BioImage, 320 N. Main, Suite 301, Ann Arbor, MI 48104 U.S.A.

Alan S. Barrett
Optronics International Inc., 7 Stuart Rd.
Chelmsford, MA 01824 U.S.A.

David W. Sammons
The Upjohn Company, Kalamazoo, MI 49001 U.S.A.


Introduction

The recent extension of traditional one-dimensional gel electrophoresis into two dimensions has vastly improved the resolution
and precision of the technique, thus greatly increasing the
amount of information that can be wrung from samples containing
complex mixtures of proteins.  This information "glut" has quite
naturally stimulated interest in the use of computers to aid in
the qualitative and quantitative analysis of proteins separated
on such gels, and considerable work has recently been done
toward this end (1-7).

Whereas the utility of two-dimensional electrophoresis lies in
the replacement of a single variable with a pair of orthogonal
variables representing the net charge and molecular weight of
component proteins, the silver stain process described by Adams
and Sammons (8) extends this pair to three:  net charge,
molecular weight, and color.  In addition to a characteristic

position (x and y coordinates) on a gel and an intensity, the
spot formed by a given protein exhibits upon silver staining a
characteristic hue as well, usually of various shades of yellow,
red, green, and blue.  As will be shown below, the implications
of this are more than merely aesthetic; the "third dimension"
will permit greater resolution, accuracy, and reliability of
either automatic or manual analysis than is attainable via
existing monochrome approaches.

This paper will concern itself chiefly with the preliminary
results of an ongoing study at BioImage Corporation of the
properties and applications of digitized multispectral images
of silver stained electrophoresis gels from automatic digital
processing and analysis of multispectral image data from remote
sensing satellites, particularly LANDSAT, as well as from other
sources.  A large body of generalized image-processing software
has evolved that is capable of performing a wide variety of
operations (e.g., contrast stretching, edge enhancement,
spectral ratioing, geometrical transforms, automatic spectral
recognition and classification, spatial feature analysis,
statistical analysis, and others) on digital images with an
arbitrary number of spectral channels (9).  The focal point of
this paper implements this existing software toward analysis of
a 2-D electrophoretogram that has previously been stained with
silver such that the polypeptides fixed therein are displayed
in their characteristic colors.


Methods


Two-dimensional electrophoretograms of protein samples were
derived from the liver of non-diabetic and diabetic animals
(Upjohn's colony of spontaneously diabetic Chinese hamsters)
and were prepared by conventional procedures for the ISO/Dalt
system (10, 11).  Two hundred milligrams of wet weight tissue
were homogenized into 1 ml of ISO urea solution (12).  After

centrifugation at 200,000 x g, 10 µl of sample were loaded onto each ISO (LKB Ampholine, pH 3-10) gel. The separated polypeptides in the 2-D gels were fixed, washed, and complexed with silver using the gel electrophoresis color development system previously described (8). After staining of the 2-D gel with silver, an actual size 8x10 inch color transparency (Kodak Ektachrome 64, 6117) was prepared from the gel by backlighting with a daylight fluorescent tube. The resulting transparency was digitized on an Optronics International P-1000 Colorscan drum scanner. Each transparency was scanned three times at 100 µm resolution through narrow bandwidth filters passing light only in one of the three additive primary colors, i.e., blue, green, and red. Three data values were thus generated for each 100 µm picture element (pixel) that range from 0 to 255 and correspond to optical densities in each primary color from 0.0 to 3.0 O.D. Each scan required 3.75 minutes for completion, or a total of 11.25 minutes per gel.

The three data sets so obtained for each gel were recorded on magnetic tape and were assimilated into single files (one per gel) to facilitate further processing and analysis.


Results and Discussion


Figure 1 illustrates 2-D patterns of liver polypeptides from a nondiabetic and diabetic Chinese hamster. In the area designated by "A" is a yellow polypeptide which is to be compared by ratio recognition criteria. The area designated by B indicate two overlapping spots, one yellow and one black. These two spots from the nondiabetic pattern is later addressed with respect to analysis of distinct spectral signatures for each of the overlapping spots.

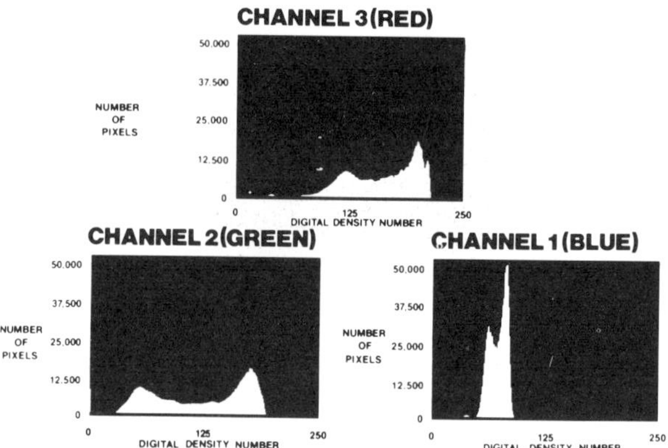



Figure 2. Histograms of spectral channels from the nondiabetic 2-D pattern

The histograms of Figure 2 illustrate the distribution of intensity levels for each of the three spectral channels (blue, green, and red) in a digital multispectral image made of the nondiabetic 2-D pattern. All three histograms are bimodal, but it is obvious that the variance in channel 1 (blue) is much smaller than in the other two channels. It is possible to ameliorate such histogram constriction either in the course of the initial scanning (by adjusting the sensitivity of the digitizer) or subsequently to digitization, by transforming data values such that their histogram is spread over the desired region. For current purposes, however, it was decided best to preserve the linear relationship between optical density and digitized value that obtains identically for each channel. If multispectral analysis as applied to color electro- phoretograms is to have any kind of compelling utility, two requirements are necessary: (1) classes of protein spots in a single gel image must be differentiable on the basis of their spectral properties, and (2) the spectral properties of the spot of a given protein must be reproducible in different gels. That the first condition holds may be confirmed by a cursory inspection of Figure 1. In practice, we have found that most of the protein spots can be grouped into four reasonable dis-

tinct spectral classes corresponding approximately to the colors
yellow, orange, red, light green, and dark blue-green.

Figure 3.    Automatic recognition map of yellow polypeptide
             spots from nondiabetic 2-D pattern

Figure 3 shows the output of a program (AUTOREC) that was used
to recognize picture elements of a gel image belonging to the
yellow spectral class.  After pixels of a small number of
representative yellow spots were examined statistically, a set
of spectral signatures were formulated that could be applied to
the entire image to recognize other yellow spots while minimizing the false inclusion of background, much of which is yellow-
orange.  Because we wished to base the classifications on hue
rather than intensity - on the presumption that hue would be
less influenced than intensity by gel-to-gel differences - the
spectral signatures were based not on the density values themselves, but rather on the ratios between optical densities in
different channels.  Excluding reciprocation, three spectral
ratios are defined by the three channels (C1, C2, C3, representing blue, green, and red, respectively):  C2/C1 (hereafter
designated as R21), C3/C2 (R32), and C3/C1 (R31).  Ordinarily,
ratios are computed by a table look-up technique (to minimize
computation time) after subtracting from the respective channels
one less than their minimum spectral density values, i.e., if
$CA_{ij}$ and $CB_{ij}$ are the values of channels A and B for a pixel at
image coordinates i, j, the corrected ratio at that point is
given by

$$RAB_{ij} = (CA_{ij} - MIN(CA) + 1) / (CB_{ij} - MIN(CB) + 1) \quad (1.)$$

where MIN(CA) represents the minimum value occurring over the entire image in channel A.  This simple correction is introduced to compensate for variable histogram offsets that may arise in the photographic and scanning processes.  For the spectral recognition map of Figure 3, these subtractive corrections were 26, 22, and 24 for channels 1, 2 and 3, respectively, and a pixel was assigned to the spectral class in question (yellow) if, after such correction, $0.29 \leq R21 \leq 0.53$, $0.27 \leq R32 \leq 0.44$, and $0.11 \leq R31 \leq 0.18$.

Table 1 shows the ranges of ratio values that were used to recognize several other spectral classes of protein spots on the same gel image.

Table 1.  Spectral Ratio Signatures

| Spot Color | R21 | R32 | R31 |
|---|---|---|---|
| Yellow | 0.29-0.53 | 0.27-0.44 | 0.11-0.18 |
| Red | 0.70-0.97 | 0.27-0.46 | 0.24-0.38 |
| Light-Green | 0.65-0.94 | 0.83-0.87 | 0.52-0.71 |
| Dark Blue-Green | 0.89-1.19 | 0.66-0.85 | 0.70-0.90 |

A recognition map made using all of these targets is presented in color-coded form as Figure 4; yellow class pixels are mapped as yellow, red, as red, light green as green, and dark blue-green as blue.  Pixels mapped in cyan (light blue) represent the intersection of the light green and dark blue-green targets. It may be noted that some background pixels were falsely recognized as red in the upper portion and yellow in the lower portion of the gel.  This reflects the fact that the background of the silver-stained gel images is not uniform across its full extent, but rather varies in broad local regions from orange-red (generally at the top) to yellow-orange (at the bottom). This presumably can be ameliorated by subtracting local background from pixel density values; at the time of this writing this has not yet been done.  At any rate, the methodology described above suffices for current purposes.

An attempt is made to apply the ratio recognition criteria for
yellow protein spots in the nondiabetic pattern to a second gel
prepared in a similar fashion, but with a liver tissue sample
from a diabetic hamster. The use of ratios rather than channel
values to define spectral classes was motivated by a desire to
minimize (although not eliminate) both quantitative effects and
spectral variations between gels. How great these variations
are is a subject of our on-going work, for example, the spectral
properties of electrophoretograms prepared from protein reference standards are currently being investigated. Figure 5a
shows the result of this spectral recognition encoded as red
and, overlaid on the central region of the spectral recognition
map of the first gel image, encoded as green; overlapped pixels
are expressed as yellow (R+G→Y). The second (red) map was subsequently corrected geometrically by the method of nearest
neighbor resampling (using the program RSAMPLG) after geometrical transform coefficients were calculated on the basis of
ten landmark spots identified on both gels (using the program
TRANSFORM). This transformation is essentially a mapping from
one arbitrary quadilateral to another, including whatever
translation, rotation, and scale changes are necessary to
minimize the mean square error between corresponding landmark
spots on the two gels. The results of this transform is displayed as Figure 5b; generally, corresponding spots are located
within 2-3 mm of each other. In general, the spectral criteria
used to recognize yellow spots on the nondiabetic pattern
worked well when applied without adjustment to the diabetic
pattern. In most instances where a spot was recognized on one
gel but not the other, inspection of the original gel showed
the unrecognized spot to be very much weaker or poorly resolved
(streaked); in all cases the hue of the "missing" spot was
similar to its counterpart, but obviously not sufficiently so
to have met the recognition criteria.

In at least one such case, this "recognition failure" points to

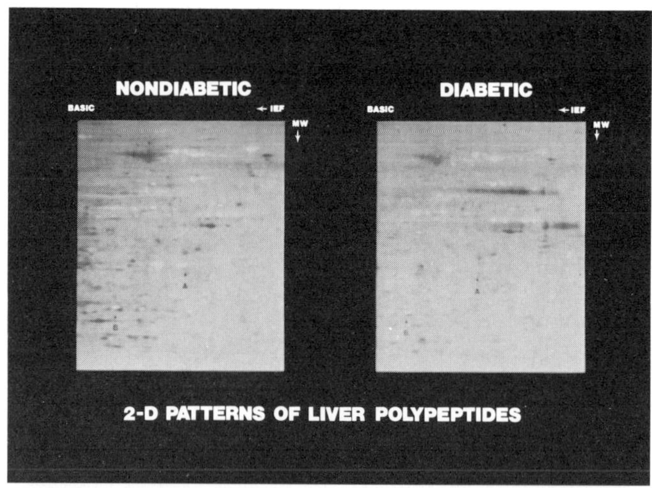



2-D PATTERNS OF LIVER POLYPEPTIDES

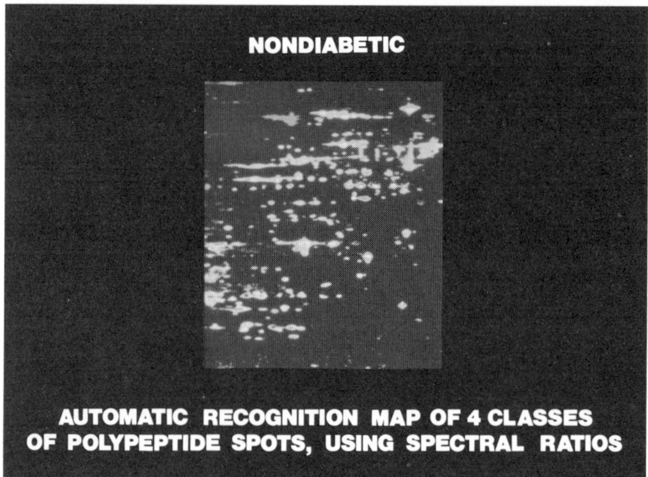



AUTOMATIC RECOGNITION MAP OF 4 CLASSES
OF POLYPEPTIDE SPOTS, USING SPECTRAL RATIOS

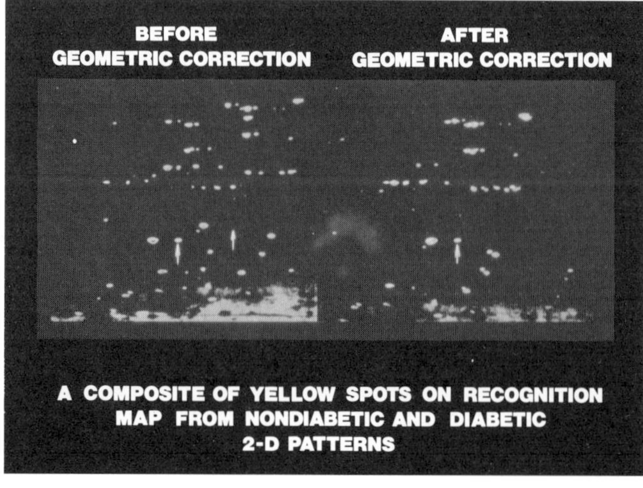



A COMPOSITE OF YELLOW SPOTS ON RECOGNITION
MAP FROM NONDIABETIC AND DIABETIC
2-D PATTERNS

what appears to be a significant quantitative difference between the two gels; this involves the spot labelled "A" in
Figure 1 and is designated by the arrow in Figure 5a and 5b.
While there is a spot in the nondiabetic gel that corresponds
to spot A in the diabetic gel, it is very much weaker and
shifted slightly in hue toward red.  It cannot, of course, be
concluded on the basis of two gels that a quantitative difference such as the one described here is somehow related to the
fact that one hamster was diabetic and the other was not.  It
is not unlikely, however, that where such differences are
indeed significant, multispectral digital analysis will be a
valuable aid in their detection and quantitation.  The spot
difference noted here, in fact, had gone unnoticed until our
attention was drawn to it by the spectral recognition/
geometric correction results.

Perhaps the most critical problem to be dealt with in the
computerized image analysis of two-dimensional gels is that
of segmentation, i.e., the accurate identification of spot
boundaries on the gel.  Considerable work on the segmentation
of monochrome gel images has been done in recent years which
has led to several computationally simple and acceptable
accurate solutions to this problem, such as peak detection
and inflection point delimiting (3) or "curvex area detection"
(4).

Thresholding, whether monochrome or - as illustrated above as
"spectral recognition" - multispectral, is generally not itself an adequate approach to the segmentation problem, since
it requires near-perfect spectral uniformity between gel
images or at least a priori knowledge of systematic non-
uniformities.  Several techniques are currently being evaluated for the segmentation of multispectral gel images that
are analogous to the various monochrome approaches in use.

380

The potential usefulness of the color dimension for segmentation is apparent on close inspection of Figure 1, where it is
common to find overlapping spots with very distinct spectral
signatures.  Such an instance is illustrated in terms of the
3-channel digital density values in Figure 6, which covers
the immediate area surrounding the two spots labelled "B" in
Figure 1; to conserve space, only every second pixel of every
second line is displayed.  Approximate outlines are drawn in
for each of the two spots.  It can be seen that the two spots,
which fall respectively within the yellow and blue-green
spectral classes, are very clearly distinguishable despite
their spatial proximity.



```
185 184 180 180 182 185 178 181 181 180 185 184 184 180 187 187 187 183 188 182 183 183 184 185 181 179 184 183
181 181 181 177 185 183 185 179 184 178 187 180 182 180 185 186 182 182 186 187 182 180 185 186 180 179 179 179
182 183 182 182 182 176 180 179 177 177 177 178 176 177 179 174 184 181 181 184 189 184 184 179 183 184 181 180
183 184 181 181 179 178 177 177 178 174 175 169 168 171 169 174 174 179 181 179 182 182 184 183 183 182 183 177
183 186 181 175 176 174 170 170 170 167 170 168 172 169 166 171 169 167 172 178 180 183 183 184 183 177 184 183
188 183 179 174 178 170 169 169 169 163 169 168 166 168 165 170 171 172 169 178 180 181 188 185 187 186 180 184
191 180 178 177 177 173 167 169 167 166 173 168 166 164 168 170 170 171 174 176 183 186 185 186 188 185 180 186
187 186 184 179 178 175 176 166 171 167 166 171 169 172 169 172 177 179 187 189 191 190 194 190 193 190 181 182
184 190 188 182 179 182 177 174 174 175 175 172 171 178 181 184 185 189 193 191 190 192 194 190 188 188 188 184
183 193 189 184 186 183 180 181 178 183 174 178 185 185 191 194 193 198 201 193 197 197 200 192 187 193 196 190
192 190 186 186 190 189 187 184 189 186 186 188 193 193 197 192 193 194 198 192 200 194 191 197 192 194 191 187
186 188 185 185 188 182 185 190 186 190 188 188 188 198 197 191 193 197 197 198 196 192 195 195 190 189 190 189
185 186 184 189 184 184 185 186 182 184 189 190 184 191 193 190 192 194 193 202 191 194 194 196 195 191 189 189
176 178 180 181 184 180 176 185 185 185 181 185 184 188 187 187 189 190 191 195 192 193 192 192 191 191 187 183
177 180 184 179 177 179 178 181 184 182 180 183 184 182 189 185 186 183 187 182 189 191 189 191 192 182 187 180
179 178 182 178 178 181 175 176 180 174 182 184 182 182 181 180 185 181 185 188 183 183 186 188 186 183 185 173
```



```
134 129 128 126 129 131 134 139 137 139 145 147 147 150 151 147 151 148 144 143 139 137 136 133 131 133 130 126
130 121 128 127 133 134 136 138 136 140 142 140 138 144 145 147 150 143 147 143 140 140 138 132 128 130 124 121
132 130 132 131 133 133 131 130 120 119 117 112 112 115 119 122 129 137 144 145 145 144 141 135 131 127 127 125
135 131 133 131 122 113 108 102  88  85  85  85  83  82  88  92  99 108 124 136 141 144 145 141 139 130 128 126
141 138 131 123 111  97  87  83  75  70  69  67  67  70  69  75  80  89 104 117 134 141 148 143 142 135 133 132
144 141 133 121 103  90  77  74  68  65  66  64  64  63  66  69  73  81  95 115 131 147 148 149 149 140 136 135
146 140 135 121 106  90  78  74  70  66  64  65  64  66  66  70  77  89 110 128 150 155 155 156 148 143 137 135
150 150 142 132 121 105  92  83  76  73  69  70  72  76  81  91 107 129 150 164 177 170 172 165 158 147 137 138
159 144 149 143 138 128 115 109 101  93  92  92  97 111 121 149 167 179 186 188 189 188 178 170 163 150 143 138
160 153 147 145 147 144 143 139 128 129 123 134 145 165 172 184 186 185 189 191 185 187 178 176 172 160 151 141
168 154 152 142 146 147 146 153 153 157 160 167 174 177 183 191 186 187 189 193 194 193 186 184 172 168 157 146
162 146 146 143 141 143 140 154 155 151 162 160 168 176 181 185 187 185 189 193 189 190 187 179 177 168 157 155
150 141 137 132 133 129 133 148 141 153 147 153 153 160 169 172 178 181 185 180 181 185 182 175 175 164 157 142
141 135 131 127 129 125 126 135 130 134 134 136 141 144 153 158 160 169 170 177 173 170 173 170 158 151 148 140
129 126 121 123 117 118 123 130 120 124 127 131 128 130 136 141 143 150 154 157 159 158 156 154 143 140 131 134
119 119 120 115 117 114 116 126 118 116 120 123 126 126 122 128 131 133 133 136 139 138 140 135 134 125 123 119
```

Figure 6.  Digital discrimination of two overlapping spots
in channels 1 and 2

Even more than segmentation, the availability of multispectral data to a computer performing automatic analysis
will greatly facilitate the comparison of one gel (perhaps
prepared for diagnostic purposes) to another (perhaps a
reference standard) through respective data bases containing
information (such as coordinate position, size, spectral
characteristics, and possibly the integrated intensity over
the spot of a 3-dimensional spectral vector) collected in the
course of the segmentation processes.  For example, one problem that often occurs in gel comparisons is the need to decide

which of several spots in close proximity on a reference gel
is to be assumed to correspond to a given spot on a gel under
analysis.  With the addition of spectral criteria, it should
be possible to make an unambiguous choice in a large majority
of cases, and to substantially reduce uncertainty in the
others.  Furthermore, it is likely that the complex, relatively
high-precision geometrical transforms on spot coordinates (on
the digital images themselves) that are required for the
reliable comparison of monochrome gels can be replaced by
simpler, computationally cheaper geometric adjustment schemes
with little sacrifice of reliability.


References


1.  Lutin, W. A., Kyle, C. F., Freeman, J. A.:   In Electro-
    phoresis 1978, ed. N. Catsimpoulas, Elsevier North
    Holland, Amsterdam (1978).

2.  Garrels, J. J.:   J. Biol. Chem. <u>254</u>, 7961-7977 (1979).

3.  Bossinger, J., Miller, M. J., Kiem-Phong, U., Nguyen-Huu,
    X.:  J. Biol. Chem. <u>254</u>, 7986-7998 (1979).

4.  Kronberg, H., Zimmer, H. G., Neuhoff, V.:   Electrophoresis
    <u>1</u>, 27-32 (1980).

5.  Lester, E. P., Lemkin, P., Lipkin, L., Cooper, H.L.:
    Clin. Chem. <u>26</u>, 1392-1402 (1980).

6.  Lipkin, L. E., Lemkin, P. F.:   Clin. Chem. <u>26</u>, 1403-1412
    (1980).

7.  Taylor, J., Anderson, N. L., Coulter, B., Scandora, A.,
    Anderson, N. G.:   "Electrophoresis 1979" Symposium,
    Munich, Germany, October (1979).

8.  Adams, L. D., Sammons, D. W.:   Electrophoresis 1981
    Symposium, Charleston, South Carolina, April (1981).

9.  Vincent, R. K.:   Proceedings of the 1975 IEEE Conference
    on Decision and Control, Houston, Texas, December (1975).

10. Anderson, N. G., Anderson, N. L.:   Anal. Biochem. <u>85</u>,
    331-341 (1978).

11. Anderson, N. L., Anderson, N. G.:   Anal. Biochem. <u>85</u>,
    341-354 (1978).

12. Anderson, N. G., Anderson, N. L., Tollaksen, S. L.:   ANL-
    BIM-79-2, Argonne, Illinois (1979).

A COMPUTERIZED SYSTEM FOR MATCHING AND STRETCHING TWO-DIMENSIONAL
GEL PATTERNS REPRESENTED BY PARAMETER LISTS


J. Taylor, N. L. Anderson, and N. G. Anderson
Molecular Anatomy Program, Division of Biological and Medical
Research
Argonne National Laboratory
Argonne, IL 60439  U.S.A.


Introduction

The analysis of high resolution two-dimensional electrophoretic
patterns requires that individual spots from one pattern be
matched with the corresponding spots in other similar patterns.
In addition, it is highly desirable to be able to overlay
patterns onto one another in such a way that analogous spots
are in registration across the entire pattern.  This allows
the easy visual comparison of any two gels and verification
of computer-made identifications.  Direct superimposition of
patterns without stretching is possible only for the simplest
cases, however, as the variability of patterns from different
electrophoretic runs is still too great.  Of particular
concern are the nonlinear distortions caused by variations in
the first-dimension pH gradient which make comparisons of
gels run by different laboratories extremely difficult.

For large studies, it is not realistic to perform pairwise
pattern matching of every gel pattern to every other gel
pattern.  For example, a study of 100 gels would require
almost 5000 matching runs.  A more feasible way would be to
pick one gel as a reference and match to it all the other
gels (1).  Thus, for the case of a 100 gel study, only 99
pattern matching operations would be needed.  However, this
method suffers from one serious defect:  a protein that

does not appear in the reference pattern cannot be matched, and information about its occurrence  in other gels is unavailable.  Unfortunately, information about a protein that appears in some samples, but not in others is often the most crucial part of an experiment.  What is needed is a <u>composite</u> pattern - one that contains every spot detected on every gel in the study.  The composite would serve as the reference pattern, and all gel patterns in the study would be matched to it.  This procedure would allow the cross-indexing of all detected spots throughout the entire set of gels.

A composite gel pattern could be produced in several ways. In some cases a pooled sample would be possible.  However, proteins of low abundance which appear in only a small fraction of the samples might be below the detection threshold in the composite pattern.  Also, all samples would have to be available at the same time, a limiting requirement in the case of progressive study in which additional samples would be added intermittently over a long period.

Another alternative is to form the composite by patching together the digital images of the individual gels.  This method requires that the images be stretched into registration with each other, a difficult (but not impossible) task.  The images are large (2 to 3 Mb) and the distortion functions appear to be complex.

Our approach is a variant of the above alternative.  Instead of stretching and modifying each digital image, we stretch a model of each image, which is composed of a set of two-dimensional Gaussian distributions, with each distribution corresponding to a protein spot (2-4). The individual distributions are characterized by five parameters - an amplitude, the x and y coordinates, and the x and y widths.  The distributions are restricted to be aligned along the x and y axes.

The present paper describes the method by which one pattern
is stretched by a computer into registration with another
pattern and how individual protein spots are matched between
the two. One of these patterns is called the "master" pattern
and may be a composite of many other patterns; it serves as a
reference against which successive new (or "object") patterns
are matched.   Each spot of the master pattern is assigned a
unique "master spot number" (MSN).   A match is established
simply by  associating the MSN of a master set spot to its
homologous object spot.   By matching spots from a series of
patterns to a unique, prenumbered master, the assigned MSN's
cross-index each spot through all the gels.

The computer algorithm described in this paper is implemented
in two basic parts.   The first part is the coordinate stretching
procedure.   The goal of this procedure is to transform the
object spot coordinates so that registration is improved in a
local region of the pattern.   The second basic part of the
implementation is the spot identification or matching
procedure, which matches object spots with master spots in a
local neighborhood.

A small number of initial matches are supplied to the system
in an interactive session, and the remaining matchings are
made automatically.   While these two fundamental parts (the
stretching and identification procedures) are invoked
iteratively according to various higher level strategies,
they are distinct operations and will be described separately.


The Stretching System


The stretching system works by applying a set of local
transformations to the positions of the object spots.   Table
I specifies the notation used in describing this procedure.

## Table I. Stretching System Notation

m – A superscript to designate that a parameter is for a spot in the master set.

o – A superscript to designate that a parameter is for a spot in the object set.

S – The set of matched pairs.  The elements of S are designated $(s_i^o, s_i^m)$. Thus match i pairs spot $s_i^o$ in the object set with spot $s_i^m$ in the master set.  To keep the notation from becoming too cumbersome, the superscripts are dropped, as the context in which the elements $s_i$ are written always clarifies which is intended.  Thus $x_{s_i^o}^o$ is written $x_{s_i}^o$ and designates the x coordinate (in the object system) of the $i^{th}$ matched pair.

N – The number of matched pairs.

(x,y) – The coordinates of a spot measured in units of 100 microns (the standard pixel spacing).  The (x,y) pair is written with superscripts m and o (see above).  Thus $x_\ell^m$ represents the x position of the $\ell^{th}$ spot in the master set.

u – The uncertainty or error associated with parameter $x^o$.

v – The uncertainty or error associated with parameter $y^o$.

' (apostrophe) – A primed parameter denotes an estimated value for that parameter based on the fitting process.  These values do not directly replace the old values, but are weighted with a factor decreasing exponentially with distance from a pivot point, as will be discussed later.  Thus $(x_\ell^{o\,\prime}, y_\ell^{o\,\prime})$ is the predicted value for the $l^{th}$ spot in the object system.  Since only the object set parameters are predicted here, the notation $x_\ell^{o\,\prime}$ is shortened to $x_\ell^{\prime}$.

The goal of the stretching procedure is to transform all object spot coordinates so that the differences between those coordinates and the coordinates of their genuine counterparts in the master system are small. Only the $(x^o, y^o)$ positions are changed. The transformations of unmatched spots must be "reasonable," so that the transformed picture still strongly resembles the original. (Although this concept of "reasonableness" sounds nebulous, it is easily quantified by holding known matches out of the matched set and comparing their final positions with their counterparts in the master set.)

Each individual transformation is based on a fit of linear, quadratic, or cubic functions of x and y using the already matched pairs as a guide. The predicted values from this fit provide the crucial information for updating the pattern. The model used is:

$$x'_\ell = \sum_{k=o}^{n} c_k^x f_k (x^o_\ell, y^o_\ell) \tag{1}$$

$$y'_\ell = \sum_{k=o}^{n} c_k^y f_k (x^o_\ell, y^o_\ell) \tag{2}$$

where $n = 2, 5,$ or $9$ depending on the order of the fit. The functions $f_k (x, y)$ are defined as follows:

$$f_o = 1, \; f_1 = x, \; f_2 = y,$$
$$f_3 = x^2, \; f_4 = xy, \; f_5 = y^2, \tag{3}$$
$$f_6 = x^3, \; f_7 = x^2 y, \; f_8 = xy^2, \; f_9 = y^3$$

388

For notational convenience, the column vector $f$ is defined to be: $f = (f_0, f_1, f_2, \ldots, f_9)^T$. (The symbol T denotes transpose.) Each transformation is intended to emphasize a particular (circular) region of the pattern. This notion is parameterized by a center position (also called a pivot point) $(P_x, P_y)$ and a "fall-off" parameter $\lambda_1 \geq 0$. The error estimates $u_{s_i}$ and $v_{s_i}$ for a spot at $x^o, y^o$ are multiplied by a factor $\exp[\lambda_1((x^m_{s_i} - P_x)^2 + (y^m_{s_i} - P_y)^2)^{1/2}]$ to form $w^x_i$ and $w^y_i$, respectively. The fitting procedure takes the error estimates into account and tends to fit regions of low error more closely than regions of higher error. For the most part, the procedure is a standard $\chi^2$ minimization:

$$\chi^2_x = \sum_{i=1}^{n} (x'_{s_i} - x^m_{s_i})^2 / w^2_{x_i} . \tag{4}$$

The $\chi^2$ sum can be rewritten:

$$\chi^2_x = \sum_{i=1}^{N} \left[ \sum_{k=0}^{n} c^x_k \, f_k \, (x^o_{s_i}, y^o_{s_i}) - x^m_{s_i} \right]^2 / (w^x_i)^2 . \tag{5}$$

As usual in linear fitting problems, we take the partial derivatives with respect to the parameters and require that they be zero at the solution:

$$\frac{\partial \chi^2_x}{\partial c^x_j} = 2 \sum_{i=1}^{N} \left(\frac{1}{w^x_i}\right)^2 \left[ \sum_{k=0}^{n} c^x_k \, f_k \, (x^o_{s_i}, y^o_{s_i}) - x^m_{s_i} \right] f_j(x^o_{s_i}, y^o_{s_i}) = 0 \tag{6}$$

for $j = 0$ to $n$.

The n by n matrix A is to be defined by specifying its elements:

$$a_{jk} = \sum_{i=1}^{N} \left(\frac{1}{w_i^x}\right)^2 f_j(x_{s_i}^o, y_{s_i}^o) \, f_k(x_{s_i}, y_{s_i}). \tag{7}$$

The column vector $\underline{d}^x$ is defined to have elements:

$$d_j^x = \sum_{i=1}^{N} \left(\frac{1}{w_i^x}\right)^2 x_{s_i}^m f_j(x_{s_i}^o, y_{s_i}^o). \tag{8}$$

The desired coefficients are represented by the column vector $c^x = (c_0, c_1, \ldots, c_n)^T$.

The system of equations is expressed in matrix notation as:

$$A \, c^x = d^x. \tag{9}$$

The system is then scaled in order to maintain as much numerical precision as possible. This produces two diagonal matrixes $D_r$ and $D_c$ for the row and columns scale matrices, respectively. Equation 9 can be written:

$$(D_r \, A \, D_c)(D_c^{-1} \, c^x) = D_r \, d^x \tag{10}$$

(The superscript -1 means the matrix inverse.) Solving for the coefficient vector $c^x$,

$$c^x = D_c \, (D_r \, A \, D_c)^{-1} \, D_r \, d^x \tag{11}$$

The predicted values are:

$$x' = f^T(x_\ell^o, y_\ell^o) \, c^x \tag{12}$$

and the error estimate (5) for x' is

$$u_{\ell}'^2 = f^T(x_{\ell}^o,\ y_{\ell}^o)D_c(D_r\ A\ D_c)^{-1}\ D_r\ f(x_{\ell}^o,\ y_{\ell}^o)\sigma^2$$

$$+\ \left[\frac{\partial f^T(x_{\ell}^o,\ y_{\ell}^o)\cdot c^x}{\partial x^o}\right]^2 \cdot\ u_{\ell}^2\ +\left[\frac{\partial f^T(x_{\ell}^o,\ y_{\ell}^o)\cdot c^x)}{\partial y^o}\right]^2 v_{\ell}^2 \tag{13}$$

where $\sigma^2$ is estimated by:

$$\sigma^2\ =\ \frac{1}{(N-n)}\ \sum_{i=1}^{N}\ [x_{s_i}^m\ -\ f^T(x_{s_i}^o,\ y_{s_i}^o)\ c^x]^2. \tag{14}$$

A condition number (6) for the matrix is monitored and if it
is too small, either  the order of the functions involved in
the fit is reduced or else no transformation is applied for
this choice of a pivot point. The process is repeated in a
similar fashion for the y coordinates (much of the work is the
same and does not need to be repeated).

The primed coordinates and error estimates do not replace the
old values directly.  Since the transformation was calculated
locally by de-emphasizing regions far from the pivot point,
the final effect of the local transformation must also be
de-emphasized progressively with distance.  There are many
ways to do this.  We have chosen simply to weight the effect of
the transformation by a factor which decreases exponentially
with distance from the pivot point:

$$x_{\ell}^o \leftarrow x_{\ell}^o\ +\ (x_{\ell}'-x_{\ell}^o)\exp\left\{-\lambda_2[(P_x-x_{\ell}^o)^2\ +\ (P_v^y-y_{\ell}^o)^2]^{1/2}\right\} \tag{15}$$

$$u_{\ell}^o \leftarrow u_{\ell}^o\ +\ (u_{\ell}'-u_{\ell}^o)\exp\left\{-\lambda_2[(P_x-x_{\ell}^o)^2\ +\ (P_v^y-y_{\ell}^o)^2]^{1/2}\right\} \tag{16}$$

Similar replacements are done for y and V.  In most cases,
the parameter $\lambda_2$ is chosen to be the same as $\lambda_1$.

The first stages of the stretching are generally done with
$\lambda_1 = \lambda_2 = 0$, so that no localized stretching is done
at all.  This procedure translates, rotates, and scales the
pattern and provides a good starting point for subsequent
stretching stages.

**Initially all uncertainties are set** to 1.0.  In general,
the uncertainties in the x and y positions tend to grow as
more transformations are made.  Periodic corrections are
performed to lower the uncertainties according to how well
the neighborhood around a spot is matched to master spots.
Those regions that have a high proportion of matched spots
are considered more certain and thus carry more weight in
subsequent transformations:

$$u_i \leftarrow 1 + (u_i - 1) \left[ 1 - \left[ \frac{\sum\limits_{i \ (\text{matched})} \exp \left\{ -\lambda_3 \left\{ (x_i^o - x_{s_j}^o)^2 + (y_i^o - y_{s_j}^o)^2 \right\}^{1/2} \right\}}{\sum\limits_{i \ (\text{all spots})} \exp \left\{ -\lambda_3 \left\{ (x_i^o - x_j^o)^2 + (y_i^o - y_i^o)^2 \right\}^{1/2} \right\}} \right]^{1/2} \right] \tag{17}$$

A corresponding reduction is made for $v_i$.

The summation in the numerator is over all identified spots,
whereas the summation in the denominator is over all spots.
Thus, the uncertainties tend to return to their original value
(1.0) as more and more close neighbors are identified.  The
choice of 1.0 as the starting uncertainty value is arbitrary,
and may be replaced at a later time with an uncertainty
derived as a by-product of the optimization procedure.

<u>Spot Identification</u>

Spots in the object set are identified by matching them
with their counterparts in the master list.  The procedure for
performing this matching is quite simple and is described below.

In addition to the definitions used in the stretching section, several others are necessary here (Table II).

### Table II. Notation for the Identification System

$\sigma_x$ - A measure of the half width of a spot along the x axis.
$\sigma_y$ - A measure of the half-width of a spot along the y direction.
A  - The amplitude or height of a spot.
V  - The volume of a spot.

Since we are primarily concerned with Gaussian forms, the volume V is simply $2\pi A\sigma_x\sigma_y$. User specified parameters used in the matching are $R_1$, $S_1$, $S_2$, $a_1$, $a_2$, $a_3$, and $a_4$.

Typical values for these are listed in Table III and are adjusted by observing the performance of the algorithm until satisfactory results are obtained over a representative set of gel patterns. The volumes of the object spots are scaled using a $\chi^2$ minimization fitting procedure on the already matched spots with a quadratic (no constant term) model. Only the spots in this object set within a distance R of the pivot point $(P_x, P_y)$ are considered. A slippage term is added for both x and y dimensions so that the already identified spot closest to $P_x, P_y$ is exactly matched. The interpoint distances to all spots in the corresponding (but slightly enlarged) region of the master set are computed and compared with the threshold parameter $S_1$. Any geometrical distance less than $S_1$ is considered a possible match. An additional check is then made using a generalized distance measure:

$$d_{ij} = [a_1(x_i^o - x_j^m)^2 + a_1(y_i^o - y_i^m)^2 + a_2(\ln V_i^o - \ln V_i^m)^2$$

$$\qquad\qquad (18)$$

$$+ a_3(\sigma_{x_i}^o - \sigma_{x_i}^m)^2 + a_4(\sigma_{y_i}^o - \sigma_{y_i}^m)^2]^{1/2}$$

This measure $d_{ij}$ must be less than threshold $S_2$ before spot i and spot j can be matched.

It is, of course, possible that a spot in the object set could satisfy the match criteria for more than one spot in the master set. This ambiguous situation is dealt with by ranking the distance measures $d_{ij}$ in descending order and matching the worst possibilities first. The identification process allows for the breaking of existing matches if necessary, so that a better match for a spot will supersede an existing match (the exception to this is the case where the match to be broken is a manually entered pair).

Upper-level Strategy

The stretching and identification procedures described above are called from a high level routine according to several strategies. We currently use four operational stages for the patterns produced in this laboratory. Table III gives the parameters for each of these stages. These parameters are not critical, as the routines appear to work over a wide range of values for each parameter.

Table III. <u>Typical Parameters for Each Stage of the Stretching-identification System</u>

| Stage | 1 | 2 | $R_1$ | $S_1$ | $a_1$ | $a_2$ | $a_3$ | $a_4$ | $S_2$ | 3 | Max order |
|---|---|---|---|---|---|---|---|---|---|---|---|
| 1 | 0 | 0 | 100 | 5.0 | 1 | 1 | 1 | 1 | 10 | .01 | 3 |
| 2 | .01 | .01 | 150 | 5.0 | 1 | 1 | 1 | 1 | 10 | .01 | 2 |
| 3 | .01 | .01 | 200 | 7.5 | 1 | 1 | 1 | 1 | 12 | .01 | 2 |
| 4 | .01 | .01 | 250 | 10.0 | 1 | 1 | 1 | 1 | 15 | .01 | 2 |

394

The first stage transforms the pattern globally to bring
it into gross alignment with the master pattern.  At this
point the only matched pairs are those entered manually by
the user, and the order of the fit is limited by their
number.  If the order is 1, then this transformation is
essentially a translation, rotation, and scaling of the entire
pattern.  The global stetching operation is followed by a
sequence of identification operations in small neighborhoods
around each of the initial matches.  In the second stage, the
regions around the best matches are processed first by the
stretching procedure and then by the identification procedure.
The primary purpose of this stage is to increase the number of
matches rapidly in order to add stability to the stretching
process.

The third stage alternately stretches and identifies local
regions in a grid-like pattern which covers the entire
image.  The fourth and final stage visits the region of the
worst matched spots to improve their alignment.  Both stretching
and identification operations are performed.  Between stages,
an operation is performed to eliminate the pairing of spots
whose mismatch exceeds $S_1$ for the previous stage.  Such
misidentifications can occur by a spot being matched early in
the operation and then shifted out of range by subsequent
stretching operations.

<u>Examples</u>

Two representative patterns were obtained from autoradiographs of different gels run from the same sample of human
lymphocytes labeled with $^{35}$S-methionine.  The patterns were
scanned at 100 micron resolution and analyzed to produce a
spot list of Gaussian parameters.  The identification and
stretching system was then run to match and align the two
patterns.  Figures 1A-1E show plots of the identified spots in

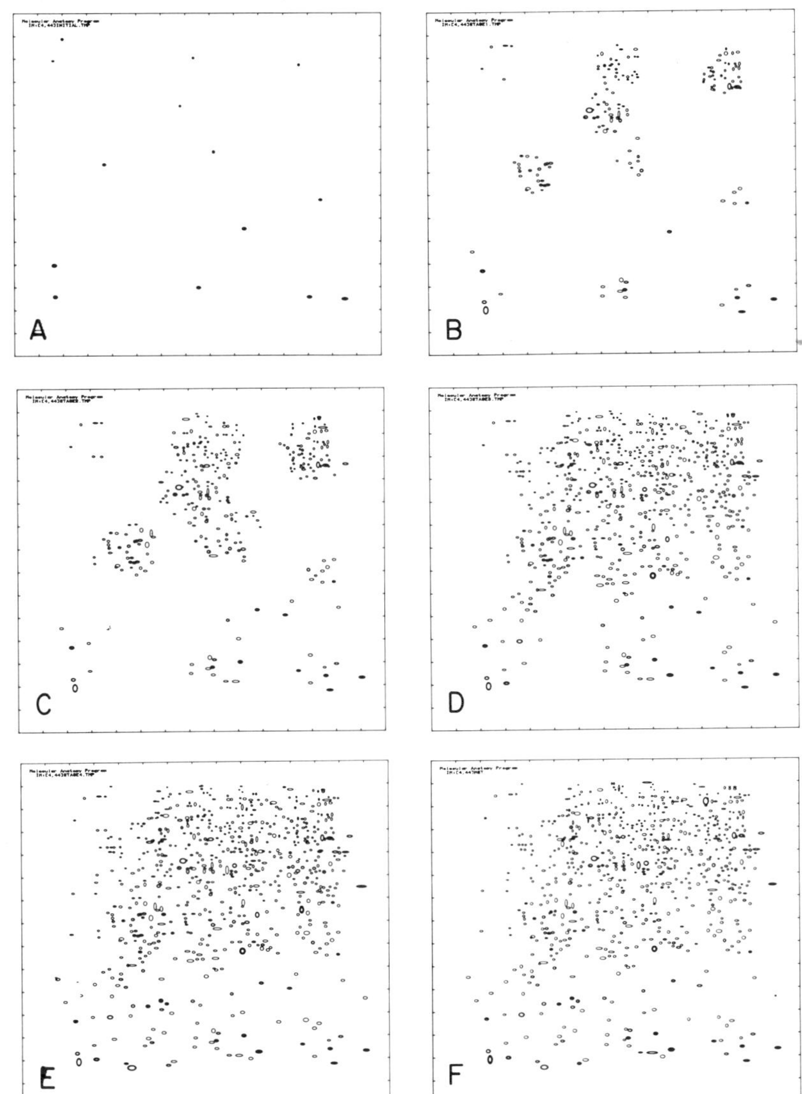

Figure 1.  Frames A through C show the identified spots in the object pattern at several stages in the analysis. Frame A shows the intitial identifications, and frames B, C, D, and E show the identifications at the end of stages 1, 2, 3, and 4, respectively. Frame F is a plot of the master pattern.  The least abundant spots have been omitted for clarity in all frames.

Figure 2. The distortion grid resulting from stretching a square grid in the same manner that the object pattern is stretched into registration with the master pattern.

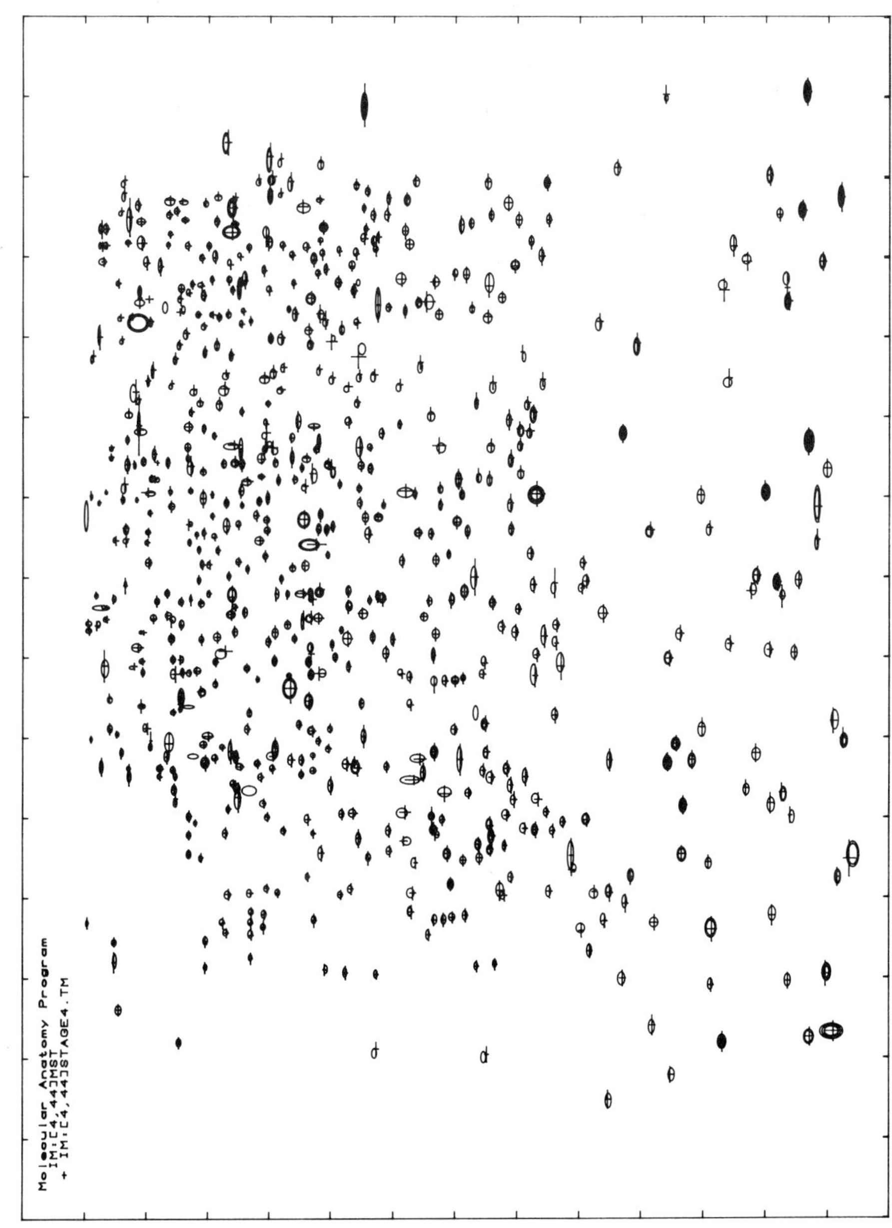


Figure 3. A plot of both the master and object patterns in the example. The master pattern spots are represented by ellipses and the object spots by the distorted crosses.

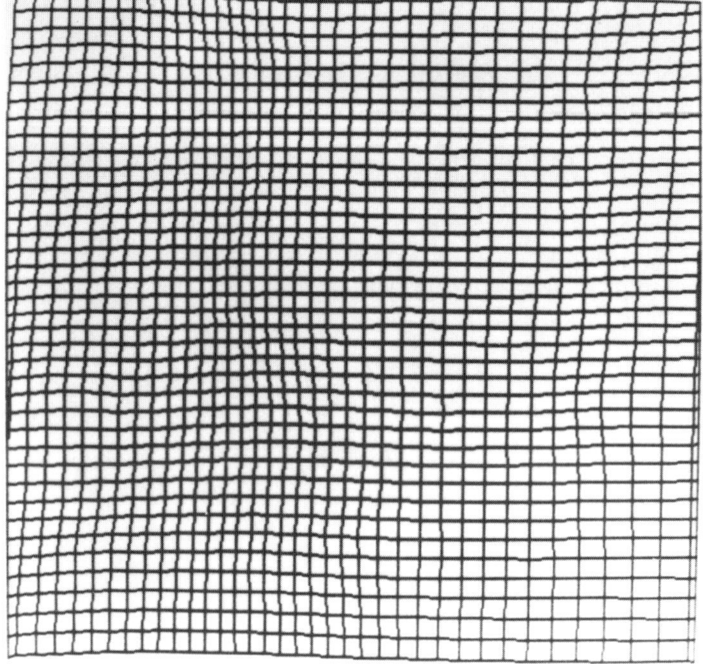

Figure 4.  A distortion grid resulting from the comparison of two gels of different samples run under non-identical conditions.

the object pattern at the end of each stage of the procedure.
Fourteen initial matches were entered and the parameters used
are the ones listed in Table III.  Figure 1F is the master
pattern.  Figure 2 shows the distortion grid, which is produced
by stretching a square 40 by 40 grid in the same exact manner
that the object pattern is stretched.  Figure 3 shows the
final registration of the master and object patterns.  A
total of 668 spots were paired in this run. Patterns for gels
produced under similar conditions are generally easy to align
and match, with only 5-15 initial matches supplied by the
user.  Attempts have also been made to align patterns produced
using grossly different experimental conditions.  These cases
are far more difficult to match and the final alignment is not
as close as that induced for similar gels.  Figure 4 shows the
severe distortion obtained in a case where different ampholyte
brands were used in the first dimension electrophoresis.
Although satisfactory results can be obtained for these cases,
the user interaction is high (as many as 50 initial matches
are sometimes necessary).  Standardization of gel running
procedures will undoubtedly make the pattern-to-pattern
comparisons easier and more accurate.


Acknowledgment

This work is is supported by the U.S. Department of Energy under
contract No. W-31-109-ENG-38.


References

1.  Lipkin, L. E., Lemkin, P. F.: Clin. Chem. 26, 1403-1412
    (1980).

2.  Garrels, J.I.: J. Biol. Chem. 254, 7961-7977, 1979.

3.  Lutin, W. A., Kyle, C. F., Freeman, F. A.:    In
    Electrophoresis  '78., N. Catsimpoolas, ed.,
    pp. 93-106..

4.  Taylor, J., Anderson, N. L., Coulter, P. B., Scandora,
    A. E., Jr., and Anderson N. G.:  In Proceedings of
    Electrophoresis '79, B. Radola, Eds., W. deGruyter,
    pp 329-339, 1980.

5.  Searle, S. R.: <u>Linear Models</u>, John Wiley and Sons, Inc.,
    New York (1971).

6.  Gilberts, S.:  Linear Algebra and Its Applications,
    Academic Press, New York (1976).

# GELLAB: MULTIPLE 2D ELECTROPHORETIC GEL ANALYSIS

P. F. Lemkin, L. E. Lipkin

Image Processing Section, DCBD, NCI, National Institutes of Health,
Bethesda, Md. 20205

## INTRODUCTION

Two-dimensional polyacrylamide gel electrophoresis (PAGE) is a rapidly
developing biochemical tool with ever increasing application to a wide variety of
problems in molecular biology, basic biochemistry, genetics and clinical research
(1).

GELLAB is a computer system for the analysis of 2D gel electrophoresis
images. It is particularly suited for the semi-automation of comparisons among
multiple gels rather than simple gel-pair analyses. The multiple gel spot data
base is constructed by first extracting spot (density,x,y) triples (using a
segmentation procedure), then pairing (n-1) gel spot lists to that of a
representitive gel using a landmark proximity algoritm. Finally corresponding
spots are merged into a single data base (DB). Interrogation of and experimentation on this multiple gel data base may then be performed. One can thus find
particular spots  and extract measurements within the set of gels. The data base
may be partitioned both by gel class and spot statistics. Various interactive
graphics facilities are available for observing intermediate results. It is
possible to search for spots in the spot data base having statistically
significant differences between classes, the data base having been partitioned
into several classes. The system may be directed toward identification of
biologically significant quantitative changes manifest as spot differences
despite complications introduced by a wide variety of gel preparative conditions.
This paper focuses on methods for handling multiple gels using the concept of the
the representative gel, R-gel. The R-gel serves as the framework within which to
link spots across a set of gels and serves as a practical substitute for a
practically unattainable canonical gel model. The primary data base consists of
the totality of the lists of corresponding spots (R-spot sets) and their
associated properties and interrelations. The multiple gel data base system is
very general in that gel segmentation and spot pairing algorithms other than
those now used in GELLAB may be substituted in the early phases of data base

construction. CGELP, the GELLAB program for constructing, partitioning, normalizing, searching, extrapolating missing spots, retrieving, and formatting the data base is discussed.

ANALYZING MANY GELS SIMULTANEOUSLY

The 2D page technique has been used to separate hundreds to several thousand polypeptide components in a two dimensional matrix (2). Given the large number of spots normally present, it is difficult to manually compare a number of gels against each other, particularly in terms of comparative densities of corresponding spots. Spot measurement (position and quantity of polypeptide), spot correspondence (between gels), and the assembling and statistical analysis of corresponding spots are some of the areas where automation can be helpful. The GELLAB system automates some of these functions (3-10). GELLAB is written in the SAIL computer language (10) for a DECSYSTEM-10 or DECSYSTEM-20.

Figure 1 illustrates the first two steps in a typical semiautomated analysis. Gels are first scaned and accessioned into our computer (a DECSYSTEM-2020) as 512x512 pixel (picture element) images. Gel images may be scanned with our TV vidicon or copied from magnetic tapes produced by an Optronics drum scanner. The neutral density standard wedge is scanned with the gel when using the vidicon camera in order to calibrate the set of images to the same neutral density, ND, standard. Figures 1c and 1d show spots extracted using a segmentation program called SG2DRV (3,7) which is based on finding the central cores of spots and then propagating out to an inflection region defined by the second 2D derivative. Figures 1e and 1f show spots paired between these two gels. The pairing algorithm, CMPGEL, uses a set of interactively defined landmark spots. These are defined by the operator flicker—aligning the two gels on our TV system and noting the correspondence of these landmark spots with an interactive tablet (4,7). Segmented spots are assigned to the nearest landmark set and then paired by best-fit within each landmark region thus greatly reducing the combinatorics of checking each spot with all other spots.

DATA BASE MANIPULATION TO EXTRACT "INTERESTING" SPOTS

The data base composed of a set of gels may be viewed as a rectangular solid constructed by stacking the gels as in Figure 2a. Ideally, one could thread a path for each spot through each gel. In those instances where a spot was missing, the thread could pass through where it "would have been". Such a spot

mapping would define a cannonical gel (C-gel) for this set of gels. In practice,
a representative gel (R-gel) is selected from the set of gels. Only those spots
appearing in the R-gel are used initially in the data base. It is possible and
reasonable to interpolate missing spots on the R-gel which are found in the other
set of gels.

A R-spot may be visualized as a spot threaded through the set of gels
starting at the R-gel. A R-spot set is the set of spots in a subset of the gels
in the DB along the thread defined by the R-spot. The number of independent
variables (and resultant gels) can be quite large for some experiments. This is
shown, for example in Fig. 2b by sorting the gels in the rectangular solid as
to class and then defining class partitions. Similarly, a subset partition of
R-spot sets in the DB can be defined based on various statistical and logical
criteria.

ALGORITHM FOR CONSTRUCTING A DATA BASE

The algorithm for constructing such a composite gel (CGEL) data base for
n-gels is presented here in a brief form. The initial data is a set of gel
pairings - Gel Comparison Files - {GCF1,...,GCFn-1} produced by spot extraction
and gel pairing programs such as (1-2,12-14).

> Process each GCFi in turn. For each R-gel spot, if it is not in the data
> base create a new R-spot and a new R-spot set and insert both spot data
> (x,y,density,etc.) into the R-spot set. Otherwise, find the R-spot set and
> insert the non-Rgel spot data from gel i into the R-spot set.

Obviously for a large number of gels with over a thousand spots each, it
would be very difficult if not impossible to keep this amount of data in the main
core memory of a computer. Thus a disk memory paged data base is used with
effectively infinite storage for a large number of gels. By paging we mean that
only part of the data base (i.e. a page of some standard size) is transfered
between the computer main memory and the disk as required. The page data base
construction and analysis program in GELLAB is called CGELP (5,7).

CGELP is an interpreter, which allows easy user access to the data base
in a wide variety of ways. It has various procedures for testing the validity of
a set of data including computing the correlation coefficients for all gels taken
two at a time, performing searches through partitions of the data base for R-spot
set meeting various statistical criteria as well as other operations to be
discussed.

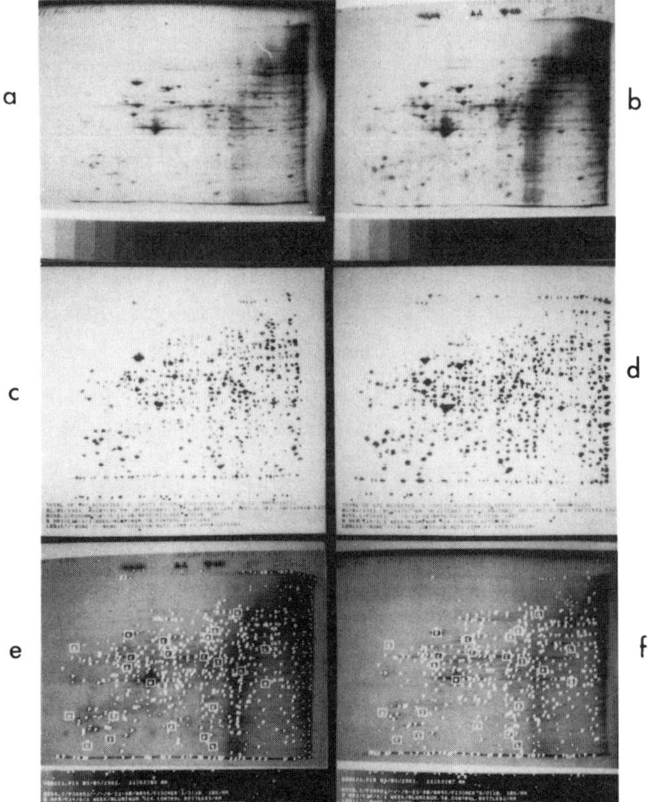

Figure 1. A composite of two typical gel images of a P388D1 mouse macrophage cell line (8-9). These gels are part of a series of 16 gels in this particular data base analyzed by GELLAB. The gels are split into two groups, one harvested 24 hours after the first) are shown in a,c,e (time T0) and b,d,f (time T24 - 24 hours later). Original gel images a) time T0 and b) time T24. Segmented spot images of c) T0 and d) T24. Labeled spot comparison images for e) T0 and f) T24. The labels are S (sure pair), P (possible pair), A (ambiguous pair), and + (unresolved spot). Landmark spots are denoted by sequential letters with squares.

## PARTITIONING AS AN AID IN ORGANIZING THE DATA BASE

The utility of a data base management system only becomes apparent when questions can be easily posed to and answered by the system. In this regard, CGELP has various facilities for manipulating the Q&A environment. Partitioning of both the space of gels and the space of R-spot sets and of individual spots is possible.

The major gel partition of the DB is defined by the data structure called

"working set of gels". This is simply a subset of all the gels in the DB. The
working set of gels may be conveniently defined in various ways including: 1)
explicit enumeration of particular gels, 2) responding Yes/No to a prompt for
each gel, 3) by being set to an equivalence with a gel subset (of which there may
be up to 10 gel subsets which themselves may be defined in a variety of ways).

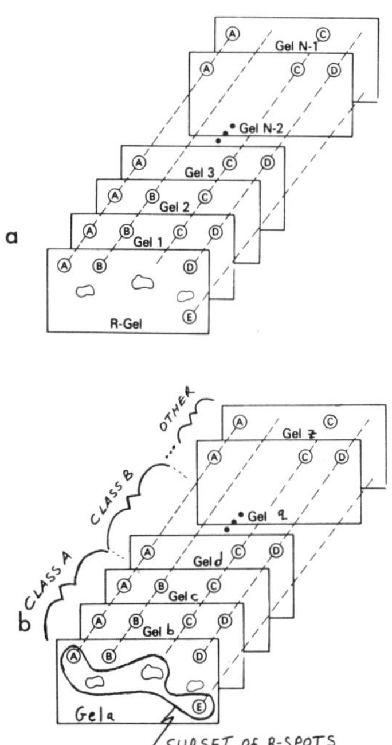


Figure 2. A multiple gel data base viewed as 2D projection on 3D rectangular
solid. a) Full set of gels X full set of R-spot sets. b) Partition of working set
of Gels X class partition of gels X subset of R-spot sets meeting partition
criteria.

The next type of gel partition is only operative when performing a
multiclass DB search. Up to 9 gel classes may be defined with gels possibly being
present in several classes. The set of all gel class subsets is itself a subset
of the working set of gels. What this means is that a classification scheme can
be applied to the DB and may be automatically updated to reflect the latest
change in the working set of gels. For small numbers of gels with little
variation in experimental design, this concept is not all that useful. Its
utility becomes apparent when the number of classes and gels in the data base
becomes large. Currently up to 128 gels may be used in a data base. With minor
extensions, this limit could be in the thousands.

In (4,7) the concept of spot pairing label is introduced. Spot labels are SP (sure pair), PP (possible pair), AP (ambiguous pair), US (unresolved spot), and EP (extrapolated pair). The system also permit_s partitions by label. Normally, SP and PP are used for investigating changes in relative spot density among gels. AP and US labels reflect either fragmentation of a spot due to segmentation error, system noise or a real spot difference. EP is defined in CGELP for gels with spots missing by an extrapolation algorithm as follows. The position where the spot would be is estimated by adding the mean offset of the spots present from their corresponding landmark spot to the landmark position in the current gel. Since landmark positions are known, this calculation can always be performed.

Various statistical and metric parameters of an individual spot or R-spot set may be used to partition spots and/or R-spot sets. The parameters must pass a set of limits in order to be considered for any CGELP operation. These include various distance, area, density metrics. Thus particular aspects of the data base can be enhanced and investigated using statistical partitioning.

Finally, there are many types of statistical searches through the DB available in CGELP. The results of a search is always a "search results list" which can be used in various ways for visual feedback of the spots found. The searches may be divided into two main groups: single-class and multi-class. Examples of single class searches include finding all landmark R-spot sets and finding R-spot sets meeting the current statistical limits. Multiclass searches include: t-test, F-test, and 2- (Wilcoxson Mann Whitney) or m-class (Kruska Wallis) rank order searches (15). The numbers of the R-spot sets meeting the statistical criteria are printed as they are found in a linear search through the data base. In addition, for the 2-class (t-, F- or rank order test) searches, a histogram of the ratio of the mean densities of each of the two classes is computed and printed at the end of the search as shown in Table I. All searches as well as explicit printing of a R-spot set may be optionally saved in disk files. Any R-spot set may be printed with density expressed as a % of total density, ratio of a subset of spots, least square normalized with a linear approximation to the density range of the R-gel, absolute measurement density or estimated gaussian volume. The R-spot sets may be ordered by any of these density format modes. This rank ordering by density is the normal mode of operation. CGELP permits any R-spot set to be printed at any time so that spots of possible interest can be checked under various partitioning conditions.

Table I. Multi-gel 2-class mean density ratio R-spot histogram. The ratio m2/ml is the ratio of the mean density of each class (2=T24 and 1=T0) for a given R-spot set. It is a useful aid in determing the clustering of patterns of spot changes. Each histogram entry is a R-spot number found by the search.

```
F-test class search at  .95 significance
Found 29 R-spots,  mean (sd/mn= .42+- .78

m2/ml|  R-spot sets
 .40| 426
 .45|
 .50|
 .55|
 .60| 252
 .65|  64   96 190
 .70|
 .75|
 .80|  97
 .85|  37
 .90|
     .
     .
     .
1.25|
1.30|  39 164 397
1.35| 117 163
1.40| 162 475
1.45|
1.50|
1.55|
1.60| 136
1.65|
1.70|
1.75| 512
1.80|
1.85|
1.90|
1.95| 106
2.00|  18 538 546
2.05| 123
2.10|  50 540
2.15|
2.20|
2.25|
2.30|  88 339
2.35| 154
2.40|
     .
     .
     .
2.75|
2.80|  44
2.85|
     .
     .
     .
4.50|
4.70|  69 393
```

Given the ability to partition the DB using a wide varity of methods, it is interesting to examine the distributions of, and correlations among these parameters of a given data base being analyzed. This is useful for example, in "tuning" statistical limits. Various types of histogram and feature-feature plots which can be generated on the user's display terminal and later plotted on a hardcopy plotter. There is also a 3D feature plot in addition to the 2D feature plot as well as the facility to perform density-density log plots between any two gels.

Essential to making the link between the non-pictorial gel DB - the set of R-spot sets - is the R-map. This is an image of the gel with the spots labeled with their R-spot numbers. A graphics approximation can be interactively drawn while in CGELP where labeled spots are circles proportional to their density. Figure 3 illustrates several R-maps generated from (CGELP produced) search result lists (SRL) files using an auxillary program called MARKGEL.

Another very useful derived image is the gel mosaic illustrated in Figures 4a-4d. This is generated from copies of the gel images as well as using a CGELP produced SRL file by extracting (and optionally zooming) the region of each corresponding gel centered at a particular R-spot set entry of interest. The

SEERSPOT program is used to generate mosaics. Using mosaics, extrapolated spots may be visualized as in Figure 4d.

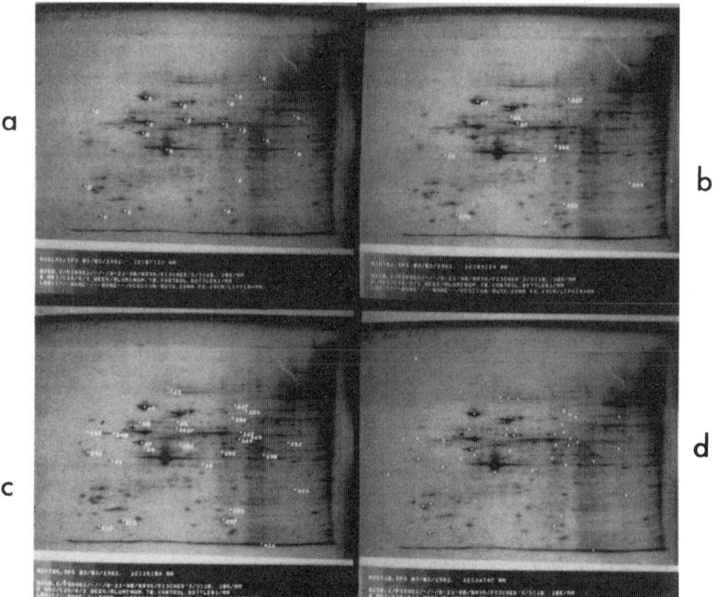


Figure 3. R-maps generated by overlaying a copy of the gel image with labels using the MARKGEL program for the 16 gel P388Dl data base. a) Landmark set of R-spots, b) results of F-test search between T0/T24 at 99% confidence limit, c) same at 95% limit, d) same at 90% limit drawn with the no-label option.

A small number of selected spots may be plotted adjacent to one another in a rank order table as shown in Table II. This facilitates checking the relative densities of spots in these sets across a set of gels.

SUMMARY OF SYSTEM CHARACTERISTICS

Given the types of tools just discussed, it is thus possible to build, partition and search a multiple gel data base to find statistically interesting spots. After normalizing and tuning the statistical parameter limits, several searches are usually performed resulting in search result lists (and associated files). These in turn are often used to produce R-map images of these SRLs. After manually checking these R-map images, a small number of particularrly interesting R-spots maybe checked using the mosaic images. This process then enables the investigator to concentrate on and verify significant density changes in a

multigel DB. This is done by effectively reducing the combinatorics of searching
the 3D gel-spot space by the method of successive partitioning.

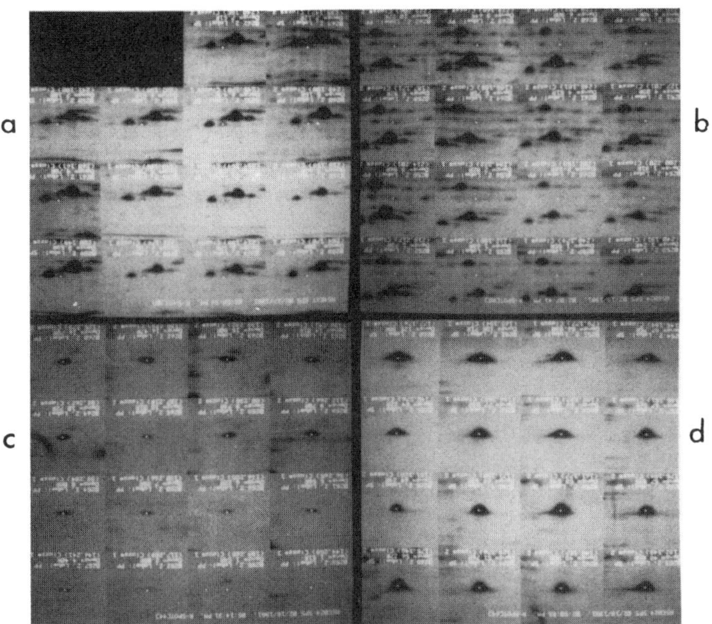

Figure 4. Mosaics of 2 classes (T0/T24) of gels with significant differences for
P388D1 set of gels. Pannels are ordered with lightest first. a) R-spot[64], b)
R-spot[44], c) R-spot[88], d) R-spot[30] with EP in 1st pannel where a very light
spot might be.

Because of the organization of the DB as an effectively infinite list of
relative small R-spot sets, each set being the size of the number of gels in the
DB, the comparison of a large number of gels is computationally feasible.

Experiments have been performed on reproducibility in the system using
E.coli gels. Multiple scans of the same gels and multiple gels of a split sample.
Multiple scans of the same gels and multiple gels of a split sample were
accessioned into the sytem. Correlation coefficients of 99% were seen for a large
number of SP+PP spots. Where there was poor correlation, it was found that many
of the spots were poorly defined or fragmented by the segmenter program, SG2DRV.
Additional work is underway to improve this peformance. Occasionally, we found
that if a landmark spot was poorly chosen (i.e. a very light or saturated spot),
the landmark position would be incorrect for some of the gels. Thus all spots in
the associated landmark set might be shifted by this amount and mispairings
occur. This was found to be avoidable by not choosing such spots as landmarks.

**Table II. Example of a rank order table of selected spots used to test hypotheses of relative change of selected R-spots.**

```
File: RNKTB1.TBL 02/26/1981,  10:17:23 AM
RANK-ORDER table: <ACC#>&<LMset>&<Class #>
Paged CGL data base file: J5.PCG[61,1]]
Using least square normalization.
User defined spot list

Density
 22.3 |                          0263.2G2
 21.8 |
 21.3 |
 20.8 |
 20.3 |                          0260.2G2
 19.7 |                          0262.2G2
 19.2 |
 18.7 |                          0261.2G2
 18.2 |
 17.7 |
 17.2 |                                                   0263.2V2
 16.7 |  0260.2B2
 16.2 |  0262.2B2               0264.2G2
 15.6 |  0263.2B2               0259.2G2                   0262.2V2
 15.1 |
 14.6 |                         0250.2G1
      |                         0265.2G2
 14.1 |                         0258.2G2
 13.6 |  0261.2B2                                          0259.2V2
 13.1 |
 12.6 |                                                   0250.2V1
 12.1 |
 11.5 |                                  0262.2P2
 11.0 |                                                   0265.2V2
      |                                                   0260.2V2
 10.5 |  0264.2B2
 10.0 |  0265.2B2 0262.2E2
  9.5 |  0259.2B2 0263.2E2
      |  0258.2B2
  9.0 |                                                   0261.2V2
  8.5 |                                  0260.2P2
  8.0 |
  7.4 |
  6.9 |  0250.2B1                        0263.2P2 0764.2V2
  6.4 |            0260.2E2              0259.2P2
  5.9 |            0250.2E1              0264.2P2
      |            0259.2E2
  5.4 |            0264.2E2                                0258.2V2
  4.9 |                                  0258.2P2
  4.4 |            0261.2E2              0265.2P2
  3.9 |            0258.2E2
  3.3 |
  2.8 |
  2.3 |
  1.8 |            0265.2E2
-----------------------------------------------------------------
R-spot |      44         88        117       393       512
Class #   1=T0, Class #   2=T24
```

The data base analysis is as good as paired spot measurements used to construct it and therefore improvements in both gel technology and spot measurement and comparison is needed to achieve optimum performance.

In the current CGELP implementation, only US spots in the R-gel are accounted for. There is a need to extrapolate new R-spot sets (eR-spots) for US found in other gels thus guarenteeing the completeness of the DB. The DB to could be extended to access protein atlases of known proteins by building a global tree or graph structure of various data bases with R-spot transformation tables mapping one atlas to another. Movement around this "library" would then be between "volumes" of individual data bases.

CONCLUSION

We have presented mechanisms for dealing with the PAGE 3D spot analysis problem, using a wide variety of user selected partitionings. Although not fully automated, the laborous part of spot measurement, pairing and congener spot assembly has been automated while the analyses of reduced data is presented to the investigator for final decisions.

REFERENCES

1. Anderson, N. G., Anderson, N. L., Behring Inst. Symposium 1977, Mitt. Behring Inst. Sympos., 63, 169 (1979).

2. O'Farrell, P. H., J. Biol. Chem., 250, 4007 (1975).

3. Lemkin, P., Lipkin, L., Comp. Biomed. Res., June, 1981.

4. Lemkin, P., Lipkin, L., Comp. Biomed. Res., August, 1981.

5. Lemkin, P., Lipkin, L., Comp. Biomed. Res., submitted to Comp. Biomed. Res.

6. Lemkin, P., Merril, C., Lipkin, L., Van Keuren, M., Oertel, W., Shapiro, B., Wade, M., Schultz, M., Smith, E., Comp. Biomed. Res., 12, 517 (1979).

7. Lipkin, L., Lemkin,P., Clin. Chem, Sept., 1403, (1980).

8. Lester, E.P., Lemkin, P., Cooper, H.L., Lipkin, L.E., Clin. Chem., Sept., 1392, (1980).

9. Lemkin, P., Lipkin, L., Merril, C., Shiffrin, S., Envir. Health. Perspect., 34, 75 (1980).

10. Lester, E.P., Lemkin, P., Lipkin, L.E., Anal. Chem., 53, 390A (1981).

11. Reiser, J.F., N.T.I.S. #AD-A045-102, Springfield, Va., 1976.

12. Garrels, J. I., J. Biol. Chem., 254, 7961 (1979).

13. Lutin, W.A., Kyle, C.F., Freeman, J.A., in <u>Electrophoresis '78</u>, N. Catsimpoolas (ed), Elsevier North Holland, Inc., pp 93-106 (1978).

14. Bossinger, J., Miller, M.J., Kiem-Phing, V., Geiduschek, P., Xuong, N., J. Biol. Chem., 254, 7986 (1979).

15. Natrella, M.G., Experimental Statistics, <u>NBS Handbook 91</u>, U.S. Govt Printing Office, Wash, D.C., Oct. (1966).

IMPLICIT MODELING OF SPOTS FOR THE EVALUATION OF TWO-DIMENSION-
AL ELECTROPHORETOGRAMS


Harald Kronberg, Hans-Georg Zimmer, Volker Neuhoff
Max-Planck-Institut für experimentelle Medizin
Forschungsstelle Neurochemie
D-3400 Göttingen, Germany


Introduction

There are two ways by which the automatic evaluation of electrophoretograms can be improved. One is to increase the reliability
of the data. The other relates to signal processing and provides
more flexibility in modeling of the spots.

For a quantitative analysis of barely-visible and rather opaque
stained protein spots in slab gels there is a need for measurements of optical densities (OD) in the range from 0.001 O.D. to
2.5 O.D. (units of optical density). Directly scanning the electrophoretogram by a photometer meets this requirement and will
be discussed in the following chapter.

If the measuring device accepts only photographs or radiographs
of the electrophoretogram, the processing of the film becomes
a critical step. Even though linearity may be achieved between
0.2 O.D. and 2.6 O.D. (1), the fog level of 0.2 O.D. limits the
dynamic range of the data to be evaluated. Because film grain
noise is signal-dependant and aperture-dependant (2), the measuring values deteriorate and yield quantitative results of protein contents only in a range of 1 : 10 (1).

If videodensitometry is used to scan gel electrophoretograms

---

*Supported by the Deutsche Forschungsgemeinschaft

414

(3,4) the quantification of well-resolved proteins stained with
Coomassie Brilliant Blue R-250 seems to be restricted to a range
of 1 : 4 (3). This is mainly due to the limited resolution of
the gray scale even of optimized TV-systems (5,6), which is a
consequence of the local variation of sensitivity, dark-current,
limited linearity and the fading of the measured signal. In
addition, the illumination of the entire image field produces
glare and thus impairs the signal.

Two approaches to the problem of finding the locations and extents of spots in electrophoretograms or their images have been
delt with in the literature. The thresholding technique (cf.7)
employs a fixed or adaptive level of the gray scale to discriminate between the spots and the background. This simple model
raises difficulties in finding the best threshold and in separating confluent spots. The fitting of 2-D Gaussian functions
(8,9,10) relies on an explicit mathematical model of the spots
which is better adapted to reality than thresholding, but may
still lead to wrong decompositions of distorted spots.

Because of the asymmetry and irregularity of the spots in real
electrophoretograms (see Fig. 1) we have not tried to model the
spots on a predeterminate pattern. Instead, an implicit, more
flexible model is used by defining clear background, kernels

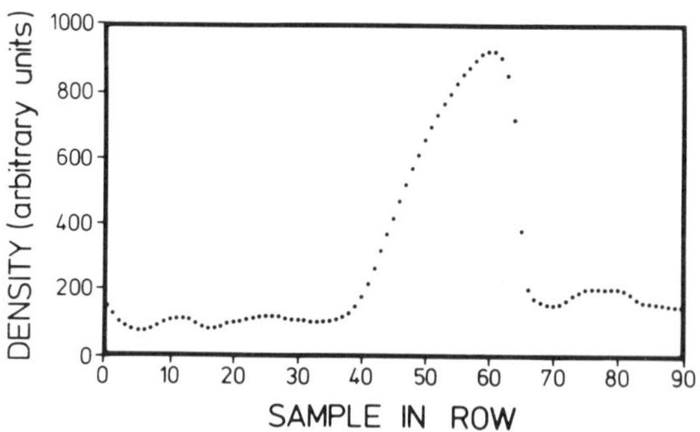


Fig. 1: Profile of a protein spot scanned in the SDS dimension

and natural boundaries of spots by their obvious properties and thereafter filling in the unclassified areas by region growing.


Data Acquisition

The electrophoretograms were prepared by J. Klose, Berlin, in his studies of mapping proteins extracted from embryonic mice (11). The size of the evaluated section is about 40 mm x 30 mm and data acquisition is performed by a scanning-microscope photometer. The technical details are described in (12,13,14) and therefore only the main aspects are summarized here. A section of a two-dimensional electrophoretogram is shown in Fig. 2a. It is 1 mm thick, stained with Coomassie Brilliant Blue R-250, mounted between glass plates, and scanned with a microscope photometer by mechanical movement of the motorized stage. The step size is 0.1 mm in both x- and y-direction, the frequency 200 steps per second. Condensor and lens are 10/0.30 micro-objectives, the illuminated measuring area in the specimen is 0.2 mm x 0.2 mm, the wavelength is 550 $\pm$ 6 nm.

The noise of the measured signal is predominantly due to the stochastic nature of the photons. Extracting the square root of the analog output of the photomultiplier transforms the signal-dependant Poisson-distributed noise of photoelectrons into a signal-independant additive noise, which can be reduced by linear filtering. Considering the square root of the intensity as the signal these measures lead to a signal-to-noise ratio of 1,100 in the blank field and 300 for an optical density of 2.4. Therefore the spot profile in Fig. 1 appears to be smooth. However, there may be disturbances e.g. caused by dust or tiny bubbles in the gel. They are attenuated by two-dimensional low-pass filtering of the digitized picture. Data are stored as a matrix of 400 x 300 square rooted intensities $f(i,j)$. Each sample $f(i,j)$ is replaced by the mean of $f(i,j)$ itself (with the weight 4) and the four nearest neighbors (with the weight 1).

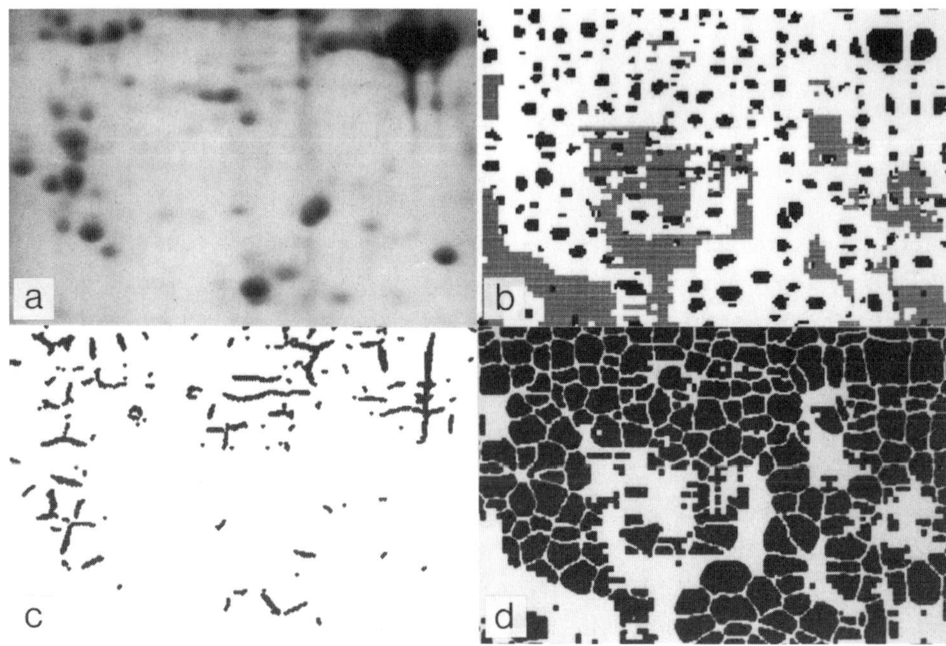

Fig. 2: (a) Section of a two-dimensional electrophoretogram
stained with Coomassie Brilliant Blue R-250. Photographed under
illumination with 550 nm, underexposed and reproduced with high
contrast to show faint spots. (b) Segmentation of (a). Gray:
flat background segmented using coarse sampling grid; black:
convex areas segmented using medium coarse sampling grid.
(c) Natural boundaries detected in fine, original sampling grid.
(d) Final result of segmentation after three coarse and one
fine expansion of convex areas.


The procedure is recursive, because the lower and left neighbors
are already smoothed while right and upper neighbors are still
unprocessed. A sketch of the filtering scheme is given in Fig. 3.
It shows sections of the sampling grid and the weights to be
applied to the raw and smoothed samples contained in the squares.
This local operation proceeds from left to right and from bottom
to top of the whole matrix. It does not affect the integrated
optical densities but just attenuates high frequency disturbances because the electrophoretograms are oversampled about four
times (12).

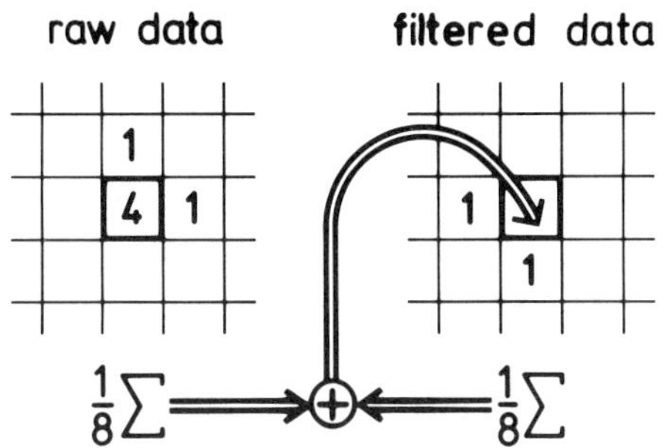



Fig. 3: Recursive smoothing in the original sampling grid. The moving average of the picture points of the raw and filtered data is calculated using the weights indicated in the diagram and is stored as the actual filtered picture point.

Finally the samples are negated to fit into existing software, so that spots appear to be elevated above the background.


Segmentation

Since electrophoretic spots asymptotically approach the background or penetrate one another, the spot boundaries are not well defined. This is why we define a spot by a flexible model, which is given implicitely by a rule for segmentation. Cone structured analysis based on the detection of obvious properties of the electrophoretogram like clear background, kernels of spots and natural boundaries between spots is a means to detect distinct spots even when they are barely visible or separated only by a slight inflection of the gray level surface. Fig. 4 illustrates this notion. The unclassified areas (shown in white) have to be filled in by a final procedure of region growing.

The clear, essentially flat background B is defined as an area of little local variation of the gray level in the following way (see Fig. 5): A coarse grid is constructed by selecting every fourth pixel in every fourth line. The center C of a 5 by 5 moving window in this coarse grid, which corresponds to a 20

418

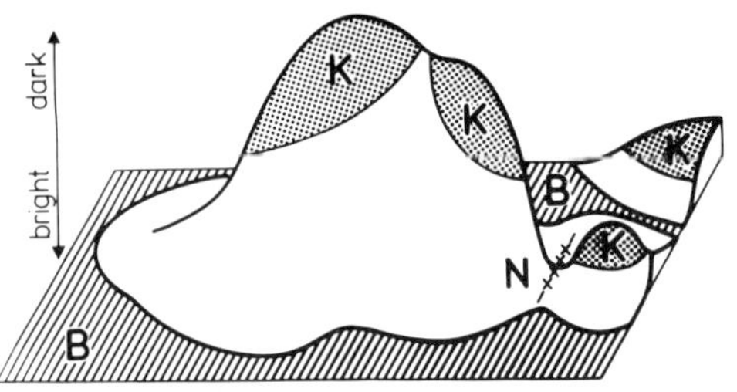

Fig. 4: Implicit modeling of electrophoretograms. B = background, K = kernel of spot, N = natural boundary.


by 20 window in the original sampling grid, is classified as
background, if the mean absolute difference of gray levels between the center C and eight picture points R on the rim of the
window does not exceed a fixed threshold. According to our experience the threshold is not critical but the size of the window is important. If a sample is classified as background all
its neighbors in a 4 by 4 window in the fine grid are also
called background. All picture points falling into this catego-
ry are excluded from further processing. The background thus
determined is shown in Fig. 2b in gray.

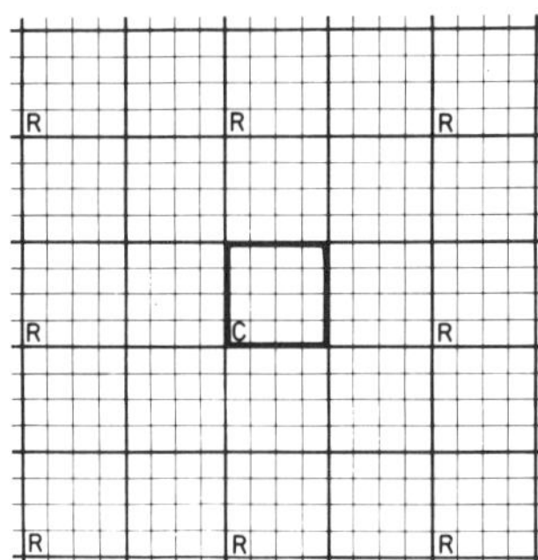

Fig. 5: Center C and rim R of
the moving window used to detect the background. Coarse
sampling grid constructed of
every fourth pixel in every
fourth line of the original
(fine) sampling grid.

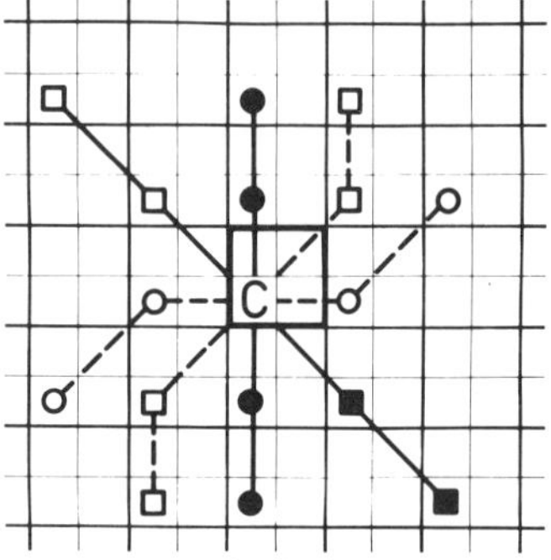

Fig. 6: Definition of convex
area detector: Center C and
four basic sequences of samples to calculate second deri
vatives in four directions
through C. Rotating the pattern by 90 degrees yields another four directions. Medium
coarse sampling grid.

The kernels K of the spots (cf. Fig. 4) are first approximations of position, size, and shape. They are determined in a medium coarse grid consisting of every second sample in lines and rows of the original image and defined as convex areas of the gray level surface in the following way (see Fig. 6): In a 5 by 5 moving window in the medium grid eight directions through the center are established. Four of them are displayed. Rotation of the diagram by 90 degrees would show the other four directions. Each direction hits five picture points. If the difference between the gray level in the center and the mean level of the remaining four samples is positive, the curvature in this direction is called convex. If this holds for all eight directions the center is classified as a sample of the kernel. The kernels thus determined are shown as light areas in Fig. 2b.

Inspecting less than eight directions failed to separate badly overlapping spots. Several local operations have been tested for segmentation of the kernels. By comparison we found that the size of the window is important and related to the spatial frequencies of the spots. The described procedure of calculating the curvature is superior to the second derivative of the usual form (15) in that it reveals a high potential to dissect spots while on the other hand it does not suppress extended weak spots.

Since the integrated optical density of a spot is measured in relation to the background, the extent of the gradual transition into the background is not critical. But the mutual delimination of adjacent spots has to be correct, especially in the case of asymmetric gray level profiles. This is achieved by drawing natural boundaries N (cf. Fig. 4), defined as pronounced local minima in the original sampling grid and determined in the following way (see Fig. 7): In a 7 by 7 moving window four directions (horizontal, vertical, and two diagonals) through the center C to opposing samples $R_i$ on the rim are tested. If for at least one direction both differences between the gray levels

420

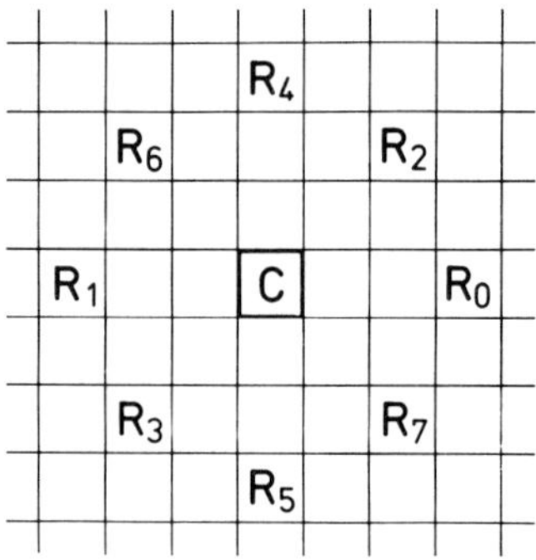

Fig. 7: Moving window used for the determination of natural
boundaries and for region growing. Original (fine) sampling
grid.


on the rim and the center are above a suitably fixed threshold,
then C belongs to a natural boundary. The threshold has to be
high in order to prevent local minima from being created just
by noise in the background. The natural boundaries of the spots
in Fig. 2a are shown in Fig. 2c.

Only short sections of spot boundaries are defined by pronounced minima. As the convex areas are good estimates of spot
shapes, it is meaningful to expand them by iterative local blow
operations as a simple method of region growing, until they
reach either the flat background, natural boundaries or each
other. The points of mutual contact determine the final boundaries. Region growing is performed using the fine grid and the
moving window shown in Fig. 7. If the center C is still unclassified, it becomes part of a kernel if this kernel and no other
coincides with at least one sample $R_i$ in the window. After three
iterations the boundaries are refined in a final run using a 5
by 5 window in a similar manner. The complete segmentation of
the image into background and protein spots is shown in Fig. 2d.

Integration of Optical Density

Prior to integration the gray values have to be converted to optical density because during segmentation they represent the negated square rooted intensity. The reference for the optical density of a spot is estimated by the trend of low optical density values within the spot area. Whenever during integration the next value $OD_n$ is fetched from memory, it is tested to check if $OD_n$ is smaller than the smallest already processed value $OD_p$ of that spot. In this case the reference $OD_r$ for the optical density is updated according to the formula $OD_r=(OD_p+ OD_n)/2$. When all optical density values of a spot are summed up, the integrated optical density is referred to the calculated background level $OD_r$. In the case of the electrophoretogram of Fig. 2a 223 spots are detected with integrated optical densities ranging from 0.5 O.D. to 1,030 O.D.


Discussion

Data acquisition with a high resolution and a large dynamic range of the gray scale is the basis of the presented method of evaluating two-dimensional electrophoretograms by implicit modeling. After analog and digital filtering gray values are available which correspond to optical densities between $5 \cdot 10^{-4}$ O.D. and 2.5 O.D.

The evaluation of a protein pattern relies on an automatic segmentation of the spots in the digital image. The described procedure detects barely-visible spots and separates even severely confluent spots while needing only occasional interactive corrections. The local operations used for segmentation are flexible and easy to modify, if required by different procedures of electrophoresis.

The reproducibility of the integrated optical density relative to the background surrounding the spot is between 5% and 10%. It is mostly at 7% for both faint and intense spots in repeated measurements and evaluations of the same electrophoretogram. This seems to depend on impurities within, or on, the gel but it will be sufficient for most biological studies of protein patterns, particularly with regard to the known problems of quantitative protein staining (16).

It is planned to extend the presented methods to the evaluation of two-dimensional electrophoretograms measuring 50 mm x 70 mm and to add an automatic classification of protein spots by matching subtemplets of a reference electrophoretogram according to a procedure (17) developed for chromatograms.


References


1.   Giebel, W., Striebel, H.M.: Microsc. Acta Suppl. 1, 133-149 (1977)

2.   Billingsley, F.C.: In: Huang, T.S. (Ed.): Picture Processing and Digital Filtering. Topics in Applied Physics 6, 250-281, Springer-Verlag, Berlin, Heidelberg, New York (1975)

3.   Kramer, J., Gusev, N.B., Friedrich, P.: Analyt. Biochem. 108, 295-298 (1980)

4.   Klose, J., Schneider, W.: In: Radola, B.J. (Ed.): Electrophorese Forum '80, 95-105, Munich (1980)

5.   Erhardt, R., Reinhardt, E.R., Schlipf, W., Bloss, W.H.: Analyt. Quant. Cytol. 2, 25-40 (1980)

6.   Erhardt, R., Reinhardt, E.R.: Forschungsbericht RV12/GfW/1, Inst. f. Physik. Elektronik, Universität Stuttgart (1975)

7.   Bossinger, J., Miller, M.J., Vo, K.-P., Geiduschek, E.P., Xuong, N.-H.: J. Biol. Chem. 254, 7986-7998 (1979)

8.   Garrels, J.I.: J. Biol. Chem. 254, 7961-7977 (1979)

9.   Lutin, W.A., Kyle, C.F., Freeman, J.A.: In: Catsimpoolas (Ed.): Electrophoresis '78, 93-106, Elsevier North Holland, Inc. (1978)

10.  Taylor, J., Anderson, N.L., Coulter, B.P., Scandora, A.E., Anderson, N.G.: In: Radola, B.J. (Ed.): Electrophoresis '79, 329-339, Walter de Gruyter, Berlin (1980)

11. Klose, J., Blohm, J., Gerner, I.: In: Neuberg, D., Merker, H.J., Kwasigroch, T.E. (Eds.):Methods in prenatal toxicology, 303-313, Georg Thieme, Stuttgart (1977)

12. Kronberg, H., Zimmer, H.-G., Neuhoff, V.: Electrophoresis 1, 27-32 (1980)

13. Zimmer, H.-G.: J. Microsc. 116, 365-372 (1979)

14. Zimmer, H.-G., Kronberg, H., Bernstein, R., Neuhoff, V.: Pattern Recognition 13, 79-82 (1981)

15. Rosenfeld, A., Kak, A.C.: Digital Picture Processing. Academic Press, p.183, New York (1976)

16. Neuhoff, V., Philipp, K., Zimmer, H.-G., Mesecke, S.: Hoppe-Seyler's Z. Physiol. Chem. 360, 1657-1670 (1979)

# EVALUATION OF IMMUNOGLOBULIN DIVERSITY

K. Felgenhauer and H. Mohrmann
Research Department, Neurologic University Clinic
5ooo Cologne 41

We have become used to the fact, that many proteins, which
appear homogeneous with other techniques split into several,
sometimes up to 2o bands on isoelectric focusing. Even more
surprising however is the reduction of the almost infinite
complexity of the antibody family to about the same number of
bands. It is often tacitly assumed, that the bands are individual antibodies, since their appearence does not differ from
unequivocally homogeneous proteins.Thus the focusing pattern
of normal immunoglobulins does not reflect the continuous
heterogeneity of the estimated $10^5$ to $10^6$ antibodies of a
healthy person. The discrepancy is the more obvious as in isoelectric focusing individual immunoglobulins as present in the
serum of myeloma patients generally exhibit 3-5 sharp bands.

According to present undisputed knowledge the diversity of
antibodies is virtually infinite. They extend over a broad
range of isoelectric points between pH 6 and 9 with a rather
symmetric charge distribution profile. All electrophoretic
techniques in non-restrictive media exhibit therefore a broad
migration zone with a more or less smooth, bell shaped distribution profile, as with a transparent upper gel modification
of the Ornstein-Davies technique (Felgenhauer, 1971). The
difference between the sharp bands of individual proteins in
the lower gel, e.g. $\beta$-Lipoprotein, $\alpha_2$-Macroglobulin and the
haptoglobin-polymers on the oneside and the polyclonal immunoglobulins at the other side becomes evident with this technique.

If a single B-cell grows beyond control as in multiple myeloma
all other lymphocyte clones are rolled back and the monoclonal
immunoglobulin dominates the whole $\gamma$-region (Felgenhauer,
Alzer and Schumacher, 1972). If a fast moving myeloma protein
reaches the lower gel its appearence does not differ from the
other well defined serum proteins, e.g. transferrin. There
are bi- and triclonal IgM-proteins in some neoplastic diseases
with a concomittant decline of all other immunoglobulins.

Meanwhile the limited number of antibodies that are produced
upon stimulation with foreign antigens can generally not been
differentiated from the vast number of other serum immunoglobulins, there are some compartments of the human body that are
set apart by rather impermeable barriers to conserve locally
synthesized antibodies. One of these privileged sites is the
cerebrospinal fluid compartment. A non-tumour type monoclonal
immunoglobulin may grow out of the otherwise polyclonal $\gamma$-re-
gion in subacute or chronic inflammatory diseases of the central nervous system, but there is a broad variety of antibody
patterns in these diseases: biclonal, oligoclonal and polyclonal. However in some cases the question can not be answered
whether there is an unresolved oligoclonal pattern or a true
polyclonal reaction with some antibodies popping out from the
background. The expectation seemed justified, that isoelectric
focusing may solve this separation problem. The introduction
of the low endosmosis agarose enabled us to make a systematic
comparison between electrophoresis and focusing in the same
separation medium.

The immunoglobulins are nicely spread over almost the whole
gel path if ampholytes pH 6-9 are used (Fig. 1).

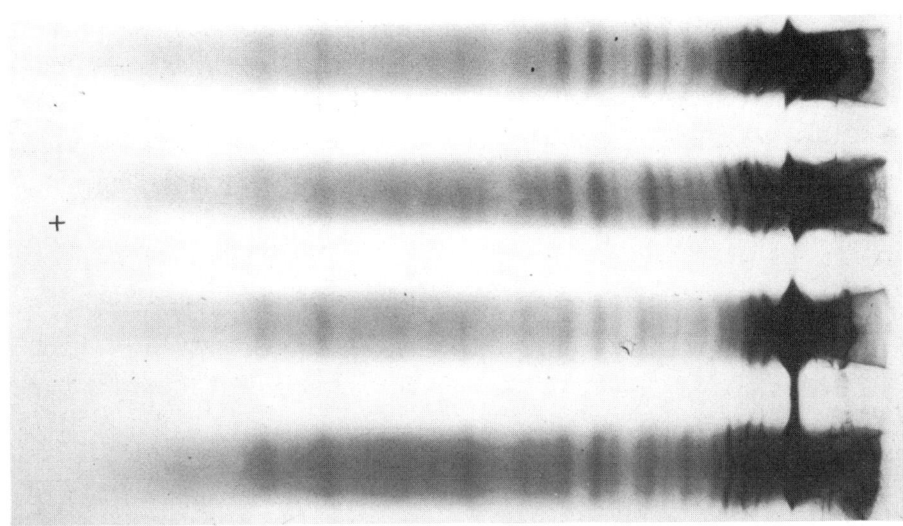


<u>Fig. 1:</u> The overall identity of the normal immunoglobulin patterns from four healthy persons. Agarose focusing with LKB-Ampholine pH 6-9.


There is in the upper pH-range a diffuse staining with some more or less broader subfractionation. In the lower range about 10 sharp bands can be counted that are also immunoglobu- lins. A comparison of the individual patients reveals com- pletely identical patterns. If these bands are individual anti- bodies, or at least defined groups of antibodies we have to conclude that all healthy persons have identical antibody po- pulations irrespective of their individual immune histories. We can only accept such antibody pattern identities if we dis- regard germ line and clonal selection theories.

The actual band patterns obtained vary with the commercial source and the question must be raised whether the ampholytes themselves as detectable for instance with the glucose-cara- mel reaction (Felgenhauer and Pak, 1973) may have an impact upon the protein pattern. Meanwhile there is no pattern iden- tity of ampholytes and immunoglobulins it is noteworthy, that

430

those pH-regions where multiple bands appear correspond
rather well. It is probable that conductivity discontinuities
are the cause for the partition of the truely polyclonal
immunoglobulins. Pseudooligoclonal partition becomes
especially evident with electrophoretically monoclonal myeloma proteins as demonstrable with all separation media yet
applied, e.g. Sephadex thin layer,polyacrylamide gel or agarose (Fig. 2).

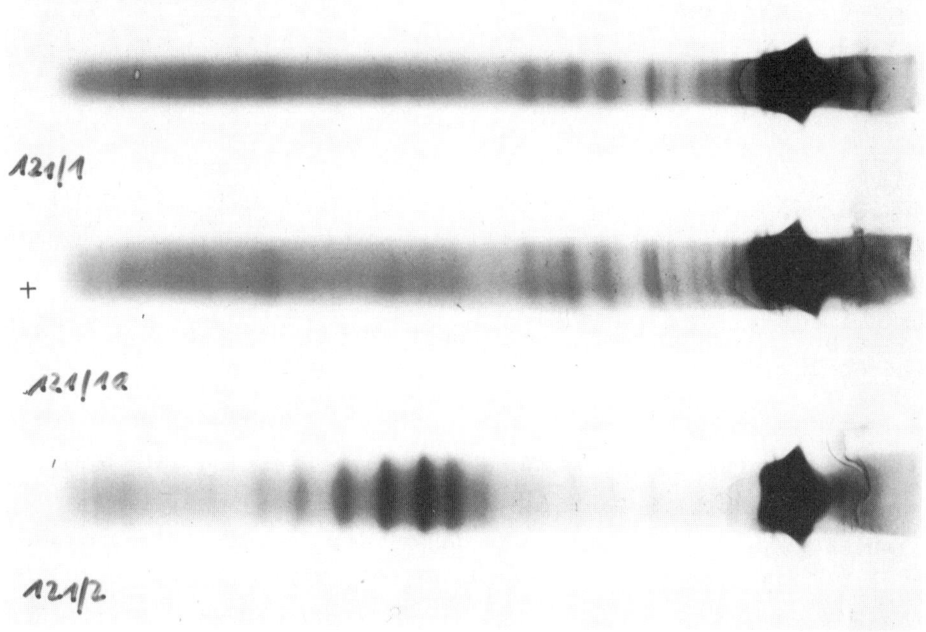


Fig. 2: Pseudooligoclonal partition (121/2) of an otherwise
monoclonal myeloma protein on isoelectric focusing. (Agarose,
LKB-Ampholine pH 6-9).

This multiplicity of all monoclonal immunoglobulins makes an
enumeration of inflammatory antibodies after isoelectric
focusing virtually impossible. In some inflammatory diseases
we see comparatively broad bands after agarose electrophoresis
and we would like to know, whether there are several unseparated antibodies or a true polyclonal population of immuno-

globulins. Isoelectric focusing reveals many bands that normally are not present (Fig. 3) but can we dare to decide which of these bands represent true individual antibodies, if we discover that several of them correspond to serum bands which clearly are not individual antibodies.

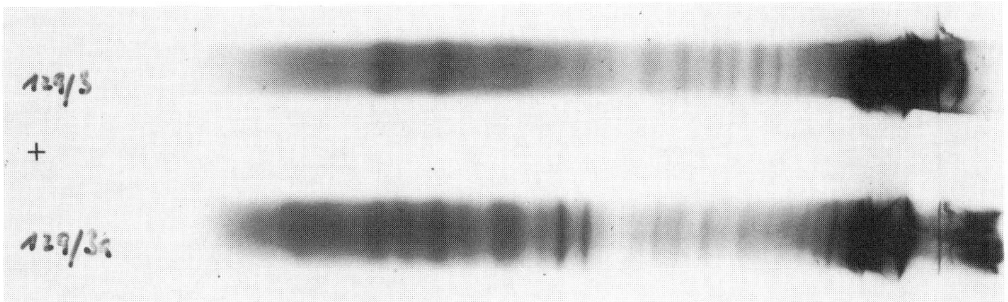


**Fig. 3:** Isoelectric focusing of serum (129/3) and CSF (129/3a) immunoglobulins from a patient suffering from chronic encephalitis. (Agarose, LKB-Ampholine pH 6-9).

At the first glance the situation may be less complicated at the far alcaline end of the $\gamma$-globulins where sharp bands normally do not appear in the serum. There are indeed several immunoactive neurological diseases where very alcaline bands can be seen exclusively in the cerebrospinal fluid. In all these cases isoelectric focusing reveals also multiple bands in the far alcaline region. Such alcaline immunoglobulins are very seldom found in the serum, but if present their patterns are similar in serum and CSF.

After all we have to admit, that all attempts to improve the separation of disease typical immunoglobulins beyond the agarose electrophoretic standards have rendered results, that are rather ambiguous for diagnostic purposes. Nevertheless we should further work on a method that combines the reliability and simplicity of agarose electrophoresis with the overall separation expectancies of isoelectric focusing. On application of immunoglobulins we should stop to count bands, but rather aim to fulfill the no-band premise that is more likely to reflect the normal state of our immune system.

References

1.  Felgenhauer, K.: Clin. Chim. Acta 39, 175-181 (1972).

2.   - , G. Alzer and K. Schumacher: Klin. Wochenchr. 5o, lo33-lo36 (1972).

3.   - and S.J. Pak: Annals N.Y. Acad. Science 2o9, 147-153 (1973).

STUDIES OF ANTIBODY CLONOTYPE PATTERNS IN RABBIT ANTIMICRO-
COCCAL SERA BY ANALYTICAL ISOELECTRIC FOCUSING IN AGAROSE
GELS[1]



Steve Binion and L. Scott Rodkey[2]
Division of Biology, Kansas State University
Manhattan, Kansas  66506



## Introduction

We recently reported a simplified method for the synthesis of
inexpensive ampholytes from pentaethylenehexamine (PEHA) and
described their use for isoelectric focusing of antibodies in
agarose gels (1).  In this paper we present preliminary results of experiments using analytical isoelectric focusing in
agarose gels to study antibody clonotype patterns in rabbit
antisera.


## Materials and Methods.

Antisera were obtained from rabbits after multiple immuni-
zations with <u>Micrococcus</u> <u>lysodeikticus</u> vaccine (2).  Ten to
twenty-five microliter volumes of antisera were focused in
gels made of 1% Isogel agarose (Marine Colloids, Rockland,
ME) bonded to Gelbond film (Marine Colloids).  Ampholytes
synthesized in our laboratory from PEHA (Columbia Organic


[1] Supported, in part, by NSF Grant PCM 79-21110.  Contribution 81-358-B, Division of Biology, Kansas Agricultural
Experiment Station, Manhattan, Kansas.

[2] Recipient of a USPHS Research Career Development Award
5 K04 AI00217.

Chemicals, Columbia, SC) covering the pH range 3.5-9.5 (1)
were incorporated in the gels at a concentration of 2%. Gel
dimensions were 25 x 11.5 x 0.2 cm.

Focusing was done in the short direction on an LKB Multiphor
(LKB, Sweden) equipped with an LKB 3371E power supply.
Coolant temperature was 8° C. The anode and cathode
solutions were 1N phosphoric acid and 1N sodium hydroxide,
respectively. The focusing run time for each gel was 3.5
hours. Initially, the voltage was set at 150V. After one
hour the filter paper sample applicators were removed and
the voltage was set at 200V. The voltage was increased in
100 volt steps at 30 minute intervals until a final voltage
of 500V was reached. This voltage was maintained for 30
minutes. The pH gradient in the gel was measured at room
temperature using a surface electrode (#9210, Lab Research
Products, Lincoln, NE). The gel then was focused for an
additional 30 minutes at 500V.

Focused antibodies were precipitated and fixed in the gel
using the method of Keck et al (3). Each gel was washed
in 500 ml of 20% sodium sulfate in borate-buffered saline
(BBS, 0.16 M borate in 0.85% sodium chloride, pH 8.0) for
3-6 hours with gentle rocking, then washed 12-16 hours
in a second 500 ml volume of the same solution. Each gel
then was washed successively in:  a) 500 ml of 18% sodium
sulfate, pH 7.0, containing 0.05% glutaraldehyde, for 3
hours b) 500 ml of BBS containing 0.1% sodium borohydride
for one hour, and 500 ml of 0.075 M phosphate buffer, pH 7.5
for 3 hours (with hourly changes of buffer). The gel then
was incubated 12-16 hours with $^{125}$I-labeled micrococcal
carbohydrate antigen (4) and washed in 500 ml of 0.075 M
phosphate buffer, pH 7.5 for 3 hours (with hourly changes of
buffer). The gel then was air dried and exposed to Kodak
Min-R X-ray film (Eastman Kodak, Rochester, NY). Each gel
was stained with 0.12% Coomassie Brilliant Blue R-250
(Eastman Kodak) in an ethanol/glacial acetic acid/distilled

water solution (25/8/67, v/v/v) and was destained in the
same solution without dye.


Results and Discussion

Analytical isoelectric focusing has proven to be an extremely
valuable technique for studying antibody heterogeneity (4, 5,
6).  When we began using isoelectric focusing in our laboratory we recognized two drawbacks to the procedure as it was
commonly employed. ˙The first drawback was the extremely high
cost of commercial ampholytes.  The second drawback was the
use of polyacrylamide gels because acrylamide is a cumulative
neurotoxin, the gels often do not polymerize, or fail to
stick to glass plates after casting, and IgM antibodies cannot be focused in polyacrylamide gels because they are too
large to enter the gel matrix (6).

We have eliminated the need for expensive commercial ampholytes by synthesizing ampholytes from PEHA (described in
reference 1).  These ampholytes are approximately seventy
times less expensive per milligram than commercial ampholytes
and give very satisfactory pH gradients.  Figure 1 shows the
pH gradient produced by PEHA ampholytes and the nearly
identical pH gradient produced by LKB Ampholines.  To overcome the problems associated with the use of polyacrylamide
gels, we substituted Isogel agarose gels bonded to Gelbond
film for polyacrylamide gels adhered to glass plates.  Isogel
agarose is non-toxic and gives gels with good mechanical
stability that remain firmly attached to the Gelbond throughout all the steps in our procedure.  In addition, IgM can be
focused in agarose gels due to the large pore size in the gel
matrix (7).

By combining PEHA ampholytes with agarose gels we have
obtained an inexpensive and dependable system for performing
isoelectric focusing that allows us to examine large numbers

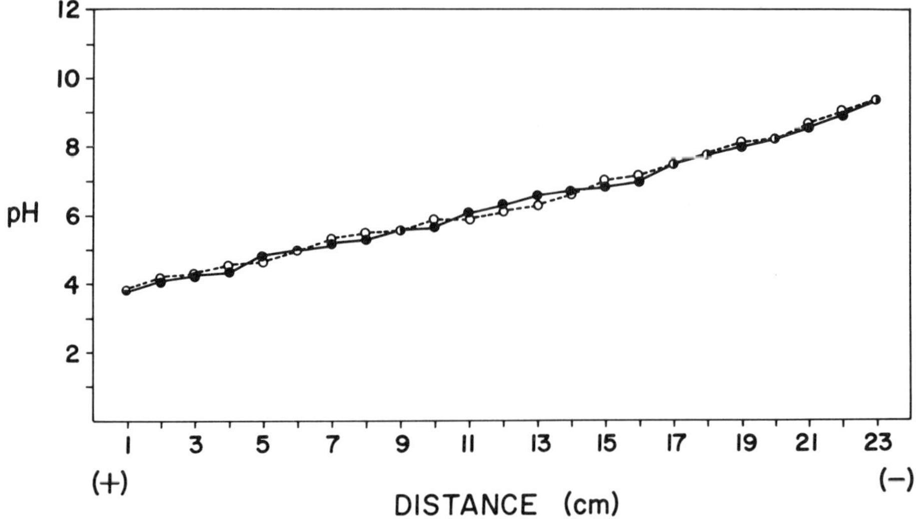

Figure 1. pH gradient obtained in agarose using PEHA ampholytes synthesized in our laboratory (open circles) and pH
gradient obtained with LKB Ampholines (closed circles).

of antisera.  Figure 2 shows the stained gel (A) and radioautograph (B) from a focusing run in which we compared antisera obtained from eight outbred rabbits during three successive rounds of immunization with Micrococcus lysodeikticus.
As seen in the radioautograph, these outbred rabbits exhibited similar antibody clonotype patterns.  Comparison of
clonotype patterns in antisera obtained from the same
individual showed apparent shifts in clonotype expression in
several instances.  These shifts included the deletion of
previously expressed bands, appearance of new bands, and
cycling of band expression.

As we have shown here, analytical isoelectric focusing in
agarose is a powerful tool for studying antibody clonotypes.
This technique provides a way to detect antibody clonotypes
which may be shared among outbred animals, and also provides
a way to study changes in antibody clonotype expression with-

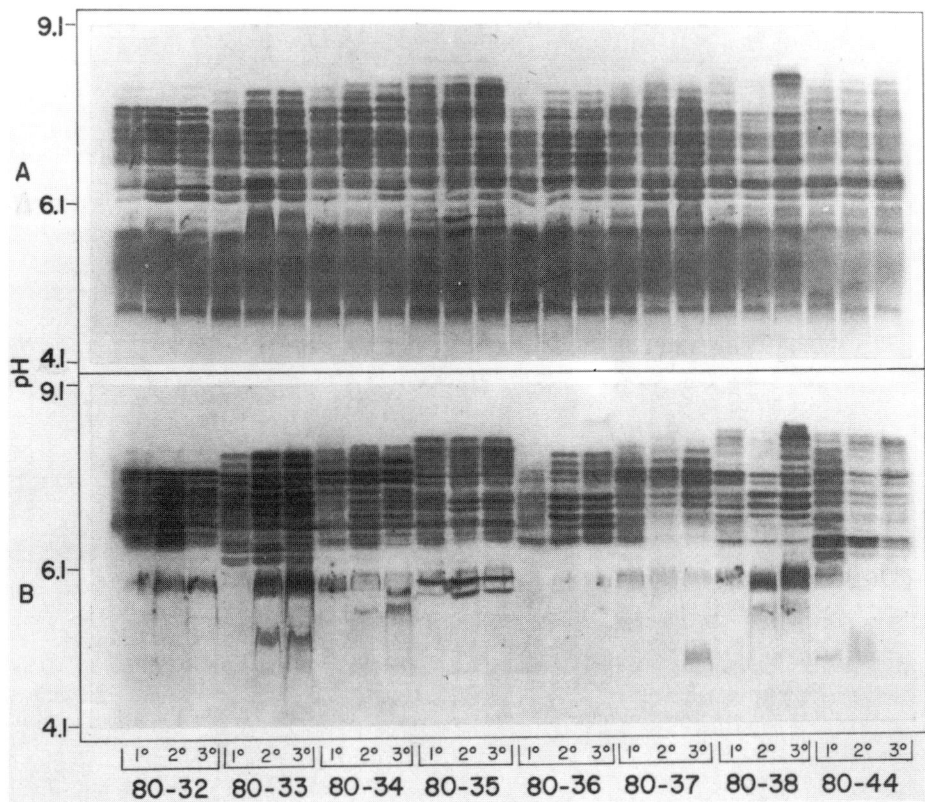



Figure 2. Stained gel (A) and radioautograph (B) showing patterns of focused antisera from eight outbred rabbits. Radioautograph shows clonotypes of antibodies specific for micrococcal carbohydrate antigen.

in an individual. The ability to monitor such changes should make it possible to investigate factors which control such changes, and current work in our laboratory involves the application of isoelectric focusing to the study of this problem and other immunological questions.


References


1.   Binion, S., Rodkey, L. S.: Anal. Biochem.   (In Press).

438

2.  Osterland, C. K., Miller, E. J., Karakawa, W. W.,
    Krause, R. M.:  J. Exp. Med. <u>123</u>, 599-614  (1966).

3.  Keck, K., Grossberg, A. L., Pressman, D.: Eur. J.
    Immunol. <u>3</u>, 99-102  (1973).

4.  Awdeh, Z. L., Williamson, A. R., Askonas, B. A.: Nature
    <u>219</u>, 66-67  (1968).

5.  Williamson, A. R,:  in Handbook of Experimental
    Immunology Vol. 1, ed. D. M. Weir. pp. 9.1-9.31  (1978)
    Blackwell Scientific Publications, Oxford

6.  Braun, D. G., Hild, K., Ziegler, A.:  in Immunological
    Methods, eds. I. Lefkovits and B. Pernis, pp. 107-120
    (1979)  Academic Press, New York.

7.  Saravis, C. A., Zamcheck, N.:  J. Immunol. Methods <u>29</u>,
    91-96  (1979).

MONOCLONAL GAMMOPATHY WITH FREE LIGHT CHAINS

Tsieh Sun and York Y. Lien

Department of Laboratories, North Shore University Hospital, Cornell

University Medical College

Manhasset, N.Y. 11030, U.S.A.


Introduction


The presence of free monoclonal light chain or Bence Jones protein (BJ) in

urine is generally considered one of the most reliable indicators for

malignant monoclonal gammopathy. However, the clinical significance of

Bence Jones proteinemia has seldom been discussed. The relationship of

the BJ in serum and in urine has also not been well studied. The present

paper reports several parameters of BJ to substantiate the controversial

statistical data in this field and to illustrate the clinical significance

of BJ in serum and its relationship to BJ in urine.


Materials and Methods


In the past 3½ years, 5,321 sera and 1, 129 urines were tested in our

laboratory by a high-resolution electrophoresis (1). When a monoclonal

band was detected, the abnormal immunoglobulin was then identified by

immunoelectrophoresis and immunofixation techniques (2, 3). The molecular

form of the free monoclonal light chains was studied by SDS-polyacrylamide

gel electrophoresis.

Results

Of the 5,321 sera, 286 showed monoclonal gammopathy, giving an incidence
of 5.3%. These 286 specimens were from 183 patients. Immunologic analyses
revealed that 55.7% of patients had monoclonal IgG, 16.4% IgA, 12.5% IgM,
10.9% kappa or lambda light chain, 2.2%IgD, and 2.2% biclonal gammopathy
(Table 1).

TABLE 1

FREQUENCY OF MONOCLONAL GAMMOPATHIES IN
NORTH SHORE UNIVERSITY HOSPITAL
(1976-1980)

| CLASSIFICATION | NUMBER OF PATIENTS | PERCENTAGE |
|---|---|---|
| IgG-kappa | 68 | 37.2% |
| IgG-lambda | 34 | 18.6% |
| IgA-kappa | 15 | 8.2% |
| IgA-lambda | 15 | 8.2% |
| IgM-kappa | 18 | 9.8% |
| IgM-lambda | 5 | 2.7% |
| kappa | 10 | 5.5% |
| Lambda | 10 | 5.5% |
| IgD | 4 | 2.2% |
| Biclonal gammopathy | 4 | 2.2% |
| Total | 183 | 100.0% |

Sera from 12 patients contained BJ in addition to the major monoclonal
immunoglobulin. All 12 patients had myelomatous disorders, including 7
IgG, 4 IgD and 1 IgA myeloma. Of the 1, 129 urines, 128 were obtained
from patients with monoclonal gammopathy. Of these 128 specimens, 73
contained BJ in urine, an overall positive rate of 57.0%. All patients
with light chain disease or BJ in serum showed BJ in urine. Patients with
lambda type monoclonal gammopathy had a higher incidence (31/51 or 60.8%)
of BJ in urine than those with kappa type (42/77 or 54.5%) (Table 2).
Only 2 BJ positive cases were not associated with monoclonal gammopathy.

All BJ in serum (9/9) had a monomeric form and most of the BJ in urine
(18/20) were dimer **or** both dimer and monomer (Table 3). The molecular
forms in urine and serum were not always consistent but the molecular
weight of BJ in urine was usually higher than that of BJ in serum.

TABLE 2
FREQUENCY OF BENCE JONES PROTEINURIA IN MONOCLONAL GAMMOPATHIES

| Monoclonal Proteins | Total | Positive | Negative |
|---|---|---|---|
| IgG-kappa | 48 | 21 (44%) | 27 (56%) |
| IgG-lambda | 27 | 12 (45%) | 15 (55%) |
| IgA-kappa | 11 | 6 (55%) | 5 (45%) |
| IgA-lambda | 9 | 5 (56%) | 4 (44%) |
| IgM-kappa | 7 | 5 (71%) | 2 (19%) |
| IgM-lambda | 3 | 2 (56%) | 1 (33%) |
| kappa | 8 | 8 (100%) | 0 (0%) |
| Lambda | 8 | 8 (100%) | 0 (0%) |
| IgD | 4 | 4 (100%) | 0 (0%) |
| Biclonal | 3 | 2 (67%) | 1 (33%) |
| Total | 128 | 73 (57%) | 55 (43%) |

TABLE 3. THE MOLECULAR FORM OF FREE LIGHT CHAINS IN SERUM AND URINE

| Patients | B.J. Type | Urine | Serum |
|---|---|---|---|
| WR | λ | D | M |
| MS | λ | D | M |
| MM | λ | M + D | M |
| BP | λ | D | M |
| SM | k | − | M |
| IP | λ | − | M |
| SJ | λ | − | M + D |
| KL | λ | − | M + D |
| TJ | λ | M + D | M + D |
| CD | λ | D | − |
| SA | λ | D | − |
| WW | λ | D | − |
| HD | λ | M + D | − |
| MN | k | M | − |
| SB | λ | M + D | − |
| KS | k | M + D | − |
| VH | k | M + D | − |
| SJ | λ | D | − |
| BJ | k | D | − |
| AC | k | M + D | − |
| TJ | λ | D | − |
| DE | k | M + D | − |
| SS | k | M | − |
| DF | λ | M + D | − |

442

DISCUSSION

While the association between BJ in urine and malignant gammopathy needs
further clinical analysis, this study shows that an intimate correlation
exists between monoclonal gammopathy in serum and BJ in urine, confirming
the specificity of the electrophoretic and immunologic techniques (4).
In addition, the presence of BJ in serum appears to be an even more
reliable indicator for malignancy than BJ in urine. It has been reported
that the molecular form of BJ in serum and in urine taken from the same
patients to be always identical (5). In this study, however, it was
observed that in the same patient when serum BJ is monomeric urine BJ
can be monomeric or dimeric. It is possible that polymerization of BJ
takes place in the kidney as well as plasma cells.

References


1. Sun, T., Lien, Y.Y., Gross, S.: Ann. Clin. Lab. Sci. $\underline{8}$: 219-227 (1978).

2. Sun, T., Lien, Y.Y., Degnan, T.: Am. J. Clin. Pathol. $\underline{72}$: 5-11 (1979).

3. Sun, T., Lien, Y.Y.: Clin. Chem. $\underline{26}$: 1763-1764 (1980).

4. Pruzanski, W., Ogryzlo, M.A.: Adv. Clin. Chem. $\underline{13}$: 336-382 (1970).

5. Zolla, S. et al.: J. Exp. Med. $\underline{132}$: 148-156 (1970).

# ISOELECTRIC FOCUSING OF IMMUNOGLOBULINS IN ULTRATHIN LAYERS OF AGAROSE


Bruno Schmidt, Gertrude Pfeifer
Ludwig Boltzmann Institut für dermato-venerologische Sero-
diagnostik, A-1130 Wien, Austria


Introduction

Isoelectric focusing (IEF) in acrylamidegel to analyse IgG
antibody heterogeneity has become the method of choice for
over 10 years. The conventional technique has some limitations
e.g. the time needed for focusing including staining and
destaining is normally more than 15 hours, or large sized
gels may detache from the glass plate during fixation or
staining and so increasing the possibility of mechanical
injuries of the gel. These disadvantages were overcome by the
recently described technique of ultrathin-layer-IEF in
polyacrylamidegels on cellophane (1) or on silanized supports
(2). The method combines high resolution, speed and reagent
economy. However, the polyacrylamidegel severely restricts
the mobility of large proteins due to molecular sieving.
Agarose should be an ideal support for IEF because of its
mechanical strength and its large pore size, however, the
high electroendosmotic flow  causes  problems. Rosen (3)
has used an agarose with blocked charged groups and Jackson
et al. (4) have tried composite acrylamid-agarose gels to
overcome electrosmotic effects. Problems with endosmosis are
potentiated by reducing the gel thickness. We are presenting
some experiences with ultrathin layers of agarose on
focusing immunoglobulins especially in the more basic region.

Results

Serum samples derived from patients infected with <u>Treponema</u>
<u>pallidum</u>, who developed IgM-anti-IgG autoantibodies,
were taken at various stages before and after treatment.
Complete sera or QAE-A50 (Pharmacia) fractions thereof
containing basic immunoglobulins were focused together with
standard marker proteins in a Multiphor (LKB) or Mediphor
(Desaga) apparatus with a cooling temperature of $10^{\circ}$C.
The "flap technique" (2) was used for casting gels, starting
at first with high porosity acrylamidegels (T3%,C20%). The
resolution in the pI region from 7 - 10 was poor and did not
markably improve when gels were cast   with 6M urea.
Preliminary experiments with agarose in 0,2 mm thin layers
showed that the good resolution of individual bands was even
better without addition of urea.
The main problem of focusing in agarose gels is the high
electroendosmotic flow. When using IsoGel Agarose (Marine
Colloids) we were forced to dry the surface of the gel every
10 minutes and voltages above 1500 volts were impossible. On
the other hand Pharmacia Agarose IEF can be run to 2500 V.h
starting with 500 volts and increasing to 2000 volts. For
better resolution in basic pH regions it is advisable to
flatten the pH gradient by adding basic ampholytes. Using a
3:1 mixture of Servalyte T 4-9 and other ampholytes 8 - 10
( or 8 - 10,5) the effects are rather controversial. The
addition of Servalyte 9 - 11 hardly improved the focusing
pattern, while Pharmalyte 8 - 10,5 elongated the separation
zone pH 7 - 10 from 1,5 cm to 3 cm. The best resolution can
be obtained with the new Agarose Ampholine pH 3,5-9,5 (LKB)
in combination with Pharmalyte pH 8 - 10,5 (Pharmacia).
Figure 1 shows the focusing pattern. Note the development of
discrete bands (indicated by bars) in sera C and D, 13 and
31 months after treatment. The zones can also be seen in the
QAE-A50 fraction thereof (E), while there are no bands

visuable before treatment (B). Samples E and D together with
marker proteins (A) are shown on the right side of the figure
(II) focused on normal polyacrylamide gels (Serva Precotes)
on the same pH range.

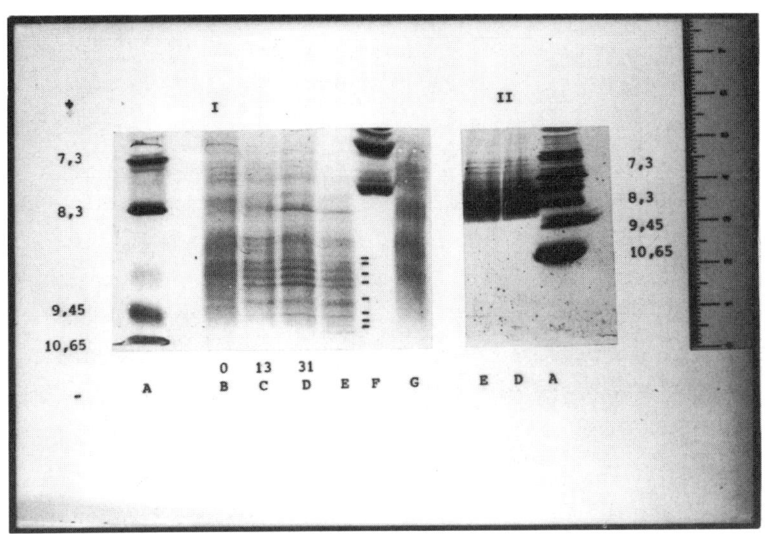

Figure 1
Comparison of IEF in ultrathin sheets of agarose and acrylamid gels: I. 0,1 mm layers of 1% Agarose IEF (Pharmacia)
with 4% ampholytes (mixture of 3 parts LKB Agarose Ampholine
pH 3,5 - 9,5 with 1 part Pharmalyte 8 - 10,5).
II. 0,1 mm acrylamide gel T5%, C3%, pH 3-10,5 (Serva Precote)
Samples: A marker protein mixture containing Cytochrom C
[ pI 10,65 ], Ribonuclease [ pI 9,45 ], whale myoglobin
[ pI 8,3 ] and horse myoglobin [ pI 7,3 ]; B,C and D are
sera of patient N.G., infected with _Treponema_ _pallidum_, before
(B), 13 and 31 months (C,D) after treatment. E is a QAE-A50
fraction of serum D, F is Con A marker and G is human IgG
(Sigma). Note the development of definite bands in the
basic region (indicated by bars) in the sera taken after
treatment. Temperature of focusing 10$^{\circ}$C, 2500 V.h.

It should be mentioned that the samples are applied near pH 6
as  no focüsing of immunoglobulins can be obtained by
application near the cathode. Studies with ultracentrifugation
and immunofluorescence techniques seem to indicate that the
discrete bands in C and D are IgM-IgG complexes with 23 S
and the IgM part is reacting with the Fc-part of IgG, while
the IgG is specific for Treponema pallidum. At the moment
we cannot exclude absorption effects. Looking for the IgM after disassociation (rheumatoid factors have pI 3,7 - 4,8 [5])
we were unable to get reasonable focusing patterns in the
acidic pH range due to endosmotic flow. In the basic part
pH 7 - 10 the resolution is remarkable better by using agarose
gels instead of acrylamide.
We found the ultrathin layer technique in agarose gels of
value in the analysis of immunoglobulins especially for
IgM-autoantibodies. The optimal conditions are summarized
in Figure 1.


References


1. Görg, A., Postel, W., Westermeier, R.: Anal.Biochem.89
   60-70 (1978)
2. Radola, B.J.: Electrophoresis 1, 43-56 (1980)
3. Rosen, A., Ek, K., Aman, P.: J.Immunol.Meth. 28,
   1-11 (1979)
4. Jackson, D.E., Skandera, C.A., Owen, J, Lally, E.T.,
   Montgomery, P.C.: J.Immunol.Meth. 36, 315-324 (1980)
5. Trieshmann, H.W., Abraham, G.N., Santucci, E.A:
   J.Immunol. 114, No.1,176-181 (1975)

ELECTROPHORETIC ABNORMALITIES OF HIGH DENSITY LIPOPROTEINS IN
LIVER DISEASES

Kazuhisa Taketa
Health Research Center, Kagawa University, Takamatsu 760, Japan

Satoru Ikeda
First Department of Internal Medicine, Okayama University
Medical School, Okayama 700, Japan

Makoto Watanabe
Okayama Saiseikai General Hospital, Okayama 700, Japan

Introduction

The electrophoretic abnormality of lipoproteins in liver
diseases is characterized by the deletion of α- and pre-β-
lipoprotein bands, the appearance of a single band migrating
between the β and pre-β position and the presence of cathodally
migrating lipoprotein X (LP-X) on agar gel depending on the
contribution of cholestatic mechanism.  The disappearance of
α band is suggested by Sabesin <u>et al.</u>(1) to be due not only to
the decreased plasma level of high density lipoproteins (HDL),
but also to their reduced mobility.  We have reported several
abnormal HDL's with altered electrophoretic mobilities appearing in liver diseases (2,3,4).
In this presentation, we report major electrophoretic abnormalities of HDL in hepatitis and cholestatic patients as revealed
by polyacrylamide-gel disc-electrophoresis (PAGE).  In patients
with parenchymal liver injury, distribution of HDL among its
subclasses, $HDL_2^e$ and $HDL_3^e$ (electrophoretically defined sub-
fractions corresponding to ultracentrifugally separated $HDL_2$
and $HDL_3$, respectively) (5) tended to shift toward the larger

448

molecular form, $HDL_2^e$, with decreased intensities of total $\alpha$-lipoprotein band. A similar molecular shift was observed in mild cholestasis and in anicteric elevation of biliary enzyme activities (non-cholestatic cholestasis)(6) with normal or high levels of HDL. An HDL migrating slower than $HDL_2^c$, termed as Slow-Migrating HDL (HDL-S)(3), also emerged frequently under such cholestatic conditions of the liver.


Materials and Methods

PAGE of lipoproteins was performed on whole serum or plasma and isolated HDL (or HDL + residue) fraction. The plasma obtained with EDTA was used for ultracentrifugal separation of HDL and the serum for removal of low and very low density lipoproteins either by dextran sulfate-$MgCl_2$ (7) or by anti-$\beta$-lipoprotein antibody (HDL-Fractionation Kit-S, Nihonshoji Co. Ltd., Osaka). Polyacrylamide gels with a monomer concentration of 3.6% (8) were used mainly for electrophoretic demonstration of HDL-S and those with a monomer concentration of 7.5% and with added lauric acid (Utermann's system) (9) for subfractionation of HDL into $HDL_2^e$ and $HDL_3^e$. Lipid concentrations were determined by routine laboratory techniques.


Results and Discussion

Among 104 patients with various liver diseases studied for PAGE pattern of serum lipoproteins with 3.6% gels, 80 cases revealed markedly reduced or virtually absent, 15 unchanged and 9 rather increased intensities of $\alpha$-lipoprotein band (Table 1). Decreased $\alpha$-lipoprotein band was found not only in patients with parenchymal liver injuries, such as acute and chronic hepatitis, liver cirrhosis with or without primary hepatoma or alcoholic liver injury, but also in cholestatic

Table 1.  Intensities of α-Lipoprotein Band in PAGE of Serum from Patients with Liver Diseases

| Liver diseases | Number of cases | | |
|---|---|---|---|
| | Intensity of α-lipoprotein band | | |
| | Decreased or absent | Unchanged | Increased |
| Intrahepatic cholestasis[*1] | 20(10) | 1 | 6(4) |
| Biliary obstruction[*2] | 22(6) | 1 | |
| Hepatocellular carcinoma[*3] | 10(4) | | |
| Acute hepatitis | 9 | 4 | |
| Chronic hepatitis | 3 | 3 | |
| Liver cirrhosis | 8 | 4 | 2 |
| Alcoholic liver injury | 8 | 2 | 1 |
| Total | 80 | 15 | 9 |

[*1], Primary biliary cirrhosis (early and advanced stages), mild cholestasis and non-cholestatic cholestasis (Taketa et al.) included.
[*2], malignant, extrahepatic.
[*3], Cases with biliary obstruction included.
The number of cases with positive HDL-S is given in parentheses.

patients.  Increased α-lipoprotein band was found in some cases of intrahepatic cholestasis but in none of biliary obstruction.  The intrahepatic cholestasis with increased HDL, as measured by its cholesterol content, was characterized by slight or no increases in serum bilirubin level in the presence of markedly increased activities of biliary enzymes (alkaline phosphatase, leucine aminopeptidase and γ-glutamyltransferase) ; cases with early stages of primary biliary cirrhosis and anicteric biliary enzyme elevation being included.

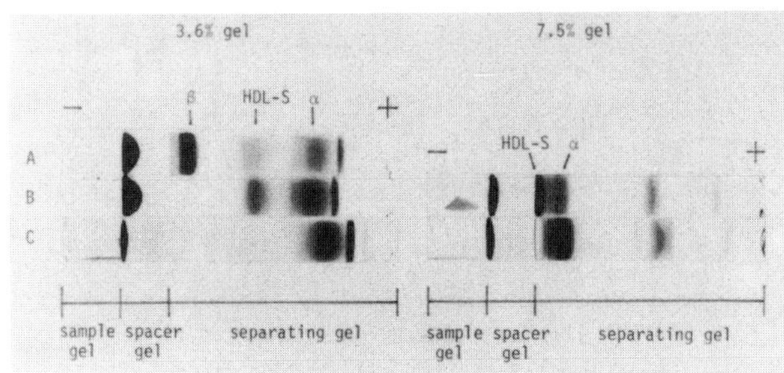


Fig. 1.  PAGE pattern of serum lipoproteins in a case of non-cholestatic cholestasis.  A, serum; B, supernatant from serum treated with anti-β-lipoprotein antibody; and C, supernatant from serum treated with dextran sulfate-$MgCl_2$.

HDL-S, an electrophoretically defined new class of cholestatic
HDL ($1.063 < d < 1.083$) migrating slower than the usual $\alpha$-lipo-
proteins in PAGE, can be readily demonstrated by differential
precipitations with anti-$\beta$-lipoprotein antibody (or concanavalin A) and dextran sulfate-$MgCl_2$ (Fig. 1). The incidence of
positive HDL-S was much higher in intrahepatic cholestasis than
in biliary obstruction and none in parenchymal liver injuries
as reported earlier (4,10). Although the appearance of HDL-S
was not necessarily associated with increased $\alpha$-lipoprotein
band, the intensity of HDL-S band roughly correlated with the
level of HDL-cholesterol. Thus, the HDL-S bands in intrahepatic cholestasis with mild jaundice were often darker than those
with marked jaundice (Table 2) opposite to the changes in LP-X
level in cholestasis (4).
The mobility and intensity of HDL-S were unaltered by the presence of sodium deoxycholate up to 1 mg/mg serum, by _in vivo_
heparin treatment (i.v., 0.1 mg/kg body weight) and incubation
of heparin-treated plasma for 30 min at 37°C, or by incubation
of a mixture of equal volumes of HDL-S positive and negative
sera for 2 h at 37°C. Thus, the liver and other tissues appear
to be required for the metabolism (production and removal) of
HDL-S.

Table 2.  Intensities of HDL-S Band in PAGE of Serum from
Patients with Liver Diseases

| Liver Diseases | Serum bilirubin (mg/dl) | Number of cases | | | | |
| --- | --- | --- | --- | --- | --- | --- |
| | | Intensities of HDL-S band | | | | |
| | | - | ± | + | ++ | +++ |
| Intrahepatic cholestasis*1 | > 5 | 5 | 4 | 1 | | |
| " | ≤ 5 | 9 | | 4 | 5 | |
| Biliary obstruction*2 | > 5 | 12 | 2 | | | 1 |
| " | ≤ 5 | 3 | | 1 | 1 | |
| Hepatocellular carcinoma*3 | | 13 | 1 | 2 | 1 | |
| Total | | 42 | 7 | 8 | 7 | 1 |

*1, Primary biliary cirrhosis (early and advanced stages), mild
cholestasis, non-cholestatic cholestasis(Taketa et al.) and each
one case of fulminant hepatitis with cholestatic feature and lupoid
hepatitis with cholestatic feature included.
*2, Malignant, extrahepatic.
*3, Cases with biliary obstruction included.

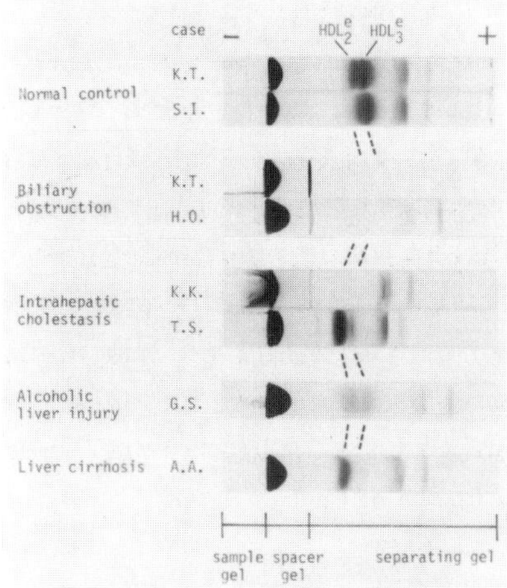



Fig. 2. PAGE patterns of HDL subclasses in patients with liver diseases. Case K.K., advanced and Case T.S., early primary biliary cirrhosis. Dextran sulfate-MgCl$_2$-treated sera were applied to 7.5% gels with Utermann's system.

Although a fine resolution of α-lipoprotein into several sub-bands was achieved by increasing the monomer concentration of polyacrylamide as in Fig. 1, characterization of each band became rather difficult. HDL-S moved as a single band least into the separating gel at a 7.5% monomer concentration. In our previous studies, the level of serum HDL, as determined by its cholesterol content, decreased significantly in both parenchymal and obstructive liver diseases, although the decrease was variable in alcoholic liver injuries and insignificant in a group of intrahepatic cholestasis (11,12), as it was demonstrated in the present study by the change in α-lipoprotein band intensity. The decreased level of serum HDL-cholesterol appeared to be primarily related to the fading of HDL$_3^e$ even though a wide individual variation (9) in the intensity of HDL$_2$ was taken into account (Fig. 2). In liver diseases, HDL$_2^e$ was generally darker than HDL$_3^e$ in contrast to the predominant proportions of HDL$_2^e$ ≤ HDL$_3^e$ as compared with

452

$HDL_2^e > HDL_3^e$ (9) in control subjects.  A group of patients with intrahepatic cholestasis presenting with unusually increased HDL-cholesterol concentrations in serum gave densely stained bands of $HDL_2^e$ (e.g. Case T.S. in Fig. 2).  In alcoholics with increased levels of serum HDL-cholesterol (cases with mild alcoholic liver injuries (12)), $HDL_2^e$ band was also pronounced. The results indicate the usefullness of electrophoretic analysis of HDL abnormalities in understanding the mechanisms of altered lipoprotein metabolism in liver diseases and also for a diagnostic purpose when particularly unusual cases of intrahepatic cholestasis are dealt with.


References


1. Sabesin, S.M., Hawkins, H.L., Kuiken, L., Ragland, J.B.: Gastroenterology 72, 510-518 (1977).

2. Watanabe, M., Taketa, K.: Proc. Jap. Soc. Clin. Biochem. Met. 15, 65-67 (1978).

3. Watanabe, M.: Acta Med. Okayama 33, 269-285 (1979).

4. Watanabe, M.: Acta Med. Okayama 33, 327-341 (1979).

5. Ikeda, S., Nagashima, H., Taketa, K., Watanabe, M.: Acta Med. Okayama 35, 141-146 (1981).

6. Taketa, K., Izumi, M., Yamamoto, H., Ikeda, S., Watanabe, M.: Am. J. Gastroenterol. 74, 96 (1980).

7. Kostner, G.M.: Clin. Chem. 22, 695 (1976).

8. Naito, H.K., Wada, M., Ehrhart, L.A., Lewis, L.A.: Clin. Chem. 19, 228-234 (1973).

9. Utermann, G.: Clin. Chim. Acta 36, 521-529 (1972).

10. Taketa, K., Watanabe, M., Nagashima, H.: Am. J. Gastroenterol. 72, 348 (1979).

11. Watanabe, M., Taketa, K., Nagashima, H., Yamamoto, Y.: Acta Med. Okayama 33, 323-326 (1979).

12. Ide, T., Taketa, K., Watanabe, M., Ikeda, S., Izumi, M., Sokabe, T., Kono, H., Yamamoto, Y.: Acta Med. Okayama 34, 293-299 (1980).

SDS-PAA-GEL-ELECTROPHORESIS (SDS-PAGE) OF URINARY PROTEINS:
CORRELATION OF DIFFERENT PROTEIN PATTERNS TO RENAL PROTEIN
CLEARANCES AND TO RENAL AND EXTRARENAL DISEASES

Wolf H.Boesken

Nephrologisches Labor,Medizinische Klinik der Universität
D-7800 Freiburg im Breisgau,Germany


Introduction

Proteinuria is one of the markers of renal diseases.Its quantitative estimation alone allows limited diagnostic conclusi-
ons.Electrophoretic procedures have been applied for further
differentiation,however,all zone electrophoreses were of little value,except in the case of paraproteinurias. - The kidneys
handle different serum proteins merely by their size,the surface charge plays a minor role.The glomerular filter operates
as a sieve:While serum albumin(Mol.wt 67000 d)is cleared at a
glomerular filtration rate of about 0.02 ml/min,the clearance
of ß-2-microglobulin(ß2M;11900 d)equals the one of endogenous
creatinine(100 ml/min)and immunoglobulin G (IgG;156000 d) is
retained completely.The glomerular ultrafiltrate contains about 5 g protein/24 h,which with the exception of 100 mg/24 h
will be resorbed completely.The resorption,too,depends on the
molecular size of the filtered small proteins (1).In disease
states the glomerular filter operates as sieve with a 'variable pore size',leading to proteinurias with all proteins('unselective')or only with the smaller serum proteins('selective').
Here the different net charge possibly is involved.Similarly
the disturbed tubular function leads to a micromolecular proteinuria with all or only some of these small proteins.So,a
cascade-like system of size-dependant glomerular and tubular
mechanisms retains the proteins filtered physiologically at

the glomerular membrane. - For these reasons electrophoretic
procedures had to be applied,which separate according to molecular weight and do not destroy the physiological structure of
urinary proteins.In general two different ways have been found
:the microzone electrophoresis with a continouus PAA concentration gradient (2)and the SDS-PAGE without mercaptoethanol
(3,4,5).The results reported here have been obtained by the
latter method.


Materials and methods


Urinary samples of about 4500 patients were analysed;diagnosis
was established in 1500 by kidney biopsy,autopsy or by clear
clinical conditions,as diabetes,acute glomerulonephritis etc.
Urine was collected for 24 h,sodium azide added to prevent
bacterial growth and samples stored at 4-6 $^{o}$C,after insoluble
material had been removed by filtration and centrifugation at
3000 rpm,10',4 $^{o}$C.The presence of blood,protein and glucose
was analysed by dipsticks.Urinary proteins were quantitated by
the TCA-biuret- and the tannin -ferrichloride-methods (6).Ali-
quots of native urine with protein concentrations above 1 g/L
were subjected to the procedure;less proteinuric samples had
to be concentrated by negative pressure ultrafiltration using
Visking tubes,by Minicon B-15 cells or by the tannin -Caffeine
-procedure (7).Total and additional relative protein losses
were calculated by total protein,albumin and retinol-binding-
protein concentrations. - 7.5%-PAA-gel rods of 5x60 mm were u-
sed;about loo ug protein in 20 -200 ul buffer(o.1 M NaH$_2$PO$_4$;
o.1% SDS;pH 7.2)with 15% glycerol and bromphenolblue were applied to the gel and electrophoresed at 10 mA/gel for about 3
h at room temperature.Gels were stained by o.6% amidoblack and
destained over night in large volumes of 7% acetic acid.Quan-
titative evaluation was carried out by a Joyce-Löbl-microzone
densitometer.Gels could be stored for several years in glas
tubes for follow up studies. - For standardization catalase,

IgG,transferrin,albumin,ovalbumin and cytochrome C were used.
For practical purpose urinary albumin,IgG and monomeric hemo-
globin(arteficially split by SDS)were taken as easily identified internal standards.The migration distance of proteins was
expressed in relation to albumin(RM-albumin = 1).


Results and Discussion


1) Technical results
By SDS-PAGE as described above urinary proteins were separated
according to their molecular weight;a linear relation between
the proteins RM and mol.wt was obtained in the range of 20. -
200.ooo d.By the addition of SDS -necessary in this technique
for micelle formation- no major serum protein was destroyed,in
contrast to the additional reduction by mercaptoethanol.The albumin dimer (8)as well as IgG migrated as complete molecules.
The only known artefact is the splitting of hemoglobin in its
mono- and dimers.This fact,however,prevented the superimposing
of Hb onto albumin,impeding the analysis of hematuric samples
on SDS-free microgradient electrophoreses. - Necessary concentration procedures of urines with less than 1 mg protein/ml
are a disadventage of this method compared to microelectropho-
reses.Beside global protein losses additional proteins might
be lost by the pore structure(i.e. ß2M or RBP)or by adhesive
properties of the device used for concentration.While by Minicon B-15 cells beside a 4o -7o% global loss small proteins are
lost selectively,good results were obtained by Amicon membrane
and by the tannine precipitation and its subsequent competitive dissolution by caffeine (7). - The colorimetric quantitation
of tannine-precipitated,ferrichloride-stained proteins (6) in
urines was proved as a sensitive method in detecting low protein concentrations as well as the micromolecular proteins,in
contrast to the TCA-biuret test.
2)Urinary protein pattern
Following SDS-PAGE different elements i.e.groups of proteins ,

| A.  Macromolecular elements | |
|---|---|
| .1: all macromol.serum proteins | 60.- $>$ 350.000 d |
| .2: only smaller proteins | 60.-  130.000 d |
| .3: postglomerular hematuria | A-1 plus hemoglob. |
| .4: postrenal serum leakage, chyluria,etc | A-1 plus a typical protein c.45.000 d |
| .5: postrenal Ig-secretion | 156/325/900.000  d |
| .6: albumin polymers | 135.000 d and more |
| .7: macromol.paraproteins | 156.000 d and more |
| B.  Micromolecular elements | |
| .1: all micromol.serum proteins | 10.-70.000 d |
| .2: only greater microproteins | 40.-70.000 d |
| .3: 'overflow' as myoglobin, micromol.paraproteins 'acute phase'-proteins etc | 17.000 d 22./45.000 d 40.-60.000 d |

<u>Tab.1 :</u> Renal and extrarenal elements in proteinuria

contributing to proteinuria,could be defined(Tab.1).First,the
combination of albumin and proteins of higher molecular weight
was classified as macromolecular or -for the high coincidence
with the disturbed renal function- glomerular proteinuria,the
one with albumin and smaller proteins as micromolecular or tubular proteinuria.Glomerular proteinurias exhibited a variable
composition:serum proteins may reach the final urine in unchanged relative concentrations('unselective')or they leak through
the glomerulus only in part('selective',i.e.only albumin and
transferrin).On the other hand macromolecules may be secreted
into the urine during the postglomerular passage. - Micromolecular proteinurias contain proteins of a mol.wt between 10.-70
ooo d.The pattern with 6 or more proteins spread over this range represents the 'classical' tubular proteinuria.A second micromolecular pattern consists of proteins mol.wt 40.-70.ooo d.
In certain clinical situations one can observe a gradual shift
from selective to unselective glomerular proteinuria and from
the complete to the partial tubular pattern. - Furthermore micromolecular proteins may be found in the urine by SDS-PAGE
independant from renal dysfunction:Paraproteins of small molecular weight i.e.mono- and/or dimer Ig-L-chains or myoglobin
or 'acute phase'-proteins are detected by their high mobility.

| Pattern | Definition | Composed from |
|---------|-----------|---------------|
| O | Physiological | - - |
| I | Unselective glomerular | A-1 (+ B-1,-2,A-3) |
| II | Medium selectivity | A-1 - A-2 (+ B-2 ) |
| III | Selective glomerular | A-2 alone or +A-6) |
| IV | Tubular | B-1 alone or +A-5) |
| V a | Glomerulo-tubular | A-1 + B-1 |
| V b | Tubulo-glomerular | B-1 + A-1 ( + A-5) |
| VI | Vascular pattern | A-1 + B-2 |

Tab.2 : The typical patterns of renal proteinurias and
their composition from various elements(s.Tab.1)

The physiological urine is composed from albumin and some macro- and micromolecular proteins,not exhibiting however one of the typical patterns.- The urinary protein excretion of an individual patient may be composed from one or more of the elements,listed in Tab.1,indicating pathogenetically different sources of proteinuria within the nephron.This reflects the situation in renal disease,which frequently is not restricted to one part of the nephron.So proteinuria indicates glomerular as well as tubular damage of different localisation as interstitial and vascular renal diseases.In Tab.2 the most common patterns of proteinuria are listed.Fig.1 shows typical samples of renal proteinurias,fig.2 extrarenal proteinurias as seen by SDS-PAGE.Beside composed renal proteinurias combinations of renal and extrarenal proteinurias occur,i.e.in active pyelonephritis (B-1 + A-5),in Bence Jones-tubulopathy or rhabdomyolytic kidney failure (B-1 + B-3),etc.

3)Correlation of SDS-PAGE-pattern and protein clearances
The main difficulty in renal pathophysiology of man is the fact,the the composition of the urine reflects the filtration of substances at the glomerulus as well as their resorption by the proximal tubulus.In discussing here electrophoretic aspects only the results are reported,which are necessary for the correct interpretation of electrophoretic findings.The different macromolecular proteinurias are distinguished by the glomerular selectivity,which correctly should be calculated by protein clearances.Comparing however the results of the latter procedure with SDS-PAGE-pattern in 5o patients a good correlation

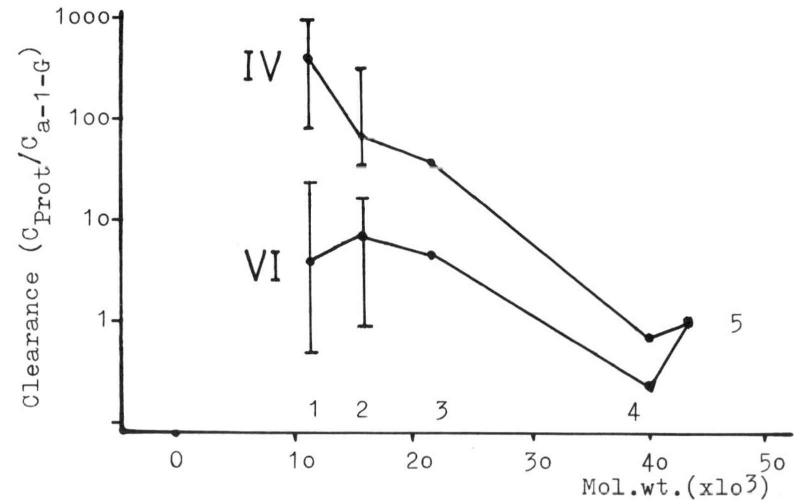



Fig.3 : Relative clearances of micromolecular proteins 1= ß-2-microglobulin,2=lysozyme(15.ooo d),3=retinol-binding protein(21.ooo d),4=ß-2-glycoprotein I(40. ooo d)in relation to 5=a-1-acid glycoprotein(44.ooo d)in 2 groups of 6 pat with proteinuria IV resp.VI.

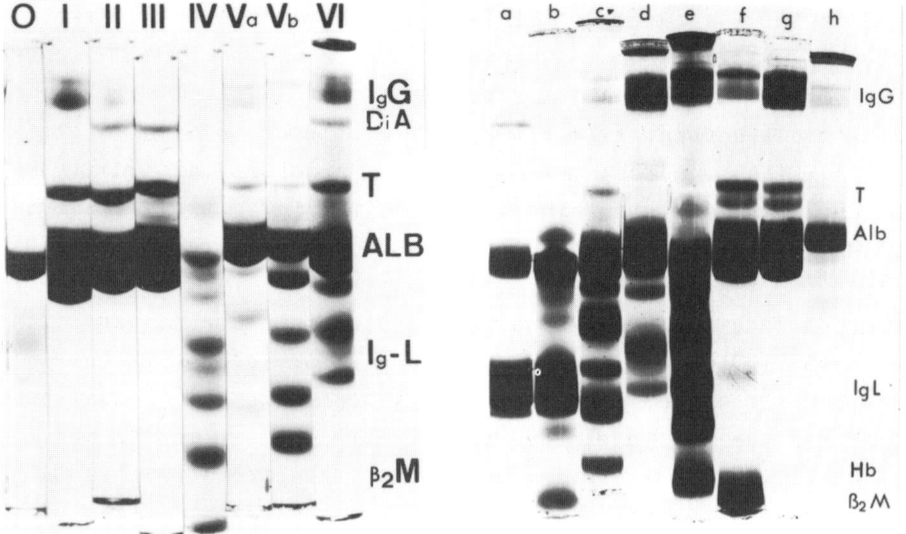



Fig.1 + 2 :Renal(I-VI)and extrarenal(a-g)proteinurias compared to the physiological one(O)as seen by SDS-PAGE.The main constituant proteins are indicated:IgG,transferrin,albumin,Ig-Light-chains,hemoglobin and ß-2-microglobulin.For explanation of pattern O-VI refer to table 2.Extrarenal proteinurias are indicated as a=paraproteinuria,b=Bence Jones-tubulopathy,d+e=postrenal Ig-secretion,f=bleeding and g=chyluria (from ref.9).

was obtained.The SDS-PAA-selectivity index,calculated by the
ratio of urinary transferrin and immunoglobulins,can be accepted therefore as a sufficient clinical parameter.- The postrenal source of Ig was proved by the discrepancy of transferrin
and Ig clearances. - Similar observations were made in analy-
sing the different tubular patterns,and defined as tubular selectivity for microproteins.The two elements B-1 and B-2(Tab.1)
resp.the patterns IV and VI(Tab.2)reflect different clearances
for microproteins when measured immunochemically.The slope of
the regression line of 5 micromolecular protein clearances was
taken as index of tubular selectivity.Fig.3 shows the clearances in relation to the one of a-1-acid-glycoprotein calculated
in two groups of 6 patients exhibiting the pattern IV and VI
respectively.Furthermore in Bence Jones-tubulopathy the Ig-L-
chain-paraprotein clearance exceeded clearly the one of a similarly sized microprotein. - From these results a hypothesis
of selective resorption of microproteins by the tubulus has
been deducted.In clinical practice the unselective tubular
pattern indicates a severe damage,the 'partial'selective one a
moderate,possibly vascular disease.
4)Correlation of SDS-PAGE-pattern to clinical data
The aim of electrophoresing urinary proteins is to come to a
clinical diagnosis without invasive procedures.This might be
achieved in a limited way,for SDS-PAGE pattern just reflect as
the kidney handles serum proteins.But taking this as pars pro
toto and referring to additional clinical information as hist-
ory,renal function,urine volume and clinical course the electrophoretic information may enable a rather detailed diagnosis.
First,it differentiates between vascular and glomerular and
tubulo-interstitial disorders.In patients with little or no
proteinuria mild diseases can be discriminated from the physiological state.This is an important aim in interstitial disorders and in slowly progressive systemic diseases as diabetes,
hypertension,lupus erythematosus etc.It is explained by the
fact that  physiological and pathological proteinurias overlap
at 40 to 400 mg protein/24 h. - Secondly,within the group of

glomerulopathies those with a selective proteinuria represent
minimal change lesions and early stages of membranous nephro-
pathy,focal glomerular sclerosis and amyloidosis,while chronic
and proliferative glomerulonephritides as well as dgenerative
glomerulopathies caused by diabetes,hypertension and myeloma
lead to an unselective glomerular proteinuria.It is not possible however to discriminate further within the two groups.-
Micromolecular proteinuria indicates tubulo-interstitial damage,the classical tubular proteinuria a severe tubular insuffi-
ciency,the partial pattern a milder damage of possibly vascular pathogenesis.A overproportional excretion of Ig-L-chains
proves a myeloma-associated tubulopathy.The different pathogenesis leading to identical damage with identical proteinuria
can not be elucidated by electrophoretic protein analysis.-
SDS-PAGE enables easily follow-up studies in patients with renal disease.Fig.4 shows the variability of proteinuria in one
patient within 4 months after cadaveric kidney transplantation

| 8000 | 1500 | 25 | 5000 | 6500 | 2600 | 2200 | 2400 | Prot. (mg/d) |
|------|------|-----|------|------|------|------|------|-------------|
| 1 | 4 | 55 | 111 | 115 | 121 | 130 | 146 | d. p. transpl. |

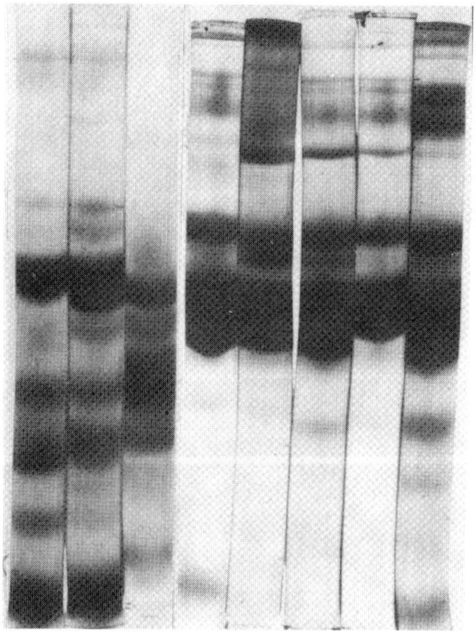

IgG
T
Alb
IgL
ß$_2$M

Fig.4:SDS-PAGE pattern
changes from tubular(1/4)
to the partial micromole-
cular pattern(55) and over
the selective glomerular
(111) to an unselective
pattern (146) within 5
months, indicating a re-
current glomerulonephritis
into the transplant (from
ref. 9).

## Summary

SDS-PAA-gel electrophoresis has been used to analyse urinary proteins of about 4500 patients suffering from various diseases,in more than 1500 the diagnosis was established by clear parameters.SDS-PAGE was carried out in 7.5%-PAA-o.1%-SDS-gels without spacer.Native samples of 24-h-urines containing about loo ug protein/gel were applied;samples containing less than 1 g/L had to be concentrated. - Different patterns of urinary protein composition could be distinguished beside the physiological one: A-renal origin: I-III glomerular proteinurias of different selectivity,grouped by the ratio transferrin/immunoglobulins; IV micromolecular tubular proteinuria; V mixed glomerular and tubular; VI partial micromolecular(mol.wt 40.-70. ooo d)plus glomerular proteinuria. B-extrarenal origin:prerenal('overflow'7i.e.myeloma proteins,myoglobin etc and postrenal,i.e.bleeding or inflammation.Proteinuria pattern correlated closely to pathological alterations of the nephron:the unselective pattern reflected diseases of membrane and mesangium the selective one alterations within the podocyte and outer membrane layer.The classical tubular proteinuria signalized an impairment of tubular microprotein resorption,while the 'selective tubular pattern'most probably was due to peritubular vascular diseases.In paraproteinurias the components could be differentiated by their size. -The various states of selectivity in glomerular as well as in tubulo-interstitial diseases were proved by protein clearances:so the decreasing glomerular selectivity in glomerulosclerosis and the improving tubular function after acute transplant rejection could be monitored easily by SDS-PAGE. -Proteinuria pattern do not replace renal biopsy,but -as a functional parameter- give additional and different answers to the clinician.SDS-PAGE has proved as valuable tool in early detection and in follow-up studies of all renal diseases in clinical and experimental situations.

References

1.  Boesken,W.H.,Wacker,B.,Kleuser,D.,Mamier,A.:
    Verh.Dt.Ges.Innere Med. $\underline{86}$,235-238 (1980)

2.  Reichel,W.,Wolfrum,D.,Weber,M.,Scheler,F.,Neuhoff,V.:
    Contrib.Nephrol. $\underline{1}$,109-118 (1975)

3.  Pesce,A.J.,Boreisha,I.,Pollack,V.E.:
    Clin.Chim.Acta $\underline{40}$,27-34 (1972)

4.  Boesken,W.H.,Kopf,K.,Schollmeyer,P.:
    Clin.Nephrology $\underline{1}$,311-318 (1973)

5.  Virella,P.,Pires,M.T.:
    Clin.Chim.Acta $\underline{50}$,63-75 (1974)

6.  Yatzidis,H.:
    Clin.Chem. $\underline{23}$,811-12 (1977)

7.  Mejbaum-Katzenellenbogen,W.,Dobryszycka,W.:
    Nature $\underline{193}$,1288-89 (1962)

8.  Boesken,W.H.,Schindera,F.,Billingham,M.,Hardwicke,J.,
    White,R.H.R.,Williams,A.:Clin.Nephrology $\underline{8}$,395-99(1977)

9.  Boesken,W.H.: J.Clin.Chem.Clin.Biochem.$\underline{16}$,199(1978)and
    Curr.Probl.Clin.Biochem. $\underline{9}$,235-247 (1979)

10. Boesken,W.H.,Rohrbach,R.,Schollmeyer,P.:Nieren-u.Hoch-
    druckkrankheiten $\underline{5}$,206-214 (1978)


Acknowledgement

This work was supported by a grant (Bo 378/11) of the
Deutsche Forschungsgemeinschaft Bonn-Bad Godesberg

HEPATIC IMPLICATION IN THE METABOLISM OF HUMAN SALIVARY
AMYLASE ISOENZYME

Atsuo Murata, Michio Ogawa, Ken-ichi Fujimoto, Takeshi
Kitahara, Yasuki Matsuda, Goro Kosaki
The Second Department of Surgery, Osaka University Medical
School, Osaka, Japan

Introduction

Elevation of serum amylase activity may occur in liver diseases
(1,2), but whether this increment to circulating amylase originates in the liver has been unclear.  The liver may possibly
contribute to the amylase activity normally present in human
and mammalian serum (3), but liver amylase has not been identified for certain in human serum, either in health or in disease.
On isoamylase analysis, elevated serum amylase in liver diseases proved to be composed essentially of salivary-type isoamylase.  Human salivary amylase is known to be seperated into
two families of isoenzymes: the isoenzymes of family A are
glycoproteins, whereas those of family B are not (4,5).  In
recent years, several works (6,7,8) showed that hepatocytes
contain the receptors which bind specifically to glycoproteins.
Family A isoamylase, therefore, could be a clue trapped by the
liver.  The purpose of this study is to clarify the hepatic
implication in the metabolism to develop hyperamylasemia in
hepatic dysfunctions.

Materials and Methods

1)  Purification of human salivary isoamylases
Two families of salivary isoamylases were purified from whole
saliva by adsorption to corn starch, gel filtration and con-

canavalin A — Sepharose affinity chromatography (5,9).  The
preperations obtained were found to be homogenous by the criteria of analytical polyacrylamide gel electrophoresis and
electrophoretic analysis in the presence of sodium dodecyl
sulfate.

2)  Polyacrylamide gel electrophoresis
Five percent polyacrylamide gel electrophoresis was performed
according to the modification of the method reported by Otsuki
et al. (10).  Electrophoresis was done horizontally at 4°C for
3 hours and after electrophoresis the gel was stained for amylase activity with iodine-starch reaction.

3)  In vivo studies
Aliquots  of family A and family B were iodinated with [131]I and
[125]I using the chloramine T method, respectively.  Clearance
studies in rabbits were performed by bolus injection of the
mixture of [131]I-labeled  family A and [125]I-labeled  family B.
Serum samples were collected at suitable intervals for approximately 60 minutes and their radioactivities were measured.  As
hepatic failured rabbits (11), 1.5 gm/kg weight of D-galactosamine was injected intraperitoneally, and 24 hours later these
rabbits were used.


Results


Typical electrophoretic patterns of isolated family A and family B isoamylases are shown in Figure 1.  Family A isoamylase
had slower mobility than family B isoamylase.  Figure 2 shows
the electrophoretic patterns of serum in the case of salivary-
type hyperamylasemia following right hepatic lobectomy for
primary liver carcinoma.  On the first post-operative day, the
marked hyperamylasemia was noticed and continued for three days.
The disappearance curves of two families in serum of normal
rabbits are shown in Figure 3.  Family A disappeared from serum
faster than family B.  The mechanisms of their clearance

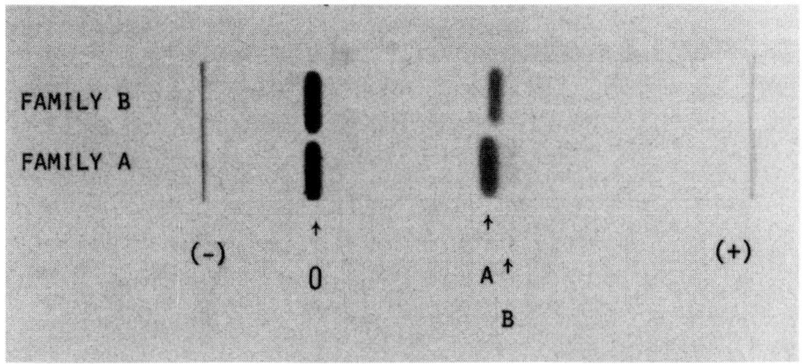



Fig. 1   Zymograms of salivary family A and family B
isoamylases.

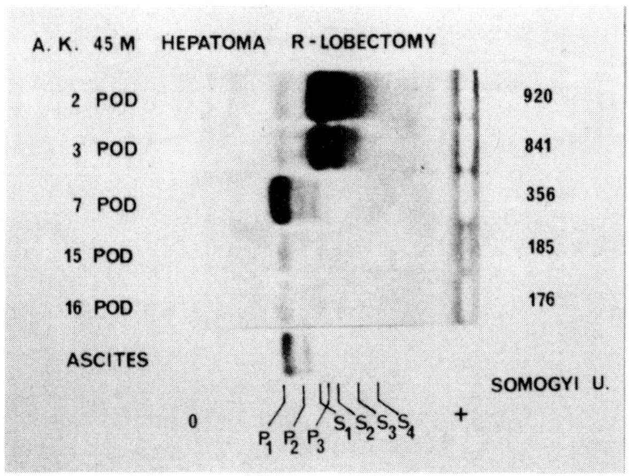



Fig. 2   Zymograms of amylase isozymes following surgery.

occurred in the different ways, though each of injected families was recognized as foreign bodies for rabbits.   To neglect
interference with renal factors, the mixture of two families
was injected to nephrectomized rabbits.   The same disappearance
curves as normal group were obtained as shown in Figure 4.
Figure 5 shows the histological examination of the liver of
rabbit sacrificed at 24 hours after the injection of D-galac-

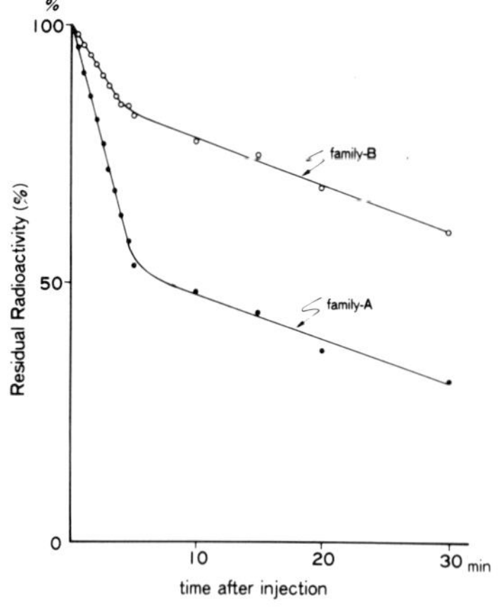

Fig. 3 The disappearance
curves of two families in
serum of normal rabbits.

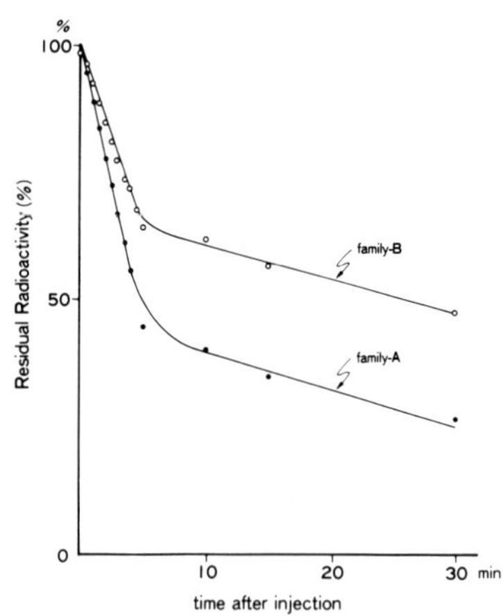

Fig. 4 The disappearance
curves of two families in
serum of nephrectomized
rabbits.

tosamine. Hepatic congestion, hepatic cord destruction, polymorphism and disappearance of hepatic nuclei were found. The
values of the typical liver enzymes GOT, GPT and Al-P were
significantly elevated. As shown in Figure 6, the clearance rate of family A in these hepatic failured rabbits was
slower than that of normal, while that of family B was the
same as normal. The relative half-times for disappearance of
two families from serum are shown in Table. Half-times of
family B were almost the same in three groups, average 45 minutes. On the other hand, half-times of family A were almost
the same in normal and nephrectomized rabbits, while, in hepatic failured rabbits, was prolonged to 20 minutes from 8
minutes.

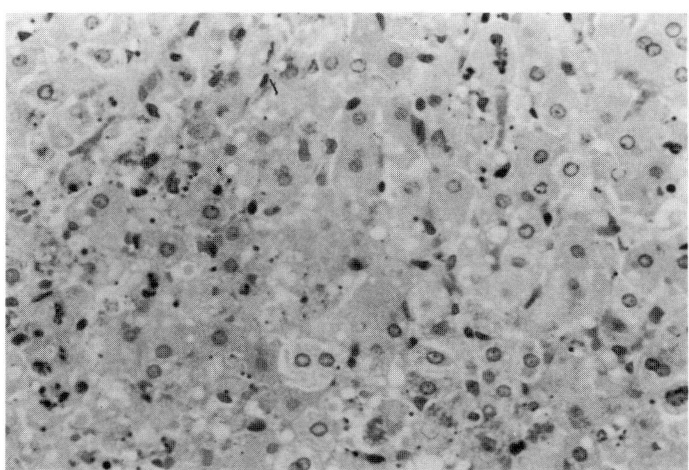

Fig. 5  Histological feature of the liver of the rabbit
injected with D-galactosamine (1.5 gm/kg).

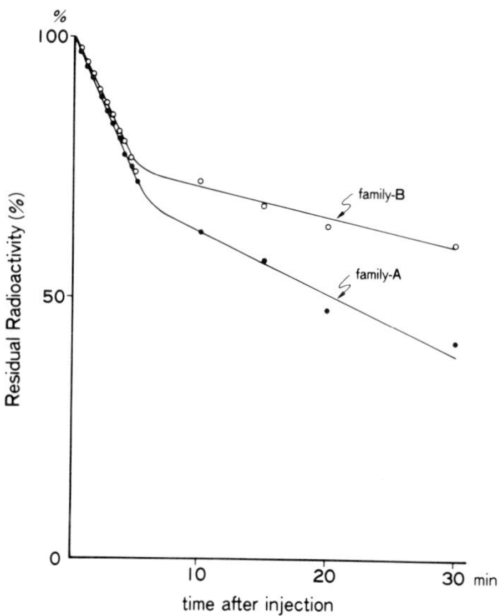


Fig. 6  The disappearance curves of two families in serum of
liver failured rabbits.

Table   Half-times for disappearance of two families from
        serum in three groups.

|                          |          | family A | family B |
|--------------------------|----------|----------|----------|
| normal group             | ( n=5 )  | 8 min    | 45 min   |
| nephrectomized group     | ( n=3 )  | 7 min    | 40 min   |
| hepatic failure group    | ( n=5 )  | 20 min   | 48 min   |

Discussion

The metabolic studies for mammalian amylases reported by Duane
et al. (12,13) showed the following characteristics: the pure
amylases isolated from pancreas and parotid glands of baboon,
an animal with similar serum amylase level and renal clearance
of amylase to man, were cleared through the kidney and serum
half-times of them were about 130 minutes.  If human salivary
amylase enter the blood stream in the same way as other exocrine enzymes, family A would be trapped and metabolized by
the liver as a glycoprotein.  Our present study revealed that,
in hepatic failured rabbits, family A isoamylase was cleared
more slowly than normal rabbits.  The metabolism of family A,
therefore, was demonstrated to be related to some functions
of the liver.
Several investigators (6,8,14,15) reported that mamalian hepatocytes have the receptors specific for serum glycoproteins
and that the binding mechanism was showed the similar ways in
all mammalians.  The receptors of mammalian liver, which are
glycoproteins themselves and identified on liver plasma membranes, are responsible for the rapid serum clearance and
lysosomal catabolism of desialylated glycoproteins through
exposed galactose residues.  Prieels et al. (7) showed another
mechanism that bound specifically to fucosyl $\alpha$1-3 N-acetylglu-

cosamine linkage found in sugar chains of glycoproteins. In
the structural studies by Yamashita et al. (16), family A isoamylase contains three types of single asparagine-linked sugar
chains in one molecule, and both exposed galactose residue and
fucosyl αl-3 N-acetylglucosamine linkage are present in their
sugar chains. Tolleshaug et al. (17) reported that the uptake
of asialo-glycoproteins in hepatocytes was one of the receptor-
mediated endocytoses.

Thus, it can be concluded that the liver takes a role in the
metabolism of human salivary family A isoamylase and that in
damaged liver the clearance rate of family A is prolonged.
Family A isoamylase in human serum would be trapped by the
liver and degraded in the hepatic lysosomes. And liver diseases would cause injury of this mechanism. Impaired hepato-
cytic uptake for family A isoamylase, therefore, could be one
of the causes of salivary-type hyperamylasemia in liver
diseases.


References


1.  Bhutta, I.H., Rahman, M.A.: Clin. Chem. 17, 1147-1149
    (1971).

2.  Warshaw, A.L., Bellini, C.A., Lee, K.-H.: Gastroenterol.
    70, 572-576 (1976).

3.  Takeuchi, T., Matsushima, T., Sugimura, T.: Biochim.
    Biophys. Acta 403, 122-130 (1975).

4.  Keller, P.J., Kauffman, D.L., Allan, B.J., Williams, B.L.:
    Biochem. 10, 4867-4874 (1971).

5.  Takeuchi, T.: Clin. Chem. 25, 1406-1410 (1979).

6.  Ashwell, G., Morell, A.G.: Adv. Enzymol. 41, 99-128 (1974).

7.  Prieels, J.-P., Pizzo, S.V., Glasgow, L.R., Paulson, J.C.,
    Hill, R.L.: Proc. Natl. Acad. Sci. U.S.A. 75, 2215-2219
    (1978).

8.  Steer, C.J., Ashwell, G.: J. Biol. Chem. 255, 3008-3013
    (1980).

9.  Matsuura, K., Ogawa, M., Kosaki, G., Minamiura, N.,
    Yamamoto, T.: J. Biochem. 83, 329-332 (1978).

10. Otsuki, M., Saeki, S., Yuu, H., Maeda, M., Baba, S.:
    Clin. Chem. 22, 439-444 (1976).

11. Keppler, D., Lesch, R., Beutter, W., Decker, K.:  Exp.
    Molecul. Pathol. 9, 279-290 (1968).

12. Duane, W.C., Frerichs, R., Levitt, M.D.: J. Clin. Invest.
    50, 156-165 (1971).

13. Duane, W.C., Frerichs, R., Levitt, M.D.: J. Clin. Invest.
    51, 1504-1513 (1972).

14. Regoeczi, E., Hatton, M.W.C., Wong, K.-L.: Can. J.
    Biochem. 52, 155-161 (1974).

15. Newman, R.A., Fricks, U., Klein, P.J., Uhlenbruck, G.,
    De Vries, A.L.: J. Clin. Chem. Clin. Biochem. 18, 31-37
    (1980).

16. Yamashita, K., Tachibana, Y., Nakayama, T., Kitamura, M.,
    Endo, Y., Kobata, A.: J. Biol. Chem. 255, 5635-5642
    (1980).

17. Tolleshaug, H., Berg, T., Holte, K.: Eur. J. Cell Biol.
    23, 104-109 (1980).

# POSTTRANSLATIONAL MODIFICATION OF PANCREATIC AMYLASE IN ACUTE PANCREATITIS: POSSIBLE ROLE OF PANCREATIC DEAMIDASE


Michio Ogawa, Naoko Saito, Ken-ichi Fujimoto, Masanobu Masuike, Atsuo Murata, Goro Kosaki
Second Department of Surgery, Osaka University Medical School, Osaka 553, Japan

Noshi Minamiura and Takehiko Yamamoto
Faculty of Science, Osaka City University, Osaka 558, Japan


## Introduction

It is known that the incubation of the major component of human pancreatic amylase transforms into many minor components of more anionic forms (1, 2). In acute pancreatitis, these minor components increase in serum, which has been used as a diagnostic aid for the pancreatic inflammation (3, 4, 5). In our previous paper, we demonstrated that the transformation was brought about by deamidation, using peptidoglutaminase I and II (2). In general, these phenomena have been considered to be due to non-enzymatic deamidation and to be useful timer of development and aging (6).
Minor components of pancreatic amylase, however, were identified in serum even at the begining, or at the time of admission, of acute pancreatitis, which is very short period for non-enzymatic deamidation. In the present study, the mechanisms of deamidation of pancreatic amylase in acute pancreatitis were investigated.

## Materials and Methods

1) Minor components of serum pancreatic amylase after surgery

<u>Serum</u>  As a model of acute pancreatitis, patients operated
for pancreatic or biliary tract disease were selected and
serum samples were taken at 7 to 12 hours after operation.
Urine was collected simultaneously.

<u>Amylase activity and creatinine concentration</u>  Serum and
urine amylase activities were determined by a chromogenic
method using blue starch (Pharmacia Fine Chemicals, Sweden)
as a substrate (7).  The assay was carried out as described
in the manufacturer's instruction.  Creatinine concentration
was determined by an autoanalyzer technique.

<u>Amylase-creatinine clearance ratio (ACCR)</u>  Amylase-creatinine
clearance ratio (ACCR) was obtained according to the formulas
(8):

$$ACCR(\%) = \frac{\dfrac{[\text{urine amylase}]}{[\text{serum amylase}]} \times \text{urine volume per unit time}}{\dfrac{[\text{urine creatinine}]}{[\text{serum creatinine}]} \times \text{urine volume per unit time}} \times 100$$

$$= \frac{[\text{urine amylase}]}{[\text{serum amylase}]} \times \frac{[\text{serum creatinine}]}{[\text{urine creatinine}]} \times 100$$

Normal range of ACCR in our laboratory was 1.1-3.0%.

<u>Electrophoresis</u>  Serum samples were subjected to electropho-
resis on a 5% polyacrylamide gel by the method reported by
Otsuki, et al. (9).  The gel plates obtained were scanned on
Two Waves TLC Scanner CS-900, Shimazu Co., Japan, and the
ratio of a minor component ($P_2$) to the major component ($P_1$)
of serum pancreatic amylase was assessed.  Normal range of
$P_2/P_1$ ratio was less than 0.2.  The patients with preopera-
tive $P_2/P_1$ ratio above 0.4 were excluded in this study, as
there was an inherited trait of pancreatitis-like isozyme
pattern (10).

2) Purification of pancreatic deamidase

<u>Treatment of human pancreatic tissue</u>  Human pancreatic tis-
sue obtained on autopsy and confirmed as normal histologi-
cally, was sliced.  It was then soaked in acetone and in
ether for several days in the cold to remove lipid materials.
The treated tissue was dried in an air current.  The defatted
tissue was subjected to purification of pancreatic deamidase.

<u>Deamidase activity</u>  Deamidase activity was measured spectro-
photometrically by the indophenol method using 4% bovine
serum albumin as a substrate (11).
<u>Molecular weight</u>  Molecular weight was estimated by the gel
chromatography on a Sephadex G-200.


Results


1) Minor components of serum pancreatic amylase after surgery

Electrophoretic pattern and designation of human amylase
were shown in Figure 1.  Serum amylase activity remained low
within 6 hours after surgery, followed by the sharp increase
in amylase activity.  Figure 2 shows serial isozyme analysis
of serum amylase in a patient underwent sphincteroplasty.
10 hours after operation, total serum amylase increased to
2,920 Somogyi units.  At this time ACCR was 8.74% and $P_2/P_1$
ratio was 0.7 (calculated from the densitometry of the
diluted sample).  As shown in this Figure, increase in $P_2/P_1$
ratio was seen till the second postoperative day, and the

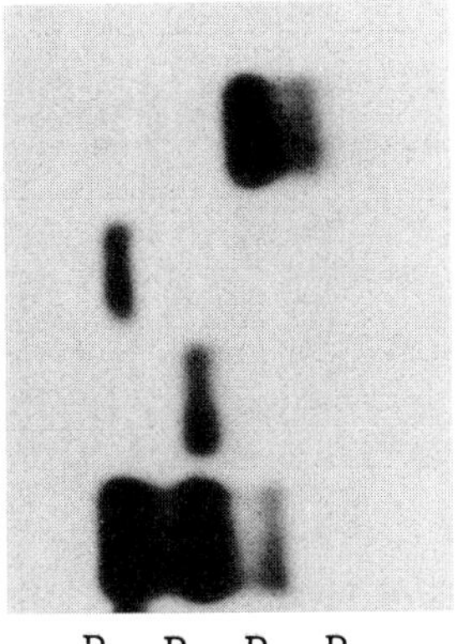


Figure 1  Electrophoretic pattern and designation of human
pancreatic and salivary amylase.

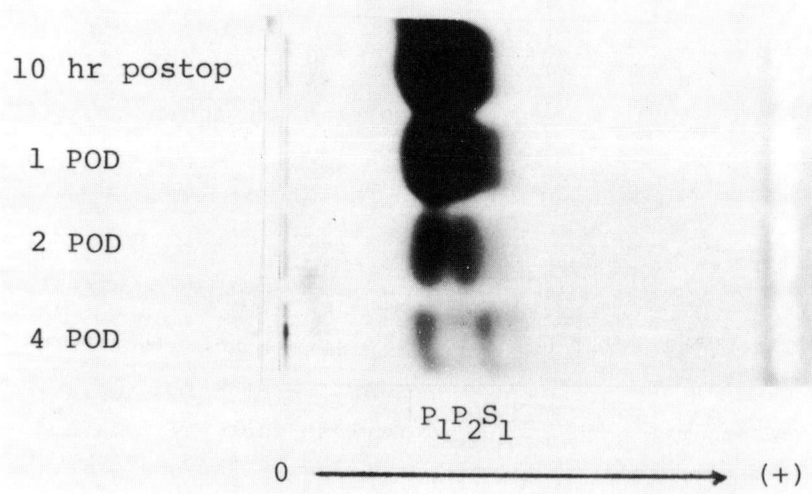


Figure 2  Serial isozyme analysis of serum amylase in a patient after sphincteroplasty

zymogram of the fourth postoperative day returned to the normal pattern with disappearance of complaints.
Table 1 shows the relationship between ACCR and $P_2/P_1$ ratio.  In the group with moderately elevated ACCR (3.1-4.8%), 25% showed the increase in minor components of pancreatic amylase in serum.  In the highly elevated ACCR

Table 1  Relationship between amylase-creatinine clearance ratio (ACCR) and $P_2/P_1$ ratio of serum pancreatic amylase

| ACCR (%) | No. | No. of $P_2/P_1 \geqq 0.4$ (%) |
|---|---|---|
| - 1.0 | 5 | 0 ( 0) |
| 1.1 - 3.0 | 27 | 1 ( 3.7) |
| 3.1 - 4.8 | 28 | 7 (25.0) |
| 4.9 - 9.9 | 11 | 8 (72.7) |
| 10.0 - | 7 | 1 (14.3) |

group (4.9-9.9%), the increase in minor components observed in 72.7%. However, the anomalous elevation of ACCR (larger than 10.0%), which was thought to be due to the change in amylase excretion and not to acute pancreatitis, was not accompanied by the increase in minor components of pancreatic amylase.

2) Purification of pancreatic deamidase

The defatted tissue was suspended in 20 mM Tris-HCl buffer, pH 7.6, containing 2 mM $CaCl_2$, and homogenized in a Waring blender. The homogenate was centrifuged at 20,000 × g, and the supernatant was applied to a DEAE-cellulose column, equilibrated with 20 mM Tris-HCl buffer, pH 7.6, containing 2 mM $CaCl_2$. As shown in Figure 3, pancreatic deamidase activity was adsorbed on the column and eluted with a linear concentration gradient of NaCl. The active fractions were combined and concentrated in a membrane filter. The concentrate was applied to a hydroxyapatite column equibrated with

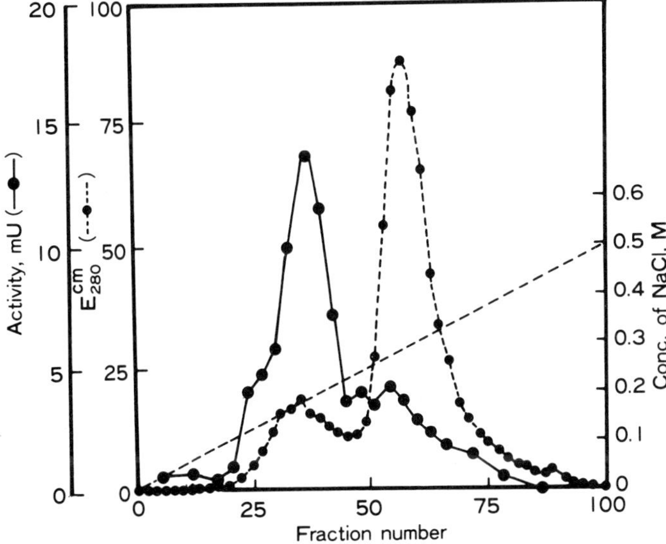


Figure 3   Chromatography of pancreatic deamidase on a column of DEAE-cellulose

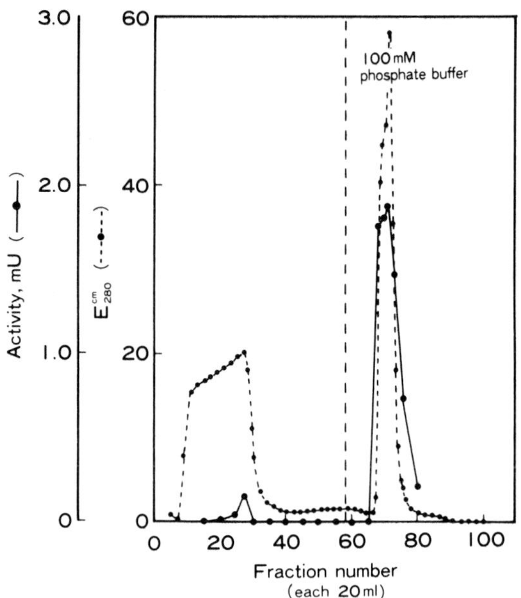



Figure 4   Chromatography of pancreatic deamidase on a column of hydroxyapatite

10 mM phosphate buffered saline, pH 7.6.  Deamidase activity was adsorbed and eluted from the column with 100 mM phosphate buffered saline, pH 7.6 (Figure 4).  The active fractions in the elute were combined and concentrated.
The purification of pancreatic deamidase was increased by 22 fold  at this step.  However, the preparation was not homogeneous by analytical polyacrylamide gel electrophoresis. The molecular weight of pancreatic deamidase estimated by the gel chromatography on a column of Sephadex G-200 were approximately $2 \times 10^4$.  The optimal pH of the activity was at 7.5.  The activity decreased by 50% at pH values below 6.9.  The enzyme activity was lost at 60°C for 30 min.

## Discussion

In acute pancreatitis, the renal clearance of amylase to
creatinine (ACCR) is elevated to levels several times higher
than normal (8).  Therefore, this ratio has been widely used
for the diagnosis of acute pancreatitis.  As shown in the
present study, deamidation of serum pancreatic amylase was
frequently observed in patients with elevated ACCR, postoperative pancreatitis.

Non-enzymatic deamidation was known to require a long time
even at alkaline pH and high temperature (2, 12, 13).  In
contrast, the deamidation in acute pancreatitis was seen in
a short period after operation.  These results together with
pH range of human blood (pH 7.35-7.45) and metabolic turnover
of amylase (14) suggested a possible existance of deamidase
in the human pancreatic tissue and activation of the enzyme
in acute pancreatitis.

The present study revealed that deamidase activity was present
in the pancreatic tissue, and pancreatic deamidase was partially purified on columns of DEAE-cellulose and hydroxyapatite.  As for related enzymatic activity, Mycek and Waelsch
(15) reported that transglutaminase from the soluble fraction
of guinea pig liver could catalyze some amide groups of
protein-bound glutamine residues.  Further study is needed to
clarify their differences or similarities.

In conclusion, deamidation by activated pancreatic deamidase
in acute pancreatitis could be taken into consideration for
the appearance of minor components in acute pancreatitis sera.


## References


1.  Takeuchi, T., Matsushima, T., Sugimura, T., Kozu, T.,
    Takeuchi, T., Takemoto, T.: Clin. Chim. Acta 60, 205-206
    (1975).

2.  Ogawa, M., Kosaki, G., Matsuura, K., Fujimoto, K.,
    Minamiura, N., Yamamoto, T., Kikuchi, M.: Clin. Chim. Acta

478

87, 17-21 (1978).

3.  Legaz, M. E., Kenny, M. A.: Clin. Chem. 22, 57-62 (1976).

4.  Otsuki, M., Yuu, H., Maeda, M., Yamasaki, T., Okano, K., Sakamoto, C., Baba, S.: Clin. Chim. Acta 79, 1-6 (1977).

5.  Frost, S. J.: Clin. Chim. Acta 87, 23-28 (1978).

6.  Robinson, A. B., McKerrow, J. H., Cary, P.: Proc. Nat. Acad. Sci. 66, 753-757 (1970).

7.  Ceska, M., Birath, K., Brown, B.: Clin. Chim. Acta 26, 437-444 (1969).

8.  Levitt, M. D., Rapoport, M., Cooperband, S. R.: Ann. Intern. Med. 71, 919-925 (1969).

9.  Otsuki, M., Saeki, S., Yuu, H., Maeda, M., Baba, S.: Clin. Chem. 22, 439-444 (1976).

10.  Otsuki, M., Maeda, M., Yuu, H., Yamasaki, T., Okano, K., Baba, S.: Gastroenterol. Jap. 13, 224-230 (1978).

11.  Bolleter, W. T., Bushman, C. J., Tidwell, P. W.: Anal. Chem. 33, 592-594 (1961).

12.  Keller, P. J., Kauffman, D. L., Allan, B. J., Williams, B. L.: Biochemistry 10, 4867-4874 (1971).

13.  Lorentz, K., Flatter, B.: Enzyme 24, 163-168 (1979).

14.  Duane, W. C., Frerichs, R., Levitt, M. D.: J. Clin. Invest. 51, 1504-1513 (1972).

15.  Mycek, M. J., Waelsch, H.: J. Biol. Chem. 235, 3513-3517 (1960).

# EFFECTS OF DIALYZABLE LEUKOCYTE EXTRACTS WITH TRANSFER FACTOR ACITIVITY ON LEUKOCYTE MIGRATION IN VITRO:
# VI.  STUDIES ON THE PRIMARY STRUCTURE OF TRANSFER FACTOR


Gary V. Paddock, Gregory B. Wilson,[2] Fu-Kuen Lin, Nicki O'Leary
and H. Hugh Fudenberg

Department of Basic and Clinical Immunology and Microbiology,
Medical University of South Carolina, Charleston, South Carolina 29425


## Introduction

Human transfer factor (TF) is a component of molecular weight (M.W.) near
2000 found in the dialyzable portion of leukocyte extracts (DLE) from immune
donors.  TF can transfer antigen-specific cell-mediated immunity (CMI) to
non-immune lymphocytes in vitro as shown by the agarose leukocyte migration
inhibition (LMI) assay (1,2).  Lymphocytes from cows, mice, and presumably
other mammals also produce TF active in the LMI assay (3,4).  Previously we
reported the partial purification and characterization of human TF specific
for PPD (5,6) and showed that after high pressure reverse phase liquid
chromatography (HPLC) of partially purified TF activity, the PPD-specific TF
activity segregated into two fractions (TF-H5 and TF-H7), both able to
transfer specific CMI to PPD.  Both TF-5 and TF-7 are oligoribonucleopeptides, but they differ structurally with respect to their nucleic acid
segments (6).  In the present communication, we report the further
purification of both TF components and confirmatory data on the primary
sequence of the ribonucleic acid segment of human TF-H7.  In addition, we
report the purification and structure of bovine TF specific for PPD.

---


[1]Publication no. 416 from the Department of Basic and Clinical Immunology
 and Microbiology, Medical University of South Carolina.

[2]Reprint requests and correspondence:  Dr. G.B. Wilson.

Methods

Human TF-H5 and TF-H7 specific for PPD were purified from crude DLE through
the HPLC step, and thin-layer chromatography (TLC) was performed as
described previously (5-7).  Further purification of TF-H7 and TF-H5, with
or without prior treatment with bacterial alkaline phosphatase (BAP) (6,8),
was accomplished using boronate affinity chromatography (Affi-gel 601;
Bio-Rad Laboratories).  The boronate gel was washed and equilibrated in
0.25 M ammonium acetate buffer (pH 8.8) with or without 0.15 M NaCl added.
Mixtures containing TF activity were fractioned using stepwise elution,
first with equilibration buffer, then with water, and then with 0.1 N
acetic acid.  At some step in the purification of TF activities specific
for PPD, there is a requirement for their separation from TF of unrelated
specificities.  This does not occur up through the HPLC step of
purification, since PPD-positive and coccidioidin-positive TF elute in
identical fractions (5,6).  To attempt this separation, we explored the use
of polystyrene coated with specific antigen, i.e., PPD (as described in
ref. 9, with modifications.

Radioiodination of human TF-H7 (and TF-H5 as a control) at cytidine residues
was attempted using the method of Prensky (10) with modifications in
temperature, incubation time, and thallic ion concentration.  Analyses of
radiolabeled components by TLC or by acrylamide gel electrophoresis (25%
acrylamide, 7 M urea) was performed using established methods (6-8,11).

Bovine TF specific for PPD (B-TF) was purified from the crude dialyzable
fraction (MW <20,000) of material released by immune lymph node cells (LNC)
incubated at 37°C for 4 hr (3,12), following our purification schemes for
human TF without modification (1,6).  Structural properties of B-TF were
deduced by evaluating the effects on TF activity of incubating B-TF with 11
different enzymes, following the same procedures employed in our structural
studies on human TF (5,6,8).  Bovine and human TF activities were detected
in vitro using the agarose LMI assay (1-3).

Results and Discussion

The purification of either B-TF or human TF entails performing the following steps sequentially: (a) Sephadex G-25 gel filtration chromatography, (b) phenol extraction and ethanol precipitation, (c) HPLC, (d) TLC, (e) boronate affinity chromatography, and (f) binding to and elution from antigen-coated polystyrene. Step (d) usually needs to be repeated after step (e) to ensure homogeneity of the preparation, depending on the TF studied. Step (f) may also be performed prior to step (a) or step (b) depending on the TF being purified. Our studies to date indicate that co-factors present in the crude starting material or specific subfractions thereof must be present for optimal binding. It may be possible to add these components to the product of step (e). Details of these steps can be found elsewhere (5-8; G.B. Wilson et al., submitted).

Two steps in TF purification (steps c and e) suggest definite differences between B-TF and human TF specific for PPD. At step (c), B-TF activity in contrast to human TF activity eluted in a single fraction with a retention time close to that of human TF-H5 (5,6). However, the requirements for binding to boronate (step e) were similar for B-TF and human TF-H7 (human TF-H5 required prior treatment with BAP for binding). Our results from enzymatic inactivation of the B-TF activity from step (a) or (c) indicated that B-TF was inactivated by Pl nuclease, RNase Tl, RNase U2, snake venom phosphodiesterase (SVP), pronase, and carboxypeptidase A. It was not inactivated by DNase I, RNase A, spleen phosphodiesterase (SPP), BAP, leucine aminopeptidase, or by sequential treatment with BAP and then SPP. With one exception, the results for enzyme inactivation of B-TF were identical to those reported for human TF-H5 (6). Human TF-H5 was not affected by SVP unless first treated with BAP to remove a phosphate, which we propose is on the 2' or 3' hydroxyl of the 3' terminal ribose (ref. 6; Fig. 1). Figure 1 presents simplest-case models for the structure of human TF-H5, TF-H7, and B-TF, which are derived from our results for enzymatic inactivation of the TF activities. Our rationale behind the selection of the models shown over other possible models is discussed in detail else-where, as are certain restrictions that must be placed on these models (6).

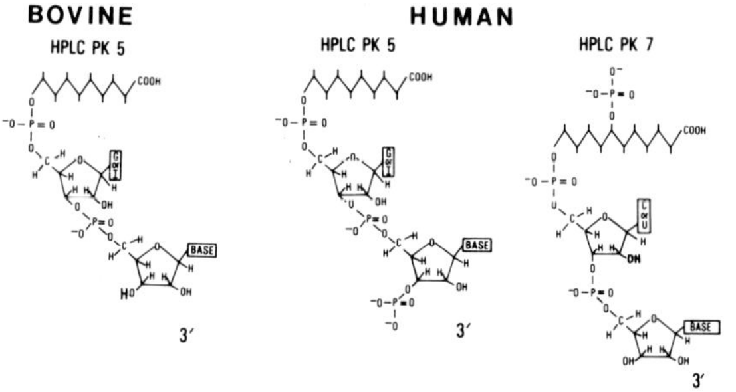



Fig. 1.  Models for the structure of human TF-5 and TF-7 and B-TF specific for PPD.

The model for B-TF was derived by following a line of reasoning similar to that used to develop the model for human TF-H5.

Our results from boronate affinity chromatography agree well with our proposed models for the structures of bovine and human TF.  B-TF and human TF-H7 are proposed to have a free co-planar cis-diol group (the 2' and 3' hydroxyls of the 3' terminal ribose).  Indeed, both activities bind to boronate with no requirement for pretreatment with BAP.  BAP treatment probably removes the 2' or 3' phosphate shown on TF-H5, which then allows it to bind to boronate.  Our structural data indicate that all three TF moieties shown in Fig. 1 are oligoribonucleopeptides; however, human TF-H7 differs from human TF-H5 and B-TF since the former TF should contain primarily pyrimidine residues (certainly an internal cytidine or uridine) whereas the latter two should contain primarily purines (ref. 6; Fig. 1).

Our results from attempts to label human TF-H5 and TF-H7 also agree with the above conclusions.  Whereas we have as yet been unable to label TF-H5,

we have unequivocally labeled TF-H7.  As Figure 2 (left) shows, additional
labeled components (of M.W. 1000-3000 by Bio-gel P-2) are present in TF-H7
as compared to a reaction mixture control without TF.  As Figure 2 (center)
shows, we can obtain an isolated labeled component from TF-H7 after TLC and

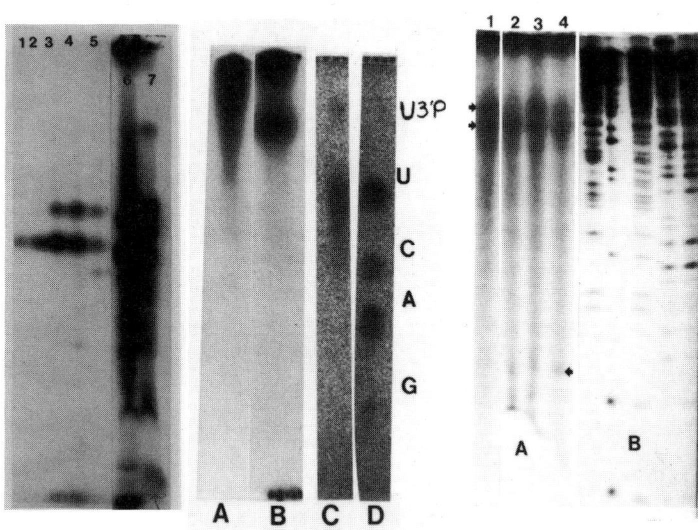

Fig. 2.  Analyses of [125]I-labeled TF-H7 by thin-layer chromatography and
polyacrylamide gel electrophoresis.  <u>Left</u>:  TF-H7 was labeled with [125]I at
cytidine residues and then either first fractionated by Bio-gel P-2
filtration (1) or immediately subjected to analysis by acrylamide gel
electrophoresis (anode at bottom).  Numbers 1 to 5 are consecutive
fractions from Bio-gel P-2 covering the M.W. range where the TF activity
elutes (1 is just after the void volume of the column).  Number 6 is the
crude reaction mixture containing TF-H7.  Number 7 is the same reaction
mixture without TF-H7.  <u>Center</u>:  TLC (cellulose, MEOH/$H_2O$/HOAc) of labeled
TF-H7 components eluted from boronate with 0.1 N acetic acid (A) or with
water (B).  Standards run on the same plate were (C) 3' nucleotide
monophosphates (U 3'P at top) and (D) free bases (top to bottom:
U, C, A, G).  (A) and (B) are results from autoradiography; (C) and (D)
were photographed while illuminated with ultraviolet light (6).  <u>Right</u>:
(A) Results of analyses of labeled TF-H7 from boronate by acrylamide gel
electrophoresis either untreated (1) or after incubation with SVP (2), SPP
(3), or pronase (4).  Changes in the relative intensities of components
(indicated by arrows) can be observed by comparing tracks 2 and 4 to 1 and
3.  (B) Deoxyribonucleotide fragments for markers obtained by established
methods (11), which entailed the digestion of a restriction fragment
labeled at the 5' end with [32]P.

boronate affinity chromatography with a mobility close to the uridine 3'
phosphate marker on TLC.  This is precisely where the TF activity in
fraction TF-H7 resides after TLC (6).  We have, in fact, eluted labeled
material and shown that it has TF activity.  To obtain further data on the
structure of TF-H7, we have begun to perform experiments in which the
effects of those enzymes used in previous structural studies involving
inactivation of the TF activity were evaluated.  As Figure 2 (right) shows
both SVP and pronase promoted changes in the intensity of labeled
components present in the TF-H7 product; SPP did not.  These results are
consistent with our previous data (6).  Further studies are in progress to
elucidate completely the structural properties of TF-H7 (and TF-H5 by
alternate methods of labeling).

In comparing B-TF to human TF, we have found two interesting differences:
first, the absence of a component similar to human TF-H7 in the crude
material from bovine LNC, and second, the lack of a 2' or 3' phosphate on
B-TF, which is otherwise structurally similar to human TF-H5 (Fig. 1).  We
have previously hypothesized that TF-H7 and TF-H5 are both present in human
DLE only because the immunocytes containing them are disrupted during
freeze-thawing (6).  TF-H7, being an oligoribonucleophospeptide (Fig. 1;
ref. 6), is thought to be strictly an intracellular product, perhaps a gene
regulator involved in the synthesis of TF-H5 (6).  Since the bovine LNC
were not disrupted to prepare the crude starting material containing B-TF
activity a human TF-H7 equivalent would not be present.  TF-H5 is thought
to be secreted by the cell, possibly becoming part of a T-cell antigen
receptor (T-AR) necessary for T-cell activation by the binding of Ia
antigen determinants presented to T cells by monocytes (macrophages).  The
fact that TF can bind to specific antigen (our results and ref. 9) lends
support to our hypothesis for this mechanism of action of TF in CMI (6).
If we assume that the 2' (or 3') phosphate of TF-H5 is normally removed
prior to its secretion, and that the final product becomes part of the
T-AR, then we can explain the absence of a 2' or 3' phosphate on B-TF
(which is probably simply released from the surface of immunocytes during
the 4-hr incubation at 37°C).  A human TF similar to the B-TF described
here (Fig. 1) presumably exists on human T cells immune to PPD; however, we
speculate that exonucleases released by leukocytes during the preparation

of DLE destroy it.  Most likely its intracellular counterpart (TF-H5) is protected from such destruction, since it still contains the 2' or 3' phosphate necessary to block the action of exonucleases similar to SVP.

Acknowledgment

We thank Charles L. Smith for excellent editorial assistance.

References

(1)  Wilson, G.B., Fudenberg, H.H.:  Lymphokines $\underline{4}$, in press (1981).

(2)  Wilson, G.B., Fudenberg, H.H., Horsmanheimo, M.:  J. Lab. Clin. Med. $\underline{93}$, 800-819 (1979).

(3)  Wilson, G.B., Newell, R.T., Burdash, N.M.:  Cell. Immunol. $\underline{47}$, 1-15 (1979).

(4)  Borkowsky, W., Lawrence, H.S.:  J. Immunol., in press (1981).

(5)  Wilson, G.B., Paddock, G.V., Fudenberg, H.H.:  Trans. Assoc. Amer. Phys. $\underline{92}$, 239-256 (1979).

(6)  Wilson, G.B., Paddock, G.V., Fudenberg, H.H.:  Thymus $\underline{2}$, 257-276 (1981).

(7)  Paddock, G.V., Wilson, G.B., Wang, A.C.:  Biochem. Biophys. Res. Commun. $\underline{87}$, 946-952 (1979).

(8)  Paddock, G.V., Wilson, G.B., Fudenberg, H.H., Wang, A.C., Lovins, R.E.:  In: Immune Regulators in Transfer Factor (A. Khan, C.H. Kirkpatrick and N.O. Hill, eds.), Academic Press, N.Y., pp. 419-431 (1979).

(9)  Borkowsky, W., Lawrence, H.S.:  J. Immunol. $\underline{126}$, 486-489 (1981).

(10) Prensky, W.:  Methods Cell Biol. $\underline{13}$, 121-152 (1976).

(11) Maxam, A.M., Gilbert, W.:  Methods Enzymol. $\underline{65}$, 499-560 (1980).

(12) Klesius, P.H., Fudenberg, H.H.:  Clin. Immunol. Immunopathol. $\underline{8}$, 238-246 (1977).

CHARACTERIZATION OF HORSE PLASMA PROTEINS BY SPECTROPHOTOMETRY
AND ISOELECTRIC FOCUSING

Charles F. Pelzer
Department of Biology, Saginaw Valley State College
University Center, Michigan, USA


Lawrence D. Aronson
Department of Medicine, Michigan State University


N. Edward Robinson
Veterinary Clinical Center, Michigan State University
East Lansing, Michigan, USA


Abstract

Spectrophotometric analysis of plasma samples from 19 normal
Arabian horses using benzoyl-l-arginine ethyl ester as substrate revealed a mean trypsin inhibitory capacity (TIC) of
0.65 ± 0.107 mg trypsin inhibited per ml of plasma.  Sex
differences were not observed.  Three other healthy Arabians
were seen to have significantly lower TIC values.

Polyacrylamide gel isoelectric focusing (IEF) at ph 3.5 - 5.0
resulted in the resolution of more than 11 anodal protein
zones from plasma of Arabians and four other horse breeds.
Based upon migration rate, staining properties, and inhibition
of activity by pre-treatment with trypsin incubation, certain
protein zones have been identified tentatively as protease
inhibitors (Pi's).  No normal horse having grossly deficient
Pi bands and an extremely low TIC value was observed in our
sample.  In addition, an examination of the IEF protein
patterns of blood plasma from 20 ill horses also failed to
identify Pi-deficient individuals.

488

Introduction

Human and horse plasma contain five major protein families with characteristic electrophoretic (or isoelectric focusing, IEF) properties: albumin, alpha$_1$, alpha$_2$, beta and gamma globulins. Each human and equine plasma fraction contains a number of specific proteins including an alpha$_2$ macroglobulin, fibrinogen and transferrin in the beta fraction and immunoglobulin G in the gamma zone. Erickson (1) has shown that the alpha$_1$ fraction of horse plasma is similar to human blood by possessing anti-tryptic or protease inhibitory (Pi) activity.

Laurell and Eriksson (2) were the first to demonstrate the association of an inherited deficiency of human plasma alpha$_1$ antitrypsin and emphysema. Subsequent investigations have correlated Pi phenotypes and a number of other human conditions (3, 4, 5). In the turkey, a deficiency of Pi's is found among birds afflicted with round heart disease (6).

In view of the association of disease states and biochemical markers in the blood, a study of equine plasma proteins including Pi's was undertaken in normal and ill horses. The present communication reports the trypsin inhibitory capacity (TIC) values from healthy Arabian horses and the isoelectric focusing (IEF) protein pattern from normal and ill horses of various breeds.


Materials and Methods

Horses. Blood samples from normal horses of the following breeds were collected: Arabians (23 horses), Standardbreds (9) Thoroughbreds (9), Quarterhorses (9), and Appaloosa (1). In addition, we examined blood from 20 ill horses displaying arthritis (5 horses), fractures (5), chronic respiratory problems (4), peritonitis (2), absence of ovaries (2), soft palate (1), and a sarcoid tumor (1). All samples were isoelectric focused.

Normal Arabian horses were also analyzed for TIC.

TIC assay.  A spectrophometric method modified from Homer and others (7) was used with the help of a Beckman Acta III.  To standardize the trypsin (E.C.  3.4.4.4), increasing amounts of trypsin (0.71, 1.06, and 1.42 x $10^{-5}$M) were added to a 0.01 M solution of p-nitrophenyl-p´-quanidinobenzoate.  The rate of the reaction read on the spectrophotometer at 402mm was measured by plotting optical density versus volume (ml.) of trypsin added. The point obrained by extrapolating the line to zero time was decreased by the O.D. value of the control containing no trypsin and the resulting number multiplied by a conversion factor of 6.025 x $10^{-5}$ (8) to give the molar amount of active trypsin.  A percent purity figure was then obtained and used to calculate the amount (mg/ml) of active trypsin present.

For the plasma assay, increasing amounts of diluted plasma (1:5) were added to trypsin (850 ug/ml).  This mixture was then pipetted into cuvets containing 0.05 mM benzoyl -L- arginine ethyl ester.  The reaction rate was linear as shown by plotting absorbance units per minute against volume (ml) of plasma added. The line was extrapolated to zero rate to determine the theoretical amount of plasma required for total inhibition of trypsin. The TIC values were calculated by dividing the trypsin concentration (derived from the stadardization procedure) by the amount (ml) of plasma needed to inhibit 1.0 ml of trypsin.

IEF procedure.  Since human plasma $alpha_1$ - antitrypsin proteins have isoelectric points in the range of pH 4.2 to 4.65 (9), IEF was performed in thin polyacrylamide gels at pH 3.5 - 5.0 for about 2.5 hours at 4°C employing the LKB Multiphor apparatus. The gel was prepared according to the method of Kueppers (10). Filter paper tabs soaked in horse plasma were applied to the cathodal end and the apparatus set at a constant current of 50ma.  For the final 2 hours without the tabs, the run was adjusted to a constant voltage of 800 mV.  Fixing and staining the gel with Coomassie Brillant Blue R-250 involved the method of Arnaud and others (11).

Incubation with trypsin.  To establish protein zones obtained by IEF as Pi's, the plasma was incubated with excess trypsin $(0.6 \times 10^{-4}M$ and $0.9 \times 10^{-4}M)$ in phosphate buffer at pH 7.4 for about 20 minutes at 37°C (12).  For comparison, control plasma plus phosphate buffer were incubated.  Samples were then submitted to IEF.

Staining for Pi's.  Protease inhibitors were also identified by the method of Uriel and Berges (13).  After IEF, the gels were incubated in a fresh solution of trypsin (0.08 mg/ml) and 0.1 M phosphate buffer at pH 7.4 for 45 minutes.  The gels were next placed in a fresh solution of acetyl - DL - Phenylalanine -$\beta$- naphyl ester prepared as follows: 5mg of the ester dissolved in 2 ml of dimethyl formamide was brought to 20 ml with 0.05 M phosphate buffer at pH 7.4 containing 10mg. of tetrazotized ortho-dianisidine dye.  After staining, the gel assumed a pink color except along the focusing path where zones exhibiting Pi activity appeared as unstained bands.

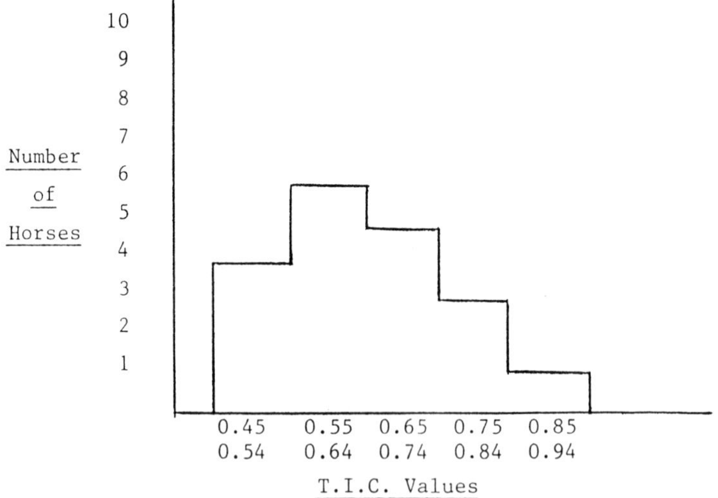


Figure 1.  Histogram of TIC values.

## Results and Discussion

Figure 1 is a histogram illustrating the TIC values of 19 normal, healthy Arabian horses. The mean value was $0.65 \pm 0.107$ mg trypsin inhibited per ml plasma with a range of 0.48 to 0.86. For stallions, the mean was $0.63 \pm 0.106$; for the mares, the average was $0.66 \pm 0.105$. Thus, there was essentially no difference between the sexes.

Three normal Arabians had lower TIC values of 0.36, 0.40 and 0.42. The mean of 0.39 $\pm$ 0.025 for these three horses is statistically significant (t = 3.897, df = 20; where t crit = $\pm$ 2.845 at $\alpha$ = 0.01). The average TIC of these three horses represents 60% of the trypsin inhibitory capacity of the other Arabians with higher TIC values. The latter finding is not unlike that observed in human plasma where Pi MZ phenotypes with intermediate levels of $alpha_1$ - antitrypsin possess about 58% of the trypsin inhibitory capacity associated with the Pi M phenotype (14).

Figure 2 reveals the plasma protein patterns obtained by IEF from 5 breeds of horses: Thoroughbred, Standardbred, Appaloosa, Quarterhorse and Arabian. At least 11 well resolved anodal protein zones are recognized. Common protein bands are observed among the 5 breeds as well as distinct differences within and between breeds. Breed specific bands were not observed. It was not possible to determine the genetic basis, if any, of the observed variation in protein zones between horses because of the unavailability of critical maternal or paternal blood samples for each of the pedigrees involved in this investigation.

Since the IEF patterns of the 3 Arabian horses with lower TIC values were not grossly different from other horse plasma profiles, an attempt to correlate TIC data and IEF patterns was unsuccessful. A similar finding of difficult discrimination among protein patterns is observed in the human Pi system. The M and MZ phenotypes are difficult to differentiate especially

492

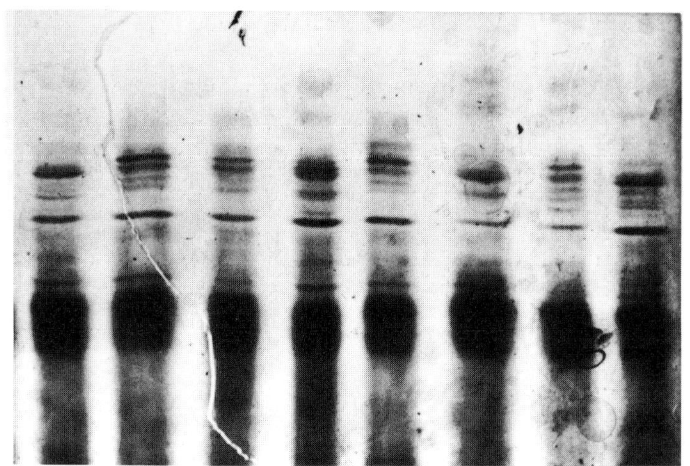

Figure 2.   IEF pattern of horse plasma proteins.   Channel 1:
Thoroughbred mare;  2,3: Standardbred stallion, mare;  4: Appaloosa mare;  5,6,7: Quarterhorse mares;  8: Arabian mare.

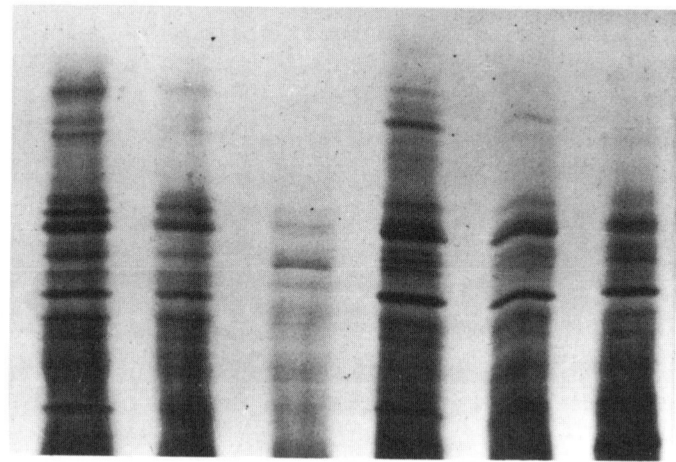

Figure 3.   Effect of trypsin pretreatment on IEF pattern.
Channel 1: Arabian mare (R20);  2: R20 plasma plus $0.6 \times 10^{-4}$M
trypsin;  3: R20 plus $0.9 \times 10^{-4}$M trypsin;  4: Arabian stallion
(R8);  5: R8 plus $0.6 \times 10^{-4}$M trypsin;  6: R8 plus $0.9 \times 10^{-4}$M
trypsin.

by starch gel electrophoresis.

Plasma from 20 horses diagnosed as having specific pathologic
ailments were also focused by IEF. Included in this sample
were 4 horses with respiratory difficulties. All samples
appeared in the normal range of protein patterns.

Incubation of horse plasma with trypsin for 20 minutes prior
to IEF indicates that certain protein zones (Fig. 3) are diminished or eliminated as a result of the binding of trypsin
and plasma antitrypsin. It would thus appear that these bands
are true protease inhibitors. In addition, the horse bands
were identified as Pi's on the basis of staining properties.
Application of the Uriel and Berges staining method resulted
in the appearance of rapidly migrating zones plus several
more slowly moving bands in the human $alpha_1$ - antitrypsin
region. The latter zone was identified by running human Pi M
types along side equine samples.

Although some normal Arabian horse plasma had reduced amounts
of Pi activity on the basis of TIC data, a horse severely deficient for Pi activity or Pi bands similar to that found in
some humans and turkeys was not found. Breeze and others (15)
also failed to discover a severely deficient animal among 44
horses without respiratory disease and 47 horses reported to
have some form of pulmonary problem. Since only about 7 per
10,000 humans (16) possess the severely deficient Pi ZZ phenotype it may not be surprising to discover that we were unable
to locate a horse drastically lacking plasma Pi's.

The excellent technical assistance of Barbara Conley is gratefully acknowledged. One of us (CFP) was supported by funds
from Michigan State University and Saginaw Valley State College
Foundation.

References

1.  Erickson, E. D.: In Kitchen, H., Krehbiel, J. (ed.), Proceedings of the First International Symposium on Equine Hematology, 152-158, Amer. Assoc. Equine Practitioners, Golden, CO, (1975).

2.  Laurell, C.- B., Ericksson, S.: Scand. J. Clin. Lab. Invest. 15, 132-140 (1963).

3.  Cox, D. W., Huber, O.: Lancet 1, 1216 (1976).

4.  Palmer, P. E., Gherardi, G. J., Baldwin, J. M., Wolfe, H. J. Amn. Int. Med. 88, 59-65 (1978).

5.  Sharp, H. L., Bridges, R. A., Krivit, W., Freier, E. F.: J. Lab. Clin. Med. 73, 934 (1969).

6.  Neumann, F., Meirom, R., Rattner, D., Trainin, Z., Klopfer, U., Nobel, T. A.: Amer. J. Path. 84, 427-430 (1976)

7.  Homer, G. M. Katchman, B. J., Zipf, R. E.: Clin. Chem. 9, 428-437 (1963).

8.  Chase, T., Shaw, E.: Biochem. Biophys. Res. Comm. 29, 508-514 (1967).

9.  Allen, R. C., Harley, R. A., Talamo, R. C.: Amer. J. Clin. Path. 62, 732-739 (1974).

10. Kueppers, F.: J. Lab. Clin. Med. 88, 151-155 (1976).

11. Arnaud, P., Creyssel, R., Chapuis - Cellier, C.: LKB Application Note 185 (1975).

12. Allen, R. C., Oulla, P. M., Arnaud, P., Baumstark, J. S.: In Radola, B. J., Graesslin, D. (ed.): Electrofocusing and Isotachophoresis, Walter de Gruyter, Berlin, New York, (1977).

13. Uriel, J., Berges, J.: Nature 218, 578-580 (1968).

14. Kueppers, F., Black, L. F.: Amer. Rev. Resp. Dis. 110, 176-193 (1974).

15. Breeze, R. G., Nicholls, J. M. Veitch, J. Selman, I. E., McPherson, E. A., Lawson, G.H.K.: Vet. Rec. 101, 146-149 (1977).

# ALPHA$_1$-ANTITRYPSIN DEFICIENCY AND DISEASE

P. Arnaud and R.C. Allen

Departments of Basic and Clinical Immunology and Microbiology, Pathology, and Laboratory Animal Medicine, Medical University of South Carolina, Charleston, South Carolina 29425

## Introduction

The ability of plasma to inhibit proteases has been recognized since the work of Camus and Gley in 1897 (1). This ability is mainly found in alpha$_1$-antitrypsin (A$_1$AT), a glycoprotein present in the biological fluids including plasma, extracellular fluids, cerebrospinal fluid, and a number of secretions such as tears, saliva, seminal fluid, and cervical mucus. Its concentration in normal plasma is 130 mg% in contrast with previous values of 200 mg%. The discovery that hereditary A$_1$AT deficiency was associated with chronic pulmonary disease (2) created greater interest in its study in association with other diseases, such as the demonstration of the association between its deficiency and liver diseases (3).

The physicochemical characteristics and biological roles of A$_1$AT (also often called alpha$_1$-protease inhibitor) have been recently reviewed by Jeppson and Laurell (4) and the genetic aspects of the Pi system have been discussed in detail by Fagerhol (5). The aim of the present article is to review the recent findings on the genetics of the Pi system of A$_1$AT with special emphasis on the alleles and phenotypes leading to A$_1$AT deficiency, and to revise the disease associations which have been reported to be associated with A$_1$AT deficiency.

---

Publication no. 431 from the Department of Basic and Clinical Immunology and Microbiology, Medical University of South Carolina. Research supported in part by USPHS Grants HD-09939 and CA-25746.

496

Genetic Polymorphism of A$_1$AT and Deficient Alleles of the Pi System

In 1965, Fagerhol and Braend (6) reported a genetic polymorphism in the
prealbumin region, evidenced by acid-starch gel electrophoresis (ASGE),
which corresponded to A$_1$AT variants. Due to the pioneer work of Fagerhol,
the Pi genetic system of A$_1$AT was shown to consist of at least 20 alleles
inherited in the autosomal codominant mode. Large population studies
demonstrated variations in the distribution of the Pi allele frequencies
according to the ethnicity. As an example, the Z deficient allele appears
to be more frequent in Caucasians, the S allele is more frequent in Spanish
populations, and Pi variants are infrequent in Blacks and Asians (5).

In 1976, the introduction of thin-layer isoelectric focusing (IEF) for Pi
typing (8,9) gave a new interest in the study of the genetics of A$_1$AT
variants. This technique was able to demonstrate more polymorphic variants
than starch gel. This is illustrated in Table 1, which compares the
classification of Pi alleles identified by ASGE and by IEF. One of the
major results of the IEF technique was to permit the dissection of the
common M allele in several subtypes, inherited in the same manner as the
other Pi alleles (10). As a result, more than 85% of the individuals in
normal populations were found homozygous for the M allele by ASGE. In
contrast, more than 50% of the individuals are now found to possess a
variant different from the most common M$_1$ subtype. This increases
dramatically the interest of the Pi system as a genetic marker, and A$_1$AT
can now be considered as the most polymorphic protein in human plasma, with
more than 50 different alleles (see Table 1) confirmed by family studies.
This is especially useful in forensic medicine, paternity tests, and has
been recently used in linkage studies.

An additional point of interest is that the basal plasma values of A$_1$AT are
genetically determined, i.e., the expression of some alleles is associated
with different levels of A$_1$AT deficiency. These (see Table 2) vary from
moderate deficiency, such as that associated with the S and I alleles, to
severe deficiency such as that present with the Z and P alleles, to almost
complete absence, which is associated with the so-called "null" or (-)
gene. In addition, A$_1$AT is an acute-phase reactant protein, and its level

Table 1.  Pi alleles identified by acid-starch gel electrophoresis (ASGE) and isoelectric focusing (IEF).  In addition, two alleles are designated by an asterisk: MUnstable and MSan Francisco which present a thermal instability.  It is not known whether BAlhambra is identical or different from B.  Data from refs. 34-36 and from unpublished data.

| ASGE | IEF | ASGE | IEF | ASGE | IEF |
|---|---|---|---|---|---|
| B | BSaskatoon | M | $M_1$ | P | PClifton |
|  | BAlhambra(?) |  | $M_2$ |  | PSaint Louis |
| C |  |  | $M_3$ |  | PBudapest |
| D |  |  | $M_4$ |  | PKyoto |
| E | ELemberg |  | $M_5$ | R | RGambia |
|  | ECincinnati |  | MSalla | S | $S_1$ |
|  | ETokyo |  | MChapel Hill |  | $S_2$ |
|  | $E_2$ |  | MMalton |  | $S_3$ |
| F |  |  | MDuarte |  | $S_4$ |
| "F" |  |  | MUnstable* | T |  |
| G | GCler |  | MSan Francisco* | V |  |
| I | $I_1$ |  |  | W | WGazak |
|  | $I_2$ (IFecamp) | N | NHampton |  | WToulouse |
| L | LVibeuf |  | NRouen |  | WConstantine |
|  |  |  | NLyon |  | WSalerno |
|  |  |  | NLetrait | X |  |
|  |  |  |  | Y |  |
|  |  |  |  | Z | ZCathodal |
|  |  |  |  |  | ZPratt |
|  |  |  |  |  | $Z_2$ |
|  |  |  |  | Null |  |

Table 2.  Alpha$_1$-antitrypsin expression associated with deficient alleles.  Mean values for the M allele were considered as 100%.  From ref. 37 and unpublished data.  For the % expression associated with MMalton and MDuarte alleles, see ref. 11.

| Pi allele | % expression | Serum $A_1AT$ |
|---|---|---|
| M | 100 | Normal |
| S | 472.7 | Moderately deficient |
| I | 68.3 | Moderately deficient |
| P | 34.9 | Severely deficient |
| Z | 26.3 | Severely deficient |
| Null | 0 (or trace) | Absent |

is increased in infections and inflammatory status. This stresses the
importance of the precise determination of the Pi phenotypes to obtain an
accurate information on possible $A_1AT$ deficiency. Other deficient alleles
include Mmalton and Mduarte (11). The basis of the polymorphism of the Pi
allele products appears to be due to point amino acid substitution(s) which
affect the net charge of the molecule, leading to differences in the
isoelectric points of the different isotypes (4).

The linkage between Pi and the Gm markers of the immunoglobulin heavy chain
was first demonstrated by Gedde-Dahl et al. (12). The important point
concerning this linkage is that it is sex dependent and allele-specific,
i.e., the recombination fractions in males are lower than in females, and,
in addition, the recombination fractions in males with Z and S (deficient)
alleles are lower than those in other males and in females. This
allele-specific effect represents a unique case in mammalian genetics. A
preferential transmission of the Z deficient allele from male parents has
been proposed by us (13,14). This was found by the observation of a
segregation distortion in children from heterozygous parents. This finding
is of special interest because it could, at least in part, explain the
so-called heterozygous advantage which is though to be the major basis for
the persistence of genetic polymophisms.

$A_1AT$ Deficiency and Diseases

The association between $A_1AT$ and disease was discovered by the finding by
C.B. Laurell of a decrease in the $alpha_1$-globulin peak in subjects with
pulmonary emphysema (2). The thesis of Eriksson demonstrated that genetic
deficiency of $A_1AT$ could be associated with precocious pulmonary emphysema.
Two major mechanisms of disease predisposition can be considered, those
apparently due to an imbalance between proteases and inhibitors, and those
apparently related to $A_1AT$ retention within the hepatocyte.

Imbalance between proteases and inhibitors

Chronic obstructive lung disease (COPD). COPD represents the first
reported and most common disorder associated with $A_1AT$ deficiency. Its

onset and severity appear related to the degree of the deficiency, i.e., all completely deficient (homozygous null) subjects appeared to present symptoms, when it represents a major risk in homozygous deficient (ZZ) subjects, although part of them do not present clinical symptoms. Finally, SZ subjects appear also to be predisposed to COPD (15). The clinical picture of the disease has been recently reviewed (16), but in its typical expression it realizes an early onset diffuse pulmonary emphysema, whose prognosis appears to be severe. A controversy has been developing for several years as to the risk of heterozygous (mostly MZ) individuals to develop COPD. This can only be answered by follow-up of individuals during a prolonged period of time, but such an answer is of critical importance, in view of the relatively high frequency (4 to 6%) of heterozygous individuals in random populations. Indeed, if there is not, for the moment, place for substitutive treatment in $A_1AT$ deficiency, preventive measures can be taken in individuals at risk which should avoid cigarette smoking and working in atmosphere containing air pollutants. It appears also that $A_1AT$ deficiency can be associated with asthma in children. A highly significant increase of heterozygous deficient phenotypes was found in nonatopic asthma in infancy (17). In addition, those asthmatic children with deficient phenotypes were characterized by an earlier onset and a greater severity of the disease.

<u>Rheumatic disorders</u>. In 1976, Cox and Huber (18) reported an association between adult rheumatoid arthritis and the possession of the Z allele in the heterozygous state. Similarly, we have shown a significant association between the heterozygosity for the Z allele and juvenile chronic polyarthritis (19). Therefore, the predisposing role of genetically-determined $A_1AT$ deficiency in rheumatic disorders appears well established.

<u>Role of $A_1AT$ as an inhibitor of protease</u>. Although named from its inhibitory activity on trypsin, it is unlikely that the inhibition of trypsin represents a physiological role for $A_1AT$, except perhaps in pancreatitis. In contrast, $A_1AT$ is mostly involved in inhibition of a series of proteases, such as elastase, lysosomal proteases, and neutral proteases, which are released by polymorphonuclear (PMN) leukocytes and can

be considerably increased during such processes as infection and
inflammation.  An additional point of interest comes from the observation
that cigarette smoke can inactivate $A_1AT$ (20), explaining the aggravating
role of smoking in $A_1AT$ deficiency.

## Retention of $A_1AT$ inside the hepatocyte

A basic interest of $A_1AT$ deficiency comes from the observation by Sharp (3)
that $A_1AT$ deficiency of the ZZ phenotype was associated with neonatal
hepatitis.  Further studies demonstrated that infants with ZZ phenotypes
could develop early progressive cirhosis of the liver (21).  Additional
studies have shown that liver damage could also be observed in individuals
heterozygous for the Z allele (22).  Another important association with the
Z allele appears to be the possibility to develop hepatoma, as first
proposed by Eriksson (23).  Finally, the distribution of the Z allele
appears dramatically increased in individuals with liver damage associated
with hepatitis B antigen (24).  The common point of these observations is
the fact that $A_1AT$ is alway retained inside the hepatocytes in association
with the Z allele (25), as demonstrated by periodic acid Schiff stain and
specific immunofluorescence.  This $A_1AT$ is very similar to the circulating
molecule, although is lacks part of the sugars of the normal glycoprotein
(4).  Recently, evidence has been presented that retention of $A_1AT$ within
the hepatocyte occurs with other severely deficient phenotypes such as
Mmalton, Mduarte (see Cox et al., ref. 11), and P.  The mechanisms by which
this retention predisposes to liver diseases is still unclear.  The
proportion of individuals wiht homozygous ZZ phenotypes and infantile
cirrhosis appears to be around 15% (26).  Other additional environmental
and/or genetic factors are certainly needed to permit the appearance of
these clinical manifestations.  When they are present, few therapeutic
resources are offered, although infantile cirhosis due to $A_1AT$ deficiency
represents one of the indications of heterotopic liver transplantation
(27).

## $A_1AT$ Deficiency and Reproduction

Apart from the segregation distortion previously mentioned (13,14), $A_1AT$

deficiency has been associated with abnormal fertilization. This involves an apparent increased fertility in women carrying deficient $A_1AT$ genes (28) as well as a higher incidence of miscarriages (29). The latter has to be discussed in view of a possible effect of deficient alleles on chromosome aberrations (5). It has to be kept in mind that the fertilization process is under the control of a delicate imbalance between proteases and its inhibitors, both present in cervical mucus and seminal fluid, and that a genetically-determined $A_1AT$ deficiency could modify this fragile equilibrium.

Other Diseases Which Could Be Associated With $A_1AT$ Deficiency

Several other diseases have been shown to be associated with $A_1AT$ deficiency (see Table 3). Two disease associations appear to be likely, pancreatitis and renal diseases. In addition, preliminary studies have implicated both Pi deficiency and an association of the MZ allele with the A, B, and AB blood groups with an increased frequency of periodontitis (30). This represents one of the first observations of two genetic determinants associated to predispose to a disease state. Systematic Pi determinations should be performed in selected groups of patients compared with controls of similar ethnicity, to increase our knowledge not only on the risk associated with $A_1AT$ deficiency, but also to delineate the exact role of this protein. Indeed, recent studies have attributed a role of $A_1AT$ as a modulator of the immune response (31).

Clinical Prognosis of $A_1AT$ Deficiency

Liver diseases. In a prospective study which can be considered as a model, Sveger (26) screened 200,000 new borns, among which 122 homozygous for Z phenotypes were individualized. Neonatal cholestasis was clinically evident in 7% of the babies, mild liver symptoms in 4%, and probable liver involvement was suspected in 6.5%. Therefore, liver symptoms appeared to affect only 18% of the children. Interestingly, Sveger (32) followed up 120 children form the initial group 4 years later. Three of them had severe symptoms of the disease. 45 children (36.5%) has abnormal liver tests, although the remainder appeared not to suffer from liver disease.

502

Table 3.  Diseases or syndromes in which $A_1AT$ deficiency has been reported

| Disease or syndrome | Type of $A_1AT$ deficiency |
| --- | --- |
| Glomerulonephritis | Pi ZZ |
| Acne urticaria and angioneurotic edema | Decreased plasma levels, Pi ZZ |
| Biliary atresia | Pi ZZ |
| Ehlers-Danlos | Pi null |
| Chronic pancreatitis | Pi MZ and ZZ |
| Weber-Christian syndrome | Pi ZZ |
| Spherocytosis | Pi MZ |
| Thyroiditis | Decreased plasma levels |
| Gastric or duodenal ulcer | Decreased plasma levels |
| Severe combined immunodeficiency | Pi ZZ |
| Coeliac disease | Decreased plasma levels |
| Psoriasis | Pi MZ |
| Familial hypercholesterolemia | Pi ZZ |
| Inherited aminoaciduria | Pi MZ and ZZ |
| Multisystem fibrosis | Pi SZ |
| Fibrosing alveolitis | Pi MZ |
| Multiple endocrine adenomatosis | Pi ZZ |
| Pyloric stenosis and hyperbilirubinemia | Pi ZZ |
| Allergic contact dermatitis | Pi MS and MZ |
| Uveitis | Pi MZ |

In another study performed in Sweden, Larsson (33) collected data from 246
adults with apparently homozygous PiZZ deficiency.  An overall 12% of the
subjects had a diagnosis of liver cirhosis which reached 19% in subjects
over 50 years of age.  Finally, liver cirrhosis was responsible for the
deaths in 13% of the patients who died.

Lung diseases.  Children with homozygous ZZ deficiency do not appear to
present more pulmonary symptoms than control children (32).  In contrast,
in adults, the prevalence of chronic obstructive lung disease is extremely
high.  Although selection of individuals can bias the exact frequency of
pulmonary symptoms, Larsson (33) found an overall of 39% of the ZZ subjects
with COPD under the age of 40, when, over this age, 85% of the subjects
suffered from COPD.  It was responsible for 59% of the death observed in
this group.  This, added to liver cirhosis, led to a reduced life
expectancy which was even more severely reduced in smokers.

## Conclusion

Alpha$_1$-antitrypsin is a model for polymorphic proteins; the multiplicity of
its alleles makes it very valuable for genetic and forensic studies.  The
association of two major (lung and liver) diseases with genetically
determined A$_1$AT deficiency is of interest.  These reduce the life
expectancy of severely deficient individuals.  In addition, A$_1$AT deficiency
appears to favor the onset of several other groups of diseases for which
large genetic and prospective studies are needed.

## References


1.  Camus, L., Gley, E.: C.R. Soc. Biol. (Paris) 49, 825-829 (1897).

2.  Laurell, C.B., Eriksson, S.: Scand. J. Clin. Lab. Invest. 15, 132-140
    (1963).

3.  Sharp, H.L., Bridges, R.A., Krivit, W., Freier, E.F.: J. Lab. Clin.
    Med. 73, 934-939 (1969).

4.  Laurell, C.B., Jeppson, J.O.: in: The Plasma Proteins, F.W. Putnam,
    Ed., Academic Press, New York, vol. 2, pp. 229-264 (1975).

5.  Fagerhol, M.K.: Postgrad. Med. J. 52, 73-79 (9176).

6.  Fagerhol, M.K., Braend, M.: Science 149, 986-987 (1965).

7.  Fagerhol, M.K., Laurell, C.B.: Clin. Chim. Acta 16, 199-203 (1967).

8.  Arnaud, P., Chapuis-Cellier, C., Creyssell, R.: C.R. Soc. Biol. (Paris)
    168, 58-61 (1974).

9.  Allen, R.C., Harley, R.A., Talamo, R.C.: Amer. J. Clin. Pathol. 62,
    632-642 (1974).

10. Frants, R.R., Eriksson, A.W.: Hum. Hered. 26, 435-440 (1976).

11. Cox, D.W., Billingsey, G.D., Smyth, S.: in:  Electrophoresis 81, Walter
    De Gruyter, Berlin, pp. 505-510 (1981).

12. Gedde-Dahl, R., Fagerhol, M.K., Cook, P.J.L., Noades, J.: Ann. Hum.
    Genet.  35, 393-399 (1972).

13. Chapuis-Cellier, C., Arnaud, P.: Science 205, 607-608 (1979).

14. Iammarino, R.M., Wagener, D.K., Allen, R.C.: Amer. J. Hum. Genet. 31:
    508-517 (1979).

15. Larsson, C., Dirksen, H., Sundstrom, G., Eriksson, S.: Scand. J. Resp.
    Dis.  57, 267-280 (1976).

504

16. Morse, J.O.: N. Engl. J. Med. 299, 1045-1048 and 1099-1105 (1978).

17. Arnaud, P., Chapuis-Cellier, C., Souillet, G., Carron, P., Wilson, G.B., Creyseel, R., Fudenberg, H.H.: Trans. Assoc. Amer. Phys. 89, 205-214 (1977).

18. Cox, D.W., Huber, O.: Lancet i, 1216-1217 (1976).

19. Arnaud, P., Galbraith, R.M., Faulk, W.P., Ansell, B.M.: J. Clin. Invest. 60, 1442-1444 (1977).

20. Gader, J.E., Fells, G.A., Crystal, R.G.: Science 206, 1315-1316 (1979).

21. Hadchouel, M., Gauthier, M.: J. Pediat. 89, 211-215 (1976).

22. Palmer, P.E., Gherardi, G.J., Baldwin, J.M., Wolfe, M.J.: Ann. Int. Med. 88, 59-60 (1978).

23. Eriksson, S., Hagerstrand, J.: Acta Med. Scand. 195, 451-458 (1976).

24. Hodges, J.R., Millward-Sadler, G.H., Barbatis, C., Wright, R.: N. Engl. J. Med. 304, 557-560 (1981).

25. Lieberman, J., Mittman, C., Gordon, H.W.: Science 175, 63-65 (1972).

26. Sveger, T.: N. Engl. J. Med. 294, 1316-1321 (1976).

27. Hood, J.M., Koep, L.J., Peters, R.L., Schroter, G.P.J., Weil, R., Redeker, A.G., Starzl, T.E.: N. Engl. J. Med. 302, 272-275 (1980).

28. Evans, H.E., Bognackins, M.S., Perrott, L.M., Glass, L.: J. Pediat. 90, 621-626 (1977).

29. Aarskog, D., Aarseth, P., Fagerhol, M.K.: Clin. Genet. 13, 81-84 (1978).

30. Allen, R.C.: in: Electrophoresis 79, B.J. Radola, Ed., Walter de Gruyter, Berlin, New York, pp. 631-645 (1979).

31. Arora, P.K., Miller, H.C., Aronson, L.D.: Nature 274, 589-590 (1978).

32. Sveger, T., Thelin, T.: Acta Pediat. Scand. 70, 171-177 (1981).

33. Larsson, C.: Acta Med. Scand. 204, 345-351 (1978).

34. Frants, R.R., Eriksson, A.W.: Hum. Hered. 30, 333-342 (1980).

35. Charlionet, R., Sesboue, R., Morcamp, C., Lefebvre, F., Martin, J.P.: Hum. Hered. 31, 104-109 (1981).

36. Cox, D.W.: Amer. J. Hum. Genet. 33, in press (1981).

37. Arnaud, P., Chapuis-Cellier, C., Vittoz, P., Fudenberg, H.H.: J. Lab. Clin. Med. 92, 177-186 (1978).

RARE TYPES OF $\alpha_1$-ANTITRYPSIN ASSOCIATED WITH DEFICIENCY


Diane Wilson Cox, Gail D. Billingsley, Shirley Smyth

The Research Institute, The Hospital for Sick Children and Departments of Paediatrics and Medical Genetics, University of Toronto, Toronto, Canada


Introduction

More than 30 genetic variants (PI types) of the plasma protease inhibitor $\alpha_1$-antitrypsin ($\alpha_1$AT), or $\alpha_1$-protease inhibitor, have been described (1,2). Only a few of these are associated with a pronounced deficiency of $\alpha_1$AT in the serum. The most common deficiency allele is *PI*Z*. In a Canadian white population of 504 adults and 523 children, we have calculated the frequency of *PI*Z* to be 0.0122, with a corresponding frequency of PI ZZ homozygotes of 1 in 6,700.

Other rare deficiency alleles have been reported. These include *PI* Mmalton* (3), *PI*Mduarte* (4), *PI*QO* (null), and an 'M-like' variant (6). We are reporting here studies of PI types in 112 individuals with $\alpha_1$-antitrypsin deficiency, and in their relatives. The frequencies of the rare deficiency types have been calculated. The variants have been compared by isoelectric focusing and by electrophoresis in starch and agarose gels.


Materials and Methods

Serum samples were available from 78 adults and 34 children with $\alpha_1$AT deficiency. The majority of the adults had chronic obstructive pulmonary disease (COPD). The majority of children were ascertained because of liver abnormalities.

Sera were reduced, prior to electrophoresis, with 20 mM dithioerythritol, or with cysteine (7). Isoelectric focusing (IEF) was carried out using LKB PAG plates, pH 4-5, as previously reported (1), except that prefocusing was carried out for 2 hours at 1,000 V; run time was 2 hours at 1,600 V. Deficiency alleles were compared following Coomassie Blue R250 staining of the IEF gel and by print immunofixation (8). Fagerhol's method of starch gel electrophoresis at pH 4.9 was carried out on selected sera, as described previously for our laboratory (9). Deficiency types of $\alpha_1$AT were visualized following immunofixation in the starch gel (10). Agarose electrophoresis was carried out at pH 8.6 (11) followed by immunofixation in the gel (12). Specific $\alpha_1$AT antiserum was obtained from Atlantic Antibodies.

The serum $\alpha_1$AT concentration was determined by electroimmunoassay (13). $\alpha_1$AT concentration was expressed as % of a normal pool, where the pool was that of 1,000 normal donors sampled through the University of Washington.

Sera from available relatives including children, parents, and sibs of patients were PI typed and $\alpha_1$AT was assayed. Only the original patients (probands) have been included in estimates of gene frequencies.


Results


All sera were compared by IEF, and $\alpha_1$AT concentration was assayed. Homozygotes for PI null were detected by their exceptionally low concentrations of serum $\alpha_1$AT (< 2% of normal). Other rare deficiency alleles were, in some cases, not initially detected by IEF with standard protein staining. In several families, the occurrence of discrepancies from the expected PI types in offspring led to detection of the rare alleles. Therefore the serum of each of the 112 individuals with $\alpha_1$AT deficiency was examined by IEF followed by print immunofixation. The deficiency types of $\alpha_1$AT could be clearly distinguished by IEF. Serum from the only Mduarte homozygote found to date was kindly provided by Dr. J. Lieberman. Initial comparisons indicated that the serum had degenerated in storage and could

not be used for comparison.  However reduction of the sample with DTE or
cysteine resulted in a clear pattern by IEF.  The position by IEF was
further confirmed by print immunofixation.  Position of the rare deficiency
types by IEF was in the M region;   isoelectric points were considerably
lower than those of the Z variant.  The relative positions, anodal to
cathodal, by IEF, are as follows:

$$M1 > M3 = Mduarte > M2 > Mmalton = Nhampton$$

The deficiency types could be observed in starch only by immunofixation.
The relative mobilities, anodal to cathodal, were as follows:

$$M2 = Mduarte = Mmalton > M3 = M1$$

Band patterns in agarose gels could be observed only after immunofixation.
Relative mobilities anodal to cathodal, were as follows:

$$Mmalton > M2 > Mduarte > M3 = M1$$

An unusual phenomenon was observed in agarose electrophoresis.   The
mobility of Mduarte $\alpha_1$AT in the homozygote (sera from proband, provided
by Dr. Lieberman) appeared to be slightly cathodal to that of the Mduarte
in the PI MduarteZ heterozygote from our series.  We have previously
observed a difference in mobility of S $\alpha_1$AT in heterozygous PI type M2S
individuals, when compared to that in sera of PI type M1S.  There appears
to be some interaction between M2 and S, perhaps because of interaction
with charged groups in the agarose.  This same effect may be occurring
with Mduarte $\alpha_1$AT in agarose.  We have carefully investigated the
possibility that these might be different variants.  A prolonged run on
IEF, which usually shows small differences in isoelectric point, failed to
reveal any differences between these two variants.
The distribution of PI types was as follows:

$$ZZ: 107, \quad MmalZ: 2, \quad MduaZ: 1, \quad Z-: 1, \quad null (--): 1$$

These numbers do not include relatives of the patients studied.  The gene
frequencies for the rare alleles are shown in Table 1.

TABLE 1

Frequency of PI Deficiency Alleles

| Allele | Frequency |
|---|---|
| Z | $1.22 \times 10^{-2}$ |
| Mmalton | $1.69 \times 10^{-4}$ |
| Mduarte | $0.56 \times 10^{-4}$ |
| null | $1.69 \times 10^{-4}$ |

The frequencies of the rare alleles have been calculated as a fraction of
the Z gene frequency. Because of the relatively small number, they must
be regarded as approximate. The concentration of $\alpha_1$AT in all sera from
patients in this study was less than 40% of normal. Concentrations
obtained for sera of deficient and partially deficient PI types are shown
in Tables 2 and 3. Results for probands and for relatives of individuals
with rare PI types have been included. In the two individuals of PI type
MduarteZ, the concentration of Mduarte is only about one-third that of the
Z component, in both individuals. The Mduarte homozygote has a relatively
low concentration of $\alpha_1$AT, 6% of normal by our method. While the Mmalton
$\alpha_1$AT does not appear to be associated with a lower concentration of $\alpha_1$AT
than that of Z, there were occassional individuals among the relatives in
whom the Mmalton component could be visualized only after increased
staining of the gels.

Among probands and relatives in this study, early onset COPD occurred in
one patient PI MmaltonZ, two PI MduarteZ brothers, one Z- patient, and
two PI null homozygotes. Three other individuals with rare deficiency
alleles were less than 25 years of age and did not yet have COPD.

Discussion

Our study indicates that other types of $\alpha_1$AT, in addition to Z, are
associated with a pronounced deficiency of $\alpha_1$AT and can be identified by
appropriate laboratory techniques. Immunofixation is particularly useful
in identifying these alleles, which could be missed by standard typing
techniques.

TABLE 2

Concentration of $\alpha_1$AT in Deficiency PI Types

| | PI types | | | | | |
|------|------|-------|------|-------|-------|------|
| | Z | QO(-) | Mmal | MmalZ | MduaZ | Z- |
| No. | 105 | 3 | 2 | 4 | 2 | 1 |
| Mean | 17.7 | <2 | 15.0 | 18.0 | 12.5 | 13.0 |
| 1 SD | 5.1 | | 1.4 | 4.6 | 2.1 | |

TABLE 3

Concentration of $\alpha_1$AT in Partially Deficient and Normal PI Types

|  | PI types | | | | |
|---|---|---|---|---|---|
|  | Partially deficient | | | | Normal |
|  | MZ | MMmal | MMdua | M– | M |
| No. | 10 | 15 | 4 | 9 | 22 |
| Mean | 67.6 | 63.3 | 48.2 | 59.4 | 109.4 |
| 1 SD | 10.8 | 9.7 | 12.3 | 6.8 | 13.4 |

The frequencies of the other deficiency alleles are considerably lower
than that of *PI*Z*. They result in as pronounced a deficiency of $\alpha_1$AT
as in PI ZZ individuals. The small amount of data on PI Mduarte indicates
that this allele may result in a more pronounced deficiency of $\alpha_1$AT than
in PI type ZZ. We have also found that in certain Mmalton heterozygotes,
the Mmalton component is less abundant than one would expect for a Z
component. However, a reduced level is not reflected in the MMmalton
heterozygotes in this study.

Individuals with the rare deficiency alleles are also at increased risk
for developing COPD. This is expected because of the low levels of
$\alpha_1$AT in the serum.

The low concentration of $\alpha_1$AT in serum, associated with PI type ZZ, has
been shown to be due to lack of secretion of $\alpha_1$AT from the liver (14).
The half-life of Z $\alpha_1$AT in the circulation is only slightly below that
of normal (15). The half-life of Mmalton $\alpha_1$AT is similar to that of the
normal M $\alpha_1$AT (16). Mmalton $\alpha_1$AT, like Z, is not secreted normally from
the liver. We have recently had the opportunity to demonstrate this
through the study of an individual of PI type MlMmalton. Typical PAS
positive granules have been found in his liver biopsy. This material
shows fluorescence with $\alpha_1$AT (17). Similar inclusions have been
reported in the Mduarte homozygote, although this was not confirmed by
studies with fluorescent $\alpha_1$AT antibody (4).

At least three types of $\alpha_1$AT which are not secreted normally from the

liver can now be recognized:  Z, Mmalton, and Mduarte.  The reason for
this lack of secretion from the liver remains an intriguing problem and
should provide insight into the normal mechanisms of excretion of
glycoproteins from the liver.


References


1.  Cox, D.W.:  Amer. J. Hum. Genet. 33, in press (1980).

2.  Cox, D.W., Johnson, A.M., Fagerhol, M.K.:  Hum. Genet. 53, 429-433
    (1980).

3.  Cox, D.W.:  Protides of the Biological Fluids, Vol. 23, pp. 375-378,
    H. Peeters, Pergamon Press, Oxford. (1976).

4.  Lieberman, J., Gaidulis, L., Klotz, S.D.:  Am. Rev. Resp. Dis. 113,
    31-36 (1976).

5.  Talamo, R.C., Langley, C.E., Reed, C.E., Makino, S.  Science 181,
    70-7; (1973).

6.  Kueppers, F., Utz, G., Simon, B.:  J. Med. Genet. 14, 183-186 (1977).

7.  Pierce, J.A., Jeppsson, J-O., Laurell, C.-B.:  Anal. Biochem. 74, 227-
    241 (1976).

8.  Arnaud, P., Wilson, G.B., Koistinen, J., Fudenberg, H.H.:  J. Immunol.
    Methods 16, 221-231 (1977).

9.  Cox, D.W., Celhoffer, L.:  Can. J. Genet. Cytol. 16, 297-303 (1974).

10. Lieberman, J., Gaidulis, L.:  J. Lab. Clin. Med. 87, 710-716 (1976).

11. Jeppsson, J.-O., Laurell, C.-B., Franzen, B.:  Clin. Chem. 25, 629-
    638 (1979).

12. Johnson, D., Travis, J.: Biochem. Biophys. Res. Commun. 72, 33-39
    (1976).

13. Laurell, C.-B.:  Scand. J. Clin. Lab. Invest. 29, suppl. 124, 21-37
    (1972).

14. Sharp, H.L., Bridges, R.A., Krivit, W., Freier, E.F.:  J. Lab. Clin.
    Med. 73, 934-939 (1969).

15. Laurell, C.-B., Nosslin, B., Jeppsson, J.-O.:  Clin. Sci. Mol. Med.
    52, 457-461 (1977).

16. Jeppsson, J.-O., Laurell, C.-B., Nosslin, B., Cox, D.W.:  Clin. Sci.
    Mol. Med. 55, 103-107 (1978 ).

17. Roberts, E., Cox, D.W., Cutz, E.:  Unpublished observations.

ISOLATION AND CHARACTERIZATION OF AN ALPHA-1-ANTITRYPSIN-RELATED
GLYCOPROTEIN FROM HUMAN LIVER

Robert H. Glew, J.L. Zidian, J.P. Chiao, T. Kuhlenschmidt

Department of Biochemistry, School of Medicine, University of Pittsburgh,
Pittsburgh, PA 15261

Richard M. Iammarino and K.P. Brooks

Departments of Pathology and Biochemistry, West Virginia University,
Morgantown, WV 26506

Introduction

Alpha-1-antitrypsin ($A_1AT$) is the major protease inhibitor in human
plasma, and as its name implies, it inhibits serine active site proteolytic enzymes including elastase, collagenase, trypsin, chymotrypsin, and
leucocyte proteases (1). $A_1AT$ is a 48,000 dalton glycoprotein (13% carbohydrate) that is synthesized exclusively by the liver. $A_1AT$ deficiency is
a human genetic disease of major importance. The most common disease
variant, designated Pi ZZ, is associated with pulmonary emphysema and
hepatic cirrhosis. These individuals, and all who carry a single Z
allele, have hepatic inclusions of PAS-positive, diastase-resistant
globules which cross-react immunologically with $A_1AT$. The $A_1AT$-related
inclusions are enclosed in membranes and they lack sialic acid. Using an
immunoperoxidase staining technique, Huff and co-workers (2) found cross-
reactive material (CRM) in normal liver cells, liver cells of individuals
with a variety of other liver diseases as well as individuals with the
classic form of Pi ZZ $A_1AT$ deficiency. Phenotype characterization of the
first two groups shows absence of the Z gene product in serum.

Larsson and Eriksson (3) isolated $A_1$AT-like material from liver of Pi ZZ diseased individuals. They described a somewhat lower molecular weight, carbohydrate-deficient protein which cross-reacted immunologically with antibodies prepared against plasma $A_1$AT. No one has, as yet, reported the isolation of a liver protein with a greater molecular weight than $A_1$AT which might be a candidate for a precursor to the plasma protease inhibitor. With this in mind, we set out to isolate the major $A_1$AT-CRM in human liver.

In this report, we describe the isolation of a 68,000 dalton glycoprotein from human liver that cross-reacts with $A_1$AT antibodies. The relationship of this hepatic glycoprotein to $A_1$AT is unclear since the two glycoproteins are substantially different from each other. These data underscore the need for cautious interpretation of results obtained using immuno-chemical methods of protein identification.

Materials and Methods

Human $A_1$AT was purified to homogeneity from pooled plasma as described elsewhere (4,5) and antibodies to the glycoprotein in Freund's adjuvant were raised in New Zealand white rabbits. The isolation of the intrahepatic $A_1$AT-CRM was initiated by homogenizing 400 grams of human liver, obtained at autopsy within five hours of death, in 2 L of distilled deionized water with the aid of a Waring blender. After filtering through cheesecloth the homogenate was centrifuged at 100,000 x g for 1 h. Unless specified otherwise, all procedures were performed at 1-4°C. The super-natant fraction was loaded onto a DEAE-cellulose column as described in

Figure 1. The fractionation of $A_1AT$-CRM was monitored by rocket immuno-electrophoresis. The various steps in the purification scheme are shown below; these include column chromatography on DEAE-cellulose, QAE-Sephadex (A-50), hydroxylapatite and Sephadex G-200 and isoelectric focusing. Protein was determined by the procedure of Lowry _et al._ (6) using bovine serum albumin as standard.

Table I

Summary of the Purification of Alpha-1-Antitrypsin Cross-Reactive Material from Human Liver

| Fractionation Step | A. Total Protein (mg) | B. $A_1AT$-related material (mg) | B/A x 100 |
|---|---|---|---|
| 1. High-speed supernatant | 21,300 | 48.0 | 0.23 |
| 2. DEAE-cellulose | 6,410 | 48.0 | 0.75 |
| 3. QAE-Sephadex | 438 | 15.8 | 3.6 |
| 4. Hydroxylapatite | 157 | 12.5 | 8.0 |
| 5. Sephadex G-200 | 90 | 7.2 | 8.0 |
| 6. Isoelectric focusing | 60 | 5.7 | 9.5 |

The purification scheme was initiated using 400 g of liver. The content of $A_1AT$-related material in human liver is 0.12-0.48 mg/g wet weight (mean 0.3 mg/ml). The highest purity preparation of intrahepatic $A_1AT$-CRM had approximately 10% of the immunoreactivity of pure human plasma $A_1AT$.

Results and Discussion

The isolation scheme yields approximately 60 mg of purified $A_1AT$-CRM representing an overall yield of approximately 10%. As indicated in Table I, the $A_1AT$-CRM has 9.5% of the immunoreactivity of plasma $A_1AT$. The purified hepatic $A_1AT$-related substance appears to be free of contaminating proteins by several criteria. The protein migrates as a single component in SDS polyacrylamide gels in the presence or absence of 2-mercaptoethanol (Figure 6); the protein migrates very close to bovine serum

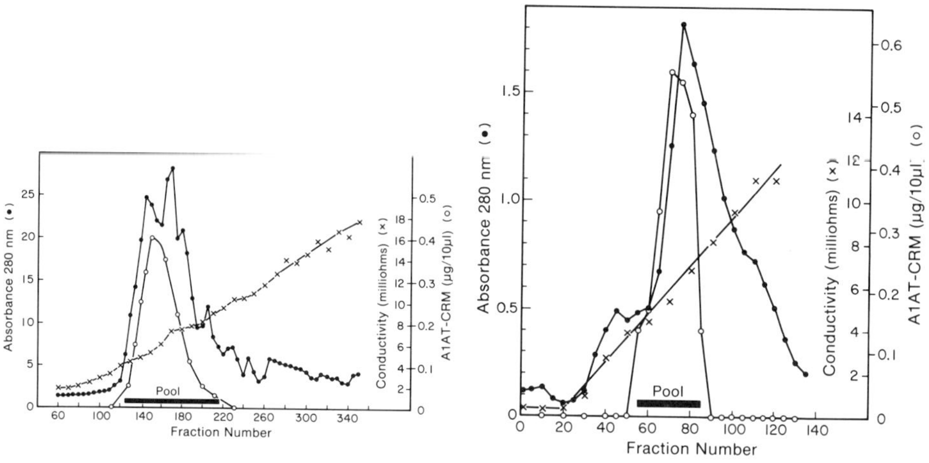

Figure 1

Figure 2

Figure 1 - DEAE-Cellulose Chromatography - After centrifugation as described in Materials and Methods, the supernatant was decanted and loaded on a DEAE-cellulose column (bed volume 2 L) equilibrated with standard Tris buffer (0.05 M Tris-HCl, pH 7.6) supplemented with 5 mM MgCl$_2$ and 5 mM 2-mercaptoethanol. The column was washed with 10 L of buffer and then developed with an 8 L linear gradient of NaCl (0-0.4 M). Fractions (18 ml) were collected and analyzed for protein (0), A$_1$AT-CRM (0), and salt concentration by conductivity (X). Fractions 125-218 were pooled and dialyzed against the standard Tris buffer.

Figure 2 - QAE-Sephadex Chromatography - The A$_1$AT-CRM from the previous column was loaded on a QAE-Sephadex column (4 x 35 cm) equilibrated in the standard Tris buffer. The column was washed exhaustively and developed with a 3 L linear gradient of NaCl (0-0.6 M) in the standard Tris buffer. Fractions (24 ml) were collected and analyzed as in Figure 1. The pooled fractions (45-85) were dialyzed against 0.01 M sodium phosphate buffer, pH 7.0, supplemented with 5 mM 2-mercaptoethanol.

albumin and considerably more slowly than plasma $A_1AT$ and the "apparent"
molecular weight of the hepatic $A_1AT$-CRM is 64,200 daltons. The purity of
the material was also indicated by the results of sedimentation equilibrium analysis according to Yphantis (15); a plot of ln Y (fringe displacement versus radius $^2$ yielded a straight line which is indicative of
size homogeneity (data not shown). Employing a partial specific volume of
0.736 ml/g estimated from the results of amino acid and carbohydrate
analysis (Table II), the molecular weight of the $A_1AT$-CRM was calculated
to be 70,200 daltons. Based on the results obtained from a number of
determinations by electrophoresis and sedimentation analysis, we estimate
the size of the $A_1AT$-CRM to be 68,000 daltons.

Despite the evidence for homogeneity, isoelectric focusing resolved the
$A_1AT$-CRM into 2 fractions (Figure 5); the more acidic species was about
twice as immunoreactive as the less acidic material.

The amino acid and carbohydrate composition of the hepatic $A_1AT$-related
glycoprotein is summarized in Table II. The $A_1AT$-CRM contains reduced
amounts of the same monosaccharides as plasma $A_1AT$. The results of
cyanogen bromide cleavage studies and ion-exchange chromatography on DEAE-
cellulose (data not shown), indicated that the hepatic glycoprotein
contains only 2 oligosaccharide chains in contrast to plasma $A_1AT$ which
contains three major glycopeptides when subjected to cyanogen bromide
fragmentation.

The possible relatedness of the 68,000 dalton hepatic glycoprotein to
plasma $A_1AT$ was investigated using fragmentation techniques involving

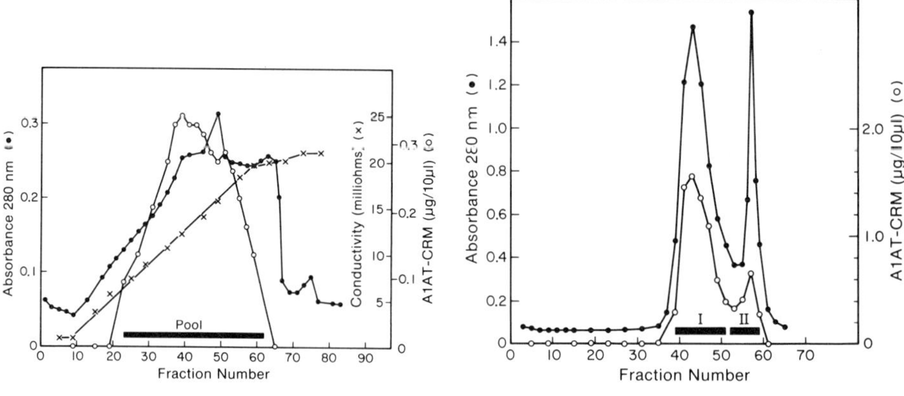

Figure 3

Figure 4

Figure 3 - Hydroxylapatite Chromatography - The $A_1$AT-CRM from the QAE-Sephadex column was chromatographed on a hydroxylapatite column (2 x 35 cm) prepared by mixing one part cellulose powder with 1 part hydroxyl-apatite powder in 0.01 M sodium phosphate buffer, pH 7.0. The column was developed with a 2 L linear gradient of ammonium sulfate (0-0.2 M) in the same phosphate buffer. Fractions (2.5 ml) were collected.

Figure 4 - Sephadex G-200 Chromatography - The $A_1$AT-CRM pool from the hydroxylapatite column was dialyzed exhaustively against sodium phosphate buffer, pH 7.6 containing 5 mM 2-mercaptoethanol, and loaded on a Sephadex G-200 column (2 x 135 cm) with a 5 cm layer of QAE-Sephadex at the top of the column to simultaneously concentrate and load the protein sample according to the method of Miller et al. (4). Fractions 39-51 were pooled (I), dialyzed against the same phosphate buffer and subjected to isoelectric focusing (Figure 5). Pool II was not studied further.

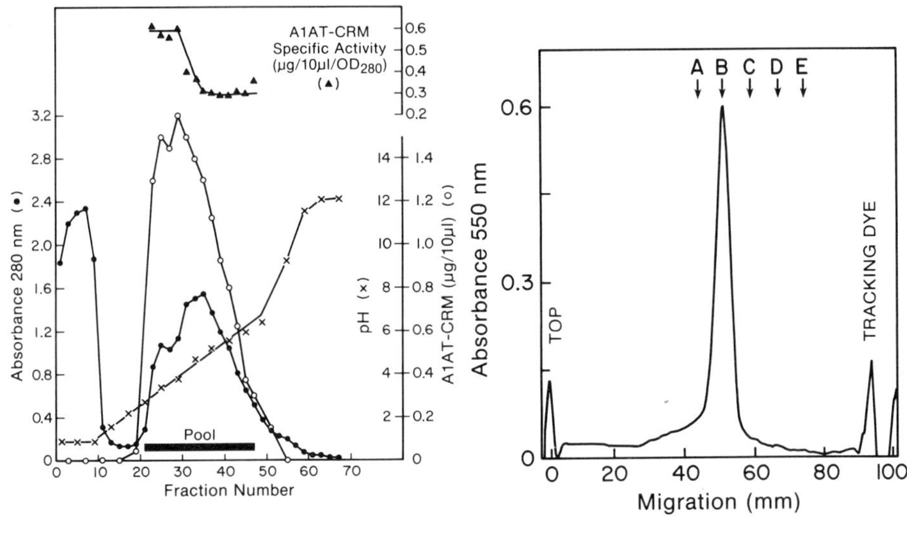



Figure 5                                         Figure 6

Figure 5 - Isoelectric Focusing - Preparative isoelectric focusing of the
$A_1$AT-CRM was performed using an LKB 8101 isoelectric focusing column
containing a 100 ml linear (0-50%, w/v) sucrose gradient and (1%, w/v)
pH 4.0-6.5 ampholynes.  The column was focused at 13 watts for 20 hours.
Fractions (1.2 ml) from the column were collected and analyzed for
$A_1$AT-CRM (●), protein (O) by absorbance at 280 nm and pH (X) using a glass
electrode.


Figure 6 - Polyacrylamide (25%) Gel Electrophoresis of the Purified
Preparation of $A_1$AT-CRM from Human Liver.  Electrophoresis was performed
in the presence of SDS and 2-mercaptoethanol according to Laemmli (7).
The standard protein markers are: A, phosphorylase B (94,000); B, bovine
serum albumin (68,000); C, ovalbumin (43,000); D, carbonic anhydrase
(30,000); and E, soybean trypsin inhibitor (21,000).  Based on the results
of 6 experiments, the "apparent" molecular weight of the hepatic $A_1$AT-CRM
was 64,200 daltons.  In the same gel, the molecular weight of human plasma
$A_1$AT was 47,000 daltons.  The results of the purification procedure are
summarized in Table I.  The $A_1$AT-CRM also runs as a single component when
electrophoresis is carried out in the absence of SDS (data not shown).

Table II

Amino Acid and Carbohydrate Composition of Intrahepatic Alpha-1-
Antitrypsin-related Material and Plasma $A_1AT$

| Amino Acid | Residues per 67,800 grams | |
|---|---|---|
| | $A_1$AT-CRM | Plasma $A_1$AT |
| Lysine | 60.0 | 33.0 |
| Histidine | 16.2 | 11.0 |
| Arginine | 27.2 | 8.0 |
| Aspartic acid | 61.2 | 29.0 |
| Threonine | 31.1 | 26.0 |
| Serine | 28.9 | 20.0 |
| Glutamic acid | 73.3 | 48.0 |
| Proline | 27.0 | 20.0 |
| Glycine | 21.3 | 23.0 |
| Alanine | 69.9 | 23.0 |
| Half-cystine | 2.6 | 2.0 |
| Valine | 36.1 | 22.0 |
| Methionine | 8.1 | 8.0 |
| Isoleucine | 16.4 | 14.0 |
| Leucine | 67.9 | 43.0 |
| Tyrosine | 17.7 | 7.0 |
| Phenylalanine | 34.0 | 24.0 |
| **Monosaccharide** | | |
| Mannose | 1.6 | 7.7 |
| Galactose | 2.1 | 6.6 |
| Glucosamine | 5.0 | 11.8 |
| Sialic acid | 3.3 | 7.3 |

The partial specific volume of the $A_1$AT-CRM from human liver was estimated
to be 0.736 ml/g based on carbohydrate and amino acid composition calculated according to Schachman (8) and Gibbons (9). Tryptophan was not
determined. The acid hydrolysates were prepared by digesting samples with
6 N HCl at 100°C in vacuo for 24 hours as described by Moore and Stein
(10). Analyses were performed on a Beckman Model 120 amino acid analyzer
according to the procedure of Spackman et al. (11). Half-cystine residues
were determined as carboxymethylcysteine.

Carbohydrate analysis was performed using an automated sugar analyzer as
previously described (12). Hydrolysis was performed in tubes sealed with
Teflon-lined caps in 2 N trifluoroacetic acid for 4 h at 100°C for neutral
sugars and in 4 N HCl for 6 h at 100°C for amino sugars. Sialic acid was
determined according to Warren (13) after hydrolysis of samples for 1 h at
80°C using 0.05 N $H_2SO_4$.

cyanogen bromide (Figure 7) and chymotrypsin fingerprinting (Figure 8).
Cyanogen bromide cleaved plasma $A_1AT$ into 4-5 discernable peptide fragments. However, many more fragments (approximately 10) were generated by
cyanogen bromide digestion of the $A_1AT$-CRM from human liver; furthermore,
few of these fragments corresponded to the cyanogen bromide-generated
peptide fragments from plasma $A_1AT$. Also, very dissimilar fingerprints
for the two glycoproteins resulted from incubation with chymotrypsin
(Figure 8). These results indicate that although the 68,000 dalton
hepatic glycoprotein reacts with antibodies directed against plasma $A_1AT$,
the two proteins do not share any significant portion of their primary
structures.

The hepatic $A_1AT$-CRM lacks antiprotease activity and, as shown above,
differs substantially from plasma $A_1AT$ with regard to molecular weight and
composition. Since the major criteria involved in the purification scheme
was immunochemical specificity employing an antiserum derived from rabbits
immunized with serum $A_1AT$ antigens, the observation of Huff _et al_. (2)
concerning the frequent finding of $A_1AT$-CRM in normal and diseased liver
may be explained by our findings. The hepatic glycoprotein which we
purified accounts for approximately 0.1% of the protein content of the
normal liver. The basis for the relatedness of the intrahepatic $A_1AT$-CRM
to plasma $A_1AT$ remains to be determined.

References

1. Morse, J.O.: N. Engl. J. Med. **299**, 1099-1105 (1978).

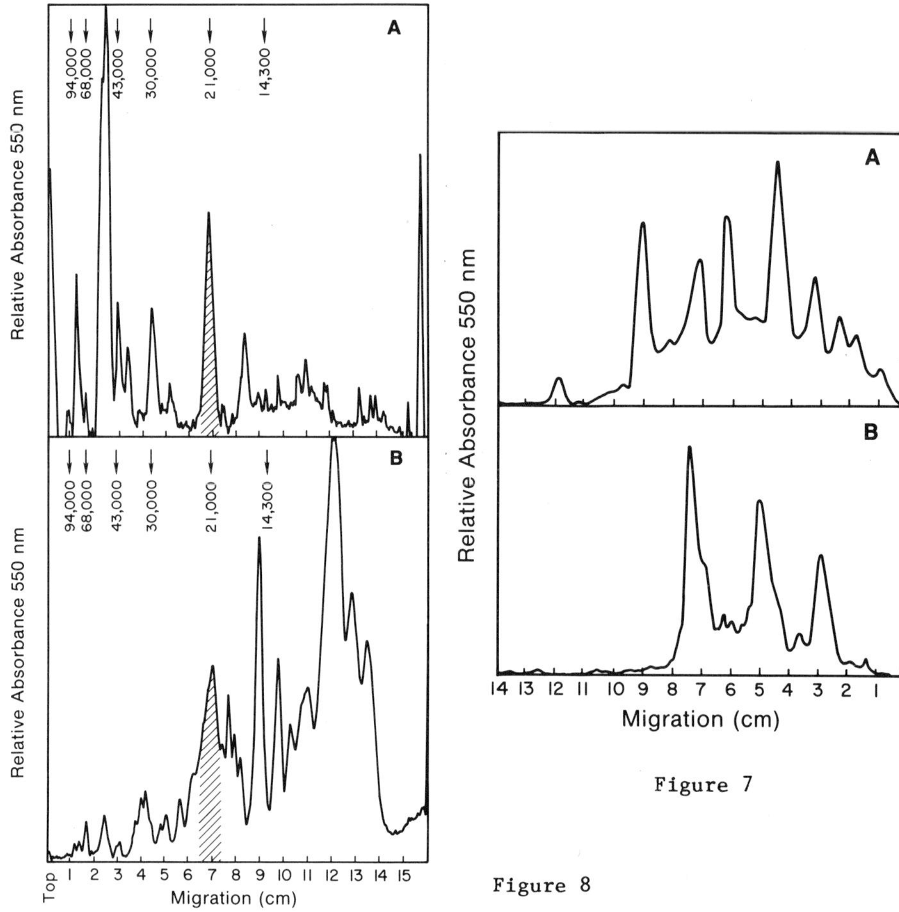

Figure 7

Figure 8

Figure 7 – Cyanogen Bromide Cleavage of Hepatic $A_1$AT–CRM and Plasma $A_1$AT – Purified preparations of the two proteins were subjected to cyanogen bromide cleavage (9) and polyacrylamide (20%) gel electrophoresis in the presence of SDS and 2-mercaptoethanol as described elsewhere (7): A, hepatic $A_1$AT–CRM; B, plasma $A_1$AT.

Figure 8 – Chymotrypsin Fingerprints of the Hepatic $A_1$AT–CRM and Plasma $A_1$AT – Purified preparations of A, $A_1$AT–CRM (20 µg) and B, plasma $A_1$AT (20 µg), were incubated with chymotrypsin (1 µg) for 16 h at 20°C and subjected to polyacrylamide (15%) slab gel electrophoresis in the presence of SDS and 2-mercaptoethanol according to the procedure of Cleveland et al. (14). Gels were stained with Coomassie blue and scanned with a recording densitometer. The gels are calibrated (arrows) with various protein standards. The shaded areas indicate the staining of chymotrypsin.

2. Huff, D.S., Punnett, H., Lischner, H. W.: Lab Invest. _38_, 389 (1978).

3. Jeppsson, J-O., Larsson, C., Eriksson, S.: N. Engl. J. Med. _293_, 576-579 (1975).

4. Miller, R. R., Peters, S.P., Kuhlenschmidt, M. S., Glew, R.H.: Anal. Biochem. _72_. 45-48 (1976).

5. Kuhlenschmidt, M. S., Coffee, C. J., Peters, S. P., Glew, R. H., Sharp, H. L.: in Proteases and Biological Control, edited by Reich _et al_., Cold Spring Harbor Symposium _2_, 415-428 (1975).

6. Lowry, O. H., Rosebrough, N. J., Farr, A. L., Randall, R. J.: J. Biol. Chem. _193_, 265-275 (1951).

7. Laemmli, U. K.: Nature _227_, 680-685 (1970).

8. Schachman, H. K.: in Methods in Enzymology (Colowick, S. and Kaplan, N., editors) Academic Press, New York _4_, 65-71 (1957).

9. Gibbons, R. A.: in Glycoproteins (Gottschalk, A., editor) Part A, Elsevier Publishing Co., New York, NY, 78 (1972).

10.Moore, S., Stein, W.H.: in Methods in Enzymology (Colowick, S. and Kaplan, N., editors) Academic Press, New York, _6_, 819-931 (1963).

11.Spackman, D. H., Stein, W.H., Moore, S.: Anal. Chem. _30_, 1190-1206 (1958).

12.Lee, Y. C., Johnson, G. S., White, B., Scocca, B.: Anal. Biochem. _43_. 640-643 (1971).

13.Warren, L.: J. Biol. Chem. _234_, 1971-1975 (1959).

14.Cleveland, D.W., Fisher, S. G., Kirschner, M.W., Laemmli, U.K.: J. Biol. Chem. _252_, 1102-1106 (1977).

15.Yphantis, D. A.: Biochem. _3_, 297-317 (1964).



Acknowledgement: This work was supported by a grant (AM-26916) to R.H.G. from the National Institutes of Health, U. S. Public Health Service.


Abbreviations:
| | |
|---|---|
| alpha-1-antitrypsin | $A_1AT$ |
| quaternary aminoethyl | QAE |
| diethyl aminoethyl | DEAE |
| cross-reacting material | CRM |
| sodium dodecyl sulfate | SDS |

CK-MB BY AN IMPROVED ELECTROPHORESIS METHOD:  A CLINICAL EVAL-
UATION OF THE METHOD


Harvey L. Kincaid

Children's Hospital National Medical Center and George Washington University
School of Medicine, Washington, D.C.


Introduction

Creatinine kinase (CK)(EC 2.7.3.2) exists in 3 different isoenzymic
forms (MM, MB and BB).  Differences in the CK isoenzyme composition of
various tissues include:  CK-MM in skeletal muscle and the myocardium;
CK-MB in the myocardium in significant amounts; CK-BB predominantly in
brain tissue and as a feature of certain neoplasms.  Serum CK-MB in the
differential diagnosis of acute myocardial infarction (AMI) is well
established (1,2,3).  There is a considerable amount of controversy,
however, over which method to use.  Chromatographic and electrophoretic
methods separate CK-MB prior to measuring its activity.  This study
compares a commercially available column method and an in-house column
method modified after the original method of Mercer (4) with an improved
electrophoretic procedure(5).

Materials and Methods

Blood specimens were drawn from hospital patients who were suspected of
having an AMI.  Other sera included those of hospital patients and
laboratory personnel having both elevated and normal total CK activity.

524

The column chromatographic procedure was carried out according to the manufacturer's instructions (Worthington Diagnostics, Freehold, NJ). CK assays were performed on an ABA 100 discrete analyzer at 37 °C (Abbott Diagnostic, Chicago, IL) using Worthington reagent. DEAE-Sephadex A-50 ion exchange columns were also prepared. To insure a complete separation of the isoenzymes, two additional "washout" fractions were collected; one in-between the CK-MM and CK-MB fractions and the other in-between the CK-MB and CK-BB fractions. Activity measurements were performed on a Beckman Model 25 recording spectrophotometer and a Rotachem II-a (Travenol Laboratories, Columbia, MD) centrifugal analyzer.

A thin film agarose support media was used for the electrophoretic separation (Coring Special Purpose Film, Corning Medical, Medfield, MA). Aliquots (2 uL) of patient sera or control were applied to the thin film agarose and the electrophoretic separation was carried out at 90V (constant voltage) for 20 minutes in a Corning casette with MOPS buffer (0.05 mol/L, pH 7.8). The agarose plate was removed and placed on a larger piece of moistened filter paper with the gel surface facing up. Numbered strips of a special grade of filter paper (Whatman Chromatography No. 542) were soaked in substrate reagent (Beckman Instruments, Berea, CA) and then placed onto the correspondingly numbered electroporesis channels of the gel. After (30 minutes, 41 °C), the strips of filter paper were removed from the gel surface, placed in an aqueous-methanol mixture (80:20) for 1-3 minutes and air dried without blotting. The strips were scanned on a Beckman CDS-100-F computing densitometer in the fluorescent mode and the amount of CK-MB reported as a percentage of the total fluorescence in the pattern.

Technical Results

The chromatographic method was capable of measuring CK activity in column eluates when a sample's total CK was less than 50 IU/L 30 °C. In contrast, the electrophoretic method proved imprecise in this range. The electrophoresis procedure became reliable when the sample's total CK reached a value of 80 IU/L (30 °C). Samples having a total CK above 450 IU/L (30 °C) required dilution because it was shown that the method will over-estimate the amount of CK-MB. The fluorescent scans obtained with the paper strips were compared with those obtained from scanning the gel itself. Problems with non-specific fluorescence (8) and imperfections in the gel were encountered when the gel was scanned directly. Any non-specific fluorescence present in a patient's sample remained fixed in the gel, however, and was not transferred onto the overlay paper. In the absence of interferences and artifacts, quantitation of CK-MB by scanning the gel compared favorably ($\pm$ 2%) with the percentage obtained by scanning the paper strips and indicating that there was a quantitative trapping of the reduced NAD in the paper. Similarly, quantitation of CK-MB by the overlay technique compared favorably (1-2%) with CK-MB by column in most cases. However, an apparent altered column separation of the isoenzymes was demonstrated with certain patient samples (see below).

Day-to-day precision for the electrophoresis method was determined using 2 different lots of lyophilized control material (Ortho Diagnostics) and a coefficient of variation of less than 10% was obtained for the quantitation of the CK-MB and CK-MM isoenzymes. The estimation of the percentage of CK-MB was shown to be linear over a range from 3 to 35%.

Diagnostic Results

The CK-MB data obtained for 36 patients were correlated with their diagnosis relative to AMI. Serum specimens from an additional 45 hospital patients were also analyzed by both methods. Using a CK-MB value of greater than 6% as being indicative of myocardial damage, the sensitivity and specificity (6) for the electrophoresis-overlay method was 91% and 84%, respectively. The figures for the column method were 82% and 84%, respectively. Some samples gave widely discrepant results. One patient had an anomalous isoenzyme that gave a false positive result by the column method.

Discussion

Electrophoresis easily distinguishes CK-MB from atypical isoenzymes because the patient's isoenzyme pattern is observed and directly compared with that of a control that is run in parallel. Three patients were encountered in the course of this study that demonstrated a CK-MB by column which was significantly lower than that obtained by electrophoresis. For 2 cases, the discrepancy was accounted for by measuring the activity eluted as CK-BB. The third case (TT) had 16% CK-MM, 0% CK-MB, and 77% CK-BB by the column method. The sum of CK-MB (21%) and CK-BB (55%) by electrophoresis was within 1% of the sum of CK-MB (0%) and CK-BB (77%) by column. It appeared that in this case as well, CK-MB was eluted from the column in the CK-BB buffer. This patient had an advanced oat cell carcinoma and the large percentage of CK-BB is explained as the product of the metastatic disease process (7).

This study demonstrates that diagnostically, reliable results are achieved by measuring CK-MB in terms of relative fluorescence. The separated isoenzymes in the gel react with the substrate in the overlay paper and gives a semi-quantitative estimate of the percentage of CK-MB when the paper is scanned. The technique has the advantage in that non-specific fluorescence remains within the gel and is not transferred onto the overlay paper. The procedure is sensitive, precise and correlates well with other evidence of myocardial damage. It can not reliably quantitate CK-MB in samples having a total CK in the lower half of the normal range, however. In the upper half of the normal range (60-100 IU/L, 30 °C), the method can reliably detect about 4 IU/L of CK-MB when careful attention is taken in the application of the samples and the overlay strips.

References


1.  Varat, M.A. and Mercer, D.W.  Cardiac specific creatine phosphokinase isoenzyme in the diagnosis of acute myocardial infarction.  Circulation 51, 855 (1975).

2.  Galen, R.S., Reiffel, J.A. and Gambino, R., Diagnosis of  acute myocardial infarction.  J. Am. Med. Assoc. 232, 145 (1975).

3.  Knottinen, A. and Somer, H., Specificity of serum cretine kinase isoenzymes in diagnosis of acute myocardial infarction. Brit. Med. J. 1, 386 (1973).

4.  Mercer, D.W., Separation of tissue and serum creatine kinase isoenzymes by ion-exchange column chromotagraphy.  Clin. Chem. 20, 36 (1974).

5.  Kincaid, H.L., A comparison of ion exchange chromatography with an improved elctrophoretic method for measuring CK-MB.  Clin. Chem. 24, 1034 (1978).

6.  Galen R.S. and Gambino, S.R., _Beyond and Normality:  The Predicative Value and Efficiency of Medical Diagnoses_.  John Wiley & Sons, 1975, pp. 10-19.

7.  Silverman, L.M., Dermer, G.B., Zweig, M., et al., Creatin kinase BB:  A new tumor associated marker.  Clin. Chem. _25_, 1432-1435 (1979).

DETECTION AND CHARACTERIZATION OF CYSTIC FIBROSIS PROTEIN EMPLOYING
ISOELECTRIC FOCUSING AND IMMUNOELECTROPHORETIC TECHNIQUES[1]


Gregory B. Wilson and Eugenia Floyd

Department of Basic and Clinical Immunology and Microbiology,
Medical University of South Carolina, Charleston, South Carolina 29425


Introduction

The cystic fibrosis protein (CFP) is a blood component with an isoelectric
point (pI) of~8.5 found specifically in sera from cystic fibrosis patients
(CF homozygotes, CFH) and in heterozygote carriers (HCF) of the gene defect
responsible for CF (1,2).  Several properties of CFP have been described
(1), and its value as a diagnostic marker for CFH and HCF subjects has been
confirmed by others (4,5).  One of us recently reported the development of
monospecific antisera to CFP raised in mice (6).  In the present
communication, we report the development of immunoassays for CFP, as well as
our progress toward determining the identity of CFP.  We have evaluated our
hypothesis (ref. 1) that the CFP may be structurally related to one or more
of the CF ciliary dyskinesia substances (CDS) which we have recently
characterized (7,8).

Methods

Counterimmunoelectrophoresis (CIE), crossed immunoelectrophoresis (CR-IEP),
and rocket immunoelectrophoresis (R-IEP) were performed following standard
procedures with minor modifications.  Except for CIE, 1.0% agarose with low
electroendostomic properties was used exclusively.  Isoelectric focusing
(IEF) for CFP detection, Bio-gel P-10 chromatography, and bioassays for the

---

[1]Publication no. 415 from the Department of Basic and Clinical Immunology
 and Microbiology, Medical University of South Carolina.

530

detection of CDS were performed as described elsewhere (2,7-9). Experiments to evaluate antibody neutralization of the CDS closely followed protocols described previously (8).

Results and Discussion

CFP occurs as a band doublet in the uppermost centimeter (pI 8.5) of the electrophoretogram resulting from focusing of serum (ref. 9 and Fig. 1A). To develop antisera to CFP, this region was excised from the unstained electrofocused gel, emulsified in phosphate-buffered saline (PBS, pH 7.4), and injected into mice. Immunogens were also prepared from material focusing in the same region which had been desorbed from protein A (3,10) or from serum components not bound to protein A (Fig. 1B). Analyses of sera from mice immunized with each of these immunogens (Fig. 2A-C) indicated that only mice which received the first two immunogens produced antibodies recognizing additional components in CFP-positive as compared to CFP-negative samples. We assume, therefore, that CFP adsorbs to protein A (directly or indirectly). In addition, analyses of the potency of the antisera produced indicated that emulsification of the excised gel in pH 7.4 buffer resulted in more potent antisera than emulsification in acetate buffer (pH 4.7) (6), everything else being identical. These results are consistent with a low molecular weight for CFP (M.W. <15,000; ref. 1); i.e., it is a better immunogen when bound to a carrier such as IgG.

From analyses of CFP-positive (CFH and HCF) samples and normal control (NC) sera by CIE using unabsorbed mouse antisera, we routinely found that CFP-positive samples produced more precipitation arcs than did CFP-negative samples (ref. 6; Fig. 2E). Absorbed antisera produced one major and at times one obvious minor arc against CFP-positive sera, whereas NC sera failed to show any precipitation arcs (Fig. 2F). Quantitative differences were at times evident between the levels of those components (Fig. 2F), in CFH and HCF sera (CFP and ?) as noted by the location of the arcs relative to the cathode (2). An R-IEP assay was therefore set up for better quantitation of the levels of these components. By CR-IEP analysis following electrofocusing (Fig. 3), we found that mouse IgGs have a pI range of about 5.0 to 7.7. R-IEP was therefore run at pH 6.5 to minimize

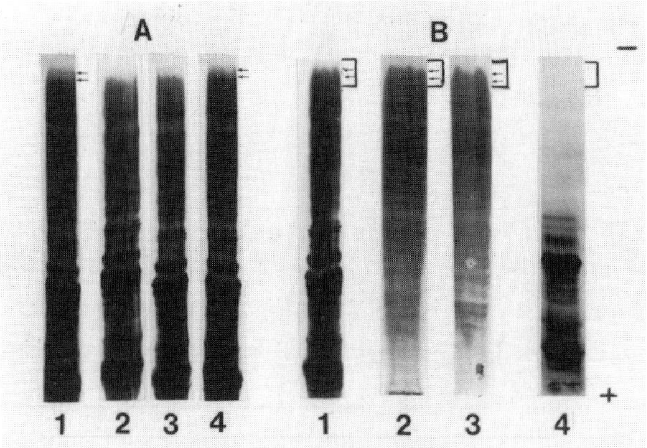

Fig. 1. IEF in thin-layer polyacrylamide gel (pH 2.5-10; 4 M urea). Anode at bottom. (A) Whole serum. CFP band doublet indicated by arrows (pI ~8.5). (1,4) CFP-positive, (2,3) CFP-negative. (B) Immunogens used to obtain anti-CFP antibodies. (1) Whole serum, (2,3) serum components adsorbed to protein A, (4) serum components not bound to protein A. The area used is enclosed in a bracket.

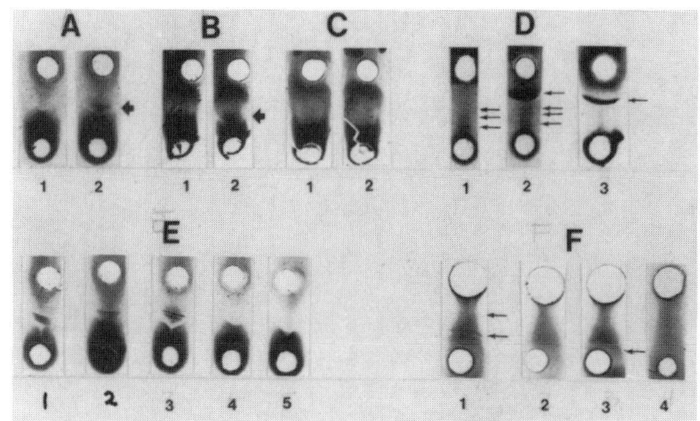

Fig. 2. Reactivity of immune mouse antisera shown by CIE. (A,B,C) Precipitation patterns obtained for mice injected with immunogens 1, 2, and 4, respectively, from Figure 1B. (1) CFP-negative, (2) CFP-positive serum. (D) Patterns obtained for reaction of mouse antisera with fractions of CFP-positive serum. (1) M.W. 3,500-15,000 components desorbed from protein A and isolated by Bio-gel P-10 chromatography. (2) Extract of CFP region of an electrofocused gel. (3) M.W. >20,000 components desorbed from protein A and isolated by Bio-gel P-10 chromatography. (E) Reactions for 3 CFP-positive (1,2,3) and 2 CFP-negative serum samples when tested versus mouse antisera (immunogen 1, Fig. 1B). (F) Reactions between absorbed mouse antisera (immunogen 1, Fig. 1B) and CFH (1), HCF (3), and normal control sera (2,4). Mouse antisera in A to E were not rendered monospecific. Barbital buffer, pH 8.6 (ionic strength 0.10).

the mobility of mouse IgG (Fig. 3). At this pH we could easily quantitate
the level of human IgG in serum (Fig. 4A), and we obtained at least
3 rockets for CFP-positive samples (Fig. 4B). However, most <u>unabsorbed</u>
mouse antisera failed to produce rockets specific for CFH and HCF samples
when whole sera were studied (even with 10% antisera). By conducting a
series of interrelated experiments (details to be published elsewhere), we
determined the major reasons for this failure:  (a) specific components
(e.g., CFP) can be obscured by the presence of rockets for unrelated
components; (b) high concentrations of antisera are required (>20%); and
(c) NC samples may generate cross-reacting components.

To counteract these problems, we have resorted to (a) using antisera
rendered selectively polyspecific by removing antibodies to specific serum
components or monospecific by using specially treated NC (CFP-negative)

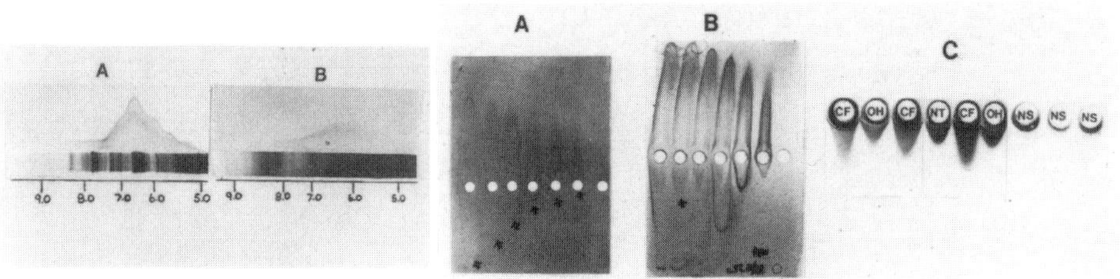

Fig. 3.  Left:  Determination of pI range for mouse IgG by IEF followed by
CR-IEP.  (A) Results when mouse serum was focused in a pH 2.5-10 gradient
containing 4 M urea.  (B) pH 2.5-10 gradient without urea.  First
dimension, anode to right (scale shows pH gradient).  Second dimension,
anode at the top.  A purified goat IgG fraction containing anti-mouse IgG
antibodies was used to obtain the patterns shown.  The effects of urea on
pI values were corrected for in A.  Fig. 4.  Right:  Analysis of reactivity
of mouse antisera by R-IEP.  (A) Plate showing quantiation of human IgG
using 1.0% mouse anti-human IgG antiserum.  Arrows show tip of cathodal
rockets.  20 µl of normal serum was cut 1/10 and serially diluted to 1/320.
(B) Unabsorbed mouse antisera (2.5%) (immunogen 2, Fig. 1B).  Wells
contained 20 µl human serum (CFP-positive) diluted 1/5 (first two wells)
then serially diluted to 1/80.  Arrow denotes a second cathodal rocket not
seen in A.  Far right well in A and B is a buffer-only control.
(C) Quantitation of material believed to contain CFP, obtained from sera of
CFH (CF), HCF (OH), normal controls (NS), and normal serum treated as
described in the text (NT).  The gel contained 25% absorbed anti-CFP mouse
antisera.  Buffer was sodium barbital - sodium acetate, pH 6.5 (ionic
strength 0.02).  Cathode at the bottom.

sera, and (b) developing methods to obtain CFP (and other relevant markers) quickly, free of contamination from other serum components, and to stabilize serum samples to prevent the production of increased amounts of CF-related markers. Similar techniques were applied to analyses by CIE. As shown in Fig. 4C, quantitative differences between CFH, HCF, and NC samples were obtained by R-IEP. Our data currently indicate that at least two components are present at higher levels in CFH and HCF samples, depending on the antisera developed and conditions used for analyses, and that at least one of them may be generated in NC samples by specific treatments (G.B. Wilson, unpublished observations) (Fig. 4C). We are currently exploring more fully the specificity of both these components (markers) for CF genotypes. It is entirely possible that each could represent one band of the CFP band doublet.

Our attempts to identify the CFP have centered in part on evaluating its possible co-identity with one of the CDS recently described by our group (7,8). Our analyses of Bio-gel P-10 fractions by CIE indicated that at least three precipitation arcs were due to components in the M.W. range of the CDS (M.W. 3,500-15,000; Fig. 2D). CD activity and at least 4 components which reacted with mouse antisera were also present in gel extracts of the CFP-containing region (Fig. 2D). These results, which are consistent with our previous findings concerning the physicochemical properties of the CDS and CFP (1), strongly indicate that most of these substances are fragments of larger serum components (1,8). If so, their hyperaccumulation in serum is undoubtedly linked to a defect in CF alpha2-macroglobulin (A2M) entailing either an abnormality in A2M-protease binding affinity or in the carbohydrate structure of A2M (the latter resulting in reduced endocytosis and therefore clearance of "macroprotease complexes" containing A2M by CF monocyte-macrophages) (3,8). To link CFP more firmly to one of the CDS, we attempted to neutralize the purified CDS activities with mouse antisera (prepared by injecting immunogen B-1 or B-2 in Fig. 1). In our initial experiments we could neutralize the CF-specific CDS activity (M.W.∼5000) and the C3-CDS (M.W. 9000) activity produced by mononuclear cells (MNC); however, we could not neutralize the C5-CDS (M.W. 15,000) activity (7,8). Although we are reluctant to conclude at this time from these experiments that the CFP doublet bands represent the first two

534

CDS activities, we are actively pursuing this as one viable possibility.
At the very least, these results suggest that the CF-specific CDS and
C3-CDS apparently share partial or complete identity with components
which focus in the same pI range as CFP, and that specific antibodies
can potentially be obtained to both of these CDS by methods similar to
those employed to produce specific antibodies to CFP.  We are currently
isolating all of the CDS from serum and cell culture medium, using IEF,
protein A, and lectin affinity chromatography, and raising antibodies to
them.

More recently, we have succeeded in increasing the sensitivity of our
immunoassays by employing $^{125}$I-labeled reagents.  Preliminary experiments
using a radio-R-IEP assay with mouse antisera of potency similar to that
employed in Fig. 4C indicate that we can use as little as 0.5% monospecific
antisera (Fig. 5).  To increase the potency (titer) of our antisera still
further, we are coupling diagnostic components to larger M.W. carriers
prior to injection in mice (or other animals) to increase their
immunogenicity.  Even at its current level of sensitivity, our radio-R-IEP
assay can be employed as an aid to detect and/or quantitate CFP and other
appropriate or potential diagnostic markers for CF in white blood cell and
skin fibroblast cultures and in amniotic fluid.

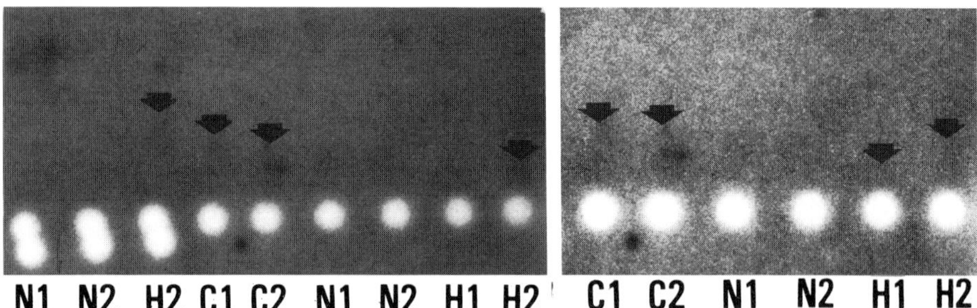



Fig. 5.  Left:  Quantitation of material containing CFP from sera of
normal controls (N), heterozygote carriers for CF (H), and CF
homozygotes (C) by radio-R-IEP.  The gel contained 0.5% mouse anti-CFP
antiserum.  The single wells contained one-third as much material as the
double wells.  Note that both normal control samples failed to react
with the mouse antiserum.  The tips of the cathodal rockets are
indicated by arrows.  Right:  Enlargment of wells 4-9.

Acknowledgments

We thank Charles L. Smith for excellent editorial assistance and
Geri Limehouse for photographic printing.

References


(1)  Wilson, G.B., Fudenberg, H.H.:  Pediat Res. 11, 317-324 (1977).
(2)  Wilson, G.B., Fudenberg, H.H., Jahn, T.L.:  Pediat Res. 9, 635-640
     (1975).
(3)  Wilson, G.B.:  Pediat. Res. 13, 1079-1081 (1979).
(4)  Nevin, G.B., Nevin, N.C., Redmond, A.O., Young, I.R., Tully, G.W.:
     Hum. Genet. 56, 387-389 (1981).
(5)  Manson, J.C., Brock, D.J.H.:  Lancet i, 330-331 (1980).
(6)  Wilson, G.B.: Lancet ii, 313-314 (1980).
(7)  Wilson, G.B., and Bahm, V.J.: J. Clin. Invest. 66, 1010-1019 (1980).
(8)  Wilson, G.B., Fudenberg, H.H., Parise, M.T., Floyd, E.:  J. Clin.
     Invest. 67, in press (1981).
(9)  Wilson, G.B., Arnaud, P., Fudenberg, H.H.:  Pediat Res. 11, 986-989
     (1977).
(10) Wilson, G.B., Fudenberg, H.H.:  J. Lab. Clin. Med. 92, 463-482 (1978).

# IMMUNOFIXATION - AGAROSE ISOELECTRIC FOCUSING TECHNIQUES FOR SCREENING GLIAL TUMOR CELL CULTURES FOR GLIAL MARKERS


Eugene A. Quindlen, Paul E. McKeever, Paul L. Kornblith
Surgical Neurology Branch, National Institute of Neurological
and Communicative Disorders and Stroke, National Institutes of
Health, Bethesda, Maryland 20205 U.S.A.


## Introduction

Characterization of tissue culture cell lines is of prime
importance in any research performed with tissue culture cells.
Some cells may produce marker proteins specific for the cell
of origin. We have performed experiments in the application
of agarose and agarose-polyacrylamide (PA) isoelectric focusing (IEF) to the analysis of glial tumor cells for glial
fibrillary acidic protein (GFAP) and S-100 protein. Direct
immunofixation after agarose IEF could not reliably detect
these protein markers because of isoelectric precipitation of
unfixed proteins. Agarose-PA IEF followed by crossed-immuno-
electrophoresis produced excellent resolution of tumor cell
extracts and quantitative identification of GFAP.

## Materials and Methods

Acrylamide and methylene Bis acrylamide were purchased from
Biorad Laboratories. LKB Ampholines were used in all experiments. Isogel and Seakem LE agarose and Gelbond polyester
films were obtained from Marine Colloids, FMC Corp. All
experiments were performed with an LKB multiphor apparatus
cooled with a Neslab RTE-4 refrigerated circulator.

538

Agarose isoelectric focusing was performed with 12 cm x 14 cm
plates of .8% Isogel agarose poured to a .8 mm thickness.  The
gel solution contained 2.5% pre-blended Ampholines 3.5 - 9.5
and 5% glycerol.  Acetic acid and sodium hydroxide .5M served
as the anolyte and catholyte respectively.  Samples were
applied at the cathodal end of the gel and IEF carried out for
1 hour at 7 watts/per plate at 10C°.  Ferritin (Horse spleen)
and hemoglobin were used as colored markers.  Direct immunofixation after agarose IEF was performed after the technique
of Ritchie and Smith (1).  The gel was then pressed and
washed for 24-48 hours in phosphate buffered saline (PBS),
pressed, dried and stained with .25% coumassie blue-R in 10%
acetic acid 30% methanol.


Agarose-polyacrylamide gels were .5% in Isogel agarose and
2.5% polyacrylamide with 3% of the cross-linker Bis.  After
a .7% hot agarose solution was made and cooled to 60°C the
solution was made 5% in glycerol, 2.5% in acrylamide, 2.5%
in Ampholines and $2 \times 10^{-5}$% in riboflavin 5- phosphate.  The
mixture was pipetted under a glass plate and a polyester foil
with spacers so that a .8 mm gel was produced.  Photopolymeri-
zation was initiated after the agarose had gelled and continued
for two hours.  After the gel was refrigerated overnight,
electrode wicks just saturated with 1M phosphoric acid and 1M
sodium hydroxide were applied.  The samples were pipetted onto
filter paper squares near the cathode and focusing carried out
at 1.0 watts/1.2cm gel length constant power for 1.5 hours.
Ferritin, hemoglobin and Pharmacia PI calibration markers 3-10
served as internal pI markers in the plate.  After focusing,
the gel was fixed in 10% trichloroacetic acid for 20 minutes,
washed in 10% acetic acid - 25% ethanol for 30 minutes and
then air dried overnight.  The dried gel was then stained with
0.25% coumassie blue-R, in 30% methanol, 10% acetic acid - 25%
ethanol and air dried at room temperature.

Human GFAP was prepared from a human fibrillary astrocytoma
K.H. and from human spinal cord using Eng's technique (2)
modified by Palfreyman (3). Anti-sera to GFAP was produced
in rabbits as described elsewhere (3). Bovine S-100 was
prepared by the technique of Levine (4) and anti S-100
produced in rabbits with the purified S-100 absorbed to
methylated bovine serum albumin. These antisera were used in
direct immunofixation at a dilution of 1:1 and in crossed
immunoelectrophoresis (CIE) at a concentration of 1-2%. CIE
plates were prepared using Seakem agarose LE 1% in a .01M
TRis-glycine buffer pH 8.6. A strip from an IEF gel was cut
out with scissors and laid directly on the lower half of the
CIE plate which contained no antibody. Electrophoresis into
the upper half of the plate containing antibody was carried
out for 14-16 hours at 1v/cm. The plates were then treated
as the agarose IEF plates above.


Tissue cultures of human glial tumors and human skin fibroblasts were established by placing 1 mm explants from fresh
biopsy material into 75 cms$^2$ plastic culture flasks with 10 ml
of Ham's F-10 medium containing 10% fetal calf serum. After
confluency was attained the cultures were passaged serially
by trypsinization and dilution with the same medium.


Cultures to be analyzed by IEF were washed x 2 with Hank's
Balanced Salt Solution, trypsinized and centrifuged at 1000 g
for 10 minutes to obtain a cell pellet of 200 ul representing
6-9 x $10^6$ cells (3, 75 cm$^2$ flasks). An equal volume of .001 M
Tris buffer pH 8.0 was added to the pellet and the pellet
sonicated for 15 seconds with a Branson sonifier with a
microtip and 600 ul of 10M urea, .5% 2-mercaptoethanol in
.001 M Tris buffer was added to the cell lysate. The mixture
was incubated at room temperature for 30 minutes and then
centrifuged in a Beckman air-fuge at 100,000  g for 15 minutes

in an A-100 rotor.   This procedure solubilized > 80% of the
total cell protein.   Protein content was determined to be
3-4 mg/ml by the Bradford assay (5).


Results


Initially, glial cells were solubilized in 0.5% Triton X-100
or sonicated and the water soluble extract used for agarose
IEF.   Although Triton solubilized 90-95% of the cell protein,
the IEF patterns obtained were smeared and showed poor
resolution.   This was not improved with the inclusion of
Triton X-100 in the gel.   Proteins and lipoprotein precipitated regardless of the application point of the sample (Fig 1).
The water soluble extract could be resolved into many distinct
bands but contained only 50% of the total cell protein and as
was found later, little of the intracellular GFA.


Direct immunofixation after agarose IEF was tested using
ovalbumin and anti-ovalbumin antisera (Miles Lab.) and found
to be quite sensitive, detecting 50 ng. of ovalbumin.   However,
when the same technique was applied to IEF patterns of glial
tumor cell Triton X-100 or 6M urea extracts, specific precipitation could not be surely detected.   Occasionally some bands
were deeper staining but there was extensive isoelectric
precipitation from pI 5-6.5 of control lanes, which received
no anti-sera (Fig 2).   Extensive washing of the plate with
0.3M saline, with or without Triton X-100 failed to remove
the precipitated proteins.   When the plate was press-blotted
with nitrocellulose membrane all of the soluble marker
proteins were completely removed from the agarose, but little
of the cell proteins.

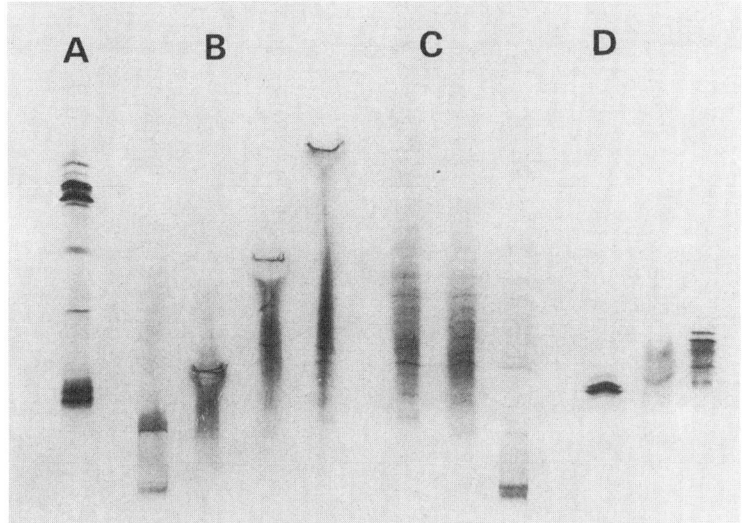

Figure 1.  Agarose IEF 3.5 - 9.5  A)  Hgb-ferritin markers
B) Triton X-100 extract of C-158 glial cells applied at
different positions, 75 µg each lane.  Note streaking and
precipitation C) "water soluble" extract of C-158 cells
75 µg D) ovalbumin marker.

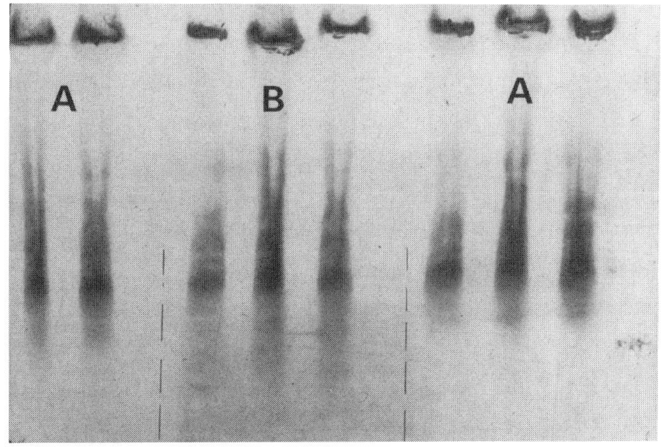

Figure 2.  Agarose IEF - Direct immunofixation of Triton X-100
extracts of C-158 cells  A) Control lanes of 100 µg extract
B) 100 µg of cell extract which received overlay of 75 µl antiserum.  Note extensive precipitation in control and test lanes,
despite 24 hour saline wash.

542

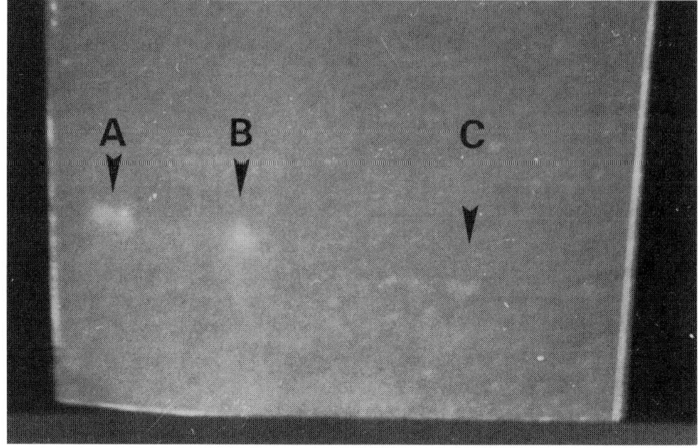


Figure 3. Agarose IEF - Indirect immunofixation with anti-
rabbit FITC anti-serum. A) 8 mg of semi-purified GFAP
B) GFAP staining of triton extract of K.H. glial tumor cells.
C) ovalbumin 25 nanograms, stained with FITC second antibody
as a marker.


When indirect immunofixation was carried out using goal anti-
rabbit IgG FITC or peroxidase conjugated anti-sera as the
second antibody, GFAP could be detected after agarose IEF
(Fig 3). However with the use of a second antibody and the
use of more extensive washing a single experiment could take
several days to complete. In addition, only one of five glial
cell cultures could be found to have GFAP by this technique
using water soluble or Triton extracts. This prompted the
search for a different technique of cell solubilization which
might improve the yield and resolution of GFA. When the cells
were treated with 6M urea 0.5% mercaptoethanol approximately
80% of the cell protein was solubilized. This urea extract
gave uniform and reproducible patterns on IEF composite gels
of 0.5% agarose - 2.5% polyacrylamide. The resolution of the
various bands was enhanced considerably when even a small
amount of polyacrylamide 1-2% was added to the agarose gel.
At higher gel concentrations > 3-5% smearing of the high
molecular weight filament proteins was found. Increasing

the urea concentration to 7-10 M, while solubilizing more of
the total cell constituents, had the disadvantage of producing
more precipitation at the sample application site in a gel
without urea.

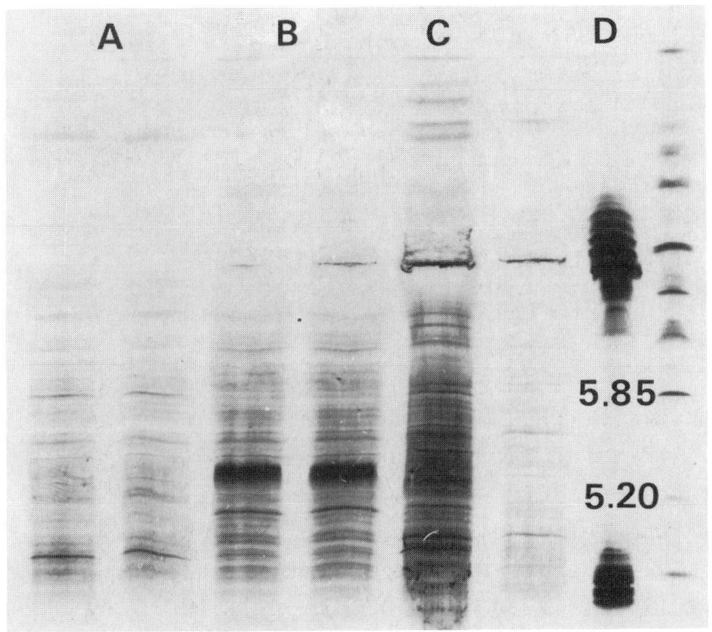


Figure 4. Agarose - PA IEF 3.5-10 of various extracts of L.M.
glial cells. A) "water"-soluble extract  B) 6M urea extract,
note dense band at pI 5.4-5.6 containing filament proteins and
little streaking or application site precipitation  C) .5%
Triton X-100 extract.  Note precipitation and streaking but
more complete solubilization of total cell proteins  D) Hgb -
ferritin markers A,B, and C represent equal aliquots of extracts
made from same number of cells % protein solubilized/total was
50%, 85%, 92% respectively.

The most striking finding of IEF of urea-cell extracts was the
presence of a broad dense band of protein at pI 5.5 - 5.6
(Fig 4).  This corresponds to the pI of GFAP but undoubtedly
represents other proteins as well.  The band could be observed

in almost every urea cell extract tested including fibroblasts
(Fig 5). However, it was more prominent in extracts from
early passage glial cells and much less prominent in fibroblasts. Pretreatment of the sonicated cells with DNase and
RNase did not affect this band.


When partially purified GFAP was analyzed by IEF-CIE, a
preciptin peak corresponding to pI 5.6 was seen representing
GFAP (Fig 6). In addition a fainter preciptin peak extending
from pI 5.3 - 5.1 was seen and probably represents tubulin a
common contaminent of most GFAP preparations (6). CIE also
demonstrated that our anti-GFAP anti-serim also contained
some antibodies directed against tubulin.


6M urea extracts of glial tumor cells and normal human
fibroblasts were likewise focused in agarose-PA gels and subjected to CIE using anti-GFAP anti-serum (Fig 7). Two tumor
cell cultures of early passage could be shown to contain GFAP
by this technique. A precipitin peak corresponding to tubulin
could also be identified in these cell lines. Four other
tumor cell lines and two fibroblasts cell lines showed only
precipitin peaks corresponding to tubulin. Be measuring the
area of the precipitin peaks of known amount of partially
purified GFAP and the GFAP peak of cell line R-19 against the
same anti-sera in identical plate, R-19 was found to have
800 µg GFAP/gm tissue. This corresponds to levels of GFAP
found in human spinal cord and glial tumors (3).


S-100 protein could also be detected by CIE after agarose-PA
IEF. Again R-19 was the only cell line found to contain S-100.
After IEF, Bovine S-100 shows 2 major bands at pI 4.15 and 4.23
with other minor components in the same range. However, IEF-
CIE of R-19 6M urea extract demonstraed a precipitin arc

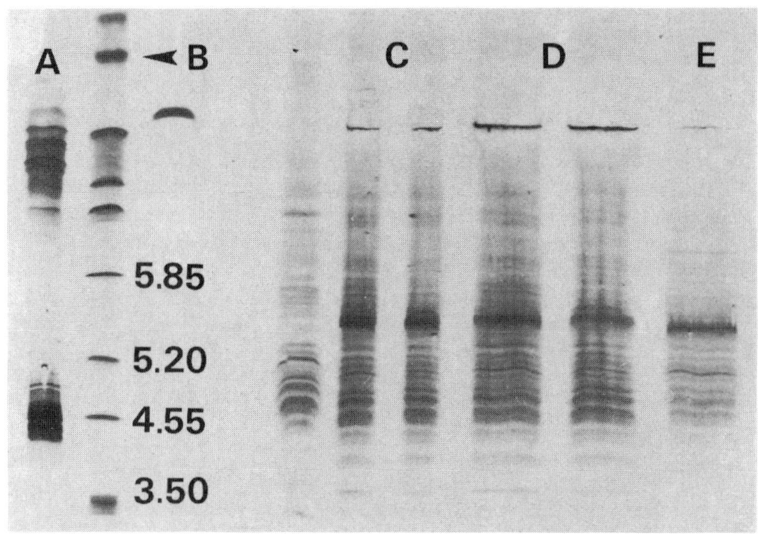

Figure 5.   Agarose - PA IEF 3.5 - 10 of 6 M urea extracts of
three different tissue culture cell lines.   125 µg wide lanes
75 µg narrow lanes.   A) Hgb-ferritin markers   B) Pharmacia
3-10 calibration markers   C) glial tumor cells R-19   D) glial
tumor cells C-158   E) normal human fibroblasts J.M.

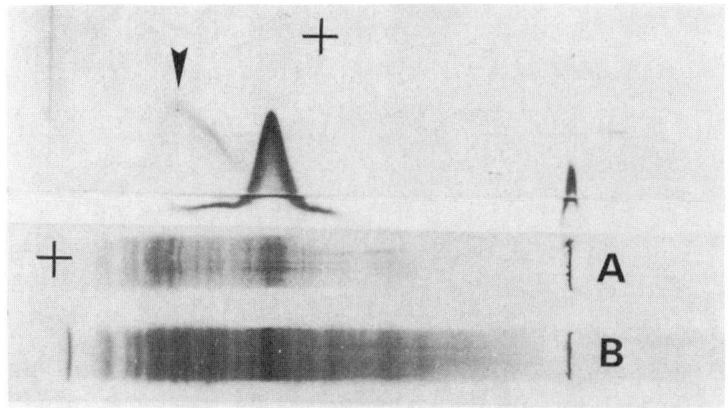

Figure 6.   Agarose - PA IEF and CIE with anti-GFAP anti-serum
of 8 µg of semi-purified GFAP lane  A); arrow indicates peak of
trace tubulin in the preparation.   B) parrallel lane of R-19
glial cell 6M urea extract run on same IEF plate shown for
comparison.

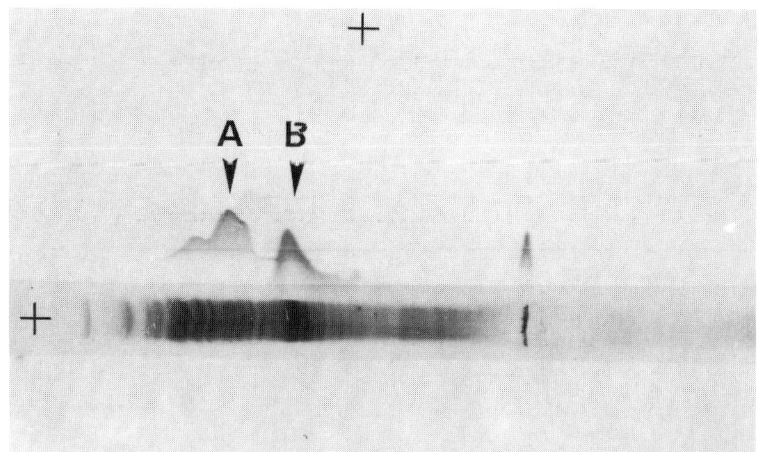


Figure 7. Agarose - PA IEF and CIE with anti-GFAP anti-serum
of 6M urea extract of R-19 glial tumor cells showing two
precipitin peaks  A) due to tubulin  B) peak due to GFAP.
Note characteristic precipitation of small amount of GFAP at
application point.

corresponding to a pI of 3.5 - 3.7.  This may indicate that
human S-100 is more acidic than the bovine S-100 but could also
be due to some carbamylation of the protein by urea.

Discussion

Despite its appealing simplicity and economy, agarose IEF
followed by direct or indirect immunofixation, was not an ideal
procedure for screening of our cell cultures for the presence
of glial markers.  The major difficulties included non-specific
isoelectric precipitation of proteins and perhaps the lack of
a suitably specific high titer anti-serum for GFAP.

Although agarose IEF and immunofixation could detect 50 nanograms of a soluble protein such as ovalbumin, we found the

greater sensitivity of CIE (7,8) was ideal for the detection of glial markers in addition to providing quantitative information.

Extraction of tissue culture cells with 6M urea, produced the most complete solubilization of the cell protein consistent with reproducible stable IEF patterns. Similar findings with tapeworm cell extracts have been reported (9). Analysis of the cell extracts by IEF in composite 0.5% agarose - 2.5% polyacrylamide gels produced optimum resolution compared to gels made from agarose of polyacrylamide alone. Such composite gels were easily dired, stained, destained and preserved and offer an ideal way to separate crude extracts of tissue containing proteins of a wide molecular weight range.

CIE after IEF detected impurities in the anti-GFAP anti-serum but produced good separation of tubulin from GFAP. The presence of anti-tubulin in the anti-serum served as a useful marker in the CIE experiment, since all of the cell lines tested contained tubulin. Two early passage (< 20) glial cell cultures were found to contain GFAP by the technique, while four other glial cell lines with extended passage in tissue culture did not. One of the cell lines that contained GFAP also had S-100 by CIE. If several different markers can be well separated by their pI on a wide range IEF plate, then CIE using mixed anti-sera could be utilized to screen for these markers in a single experiment.

548

References

1.  Ritchie, R. F., Smith, R.:  Clin. Chem. 22, 497-499 (1976).

2.  Eng, L. F., Vanderhaegen, J. J., Bignami, A., Gesstl, B.:
    Brain Res. 28, 351-354 (1971).

3.  Paffreyman, J. W., Thomas, D. G., Ratcliff, J. G.,
    Graham, D.I.:  J. Neurol. Sci. 41, 101-113 (1979).

4.  Dannies, P. S., Levine, L.:  J. Biol. Chem. 246, 6276-
    6283 (1971).

5.  Bradford, M.:  Anal. Biochem. 72, 248-252 (1976).

6.  Eng, L. F., p. 87.  Proteins of the Nervous System ed.
    Bradshaw, R. A., Schneider, D. M., Raven Press, N.Y., 1980.

7.  Johnson, A. M.:  Ann. Clin. Lab. Sci. 8, 195-201, (1978).

8.  Weeke, B.:  Scand. J. Immunol. 2, (Suppl. 1) 15-37 (1973).

9.  Kumaratilake, L. M., Thompson, R.C.:  Science Tools 26,
    21-24 (1979)

AGAROSE ISOELECTRIC FOCUSING AND RELATED IMMUNOCHEMICAL
TECHNIQUES IN THE ANALYSIS OF SERUM Gc-GLOBULIN


Jorge Lizana, Anthony Savill and Ingmar Olsson.
The Electrophoresis Group, Pharmacia Fine Chemicals, Box 175,
S-751 04, Uppsala, Sweden.


INTRODUCTION

The group-specific component (Gc-globulin), also known as the
vitamin D-binding protein, shows three common phenotypes (Gc
2-2, Gc 2-1 and Gc 1-1) reported to focus in a pH range of
about 4.8 to 5.1 (for review see ref. 1). Analysis of these
phenotypes is usually done by isoelectric focusing (IEF) in
polyacrylamide gels (2-7). Under these circumstances the Gc 1
allele is further resolved into two bands, known as Gc-1 Fast
and Gc-1 Slow. Thus, Gc-globulin can give six subtypes.

The easy of handling and casting gels of agarose, in addition
to its compatibility with subsequent immunological detection
make the use of this matrix a very attractive possibility in
routine analysis by isoelectric focusing. Such material is now
available (13), and this prompted us to use Agarose-IEF to
develop a routine isoelectric focusing method for the analysis
of Gc-globulin.

The resolving power of analytical IEF is very high (8), however
there are cases where the resolution obtainable with the
standard carrier ampholyte pH intervals commercially available
is not sufficient. Gc-globulin proteins which focus in a very
narrow pH interval constitute an example of this. The addition
of a "spacer ampholyte" (9) to the IEF gel has proven
successful for the separation of hemoglobin $A_1$ ($HbA_1$) and

glycosylated hemoglobin ($HbA_{1c}$). We decided to take a similar approach and in this report we present experiments to flatten the pH gradient by the addition of different chemical spacers into an agarose matrix. With this improved pH gradient we have developed a rapid and convenient Agarose-IEF method for the routine phenotyping of serum Gc-globulin.

## MATERIAL AND METHODS

Reagents. Carrier ampholytes Pharmalyte™ 2.5-5, 4-6.5 and 3-10, pI calibration kit 2.5-6.5, Agarose-IEF, Agarose type A (all from Pharmacia Fine Chemicals AB, Uppsala, Sweden). Coomassie Brilliant Blue R-250, TES:N-[2-Hydroxy-1,1-bis (hydroxymethyl) methyl] taurine, and MOPS:3-Morpholinopropane sulphonic acid (Fluka AG, Buchs, Switzerland). Sorbitol p.a (Merck, Darmstad, W.Germany). All other chemicals used were of analytical grade.

Serum Samples were obtained from healthy volunteers. Some typified sera were kindly provided by Mr. K. Nye, Dept. of Forensic Science, London Hospital Medical College, England. When not immediately used the sera samples were kept at $-20°C$.

Albumin-free serum  was prepared by passage of fresh serum samples (1 ml) through a column packed with Blue-Sepharose Cl-4B (14 ml) equilibrated with 0.05 M TRIS-HCl buffer pH 7.0 containing 0.1 M KCl as described elsewhere (10).

Apparatus. Isoelectric focusing was performed on a Flat Bed Apparatus FBE-3000 using the Electrophoresis Constant Power Supply ECPS 3000/150 connected to a Volthour Integrator VH-1 (all from Pharmacia). The pH measurements were done with an Ingold surface electrode.

<u>Gel casting and pH gradient</u>. Agarose-IEF gels (114x225 mm) were cast on polyester films (GelBond$^{TM}$, Marine Colloids, USA) following the instructions provided by the manufacturer: 0.3 g of Agarose-IEF, 3.6 g of sorbitol, 1.9 ml of Pharmalyte and 27 ml of distilled water. This gives a 1% Agarose gel 1 mm thick.

Preliminary experiments with different carrier ampholyte intervals showed that a mixture of three parts of Pharmalyte 4-6.5 (1.45 ml) and one part of Pharmalyte 2.5-5 (0.45 ml) proved to be the most adequate in that such a mixture gave a gradient with the range of interest in the middle. To this mixture different amounts (up to 1000 mg) of chemical spacers were added. The chemical spacers were chosen according to theoretical considerations of their pI and/or pK. Theoretically TES (pK 7.5, pI 5.5) and MOPS (pK 7.2, pI 5.0) appeared to be the most promising, and were tested at a final concentration of 34.48 mg/ml of gel.

<u>Isoelectric focusing</u> was performed at constant power (15 W) with maximum set voltage and current of 1500 V and 150 mA respectively. Cooling water at 10°C was circulated through the apparatus. The agarose gels were prefocused for 500 Volthours (around 30 min), followed by application of sample and pI markers (5 µl or less) on filter paper pieces in the middle of the gel. Thereafter focusing was continued for up to 2000 Volthours (around 90 min.). The pH gradient was then measured. A typical set up is shown in Fig. 1. After focusing, one part of the gel was fixed, washed, and stained for proteins with Coomassie Brilliant Blue. In the other part, Gc-globulin was identified by immunofixation or crossed IEF-immunoelectrophoresis.

Fig. 1  Set up for running isoelectric focusing in Agarose-IEF.

Immunofixation was carried out with cellulose acetate strips
(Sephaphore III, Gelman Instruments Co., Michigan, USA) soaked
with rabbit anti Gc-globulin (Dakopatts AS, Copenhagen,
Denmark) diluted 1:3 with TRIS-barbiturate buffer pH 8.6, ionic
strength 0.025, according to Ritchie and Smith (11) with the
exception that the strips were placed on the gel for 5 minutes.

Crossed IEF-immunoelectrophoresis was performed by laying a
lane of Agarose-IEF containing the focused serum sample on top
of an agarose-antibody gel (100x114 mm) as described earlier
(12).

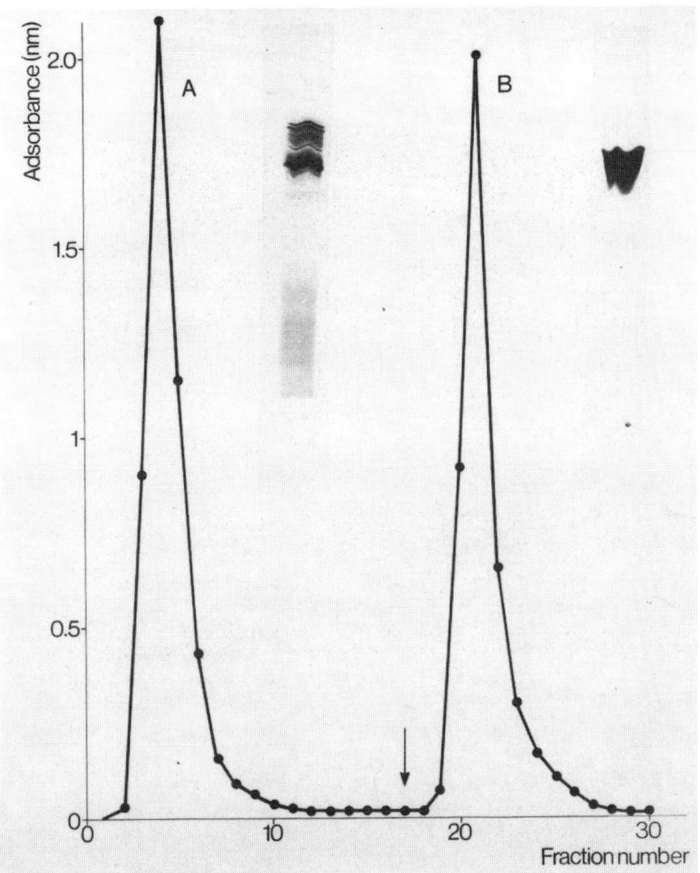

Fig 2   Chromatogram obtained with a column of Blue Sepharose CL-4B.

__Albumin-free serum.__   Fig. 2 shows the chromatogram obtained from the
column of Blue-Sepharose.   Peak A contains the various serum proteins
including Gc-globulin, as shown by isoelectric focusing.   The albumin
bound to the column was eluted by incorporation of 1.5 M KCl (arrowed)
to the equilibration buffer (peak b).   The peaks A and B were concentrated by ultra-filtration using an Amicon UM-2 membrane and run in
Agarose-IEF (Pharmalyte 3-10).

554

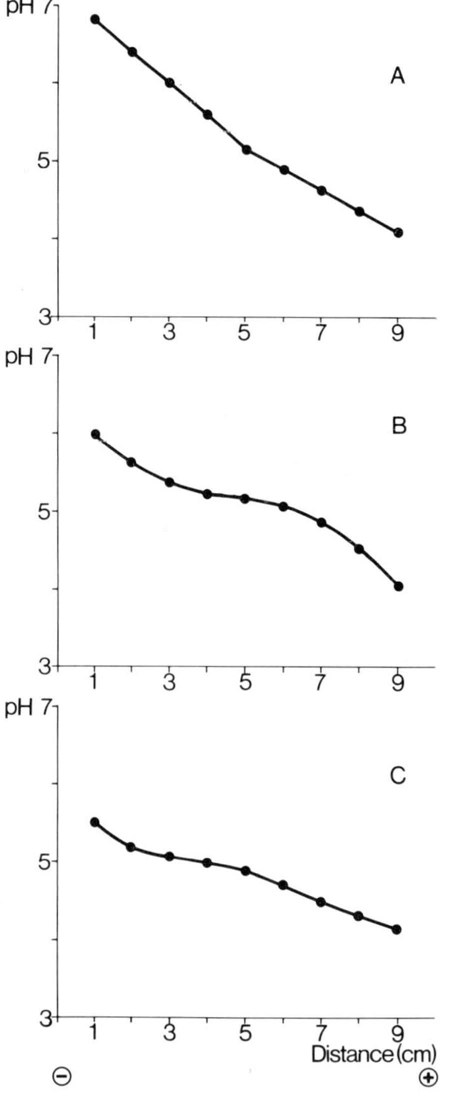



Flattening of the pH
gradient. Fig 3A shows the
pH gradient obtained with
the mixture of Pharmalyte.
The effect of TES (34.48
mg/ml) in an Agarose-IEF
gel is shown in Fig 3B, the
pH gradient is flattened in
the range 4.95-5.10. The
use of MOPS (34.48 mg/ml)
yielded the best result
with a distance of 25 mm
between pH 4.80 and 5.10
(Fig 3C).

Analysis of Gc-globulin in
Agarose-IEF. Fig. 4 shows a
section of an Agarose-IEF
gel using the modified pH
gradient. An albumin-free
serum its Gc-globulin
immunoprint (phenotype Gc
1-1) and crossed IEF-
immunoelectrophoresis is
presented in Fig. 5. A
Gc 2-1 phenotype serum with
its immunoprint and crossed
IEF-immunoelectrophoresis
is shown in Fig. 6.

Fig 3. pH gradient profiles. Measurements
     were made after 2000 Volthours (1500V,
     15W, 150 mA) in an Agarose-IEF gel.
     A. Mixture of Pharmalyte 4-6.5 and 2.5-5
     (ratio 3:1). B. Pharmalyte mixture plus
     1000 mg of TES. C. Pharmalyte mixture plus
     1000 mg of MOPS.

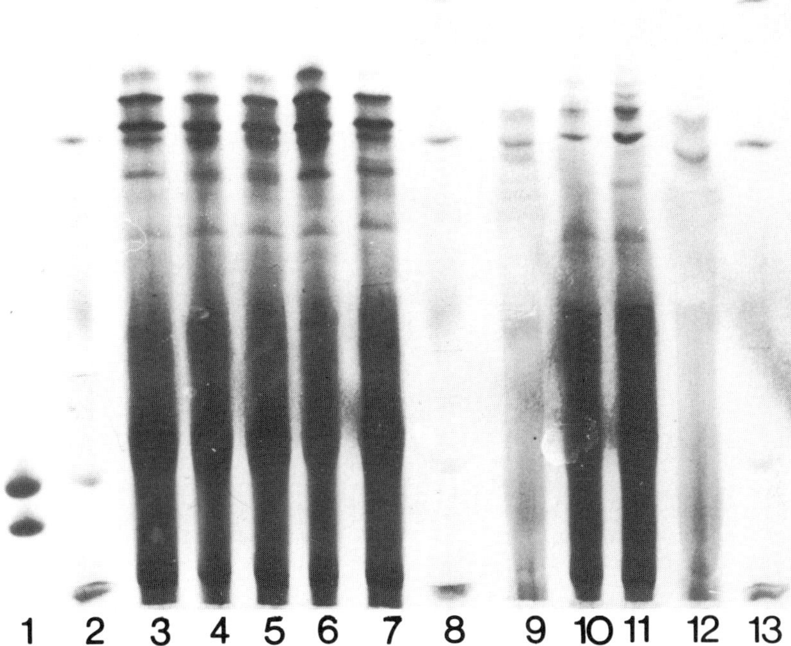



Fig. 4. Agarose-IEF run using the modified pH gradient. Anode at the top. Sample 1 was β-lacto-globulin; samples 2,8 and 13 correspond to the pI markers - the proteins are from top to bottom: amyloglucosidase (pI 3.50), glucose oxidase (pI 4.15), soybean trypsin inhibitor (pI 4.55), β-lactoglobulin (pI 5.20) and bovine carbonic anhydrase B (pI 5.85). Samples 3-8, 9-12 are unselected fresh serum samples (5 µl). Near the anode the alpha-1-antrypsin heterogenicity is well observed, the pattern of the Gc-globulin and other proteins is blurred by the focusing of albumin.

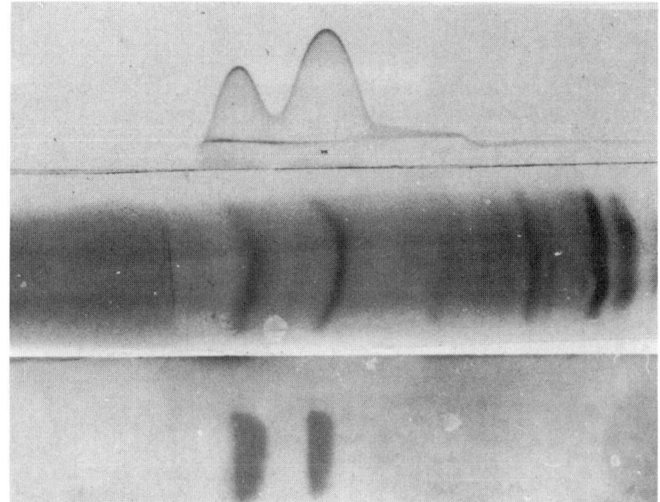

Fig. 5. Crossed IEF-immunoelectrophoresis of an albumin-free
serum Gc-globulin type Gc1-1. The immunoprint (5 min on the
gel) is also included.

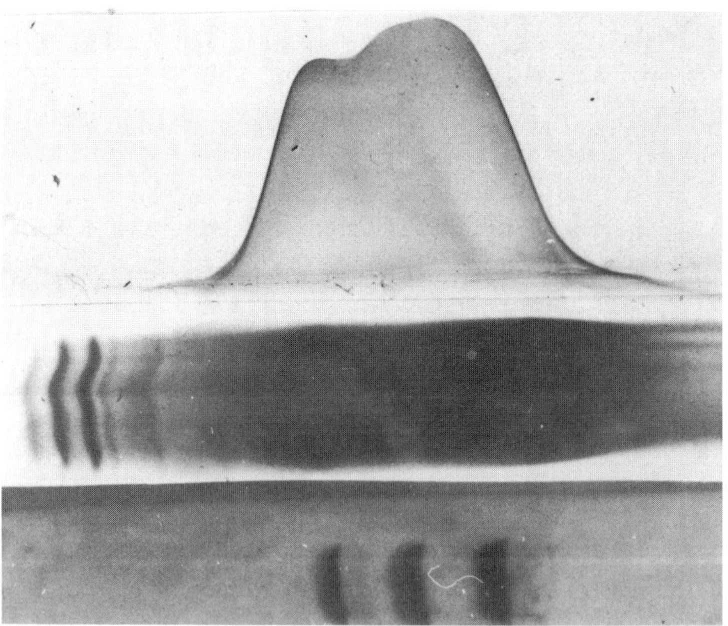

Fig. 6. Crossed IEF-immunoelectrophoresis of a serum with
Gc-globulin type Gc 2-1. The immunoprint (5 min on the gel)
is also included.

<u>pI measurements</u>. An albumin-free serum type Gc 1s-1s was run in an Agarose-IEF gel with an inter-electrode distance of 160 mm. The pH was measured at 10°C. As can be seen in Fig. 7 the separation of the two bands makes the pI measurement very easy. The recorded values were pI 4.92 for the cathodal band and pI 4.85 for the anodal band.

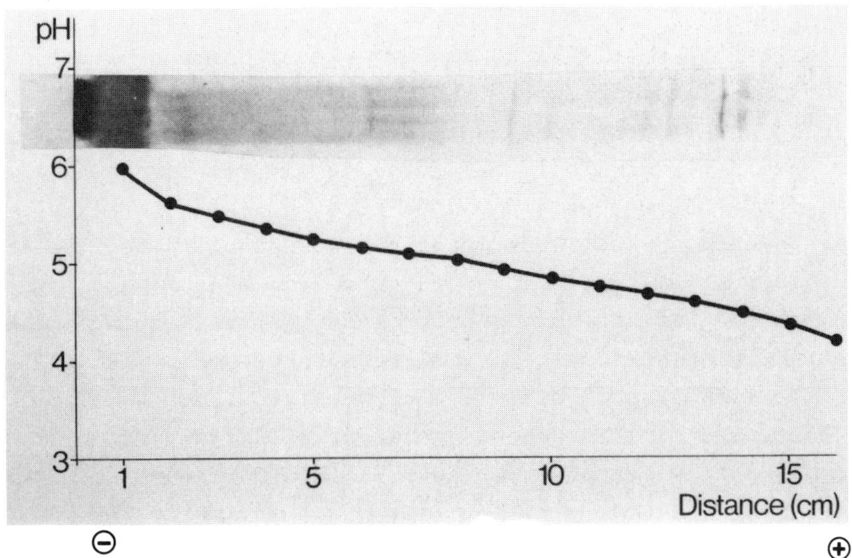


Fig. 7. pI measurements. An albumin-free serum with Gc-globulin type Gc1-1 was run in Agarose-IEF with the expanded gradient (see Methods). Inter-electrode distance 160 mm. Settings: 1500V, 15W, 150mA. Running time 5000 Volthours (around 4 hours). pI measurements were made at 10°C.


DISCUSSION


Classification of genetic markers for population studies is becoming a very important tool in human genetics, forensic science and related disciplines. One such marker, serum Gc-globulin has proved to be of great value, and guidelines for its analysis by IEF in polyacrylamide gels have been made in a recent International Workshop (6). The possibility of using agarose as the matrix for IEF is very interesting because

558

unlike polyacrylamide, agarose gels are easy to prepare
(especially important in routine work), optically clear for
scanning purposes, chemically unreactive, better suited for
immunochemical detection techniques and non-toxic. A charge
balanced agarose optimised for use in isoelectric focusing
(Agarose-IEF) is now available (13).

Indepedent of the matrix used, the normally recommended carrier
ampholyte intervals (pH 4-6 or 4-6.5) for the study of
Gc-globulin  are not optimal since Gc-globulin proteins focus
between pH 4.8-5.1 (15,16,17,18). Viau et al. (7) have used the
cathodic drift of carrier ampholytes after 4.5 hours of
focusing to obtain a suitable gradient expansion. We decided to
tackle this problem by looking for a chemical spacer that would
flatten the pH gradient in a shorter running time and in a more
reliable and stable manner. Optimal results were obtained with
MOPS (Fig.3C) added to a mixture of Pharmalyte 4-6.5 and 2.5-5
(ratio 3:1). Focusing in agarose gels (225x114x1 mm) was
carried out at 1500 V, 15 W for 2000 Volthours (around 2 h).

Thymann (14) has reported the use of agarose as a support
medium for Gc-phenotyping using carrier ampholytes (pH 4-6)
and a focusing time of 100 min (1200 V, 12 W). Sufficient
resolution was obtained by allowing the cathodic gradient
drift, due to electroendosmosis, to flatten the pH gradient.
However the electroendosmosis produced flooding which was
controlled by placing a soft absorbent paper on the gel near
the cathodic electrode strip to drain off the excess water. In
our experience this means that the gel become thinner in some
areas, with increasing risk of overheating and eventual wavy
bands. With the charge balanced agarose used by us we did not
observe flooding. Besides, the addition of the chemical spacer
(MOPS) makes it unnecessary to rely on gradient drift to
achieve a flat gradient, and allows a better control over the
pH gradient.

Serum albumin focuses in the same region as Gc-globulin and overshadows other proteins when the plates are stained with protein dyes. This necessitates the use of immunochemical detection methods to type the Gc-globulin proteins. The ease of combining Agarose-IEF with immunofixation and/or crossed immunoelectrophoresis is illustrated in Figs. 5 and 6, and makes the routine analysis of Gc-globulin a simple, fast and reliable procedure. Removal of the albumin from serum by affinity chromatography and subsequent focusing of the albumin-free fraction reveals the diversity of proteins, amongst them the Gc-globulin proteins which focus in the pH 4.5-5.0 region (see Fig. 2). However, this approach is not feasible in routine conditions.

The isoelectric point of the different Gc-globulin proteins is usually given as a pH range between pH 4.85 and 5.10 at 12°C (15). Others have reported pI values of 4.8 (16) and 4.9 (17), but the temperature of the measurements was not given. Van Baelen et al (18) cooled the flat bed apparatus at 4°C and measured isoelectric points of 5.10 (Gc 2), 5.03 (Gc 1 Slow) and 4.95 (Gc 1 Fast). A constant difference of 0.07 pH units between Gc-bands was observed. Our data measured at 10°C (Fig.7) were pI 4.85 and pI 4.92 for the anodal and cathodal bands of a Gc 1s-1s phenotype. In experiments using column isoelectric focusing in 5-50% sucrose (ampholytes 4-6) at 4°C it was reported (19) that a mixture of Gc 1 Fast and Gc 1 Slow focused in the pH range of 4.52 to 4.87. The addition of an excess of Vitamin D derivatives results in an anodic shift of the pIs of all Gc components of 0.07 (18) to 0.10 (19) pH units. Under physiological conditions the contribution to the isoelectric point, of Vitamin D binding to plasma Gc-globulin, is likely to be small and difficult to detect.

Experiments with polyacrylamide gels T5C3 (data not shown) showed that the addition of MOPS to Pharmalyte 4-6.5 yielded a similar resolution of the Gc-globulin as obtained with

Agarose-IEF. The addition of Pharmalyte 2.5-5 is not necessary
when running PAA gels because of the different degree of
gradient drift that occurs with the two matrices.

We conclude that phenotyping of Gc-globulin can be done by
Agarose isoelectric focusing, using a spacer modified pH
gradient, combined with immunofixation in a simple, fast and
reliable manner in a routine laboratory.


REFERENCES

1.  Putnam, F.W. In "The plasma proteins". (Ed. F.W.Putnam)
    vol.3, chapter 5. Acad. Press, N.Y., 1977.
2.  Cleve, H. and Patutschnick, W.: Hum. Genet. $\underline{47}$, 193-198
    (1979).
3.  Kühnl, P., Spielmann, W. and Loa, M.: Vox Sang. $\underline{35}$,
    401-404 (1978).
4.  Hoste, B.: Hum. Genet. $\underline{50}$, 75-79 (1979)
5.  Kueppers, F. and Harpel, B.: Hum. Hered. $\underline{29}$, 242-249
    (1979).
6.  Constants, J. and Cleve, H.: Hum. Genet. $\underline{48}$. 143-149
    (1979).
7.  Viau, M., Constans, J. and Bouissou, C.: Sci. Tools $\underline{24}$,
    25-26 (1977).
8.  Låås, T., Olsson, I. and Söderberg, L.: Anal. Biochem.
    $\underline{101}$, 449-461 (1980).
9.  Caspers, M., Possey, Y. and Brown, R.K.: Anal. Biochem.
    $\underline{79}$, 166-180 (1977).
10. Travis, J., Bowen, J. and Tewksbury, C.: Biochem. J. $\underline{157}$,
    301-306 (1976).
11. Ritchie, R.F. and Smith, R.: Clin. Chem. $\underline{22}$, 497-499
    (1976).
12. Söderholm, H., Smyth, C.J. and Wadström, T.A.: Scand. J.
    Immunol. $\underline{4}$, suppl. 2, 107-113 (1975).
13. Hansson, H. and Kågedal, L.: 13th FEBS Meeting, Jerusalem,
    1980.
14. Thymann, M.: In "Electrophoresis 79", (Ed. B.J.Radola),
    Walter de Gruyter & Co., Berlin, 123-128 (1980).

15. Constants, J. and Viau, M.: C.R. Acad. Sc. Paris 287, 809-812 (1978).

16. Inawari, M. and Boodman, D.S.: J. Clin. Invest. 59, 432-442 (1977).

17. Lawson, D.E.M. and Davie, M.: Vitamins & Hormones 37, 2-61 (1979).

18. Van Baelen, H., Bouillon, R. and De Moor, P.: J. Biol. Chem. 253, 6344-6345 (1978).

19. Svasti, J. and Bowman, B.H.: J. Biol. Chem. 253, 4188-4194 (1978).

ISOELECTRIC FOCUSING IN FORENSIC SEROLOGY


P.Kühnl


Institut für Immunhämatologie der Universität,
Sandhofstr.1,D-6000 Frankfurt/Main, FRG


The technique of isoelectric focusing(IEF)in flat bed
gels of polyacrylamide or agarose gels recently has been
used for an increasing number of application areas.Be-
sides the analysis of protein purity and microheteroge-
neity IEF serves as a preparative technique for the purification of proteins.The resolution of proteins which
differ in pI by only 0.0025 units, short separation times,
good reproducibility of protein patterns and practicability
are only some of the factors, that render IEF an excellent
tool in research and routine(1).In Table 1 some of the major
application areas are listed.

Table 1

<u>MAJOR APPLICATION AREAS IN ISOELECTRIC FOCUSING(IEF)</u>

1) CLINICAL CHEMISTRY (PI,Hb,Lp)

2) BLOOD GROUP GENETICS (PATERNITY TESTING)

3) FORENSIC MEDICINE (BLOOD STAINS)

4) HUMAN GENETICS (GENE MAPPING)

5) BOTANY (PLANT PROTEINS:POTATO,WHEAT..)

6) ZOOLOGY (ISOZYMES &ISOPROTEINS OF VARIOUS SPECIES)

7) IMMUNOLOGY (IMMUNOGLOBULINS,HLA MOLECULES)

8) CANCER RESEARCH (LDH ISOZYMES OF MALIGNANT TUMORS)

9) FOOD ANALYSIS(MEAT,MILK,SOYBEAN)

In the field of forensic medicine the investigation of
so-called genetic markers, i.e. isozymes and isoproteins
plays an important role in the identification of blood
stains and the analysis of various tissues.In modern
blood group genetics IEF is about to pass the threshold
and become a standard technique in the routine of paternity
testing.The fact that courts in the Federal Republic of
Germany request over 5000 expertises in affiliation cases
per year contributes considerably to this progress.In nearly
all paternity cases a basic blood group expertise is deman-
ded in the first instance.Three other kinds of expert opinions
are required only exceptionally, namely a) the anthropologi-
cal expertise(about 5% of all cases),b)the gynecological
expertise(estimation of gestation time, about 1% of all cases),
c)the  andrological expertise (fertility of the alleged man,
less than 1%).
The basic blood group expertise is called "Normgutachten" or
"Basisgutachten" and comprises a fixed list of the following
blood group systems, according to the guide-lines of our
Federal Board of Health(Bundesgesundheitsamt,Bundesgesundheits-
blatt 20,No.22,1977): ABO including $A_1$ and $A_2$,MNSs,$DCcC^WEe$,K,
$Fy^a$,Hp,Gc,$Gm^1$,$Gm^2$,$Km^1$,ACP,$PGM_1$,$AK_1$,ADA,6-PGD and ESD.If the
investigation of these systems is considered insufficient, e.g.
inexclusions based on opposite homozygozity or because of low
biostatistical values, further investigations are requested by
the court.The analysis of additional new blood group systems,
mostly including HLA and isozymes(GPT,$PGM_3$,UMPK) as well as
isoproteins(C3,Bf) and, more recently the 'isofocusing systems'
is necessary for clarifying the question of fatherhood of the
alleged man.Only those genetic markers can be applied in an
extended expertise which can be re-examined by a second independent expert.In Table 2 a number of 'IEF systems' is listed,

which are of particular interest for blood group geneticists.
In chronological sequence the most informative systems $alpha_1$-
antitrypsin (Pi),phosphoglucomutase$_1$(PGM$_1$),transferrin(Tf)
and group-specific component(Gc) are followed by esterase D

Table 2

## " IEF SYSTEMS "

| Splits of "old" systems | Material | Alleles |
|---|---|---|
| Pi | Serum | 1975(M1,M2)  1977(M3)  1980(M4,M5) |
| PGM$_1^a$ | Hemolysate White cell lysate | 1976(PGM$_1^{a1-a4}$) |
| Tf | Serum | 1977(C1,C2)  1979(C3)  1980(C4-C7) |
| Gc | Serum | 1977(1F,1S) |
| ESD | Hemolysate | 1979(ESD$^2$,ESD$^5$) |
| AMY$_1$ | Saliva,Serum | 1977(AMY$_1^1$,AMY$_1^2$)  1978(AMY$_1^3$) |

| Further systems | Material | Alleles |
|---|---|---|
| FUCA | White cell lysate | 1975(FUCA$^1$,FUCA$^2$) |
| C2 | Serum | 1976(C2$^1$,C2$^2$,C2$^0$) |
| C6 | Serum | 1976(C6$^A$,C6$^B$) |
| C8 | Serum | 1979(C8$^A$,C8$^B$) |
| Apo E | Serum | 1977(Apo E$^n$,Apo E$^d$)  1980(Apo E$^4$) |

(ESD) and amylase$_1$(AMY$_1$),where the split of one allele(Pi$^M$, Gc$^1$, Tf$^C$, ESD$^2$,AMY$_1^A$) or two alleles(PGM$_1^1$,PGM$_1^2$) led to an extension of the original formal genetic model.Whereas in the PGM$_1$ and Gc system no additional common alleles were identified since 1976/1977(except of numerous rare variants)the process of splitting goes on in the Pi (M3,M4,M5?) and Tf system(Tf C3,C4,C5,C6)(2-17).In the lower part of Table 2 five systems are shown, which have been investigated primarily by IEF, but still offer difficulties in routine work such as a) availability of sufficient quantities of white cells from small children(FUCA typing),weak intensity of the zymogram when using fluorogenic substrates as 4-methyl-umbelliferylfucopyranoside(FUCA), b)availability of complement component-deficient plasma(C2$^0$,C6$^0$,C8$^0$) and the tedious preparation of the functional(hemolytic) overlay for the detection of specific activity, and c) time-consuming preparation of sera for Apo E typing, difficulty of revealing the Apo E polymorphism in flat-bed gels.

Other limitations for routine application of IEF in forensic serology are shown in Table 3. If the experiences of our wor-

Table 3

LIMITATIONS FOR ROUTINE APPLICATION OF IEF IN FORENSIC SEROLOGY
(ISOPROTEIN AND ISOZYME SYSTEMS)

---

a) DETECTION OF POLYMORPHIC BANDS AFTER IEF IS POSSIBLE,THE SAME
   DEGREE OF SEPARATION AND INFORMATION IS OBTAINED BY CONVENTIONAL
   ELECTROPHORESIS
b) NO POLYMORPHISM IN ONE OF THE THREE MAJOR RACES (FREQUENCY OF
   THE RARE ALLELE(S) < 0.01)
c) FEW IEF POPULATION DATA CONCERNING'SPLITS' OF COMMON ALLELES,
   FORMAL-GENETIC SITUATION NOT COMPLETELY CLEAR
d) POLYMORPHISM DETECTABLE IN TISSUES WHICH ARE UNAVAILBLE FOR
   USUAL BLOOD GROUPING (LIVER,PLACENTA,GONADAL TISSUE)

---

Table 4

ESSENTIAL PREREQUISITES FOR THE INTRODUCTION OF NEW SYSTEMS

| System | Proof of specific activity | Investigation of purified protein | Comparison with other techniques | Hardy-Weinberg equilibrium given | Family investigations | Polymorphism detectable in infants | Indiv. reproducibility | Stability after storage | Modifying genes | High informative value | Silent genes, quantitative variants | Formal genetics clarified | Practicability |
|---|---|---|---|---|---|---|---|---|---|---|---|---|---|
| | 1 | 2 | 3 | 4 | 5 | 6 | 7 | 8 | 9 | 10 | 11 | 12 | 13 |
| Tf | + | + | + | + | + | + | + | + | + | + | + | + | + |
| Pi | + | + | + | + | + | + | + | + | + | + | + | + | (+) |
| Gc | + | + | + | + | + | + | + | (+) | + | + | + | + | + |
| PGM$_1$[a] | + | + | + | + | + | + | + | + | + | + | + | + | + |
| DIA$_3$ | + | ∅ | + | + | ∅ | ∅ | + | (+) | ∅ | + | ∅ | (+) | + |
| FUCA | + | ∅ | ∅ | + | + | (+) | + | ∅ | ∅ | + | (+) | ∅ | (+) |
| PEP A | + | ∅ | + | (+) | (+) | (+) | + | + | ∅ | + | + | ∅ | + |
| AMY$_1$ | + | + | + | + | + | (+) | + | ∅ | (+) | + | ∅ | (+) | + |
| C6 | + | + | + | + | + | + | + | + | + | + | + | (+) | (+) |
| C8 | + | + | ∅ | + | + | (+) | + | (+) | (+) | + | (+) | (+) | (+) |

king group and those of other investigators are summarized
and checked, if the essential prerequisites for the introduction of new systems are fulfilled(Table 4), we may consider Tf,Pi,Gc,$PGM_1^a$ and C6 as promising candidates for an
acceptance as "basic" (i.e. quite well established)systems
by the Federal Board of Health.

Table 5

EXCLUSION CHANCES FOR NON-FATHERS IN VARIOUS BLOOD GROUP SYSTEMS
INCLUDING ELECTROPHORESIS(LEFT) AND ISOELECTRIC FOCUSING(RIGHT)[+]

| SYSTEM | EXCLUSION CHANCE % | | SYSTEM | EXCLUSION CHANCE % | |
|---|---|---|---|---|---|
| | SINGLE | COMBINED | | SINGLE | COMBINED |
| ABO | 20 | | ABO | 20 | |
| MNSs | 32 | 45.6 | MNSs | 32 | 45.6 |
| Rh | 29 | 61.4 | Rh | 29 | 61.4 |
| HLA | 94 | 97.7 | HLA | 94 | 97.7 |
| K | 4 | 97.8 | K | 4 | 97.8 |
| P | 3 | 97.8 | P | 3 | 97.8 |
| Fy | 18 | 98.2 | Fy | 18 | 98.2 |
| Jk | 19 | 98.6 | Jk | 19 | 98.6 |
| Lu | 3 | 98.6 | Lu | 3 | 98.6 |
| Xg | 5 | 98.7 | Xg | 5 | 98.7 |
| Se | 2 | 98.7 | Se | 2 | 98.7 |
| ACP | 25 | 99.0 | ACP | 25 | 99.0 |
| AK | 3 | 99.1 | AK | 3 | 99.1 |
| ADA | 5 | 99.1 | ADA | 5 | 99.1 |
| $PGM_1$ | 16 | 99.2 | [+]$PGM_1^a$ | 25.3 | 99.33 |
| GPT | 19 | 99.4 | GPT | 19 | 99.46 |
| ESD | 9 | 99.4 | ESD | 9 | 99.50 |
| GLO | 18 | 99.5 | GLO | 18 | 99.59 |
| $PGM_3$ | 15 | 99.6 | $PGM_3$ | 15 | 99.65 |
| PGD | 2 | 99.6 | PGD | 2 | 99.66 |
| Hp | 18 | 99.7 | [+]FUC | 15.4 | 99.71 |
| Gc | 16 | 99.7 | [+]Gc/Sub | 24.7 | 99.78 |
| Gm | 20 | 99.8 | [+]Pi/Sub | 22.1 | 99.83 |
| Km | 6 | 99.8 | Gm | 20 | 99.86 |
| C3 | 14 | 99.8 | [+]C6 | 18.1 | 99.89 |
| Bf | 13 | 99.85 | Hp | 18 | 99.91 |
| | | | [+]Tf/Sub | 15.5 | 99.92 |
| | | | C3 | 14 | 99.93 |
| | | | Bf | 13 | 99.94 |
| | | | [+]Apo E | 7.5 | 99.944 |
| | | | Km | 6 | 99.947 |
| | | | [+]C2 | 3.1 | 99.948 |
| | | | [+]C8 | 18.1 | 99.957 |

The addition of eight IEF systems, $PGM_1^a$, FUCA, Gc, Pi, C6, Tf, Apo E, C2 and C8 would raise the combined exclusion chance for 'non-fathers' in affiliation cases from 99.85% to 99.957%, if the sequence ABO, MNSs, Rh and HLA is chosen for the first systems(18-20). The figures in Table 5 clearly reveal, that only one out of 2000 'non-fathers' cannot be identified by a complete serological expertise(99.95 vs. 0.05%). Although the percent increase from 99.85 to 99.957 is not very impressive if HLA is tested among the first systems, the true diagnostic value of IEF becomes apparent

Figure 1

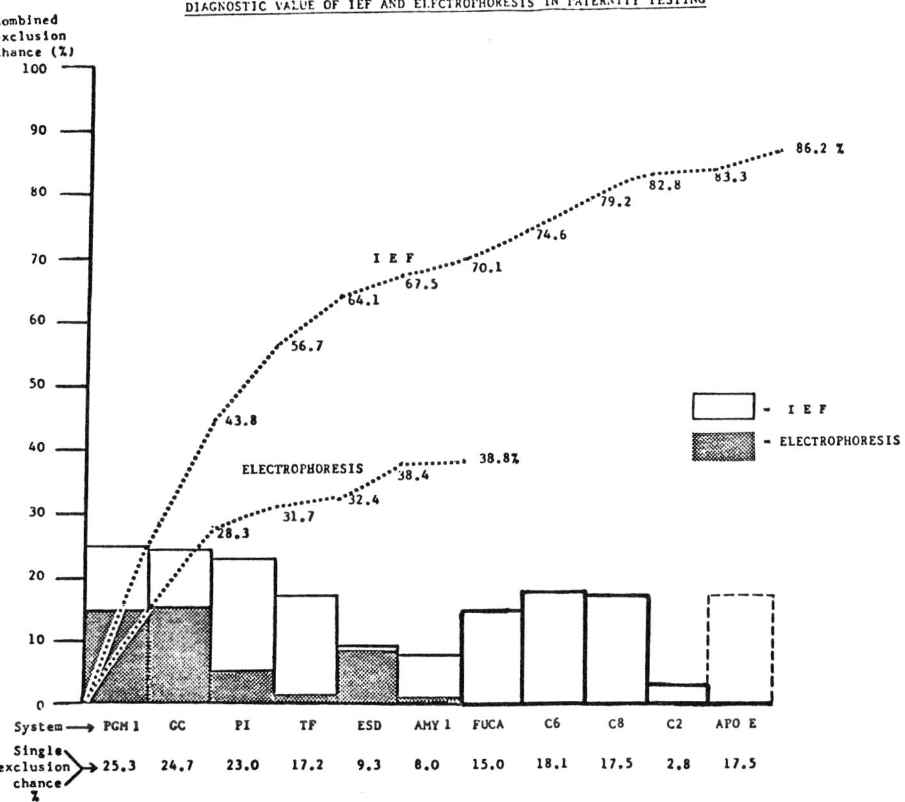

by a comparison with electrophoresis(Figure 1).After subtraction of the basic electrophoretic information the discrepancy between the two separation systems exceeds 30%
after six systems(black areas of the columns in Fig.1)
and totals 86.2% of combined exclusion chance for 'non-
fathers'.This figure suggests, that isofocusing might become
a method similarly efficient as the microlymphocytotoxicity
test in the detection of the HLA system, whose polymorphism
is still second to none of any known blood group.

## References


1) Bishop,R.:Current major application areas in electrofocusing.
Sci.Tools $\underline{26}$,2-8(1979)

2) Frants,R.R.,Eriksson,A.W.:Alpha$_1$-antitrypsin:Common subtypes
of Pi M.Hum.Hered.$\underline{26}$,435-440(1976)

3) Klasen,E.C.,Franken,C.,Volkers,W.S.,Bernini,S.:Population genetics of $\alpha_1$-antitrypsin in the Netherlands.Hum.Genet.$\underline{37}$,
303-313(1977)

4) Bark,J.E.,Harris,M.J.,Firth,M.:Typing of the common phosphoglucomutase variants using isoelectric focusing- a new
interpretation of the phosphoglucomutase system.J.Forens.
Sci.Soc.$\underline{16}$,115-120(1976)

5) Kühnl,P.,Schmidtmann,U.,Spielmann,W.:Hinweis für zwei weitere
häufige Allele am PGM$_1$-Locus.Ärztl.Lab.$\underline{23}$,229-232(1977);Ref.
Symposium Arbeitsgemeinschaft der Blutgruppensachverständi-
gen,Timmendorferstrand 1976

6) Kühnl,P.,Schmidtmann,U.,Spielmann,W.:Evidence for two common
alleles at the PGM$_1$ locus(phosphoglucomutase- E.C.:2.7.5.1).
A comparison by three different techniques.Hum.Genet.$\underline{35}$,
219-223(1977 )

7) Kühnl,P.,Spielmann,W.:Transferrin:Evidence for two common
subtypes of the Tf$^C$ allele.Hum.Genet.$\underline{43}$,91-95(1978 )

8) Kühnl,P.,Spielmann,W.:A third common allele in the transferrin
system,Tf$^{C3}$,detected by isoelectric focusing.Hum.Genet.$\underline{50}$,
193-198(1979$_b$)

9) Constans,J.,Viau,M.:Group-specific component:Evidence for two
subtypes of the Gc$^1$ gene.Science $\underline{198}$,1070-1071(1977)

10) Constans,J.,Cleve,H.,Bennett,A.,Bouillon,R.,Cox,D.W.,Daiger,S.P.
Ehnholm,C.,Fujiki,N.,Johnson,A.M.,Kirk,R.L.,Kühnl,P.,Mar-
tin,W.,Matsumoto,H.,Mayr,W.R.,Miyake,K.,Omoto,K.,Porck,H.J.
Seger,J.,Thymann,M.,Tills,D.,Toyomasu,M.,vanBaelen,H.,
Vavrusa,B.,Viau,M.:Group-specific component.Report on the
First International Workshop 27.-28.7.1978,Paris. Hum.
Genet.48,143-149(1979)

11) Martin,W.:Neue Elektrophoresemethoden zur Darstellung von Serum-
und Enzympolymorphismen:technische Verbesserungen,Hinweis
auf ein weiteres EsD-Allel.Ärztl.Lab.25,65-67(1979)

12) Pronk,J.C.:A genetic variant of amylase from human parotid saliva detected by isoelectric focusing.In:Electrofocusing
and Isotachophoresis.S.359-366.Hrsg.:B.J.Radola und ·D.
Graesslin.Berlin-New York:deGruyter 1977

13) DeSoyza,K.:Polymorphism of human salivary amylase.Hum.Genet.
45,189-192(1978)

14) Turner,B.M.,Turner,V.S.,Beratis,N.G.;Hirschhorn,N.K.:Poly-
morphism of human $\alpha$-fucosidase.Am.J.Hum.Genet.27,651-661
(1975)

15) Meo,T.,Atkinson,J.,Bernoco,M.,Ceppelini,R.:Structural heterogeneity of C2 complement protein and its genetic variants
in man:A new polymorphism in the HLA region.Proc.Natl.Acad.
Sci.(USA) 74,1672-1675(1977)

16) Alper,C.A.,Hobart,M.J.,Lachmann,P.J.:Polymorphism of the sixth
component of complement.In:Isoelectric focusing.Hrsg.:J.P.
Arbuthnott und J.A.Beely.London-Boston:Butterworths 1975

17) Utermann,G.,Hees,M.,Steinmetz,A.:Polymorphism of apolipoprotein E and occurrence of dysbetalipoproteinaemia in man.
Nature 269,604-607(1977 )

18) Mayr,W.R.:Probleme bei der Anwendung des HLA-Systems im Vater-
schaftsgutachten.Ref.Bd.S.417-441. 7th Int.Congr.Soc.
Forensic Haematogenetics, Hamburg 1977

19) Spielmann,W.,Kühnl,P.:The effic acy of modern blood group genetics with regard to a case of probable superfecundation.
Haematologia,13,75-85(1980)

20) Spielmann,W.,Kühnl,P.:Blutgruppenkunde.Stuttgart:Thieme 1981
(in press)

ISOELECTRIC FOCUSING OF THE COMMON TRANSFERRIN$^C$ ALLELE
SUBTYPES


P.Kühnl[1], J.Constans[2], M.Viau[2], W.Spielmann[1]

[1]Institut für Immunhämatologie der Universität,Sandhofstr.1
D-6000 Frankfurt/Main,FRG
[2]CNRS Centre d'Hemotypologie,CHU de Purpan,F-31300 Toulouse,
France



Abstract

We previously reported an extended polymorphism of human
transferrin(Tf) in terms of splitting the common $Tf^C$ alle-
le.An improved method of isoelectric focusing of human
sera on polyacrylamide gels(IEF) was used to investigate
further genetic heterogeneities in the Tf system.Prior to
IEF the sera were diluted 1/5 in 0.25% ferrous ammonium
sulphate x 6 $H_2O$ solution for 18 h at $+4^{o}C$. Isofocusing
on 0.5 mm thin-layers of polyacrylamide at pH 5-7 for
5 h at a cooling temperature of $+10^{o}C$ permitted an excellent resolution of the rare $Tf^B$ and $Tf^D$ variants as
well as seven different subtypes(alleles) of $Tf^C$.Where-
as $Tf^{C1}$,$Tf^{C2}$ and $Tf^{C3}$ were found at polymorphic frequencies in Whites,$Tf^{C4}$ in American Indians and $Tf^{C5}$ in Black
Americans,the two new subtypes $Tf^{C6}$ and $Tf^{C7}$ were encountered in North Africa and Southern France,respectively.The
order of the pI of the seven subtypes (from anode to catho-
de:C4,C1,C3,C5,C2,C6 and C7)reflects the sequence of their
detection and might be subject to a revision of the Tf nomenclature.

574

## Introduction

At least 22 different alleles have been described in the
Tf system since the discovery of a genetically determined
polymorphism by means of starch gel electrophoresis(1).
Besides starch, agarose and polyacrylamide gel electrophoresis isoelectric focusing was introduced as a very
useful tool for the studies of genetic heterogeneity of
Tf.The high resolution capacity of this technique and
the reproducibility of protein patterns led to the discovery of three common phenotypes among Tf C individuals,
which were explained in terms of two allelic genes,$Tf^{C1}$
and $Tf^{C2}$(2).Another common variant,$Tf^{C3}$ was recognized
by a modified procedure of IEF in a narrow pH range of
4 - 6.5 (3).Genetic studies revealed,that those new subtypes were split off from the common $Tf^{C}$ allele, which
reached its highest frequencies(between 0.98 and 0.998)
in Caucasoid populations.After 1979 a standardized procedure of pretreatment of sera with iron-solution and
IEF in a narrower pH range of 5-7 was developped, which
yielded two additional subtypes $Tf^{C4}$ and $Tf^{C5}$; for both
of them an autosomal codominant trait was established(4).
In this paper we report modifications of our standard
technique,which permitted the confirmation of another
two new alleles, $Tf^{C6}$ and $Tf^{C7}$ (5,6).

## Materials and Methods

Serum samples were obtained from blood donors and families
from Hessen,Germany and Toulouse,France.The sera were
freshly collected and used immediately or kept frozen at
$-20^{\circ}C$ until used for IEF.Prior to IEF 4 parts of an 0.25%
ferrous ammonium sulfate x 6 $H_2O$ solution were added to

1 part of serum.This mixture was stored at $+4^{\circ}$C for 18 h
(overnight) to provide an iron saturation of the Tf molecules.

IEF was performed in the pH range 5-7(Ampholine 1809-121,
LKB,Bromma.Sweden).For casting 0.5mm gels in the LKB polymerization set(245 x 115 x 0.5 mm) a thin rubber gasket
was used.A mixture of the following composition(total
volume 14.12 ml)was prepared:2.9 ml acrylamide solution
29.1%(w/v),2.9 ml N,N'-methylene-bisacrylamide solution
0.9%(w/v),7.0 ml sucrose solution 20.5%(w/v),0.7 ml
carrier ampholytes(Ampholine pH 5-7), 20 $\mu$l TEMED(N,N,N',N'-
tetramethylethylenediamine) and 0.6 ml ammonium persulfate
1%(w/v).After polymerization at room temperature (20 min)
IEF was started.A 1 M $H_3PO_4$ solution served as anolyte,a
1M ethanolamine solution as catholyte. 7 $\mu$l serum samples
were applied on Whatman filter papers(No.1) 5cm from the
cathodal electrode strip.At a cooling temperature of $+10^{\circ}$C
the following maximum settings were chosen(Power supply
2103,LKB):1500 V,15mA,10W.After prefocusing the gel for
15 min, and 60 min with the sample filter papers on the
gel the IEF was completed after another 3h 45 min(total
focusing time 5 h).The gel was then fixed and stained in
a 0.1% Coomassie Brillant Blue R 250 (Merck 12553)solution
(500 ml ethanol,160 ml acetic acid,aqua dest. ad 2000 ml)
for 1 h at $+37^{\circ}$C, and destained subsequently.

A control of Tf patterns obtained  by IEF was performed with
print immunofixation(contact time 2 min) using cellulose
acetate strips soaked with a 1/4 dilution of a monospecific
anti-Tf-immunoglobulin (TRF-IG 016-02,Atlantic Antibodies).

## Results and Discussion

After IEF in the pH range 5-7 the separation patterns of
human serum transferrins are clearly visible on the gel in
the zone located anodally to the sample application area
(pI values of 5.5 to 5.9, depending on the Tf variant phenotype).The fact,that after staining with Coomassie Brillant
Blue R 250, Tf seems to be the only major protein fraction
in this area may be explained by the preceeding treatment
of the sera with a fivefold volume of iron solution(Fig.1
and 2).One or two major bands in homo- or heterozygous
carriers provide an easy discrimination of Tf phenotypes.
The main bands observed at pH 5.5 presumably reflect the
monoferric form of the Tf molecule($Fe_1$-Tf),which appears
to be quite stable under IEF conditions(7).Minor bands ,
which correspond to the number of main bands are present
in the adjacent pH zones at 5.2 and 6.5, as confirmed by
print immunofixation.These fractions represent $Fe_2$-Tf mole-
cules(two atoms of iron) which are less stable at their
isoelectric point(5.2) and lose the second iron atom(4),
as well as iron-free Tf (apo-Tf-fraction), which remained
unsaturated even after Fe-pretreatment.A confusion between
these three molecular forms($Fe_2$-Tf,$Fe_1$-Tf and apo-Tf) is
unlikely, as the relative intensity of $Fe_1$-Tf is clearly
prevailing(8).
The phenotypes Tf C5-1 (Fig.1),Tf C6-1 and Tf C7-1(Fig.2)
were identified in US Blacks(frequency 0.02), North Afri-
cans from Algeria(0.03) and individuals from Southern France,
where the $Tf^{C7}$ allele appears to be non-polymorphic(Fig.3).
As the "slow" C-subtypes cannot be discerned from the regular C1 carriers after agarose gel electrophoresis, an identity with $D_0$ or $D_{Adelaide}$ can be excluded.

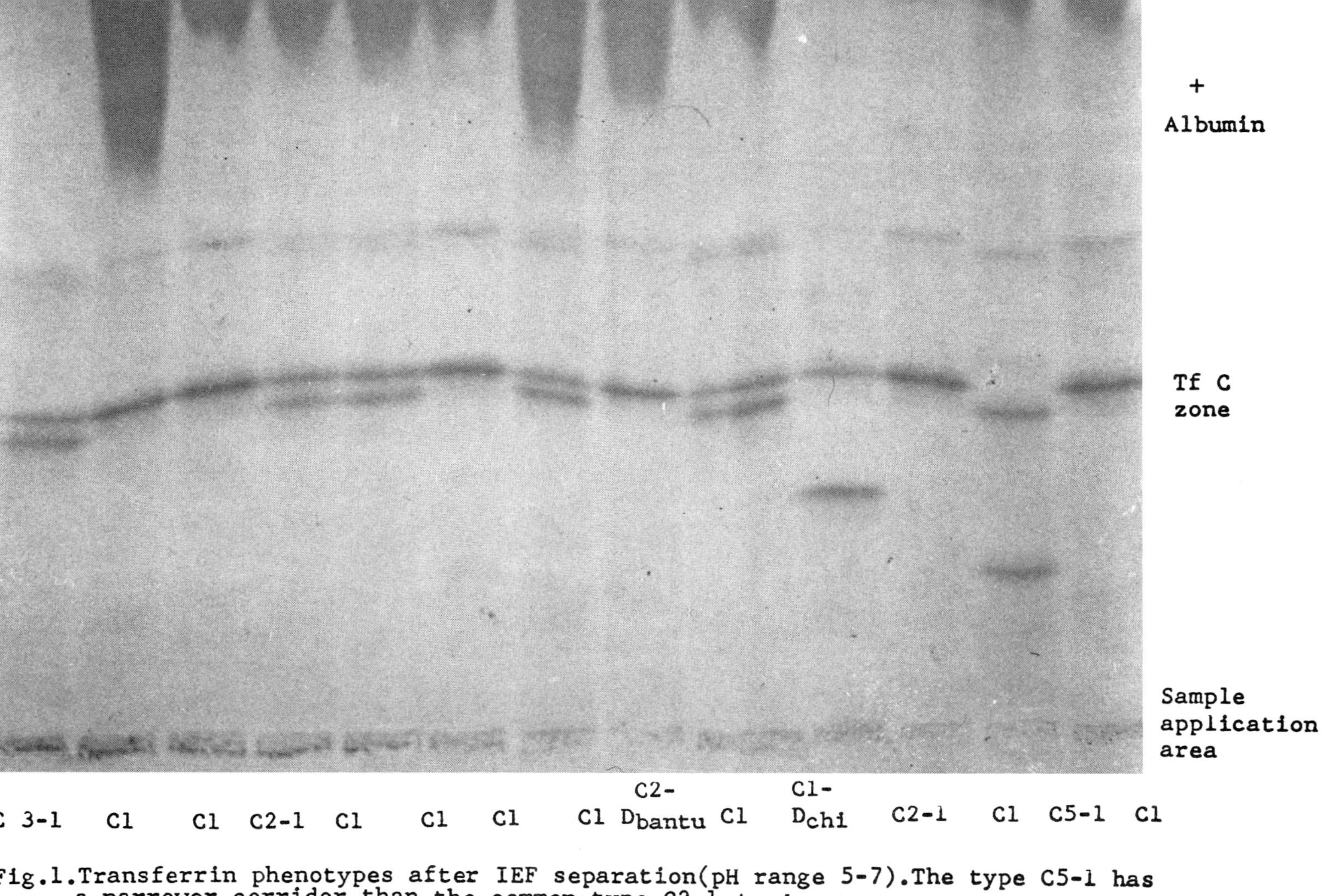



Fig.1.Transferrin phenotypes after IEF separation(pH range 5-7).The type C5-1 has
a narrower corridor than the common type C2-1.Anode at top.

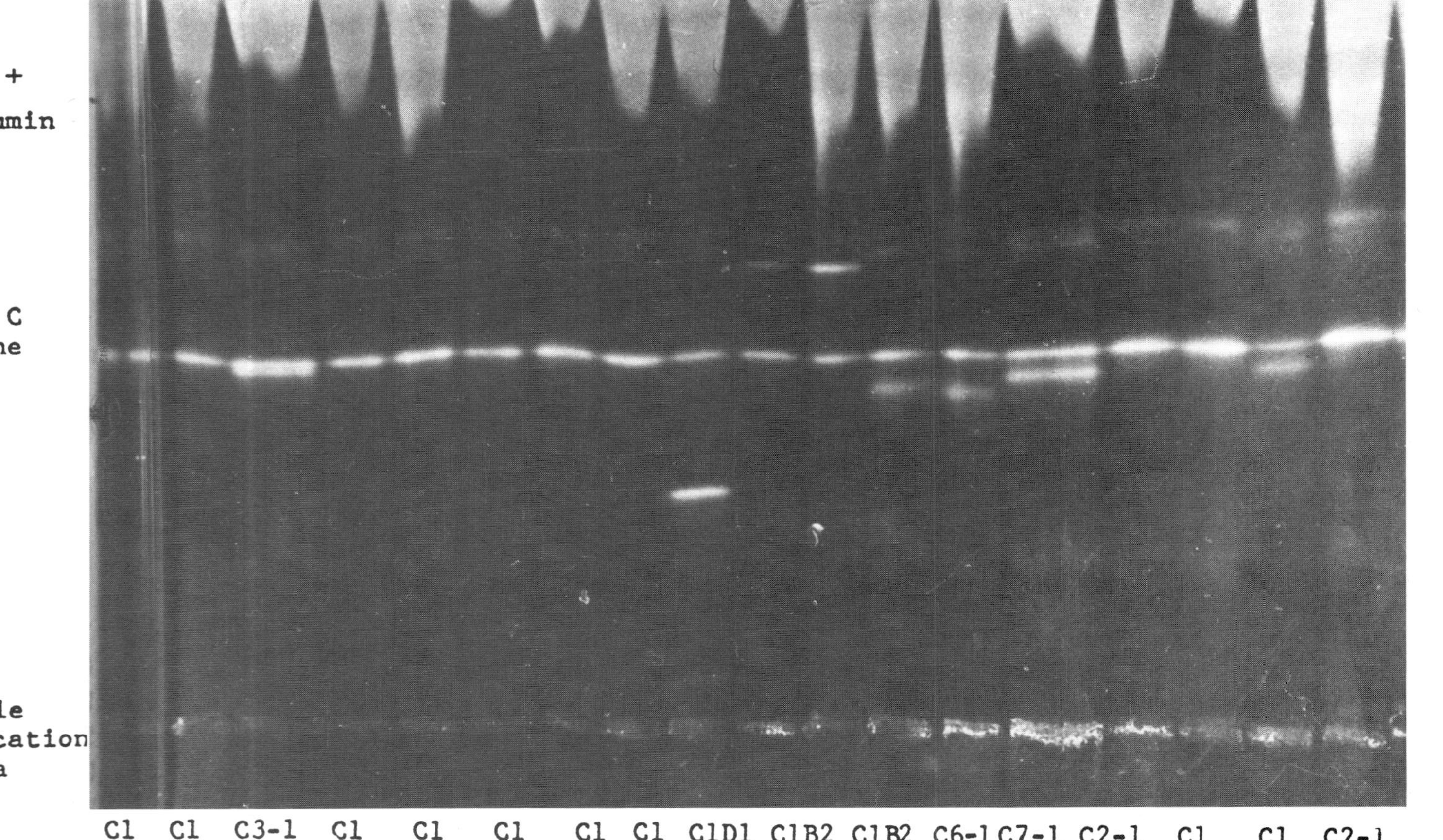


Fig.2. Transferrin phenotypes after IEF separation(pH range 5-7).Note that the corridor of C6-1 and C7-1 is markedly wider than in the common type C2-1.Anode at top

Tf^C - Subtypes
TfC1 3-1 3 2-1 3-2 2 4-1 5-1 6-1 7-1
C4
C1
C3
C5
C2
C6
C7

Tf^C - Subtypes/Frequencies
C4 ←Indians 0.04
0.77→C1
0.06→C3
Caucas.—
C5 ←Negroes 0.02
0.16→C2
C6 ←N.Africa 0.03
C7 ←S.France

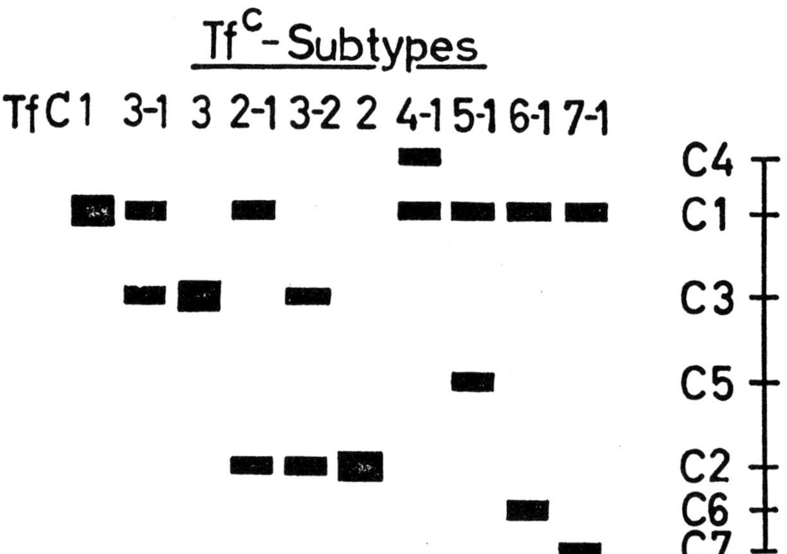

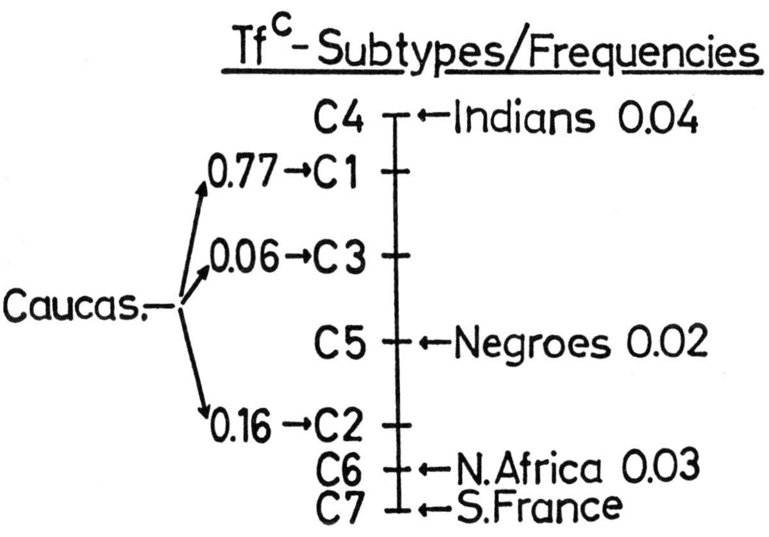

Fig.3.

Table 1

Comparison of Tf phenotypes and allele frequencies in various populations

| Authors | Population | n | Phenotypes | | | | | | | | | | Allele frequencies | | | | | |
| --- | --- | --- | --- | --- | --- | --- | --- | --- | --- | --- | --- | --- | --- | --- | --- | --- | --- | --- |
| | | | C1 | C2-1 | C2 | C3-1 | C3-2 | C3 | CB[a] | CD | C4-1 | C5-1 | $TfC1$ | $TfC2$ | $TfC3$ | $TfC4$ | $TfC5$ | $TfVar$[c] |
| Kühnl and Spielmann 1978 | FRG ( Hesse ) | 942 | 631 | 269 | 27 | - | - | - | 14 | 1[b] | - | - | 0.8195 | 0.1720 | - | - | - | 0.0085 |
| Thymann 1978 | Denmark | 132 | 90 | 35 | 7 | - | - | - | - | - | - | - | 0.8144 | 0.1856 | - | - | - | - |
| Czakanski 1979 | FRG ( Hesse ) | 294 | 209 | 73 | 8 | - | - | - | 4 | - | - | - | 0.8401 | 0.1530 | - | - | - | 0.0068 |
| Kühnl and Spielmann 1979 | FRG ( Hesse ) | 252 | 158 | 64 | 6 | 17 | 2 | 1 | 4 | - | - | - | 0.795 | 0.155 | 0.042 | - | - | 0.008 |
| Hoste 1979 | Belgium | 253 | 160 | 81 | 12 | - | - | - | - | - | - | - | 0.792 | 0.208 | - | - | - | _[d] |
| Stibler et al. 1979 | Sweden 1) Stockholm | 100 | 81 | 18 | - | - | - | - | 1 | - | - | - | 0.905 | 0.090 | - | - | - | 0.005 |
| | 2) Environs(Stockholm) | 100 | 71 | 25 | 3 | - | - | - | 1 | - | - | - | 0.840 | 0.155 | - | - | - | 0.005 |
| | 3) Lapps | 100 | 68 | 29 | 2 | - | - | - | 1 | - | - | - | 0.830 | 0.165 | - | - | - | 0.005 |
| Martin 1979 | Berlin (West) | 121 | 74 | 42 | 5 | - | - | - | - | - | - | - | 0.7851 | 0.2149 | - | - | - | - |
| Constans et al. 1980 | France(Pyreneans)[e] | - | - | - | - | - | - | - | - | - | - | - | 0.7880 | 0.1320 | 0.053 | - | - | 0.0270 |
| | North and South American Indians | - | - | - | - | - | - | - | - | - | + | - | - | - | - | 0.035 | - | - |
| | US - Negroes | - | - | - | - | - | - | - | - | - | - | + | - | - | - | - | 0.015 | - |
| Altland et al. 1980 | FRG ( Hesse and Baden-Württemberg )[e] | 352 | - | - | - | - | - | - | - | - | - | - | 0.750 | 0.165 | 0.072 | - | - | 0.0088 |
| Driessel et al. 1980 | FRG ( Western Germany ) | 380 | 229 | 86 | 5 | 43 | 7 | 2 | 8[f] | - | - | - | 0.7816 | 0.1355 | 0.0711 | - | - | 0.0118 |
| Kühnl 1980 | FRG ( Hesse ) | 876 | 519 | 213 | 24 | 86 | 14 | 3 | 15 | 2 | - | - | 0.7711 | 0.1581 | 0.0611 | - | - | 0.0096 |

Note : a) phenotypes CB2 and CB1-2 are listed together as CB
    b) a B2D1 heterozygote
    c) frequencies of $Tf^B$ and $Tf^D$ are listed together as "Var"
    d) calculation of frequencies without B- and D heterozygotes
    e) publication of allele frequencies without phenotype distribution
    f) one B2 homozygote included

In Fig.3 a survey of the hitherto identified $Tf^C$ subtypes
C1 to C7 is given together with the frequencies of the
common alleles C1,C2 and C3 in Caucasians, the apparently
race-specific allele C4(North American,Bolivian and Brazi-
lian Indians(4),C5 and C6.As the numerical nomenclature
of the seven subtypes has become somewhat confusing(sequence
from anode to cathode:C4,C1,C3,C5,C2,C6 and C7) a redesigna-
tion might be discussed, as this has been possible in the
system of the group-specific component(Gc-globulin)(9).
The comparison of Tf phenotypes and allele frequencies in
various populations shows(Table 1),that there is generally
good agreement between the data of different authors.Only
the $Tf^{C2}$ frequency calculated for Stockholm(Stibler et al.,
1979, n=100) is surprisingly low(0.09).
In conclusion we feel, that the extended Tf polymorphism
with three common alleles in Whites and three more common
$Tf^C$ subtypes in the two other major races may be useful for
gene mapping and forensic serology(identification of blood
stains, paternity testing).In affiliation cases the so-called
single exclusion chance for non-fathers is raised from about
1% in the conventional electrophoretic separation systems
(SGE,AGE) to 17.2% by the application of the IEF technique(10).


## References


1) Smithies,O.: Variations in human serum $\beta$-globulins.Nature
   180,1482(1957).
2) Kühnl,P. and Spielmann,W.:Transferrin:Evidence for two common subtypes of the $Tf^C$ allele.Hum.Genet.43,91-95(1978).
3) Kühnl,P.and Spielmann,W.:A third common allele in the transferrin system,$Tf^{C3}$,detected by isoelectric focusing.Hum.Genet.
   50,193-198(1979).

582

4) Constans,J.,Kühnl,P.,Viau,M.and Spielmann,W.:A new procedure for the determination of transferrin$^C$(Tf$^C$) subtypes by isoelectric focusing.Existence of two additional
   alleles,Tf$^{C4}$ and Tf$^{C5}$.Hum.Genet.__55__,111-114(1980)

5) Kühnl,P.,Constans,J.,Spielmann,W.,Viau,M.and Kaltwasser,
   J.P.:Analyse von fünf Subtypen des menschlichen Trans-
   ferrin$^C$-Gens. Forschungsergebnisse der Transfusionsmedi-
   zin und Immunhämatologie __7__,637-648(1980)

6) Constans,J.,Kühnl,P.,Viau,M.,Spielmann,W.and Kaltwasser,J.P.:
   Transferrin:Five common subtypes of the Tf$^C$ allele detected
   by isoelectric focusing.Abstr.Vol. 276, 18$^{th}$ Congress Int.
   Soc.Hematology & 16$^{th}$ Congress Int.Soc.Blood Transfusion,
   Montreal,1980

7) Lestas,A.N.:The effect of pH upon human transferrin:Selec-
   tive labelling of the two iron binding sites.Br.J.Hematol.
   __32__,341-350(1976)

8) Kühnl,P.:Elektrofokussierung in der Blutgruppenkunde.Habil.-
   Schrift.Universität Frankfurt/Main ,1980

9) Constans,J.et al.:Group-specific component.Report on the
   first international workshop, 27.-28.7.1978,Paris. Hum.
   Genet.__48__,143-149(1979)

10)Spielmann,W. and Kühnl,P.:The efficacy of modern blood
   group genetics with regard to a case of probable superfe-
   cundation.Haematologia __13__,75-85(1980)

11)Thymann,M.:Identification of a new serum protein polymorphism as transferrin.Hum.Genet.__43__,225-229(1978)

12)Czakanski,C.:Untersuchungen zum erweiterten Polymorphismus
   des Transferrinsystems.Med.Inaug.-Diss.Frankfurt/Main 1978.

13)Hoste,B.:Group-specific component(Gc) and transferrin(Tf)
   subtypes ascertained by isoelectric focusing.Hum.Genet.__50__,
   75-79(1979)

14)Stibler,H.,Beckman,G.,and C.Sikström:Subtypes of transferrin
   C.Hum.Hered.__29__,320-324(1979)

15)Martin,W.:Zur Gc- und Tf-Typisierung mit Hilfe der isoelektri-
   schen Fokussierung.Forensic Sci.Int.14,152(1979)
16)Altland,K.,Hackler,R.and W.Knoche:Double one-dimensional electrophoresis of human serum transferrin:A new high-resolution
   screening method for genetically determined variation.Hum.
   Genet.54,221-231(1980)
17)Driesel,A.J.,Scheil,H.G.,Pfeiffer,I.M.,and Röhrborn,G.:Gene
   frequencies of transferrin(Tf$^C$)subtypes in Western Germany
   (Düsseldorf region).Z.Rechtsmed.86,133-135(1981).

ISOELECTRIC FOCUSING OF HAIR PROTEINS


Bruce Budowle[1,3], Rodney C.P. Go[2,3], Ronald T. Acton[1,2,3]
Departments of Microbiology[1], Public Health[2], and Diabetes
Research and Training Center[3], University of Alabama in
Birmingham, Birmingham, Alabama, USA


Introduction

Properties of hair, such as color, size, shape, and texture,
are limited in their use for genetic identification.
Analysis of polypeptide components of hair may be more
informative because of greater specificity than the previously mentioned characteristics.  The principal protein
component of hair is keratin, which can be divided into two
groups--fibrous and matrix proteins (1,2).  The fibrous proteins have molecular weights of 40,000 to 58,000 daltons and
a ½ cystine content of 5 to 7 residues per 100 residues
(2,3), while the matrix proteins have molecular weights of
10,000 to 28,000 daltons and a ½ cystine content of 25 to 30
residues per 100 residues (4,5,6).  Because of the numerous
disulfide bonds between polypeptide chains of keratin, these
proteins are relatively insoluble and, therefore, difficult
to analyze electrophoretically.  However, two similar
procedures have been successful in extracting the fibrous
and matrix proteins of hair.  Shechter, et al. (1) solubilized these proteins by a method using thioglycolate reduction and iodoacetate alkylation.  The method yields S-
carboxymethyl-kerateines which when analyzed by discontinuous electrophoresis in 7.5% polyacrylamide gels containing
6 M urea reveals a total of 10 polypeptide fractions.  The
other method reported by Baden, et al. (7), uses 0.2 M Tris
containing 6 M urea and 0.2 M Mercaptoethanol at pH 9.5 to

extract the hair proteins which are then treated with
iodacetate to give the S-carboxymethyl derivative. Discontinuous electrophoresis of this extract in 17.5% polyacrylamide gels containing 6 M urea yields, including
variants, 11 different polypeptide fractions (2,7). Before
hair proteins can be used for various genetic studies, a
method is needed to increase the number of detectable
polypeptides, so more variability, if it exists, can be
demonstrated. This paper reports a highly reproducible
method which yields an increased number of polypeptide
fractions from hair.

Materials and Methods

The hair was obtained from a normal, adult, caucasian male.
Extraction and alkylation of the fibrous and matrix proteins
of hair was performed according to the method of Baden, et
al. (7). The hair was extracted in 0.2 M Tris, containing 6
M urea and 0.2 M mercaptoethanol, pH 9.5, under nitrogen at
$50^{o}C$ for three hours. Ten mg of chopped hair were used per
ml of buffer. The supernatant was treated with iodoacetate
and dialyzed against three changes of distilled water. The
extract was lyophilyzed and stored below $0^{o}C$.

Isoelectric focusing was performed in slab gels (110 mm x 95
mm x 0.8 mm) containing urea with a pH range of 2.5 - 5.0.
The gels were prepared from a solution of 6.9 g urea, 1.9 ml
of 29.1% acrylamide and 0.9% bisacrylamide, 4.9 ml of $H_2O$,
0.31 ml of 2.5 - 4.0 ampholines, 0.31 ml of 3.5 - 5.0 ampholines, and 25 µl of 10% ammonium persulfate. The solution
was deaerated and 15 µl of N, N, $N^1$, $N^1$-tetramethylethylene-
diamine was added. After polymerization for one hour on a
Bio-Rad casting tray, the gel was wrapped in Saran Wrap and
allowed to stand overnight at room temperature. For the
focusing run 0.01 M phosphoric acid was used for the

electrolyte at the anodal end and 0.02 M NaOH was the electrolyte at the cathodal end. The circulating water bath was set at 10°C. The power supply was set at 600 V and 50 mA and prerun for 30 minutes. Rectangles of Whatman #3 MM chromatography paper (5 mm x 6 mm), soaked in protein samples solubilized with 6 M urea were placed one centimeter from the cathodal end, and the run was continued for 30 minutes. The filter papers were removed and the run was continued for another 30 minutes at 600 V. The voltage was then increased to 800 V for 2 hours and finally 1000 V for 90 minutes. After the run, the gel was fixed in 12.5% trichloroacetic acid overnight, washed in a solution of 45% ethanol and 10% acetic acid for one hour and stained in 0.1% Coomassie brilliant blue R250. The gel was destained in a solution of 45% ethanol and 10% acetic acid, scanned with a tungsten lamp densitometer, and photographed.

Results and Discussion

This report describes an isoelectric focusing electrophoretic system which readily displays the protein components of hair. Using this system, there are 47 observable polypeptides (Figures 1,2), thus the resolution is more than four times greater than any previous report. Moreover, this procedure has been reproduced on four separate occasions. The results indicate that the number of different matrix and fibrous protein components of hair was greatly underestimated and still may be so. Since 17.6% of the amino acid composition of hair is comprised of glutamic and aspartic

Figure 1.   Isoelectrofocusing gel containing 9 M urea showing the band pattern of hair proteins.   The anode is to the right.

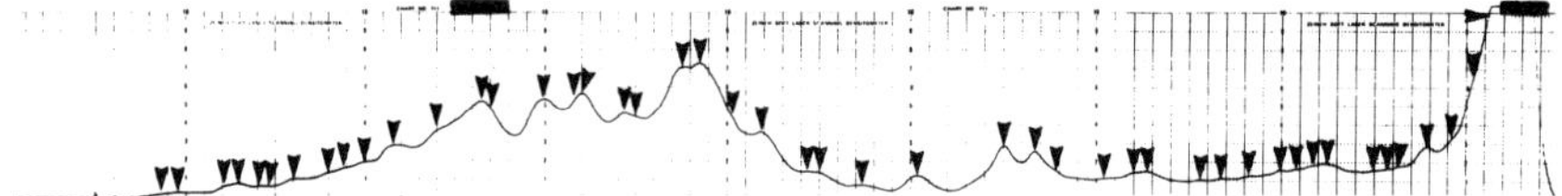

Figure 2. A densitometer scan of a gel containing isoelec-
trofocused hair proteins. Each arrow points to an individual
band. The anode is to the right.

acid (8), those proteins are acidic, and therefore, a substantial portion of these acidic proteins are stacked up at
the anode and are not identifiable. Hair proteins with
isoelectric points below 2.5 cannot be separated by the
present state of the art. To alleviate this problem and
identify an even greater number of polypeptides, we are
presently investigating a carbodiimide treatment (9) to
neutralize the carboxyl groups of glutamic and aspartic acid
by adding on isopropyl groups. Therefore, the more acidic
hair proteins will not be as acidic and possibly more
polypeptides can be identified within the established pH
range of ampholines.

Another difficulty encountered with this system is polymerization time of the IEF gel. Gels allowed to polymerize
on the casting tray for one hour gave the best results. The
protein bands were more diffuse when a gel was set on the
tray for less than one hour. If a gel sets for longer
than one hour, it has a tendency to adhere to the casting
tray.

Avoidance of excessive cooling of these gels is required;
otherwise, the urea will crystallize. Urea crystals can
tear a gel, so it will no longer be usable. Therefore,
cooling and storage of gels should be done at $10^{o}C$ or warmer.

This electrophoretic system resolves an increased number of
fibrous and matrix proteins of hair. Since, Baden et al.
(7) have shown that genetic variation exists in 5% of their

Caucasian sample, it is possible, within this greater number of polypeptides, that greater variability, if it exists, can be demonstrated. We are presently investigating different racial groups and families for genetic differences. In addition, extraction procedures, using zwittergent, cyanogen bromide, and others, are being evaluated to possibly simplify the hair extraction and solubilization procedures and thereby facilitate the analysis.

References


1.  Shechter, Y., Landau, J.W., Newcomer, V.D.:  J. Invest. Derm. 52, 57-62 (1969).

2.  Lee, L.D., Ludwig, K., Baden, P.:  Forensic Science 11, 115-121 (1978).

3.  Thompson, E., O'Donnell, I.J.:  Aust. J. Biol. Sci. 18, 1207-1225 (1965).

4.  Gillespie, J.M., Inglis, A.S.:  Comp. Biochem. Physiol. 15, 175-185 (1965).

5.  Harrap, B.S.:  Aust. J. Biol. Sci. 15, 596-597 (1962).

6.  Gillespie, J.M, Harrap, B.S.:  Aust. J. Biol. Sci. 16, 252-258 (1963).

7.  Baden, H.P., Lee, L.D., Kubilus, J.:  Amer. J. Hum. Genet. 27, 472-477 (1975).

8.  Baden, H.P., Freedberg, I.M.:  Dermatology in General Medicine, McGraw-Hill Co., New York, pp. 96-101 (1979).

9.  Rebek, J., Feitler, D.:  J. Amer. Chem. Soc. 96, 1606-1607 (1974).

ISOENZYME TYPING OF DRIED BLOODSTAINS USING MYLAR BACKED CELL-
ULOSE ACETATE MEMBRANES

Robert Briner, Robert Longwell, Yvonne Moll
Southeast Missouri Regional Crime Laboratory
Cape Girardeau, Missouri

Retha Matthews
Illinois Division of Investigation Laboratory
Fairview Heights, Illinois

Alice Abbot
St. John's Mercy Hospital Clinical Laboratory
St. Louis, Missouri

Randall Webster
Saint Louis Police Department Laboratory
Saint Louis, Missouri

Introduction

In the past decade, progress in the isoenzyme typing of blood
factors has made blood stain evidence one of the most important
investigative tools of the forensic scientist. Originally all
work in this area was carried out using starch gels. At the
present time there are four major types of electrophoretic support media being utilized for these determinations: starch,
agarose, polyacryamide, and cellulose acetate. This presentation represents the work carried out in our laboratories to
adapt the methodology to mylar backed cellulose acetate membranes (CAM).

Traditionally, the only method used in identifying human blood
in the Crime Lab was the ABO system. The rarest blood type

using the ABO system is type AB, being only 3%.  This means
that 3 out of every 100 people may have this blood type, while
an almost 50% of the human population is type 0.  It is apparent from these statistics that in order to lower the percentage
of an individual's blood type, a more detailed identification
method was needed.

In the early 1970's, research was initiated to improve the in-
dividualization of human blood on dried blood stains.  In addition to ABO, other systems were utilized in the identification
of blood.  Red cell antigens used were the MN and Rh systems
(mainly $Rh_D$).  Protein groups involved were hemoglobin and
haptoglobin.  Several enzyme systems were utilized; the more
useful of these being phosphoglucomutase (PGM), erythrocyte
acid phosphotase (EAP), adenylate kinase (AK), esterase D (ESD)
and adenosine deaminase (ADA).  Initial work was carried out by
Bryan J. Culliford of the Metropolitan Police Forensic Laboratory of London, England (1).  Using these different systems,
the frequency of occurance can be lowered considerably.  Starch
gel was used as the supporting medium for the electrophoresis.
However, another problem arose; the time factor.  When using
starch as the supporting medium, one particular enzyme system
might take anywhere from 15 to 35 hours to complete.

In the mid 1970's, workers at the University of Pittsburgh
proposed a method of multiple enzyme separation on a single
plate.  This work has been carried to completion under the direction of B. Wraxall (2).  By covering a certain portion of
the starch gel and developing each portion specifically for
that enzyme present, it was possible to determine several enzyme systems at the same time.  However, by covering a certain
portion of the starch gel, rarer isozymes of another enzyme
could have been covered up.

Substantial research has been carried out by Dr. B. W. Grunbaum
and P. L. Zajac using the Beckmann Micro-zone® system and ad-

apting the procedures of starch gel to Sartorious® cellulose
acetate membranes; the time factor was reduced tremendously
(3, 4).

An alternate system is the Helena system using cellulose acetate plates with a mylar back as the supporting medium.  This
system is now being used at the Southeast Missouri Regional
Crime Laboratory to identify genetic markers in human blood.
The results are discernable and readily reproducible.  All the
enzyme systems can be completed in less than 8 hours total.


## Methods

Certain variables may influence the outcome of the procedure.
These variables include sample preparation, cellulose acetate
plate preparation, sample applicator, the pH of the buffers,
ionic strength, voltage, amperage and development.

The dried blood stain should be stored in the freezer to preserve its enzymatic activity for as long as possible.  A .5 cm
x .5 cm piece of stained cloth is cut from the original sample
placed in a well and 2-3 drops of solvent is added.  The solvent is usually distilled water, however, Cleland's reagent,
pH 8.0, is used for EAP and mercaproethanol is used for ADA.
The samples can be soaked overnight in a humidity chamber at
$4^{\circ}$C.  Samples extracted in Cleland's reagent or mercaproethanol
are stable for approximately 8 hours.

Twenty minutes prior to applying the samples, the cellulose
acetate plates are soaked in the individual system membrane
buffer, which usually is a dilution of the tank buffer.  The
correct dilution of membrane buffer is essential for good resolution, to prevent high amperage and excess heat.  The plates
must be soaked from the bottom up to insure saturation without
causing air bubbles.  Three techniques may be used:

594

1. The plates may be placed in a beaker and membrane buffer
   allowed to flow slowly down the side from a separatory fun-
   nel.
2. The same technique as above may be used by pouring the buffer directly from the bottle carefully down the side of the
   beaker.
3. A beaker may be filled with buffer and the plate slowly and
   steadily immersed in the buffer.

Air bubbles in the cellulose acetate impede the electric current through the plate. They cannot be removed and the plate
must be discarded.

The pH of the tank buffers, membrane buffers and reaction buffers is extremely important. All enzymes have a certain pH
range for optimal activity. Extremes of pH tend to denature
the enzymes, as does temperature extremes. The ionic strength
of the buffer also is a very important variable.

The voltage and amperage play a significant part in good resolution. The voltage should remain constant, allowing the amperage to start low (about 2 milliamperes per plate) and slowly
rise. The amperage should never be allowed to rise above 10mA
per plate in order to avoid overheating of the system; as the
amperage rises, so does the temperature. If the temperature
rises over the optimal range, the enzyme will invariably be
denatured.

The length of the electrophoretic runs varies with the particular enzyme, none of which require more than one hour.

Development of the cellulose acetate plate after the electrophoretic run in critical. There are two basic types of development: visual and fluorescence. The results should be recorded and kept for future reference. The plates are photographed
with black and white Polaroid film in all cases.

## Electrophoresis

A.  Preparation of the tank:  The inner chambers are filled
    halfway with water and kept frozen until the tank in need-
    ed.  The tank is prepared by pouring 50 mls of tank buffer
    into each outer chamber.  Two wicks (2" x 5") made from
    filter paper are submerged in the buffer of the outer cham-
    bers and draped over the edges of each chamber to make el-
    ectrical contact with membrane being careful not to touch
    the ice in the inner chambers.  Cover the tank until ready
    to use.

B.  Application of the sample:  After the cellulose acetate
    plates are saturated with the membrane buffer, one plate is
    taken from the buffer and blotted dry.  With a magic marker
    place a mark on the right side of the mylar surface.  This
    marks sample number one.  Place the cellulose acetate plate
    on the aligning base mylar side down.  Most procedures call
    for a cathodic application.  One exception is PGM which
    calls for an extreme cathodic application.  For PGM, place
    the application 1 cm from the side of the plate.  To make
    an application, first clean the applicator by wetting it,
    brushing lightly with a toothbrush, rinsing and drying.
    Depress the applicator several times into the sample wells
    and apply sample to a blotter.  This helps prime the appli-
    cator, so that even spotting will be placed on the plate.
    Again depress the applicator 3 or 4 times in the sample
    wells and apply to the cellulose acetate.  Hold firmly
    on acetate for 5 seconds.  Three applications of 0.25 ml
    for each application is sufficient for most enzymes.  The
    plates are then transferred to the electrophoresis tank,
    membrane side down, and a plastic rod is placed on the back
    side of each electrode to insure good electrical contact
    with the filter paper wicks (Figure 1).  Electrophoresis
    is then carried out according to each procedure.

C.  Development:  The same number of plates used for the run
    must be soaked in reaction buffer, using the method des-

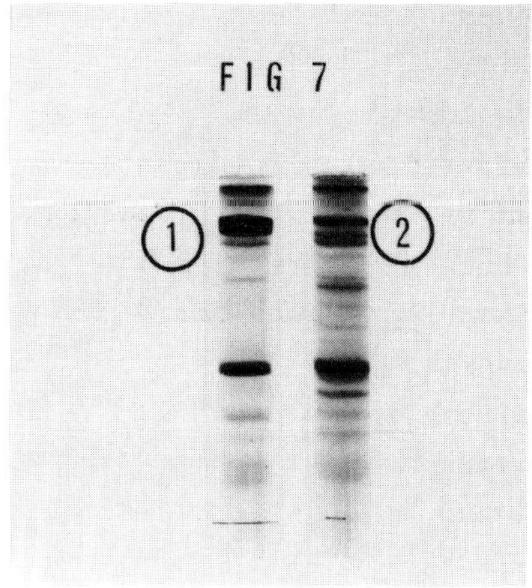



Figure 7. SDS polyacrylamide electrophoresis gels of the myofibrillar proteins of halibut muscle. Gel 1., fresh halibut muscle. Gel 2 stored frozen (-40°C) halibut muscle. Densitometric tracings A and B are the respective tracings of gels 1 and 2.

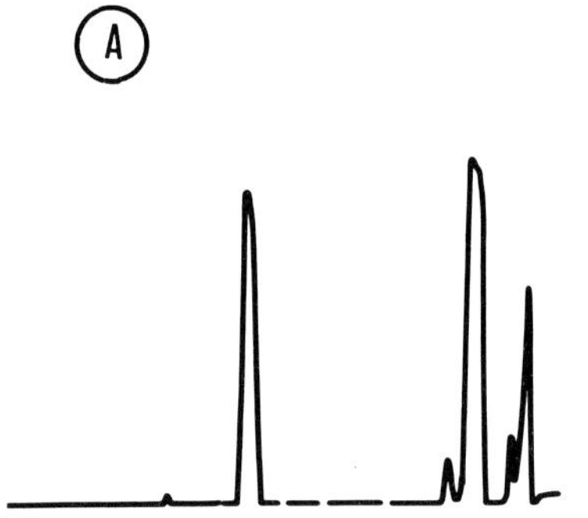



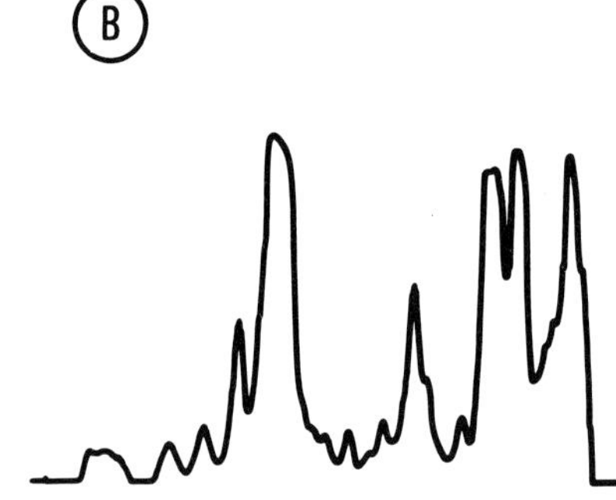

References

1.  Olson, D.G., Parrish, F.C. and Stromer, M.H. : J. Food Sci.
    41, 1036-1042 (1976).

2.  Gill, T.A., Keith, R.A. and Smith Lall, B.: J. Food Sci.
    44, 661-667. (1979).

3.  Clayton, J. W. and Tretisk, D.N. : J. Fish Res. Bd. Canada,
    29, 1169-1172.

4.  Odense, P.H., Allen, T.M., and Leung, T.C. : Can. J. of
    Biochem. 44, 1319-1326. (1966).

5.  Syner, E.N. and Goodman, M. : Science 151, 206-208. (1966).

6.  Shaw, C.R. and Koen, A.L. : In Chromatographic and Electro-
    phoretic Techniques ed. Smith, J., William Hinemann Medical
    Books Ltd., London. (1968).

7.  Porizo, M.A. and Pearson, A.M. : Biochem. Biophys. Acta 490,
    27-34. (1977).

8.  Weber, K. and Osborn, M. :  J. Biol. Chem. 244, 4406. (1969).

9.  Whitt, G.S. : J. Expt'l. Zool. 175, 1-36. (1970).

10. Sikorski, Z., Olley, J. and Kostuch, S. : Crit. Rev. in Fd.
    Sci. and Nutr. 8 (1), 97-129. (1970).

11. Fennema, O. : In Water Relations of Foods. Ed. Duckworth,
    R.B., Academic Press, London. (1975).

12. Odense, P.H., Leung, T.C. and Annand, C. : I.C.E.S.,
    C.M. H:21. (1973).

13. Kim, H.K., Robertson, I. and Love, R.M. : J. Sci. Fd.
    Agric. 28, 699-700. (1977).

# ELECTROPHORETIC BEHAVIOUR OF CROSSLINKED DNA

Eckart W. Wunder

Institut für Anthropologie und Humangenetik
Universität Heidelberg
D-6900 Heidelberg,Germany(G.F.R.)


## Introduction

Intrastrand crosslinks change some of the physicochemical properties of DNA;for example,they cause fast renaturation by
keeping the sequences in register during denaturation,which
allows snap-back.Cole[1]has demonstrated,that crosslinks change
the sedimentation behavior in alkaline sucrose gradients of
phage $\lambda$-DNA.We describe here the effect of DNA-crosslinks  on
the electrophoretic migration in gels.


## Materials and Methods

<u>Nondenaturing agarose gel</u>.A 0,8% agarose slab gel was cast in
3mm thickness in a horizontal electrophoretic apparatus as described before(this volume).Buffer conditions:Tris-HCl 40mM,pH8.0,
EDTA 2mM,Sodiumacetate 20mM.Samples were layered with 10% sucrose
and bromphenol blue plus xylenecyanine as electrophoretic markers.
 $\lambda$-DNA was digested with EcoR 1(from Boehringer,Mannheim)to completion.2$\mu$g of the fragments were treated with 8-methoxypsoralen
(Elder Comp.,Ohio)at 10mg/ml and irradiation with a Sylvania lamp.
(Blacklite Blue,FIST 8)at 2 Watt/cm$^2$for 15 min.0,7$\mu$g  of treated
and untreated DNA were run on the gel for 7$^h$ at 1,2V/cm.

<u>Denaturing agarose gel.</u>0,8% agarose horizontal slab gels were prepared as above,except the denaturing buffer conditions,according to Studier(2):NaOH 30mM,EDTA 10mM,pH 12,4.$\phi$X174DNA was completely linearized by digestion with XhoI.2$\mu$g of this form3-DNA was modified by 8-methoxypsoralen and UVa as above.0,3$\mu$g of untreated and treated DNA at concentrations of psoralen between $5 \times 10^{-3}$ and $2 \times 10^{-1}$ mg/ml in a volume of 10$\mu$l were denatured by adding NaOH to adjust pH to 12,6,and layered in the wells as described. Electrophoresis was at 1,4V/cm for $16^{h}$.

<u>Sequencing polyacrylamide gel.</u>Vertical slab gels were cast with 20% polyacrylamid(2xpurified)and 0,6% bisacrylamide in a size of 200x400mm;thickness 1,5mm.Buffer conditions:Tris-borate 45mM, pH8,3;EDTA 2,5mM.As substrate,pBR322-DNA was used.After complete digest with AvaI,the E-fragment was isolated and 5'endlabeled with $^{32}$Phosphate by polynucleotidkinase.This fragment is 249 basepairs long.After a further digest with AluI,the shorter fragment,containing 47 base pairs was isolated.For modification,10$\mu$g of trimethyl-psoralen was added to the DNA(about 0,2$\mu$g)in 40$\mu$l volume and irradiated for 30 min.using a 5mm thick glassplate as additional filter;this was repeated twice.The chemical treatment of modified and unmodified DNA was performed according to the sequencing procedure of Maxam and Gilbert(3).After formamide treatment,the different samples were layered with 10% sucrose and bromphenol blue plus xylenecyanine on the gel.Radioactivity was 5000 cpm per lane;the gel was run for $3^{h}$ at 1600 V.After the run,the gel was dried and autoradiography was made with Kodak 3X high speed X-ray film;exposure was for one week.


Results and Discussion


The nondenaturing gel shows,that heavily crosslinked DNA has an anomalous electrophoretic migration behaviour.The proportion of the distances between the restriction fragments is unchanged,
(see Fig.1)

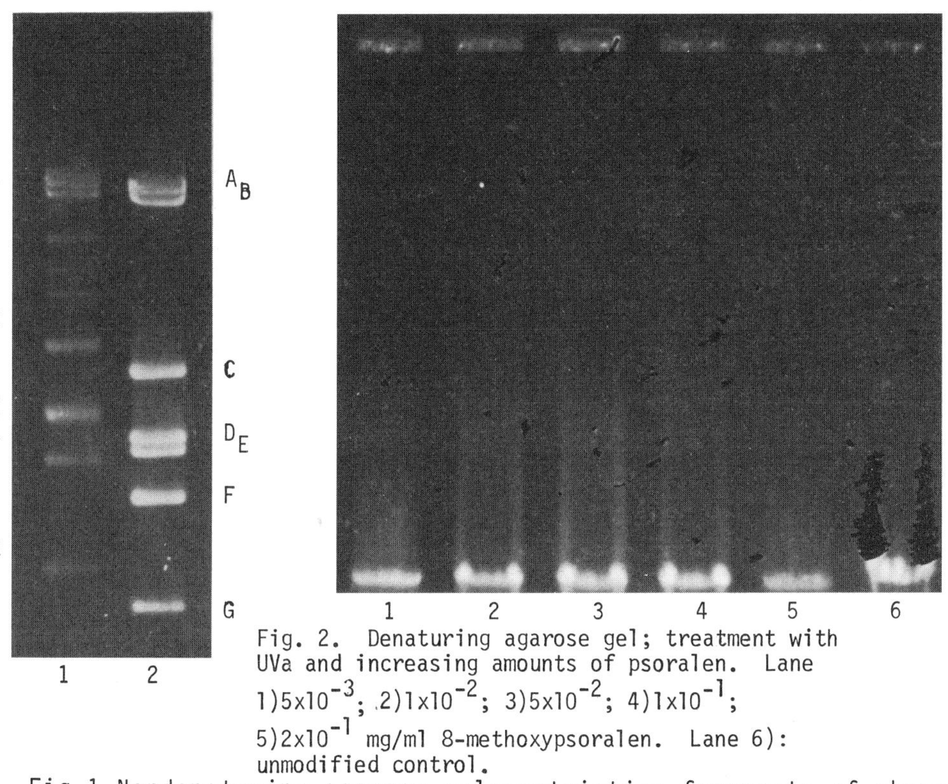

Fig. 2. Denaturing agarose gel; treatment with
UVa and increasing amounts of psoralen. Lane
1)$5x10^{-3}$; 2)$1x10^{-2}$; 3)$5x10^{-2}$; 4)$1x10^{-1}$;
5)$2x10^{-1}$ mg/ml 8-methoxypsoralen. Lane 6):
unmodified control.

Fig 1.Nondenaturing agarose gel;restriction fragments of phage $\lambda$-
DNA after EcoRI-digest.Lane 1)fragments treated with 8-methoxy-
psoralen and UVa.Lane 2)unmodified fragments.

whereas the $R_f$ of the modified fragments indicates about 8%
slower migration. If the modified DNA is electrophoresed under
denaturing
with strongly decreased migration.As can be seen from Fig. 2,the
amount of slower migrating molecules as well as the degree of
migration inhibition depends on the degree of modification.The
smears migrating slower than original molecule size,first increase with increasing modification;this can be explained by
considering,that an increasing number of singlestrand molecules
is covalently bound to their complementary strand by at least
one DNA-DNA crosslink.This way,composite molecules with double
molecular weight and decreased migration are present,which can

unfold to different degrees.With increasing numbers of crosslinks
both complementary strands are bound together more tightly and
the degree of unfolding gets more and more limited;consequently,
migration is reversed towards **the** distance of unmodified DNA with
further increased psoralen concentration.
In sequencing gels,**heavily** modified DNA shows a very similar
pattern as unmodified DNA.It is interesting to note,however,that
the longer molecules,which are found closer to the start,again
migrate slower than the corresponding unmodified molecules(**Fig.3**).
Also,the sharpness of the bands decreases towards the start.  This change
might be caused by decreased flexibility of longer molecules, which are
quasi-doublestranded, since the complementary strands are kept together
by high crosslink density.
It can be summarized,that **heavily** crosslinked DNA shows decreased
electrophoretic mobility,which is a constant factor for longer
DNA in nondenaturing gels,increases with molecule length in seque
cing gels,and has a complex dose-effect relation in denaturing
gels.

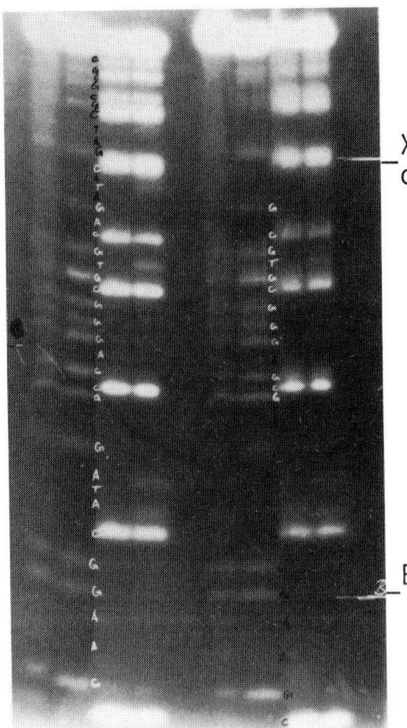


Fig. 3.  Sequencing gel of unmodified and heavily crosslinked DAN.  Lanes: 1) G; 2) A+G;3)
T+C; 4) C (=base specific nick-
ing)
  left 4 lanes:  unmodified,
  right 4 lanes: modified DNA.


References:1)Cole,R.S.,Zusman,D.:Biochim.Biophys.Acta 224, 660(1970
          2)Studier,F.W.:J.Mol.Biol.11,372    (1965)
          3)Gilbert,W.,Maxam,A.M.:Meth.Enzym. 65,499-560(1980)

# ELECTROPHORETIC MIGRATION OF FORM IV-DNA AND TOPOISOMERASE MONITORING

Eckart W. Wunder, Uwe Burghardt
Institut für Anthropologie und Humangenetik
Universität Heidelberg
D6900 Heidelberg, Germany (G.F.R.)

## Summary

An electrophoretic method is described which allows detection of form IV-DNA (covalently closed, completely relaxed rings) separately from form II-DNA by using horizontal agarose slab gels containing ethidiumbromide. This method is especially useful for topoisomerase monitoring in presence of unspecific nucleases during early steps of purification.

## Introduction

Topoisomerases can be classified as type I topoisomerase, which relaxes superhelical tension in DNA ring molecules in steps of one turn, and type II topoisomerase, which relaxes by steps of two turns, with or without ATP and $Mg^{++}$; both types of enzymes finally produce completely relaxed, covalently closed formIV-DNA from native, negatively supercoiled, replicative plasmid-or ringphage DNA (RFI). Along with the topological change, some physicochemical properties of the ring molecule change, thus making detection and quantification of the topoisomerases possible. In $CsCl_2$-gradients, differential uptake of ethidiumbromide by intercalation results in equilibration of form IV-DNA at a denser position than RFI-DNA. Using nondenaturing horizontal agarose slab gel electrophoresis with ethidiumbromide added to the

agarose gel,we could assay topoisomerase activity conveniently,
fast and at low cost in human placental extracts.


Materials and Methods


DNA substrate.PM2-DNA,containing about 93% RFI was used as substrate;150ng was used per assay.
Reaction conditions.Extracts of human placenta were prepared as
described previously(1);1µl was used per assay.Reaction volume
is 10µl.Incubation was for 1$^h$ at 37$^o$ C.Reaction buffer:Tris HCl
10mM,pH 8,0;NaCl 50mM or 150mM,resp.;EDTA 2mM or,alternatively,
Mg$^{++}$ 4mM.
Electrophoresis.Agarose(Sigma,type II)was heated with deionized
water in a waterbath until completely dissolved and adjusted to
0,8% concentration and ionic condition of the buffer at 70$^o$C;
ethidium bromide was added at 15µg/ml and well mixed;the same
concentration of ethidiumbromide was in the cathodic buffer.
Electrophoresis buffer:Tris HCl 40mM,pH 8,0;sodium acetate 20mM;
EDTA 2mM.Horizontal slab gels were run in an apparatus of our
design,which is installed over a transilluminator with light of
302nm(Ultraviolett Products,Chromato-Vue C63);the base plate is
made from UV-transparent Plexiglas.This allows checking the run
as well as photographing without interrupting the electrophoresis
The single pour construction principle is modified after Herrick(
with adjustable width of the slit.Thickness of the gels was 3mm;
running time was 12$^h$,at 1,6V/cm.The setup is seen in Fig.1.


Results and Discussion


Without ethidiumbromide being present in the agarose,migration
of the topoisomers resulting from topoisomerase action,decreases
with decreasing number of superhelical twists left in the ring
molecules.If the reaction is carried out at low temperature,
reaction intermediates can be demonstrated;incompletely relaxed

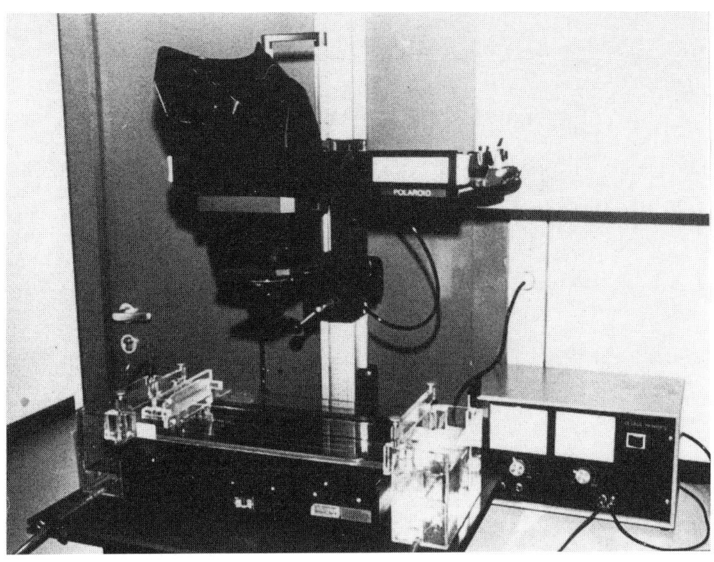

Fig. 1.Overall view of the horizontal one-pour gel electrophoresis apparatus installed immediately over a UV-transilluminator,
allowing observation and photography during the run without interruption of the electrophoresis.

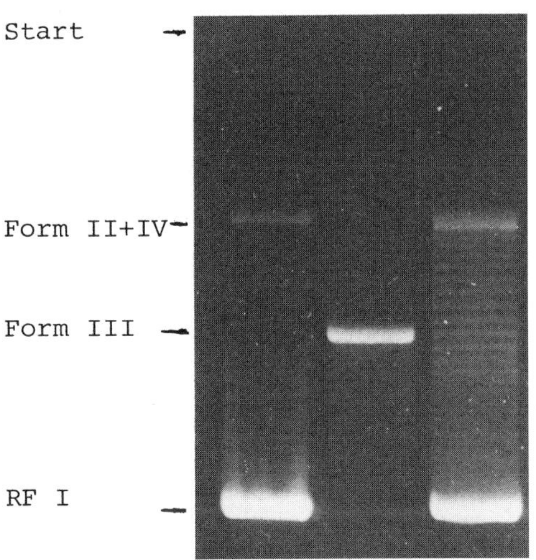


Fig. 2.Demonstration of
the decreased migration
of topoisomer rings with
decreasing number of super-
twists on an agarose gel
not containing ethidium-
bromide during electrophoresis.

Reaction intermediates
(covalently closed,partially
relaxed rings)

topoisomers,differing from each other by multiples of full twists,
migrate as a characteristic series of bands(Fig. 2);fully relaxed
rings(form IV-DNA )comigrate with form II-DNA(nicked rings).

614

Therefore,the final product of topoisomerase action cannot be
distinguished from the result of endonuclease action(form II-DNA)

By contrast,in gels containing ethidium bromide form IV-DNA migrates faster than RFI-DNA,whereas form II-DNA still migrates
slower than RFI-DNA.This has several advantages;1)topoisomerase
activity can be shown unequivocally in presence of endonucleases;
this is especially useful  in testing of unpurified extracts or
little purified fractions for $Mg^{++}$-dependant topoisomerase activity,since unspecific endonucleases are active in the presence of
bivalent cations and compete for the substrate.2)If the 'ladders'
are to be demonstrated in conventional gels,the reaction must not
be carried to completion,because form IV-DNA is indistinguishable
from form II-DNA.If the amount of topoisomerase activity is not
known,this requires several assays at different dilutions of the
extract or fraction.This problem is eliminated by the ethidiumbromide method,which at the same time requires much less DNA sub-
strate.3)Quantification of topoisomerase can be achieved by
visual or densitometric determination of 50% transition from
RFI to formIV-DNA,since the two resulting bands are easy to com-
pare.In serial dilutions of the extract,topoisomerase content
can be estimated fairly well,if endonucleases are suppressed by
EDTA and high salt concentration.
This method allowes convenient monitoring during purification
procedures like column chromatography.All fractions can be checke
in one electrophoretic run,if the gel apparatus shown above is
used,since many wells can be made in the long gel.As an example,
elution of topoisomerase activity from a hydroxyl apatite column
by phosphate gradient during purification of nuclear extracts of
human placenta is shown.

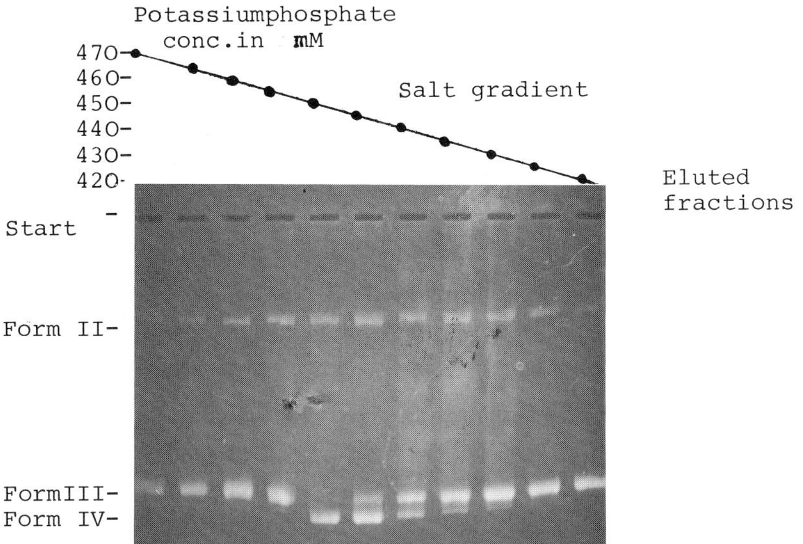


Fig. 3.Monitoring of chromatographic fractions for topoisomerase activity.Nuclear extract from human placenta was eluted
from a hydroxylapatite column by potassium phosphate gradient
as described.All fractions were reacted as in 'Methods' and
run on a horizontal agarose gel(0,8%)in presence of ethidiumbromide (not all fractions are shown).


References



1.  Wunder,E.,Burghardt,U.,Lang,B.,Hamilton,L.: Hum.Genet.,in
    press(1981)

2.  Herrick,G.: Analyt.Biochem. 108,346-347 (1980)

# PROPERTIES OF NUCLEIC ACIDS PURIFIED FROM AGAROSE


Joseph Locker
Departments of Pathology and Biochemistry
The University of Chicago
950 East 59th Street
Chicago, Illinois 60637


## Introduction

Agarose gels are indispensable to the contemporary molecular biology
laboratory.  Many formulations are widely used for electrophoresis: double-stranded DNA is generally resolved in agarose alone (1, 2), while
single-stranded molecules are best resolved in agarose gels containing a
denaturing agent (3, 4, 5).  Agarose gels are not only easy to set up and
run but afford high resolution of nucleic acids as long as 300,000 nucleotides, compared to an upper resolution limit of a few thousand nucleotides for acrylamide gels.  Since it is fairly easy to extract nucleic
acids from agarose after electrophoresis, agarose gels are often used for
high-resolution purification of individual nucleic acid species.  However,
many investigators have found that DNA or RNA purified from agarose shows
altered behavior in their experimental systems.  Particular complaints are
that this extracted nucleic acid does not digest well with restriction
enzymes, that it is a poor substrate for in vitro enzymatic labelling, and
that resolution is reduced upon re-electrophoresis.  On the other hand,
extracted nucleic acids show normal behavior in hybridization (3),  cell-
free translation (3), and bacterial transformation (6) experiments.

The deleterious effects of agarose on nucleic acids are usually attributed to agaropectins (7), highly sulfated polysaccharide precursors of
agarose.  Since agaropectins are relatively small, hydrophillic, and
highly charged, they are extracted from the gel under the conditions that
extract nucleic acids, and subsequently co-purify with them.  A variety
of different kinds of agarose are available, with significantly different

Figure 3.  <u>Inhibition of T4 Polynucleotide Kinase by Repurification of RNA from Agarose.</u>  Total yeast RNA was repurified from 0.6% agarose-6M urea gels.  Each reaction contained 5 µg RNA, a large excess, so that RNA concentration was non-limiting and the inhibitor concentration was high. The RNA was degraded with 0.2 N NaOH for 30 min at 0°C; the solution was then neutralized with HCl and Tris, and the kinase reaction was carried out with 1 unit of polynucleotide kinase in 50 mM Tris, pH 9.5, 10 mM MgCl$_2$, 67 mM Na$^+$, 5 mM DTT, 1 mM spermidine, and 0.1 mM EDTA; the reaction was carried out in 60 µl containing 60 pmole ATP (including 10 µCi $\gamma$-$^{32}$P-ATP).  These conditions are modified from Donis-Keller, Maxam, and Gilbert (9).  Duplicate aliquots were taken at 0, 30, and 120 min, precipitated and washed with trichloroacetic acid, and counted as Cerenkov radiation.

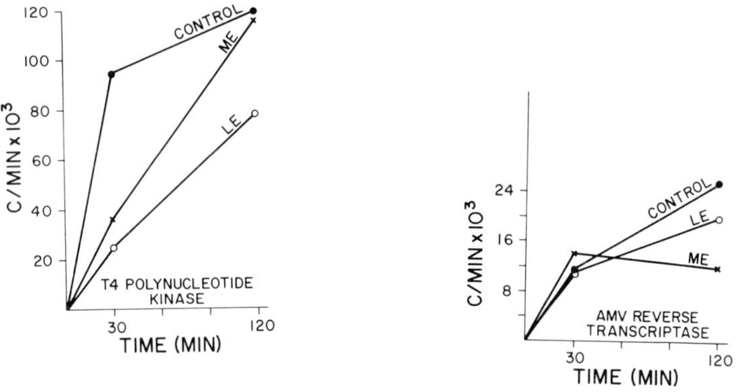


Figure 4.  <u>Inhibition of AMV Reverse Transcriptase by Repurification of RNA from Agarose.</u>  5 µg of RNA was used for each reaction mixture as described for Figure 3.  The reaction conditions used are according to Stavnezer and Bishop (10): 80 mM Tris, pH 8.1, 12.5 mM MgCl$_2$, 0.2% $\beta$-mercaptoethanol, 0.1% NP-40, 1.5 mM dATP, 1.5 mM dGTP, 1.5 mM dTTP, 6.25 µM $\alpha$-$^{32}$P-dCTP (5 µCi), 100 µg/ml actinomycin D, 700 µg/ml DNA oligonucleotide primer (boiled DNase-treated calf thymus DNA), and 1 unit AMV reverse transcriptase.  Duplicate aliquots were taken at 0, 30, and 120 min, precipitated and washed with trichloroacetic acid, and counted.

metric analysis of fluorescence in gels, and analyzed to insure lack of degradation.  Our previous experience had indicated that polynucleotide kinase showed minimal inhibition, while reverse transcriptase was difficult to work with.  Sea-Plaque agarose was not used, since it will not gel in 6M urea.  The ratio of RNA to agarose simulated normal preparative conditions.

The results of polynucleotide kinase labelling are shown in Figure 3. LE agarose is more inhibitory than ME, and, for both, the inhibition is much more pronounced at 30 than at 120 min.  Under these experimental conditions, the reaction plateaus at about 4% incorporation of labelled

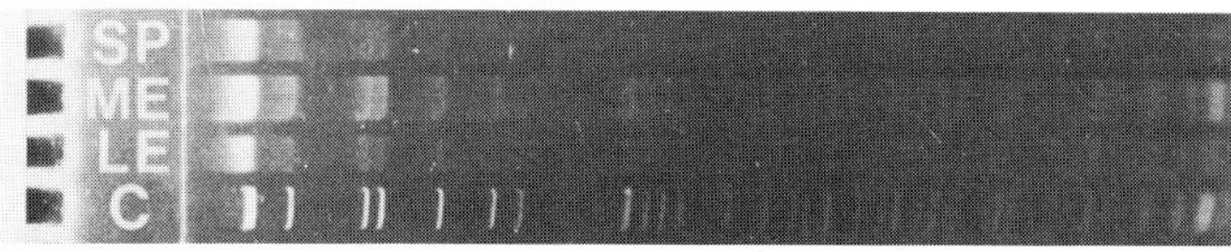

Figure 5. <u>Re-Electrophoresis in a High-Density Gel of Agarose Purified DNA</u>. One μg of control (C) or agarose repurified (LE, ME, SP) DNA was digested with Hpa II and resolved in a composite 7% acrylamide-DATD-0.7% agarose gel (11). The bands of the repurified DNA appear fainter than that of the control because they are more diffuse.

ATP. At 2 hr, ME agarose caused minimal inhibition, and LE agarose showed only 35% inhibition. This pattern suggests moderate amounts of reversible inhibition.

The results of reverse transcriptase labelling are shown in Figure 4. This is the only enzyme tested in which ME agarose was more inhibitory than LE. There is virtually no inhibition at 30 min but definite inhibition at 2 hr (23% for LE and 54% for ME). The different pattern of inhibition may indicate that a different inhibitory substance is involved. Since the initial and early reaction rates are the same, either the enzyme dies more rapidly in the agarose-treated preparations, or it dissociates from the template sooner and cannot reinitiate.

<u>Altered resolution of repurified DNA in high-density gels</u>. When repurified DNA is digested and resolved on agarose gels (up to 2% agarose), the result is indistinguishable from a digest of untreated DNA (see Figures 1 and 2). However, if the repurified DNA is analyzed on a high-density acrylamide gel (Figure 5), such as a sequencing gel, then there is a distinct loss of resolution, found equally for all varieties of agarose. This effect does not preclude use of repurified DNA for acrylamide gels and sequencing, but it makes analysis less satisfactory. Detection of bands is decreased, not because they contain less DNA, but because they are more diffuse. Loss of resolution is independent of the enzyme inhibition observed, and we cannot simulate it by mixing and incubating DNA with total agarose, the soluble agarose extract, or heparin before loading onto acrylamide gels. Heating the DNA to 80°C for 10 min or digesting with 100 μg/ml agarase at 37°C for 1 hr before loading digests on the gel does not improve resolution.

624

Discussion

Loss of resolution in acrylamide gels. The decreased resolution of
repurified DNA in acrylamide gels remains an unexplained problem.  It
seems to result from use of all types of agarose, but the effect is observed only if DNA is electrophoresed in agarose, not if DNA is merely
mixed with agarose and heated.  It seems likely that this effect is caused
by microaggregates of agarose that adhere to the DNA, possibly by wrapping
around the DNA molecules, causing frictional retardation within the acrylamide matrix.  However, if these microaggregates are ordinary agarose,
then agarase digestion under vigorous conditions, or heating, should remove them from the DNA, a result that we did not observe.  It is possible
that the loss of resolution represents a covalent modification of DNA
linking it to the agarose matrix.

The loss of resolution does not preclude use of acrylamide gels with
this material.  The effect on transfer hybridization experiments is negligible.  In experiments in our laboratory, we derive short end-sequences
from RNA molecules (about 20 nucleotides) to locate the transcripts  within long DNA sequences;  we have no difficulty using agarose-purified RNA
for this purpose.  However, the amount of information that can be obtained
from a sequencing gel of such DNA or RNA is definitely reduced.

Preparative methods.  Apart from the resolution problem, agarose purification is a desirable way to purify individual nucleic acid species for
most other experimental purposes.  The levels of enzyme inhibition that
we observe are moderate to insignificant.  The disappointing labelling of
an agarose-purified substrate observed by many investigators probably results from a combination of factors: (1) the yield of extracted nucleic
acid is generally not quantitated, and is often less than expected; (2)
there is a moderate reduction in enzyme efficiency; (3) the optimum enzyme conditions may be altered after agarose purification of substrate.
The combination of 30% recovery of nucleic acid plus 30% efficiency of
labelling would result in a 10% yield compared to control labelling values.  However, most molecular biologists are not interested in strict
quantitation of incorporation.  A labelled substrated is used for further
experiments, and the difference between 3 or 6 x $10^6$ counts in a labelled
substrate would virtually never modify an experimental design, although

a 10-fold difference might. For preparative experiments:

1) Use as low a percentage of agarose as possible. In our laboratory, DNA is isolated from 0.3% LE and ME agarose, and 0.4% Sea-Plaque, in large stable cylinder gels (1.2 cm diameter).

2) Overload the preparative gel as much as possible; this increases yield and decreases the amount of agarose to be removed.

3) Modify enzymatic reaction conditions. In particular, increase incubation times to more than 2 hr. Add more enzyme when appropriate. We routinely double the units of kinase, reverse transcriptase, and restriction enzymes for redigestion.

## References


1. Hayward, G. S., Smith, M. G.: J. Mol. Biol. 63, 383-395 (1972).
2. Sugden, B., DeTroy, B., Roberts, R. J., Sambrook, J.: Anal. Biochem. 68, 36-46 (1975).
3. Locker, J.: Anal. Biochem. 98, 358-367 (1979).
4. Bailey, J. M., Davidson, N.: Anal. Biochem. 70, 75-85 (1976).
5. Lehrach, H., Diamond, D., Wonzney, J. M., Boedtker, H.: Biochemistry 16, 4743-4751 (1977).
6. Finkelstein, M., Rownd, R. H.: Plasmid 1, 557-562 (1978).
7. Arnott, S., Fulmer, A., Scott, W. E., Dea, I. C. M., Moorhouse, R., Rees, D. A.: J. Mol. Biol. 90, 269-284 (1974).
8. Weislander, L.: Anal. Biochem. 98, 305-309 (1979).
9. Donis-Keller, H., Maxam, A., Gilbert, W.: Nucleic Acids Res. 4, 2527-2538 (1977).
10. Stavnezer, J., Bishop, J. M.: Biochemistry 16, 4224-4231 (1977).
11. Alwine, J. C., Kemp, D. J., Parker, B. A., Reiser, J., Renart, J., Stark, G., Wahl, G. M.: Methods in Enzymol. 68, 220-242 (1979).


## Acknowledgments


I would like to thank Richard Cook and the Marine Colloids Division of the FMC Corporation for information, ideas, and experimental samples of agarose. I would also like to thank Mrs. Judy Scheiner for excellent technical help. The work was supported by Grant # GM-27795 of the National Institutes of Health.

FRACTIONATION OF CONCATEMERIC DNA OF BACTERIOPHAGE T7 BY AGAROSE GEL
ELECTROPHORESIS


Philip Serwer and Gisele A. Greenhaw

Department of Biochemistry, The University of Texas Health Science Center
at San Antonio, San Antonio, Texas  78284



Introduction

Bacteriophage T7 consists of:  (a) a roughly spherical, proteinaceous capsid with a comparatively short tail, (b) a genome of linear, duplex DNA
which has a molecular weight (Mr) of $26.5 \times 10^6$ and which is packaged within the capsid (1,2).  Linear aggregates (concatemers) of the mature DNA of
bacteriophage T7 are produced after infection of *Escherichia coli* with this
bacteriophage (3-5).  Sedimentation in sucrose gradients, followed by fractionation of the gradients and assay for radioactive DNA, has been used in
the past to detect T7 concatemers (32-60S).  At least some T7 concatemers
appear to be attached to an even more rapidly sedimenting, replicating T7
DNA ($100S^+$ DNA) after comparatively gentle procedures of cellular lysis (6)
(in subsequent work, $100S^+$ DNA was renamed "flowers" [7,8]).

Sedimentation profiles of fragments of DNA released from $100S^+$ DNA by
nucleases specific for single-stranded DNA suggest that some of these fragments are concatemers heterogeneous in length (6,8).  To further analyze
these fragments, it is desirable to eliminate the loss in resolution and
sensitivity that occurs during the collection of fractions in the procedure
cited above.  Electrophoresis in agarose gels is a procedure capable of a
higher resolution by Mr for T7 concatemers than sedimentation in sucrose
gradients; DNA in such gels can be detected by staining with ethidium bromide, an optical procedure which does not require the collection of fractions (9).  In addition, if slab gels are used, several sample profiles obtained in the same gel can be compared.  Therefore, electrophoresis in agarose slab gels, followed by staining with ethidium bromide, is used here to

analyze the fragments of 100S[+] T7 DNA produced by nuclease S1, an endonuclease specific for single-stranded DNA.


Methods

1) <u>Strains</u>. Bacteriophage T7 was received from Dr. F. W. Studier. The
host for T7 was *Escherichia coli* BB/1.

2) <u>Buffers and reagents</u>. DNAs were stored in Tris/EDTA buffer: 0.10 M
NaCl, 0.01 M Tris-Cl, pH 7.4, 0.001 M EDTA. The electrophoresis buffer was 0.05 M sodium phosphate, pH 7.4, 0.001 M EDTA. For digestion
with nuclease S1, samples were added to 5 X S1 buffer: 1.15 M NaCl,
0.50 M sodium acetate, pH 4.6, 0.03 M $ZnCl_2$. Nuclease S1 was purchased from Sigma Biochemicals. HGT[P] agarose used for gels was
manufactured by Marine Colloids, Rockland, Maine.

3) <u>Isolation of 100S[+] DNA</u>. A 20 $cm^3$ culture of *Escherichia coli* BB/1 was
grown with aeration to 4.0 x $10^8$/$cm^3$ in M9 medium (10) at 30.0°C. The
culture was infected with bacteriophage T7 at a multiplicity of 15; the
culture was quenched at 22.0 min after infection, pelleted and lysed
as previously described (6), except that Sarkosyl NL97 was used for
lysis instead of Brij 58. This change was made to inactivate nucleases
that co-purify with 100S[+] DNA during fractionation of lysates made with
Brij 58. Lysates were layered on 10.6 $cm^3$, 5-25% sucrose gradients in
Tris/EDTA buffer, poured over a layer of metrizamide (11) in Tris/EDTA
buffer (density = 1.161 g/$cm^3$; volume = 1.0 $cm^3$). The lysates were
then centrifuged at 29K, 18°C for 110 min. During this centrifugation,
100S[+] DNA sediments through the sucrose gradient and is buoyed at the
top of the layer of metrizamide. Gradients were fractionated by slow
pipeting of successive layers from the top of the gradient (inner
diameter of the pipet = 2.5 mm). During fractionation, the fraction
with 100S[+] DNA was determined visually by observing light scattered by
100S[+] DNA (or something on this DNA). The 100S[+] DNA was stored at 4°C
and was used without removal of sucrose or metrizamide. During
storage, 100S[+] DNA aggregates and sinks to the bottom of tubes used for
storage; the aggregated DNA was dispersed prior to use by gentle shaking of the tube.

4) <u>Digestion with nuclease S1</u>. For the digestion of single-stranded regions of 100S[+] DNA without digestion of double-stranded regions, nuclease S1 from *Aspergillus oryzae* was used (13). To 10 µl of sample containing 100S[+] DNA was added 2.5 µl of 5 x S1 buffer. To this mixture was added nuclease S1 (1-3 µl) and incubation was continued for 15.0 min. at 30.0°C. Digestion was terminated with the addition of 4.0 µl of 0.02 M sodium EDTA. pH 7.4, followed by 30 µl of a buffer containing: 0.005 M sodium phosphate, pH 7.4, 0.001 M EDTA, 2% sucrose and 400 µg/cm$^3$ bromophenol blue. All pipeting of 100S[+] DNA was done with marked 100 µl micropipets, 1 mm inner diameter, to reduce shear-induced degradation, although some accuracy in pipeting was lost with this procedure.

5) <u>Electrophoresis</u>. Concatemers were fractionated by electrophoresis in horizontal 0.15% slabs of agarose submerged beneath 4-6 mm of electrophoresis buffer, using an apparatus and a procedure for dissolving agarose previously described (12). Samples, prepared as described below, were layered beneath the electrophoresis buffer in sample wells and were subjected to electrophoresis at 0.34 ± .03 volts/cm, room temperature (25 ± 3°C) for 20 hr. Semilogarithmic plots of the Mr of T7 concatemer-sized duplex DNAs as a function of distance migrated are linear, using the above conditions of electrophoresis (9).

6) <u>Markers</u>. The mature DNAs from bacteriophages T4, T5, and T7 were used as markers for Mr. These DNAs were prepared as previously described (9). The Mr of T4 DNA is assumed to be 110 x 10$^6$; the Mr of T5 DNA is assumed to be 77 x 10$^6$ (9).


Results


To analyze the products of the digestion of 100S[+] T7 DNA with nuclease S1, 100S[+] DNA has been subjected to electrophoresis in an agarose gel after

630

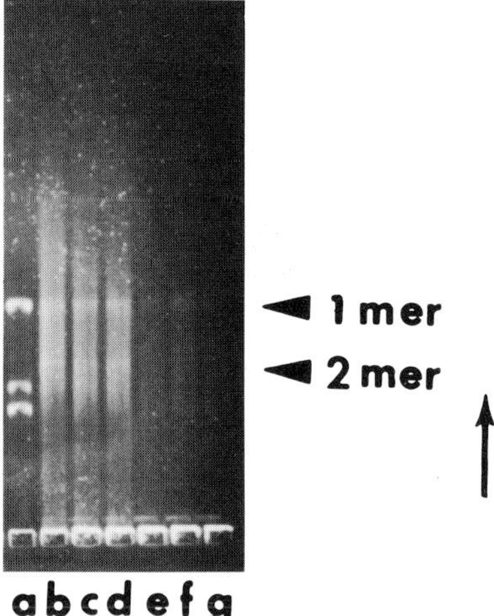


Figure 1. Digestion of 100S$^+$ DNA with nuclease S1. T7 100S$^+$ DNA from 8 x 10$^7$ cells was subjected to electrophoresis in an agarose gel after digestion with: (b) 370, (c) 37.0, (d) 9.6, (e) 3.7, (f) 0.0 units/cm$^3$ of nuclease S1. Channel a has mature DNAs of bacteriophage T4 (slowest migrating), T7 (fastest migrating) and T5. Channel g has 100S$^+$ DNA without 5 x S1 buffer added. The arrow indicates the direction of electrophoresis.

digestion with 370, 37.0, 9.6, 3.7, and 0.0 units/cm$^3$ of nuclease S1 (Figure 1, channels b-f; channel g has an undigested sample to which 5 x S1 buffer was not added). Digestion with 9.6-370 units of nuclease S1 released fragments from the 100S$^+$ DNA, most of which were heterogeneous in mobility and some of which migrated more slowly than mature T7 DNA (channel a of Figure 1 has a mixture of the mature DNAs of bacteriophages T4, T5, and T7). However, some of the DNA released formed sharp bands at: (a) the position of mature T7 DNA (some DNA migrating as mature T7 DNA was present without digestion), and (b) the position expected of a linear DNA with a length 2.13 times that of mature T7 DNA. Digestion of the sample used for Figure 1 with restriction enzymes HpaI and BglII (14) produced only fragments expected of T7 DNA (not shown); no host fragments were found (although host fragments were present in 100S$^+$ isolated at 12.0 and 15.0 min. after infection), indicating that the DNA of Figure 1 is T7 DNA. This is expected because T7 degrades the DNA of its host (15) and most DNA co-purifying with 100S$^+$ DNA from cultures terminated late in the infectious cycle is T7 DNA (6).

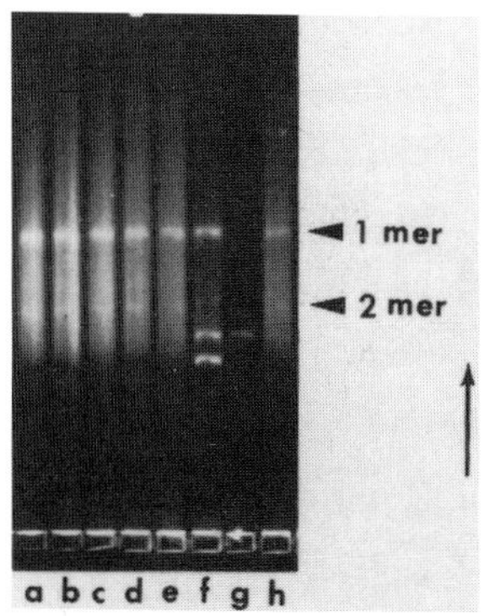



Figure 2.  Effect of the concentration of sample.  Samples of 100S+ DNA were subjected to digestion with 37.0 units/cm$^3$ nuclease S1 using DNA from the following numbers of cells:  (a) 1.4 x 10$^8$, (b) 9.2 x 10$^7$, (c) 4.8 x 10$^7$, (d) 3.4 x 10$^7$, (e) 2.4 x 10$^7$, and the digests were subjected to electrophoresis.  Other samples were (f) T4, T5, and T7 DNAs, (g) T5 DNA (15 ng), (h) T5 DNA (15 ng) + 100S+ DNA, nuclease S1-digested, from 4.8 x 10$^7$ cells.  An extra band in channel (f) is T5 DNA (possibly broken).  The arrow indicates the direction of electrophoresis.

It has been shown that the distance migrated by mature T7 DNA decreases as the concentration of DNA increases above 150 µg/cm$^2$ (9), probably because of DNA-DNA interactions.  To determine whether DNA-DNA interaction was altering the positions of the bands in channels b-d of Figure 1, 100S+ DNA digested with nuclease S1 was subjected to electrophoresis, using DNA from 1.4 x 10$^8$, 9.2 x 10$^7$, 4.8 x 10$^7$, 3.4 x 10$^7$, and 2.4 x 10$^7$ cells (Figure 2, channels a–e).  The distance migrated by DNA forming the two sharp bands increased progressively by 2.6 ± 0.9% as the concentration of sample decreased suggesting that DNA-DNA interaction was having an effect (though slight) on the distance migrated.  However, when DNA from 2.4 x 10$^7$ cells was used, the faster-migrating of the two bands was at a position indistinguishable from the position of mature T7 DNA.

As an additional test for the effect of sample concentration on the distance migrated, DNA of bacteriophage T5 (90 ng/cm$^2$) was subjected to electrophoresis without and with the addition of nuclease S1-digested 100S+ DNA from 4.8 x 10$^7$ cells (Figure 2, channels h, g, respectively).  A 1.4 ± 0.7% slowing of the T5 DNA in the presence of digested 100S+ DNA was detected.  If the position of the slower band of 100S+ DNA in Figure 2, channel h, is corrected by this 1.4%, the mobility of nuclease S1-released fragments of $100S^+$ DNA forming the band closer to the origin indicates a length 2.08 ± 0.10 times the length of mature T7 DNA, if these fragments are duplex and linear. This is equal within experimental error to the length of a linear dimer of mature T7 DNA (i.e., a dimeric concatemer).

Discussion

The DNA migrating as T7 concatemers in agarose gels is not proven by the data presented here to be linear and concatemeric. Circular and branched molecules may comigrate with concatemers in agarose gels. In additional studies we have fractionated nuclease S1 digests of $100S^+$ DNA by sedimentation in sucrose gradients and have found that: (a) most DNA sedimenting as a concatemer migrates as a concatemer of the same length during subsequent electrophoresis in an agarose gel; (b) most DNA sedimenting as a concatemer appears linear in electron micrographs; (c) agarose gel electrophoresis of restriction enzyme digests of concatemer-sedimenting fragments reveals a pattern of fragments expected of linear concatemers. (These data will be presented in a future communication.) Thus, probably most fragments of $100S^+$ DNA migrating more slowly than mature T7 DNA in Figures 1 and 2, including DNA forming the band closer to the origin, are linear and concatemeric.

The failure to detect the class of apparently dimeric T7 concatemers in previous studies (6,8) probably results from the lower resolution and sensitivity of the techniques previously used. The data presented here confirm the conclusion (8) that most concatemers released from $100S^+$ DNA by nuclease S1 are heterogeneous in length.

The presence of nuclease S1-releasable dimeric concatemers and mature-length DNA in T7-infected *Escherichia coli* has been predicted theoretically (16). However, a more detailed analysis of: (a) the structure of the apparent single-stranded regions holding these DNA's to $100S^+$ DNA, and (b) the position in the T7 assembly pathway of these DNA's is needed to know whether or not detailed predictions of this theory are correct.

## Acknowledgements

For assistance in the early stages of this work, we thank Nancy L. Smith. For technical assistance, we thank Elena S.T. Moreno. For typing this manuscript, we thank Brenda Moylan. Support was received from Grant GM 24365 of the National Institutes of Health.

## References

1.  Serwer, P.: J. Mol. Biol. 107, 271-291 (1976).

2.  McDonell, M.W., Simon, M.N., and Studier, F.W.: J. Mol. Biol. 110, 119-146 (1977).

3.  Kelly, T.J. and Thomas, C.A.: J. Mol. Biol. 44, 459-475 (1969).

4.  Schlegel, R.A. and Thomas, C.A.: J. Mol. Biol. 68, 319-345 (1972).

5.  Strätling, W., Krause, E., and Knippers, R.: Virology 51, 109-119 (1973).

6.  Serwer, P.: Virology 59, 70-88 (1974).

7.  Paetkau, V., Langman, L., Bradley, R., Scraba, D., and Miller, R.C., Jr.: J. Virol. 22, 130-141 (1977).

8.  Langman, L., Paetkau, V., Scraba, D., Miller, R.C., Jr., Roeder, G.S., and Sadowski, P.D.: Can. J. Biochem. 56, 508-516 (1978).

9.  Serwer, P.: Biochemistry 19, 3001-3004 (1980).

10.  Kellenberger, E. and Sechaud, J.: Virology 3, 256-274 (1957).

11.  Serwer, P.: J. Mol. Biol. 138, 65-91 (1980).

12.  Serwer, P.: Anal. Biochem. 101, 154-159 (1980).

13.  Vogt, V.M.: Euro. J. Biochem. 33, 192-200 (1973).

14.  Rosenberg, A.H., Simon, M.N., Studier, F.W., and Roberts, R.J.: J. Mol. Biol. 135, 907-915 (1979).

15.  Sadowski, P.E. and Kerr, C.: J. Virol. 6, 149-155 (1970).

16.  Watson, J.D.: Nature (London) New Biol. 239, 197-201 (1972).

DNA SEQUENCING ON VERY THIN (0.2 mm) GELS


Wilnelm Ansorge[*], Henrik Garoff
European Molecular Biology Laboratory,
Postfach 10.2209, 6900 Heidelberg, FRG


Introduction

Samples of oligonucleotides for DNA sequencing are produced
either by primed DNA synthesis on a single stranded DNA template
in the presence of chain elongation terminators (2',3'-dideoxy-
ribonucleoside triphosphates) or by base specific cleavages of
endlabelled restriction fragments of DNA (1,2). Subsequently they
are applied to denaturing polyacrylamide gels. As the pattern of
radioactive oligonucleotides in the gel is visualized by autoradiography, there will be some spreading of the bands due to
radiation scattering within the gel. This effect can be minimized
by reducing the gel thickness as beautifully demonstrated by
Sanger and Coulson (3) who introduced the use of thin (0.4 mm
thick) acrylamide gels for DNA sequencing. Difficulties pouring gels free of bubbles and in handling them after electrophoresis has hindered the use of even thinner gels.

Preparation of very thin ($\leq$ 0.2 mm) gels, their handling, thermostatting and application to protein separation and DNA sequencing
have recently been reported by us (6-8). The DNA sequencing gels
are covalently linked to one of the glass plates, thus facilitating gel drying after electrophoresis. The use of 0.2 mm thick
gels and their novel straight forward drying procedure are shown
to improve the resolution of the sequencing gels.

In order to ensure a uniform and constant temperature (up to
70$^{\circ}$C) throughout the gel during electrophoresis, one of the two

glass plates supporting the gel was replaced with a thermostatting water-heated plate. The elevated temperature permits the resolution of oligonucleotides that are compressed together in conventional gel electrophoresis systems. By avoiding temperature differences within the gel, the thermostatting plate creates equal electrophoresis conditions for all samples applied to the gel. Therefore the resulting autoradiograph shows straight bands in all parts of the gel.


Materials and Methods


Materials. Acrylamide was from Eastman Kodak, N,N'-methylene-bisacrylamide from Serva and urea (ultra pure) was from Schwarz & Mann. REPELCOAT was obtained from Hopkin & Williams Ltd., Rumford.


The electrophoresis apparatus was essentially as described in (4), and all buffers and solutions as described in (5). The gels were formed between two glass plates (200 x 400 x 4 mm) one of which had a notch (165 x 20 mm) cut out from the top to allow contact between the gel and the buffer of the upper electrophoresis chamber. All glass plates used were cut from floated glass. Spacers (900 x 15 mm) and the well formers for 4 mm wide and 8 mm deep slots were cut from the same piece of 0.2 mm thick PTFE. A thermoregulator plate was designed and constructed which meets the high requirements on surface planarity and heat transfer capability. The water was heated and circulated using a water thermostat, e.g. from LKB. Application of samples into the slots was done with thin-walled glass capillaries (inner diameter = 0.1 mm) for X-ray analyses (A. Müller, Glas- und Vakuumtechnik, Berlin). An LKB 2103 unit (2000 V) power supply was used.


Treatment of glass plates. To bind the gel onto the glass plates, they were treated at room temperature as described in (6) and (8). REPELCOAT was used to repel gels from glass surfaces, where desired.

Gel moulding. Gels are cast  by the novel sliding and piston techniques (6-8).

For comparative analyses 0.4 mm thick gels were also made by pouring the gel solution directly in between preassembled glass plates as described in (3).

The 0.2 mm thick gel is cast directly between the glass cover plate and thermo regulating plate.  This assembly provides isothermal conditions in the gel layer during electrophoresis.

We have used the sliding technique in a gel casting device to prepare gels in  a  thickness range of 0.01-2 mm up to about 1 m length, but even longer gels may be prepared. The technique is very easy and fast, even for an unexperienced operator.

Operation thermo regulator   plate. During electrophoresis at 2000 V the power dissipated on one plate (400 x 200 mm) was limited to 20-30 Watts, while during pre-electrophoresis the power may be as high as 100 Watts. The plates were tested at 10.000 V potential difference across the gel and electrode buffers, without failure. The water was circulated through the plate under pressure of 160-240 mbar, as delivered by the thermostat  pump. The plate was warmed up slowly at a rate of about $2^{O}$C/min during pre-electrophoresis to the desired temperature. Thermal shock should be avoided.

Electrophoresis. The glass plates with the polymerized gel were mounted on the electrophoresis apparatus and the chambers were filled with electrode buffer before the well former was removed. Unpolymerized gel solution was carefully washed out from the wells through injection of a buffer stream with a Pasteur pipette. This had to be done immediately after removing the well

638

former. After pre-electrophoresis at 2000 V for 60 min, about
1.5 μl of a DNA sample in formamide-dye    mixture (5) was loaded into the wells. The DNA samples were 5' endlabelled ($^{32}$P)-
subclones from PBR-SFV 2 or the 947 bp Eco Rl-Xho I fragment
from PBR-SFV 3 (8) that had been subjected to base specific
cleavage reactions (G, A+G, C+T, C, C+A) as described by Maxam
and Gilbert (2). Electrophoresis was done at maximum voltage
available (2000 V), or, as indicated in the legends to figures
1-4.


Results


<u>Separation of oligonucleotides on 0.2 mm thick gels</u>. The sepa-
ration of DNA molecules on the 0.2 mm thick gels was clearly
better than that which had been obtained earlier on 0.4 mm thick
gels in our laboratory. This is shown in figure 1, where the
same oligonucleotide mixture has been analysed on 0.4 mm and
0.2 mm thick gels. In both cases the film was exposed to the wet
gel covered with a polyethylene foil. The bands are much sharper
on the thinner gel and many more bases can be read from this gel
than from the thicker one.

<u>The effect of gel drying on band sharpness on autoradiograms.</u>
A DNA sample was run on two 0.2 mm thick gels under identical
conditions, and one of the gels was fixed and dried before autoradiography. The autoradiogram of the dried gel (figure 2, right)
showed much sharper and better resolved bands than that of the
wet gel covered with a polyethylene foil (figure 2, left).

<u>The usefulness of a thermoregulating plate.</u>In our earlier sequen-
cing of the Semliki Forest virus (SFV) genome using 0.4 mm thick
gels as described,    we encountered difficulties in reading
the bases 75-116 from a 947 bp Eco Rl - Xho I fragment labelled
at the Eco Rl site. The gel showed in this region severe com-
pressions and gaps (figure 3, right). When the sample was run at

Figure 1. Separation of the same sets of oligonucleotides on a
0.4 mm (left) and a 0.2 mm (right) thick 8% acrylamide gel. The
sample was end-labelled subclone 3 of PBR-SFV 2 cleaved at various bases which are indicated at the top of the gel using the
Maxam and Gilbert cleavage reactions. Three μl of sample was
applied into the slots of the 0.4 mm thick gel and 1.5 μl into
those of the thinner gel. Electrophoresis was performed using a
constant power of about 50 Watts (1800 Volts) for the thicker
gel and about 25 Watts (2000 Volts) for the thinner one. The
thin gel was run for 125 minutes and the thick one for 180 minutes after which the gels were covered with polyethylene film for
autoradiography with screen, at -30°C for 3 days. The number
100 indicates the base number from the labelled end.

Figure 2. The effect of drying of the gel on autoradiography.The
same sample has been analysed on two 0.2 mm thick 8% acrylamide
gels at 25 Watts for 90 minutes. The gel shown at the left was
subjected to autoradiography as wet for 3 days with a screen,
and the right one was first fixed, dried in an oven for 120 minutes and then autoradiographed also for 3 days with a screen.
The sample was end-labelled subclone 4 of PBR-SFV 2 cleaved at
the various bases indicated at the top of the gel.

Figure 3. Sample Eco Rl-Xho I from clone PBR-SFV 3 analysed for
base sequence on a 0.4 mm thick 8% acrylamide gel (right) and on
a 0.2 mm thick acrylamide gel with a thermostatting plate. A
constant power of 25 Watts (2000 Volts) was used during electrophoresis on the 0.2 mm thick gel and about 50 Watts (1800 Volts)
when running the 0.4 mm thick gel. The running time was about
120 minutes for the thicker gel and about 80 minutes for the
thinner gel. The measured temperature in the middle of the 0.4
mm gel was about 50°C during the run, whereas the 0.2 mm gel
was kept at 70°C with the aid of warm water circulating through
the thermostatting plate. The thicker gel has been autoradiographed wet, covered with polyethylene foil, for 4 days with a
screen, and the thinner one has been fixed and dried before
autoradiography, also for 4 days with a screen.

Figure 4. The same sample as in figure 3, run for 90 min (right),
120 min (centre) and 160 min (left).

Fig. 1

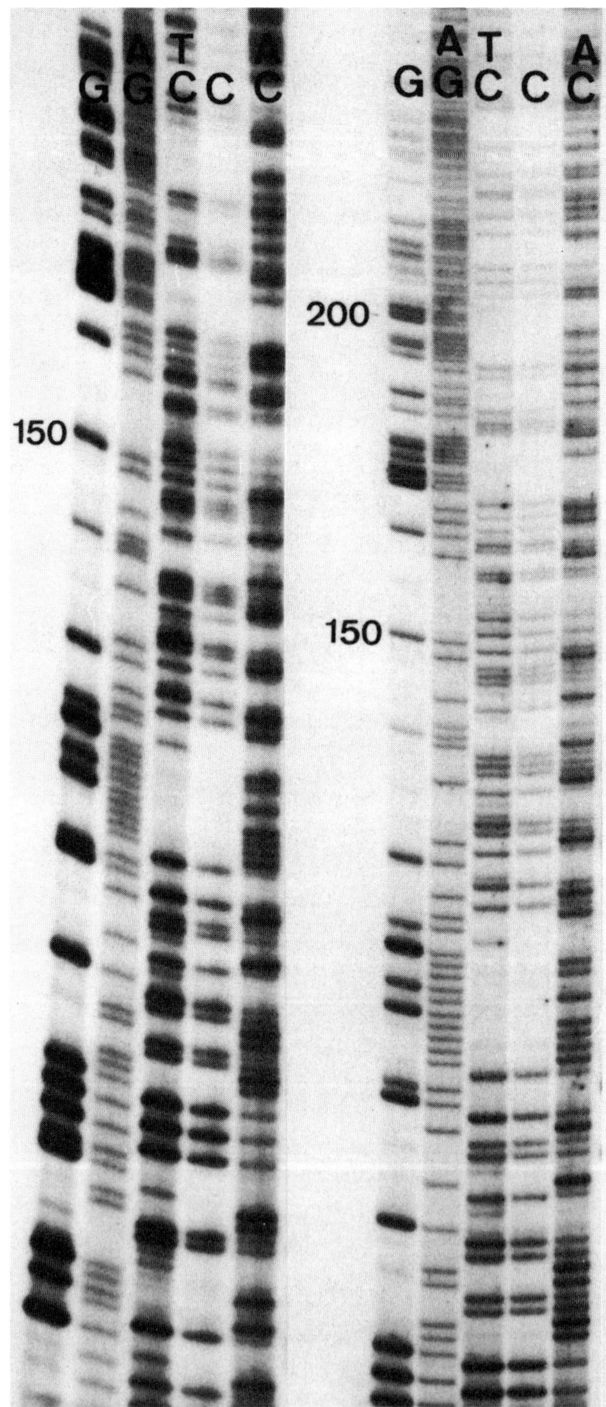

Fig. 2

100

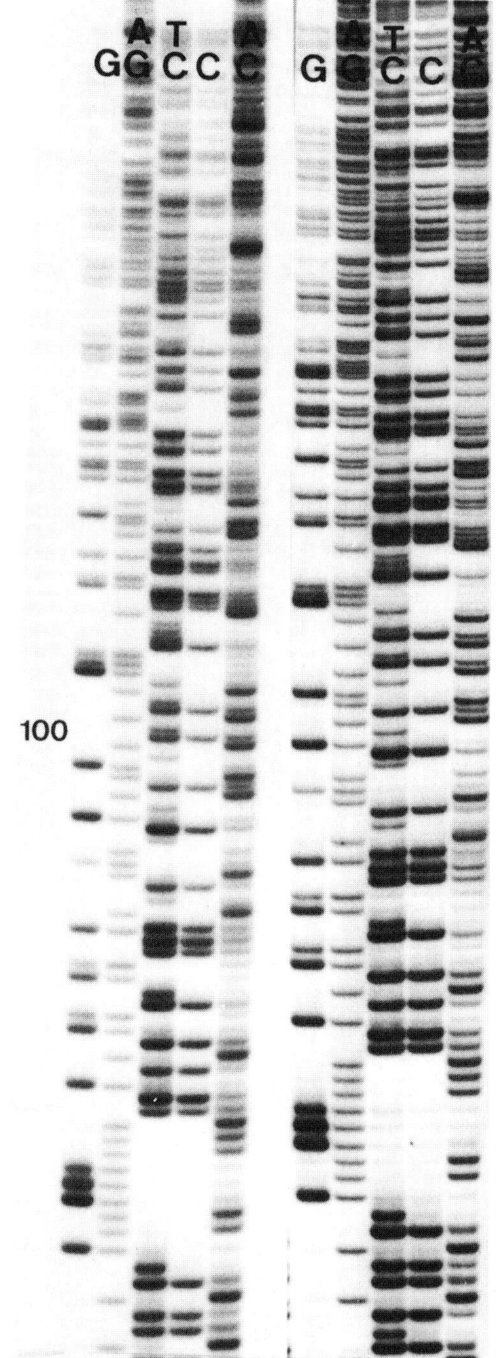

Fig. 3

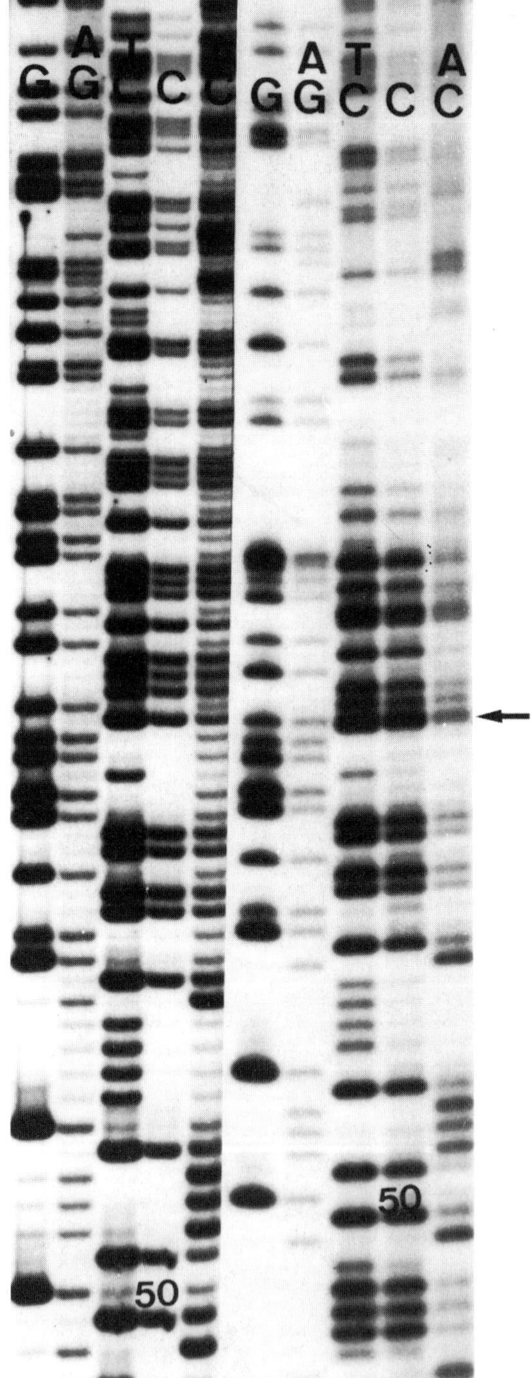

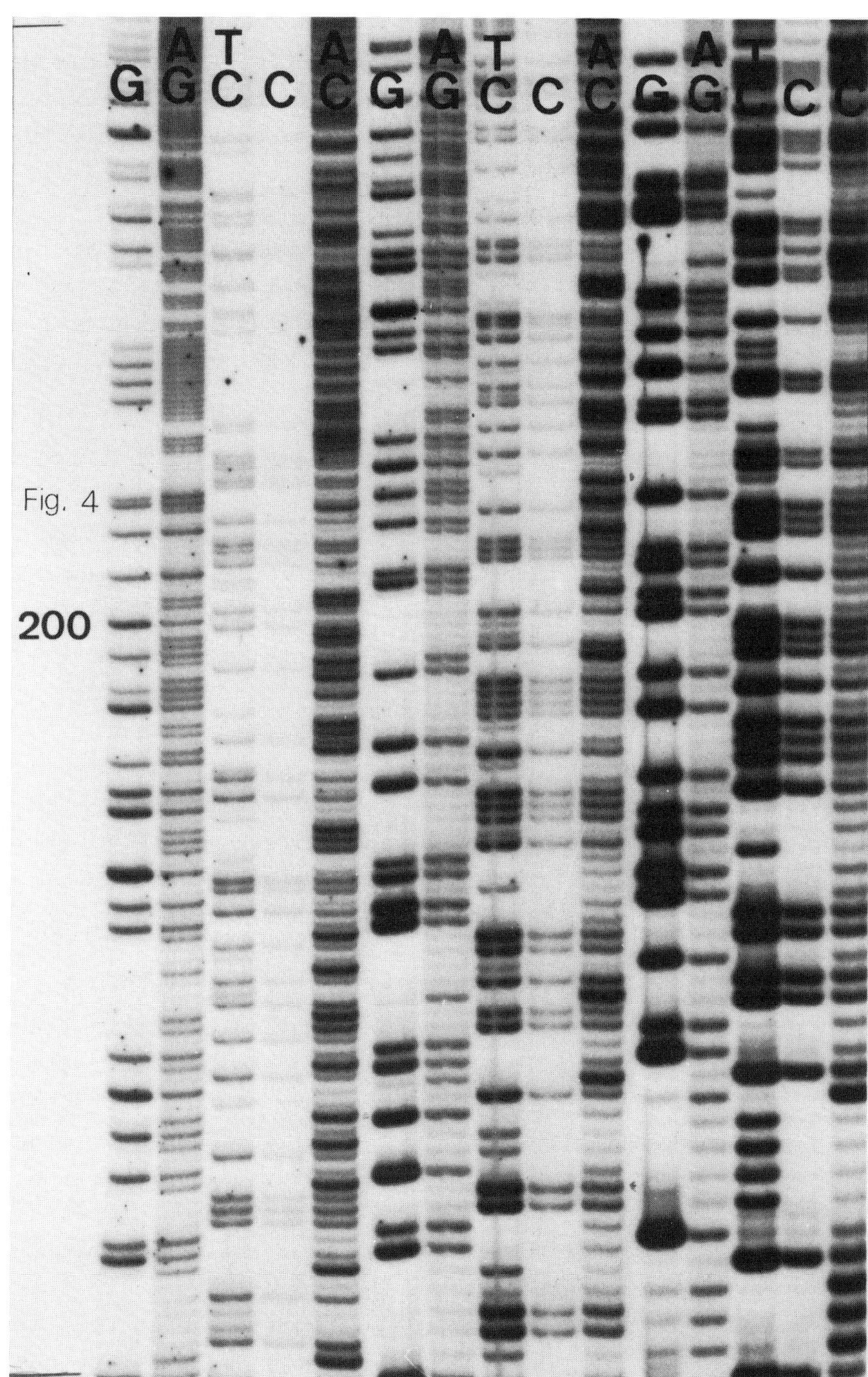
Fig. 4
200

high temperature (70$^{O}$C) with a thermoregulating  plate the sequence C-G-C-C-C-G-C-G-C-C-C-G-G-C-G-G-C-C-C-G-T-C-C-T-T-G-G-C-C-G-T-T-G-C-A-G-G-C-C-A-C-T could be read without difficulties (figure 3, left). The use of the high temperature does not increase diffusion of the bands, as shown in figure 4 (top left) where oligonucleotides containing much over 200 bases are still seen as sharp and well resolved bands. Figure 4 also shows that the use of the thermoheating  plate results in vertically and horizontally straight band patterns. There is no bending of the off-centre bands ("smiling" effect) which is often a problem in gels run without a thermoregulating plate.


Discussion


The covalent fixation of the gel to one of the glass plates offers several advantages in handling the gel both during and after electrophoresis. The sample slots are stable in their dimensions during the whole electrophoresis run; in the conventional electrophoresis system the bottoms of the slots may distort with curved bands as a result. After electrophoresis it is possible to fix the oligonucleotides and remove the urea from the gel by washing in 10% acetic acid, and dry the gel subsequently in an oven at 80$^{O}$ for 1 hour without any breakage of the gel. The dried gel layer is only 20 μm thick and can be autoradiographed directly without a polyethylene foil. We have dried in this way 4%, 6%, 8%, 12%, and 20% acrylamide gels. Gels thicker than 0.2 mm tend to crack when dried in the simple way described above

In the conventional DNA sequencing procedure (3) the gel is warmed up by the joule  heat developed by the current passing through the gel. This leads to non-uniform temperature distribution within the gel which, in turn, causes thermal distortions in the band pattern, and sometimes even cracking of the glass plates. These inconveniences can be avoided by separating the gel heating function from the applied field. We have achieved this by

constructing a simple thermoheating  plate with water circulating from a thermostat.

The modifications of the DNA sequencing gels described above
(gel thickness 0.2 mm, covalent attachment of the gel to the
glass, and   use of the thermoregulating plate)all contributed
to an improved resolution of oligonucleotides in the gel. Recent
experiments with samples containing more radioactivity have demonstrated that further improvement of the band resolution in
the autoradiogram is achieved when the intensifying screen is
omitted during autoradiography. Using 6% acrylamide we have
been consistently able to read   350 nucleotides from end-
labelled samples using these modifications. We expect that even
more bases can be read if the sample is run for a very long distance e.g. on a 900 mm long gel to maximize the separation of
the larger oligonucleotides. Such long gels have been cast on a
thermoregulating plate using the sliding technique.

Obviously the sequencing result is much dependent on the quality
of the sample. In our present work we have only analysed end-
labelled DNA samples that have been subjected to the base specific cleavage reactions described by Maxam and Gilbert (2). Although we have very limited experience with samples produced by
the chain terminator technique, these appear to give even better oligonucleotide resolution. This might be due to the fact
that very few manipulations are needed to produce the samples
and that these usually contain more radioactivity and display
a more even size distribution than the samples made by base specific cleavages.


Acknowledgements

We thank Evelyn Kiko and Richard Barker for technical assistance,
Hans Lehrach and Kai Simons for advice and support throughout
this work.

electrophoresis is that the inclusion of substrates within the gel matrix permits classes of enzyme activities to be distinguished from all other proteins. Examination of the enzymes is then possible without having to purify activities away from contaminating proteins. Many studies routinely done using spectrophotometic procedures, such as measurements of pH and ionic strength optima, may be accomplished easily for many activities simultaneously using PPAGE. Also, methods of gel technology allow conclusions regarding relationships among enzyme activities to be reached.

By carefully defining electrophoresis conditions, the use of denaturants such as sodium dodecyl sulfate or urea (8) can be avoided. This eliminates the need to renature enzymes. Also, in the absence of sodium dodecyl sulfate, information on charge heterogeneity can be obtained.

Although our interest in nucleolytic enzymes dictated the use of DNA/polyacrylamide gels, PPAGE can be extended to any enzyme:substrate system as long as the following conditions are satisfied. During electrophoresis the substrate must not migrate or be converted to product by the enzymes. It must be possible to induce enzyme activity in the gel after electrophoresis. Finally, a means to distinguish product from substrate in the gel is necessary to visualize activities. It is interesting to consider that the covalent linkage of compounds to acrylamide might provide a means to study enzymes using low molecular weight substrates.


References


1.  Karpetsky, T.P., Davies, G.E., Shriver, K.K., Levy, C.C.: Biochem. J. 189, 277-284 (1980).

2.  van Loon, L.C.: FEBS Lett. 51, 266-269 (1975).

3.  Wilson, C.: Anal. Biochem. 31, 506-511 (1969).

4.  Wilson, C.: Plant Physiol. 48, 64-68 (1971).

5.  Fairbanks, G., Steck, T.L., Wallach, D.F.H.: Biochemistry 10, 2606-2617 (1971).

6.  Maurer, H.R.: Disc Electrophoresis and Related Techniques of Polyacrylamide Gel Electrophoresis, Second revised and expanded edition (1971), pp. 8-9/ Walter de Gruyter, New York.

7.  Hedrick, J.L., Smith, A.J.: Arch Biochem. Biophys. 126, 155-164 (1968)

8.  Rosenthal, A.L., Lacks, S.A.: Anal. Biochem. 80, 76-90 (1977).

RECENT DEVELOPMENTS IN TITRATION CURVES OF PROTEINS BY ISOELECTRIC FOCUSING-
ELECTROPHORESIS

Pier Giorgio Righetti and Elisabetta Gianazza

Department of Biochemistry, University of Milano, Via Celoria 2, Milano
20133, Italy

## Introduction

By exploiting an original idea of Bjellqvist (1), we have recently demonstrated the possibility of displaying a protein titration (pH-mobility) curve
in a polyacrylamide gel slab (2) by moving the sample electrophoretically
perpendicular to a pH gradient, generated by a stack of stationary, isoelectric carrier ampholytes. The sigmoidal pH-mobility curves thus generated were indeed próportional to titration curves (3, 4). This technique can be exploited in a number of ways:

(1) by running a protein and its genetic mutant in a mixture, it is possible
to reveal the charged amino acid substitution in the variant phenotype (differential titration curves) (2);

(2) protein-ligand interactions can be demonstrated (5, 6);

(3) by the same token, macromolecule-macromolecule interactions can be detected (7, 8);

(4) by developing zymograms and/or immunoprints after titration curves of unpurified samples, it is possible to obtain proper information for a subsequent purification strategy by charge-dependent methods such as zone- and
disc-electrophoresis, isotachophoresis and ion-exchange chromatography (9);

All the data reported so far have described only qualitative or semiquantitative aspects of the technique. However, we have demonstrated recently that
this method can be used to generate physico-chemical parameters pertaining
to the molecule under investigation. They include:

(a) determination of ionization constants (pK values) for weak acids or bases or for simple, uni- uni-valent amphoteric species, from mobility values

656

in a pH gradient (10, 11);

(b) measurement of total charged amino acid differences in homologous proteins (12);

(c) assessment of buried and exposed charged groups in proteins by performing titration curves in native and denaturing conditions (13);

(d) determination of the dissociation constant ($K_d$) of a protein and its ligand from the measurement of decreasing mobilities in gels containing increasing ligand concentration. Titration curves allow also the measurement of the pH-dependence of $K_d$ (14, 15);

(e) determination of mobility data approaching free mobility values, at any pH, for proteins having a molecular mass up to $0.5 \times 10^6$ daltons, by running titration curves in highly porous gel matrices (16).

We shall review here these quantitative aspects of pH-mobility curves. Since the technique has been already described in detail in several publications (2, 9, 13, 15), the reader is referred to these articles for the necessary background.


Results

(a) pK determinations.

If the pH-mobility plots obtained by our two-dimensional IEF-electrophoresis technique are indeed titration curves, then these data could be used to measure the association constant for the binding of a proton to the titrated molecule ($pK_c$ or $pK_a$ for cationic or anionic species, respectively). Indeed, in the case of simple, weak cations and anions, the pKs can be derived in a pH-mobility plot by simply measuring the pH of $\frac{1}{2}$ mobility in the cationic or anionic directions, respectively (11) (see, in Fig. 1, the upper and lower curves which do not cross the pH axis). In the case of mono- mono-valent amphoteric molecules, two equilibrium pathways should be described: the loss of protons from the cation to the anion could go *via* a neutral molecule ($pK_c < pK_a$, subscript 1) or *via* a charged molecule ($pK_c > pK_a$, subscript 2). Most molecules of biological interest (e.g. amino acids, nucleotides) follow pathway 2 ($pK_c > pK_a$) so that the isoelectric species is a zwitterion. But there are also examples of the first case (e.g. doxorubicin, which is isoe-

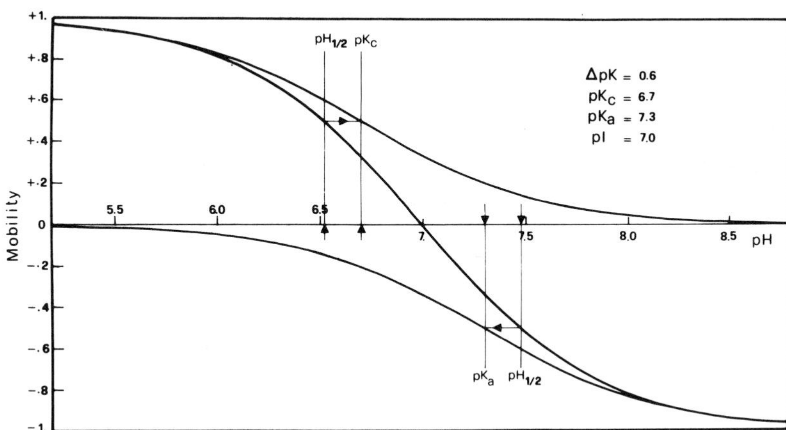

Fig. 1. Theoretical titration curve of a univalent cationic dissociating group (pK$_c$=6.7, upper), of a univalent anion (pK$_a$=7.3, lower) and of an amphoteric molecule with the same ionizable functions, pI=7, pK0.6 (intermediate tracing). Notice that the ampholyte behaviour results from the addition of the anionic and cationic segments of the titration curve. Moreover, for the simple univalent ions pK equals pH$_\frac{1}{2}$; whereas for the amphoteric species the influence of the opposite charge dissociation should be accounted for by adding (for pK$_c$) or subtracting (for pK$_a$) to pH$_\frac{1}{2}$ a correction factor (arrows). On the contrary, the pI simply corresponds to the pH of zero mobility (cross-over point) (from Valentini *et al.*) (11).

lectric between a phenolic –OH and a sugar –NH$_2$, pK$_c$< pK$_a$, thereby being in equilibrium, at the pI, with a neutral molecule). In the first case, the equation linking the total mobility (M$_t$) to the two pKs will be:

$$M_t = h \; \frac{10^{pK_{c1}-pH} - 10^{pH-pK_{a1}}}{1 + 10^{pK_{c1}-pH} + 10^{pH-pK_{a1}}} \qquad (1)$$

while in the second case (pK$_c$ > pK$_a$) it will be:

$$M_t = h \; \frac{10^{pK_{a2}-pH} - 10^{pH-pK_{c2}}}{1 + 10^{pK_{a2}-pH} + 10^{pH-pK_{c2}}} \qquad (2)$$

where the proportionality coefficient (h) represents the limiting value of M$_t$ at the two extremes of the pH scale. In fact, when pH = 0, then M$_t$ ≃ h and when pH = 14, then M$_t$ ≃ –h (see Eqns. (1) and (2)).

For the determination of pK values from mobility data, it is considered the pH (pH$_\frac{1}{2}$) corresponding to $\frac{1}{2}$ mobility in the cathodic or anodic directions, respectively, and inserted into Eqns. (1) and (2). Thus, by substituting in these Eqns. the values M$_t$ = h/2, pH = pH$_\frac{1}{2}$, and considering that pI is the arithmetical mean of the two pKs of a zwitterion one could derive two Eqns.

658

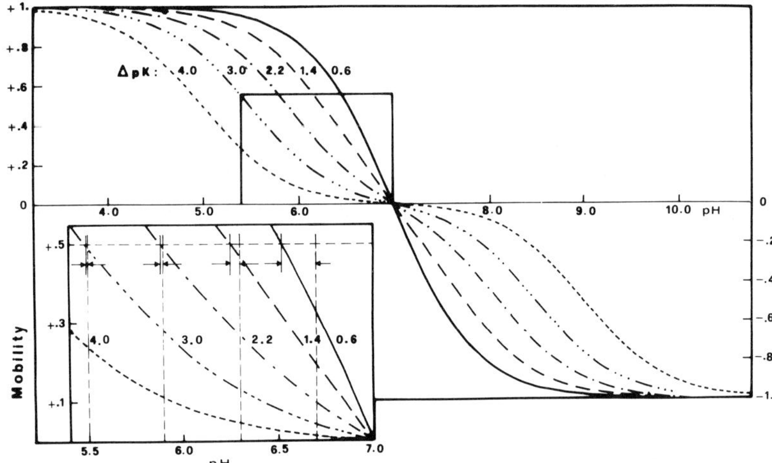


Fig. 2. Theoretical titration curves of a series of amphoteric species with pI=7 and ΔpKs of: 0.6 (——), 1.4 (-----), 2.2 (— · —), 3.0 (— ···—) and 4.0 (- - -). The discrepancy between pK and $pH_{\frac{1}{2}}$ decreases as the ΔpKs increase so that above a ΔpK value of 1.5 the influence of the opposite charge fraction on the mobility of the ion being measured becomes negligible. The insert corresponds to a magnification of the area boxed, spanning a mobility region between zero and 0.5. (from Valentini *et al.*) (11).

for $pK_c$ or $pK_a$ measurements, or, more simply, by considering the absolute value of the difference $|pI-pH_{\frac{1}{2}}|$, a general equation:

$$pK = pH_{\frac{1}{2}} \pm C_t \tag{3}$$

where

$$C_t = -\log (1 - 3\times10^{-2} |pI-pH_{\frac{1}{2}}|) \tag{4}$$

is the correction term to be added to $pH_{\frac{1}{2}}$ when the pK value is lower than pI and to be subtracted from $pH_{\frac{1}{2}}$ when the pK is higher than pI. $C_t$ is always greater than zero ($C_t > 0$) and it increases with decreasing distance between the pKs and the isoelectric point of the molecule. In Fig. 1 we have drawn the hypothetical case of an ampholyte with pI=7 and ΔpK=0.6, which, according to Rilbe (17) represents a limiting value, i.e. the minimal distance between the $pK_c$ and $pK_a$ in a zwitterion. In this limiting case, the correction term, for a proper assessment of pK from pH-mobility curves, is substantial, and it amounts to 0.18 pH units, to be added to $pH_{\frac{1}{2}}$, in the case of cations, or to be subtracted from $pH_{\frac{1}{2}}$, in the case of anions, in order to obtain the true $pK_c$ and $pK_a$ values, respectively.

It is of interest to know which is the minimum ΔpK above which the correction factor becomes negligible. In this case, the measurement of $pH_{\frac{1}{2}}$ in the catio-

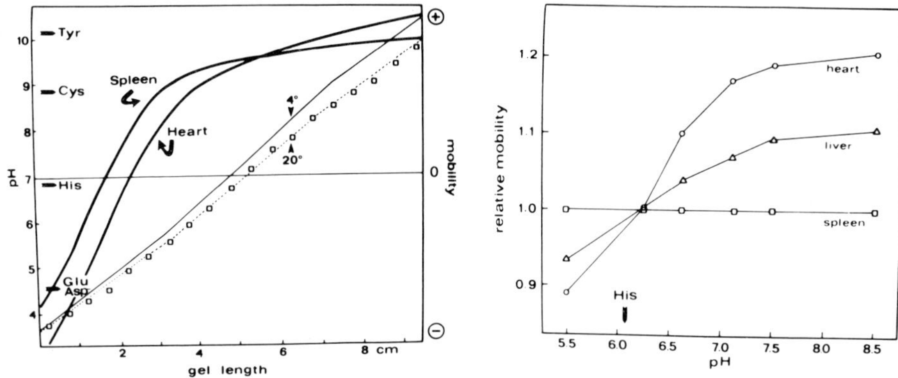

Fig. 3. A (left): drawing of the electrophoretic titration curves of ferritins from horse heart and spleen (average of 3 experiments). The pH gradient, read at room temperature (20°C), is also shown corrected at 4°C. The pK values of several protein ionizable groups are also shown. B (right): graph showing the mobilities at different pH values of heart and liver ferritins relative to spleen, taken as reference (from Gianazza and Arosio) (12).

nic or anionic directions will give direct pK values. In Fig. 2 we have drawn theoretical curves for hypothetical ampholytes, all having pI=7.0, but with increasing ΔpKs, from a minimum of ΔpK=0.6 up to ΔpK=4.0. Practically, above ΔpK=1.5, $C_t$ is already negligible, as it amounts to only 0.03 pH units. With progressively increasing ΔpKs, the pH-mobility curves flatten around the pI value so that these compounds are "isoelectric" over a progressively wider pH range. This is in excellent agreement with the titration curves of "good" and "poor" ampholytes given by Rilbe (18) in terms of net charge vs. pH plots. These theoretical curves have been experimentally verified by running titration curves of simple compounds, such as methyl red, neutral red and doxorubicin (11).

(b) Determination of amino acid differences in homologous proteins

We have previously seen that it is possible to perform "differential titration curves" by running a protein and its genetic mutants in a mixture. The shape of the respective titration curves reveals which charged amino acid has been substituted in the mutant phenotype (2). This case is quite simple, since most of the times one deals with "spot" mutations, involving one or at most two charge differences from the "wild type". We have extended these data by trying to see if our technique can probe more extensive charges in the polypeptide chain, embracing several residues of the same amino acid and mo-

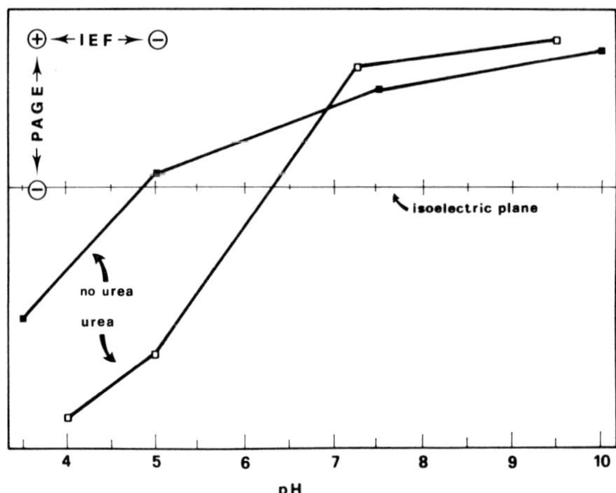



Fig. 4. Drawing of the titration curves of HSA in the folded (no urea) and unfolded (8M urea) states. The directions and polarity of the first (IEF) and of the second (PAGE) dimensions are indicated. The two curves have been run in iso-viscous gels, and highly porous matrices to correct for hydrodynamic volume differences in the unfolded state (Gianazza and Righetti, unpublished).

re than one amino acid type. We have used as a model ferritins isolated from different organs from the same mammal. Fig. 3A shows pH–mobility curves of horse heart and spleen ferritins. The two curves cross the application trench (zero mobility or isoelectric plane) at pH 5.0 and 4.5, respectively, in good agreement with the known pI values of the two proteins. Below pH 5.5, the two lines run almost parallel, with heart bearing a higher number of positive charges than spleen. However, in the anionic portion of the titration (above the pI value) the shape of the two curves differs: the mobility of heart ferritin keeps increasing with the pH, while the mobility of spleen ferritin tends to plateau at progressively higher pHs, until, at pH≈7.5, the two curves cross-over. In order to discuss these data in quantitative terms, we have re-plotted the graph in terms of differential mobilities, taking as a reference (unit) mobility throughout the pH gradient the migration of spleen ferritin. The data are shown in Fig. 3B, which also includes liver ferritin. It appears that all three molecules cross-over at the same pH, and that at any other pH liver has a mobility intermediate between spleen and heart. By assuming that the molecular volume and conformation of the protein shells do not change along the pH gradient studied, and thus that the mobility differences at any given pH represent indeed surface charge differences, we can draw the following conclusions:

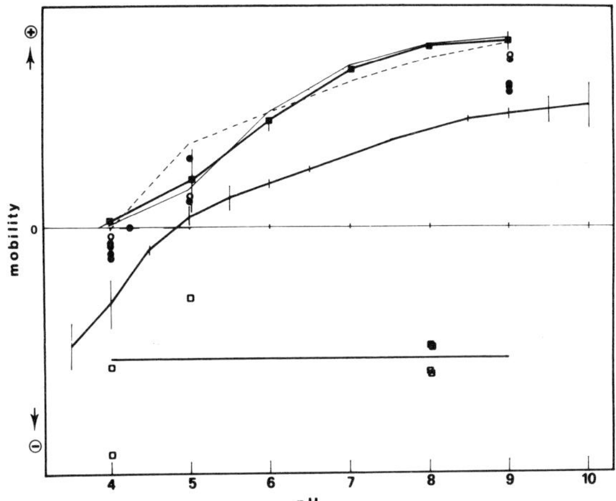

Fig. 5. Titration curves of native and chemically modified HSA. Lower box
(□-□): carboxyl-blocked albumin; solid curve with vertical standard deviation bars (pI 4.85 species): native albumin; (■-■): Lys-blocked HSA; broken
line (- - -): theoretical curve, running parallel to the control (from Gianazza and Righetti, unpublished).

(1): since heart and spleen curves run parallel between pH 3.5 and pH 5.5,
they should have a similar, if not identical, number of Glu and Asp residues;

(2): since the change in mobility around the imidazole pK is twice as large
for heart as for spleen, this suggests that there are twice as many histidi-
nes on heart as on spleen ferritin surfaces;

(3): since the anodic mobility of heart at alkaline pH keeps increasing as
compared to spleen ferritin, once the contribution of His is substracted,
this leaves about 15% more Lys and Arg in the former than in the latter.

(c) Determination of buried and exposed groups in proteins

We have previously shown that it is possible to perform titration curves of
denatured polypeptide chains in 8M urea (13) or in urea-detergent gels (4)
and that this technique can be used to probe extensive regions and overall
amino acid compositions of phylogenetically related protein species. This
was applied to α and β globin chains in normal human hemoglobin. However, if
a protein consists of a single polypeptide chain, its titration curve under
native and denaturing conditions can be used to assess changes (if any) in
the pKs of ionizable groups and the existence of buried, charged amino acid
residues. We have applied this method to the structural study of native and
denatured human serum albumin (HSA). A drawing of the two respective titra-

tion curves is shown in Fig. 4. In order to correct for the differential mobilities under the two experimental conditions, the runs have been performed in highly porous matrices (5%T, 30%$C_{Bis}$). This should ensure very similar steric hindrances for HSA in the folded and unfolded states. A correction has also been made for different viscosities by adding sucrose to the
non-urea gels, so that the two gel media should be iso-viscous. After over-
imposing the two "corrected" curves on the same graph (Fig. 4), we can see
that the two titration curves of the same protein differ markedly in shape.
Also the $\Delta pI$ between the two species is quite striking, amounting to ca. 1.6
pH units. After subtracting the pH increments due to the presence of urea in
the denatured HSA (19), this still leaves a $\Delta pI=1.26$ between native and unfolded HSA. If we knew the ratio $\Delta pI/\Delta H^{+}$, i.e. the pI variation per charge
unit variation, in the present system, we could calculate the total charge
difference, or total number of protons bound or released by the protein in
the folded/unfolded transition. This value is $(\Delta pI/\Delta H^{+})=0.1$ pH units in the
hemoglobin system (20) but varies among different classes of proteins. In order to obtain this yard-stick in our system, HSA has been chemically modified and then titrated in our 2-D technique. The results are shown in Fig.
5. When all the free $-COOH$ groups were blocked with glycine methyl ester and
the modified HSA run in IEF-electrophoresis, it moved, between pH 4and 8,
as a line practically parallel to the application trench (Fig. 5, lower part)
in the cathodic direction. When HSA was selectively $-NH_2$ acetylated in the
59 Lys residues, and run in 8M urea gels (Fig. 5, upper box) it exhibited a
pI of 4.58, which is 1.52 pH units lower than the pI of untreated, urea-de-
natured HSA. If we assume a linear relationship between $\Delta pI$ and number of
protons bound or released, we can calculate a $\Delta pI$ of 0.026 pH units per u-
nit of $\Delta H^{+}$. This means that the charge difference between folded and unfolded states in HSA is 48 protons ($\Delta pI=1.26$). This could be due to exposure of
buried charged groups, as well as to intrinsic pK variations of ionizable
groups due to the unfolding process. Work is in progress to elucidate these
aspects.

(d) Affinity titration curves

One of the most recent extensions of pH-mobility curves is the possibility
of determining dissociation constants ($K_d$) of ligands to proteins, and their
pH dependence. This is achieved by techniques developed for affinity elec-

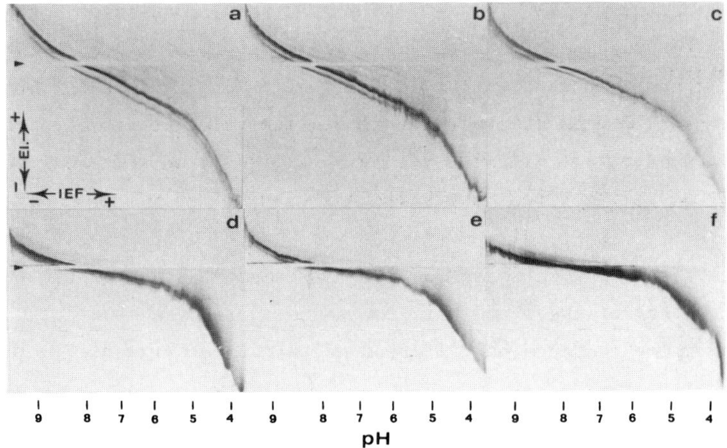



Fig.6. Affinity titration curves of lectin from *L. culinaris. The* amount of
0-α-D-mannosyl polyacrylamide incorporated in the gel was: (a) control, no
ligand; (b) 5.10⁻⁵ M; (c) 10.10⁻⁵ M; (d) 20.10⁻⁵ M; (e) 30.10⁻⁵ M and (f)
40.10⁻⁵ M. The gel contained 6%T, 4%C$_{Bis}$, 2% Ampholine pH 3.5-10 and 2 mM
each glutamic acid, aspartic acid, lysine and arginine. The amount of protein loaded in all cases was ca. 200 μg in 100 μl volume. First dimension,
80 min at 10W constant. Second dimension, 20 min at 700V constant. In both
dimensions the electrolytes were 1M H$_3$PO$_4$ at the anode and 1M NaOH at the
cathode. The gel was cooled at 4°C with a Lauda thermostat. The two arrow-
heads in (a) and (d) indicate the sample aplication trench (zero mobility
plane) (from Ek *et al.*) (14).

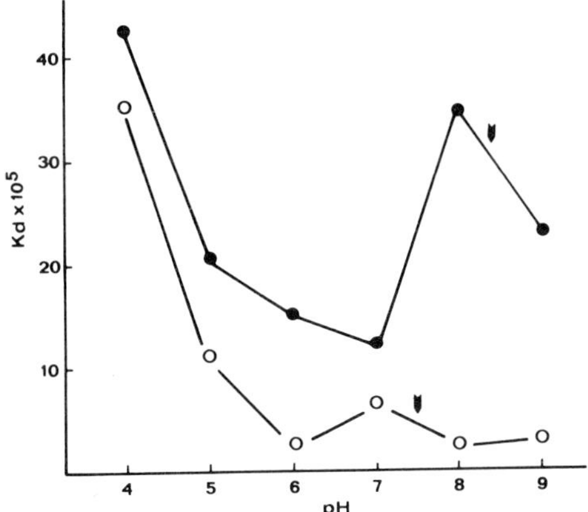



Fig. 7. Variation of the dissociation constants (K$_d$) as a function of pH for
*Ricinus communis* (O) and for *Lens culinaris* (●) lectins. The K$_d$ values reported here have been calculated from the retarded mobilities, at different pH
values, in affinity titration curves, such as the ones shown in Fig. 6. The
two arrows indicate the pI of each lectin (from Ek *et al.*) (14).

trophoresis (21). If the ligand is a macromolecule, it is simply entrapped

664

in the gel matrix, if it is a small molecule, it is covalently bound to the
gel fibers. In presence of increasing concentrations of ligand the titration
curve of the protein is progressively retarded, in a pH-dependent fashion,
and the mobility decrements, when plotted against the ligand molarity in the
gel, can be used to calculate $K_d$ values at any pH value. For this purpose,
we use the Eqn. of Tichà *et al.* (22):

$$d_o/d = 1 + C_i/K_d \qquad (5)$$

where $d_o$ is the electrophoretic mobility (at any pH, since we are running
titration curves) of the protein in the absence of ligand and d the retarded
mobilities in the presence of different molarities of ligand ($C_i$) in the gel
matrix. By plotting the ratio $d_o/d$ vs. $C_i$ straight lines are obtained, the
slopes of which are given by $1/K_d$. Thus, the dissociation constant is easily
calculated as the reciprocal of the slope. Ek *et al.* (14; 15) have developed
this technique for studying the binding of glycogen to phosphorylases a and
b, of blue dextran to several dehydrogenases and of sugars to lectins. An
example of these affinity titration curves is given in Fig. 6, which shows
binding of *L. culinaris* seed lectins to O-$\alpha$-D-mannosyl polyacrylamide entrapped in the gel matrix. $K_d$ values can be determined at any pH (usually at 6
different pH values, from pH 4 to 9, in 1 pH unit increments) and then plotted vs. pH (see Fig. 7) in order to evaluate the pH-dependence of $K_d$.

As a general conclusion, it can be stated that titration curves hold a great
potential for investigating several physico-chemical parameters of proteins
and dynamical aspects of their interaction with ligands.

Acknowledgements

Supported in part by grants from Consiglio Nazionale delle Ricerche (CNR)
and Ministero della Pubblica Istruzione (MPI, Roma).

References

1.   Rosengren, A., Bjellqvist, B., Gasparic, V.: in Radola, B.J., Graesslin,
     D. (eds.) Electrofocusing and Isotachophoresis, pp. 165-171, de Gruyter,
     Berlin, 1977

2.   Righetti, P.G., Krishnamoorthy, R., Gianazza, E., Labie, D.: J. Chroma-
     togr. 166, 455-460 (1978)

3.   Righetti, P.G., Gianazza, E.: in Peeters, H. (ed.) Protides of the Bio-
     logican Fluids, pp. 711-71(, Pergamon Press, Oxford, 1979

4.   Righetti, P.G., Gianazza, E.: in Radola, B.J. (ed.) Electrophoresis '79,

pp. 23-38, de Gruyter, Berlin, 1980

5. Krishnamoorthy, R., Bianchi Bosisio, A., Labie, D., Righetti, P.G.: FEBS Letters 94, 319-323 (1978)

6. Constans, J., Viau, M., Gouaillard, C., Bouissou, C. Clerc, A.: in Radola, B.J. (ed.) Electrophoresis '79, pp. 701-709, de Gruyter, Berlin, 1980

7. Righetti, P.G., Gacon, G., Gianazza, E., Lostanlen, D., Kaplan, J.C.: Biochem. Biophys. Res. Commun. 85, 1575-1581 (1978)

8. Lostanlen, D., Gacon, G., Kaplan, J.C.: Eur. J. Biochem. 112, 179-183 (1980)

9. Gianazza, E., Gelfi, C., Righetti, P.G.: J. Biochem. Biophys. Methods 3, 65-75 (1980)

10. Righetti, P.G., Menozzi, M., Gianazza, E., Valentini, L.: FEBS Letters 101, 51-55 (1979)

11. Valentini, L., Gianazza, E., Righetti, P.G.: J. Biochem. Biophys. Methods 3, 323-338 (1980)

12. Gianazza, E., Arosio, P.: Biochim. Biophys. Acta 625, 310-317 (1980)

13. Righetti, P.G., Krishnamoorthy, R., Lapoumeroulie, C., Labie, D.: J. Chromatogr. 177, 219-225 (1979)

14. Ek, K., Gianazza, E., Righetti, P.G.: Biochim. Biophys. Acta 626, 356-365 (1980)

15. Ek, K., Righetti, P.G.: Electrophoresis 1, 137-140 (1980)

16. Bianchi Bosisio, A., Loehrlein,C., Snyder, R.S., Righetti, P.G.: J. Chromatogr. 189, 317-330 (1980)

17. Rilbe, H.: Acta Chem. Scand. 25, 2768-2769 (1971)

18. Svensson, H.: Acta Chem. Scand. 16, 456-466 (1962)

19. Gianazza, E., Righetti, P.G., Bordi, S., Papeschi, G.: in Radola, B.J., Graesslin, D. (eds.) Electrofocusing and Isotachophoresis, pp. 173-179, de Gruyter, Berlin, 1977

20. Righetti, P.G.: J. Chromatogr. 173, 1-5 (1979)

21. Horejsì, V.: J. Chromatogr. 178, 1-13 (1979)

22. Tichà, M., Horejsì, V., Barthovà, J.: Biochim. Biophys. Acta 534, 58-63 (1978)

# A STUDY OF THE STRUCTURAL CONVERSIONS OF TWO CARBOXYL PROTEINASES EMPLOYING ELECTROPHORESIS ACROSS A pH-GRADIENT


Reinhard Rüchel and Manfred Trost
Hygiene-Institut der Universität Göttingen
D-3400 Göttingen, Germany


## Introduction

Pepsin, which belongs to the group of carboxyl proteinases
(E.C.3.4.23) has been among the first enzymes to be investigated at the onset of modern biochemistry (1). In spite of the
fact that the structure of the protein moiety, the function
of the catalytic site and variations between species and subclasses of the enzyme have been widely elucidated (2-4),
apparently no reliable data are available on the phenomenon
of alkaline denaturation, that is responsible for the inactivation of pepsin in the gut. Alkaline denaturation is considered to be irreversible and may include autolytic processes
(1,5), it is possibly common to all carboxyl proteinases.
Other carboxyl proteinases have been found for instance in a
number of fungi (6,7) and those that are secreted by C.albicans may be related to the pathogenicity of certain strains
of this yeast-like fungus (8). Such enzymes are strain-specific, they cleave immunoglobulins and can be differentiated
according to their pH of alkaline denaturation. Enzymes that
denature already at neutrality cannot exert effects beyond
the acidic milieu in the vicinity of the yeast cell. Proteinases of other strains, that denature only above pH 8 can act
independently of the yeast cell and may affect pathways of
biological regulation in the host (8).

668

## Results and Discussion

Employing electrophoretic methods we investigated some molecular aspects of alkaline denaturation of pepsin and two secretory proteinases from Candida albicans. With porcine pepsin ($M_r$ 33.000), that underwent alkaline denaturation already at pH 6.7, we found a decrease of anodic mobility upon gel electrophoresis at pH 5.5, that was accompanied by a shift of the isoelectric point from below pH 3 to a value above pH 4 (Fig. 1a,b) and which may correspond to the liberation of acidic groups that has been described earlier (1). Additionally a gain in molecular size was observed which was confirmed by SDS-electrophoresis of denatured pepsin that had been crosslinked covalently by glutaraldehyde (9). This evidence is supported by sedimentation equilibrium experiments, indicating a molecular weight of 50.000 for denatured pepsin at a partial specific volume (v) of 0.74 ml/g. If the assumption holds that upon denaturation the pepsin molecule swells (1), increase of v results in an even higher estimate of the true molecular weight.

The alkaline denaturation of two Candida proteinases, that had been purified in our laboratory, took place at pH 7.0 and 8.4 respectively (8,10). As with pepsin, a loss of anodic mobility could be observed upon polyacrylamide gel electrophoresis, but on contrast, this shift could be ascribed solely to a gain in hydrodynamic size, since a change of the isoelectric point could be excluded (Fig. 1c,d). Disc electrophoresis in polyacrylamide gel gradients revealed a shift from the ovalbumin position (45.000 $M_r$) to the BSA position ($M_r$ 68.000) upon alkaline denaturation (8). The shift could either be due to an expansion of the molecule or to dimerization. Since SDS-electrophoresis of covalently crosslinked, denatured enzyme yielded molecular weight estimates well above the weight of the monomer ($M_r$ 45.000) (8,9), formation of a dimer was assumed. The difference between the calculated molecular weight of

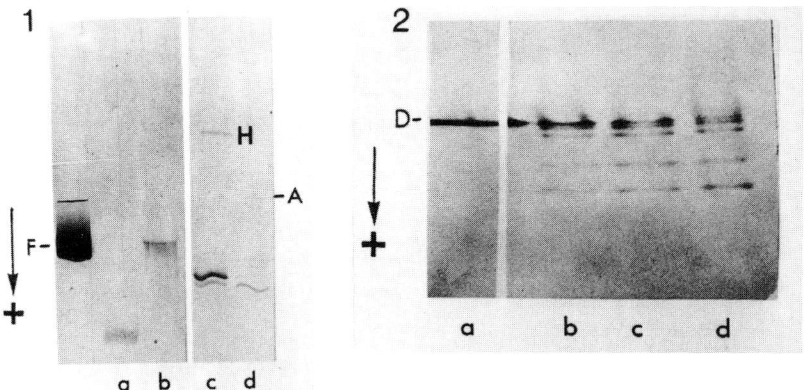


## Fig. 1

Electrofocusing of porcine pepsin (a,b) and <u>Candida</u> protease
(c,d) in a polyacrylamide slab (5% T, 3% C) of 0.1 mm thickness, containing 3% ampholytes (pH 3-10) (Serva, Heidelberg).
A section of the gel is shown after Coomassie blue staining.
Following a prerun, the sample was applied at subneutral pH
(A) to prevent alkaline denaturation. Self digestion of the
enzymes in the proximity of their pI required early termination of the run, minimal focusing time was monitored by the
dye xylene cyanole FF (Eastman, Rochester).
Native pepsin (a) and denatured pepsin (b) differ widely in
their isoelectric points, while both native (c) and denatured
(d) Candida protease focus in the same range. Secondary peaks
in c and d are due to autolysis. Horse ferritin (F) (pI 4.1-
4.6) and hemoglobin residues (H) (pI ca. 7) serve as markers.

## Fig. 2

Electrophoresis of denatured <u>Candida</u> protease in a continuous
polyacrylamide gel gradient (2.5-30% T, 2.5% C). A discontinuous buffer system was employed: 50 mM imidazole-HCl, pH 6.8
as gelbuffer and 50 mM imidazole-glycyl glycine, pH 7.2 as
running buffer. After denaturation by exposure to pH >9, a
single peak is revealed (a). Aliquots of the same sample were
titrated back to pH 7 (b), pH 6.2 (c) and pH 5 (d). Autolysis
of denatured enzyme increases with decreasing pH.

a dimer (90.000) and the apparent value obtained by gel gradient electrophoresis (68.000) may point to an abnormally
elongated shape of the aggregate, that can undergo alignment
upon clcctrophoresis (11). Such elongation has been found
with pepsin, whose axial ratio of 1:3 increases to 1:16 upon
alkaline denaturation (1,5).

Aggregation as a cause of the electrophoretic shift associated with alkaline denaturation leads to autolytic decay of
the denatured enzymes. Autolysis can be observed in the vicinity of the critical pH where denaturation occurs and which
is way above the pH-optimum of the enzymes versus protein
substrates (pH 3-4) (8,10). Autolysis of the aggregate becomes pronounced upon titration back to acidic pH and thus
clearly reflects acidic proteolytic activity (Fig. 2).
Autolysis can be suppressed by the group-specific inhibitor
pepstatin, which proves that the active site of the enzyme
is not basically affected by the denaturation process;
external substrates, though, are prohibited from access and
no binding of 14-C pepstatin methylester could be observed
(9). Since no higher oligomers have been observed, the following interpretation may be warranted: Two identical sites,
that exist only once in every monomer and thus exclude chain
formation, fuse to form a dimer. Dimerization can explain the
dependence of the rate of alkaline denaturation on enzyme concentration (1). The bonds linking the monomers are primarily
hydrophobic, they are not to be split by high salt nor by
manipulation of the pH but were cleaved by SDS alone. Binding
of substrates to carboxyl proteinases is primarily hydrophobic (1,3,12), thus aggregation may involve the hydrophobic
substrate binding sites. This would facilitate also the autolytic decay of the aggregate (Fig. 2).

The pH-dependent conformational change in porcine pepsin and in a secretory protease from Candida albicans has been monitored by a novel technique that allows for electrophoresis at right angles to a continuous pH-gradient. In contrast to the procedures of Rosengren et al. (13,14) and Borowsky et al. (15), our technique does not require use of carrier ampholytes nor an electrophoretic prerun, rather it is a single step procedure.

A homogeneous polyacrylamide gel is cast in a chamber made out of lucite. The chamber consists of two corresponding pieces sharing two drilled channels. Both pieces are kept apart by a U-shaped rubber gasket (Fig. 3a). The chamber is clamped together and is sealed by tape, it serves as an electrophoresis apparatus as well. Therefore a template of two plastic rods is inserted, that has to be removed after polymerization of the gel and thus yields the electrode channels (Fig. 3b). Employing a common gradient mixer, a pH-gradient is formed out of two solutions that differ in pH, density and catalyst concentration, but are identical with respect to acrylamide concentrations.

The recipe given below applies to a gel chamber as shown in Fig. 3a, with cylindrical reservoirs of 10 mm diameter and a flat portion of the gel measuring 35 x 15 x 0.5 mm. Thus the volume of each cylindrical reservoir exceeds the volume of the flat portion of the gel more than 13 times, providing for sufficient buffer capacity to maintain the pH-gradient during the run. The equipment can easily be scaled up for larger sample volumes.

dense solution:

0,9 ml acrylamide stock sol.
  (50% T, 2,5% C)

0,6 ml Tcmcd 1%

0,3 ml buffer: 0,6 M citric
  acid+ 1 M Tris, pH 3,2

0,7 ml sucrose 50%

0,9 ml water

0,6 ml persulfate 0,25%

light solution:

0,9 ml acrylamide stock sol.
  (50% T, 2,5% C)

0,4 ml Temcd 1%

0,3 ml  1 M Tris

0,15 ml sucrose 50%

0,05 ml bromophenol blue 0,1%

1,7 ml water

0,5 ml persulfate 0,25%

The solutions are mixed on ice and are degassed subsequently;
3,4 ml of each solution are used for the gradient which is
filled simultaneously into the two reservoirs (R) of the
lucite chamber (Fig. 3). A straight upper edge of the gel is
produced by an overlay solution (persulfate in 80 mM Tris).
Gelation is accomplished within 20 min. at $37^{O}$C, it proceeds
from top to bottom. The gradient covers the range from pH 4.5
to 9.5, it can be used right after gelation or may be stored
overnight at $4^{O}$C. The protein sample (5-15μl) has to be made
up in 10% sucrose, it is applied into the channel (C) (Fig. 3)
with a narrow glass pipet; the electrode channels are filled
with PBS. Electrophoresis is performed at 3 mA for ca. 15 min.
with bromophenol blue and xylene cyanole as tracking dyes;
the latter dye approximately marks the position of anionic
protein. The flat separation section of the gel is dissected
afterwards for subsequent staining or use in zymogram techniques, the cylindrical parts of the gel allow for measurement of the pH-gradient with a microelectrode (for instance
micro combination pH-probe type MI 410, Microelectrodes Inc.,
Londonderry, NH).

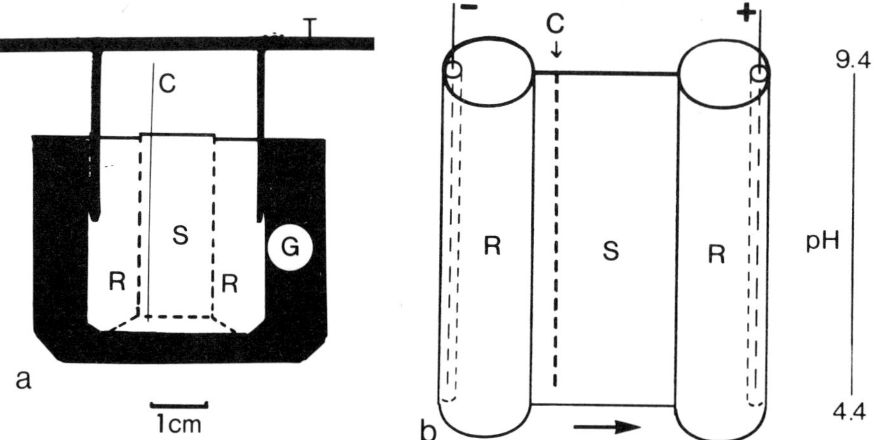



Fig. 3

a) Side view of the gel chamber with template (T) for electrode channels halfway inserted. R:cylindrical reservoirs,
S:cover slips that have been pasted onto the lucite faces for
enhanced adhesion of the polyacrylamide gel in the separation
section, G:rubber gasket, C:siliconized glass rod to form the
sample. channel.

b) Schematic view of the gel with two cylindrical reservoirs
(R) holding the electrodes at the distant edges. The separation section (S) contains the narrow channel (C) for application of the sample. The position of channel C suits anodic
migration of the sample, it can be shifted freely.


The technique described above was found to be suitable to monitor the pH-dependent conformational shift of both pepsin and
Candida protease as described in the first paragraph. Fig. 4a
illustrates that the conversion hardly involves intermediates
and thus complies with the assumption of dimerization. The pH-
profiles of such gels confirmed the critical pH-range where
denaturation of the proteases occurs and which had been identified previously by activity measurements. On contrast,
enzyme that had been denatured prior to electrophoresis migrated in a single continuous front representing a homogeneous
molecular species (Fig.4b).

674

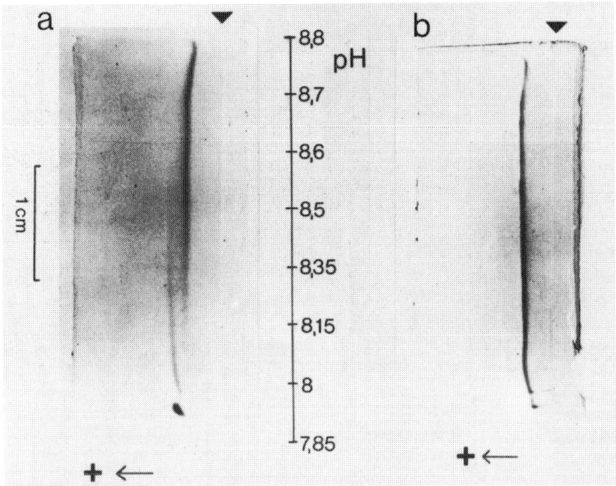



Fig. 4

a) Electrophoresis of <u>Candida</u> protease across a Tris-HCl gradient, Coomassie blue staining. The fast migrating species
   represents active enzyme, the slower species is denatured
   enzyme ( ▼ marks the origin).

b) Pattern of Candida protease that had been denatured prior
   to electrophoresis across a pH-gradient as in figure 4a.

As has been suggested previously (15), electrophoresis across
a pH-gradient should be suitable to monitor the autocatalytic
conversion of protease zymogens into active enzymes (16).
This we have demonstrated with porcine pepsinogen (Fig. 5).
After electrophoresis, the gel slab containing both pepsin
and pepsinogen was put between two layers of protein agar at
pH 3 and was incubated at 37°C; subsequent staining with amidoblack (Fig. 5c) revealed proteolytic activity in a position
corresponding to the pepsin front in figure 5b.

Summary


A one-step procedure for electrophoresis of macroions across
a pH-gradient is described that employs only conventional
buffers. Due to limitations of the polyacrylamide catalyst
system, the technique is restricted to the pH-range of 4 to

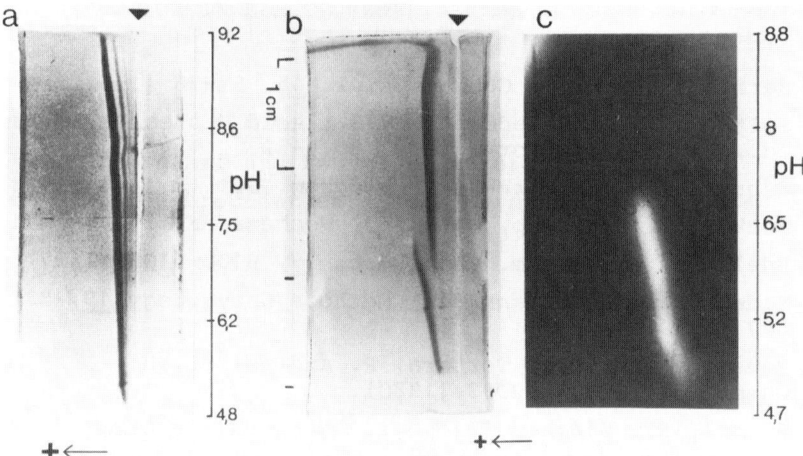

Fig. 5

a) Electrophoresis of freshly dissolved porcine pepsinogen
   across a Tris-citric acid gradient, Coomassie blue staining ( ▼ marks the origin). With its pI at pH 3.7, mobility of pepsinogen increases with increasing pH. No traces
   of pepsin have been recognized.

b) Electrophoresis of pepsinogen after 24 h of storage at
   4°C, pH 5.5, experimental conditions as in fig. 5a. Pepsinogen has been converted quantitatively into pepsin at the
   acidic side of the gradient. The conversion is due to a
   few pepsin molecules present in the aged pepsinogen solution. Prior to Coomassie blue staining the gel was passed
   through 10% TCA that denatures pepsin. Native pepsin undergoes autolysis even upon staining by the ordinary Coomassie
   blue, acetic acid, methanol procedures.'

c) Zymogram of pepsin after conversion of pepsinogen in the
   course of electrophoresis across a pH-gradient as in
   figure 4b. The overlay consisted of 1% agar containing
   2 mg/ml BSA in 0.1 M citrate buffer pH 3.0; 5 h incubation
   at 37°C and subsequent staining with amidoblack.

10. The technique has been employed at a micro scale to monitor the autocatalytic conversion of pepsinogen into pepsin
and to demonstrate the conformational shift inflicted on carboxyl proteases upon alkaline denaturation. By aid of other
electrophoretic methods evidence has been gathered that alkaline denaturation is due to dimerization and subsequent autolysis of such enzymes.

electrophoretograms of glycoproteins (3,4) it was considered important to extend the finding for elastase to other structurally well-characterized proteolytic enzymes and non-glycoproteins and also to determine the extent to which the proteins are oxidized by periodic acid under the conditions used in the PAS staining procedure.


Materials and Methods


Bovine trypsin, $\alpha$-chymotrypsin, egg white lysozyme, pepsin A, carboxypeptidase B and ovalbumin (a glycoprotein) were obtained from Worthington Biochemicals, Freehold, NJ. $\beta$-lactoglobulin was a product of the Sigma Chemical Company, St. Louis, MO. Elastase was isolated from porcine pancreas as previously described (5). Albumin was separated from human serum using previously described methods (1). Para-periodic acid (reagent) was obtained from the G. Frederick Smith Chemical Company, Columbus, Ohio. All reagents for electrophoresis were obtained from Bio-Rad Laboratories, Richmond, California.

Electrophoresis. Anionic polyacrylamide gel electrophoresis (PAGE) was carried out according to Davis (6). Cationic PAGE (7) was used for the separation of the basic proteins.

Staining. Following electrophoresis the gels were stained with Coomassie Brilliant Blue R250 and with PAS reagent using the method of Zacharius et al. (4). Briefly, the method of Zacharius et al. (4) consists of precipitating the electrophoresed proteins with 12.5% trichloroacetic acid, immersion in 1% periodic acid, washing the gels in distilled water until free of periodic acid, immersion in Schiff's reagent, treatment with 0.5% sodium metabisulfite and destaining in distilled water.

Periodic acid oxidation. It was also of interest to determine the extent to which the proteins were oxidized by periodic acid under the conditions used in the PAS staining

procedure.  The proteins (2.5 mg) were dissolved in 1.0 ml of
3% acetic acid (HOAc) followed by the addition of 1.0 ml of
10mM periodic acid (in 3% HOAc).  The reaction mixture was
incubated for 50 min in the dark and residual periodic acid
determined using the method of Knowles (8).


Results and Discussion

The results obtained when proteolytic enzymes, lysozyme, and
albumin were separated by PAGE and stained with PAS reagent
using the method of Zacharius et al. (4) are presented in
Fig. 1.  It is seen that, with the exception of albumin, all
of the proteins stain positively with PAS reagent.
$\beta$-lactoglobulin, a protein used as a negative control by
Zacharius et al. (4), was also negative in the present study.
Negative results were also obtained with prealbumin (most
anodal zone, Fig. 1, gel 10), a protein containing 0.5%
carbohydrate (9).
When the periodic acid step was omitted from the procedure
(4) the results presented in Table 1 were obtained.  Only
light staining was observed with elastase and $\alpha$-chymotrypsin.
Albumin, a non-glycoprotein, was routinely negative when
periodic acid was included in the staining procedure (Fig. 1)
but produced moderate to heavy staining when the step was
omitted (Table 1).  Electrophoretograms of serum showed
light, diffuse PAS staining in the $\alpha_0$, $\beta$ and $\gamma$ fractions when
the periodic acid step was omitted from the procedure (4).
When several of the proteins (Fig. 1) and a glycoprotein
control were subjected to periodic acid oxidation and residual
periodic acid determined (8) the results shown in Table 2 were
obtained.  It is apparent that the proteins are indeed
oxidized and that the PAS staining intensity of the proteins
in Fig. 1 are roughly in accord with the amount of periodate
consumed (Table 2).  The fact that PAS staining was positive
in the non-glycoproteins, with the exception of albumin

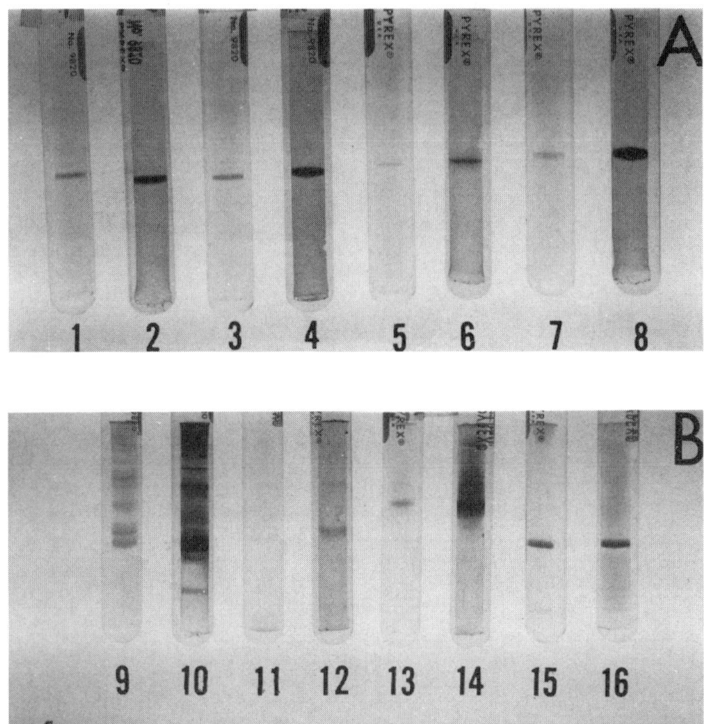

Fig. 1 (A) Separation of α-chymotrypsin (1,2), trypsin (3,4), elastase (5,6), and lysozyme (7,8) in cationic polyacrylamide gel; 12 μg protein/gel; 3.5 ma/gel for 0.5 hr.  The cathode is at the bottom of the figure.  (B) Separation of human serum (9,10), albumin (11,12), carboxypeptidase B (13,14) and pepsin A (15,16) in anionic polyacrylamide gel.  Serum, 2.5 μl; 12 μg protein/gel; 2.5 ma/gel for 2.5 hr.  The anode is at the bottom of the figure.  Odd numbers stained with PAS reagent; even numbers with Coomassie Brilliant Blue R250.

(Fig. 1B), and the fact that periodate was consumed (Table 2) strongly suggests that periodate reaction products are present on the molecules which when reacted with Schiff's reagent give the characteristic color commonly associated

TABLE 1  PERIODIC ACID-SCHIFF STAINING OF ALBUMIN, LYSOZYME,
         AND VARIOUS PROTEOLYTIC ENZYMES FOLLOWING ZONE
         ELECTROPHORESIS IN POLYACRYLAMIDE GEL[a]

| PROTEIN[b] | PERIODIC ACID OXIDATION | |
|---|---|---|
| | With | Without |
| Elastase | + | + (L)[c] |
| Trypsin | + | − |
| α-chymotrypsin | + | + (L) |
| Lysozyme | + | − |
| Albumin | − | + |
| Pepsin A | + | − |
| Carboxypeptidase B | + | N.D.[d] |

[a]Method of Zacharius et al. (4).
[b]12 μg/gel
[c](L), light staining
[d]Not determined


TABLE 2  PERIODIC ACID OXIDATION OF OVALBUMIN (A GLYCOPROTEIN)
         VARIOUS PROTEOLYTIC ENZYMES, AND LYSOZYME

| PROTEIN | MOLES PERIODATE/MOLE PROTEIN[a] |
|---|---|
| Ovalbumin | 39.3 |
| α-chymotrypsin | 13.0 |
| Elastase | 5.2 |
| Trypsin | 8.8 |
| Pepsin A | 10.5 |
| Lysozyme | 5.1 |

[a]Method of Knowles (8).

690

with the presence of carbohydrate.  The amino acids leucine,
serine,  tyrosine, arginine, tryptophan and lysine, were all
PAS positive following periodic acid oxidation, and reduction
with sodium metabisulfite.
Clamp and Hough (10) have suggested that cysteine, cystine,
methionine, tryptophan, tyrosine  and histidine would be
oxidized by periodic acid wherever  they occurred in the
peptide chain and serine and threonine only if they occurred
as N-terminal residues.  This suggestion (10) was confirmed
by Atassi (11) using sperm whale myoglobin.  Peptide bond
cleavage of the protein did not occur and serine and
threonine were apparently unchanged (11).  On the other hand
Kennedy and Butt (12) found that approximately 20% of the
serine and threonine residues were lost as a result of
periodate oxidation of human pituitary follicle-stimulating
hormone.  The question arises, are the observations made here
fact or artifact?  In the sense that the PAS procedure is
"specific" for glycoprotein the observations are artifactual.
On the other hand the _fact_ remains that the periodate-
oxidized proteins did indeed react with Schiff's reagent to
produce colored complexes characteristic of those obtained
with glycoproteins.  Obviously, interpretation of electrophoretograms of supposed glycoproteins stained by the PAS
procedure should be viewed with caution.


Acknowledgements

This work was supported by Grant HL15979 from the National
Institutes of Health, Bethesda, Maryland.  The author is
indebted to Ms. Kailas Patel, Ms. Geri Doran, Ms. Jeanette
O'Hare, and Ms. Cecelia Stodd for excellent technical
assistance and to Ms. Jan Boggs for typing the manuscript.

References

1. Baumstark, J.S., Lee, C.T. and Luby, R.J.: Biochim. Biophys. Acta 482, 400-411 (1977).

2. Hartley, B.S., Brown, J.R., Kaufman, S.L. and Smillie, L.R.: Nature 207, 1157-1159 (1965).

3. Glossman, H. and Neville, D.M. Jr.: J. Biol. Chem. 246, 6339-6346 (1971).

4. Zacharius, R.M., Zell, T.E., Morrison, J.H. and Woodlock, J.J.: Anal. Biochem. 30, 148-152 (1969).

5. Baumstark, J.S.: Biochim. Biophys. Acta 220, 534-551 (1970).

6. Davis, B.J.: Ann. N.Y. Acad. Sci. 121, 404-427 (1964).

7. Reisfeld, R.A., Lewis, U.J. and Williams, D.E.: Nature 195, 281-283 (1962).

8. Knowles, J.R.: Biochem. J. 95, 180-190 (1965).

9. Heimburger, N., Heide, K., Haupt, H. and Schultze, H.E.: Clin. Chim. Acta 10, 293-307 (1964).

10. Clamp, J.R. and Hough, L.: Biochem. J. 94, 17-24 (1965).

11. Atassi, M.Z.: Biochem. J. 102, 478-487 (1967).

12. Kennedy, J.F. and Butt, W.R.: Biochem. J. 115, 225-229 (1969).

ELECTROPHORETIC AND STAIN METHOD FOR PYRUVATE KINASE ISOENZYME
PATTERNS

Enrique Meléndez-Hevia, Javier Corzo and José Pérez
Departamento de Bioquímica, Facultad de Biología
Universidad de La Laguna, Tenerife, Spain.


Introduction

The pattern of pyruvate kinase isoenzymes is altered in some
pathological processes (1), including malignancy (2). The employment of these alterations as a tool in Clinical Biochemistry is difficult, because of the lack of a reliable method for
the development of zymograms. In fact, the most widely employed
method is based on the NADH fluorescence decay after the lactate dehydrogenase coupled reaction with the pyruvate produced by
the pyruvate kinase from phosphoenolpyruvate (3). The zymogram
must be recorded onto a photographic plate, and the relative
activity determination of the isoenzymes is not accurate. An
alternate method is the coupling with the glucose-6-phosphate
dehydrogenase-NADP reduction after the hexokinase coupling reaction, but a contaminating activity, presumibly adenylate kinase,
appears. A modification of the last procedure is presented. The
AMP addition readly blocks the ATP synthesis mediated by adenylate kinase, avoiding the pyruvate kinase purification step cons-
traint prior analysis (4).

Materials and methods

All the products were reagent grade. The samples were 35,000xg, 20 minutes centrifugation supernatants from rat liver, brain, kidney, testes and skeletal muscle homogenates. The homogenation medium contains 0.25 M sucrose; 0.25 mM fructose 1,6 diphospha- te; 0.01 M $MgCl_2$; 0.06 M KCl; 0.01 M $\beta$-mercaptoethanol, and 0.1 M tris, pH 7.4. The 26 x 12,5 cm gel mould assembly was perfor- med with a 3 mm thick glass plate, with 23 glued plastic moulds (5 microliters volume) for sample application, joined to a rec- tangular three-arm aluminium frame coated with Parafilm (Ameri- can Can Co.), a 1 mm thic glass plate, and a second 3 mm plate; the apparatus was fixed with clamps, and was heated before use to 42º C. The acrylamide/agarose solution contains 0.8% agarose (type I, Sigma Chem.Co.); 3.5% acrylamide; 0.1% NN-methylenebis- acrylamide; 0.25% TEMED; 0.025% w/v ammonium persulfate; 0.062 M glycine, and 0.074 M tris, pH 8.3 at 15º C. This solution was heated to 42º C and poured between the plates separated by the frame. After 30 minutes the polimerization was achieved and the temperature was lowered to 25º C, the agarose was allowed to gel, and the gel, supported on the thinner glass plate, was ready to use after dismounting the mould. The electrophoresis was perfor- med on a Multiphor apparatus (LKB, Sweden); the vessel buffer was 0.01 M tris, 0.33 M glycine. Time of electrophoresis was 3 to 4 hours, with an electric field strength of 10 to 12 V/cm.

The stain gel solution contains 100 mM tris-HCl, pH 7.4; 1 mM NADP; 3 mM phosphoenolpyruvate; 2 mM ADP; 12 mM AMP; 0.1 mM fruc- tose 1,6 diphosphate; 4 mM glucose; 60 mM KCl; 10 mM $MgCl_2$; 0.25 mg/ml nitro blue tetrazolium (Sigma); 0.1 mg/ml phenazine metho- sulfate;1.3 units/ml hexokinase (type f-300, Sigma); 1 unit/ml

glucose 6-phosphate dehydrogenase (type V, Sigma), and 0.6% a-
garose. This solution was poured at 42º C onto a glass plate with
a 20 x 6 cm mastic frame; the solution was allowed to gel at room
temperature. Then, the stain gel was applied to the electrophoresis gel and incubated 60 minutes at 37º C. The stain reaction
was stopped with inmersion in cold water or cold 5% acetic acid,
and was washed  until the yellow colour of nitro blue tetrazolium disappeared. The stain gel can be scanned after or before
its dessication. We employed for this purpose an Auto-scanner
Flur-Vis Helena (Helena Laboratories).


Results and discussion


The detection method presented is selective for the pyruvate
kinase activity. Without AMP added, a phosphoenolpyruvate independent spot appears that is readily removed by AMP addition at
concentrations up to 10 mM (figure 1). However, AMP slows the

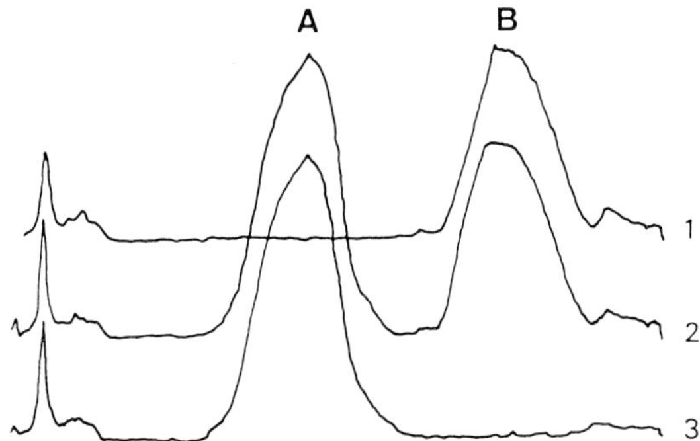


Fig. 1. Scanning tracings of rat muscle samples stained with:
1) Stain solution (see text) without phosphoenolpyruvate and
AMP. 2) The same, with phosphoenolpyruvate added. 3) With both
substrates added. A commercial sample of pyruvate kinase from
rabbit muscle (type III, Sigma) gives only peak A. Peak B, contaminating activity. AMP addition (3) remove the B contaminatig
peak.

stain reaction increasing the developing time. For this reason
the optimal AMP concentration was found to be between 10-14 mM.
All the pyruvate kinase isoenzymes give spots, as is shown in
figure 2. Their location and relative intensities are consistent
with those found with the Susor and Rutter method (3)(data not
shown). The rate of colour appearing is different for each isoenzyme, the L-type being the slowest. Thus, an incubation time of
at least one hour at 37º C is necessary in order to saturate all
isoenzymes and to obtain order zero kinetics. The formazan is
precipitated in both electrophoresis and stain gels, but the
first is more liable to crack, and its spots are generally more
diffuse than those of the second. Because of this effect, it is
more useful to remove the stain gel and process it alone. The
gel can be scanned with any conventional densitometer. The records (figure 3) show a clear peak for each isoenzyme and provide an easy method for the recognition of the pyruvate kinase i-
soenzyme pattern of any sample.

Figure 4 shows the correlation between spot intensity, as evaluated directly from the densitometer integration tracing, and
the pyruvate kinase activity applied. A good correlation ( r=
0.993) has been encountered between $2-20 \times 10^{-3}$ units of pyruvate kinase and densitometer tracing peak area.

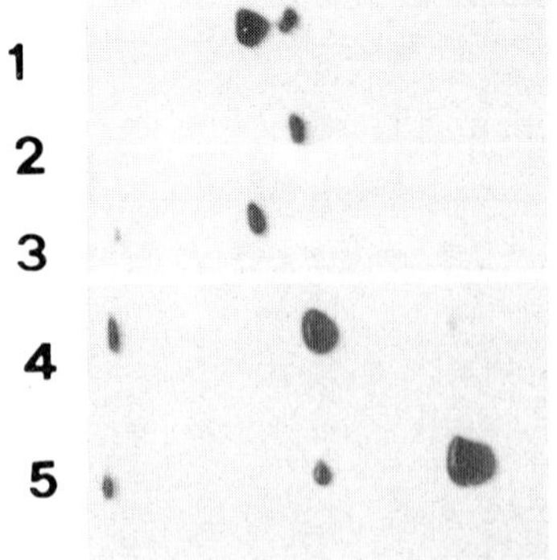


Figure 2.- Electrophoretic separation of pyruvate kinase i-
soenzymes from various tissues
of a 60 days old male rat. 1)
brain. 2) testes. 3) skeletal
muscle. 4) kidney. 5) liver.
From left to right: origin, M
or $M_1$-type, K or $M_2$-type and
L-type isoenzyme. Anode at right.

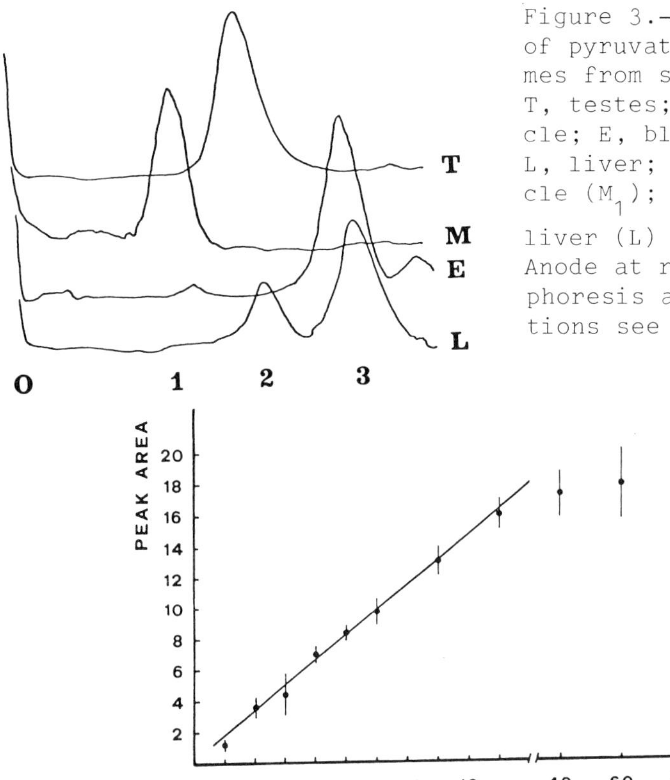

Figure 3.- Scanning tracings of pyruvate kinase isoenzymes from some rat tissues. T, testes; M, skeletal muscle; E, blood hemolysate; L, liver; O, origin. 1, muscle ($M_1$); 2, kidney ($M_2$); 3, liver (L) type isoenzyme. Anode at rigth. For electrophoresis and stain conditions see text.

Figure 4.- Correlation between densitometer peak areas and units of rabitt muscle pyruvate kinase (type III, Sigma). Four samples for each activity were assayed.

As application to a biological problem the percentage of activity of L- and K-types of isoenzymes from liver and kidney of two differents rats is shown in table 1. The method presents a remarkably degree of consistency in different samples of the same animal. As L-type isoenzyme is inducible, and $M_2$-type is not, the differences between rats can be explained as a result of their different physiological state.

This selective staining method can be employed with any other electrophoretic technique and supporting media, as cellulose

| Isoenzyme | LIVER | | KIDNEY | |
| --- | --- | --- | --- | --- |
|  | L | $M_2$ | L | $M_2$ |
| Rat 1 | $78.5 \pm 1.4$ | $21.5 \pm 1.4$ | $11.6 \pm 1.4$ | $88.4 \pm 1.4$ |
| Rat 2 | $64.4 \pm 2.6$ | $35.6 \pm 2.6$ | $15.8 \pm 1.9$ | $84.2 \pm 1.9$ |

Table 1.- Rat pyruvate kinase isoenzymes. Relative activity calculated by scanning of electrophoresis gels stained according
to the experimental procedure (see text). 1, 60 days old male;
2, 60 days old pregnant female. Three samples for each organ
were assayed.

acetate strips or agarose and starch gels. It is feasible also
to use a staining solution, without agarose, for cylindrical e-
lectrophoresis gels.

Acknowledgements.

This work was supported in part by a grant from the  Comisión
Asesora para la Investigación Científica y Técnica nº 4119-79.

References.

1. Nakashima, K., Miwa, S., Oda, S., Tanaka, T., Imamura, K.,
   Nishima, T. : Blood 43, 537 (1974)

2. Tanaka, T., Imamura, K., Ann, T., Taniuchi, K.,: GANN Mono-
   graph on Cancer Research 13, 219-234 (1972)

3. Sussor, W.A., Rutter, W.J.,: Anal. Biochem. 43, 147-155 (19
   (1971)

4. Kahn, A., Marie, J., Galand, C., Boivin, P.,: Humangenetik
   29, 271-280 (1975)

ISOZYMES OF SUPEROXIDE DISMUTASE IN <u>SERRATIA</u> <u>MARCESCENS</u>

Z. González-Lama, O.E. Santana, P. Betancor
Colegio Universitario de Las Palmas (División de Medicina)
Las Palmas de Gran Canaria, Spain

Introduction

Superoxide anion, $O_2^-$ , which is one of the active species of
oxygen offering toxic effect on living organisms is generated
biologically in living organisms themselves in the presence of
molecular oxygen (1). Superoxide dismutases are a group of me-
talloenzymes, whose function is to scavenge the superoxide radical anión $O_2^-$ by catalyzing its dismutation to hydrogen pero-
xidase and dioxygen: $2\ O_2^- + 2\ H^+ \longrightarrow H_2O_2 + O_2$, thus offering the organism protection against the toxic effect of this free radi-
cal.
Superoxide dismutase has been extensively studied in recent
years and has been shown to exist in all aerobic (2,3,4) or fa-
cultatively anaerobic microorganism (5,6) some microaerophilic
(7) and in some anaerobes (8,9,10).
We wish to report here the existence of two superoxide dismutases isozymes in the crude cell-free extract from <u>Serratia</u> <u>mar-
cescens</u> by isoelectric focusing in polyacrylamide gel slabs.

Material and methods

<u>Serratia</u> <u>marcescens</u> was isolated from clinical specimens, was
grown in Brain Heart Infusion (difco lab. Detroit, Michigan,
USA) at 37°C under aeration. Cells of <u>Serratia</u> <u>marcescens</u> were harvested by centrifugation at 800 xg for 10 min. and washed several times with cold distilled water. Cell-free extracts

were prepared by disrupting with a MSE ultrasonic desintegrator for five 1 min. periods. Cell debris was removed by centrifugation at 48,ooo xg for 1 h (4°C), in a refrigerated centrifuge Sorvall RC-5B (Dupont instrument, Newton, Conneticut USA). The supernatant fraction was used for isoelectric focusing in polyacrylamide gel.

Isoelectric focusing in polyacrylamide gel was performed using the multiphor apparatus (LKB Produkter, Bromma, Sweden) pH range 3.5-9.5. Superoxide dismutase activity in the gel was located by the photochemical method of Beauchamp and Fridovich (11). The prosthetic metal ion of superoxide dismutase is determined by two methods:

1) Inhibition by $H_2O_2$: soaking the slabs of gel after isoelectric focusing in a solution of 5 mM $H_2O_2$, 0.1 mM EDTA and 1 mM KCN prior to staining with NBT (12). Cyanide is included to inhibit catalase.

2) Inhibition by KCN: soaking the slabs of gel prior to staining with 20 mM KCN (13).


Results and discussion


Figure 1, is a photograph demonstrating superoxide activity in the cell-free crude extracts of _Serratia_ _marcescens_ by isoelectric focusing in polyacrylamide gel. We have found two distinct superoxide dismutases visualised in the anodal zone of the gel with a pI:4.8 and 5.3.

Britton and co-workers (14) found, by electrophoresis in 10% gels, in _Serratia_ _marcescens_ two isozymes with a relative mobility of 0.28, 0.48. The band of 0.48 is inhibited by $H_2O_2$; but our bands were neither inhibited by $H_2O_2$ or KCN. Since cyanide inhibits Cu,Zn-SOD but does nor affect either Fe-SOD or Mn-SOD (15); and Fe-SOD was rapidly inactivated by $H_2O_2$ and Mn-SOD remained fully active (14).

We suggest that our isozymes must been Mn-SOD; the discrepancy

between the present results and those other results lies in the
metal composition of the medium where the microorganisms were
grown, because this one changes the pattern of bands corresponding to superoxide dismutase on electrophoresis gel (16).

Figure 1: Isoelectric focusing of SOD isozymes from _Serratia_
 _marcescens_ cell-free extract.

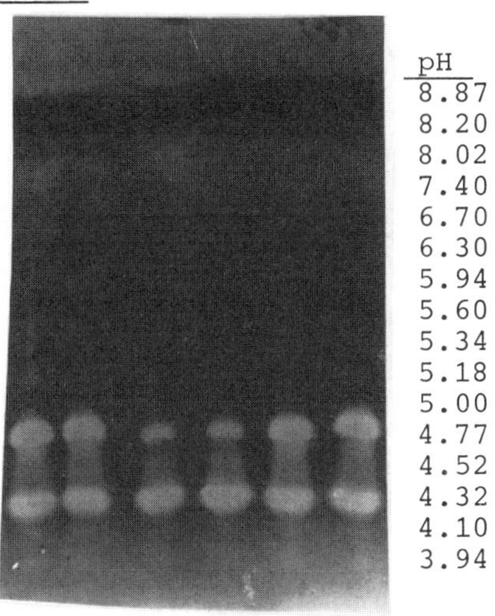



## References


1.  Fridovich, I.: Ann. Rev. Biochem. 44, 147-159 (1975).

2.  Yamakura, F.: J. Biochem. 83, 849-857 (1978).

3.  Anastasi, A., Bannister, J.V. & Bannister, W.H.: Int. J.
    Biochem. 7, 541-546 (1976).

4.  González-Lama, Z., & Cutillas, M.J.: I Meeting of the Luso-
    -Spanish Biochemical Society, Coimbra, Abstract n°P 271(1980)

5.  Steinman, H.M.: J. Biol. Chem. 253, 8708-8720 (1978).

6.  Yano, K. & Nishie, H.: J. Gen. Appl. Microbiol. 24, 333-339
    (1978).

7.  Norrod, P. & Morse, S.A.: Biochem. Biophys. Res. Com. 90,
    1287-1294 (1979).

8.  Hatchikian, E.C. & Henry, Y.A.: Biochimie 59, 153-161 (1977).

9.  Kanematsu, S. & Asada, K.: Arch. Biochem. Biophys. 185, 473-482 (1978).

10. Privalle, C.T. & Gregory, E.M.: J. Bacteriol. 138, 139-145 (1979).

11. Beauchamp, C.O. & Fridovich, F.: Anal. Biochem. 44, 276--287 (1971).

12. Mc Euen, A.R., Hill, H.A.D., Dring, G.J. & Ingram, G.S.: in Chemical and Biochemical Aspects of Superoxide and Superoxide Dismutase, Bannister, J.V. and Hill, H.A.O., eds. Elsevier/North-Holland, New York, pp. 272-283 (1980).

13. Lonnerdal, B., Keen, L.C. & Hurley, L.S.: FEBS Letters 108, 51-55 (1979).

14. Britton, L., Malinowski, D.P. & Fridovich, I.: J. Bacteriol. 134, 229-236 (1978).

15. Mc Cord, J.M.: Adv. Exp. Med. Biol. 74, 540-550 (1976).

16. Yamakura, F.: Biochim. Biophys. Acta 422, 280-294 (1978).

ISOZYMES OF SUPEROXIDE DISMUTASE IN LIVER FROM <u>LACERTA STEHLINII</u>


Z. González-Lama, A. de Armas, P. Betancor
Colegio Universitario de Las Palmas (División de Medicina)
Las Palmas de Gran Canaria, Spain.


E.M. Hevia
Departamento de Bioquímica, Facultad de Biología
La Laguna, Spain.



Introduction

One minor product of the biological reduction of molecular oxygen is a free radical nominated superoxide anion, $O_2^-$, formed by
the univalent reduction of oxygen. Superoxide dismutase (SOD)
catalyzes the dismutation of the $O_2^-$, free radicals by the following reaction: (1)

$$O_2^- + O_2^- + 2\ H^+ \longrightarrow O_2 + H_2O_2$$

SOD is an important factor in the protection against free radical damage. (2,3) This metalloenzyme has been shown to exist in
two forms in animal tissues, one containing manganese as cation
(Mn-SOD) and other containing copper and zinc as cations (Cu,Zn-
-SOD). These $SOD_s$ have been isolated from chicken liver (4) rat
(15) and mouse liver, (6) horse liver (7,8) and human liver (9).
Electrophoresis in polyacrylamide gels followed by a specific
staining technique (10) has been employed by several investigators in the study of SOD (11). Isoelectric focusing has the advantage of concentrating the protein zones as compared to electrophoresis in zone-spreading occurs. Thus, much lower quantities of enzyme can be detected by isoelectric focusing and also a much higher degree of resolution is achieved. This paper
describes the application of isoelectric focusing to the study
of SOD isoenzymes, using liver from <u>Lacerta stehlinii</u>, with the

successful separation of isoenzyme bands. The separation of
other proteins in this animal by isoelectric focusing has been
described (12).


Material and methods

_Lacerta stehlinii_ is a primitive lizard originating from the
Canary Islands less evolved than other lizards. The animals
used were adults from _Lacerta stehlinii_: no sex-dependent differences in SOD activity were noted. On the day of the experiment animals were anesthetized with ether and opened along the
ventral midline. In order to remove blood from the liver, the
superior vena cava was isolated and severed to allow circula-
tory drainage. Isotonic saline (0'9%_, made with deionized-dis-
tilled water, was perfused into the left atrium until the exudate was clear. Liver was quickly removed, washed briskly in
cold distilled water, blotted on filter paper, homogenized in
glass-Teflon Potter-Elvehjem homogenizers with 9 vol. of cold
0.25 M Sucrose. The homogenate was centrifuged at 800 xg 15
min., (4°C) in a refrigerated centrifuge Sorvall RC-5B (Dupont
instrument, Newton, Conneticut, USA) and the pellet was discarded. The supernatant was centrifuged at 10,000 xg for 10 min:

1.  The pellet was washed twice by resuspension in the origi-
    nal volume of cold 0'25 M sucrose followed by centrifuga-
    tion at 10,000 xg for 10 min. (are the mitochondria).
2.  The supernatant was centrifuged at 48,000 xg for 1 h, and
    the pellet was discarded, and the supernatant was used for
    isoelectric focusing. (soluble SOD)
The final mitochondrial pellets were suspended in one-half of
the original volume of cold 0'25 M sucrose followed by disrupting with MSE ultrasonic desintegrator for five periods 1 min.
Mitochondrial debris was removed by centrifugation at 48,000 xg
for 1 h. and the supernatant was used for assay SOD (particle

fraction). Before to assay SOD, we have obtained a chloroform-
-ethanol extract.
Isoelectric focusing in polyacrylamide gel was performed using
the multiphor apparatus (LKB Produkter, Bromma, Sweden) pH range 3.5-9.5. Superoxide dismutase activity in the gel was located by the photochemical method of  Beauchamp and Fridovich
(10).


Results and discussion

The results of electrofocusing in the pH-range of 3.5-9.5. are
shown in Fig.1.

Fig. 1.  Isoelectric focusing of <u>Lacerta</u> <u>stehlinii</u> liver iso-
         zymes (Fractions: Soluble and particle).

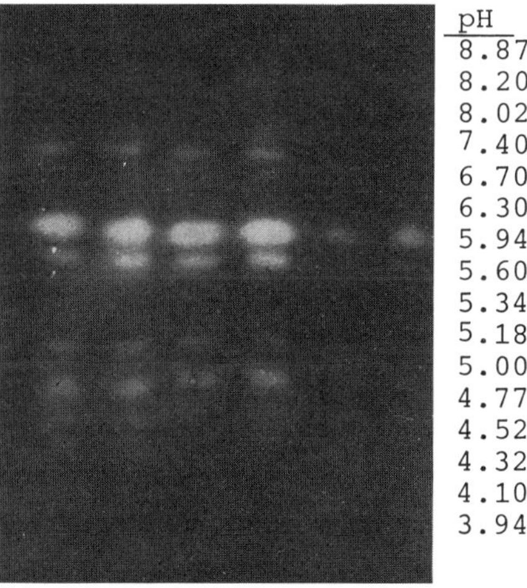


As can be seen, lizard liver contained 7 SOD isoenzymes in the
soluble fraction and 2 isoenzymes in the particle fraction. We
do not find differences in SOD activity between crude extracts
and chloroform-ethanol extracts. The pI in the soluble fraction
and the particle fraction are shown in the table 1. As can be
seen the isozymes in the particle fraction are found in the soluble fraction also. The isoelectric points of chicken liver

(5) are in close agreement to those we found for lizard liver.
The isoenzymes from bovine, rat, mouse and man liver (5,6,9)
are very different to those we found for Lacerta stehlinii liver isoenzymes. These isoenzymes are numerons as well as wides-
pread in their isoelectric points, and need further characterization.

Table I.  Isoelectric points of SOD isozymes in Lacerta stehlinii liver.

| Soluble fraction | Particle fraction |
| --- | --- |
| 4.80 | |
| 5.00 | |
| 5.15 | |
| 5.45 | |
| 6.45 | 6.45 |
| 7.05 | 7.05 |
| 8.00 | |

Values shown here are the means of at least five determinations and are given to the nearest 0'05 pH
units.


References

1.  Mc Cord, J.M. & Fridovich, I.: J. Biol. Chem. 244, 6049-
    -6055 (1969).

2.  Fridovich, I.: Annu. Rev. Biochem. 44, 147-159 (1975).

3.  Fridovich, I.: in Frontiens in Physicochemical Biology,
    Pullman, B. ed. Academic Press, New York, pp.269-281 (1978).

4.  Weisiger, R.A. & Fridovich, I.: J. Biol. Chem. 248, 3582-
    -3592 (1973).

5.  Lonnerdal, B., Keen, L.C. & Hurley, L.S.: FEBS Lett. 108
    51-55 (1979).

6.  Novak, R., Matkovics, B., Morik, M. & Fachet, J.: Experien-
    tia, 34, 1134-1135 (1978).

7.  Albergoni, V. & Cassini, A.: Comp. Biochem. Physiol. 47 B,
    767-777 (1974).

8. Ammer, D. & Learch, K.: in Chemical and Biochemical Aspects of Superoxide and Superoxide dismutase, Bannister, J.V. and Hill, H.A.O. ed. Elservier/North-Holland, New York, pp.230-236 (1980).

9. Mc Cord, J.M., Boyle, J.A., Dry, E.D.Jr., Rizzolo, L.J. & Salin, M.L.: in superoxide and superoxide dismutases, Michelson, A.M., Mc Cord, J.M. and Fridovich, I. ed. Academic Press, New York, pp. 129-138 (1977).

10. Beauchamp, C.& Fridovich, I.: Anal. Biochem. $\underline{44}$, 276-287 (1971).

11. de Rosa, G., Duncan, D.S., Keen, C.L. & Hurley, L.S.: Biochim. Biophys. Acta $\underline{566}$, 32-39 (1979).

12. González-Lama, Z., López-Orge, R.H. & Lamas, A.M.: in Electrophoresis '79, Radola, B.J. ed. Walter de Gruyter and Co. Berlin. New York, pp. 841-845.(1980)

ELECTROPHORETIC MULTIPLICITIES OF γ-GLUTAMYLTRANSFERASE IN
HUMAN SERUM AND THEIR CHARACTERIZATION BY AFFINITY CHROMATO-
GRAPHY

Masaki Izumi
First Department of Internal Medicine, Okayama University
Medical School, Okayama, Japan


Kazuhisa Taketa
Health Research Center, Kagawa University, Takamatsu, Japan


Introduction

γ-Glutamyltransferase (GGT)(EC 2.3.2.2) in human serum (or
plasma) has been separated into several bands (I, I', I", II,
II' and higher molecular species up to X) by polyacrylamide-
gradient (4-30%) gel-slab electrophoresis (PAGGSE) (1).
The distribution of GGT isozyme among the multiple forms varies
depending on the type of liver injuries involved; i.e., increased activities of III to X in alcoholic liver injury, intrahepatic
cholestasis and biliary obstruction and **of I', I" and II'**
in hepatocellular carcinoma (HCC).  In order to characterize the
multiple forms of GGT, particularly those frequently observed in
HCC, lipoprotein fractions separated by affinity chromatography
with Affi-Gel Blue and concanavalin A (Con A) were studied for
GGT isozyme by PAGGSE.


Materials and Methods

Sera or ascites obtained from patients with alcoholic liver
injury, biliary obstruction, intrahepatic cholestasis and HCC

were fractionated into each lipoprotein class (VLDL, LDL, HDL and residue) by preparative ultracentrifugation according to the method of Yasugi and Homma (2).
Affi-Gel Blue (BIO·RAD, California) affinity chromatography was performed by applying 0.4 ml of serum to a column (bed volume, 6 ml) and eluting with 0.02 M sodium phosphate buffer, ph 7.1, (Fr.I), 1.4 M NaCl in the buffer (Fr.II) and 6 M **urea in the** buffer (Fr.Ⅲ). The same amount of serum was applied to a Con A-Sepharose (Pharmacia, Uppsala) column (bed volume, 15 ml) and eluted in two steps, first with 0.05 M Tris-HCl buffer, pH 7.5, containing 1 mM $MgCl_2$, 1 mM $MnCl_2$, 1 mM $CaCl_2$ and 0.5 M NaCl and then with 0.2 M α-methyl-D-mannoside **in the buffer**.
GGT isozymes were separated by PAGGSE with a buffer system of 3 mM Tris-50 mM glycine, pH 8.5, and a running time of 24 hr at 125 V. GGT activity was detected by a color reaction with Fast Garnet GBC (2 hr at 4°C) after incubating the gels in a reaction mixture containing N-γ-L-glutamyl-α-naphthylamide as a substrate for 2 hr at 37°C (3). Lipoprotein on PAGGSE was visualized by pre-staining of samples with Sudan Black B.


Results


Zymograms of GGT resembled those of lipoproteins. GGT activity was found mainly in the HDL for cases with alcoholic liver injury and intrahepatic cholestasis and in the VLDL and LDL for cases with biliary obstruction. The GGT activity in the residue was associated with GGT-I and a high-molecular lipid complex (GGT-X) (Fig.1a). The additional GGT bands, GGT-I', I" and Ⅱ', which are frequently detected in patients with HCC, were not associated with lipoproteins. They were present in the residue (Fig.1b).
Upon Affi-Gel Blue affinity chromatography, the GGT in lipoprotein fractions (lipoprotein-associated GGT) was adsorbed on the column and eluted with 6 M urea (Fr.Ⅲ). GGT-I, I', Ⅱ' and

and X, which were not associated with lipoproteins, were recovered in Fr.I, and GGT-I" and II' in Fr.II (Fig. 2).

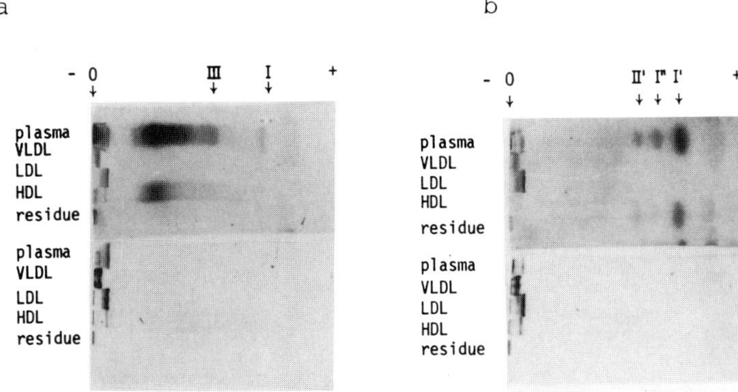



Fig. 1.  PAGGSE of ultracentrifugally separated lipoprotein
fractions.  On each panel, the upper plate was stained for GGT
activity and the lower for lipoproteins with Sudan Black B. The
results with plasma from alcoholic liver injury are shown in the
left panel (a) and those with plasma from hepatocellular
carcinoma in the right (b).

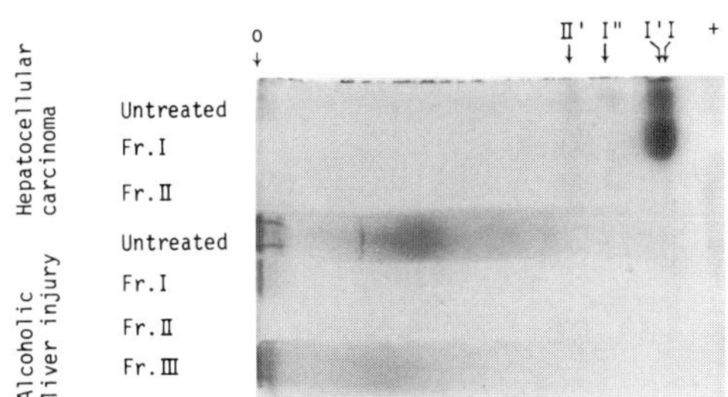



Fig. 2.  GGT staining after PAGGSE of fractions separated by
Affe-Gel Blue affinity chromatography.  The samples obtained
from serum of hepatocellular carcinoma with GGT-I, I', I" and
II' were applied in the upper three tracks and those of
alcoholic liver injury in the lower four tracks.

On the other hand, GGT was separated into two types by Con A

affinity chromatography, the unbound (eluted with Tris-HCl

buffer alone) and the bound (eluted with the additional α–

methyl-D-mannoside. They gave different electrophoretic patterns. The GGT bands associated with VLDL, LDL and high-molecular lipid complex were present in the bound fraction. Although HDL was present in the unbound fraction, GGT activity was not found in this fraction. Among the GGT isozymes unrelated to lipoproteins, the hepatoma-associated GGT-I' was in the unbound fraction and hepatoma-associated GGT-I" and non-associated GGT-I in the bound fraction (Fig. 3).

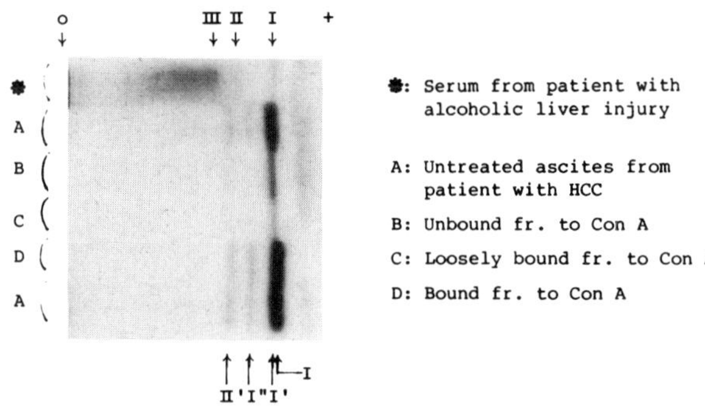


Fig. 3. GGT staining after PAGGSE of fractions separated by Con A affinity chromatography. Unfractionated ascites from a hepatocellular carcinoma patient was used as a GGT source.

The hepatoma-associated GGT-Ⅱ', which did not bind to Con A, migrated slightly cathodic to the GGT-Ⅱ', which bound to Con A. The GGT isozymes recovered in the unbound fraction was also seen in the loosely bound fraction, which was obtained by washing the Con A column with the Tris-HCl buffer alone after the unbound GGT was recovered in the void volume.

## Discussion

The GGT isozymes frequently found in hepatoma patients, GGT-I',
I" and II', which were separated by PAGGSE, were present in the
residue fraction in ultracentrifugation, unlike the other GGT
isozymes associated with lipoproteins.  The GGT associated with
lipoproteins were adsorbed on the Affi-Gel Blue column; thus
leaving GGT-I, I', II' and the GGT in high-molecular lipid
complex, which was also present in the residue, in the pass-
through fraction.  The results may be explained by the affinity
of lipoproteins to Affi-Gel Blue (4). Although GGT isozymes have
been separated on PAGGSE, it is difficult to distinguish GGT-I'
from GGT-I on the zymogram.  Since Con A chromatography readily
separates GGT-I' from GGT-I as demonstrated by the resulting
PAGGSE, GGT-I' and II', which have most diagnostic values among
other GGT isozymes for HCC, can be separated by two step affinity chromatography, Affi-Gel Blue combined with Con A columns.
This method may prove to be a useful technique to identify the
hepatoma markers of GGT.


Referance


1.  Kojima, J., Kanatani, M., Nakamura, N.: Clin. Chim. Acta
    106, 165-172 (1980)

2.  Yasugi, T., Homma, R.: Phys.-Chemi. Biol. (Chiba) 6, 119-
    122 (1959) (in Japanese)

3.  Sawabu, N., Nakagen, M., Yoneda, M., Makino, H., kameda, S.,
    Kobayashi, K., Hattori, N., Ishi, M.: Gann 69, 601-605
    (1978)

4.  Wille, L.E.: Clin. Chim. Acta 71, 355-357 (1976)

ISOAMYLASE BY ISOELECTRIC FOCUSING ON MODIFIED CELLULOSE ACETATE

Robert L. Gilman
Veterans Administration Medical Center, Gainesville, Florida.

## Introduction

Since many tissues contain amylase, identification of the
source of a patient's elevated blood amylase is important for
proper clinical diagnosis and treatment.  Methods of identification include electrophoresis, isoelectric focusing in
column, and isoelectric focusing in thin gel plates.  I have
endeavored to combine the advantages of the high resolution
of fractions offered by isoelectric focusing with the convenience and low cost of cellulose acetate film.

## Equipment

The LKB Multiphor and 2103 power supply were used.  The Multiphor was set up with the electrophoresis electrodes in place.
The buffer chambers were filled with the appropriate electrode
solution, rather than with buffer.  A piece of moist filter
paper was fitted to the inside of the electrofocusing lid.
The lid was placed in its usual position, thus affording humidity control in the space above the platten.  A $37^{\circ}$C water
bath was used for incubating the strips with substrate.  An
Ultra-Violet Products Co. view box was used to examine the
fluorescent bands produced on the strips.  A fluorescent mode
densitometer was used to scan the strips.

## Modified cellulose acetate strips

716

Cellogel IEF strips 5.7 x 14 cm were purchased from Kalex Scientific, Manhasset, N.Y.  The designation IEF indicates that these cellulose acetate gel strips are especially modified for isoelectric focusing use.


Solutions and Reagents

The anode solution was 0.5 M ethanolamine and the cathode solution was 0.2 M citric acid.  The carrier ampholyte soaking solution was made by diluting the 40% ampholyte solution with a stock solution containing 0.3 g beta-alanine and 2 mL of glycerine in 10 mL deionized water.  0.2 mL of 40% carrier ampholyte solution with pH range of 3.5 to 10 was added to 2 mL of this stock solution to make a soaking solution for individual Cellogel IEF strips.  Amylase reagent was prepared by dissolving the contents of a vial of Beckman Enzymatic Amylase-DS reagent in 0.5 mL saturated sucrose solution.


Sample and Marker Preparation

Salivary amylase marker was prepared by centrifuging fresh saliva and diluting it 1 + 10 with deionized water.  Aliquots were kept frozen until used.  At that time they were further diluted.  Pancreatic marker was prepared by diluting aspirated duodenal juice obtained post-stimulation with Pancreozymin. This juice was diluted 1 + 20 with deionized water and stored frozen.  Serum samples were used directly.


Procedure

The Cellogel IEF strips were prepared by soaking at least 2 hours, and preferably overnight, in diluted carrier ampholyte solution. The Cellogel wicks were prepared by soaking one piece of Cellogel 250 for about 20 minutes in a tray containing 50 mL electrode solution and 5 mL glycerin. The wicks were placed in position with the permeable side up, and the portion on the glass platten was blotted before the IEF strip was laid on the platten. The wick and IEF strip overlapped about 1 cm. Care was taken to avoid trapping air bubbles under the IEF strip. When the strip was in place it was blotted thoroughly. The samples were applied with a standard electrophoresis sample applicator. The platten was covered with the electrofocusing lid, and the electrophoresis electrodes were connected. Focusing time was 2½ hours with an initial voltage of 200 volts. Voltage was gradually increased to 1,000 volts. Power was not allowed to exceed 2 watts. At the end of the focusing period, the IEF strip was removed from the cell and placed on a glass lantern slide measuring 84 mm x 94 mm (94 mm is the distance between the ends of the strips). The ends of the strip were folded under the glass slide to hold it in place. Approximately 200 microliters of substrate was streaked across one end of the strip. The substrate was evenly distributed by rolling a Meyer rod back and forth across the strip a few times. The strip was then incubated for 30 min. in a closed chamber supported inside a water bath at $37^{O}$C. After the incubation period the strip was removed, dried with a hair drier, and observed in a U.V. view box. Strips can be photographed under U.V. light.

## Results

Figure 1 shows the pattern of amylase isoenzymes produced by isoelectric focusing on Cellogel IEF cellulose acetate gel strips. It is clearly seen that the isoelectric points at which the salivary amylase isoenzyme fractions focus are different from those of the pancreatic fractions.

718

Figure 2 shows a comparison between a serum sample pattern and
the two marker patterns.

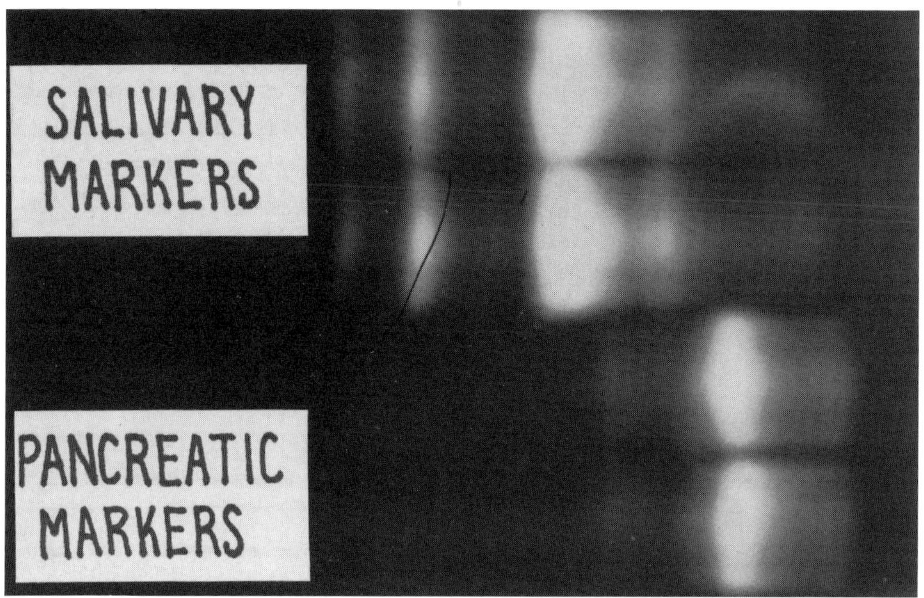


Fig. 1.  Isoelectric focusing pattern of amylase isoenzymes.

Discussion


The results by this method compare favorably with those produced by my earlier method using isoelectric focusing on
acrylamide gel plates.  This method is simpler and much less
expensive.  Shorter range pH gradients can be used to achieve
greater resolution.

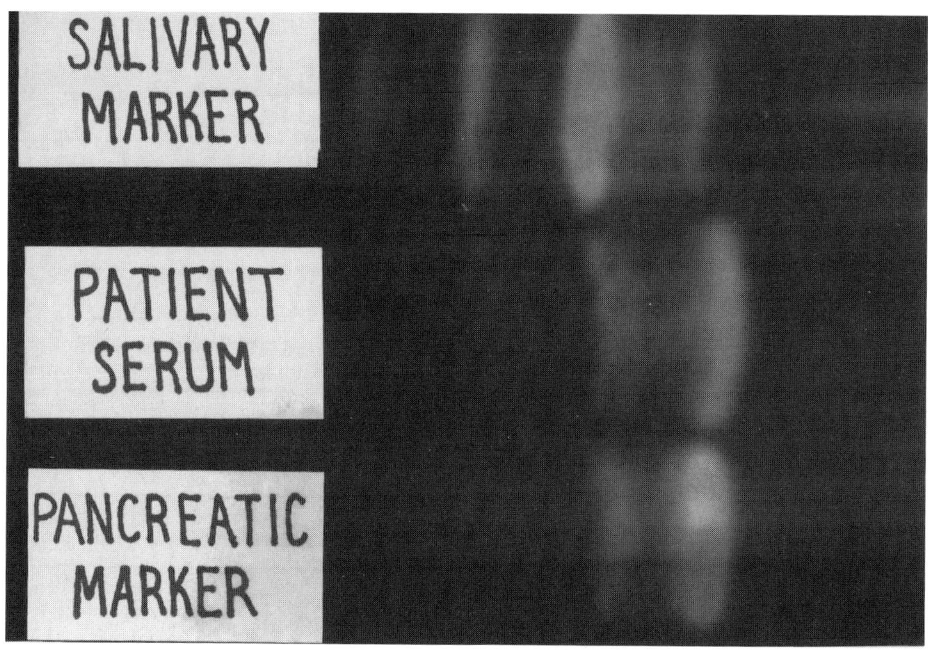


Fig. 2.  Isoelectric focusing pattern of amylase isoenzymes.


References


1.  Ambler, J.: Clin. Chem. Acta 85, 183-191 (1978).

2.  Ambler, J.: Clin. Chem. Acta 88, 63-70 (1978).

3.  Ambler, J.: Clin. Chem. 25/7, 1320-1322 (1979).

4.  Bossuyt, P.J., R. Van den Bogaert, S.L. Scharpe, and Y.
    Van Maercke: Clin. Chem. 27/3, 451-454 (1981).

5.  Cassard, G., E. Gianazza, and P. Girighetti: J.
    Chromatogr. 221(2), 279-291 (1980).

6.  Gilman, R.L.: Developments in Biochemistry Vol. 7, 31-33.
    Elsevier North Holland, Inc. (1979).

7.  Pierre, K.J., K.K. Tung, and H. Nadj: Clin. Chem. 22, 1219
    (1976).

MULTIPLE FORMS OF TREHALASE FROM AGING SOROCARPS OF THE CELLULAR SLIME MOLD, _DICTYOSTELIUM DISCOIDEUM_


Kathleen A. Killick
Boston Biomedical Research Institute
Department of Developmental Biology
Boston, MA   02114


Introduction

Differentiation in the cellular slime mold, _Dictyostelium dis-coideum_ is accompanied by the accumulation of three polysaccharides (i.e., cellulose, an acid mucopolysaccharide and cell wall glycogen) and the nonreducing disaccharide $\alpha$-$\alpha'$ trehalose. Although trehalose is detectable throughout the life cycle of _Dictyostelium_, dramatic increases in its rate of accumulation occur late in morphogenesis (i.e., 20-24 hr after initiation of development) resulting in an approximate 10-fold increase in trehalose content (1).  As such, trehalose is a major carbohydrate storage reserve of the dormant spore, where it serves as a principal energy and carbon source during germination, as a consequence of its hydrolysis to glucose, the latter reaction being catalyzed by the enzyme, trehalase (2,3).  Trehalose catabolism is absent in spore cells prior to germination due to the _in vivo_ latency of trehalase.  In stalk cells however, trehalose turnover is extensive (4).

Although most enzyme regulatory studies conducted during morphogenesis in _Dictyostelium_ have been concerned with development, little attention has been paid to the regulation of enzymatic activity in the aging organism.  This situation is unusual when one considers that about 20 to 30% of the total cell population (i.e., the stalk cells) undergoes senescence and subsequent death during development. To date, trehalase is the only enzyme, which has been shown to increase in activity dramatically during slime mold development and aging and also to undergo cell type-specific latency _in vivo_. Thus, trehalase constitutes an ideal model system within which a systematic investigation of probable _in vivo_ regulatory mechanisms, which control the establishment, maintenance and reversal of enzyme-specific latency during cellular differentiation and aging may be carried out. In order to provide a framework within which these diverse regulatory mechanisms may be studied, enzyme from the aging mature sorocarp was chosen for initial investigation. Results from these studies form the basis of this report.


Results and Discussion

1) Changes in trehalase specific activity during development.

Examination of enzyme specific activity (S.A.) during morphogenesis in _Dictyostelium_ indicated that during early aggregation (i.e., 5 hr), the S.A. decreased linearly reaching a minimal value at the pseudoplasmodial stage (i.e., 12-14 hr) (Fig. 1). The net decrease in enzyme S.A. over this period was 7 to 14-fold. During culmination as stalk and spore cell differentiation progressed co-incident with cellulose and trehalose accumulation, enzyme S.A. increased 10 to 15-fold achieving a maximal value in mature sorocarps (i.e., fruiting bodies).

2) General properties of trehalase from sorocarps.

Trehalase activity was stable to successive freeze-thaw cycles, incubation at $6^{\circ}C$ for at least 5 days and a 5 min incubation at $50^{\circ}C$. Calculation of the $Q_{10}$ and energy of activation (i.e., Ea) indicated values of 1.98 and 12.8 KCal/mole (between $30^{\circ}C$ and $40^{\circ}C$). After partial purification of trehalase by strepto-

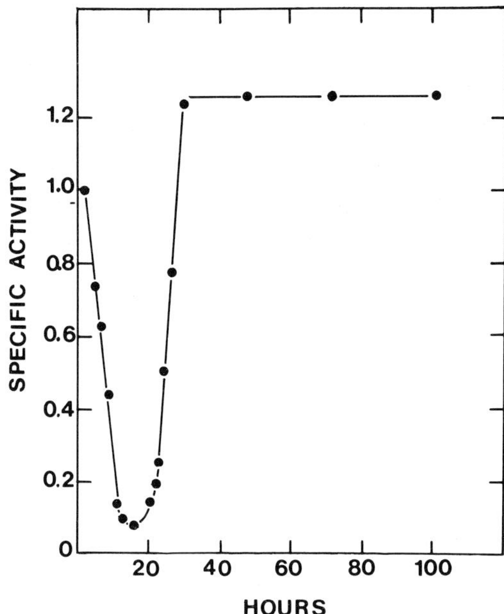

Fig. 1 .    Trehalase specific activity during development in Dictyostelium.

mycin-$SO_4$ and $(NH_4)_2SO_4$ fractionation, aliquots of the enzyme were subjected to chromatography on DEAE-cellulose. Following elution of the enzyme with a linear NaCl gradient (0-1 M), the majority of recovered trehalase activity was eluted as a single symmetrical peak at about 0.25 M NaCl with the remainder emerging as a second peak at about 0.32 M NaCl (Fig. 2). As these observations suggested the existence of multiple trehalase activities differing in charge, further studies were conducted to explore this possibility.

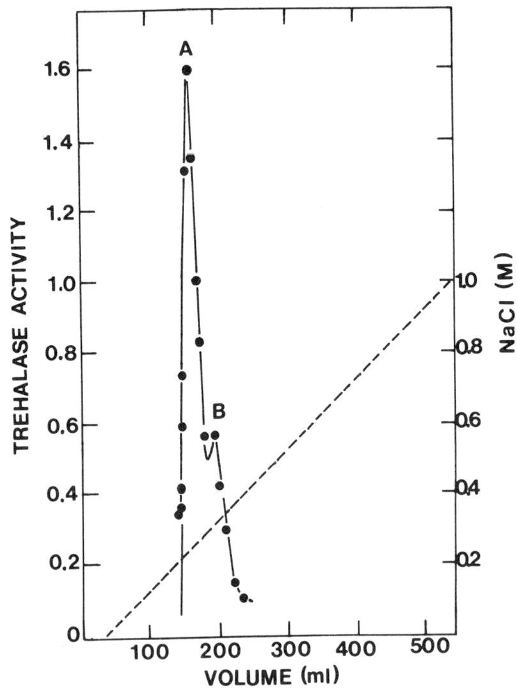

Fig. 2. Chromatography of sorocarp trehalase on DEAE-cellulose. Samples of partially-purified enzyme were subjected to ion exchange chromatography (as described in the text). Enzymatic activity was eluted from the resin with a linear NaCl (0 to 1 M) gradient in 25 mM tris-maleate buffer (pH 6.5).

3) Electrophoretic evidence for multiple trehalase activities.

Samples of trehalase, partially-purified from mature Dictyostelium sorocarps were subjected to analytical polyacrylamide gel electrophoresis. Following electrophoresis, the gels were sectioned into 2 mm slices and each segment was then transferred to and subsequently emsulsified in 25 mM tris-maleate buffer (pH 6.5). Following assay of aliquots of gel eluate for trehalase activity, the results indicated the presence of 3 peaks of enzymatic activity (Fig. 3). Recovery of trehalase activity following electrophoresis was 70-80% of the applied level and glucose production by each of the above fractions showed an absolute dependency on the presence of trehalose in

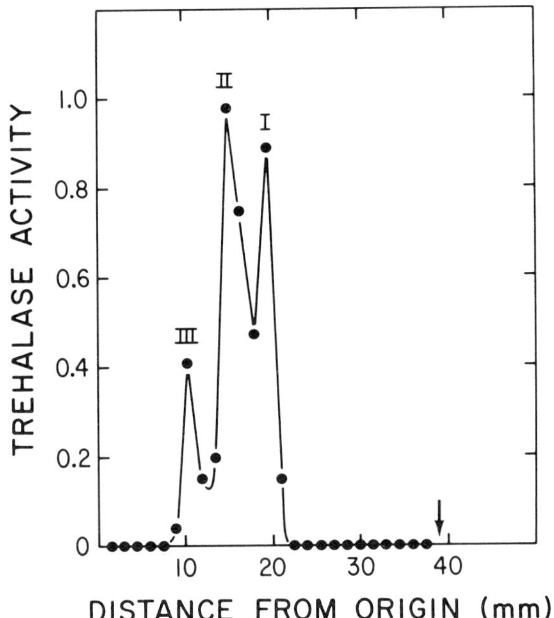

Fig. 3.    Electrophoretic profile for trehalase from Dictyostelium sorocarps.  Electrophoresis was conducted by a
modification of previously described methods (5,6).

the assay system.  Comparable results were routinely obtained
when a 10-fold range in protein concentration was subjected to
electrophoresis.  Parallel studies with trehalase purified from
vegetative Dictyostelium cells demonstrated the presence of a
single electrophoretic form of the enzyme.  When mixtures of
sorocarp and vegetative enzymes were analyzed by the same electrophoretic methods, isozyme I from sorocarps was not resolvable from vegetative trehalase.  Subsequent studies examining
a variety of parameters have led to the conclusion that sorocarp isozyme I is identical to trehalase from vegetative cells
(manuscript in preparation).  As the electrophoretic profile
observed with sorocarp trehalase preparations could be the consequence of differences in either charge and/or molecular size
between the 3 trehalose activities, preparations of enzyme were
analyzed by molecular sieve chromatography and isoelectric
focusing.

4) Molecular sieve chromatography of sorocarp trehalase.

Columns of Sephadex G-150 were equilibrated with 25 mM potassium phosphate buffer (pH 6.5) that contained 0.15 M KCl. Blue dextran 2000 and p-nitrophenol were used to determine the values for Vo and Vt respectively. The column was calibrated with standard proteins, e.g., ferritin, catalase, aldolase, bovine serum albumin, hen egg albumin and myoglobin. Standard curves were constructed by plotting the log (molecular weight) versus either the corresponding Kav or Ve/Vo value. Following application of the sorocarp enzyme to the Sephadex G-150 column, trehalase activity was eluted from the gel at a flow rate of 25 ml/hr and 4 ml fractions were collected. A single peak of trehalase activity was detectable, eluting at a molecular weight corresponding to about 90,000 (Fig. 4). Parallel studies with trehalase purified from vegetative _Dictyostelium_ demonstrated the presence of a single species of enzyme, eluting at a molecular weight of about 110,000 (Fig. 4). The results demonstrated that isozymes I and II of sorocarp trehalase were not resolvable by molecular sieve chromatography.

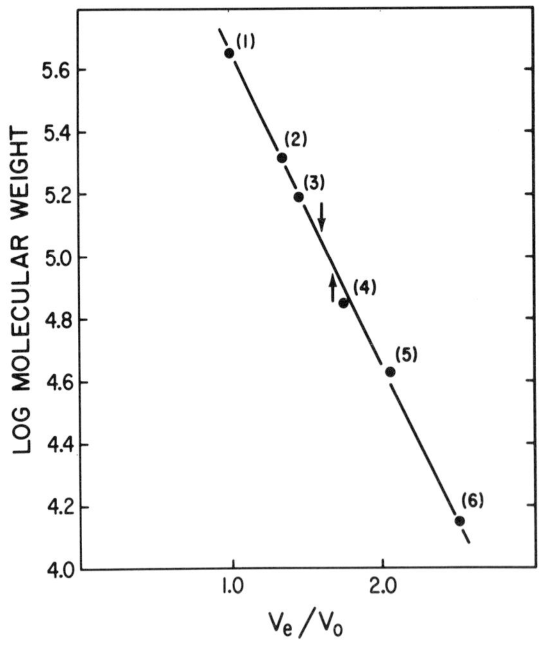


Fig. 4.

Estimation of the molecular weight of trehalase from _Dictyostelium_ sorocarps by molecular sieve chromatography. The arrows designate the trehalases from vegetative cells ( ⅄ ) and sorocarps ( ⅄ ). Standard proteins used for column calibration were: (1) ferritin, (2) catalase, (3) aldolase, (4) bovine serum (5) hen egg albumin and (6) myoglobin.

Although isozymes I and II appear to differ slightly in molecular size, experiments in progress (7) indicate that these activities differ in their pI values by about 0.5 pH units and suggest that the multiple sorocarp activities may be resolvable by the techniques of either isoelectric focusing and/or chromatofocusing.

Acknowledgements

This investigation was supported by Research Grants AG00922 and GM 25534 from the National Institutes of Health.

References

1.   Rosness, P.A. and B.E. Wright:  Arch. Biochem.  Biophys. 164, 60-72 (1974).

2.   Ceccarini, C. and M. Filosa:  J. Cell. Comp.  Physiol. 66, 135-140 (1965).

3.   Ceccarini, C.:   Biochim.  Biophys.  Acta 148,  114-124 (1967).

4.   Jefferson, B.L. and C.L. Rutherford:  Exptl.  Cell Res. 103, 127-134 (1976).

5.   Davis, B.J.:  Ann.  New  York  Acad.  Sci. 121,  404-427 (1964).

6.   Ornstein, L.:  Ann.  New  York  Acad.  Sci. 121,  321-349 (1964).

7.   Killick, K.A.:  Fed.  Proc. (St.  Louis, Mo.) (1981).

PURIFICATION OF GRANULOCYTE COLONY STIMULATING FACTOR MONITORED BY
ULTRATHIN LAYER ISOELECTRIC FOCUSING

R. Neumeier and H.R. Maurer

Pharmazeutisches Institut der Freien Universität Berlin,
Königin-Luise-Straße 2+4, D-1000 Berlin 33, Germany

## Introduction

In vitro granulocyte/macrophage colony formation requires the presence
of colony stimulating factors (CSF). CSF is found in serum, urine, certain tissues and in medium conditioned by certain cells (1,2). So far
all CSF species have been isolated from human and murine sources. Recently, we found a new source which is well suited for CSF isolation on a
large scale (3,4). Bovine lung conditioned medium contains two CSF species having molecular weights of approximately 68,000 and 29,000 d (4).
Here, we report the use of isoelectric focusing on ultrathin layer polyacrylamide gels as a tool to monitor the progress and to predict the optimum method for the isolation and purification of the high molecular
weight CSF-F from bovine lung conditioned medium (BLCM).

## Material and Methods

Ultrathin layer polyacrylamide gels were purchased from Serva (Heidelberg, Germany): Servalyt precotes 3-10, 125 x 125 mm. Anode solution: 0.83 g L-aspartate and 0.92 g L-glutamate dissolved in 250 ml water. Cathode solution: 30 ml ethylene diamine, 1.09 g L-arginine and 0.91 g L-lysine dissolved in 250 ml water. The electrode wicks were soaked with the electrode solutions and put as double layers onto the gel. The applicator strip was placed 3 cm from the anode and each hole was filled with 3-5 µl of sample solution. A Multiphor focusing apparatus and the LKB 2103 power supply were used (LKB, Bromma, Sweden). Focusing at $2^{o}$C, started with 2 watts and was increased after 1 hr to 2,5 watts for another hr. The gels were fixed for 10 min in 20 % TCA and washed with destaining solution (ethanol 200 ml/water 520 ml/glacial acid 80 ml). The gels were stained for 10 min in staining solution (2 g Serva Violett 49 dissolved in a mixture of ethanol 450 ml/water 450 ml/glacial acid 100 ml) with moderate shaking and destained in destaining solution. Finally the gels were dried overnight at room temperature.

To determine the pH gradient the focused gel was cut in two pieces. The first was fixed, stained and destained as described above. The second was further cut in 1 cm strips. From each strip the gel was removed from the supporting foil and eluted with 1 ml water. The pH of each fraction was measured. Comparing the stained gel piece with the scaled pH gradient enabled us to determine the pI of each protein band.
For graphical documentation the dried gels were fixed onto white paper and scanned at 580 nm by means of a Zeiss thinlayer scanner (M4QII).
For preparative focusing 1 mm gels were purchased from LKB: Ampholine PAGplates, pH 3.5 - 9.5, 120 x 240 mm, (LKB, Bromma, Sweden). The same electrode solutions and focusing devices as described above were used. 1 ml of partially purified CSF solution (7.0 x $10^{3}$ col/mg, 9.0 mg/ml protein) was distributed over the gel. The distance between the anode and cathode was 240 mm. The focusing was carried out at 10, 12 and 15 watts for 1 hr each. Then the gel was cut in 1 cm strips and to each fraction 1 ml 0.9 % saline was added. After determining the pH 1 ml 40 mM Tris-HCl

buffer, pH 7.0 at 23$^{\circ}$C, was added. The fractions were thoroughly mixed
and incubated forhalf an hr. After centrifuging at 4000 g the supernatants
were removed and dialysed against 20 mM Tris-HCl buffer, pH 7.5 at 4$^{\circ}$C.
The dialysed fractions were tested for biological activity in the mouse
bone marrow assay (3). The pH of the active fractions corresponded to the
pI of CSF-F.


## Results


Fig. 1 shows the stained  gel with the four consecutive purification steps
of CSF-F: a) crude bovine lung conditioned medium (BLCM), b) ammonium sulfate precipitate, c) pool of active fractions from phenyl-sepharose chromatography and d) DEAE chromatography. The in- and decrease, respectively,
of distinct protein bands can be recognized. More clearly figs. 2 a-d
show the different protein distributions using optical scanning of the
stained gel.
Analytical focusing of dialysed BLCM showed a characteristic pattern of
proteins over the pH range between 4 - 8 (Fig. 2a), which is not much
changed by ammonium sulfate precipitation (at 60 - 75 % saturation),
though a slight concentration of more acid (4,0 -5,3) and more basic
(7,0 - 8,0) proteins could be detected (Fig. 2b). By hydrophobic interaction chromatography on phenyl-sepharose only some peaks in the albumin
(BSA) range (4,0 - 5,2) and bovine hemoglobin (7,4 - 7,8) are left (Fig.
2c). Hemoglobin as a basic protein did not bind to DEAE cellulose and was
so separated, as shown in Fig. 2d. Not separated from CSF-F was BSA, being
the main contaminating protein at this stage of purification. By means
of preparative focusing the pI of CSF-F was determined as 4,0 - 5,2. The
broad region was not unexpected since other CSF species have also been
focused in similar broad peaks (5,6).

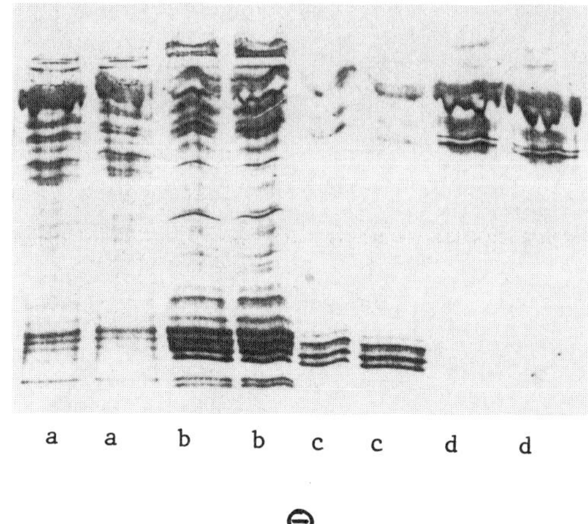



Fig. 1 : Stained ultrathin layer polyacrylamide gel. a) dialysed bovine
lung conditioned medium (4 µl; 6,3 mg/ml protein). b) Ammonium sulfate
precipitate (at 60-75 % saturation) (3 µl; 6,7 mg/ml protein). c) pool
of phenyl sepharose chromatography (4 µl; 5,0 mg/ml). d) pool of DEAE
chromatography (3 µl; 8,5 mg/ml).

## Discussion

Our studies show that ultrathin layer isoelectric focusing is a rapid,
simple and inexpensive method to monitor the progress of purification of
hormones such as CSF using different biochemical procedures. Thus the
high molecular weight species of CSF from bovine lung conditioned medium
(BLCM) was subjected to three consecutive purification steps: Ammonium
sulfate precipitation, hydrophobic interaction chromatography on phenyl
sepharose and ionexchange chromatography on DEAE cellulose. By means of
marker proteins the main contaminating substances could be identified.
Whereas the choice of the first steps is generally empirical, the choice
of the following isolation methods is most effective if the properties
of the desired and contaminating proteins are known. Thus we routinely
analyzed each fraction using ultrathin layer isoelectric focusing in
addition to determining the biological activity and protein concentration.
With separation methods resulting in many fractions 3 - 5 µl of each fraction were applied without using an applicator strip. Thus up to 15 fractions could be focused on one gel (125 x 125 mm) demonstrating the resolving capacity of the method.

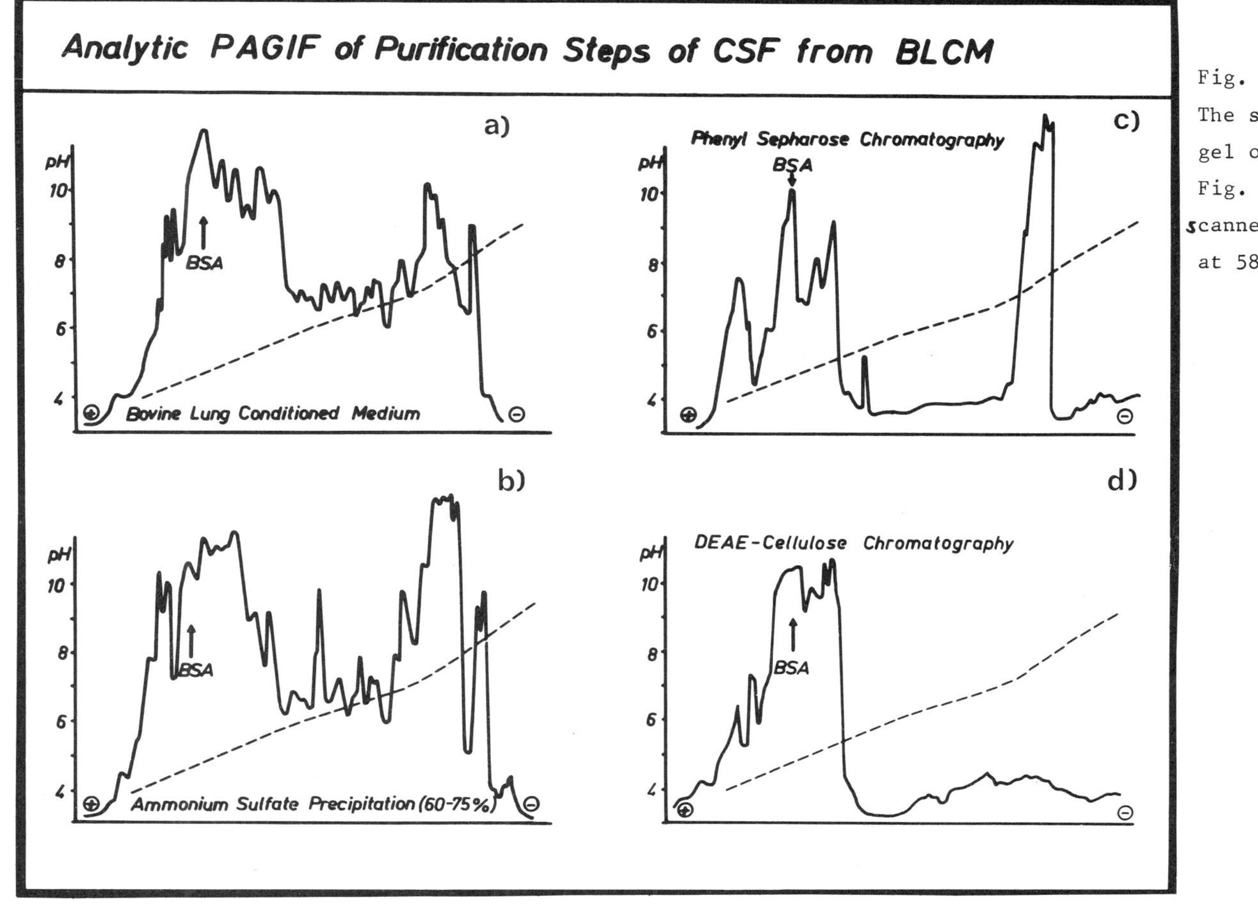



Fig. 2: The stained gel of Fig. 1 scanned at 580 nm

734

A practical application of isoelectric focusing for the separation of
CSF-F was found when the fourth purification step had to be selected.
Based on the results of analytical and preparative focusing as described
above separation methods which separate proteins due to their charge properties could be excluded. Moreover separations based on molecular weight
differences could be excluded since the main contaminating protein of
CSF-F preparations, partially purified up to the third step, was identified as bovine serum albumin (BSA), which has the same molecular weight
as CSF-F (4). These theoretical considerations were confirmed by attempting further purification with CM cellulose and gel filtration. Thus as
the last purification step an affinity chromatography was advised, which
would bind either BSA or CSF-F. It could be shown that affinity chromatography on ConA sepharose, which binds CSF-F but not BSA, was a suitable method (to be published).



References

1. Metcalf, D.; The control of neutrophil and macrophage production at
   the progenitor cell level. Exp. Hemat. Today 1978, pp. 35-46
   Springer Verlag, Berlin, Heidelberg, New York (1978).

2. Hays, EF.; Craddock, CG.; Colony-stimulating activity. Arch. Pathol.
   Lab. Med. 102; 165-171 (1978).

3. Neumeier, R.; Maurer, H.R.; Bovine lung conditioned medium as a
   source of both human and murine colony-stimulating factor.
   Exp. Hemat. 8: 728-736 (1980).

4. Neumeier, R.; Maurer, H.R.; Two different types of granulocyte colony
   stimulating factors produced by bovine lung tissue.
   Blut, in press (1981).

5. Stanley, E.R.; Heard, P.M.; Factors regulating macrophage production
   and growth: Purification and some properties of the colony stimulating
   factor from medium conditioned by mouse L cells. J. Biol. Chem. 252:
   4305-4312 (1977).

6. Tsuneoka, K.; Shikita, M.; A sialoglycoprotein stimulating proliferation of granulocyte-macrophage progenitors in mouse bone marrow cell
   cultures. FEBS letters 77: 243-246 (1977).

PROTEOLYTIC DIGESTION OF THE M PROTEIN OF INFLUENZA A VIRUSES AND
ISOELECTROFOCUSING OF ITS PEPTIDES


André Darveau, Monique Lambert and Jacqueline Lecomte

Centre de recherche en virologie, Institut Armand-Frappier
Laval, Québec, Canada


Introduction

Previous studies have shown that the two major internal proteins of influenza A
viruses, the nucleoprotein (NP) and the membrane (M) protein are very basic and
could only be resolved in the high resolution two-dimensional electrophoresis procedure of O'Farrell (1) in a non-equilibrium pH gradient (2, 3). We report here that
analytical isoelectrofocusing in thin layers of polyacrylamide gels (PAG) has permitted the separation of all the major polypeptides of influenza virus. These include
the NP, M protein and the viral surface glycoproteins, the neuraminidase (NA) and
both HA1 and HA2 subunits of the hemagglutinin. We also report that this same
procedure is readily applicable for peptide analysis of low molecular weight proteins,
using as a model the 26 000 dalton M protein.


Results and Discussion

The sucrose density purified A/PR/8/34 (H0N1) and the avian A/chick/Germ/49 (Hav2
Neq1), which will be referred here as PR8 and N virus respectively, were disrupted
with 1% SDS and reduced with 4% $\beta$-mercaptoethanol, in order to cleave the HA into
its two subunits HA1 and HA2 and applied to a 5% PAG containing 2% NP-40, 6M
urea and Pharmalytes pH 3-10 and pH 5-7 in a ratio of 2:3. Excellent resolution (Fig.
1) was obtained for all the major polypeptides of influenza virus in the range of pH
4.7 to 9.6 when gels were stained with Coomassie brilliant blue. Both glycoproteins,
the HA and NA showed a great heterogeneity of charge as reported previously (3)
which has been attributed to the extent of glycosylation (4). The M protein migrated

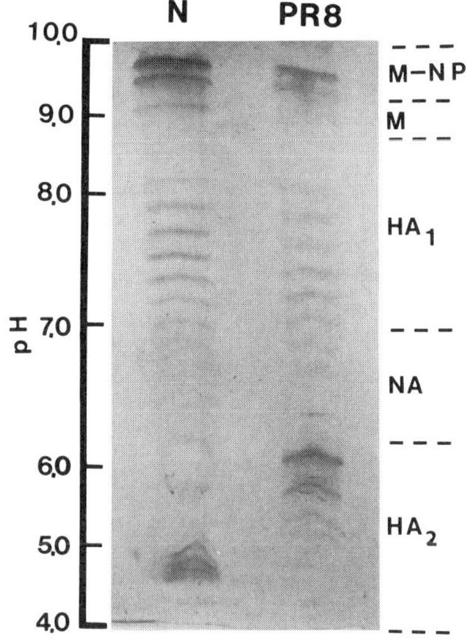



Fig. 1. IEF of purified PR8 and N virus: Purified viruses (50 μg) were disrupted with 1% SDS, 4% β-mercaptoethanol and 6M urea for 10 min at room temperature; NP-40 was then added at a final concentration of 8% v/v followed by 6 consecutive freezing and thawing. IEF was carried out in a 1 mm, 5% PAG containing 2% NP-40, 6M urea and 2% Pharmalytes of pH 3-10 and pH 5-7 in a ratio of 2:3. Electrolytes were NaOH 0.1M and $H_3PO_4$ 0.04M. Electrophoresis was run at 10W for 3 h preceded by a pre-electrophoresis at 10W for 30 min. Gels were fixed in 12.5% TCA and stained with Coomassie Brilliant Blue.

as a single band at pI value of 9.0 and co-migrated as a NP-M protein complex in two other bands at pI value of 9.3 and 9.5. The heterogeneity of charge in these latter two non-glycosylated proteins does not appear to be artefactual. Similar results were obtained when virus proteins were solubilized with the non-ionic detergent NP-40 (not shown). Comparative examination of the IEF patterns of the polypeptides of the two antigenically unrelated influenza A virus has demonstrated an overall difference in pI values for both HA1 and NA. Greater differences were readily apparent for the HA2 subunits. No significant differences were observed between the overall charge for both the NP-M protein complex and M protein.

To gain more information on the structural homology of the influenza M proteins, we have developed a rapid and convenient method for peptide mapping of this small molecular weight polypeptide. Our previous studies had shown that analysis of the cleavage products after proteolytic digestion of the M protein according to the procedure described by Cleveland et al. (5), was not readily amenable for peptides smaller than 5 000 molecular weight in 15% acrylamide SDS-gels. We have therefore exploited the sensitivity of IEF in thin layer gels. In order to overcome long extraction procedures involved in separation of M protein from the other viral

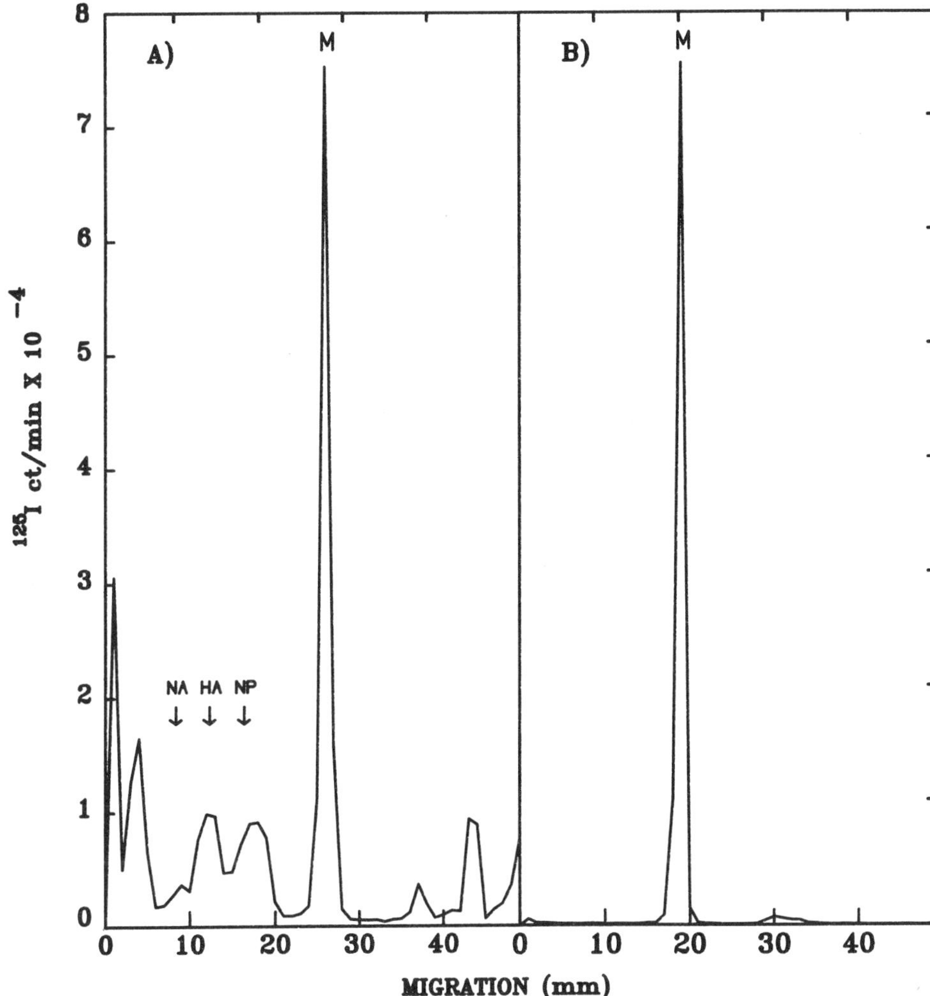



Fig. 2. Polyacrylamide gel electrophoresis in non reduced conditions: A) purified N virus in a 10% acrylamide gel; B) isolated M protein in a 15% acrylamide gel: Purified virus was labelled with $^{125}$I as described in the text. Labelled M protein was separated from other viral proteins in a 10% acrylamide gel. The homogeneity of the M protein, extracted by passive elution at 4°C for 36 h, was assessed by reelectrophoresis in a 15% acrylamide gel.

polypeptides, whole virus was labelled with $^{125}$I by a modification of the iodine monochloride method (6) and labelled virus proteins were separated by electrophoresis in a 10% acrylamide gel (7). The M protein under non reduced conditions migrates

738

as a single band with an apparent molecular weight of 26 000 (Fig. 2A). M protein
was extracted from the gel by passive elution and homogeneity of the extract was
assessed by re-electrophoresis in a 15% PAG (Fig. 2B). Before digestion with
different proteases, SDS was removed by ion-pair extraction (8) and complete
removal of SDS was monitored by incorporating $^{35}$S-SDS in the labelled $^{125}$I whole
virus.

The peptide patterns produced by complete digestion of M protein with bromelain, $\alpha$-
chymotrypsin, papain, S. aureus V8 protease and trypsin are shown in Fig. 3. The IEF
profiles were characteristic of the proteolytic enzyme used for digestion and were
highly reproducible. Although the overall charge of the M protein is very basic and
independently of the protease employed, an equal number of acidic and basic peptides

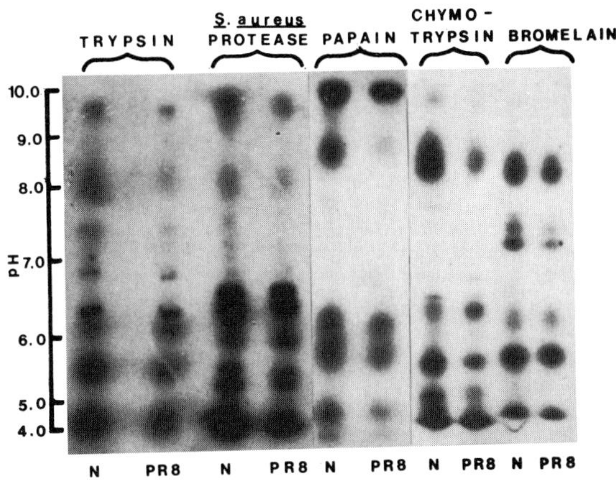


Fig. 3. IEF of digested M proteins from PR8 and N virus: SDS was removed by ion-
pair extraction with triethylamine (8) and M protein was digested in 0.01M tris-HCl
pH 6.8 at 37°C for 2 h in an enzyme: substrate ratio of 1:5. IEF was carried out in a
1 mm, 7% PAG containing 6M urea, 2% Pharmalytes of pH 3-10 and pH 5-7 in a ratio
of 2:3. Electrolytes was NaOH 1M and $H_3PO_4$ 1M. Electrophoresis was run at 20W
for 2 h preceded by a pre-electrophoresis at 20W for 15 min. Gels were fixed for
15-30 min in Coomassie Brilliant Blue G-250 dissolved in 1N $H_2SO_4$ -12% TCA
solution (9) and dried. Autoradiography was made with Kodak X-Omat R film and a
DuPont Lightning Plus screen for 24-36 h at -70°C. S. aureus protease was a gift
from Dr. G. Drapeau and all other enzymes were from Sigma Chemical Co., St. Louis,
Mo.

were generated. By comparison of the peptides generated from PR8 and N virus M proteins by each enzyme, it can be seen that no significant differences were observed among the peptide patterns. It thus appears that both these M proteins were highly conserved during evolution in a virus isolated from man (PR8) and a chicken (N).

These studies demonstrate the utility of analytical IEF in thin layer gels for achieving greater resolution of all influenza virus proteins in the pH range of 4 to 10. Enzymatic digestion of viral protein extracted from polyacrylamide gels prior to IEF offer supplementary means for peptide analysis. The large choice of proteolytic enzyme combined with the use of other radioisotopes should widen the applicability of the method in the analysis of structural homology of viral proteins.

## References


1. O'Farrell, P.H.: J. Biol. Chem. 250, 4007-4021 (1975).
2. Privalsky, M.L., Penhoet, E.E.: Proc. Nat. Acad. Sci. 75, 3625-3629 (1978).
3. Leavitt, J.C., Phelan, M.A., Laevitt, A.H., Mayner, R.E., Ennis, F.A.: Virology 99, 340-348 (1979).
4. Raghow, R., Portner, A., Hau, C.H., Clark, S.B., Kingsbury, D.W.: Virology 90, 214-225 (1978).
5. Cleveland, D.W., Fischer, S.G., Kirschner, M.W., Laemmli, U.K.: J. Biol. Chem. 252, 1102-1106 (1977).
6. Montelaro, R., Bolognesi, D.P.: Anal. Biochem. 99, 92-96 (1979).
7. Laemmli, U.K.: Nature 227, 680-685 (1970).
8. Henderson, L.E., Orozylan, S., Konigsberg, W.: Anal. Biochem. 93, 153-157 (1979).
9. Righetti, P.G., Chillemi, F.: J. Chromatogr. 157, 243-251 (1978).

# SECTION IV

## Isotachophoresis and Free Flow Electrophoresis

QUANTITATIVE ANALYSIS IN ISOTACHOPHORESIS:  THE INFLUENCE OF
VARIOUS OPERATIONAL PARAMETERS


Frans Everaerts, Nils Groot, Frans Mikkers
Laboratory of Instrumental Analysis, Eindhoven University of
Technology, 5600 MB Eindhoven, The Netherlands


Introduction

Isotachophoretic steady state configurations can be generated in
various ways, but always the choice of a discontinuous electrolyte system is required, both in terms of electrolyte constituents
and their concentrations.
Depending on the actual conditions a separand, i.e. a compound
to be separated, can migrate in an "in stack" configuration or
an "out of stack" configuration (1). In the latter case the separands (or a part of these) migrate either zone electrophoretically or according the moving boundary principle. The most simple
isotachophoretic systems are obtained in choosing only one leading constituent, one terminating constituent and one counter
constituent (2).
Commonly we may recognize still a few possibilities :

i   The constituents all are present as completely charged par-
    ticles. The "in stack" configuration simply can be given :

$$m_{terminator} < m_{separands} < m_{leading\ ion} \qquad (1)$$

ii  If actual (effective) mobilities are concerned, i.e. disso-
    ciation and association equilibria are involved, the above
    given equation will not hold anymore. It will depend on the
    actual values of the parameters, whether an "in stack" con-
    figuration is possible (1).

iii Another modification of the electrolyte systems (3,4) comprises the use of multiple counter constituents. Whereas the application of only one counter constituent will *restrict* the pH applicability of a given electrolyte system, the use of more than one counter constituent, having different physico-chemical constants, can enlarge the pH region.

The importance of system-parameters, such as pH and concentration of leading electrolyte, sample and terminating electrolyte (5,6), electric driving current and column diameter (7,8) are reasonably understood and in fact used,in practice,for optimization of separation parameters.

Using these parameters qualitative evaluation of compounds, present in the isotachophoretic stack, has been studied often. Thermal, potentiometric and conductimetric stepheights (or reduced stepheights, stepheig ht-unit values or reference-unit values) has been introduced, all having advantages and disadvantages. The values all are related to certain operational conditions chosen. Besides of signals derived from these socalled universal detectors, signals from specific (UV-absorption) detectors are available in common analytical isotachophoretic equipment. Especially the UV-absorption ratio ( or the UV-extinction ratio ) is used for the precise determination of separands, even in *one* operational system. *Enzymatic identification* (11,12) still enlarges the specificity of the qualitative evaluation in isotachophoresis.

Little research has been made recently about the influence of operational parameters on the quantitative aspects, i.e. zone-lengths in isotachophoresis. For accurate determination of (drug) metabolites in complex mixtures, such as serum and urine, this influence must be known, because a variety of operational conditions are applied for an almost "complete" survey of the separands of interest.

In this paper we will report about the tolerances in the concentration of the leading electrolyte, the tolerances in pH, supposed the electric driving current is kept constant.

Quantitative aspects in Isotachophoresis

Commonly in isotachophoretic analyses calibration curves are applied for the quantitative evaluation, i.e. injected amount is plotted versus zone-length. This has proved to be a reliable method (2). If computer facilities are available a calibration constant (2,9) can be used, having the advantage over the calibration curve  that such a constant is valid for all separands in a given operational system. A disadvantage, besides of the dependence on the operational conditions, is the knowledge of the separand concentration in its zone. The concept of relative correction factors has been described as an alternative (10). Via an internal standard these correction factors can be determined, whereas the pH of the leading electrolyte must be kept constant and in such a region that the influence of $H^+$ or $OH^-$ can be neglected.

One of the important features of isotachophoretic analyses is the constant velocity of all separand zones, which have all its own specific concentration, adapted to the operational conditions chosen for the leading electrolyte.

Zonelength in isotachophoresis can be interpreted on two different ways : a metric value (mm) derived from the recorder paper or a time-based value (sec) derived from  data-aquisition equipment. The relation between these two values are given by the isotachophoretic velocity and the paperspeed of the potentiometric recorder. It has been shown (13) that simply the zonelength of a separand ($\Delta t_x$) can be expressed as :

$$\Delta t_x = \frac{n_x F}{I} \cdot \frac{1}{T_x} \qquad (2)$$

where $n_x$ is the amount of separand x (mol), F is the Faraday constant, I is the electric driving current, $T_x$ is the transference or transport number of separand x, which depends on both the separand and various operational parameters of the electrolyte (13). A simple system (one counter-ion and one separand per zone) results in a simple relation for $T_x$. The transport number will comprise concentration (and charge) terms for systems with more mono and (or) polyvalent ions.

With a relation, as given in Eqn. 2, the influence of the pH and concentration of the leading electrolyte on the zonelength, both theoretically and practically (13) has been studied. The transport number $T_x$ has been written in a more complicated form, depending on the various system parameters (13), which are given in the Table I. It is beyond the scope of this paper to go in details, concerning the transport number.

Of course similar conditions can be given for cationic analyses. The experiment has been carried out in teflon (PTFE) capillaries with inner diameters of 0.2 mm and 0.45 mm and various current densities, which both did *not* influence the final quantitative result (7), which again proves that the influence of temperature is marginal if conditions are well chosen.

From a mixture of chlorate, formate, acetate and benzoate zone-

Table I

Operational conditions

| | ion | relative mobility | pK |
|---|---|---|---|
| Leading ion | Chloride | 1.00 | - 2 |
| Terminator | Caproate | < 0.20 | 4.83 |
| Counter-ions | β-Alanine | - 0.54 | 3.60 |
| | γ-aminobutyric acid | - 0.43 | 4.03 |
| | Creatinine | - 0.45 | 4.83 |
| Separands | Chlorate | 0.76 | - 1 |
| | Formate | 0.74 | 3.75 |
| | Acetate | 0.55 | 4.75 |
| | Benzoate | 0.43 | 4.19 |
| " Disturbing ions" | $H^+$ | - 4.38 | 7.00 |
| | $OH^-$ | 2.50 | 7.00 |

lengths ($\Delta t_x$) were measured and with Eqn. 2 the transport number ($T_x$) has been determined, while I and $n_x$ has been varied.
The results are listed in the Table II. Using GABA ,$\gamma$-aminobutyric acid,as counterion often acetate and benzoate forms mixed zones (this means they were not completely separated). By this no. accurate zonelengths could be measured for these zones. Confirming that the transport numbers approach a constant at neutral pH and at "high" concentrations. The experimentally determined transport numbers deviate approximately not more than 10 % from the calculated transport numbers, especially at low pH, which has to be ascribed to the operational conditions. At neutral pH deviations of 2% were normal.
An experimental relation, by which the separand is better characterised, has been given elsewhere (13). From the data, listed in Table II, tolerances in various operational parameters could be derived. A few of these tolerances will be presented in the following graphs.
In the Figure 1 the tolerance in $pH_{leading\ electrolyte}$ in various operational systems (Table I) is given.This tolerance indicates

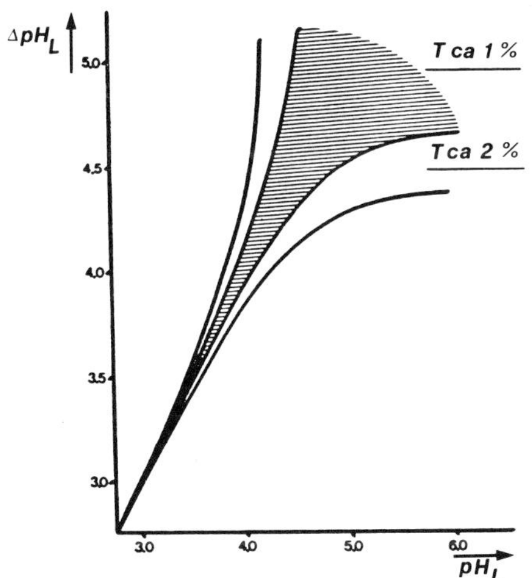


*Fig.1 Tolerancediagram for the pH of the leading electrolyte in various operational systems, for divergences less than 1 % and 2 % in $T_x$. The concentration of the leading ion : 0.010 mol $l^{-1}$.*

Table II  Transportnumbers of various separands under operational conditions listed in Table I

| $pH_L$ | Buffer | BALA | GABA | CREAT | BALA | GABA | CREAT | BALA | GABA | CREAT | BALA | GABA | CREAT |
|---|---|---|---|---|---|---|---|---|---|---|---|---|---|
| | $\bar{c}_L$ | 0.005 mol $l^{-1}$ | | | 0.010 mol $l^{-1}$ | | | 0.015 mol $l^{-1}$ | | | 0.020 mol $l^{-1}$ | | |
| | chlorate | 0.32 | 0.29 | 0.32 | 0.39 | 0.43 | 0.44 | 0.46 | 0.48 | 0.48 | 0.47 | 0.52 | 0.50 |
| | formate | 0.36 | 0.37 | 0.41 | 0.44 | 0.51 | 0.52 | 0.50 | 0.54 | 0.57 | 0.49 | 0.59 | 0.52 |
| 3.0 | acetate | 0.31 | -- | 0.38 | 0.47 | -- | 0.51 | 0.45 | -- | 0.52 | 0.42 | -- | 0.51 |
| | benzoate | 0.28 | -- | 0.29 | 0.35 | -- | 0.42 | 0.44 | -- | 0.43 | 0.40 | -- | 0.44 |
| | chlorate | 0.47 | 0.51 | 0.50 | 0.51 | 0.56 | 0.54 | 0.52 | 0.60 | 0.56 | -- | 0.63 | 0.57 |
| | formate | 0.48 | 0.52 | -- | 0.52 | 0.59 | -- | 0.52 | 0.60 | -- | -- | 0.60 | -- |
| 3.5 | acetate | 0.42 | -- | 0.50 | 0.44 | -- | 0.51 | 0.47 | -- | 0.53 | -- | -- | 0.51 |
| | benzoate | 0.39 | -- | 0.43 | 0.39 | -- | 0.45 | 0.45 | -- | 0.45 | -- | -- | 0.45 |
| | chlorate | 0.51 | 0.58 | 0.58 | 0.54 | 0.62 | 0.59 | 0.56 | 0.62 | 0.60 | 0.56 | 0.62 | 0.61 |
| | formate | 0.53 | 0.56 | 0.60 | 0.53 | -- | 0.62 | 0.55 | 0.59 | 0.64 | 0.53 | 0.59 | 0.64 |
| 4.0 | acetate | 0.49 | -- | 0.53 | 0.48 | -- | 0.53 | 0.50 | -- | 0.54 | 0.47 | -- | 0.54 |
| | benzoate | 0.42 | -- | 0.49 | 0.39 | -- | 0.46 | 0.46 | -- | 0.46 | 0.42 | -- | 0.45 |
| | chlorate | 0.55 | 0.63 | 0.60 | 0.57 | 0.62 | 0.60 | 0.54 | 0.64 | 0.62 | 0.55 | 0.64 | 0.62 |
| | formate | 0.55 | 0.61 | 0.60 | 0.54 | 0.62 | 0.61 | 0.53 | 0.63 | 0.62 | 0.54 | 0.63 | 0.62 |
| 4.5 | acetate | 0.48 | -- | -- | 0.47 | -- | 0.53 | 0.51 | -- | 0.53 | 0.47 | -- | -- |
| | benzoate | 0.42 | -- | -- | 0.44 | -- | 0.47 | 0.46 | -- | 0.46 | 0.43 | -- | -- |
| | chlorate | 0.58 | 0.66 | 0.62 | 0.56 | 0.62 | 0.61 | 0.56 | 0.65 | 0.62 | 0.53 | 0.65 | 0.62 |
| | formate | 0.58 | 0.62 | 0.60 | 0.55 | 0.62 | 0.61 | 0.53 | 0.63 | 0.62 | 0.52 | 0.62 | 0.62 |
| 5.0 | acetate | 0.53 | -- | -- | 0.56 | -- | 0.56 | 0.52 | -- | 0.55 | 0.47 | -- | 0.53 |
| | benzoate | 0.46 | -- | -- | 0.51 | -- | 0.45 | 0.44 | -- | 0.47 | 0.43 | -- | 0.45 |
| | chlorate | | | 0.62 | | | 0.62 | | | 0.61 | | | 0.61 |
| | formate | | | 0.60 | | | 0.62 | | | 0.62 | | | 0.62 |
| 5.5 | acetate | | | 0.56 | | | 0.55 | | | 0.54 | | | 0.53 |
| | benzoate | | | -- | | | 0.47 | | | 0.47 | | | 0.47 |

$pH_L$ = pH of the leading electrolyte; $\bar{c}_L$ = total concentration of the leading ion.

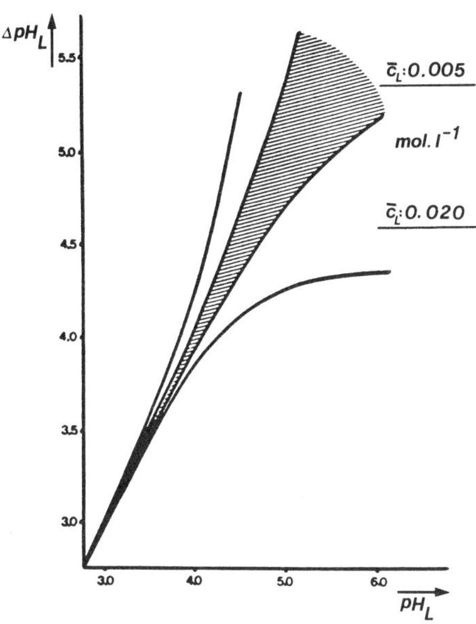


*Fig.2 Tolerancediagram for the pH of the leading electrolyte in various operational systems for a divergence less than 1 % in $T_x$. The concentration of the leading ion has been varied between 0.02 and 0.005 mol $l^{-1}$.*

a divergence of less than 1 % (2 % respectively) in the determination of the transport number $T_x$ and thus in the quantitative evaluation of isotachopherograms obtained in these systems. The concentration of the leading electrolyte (chloride + buffer) was 0.01 mol $l^{-1}$.

In the Figure 2 the tolerance in the pH of the leading electrolyte in various operational systems is given as a function of the concentration of the leading ion. This tolerance indicates a divergence of less than 1 % in the determination of the transport number $T_x$. At "high" concentrations a relatively big tolerance is permitted in the pH of the leading electrolyte.

In the Figure 3 the tolerance in the concentration of the leading ion is given at various pH-values. The tolerance again indicates the divergence less than 1 % in the determination of the transportnumber $T_x$.

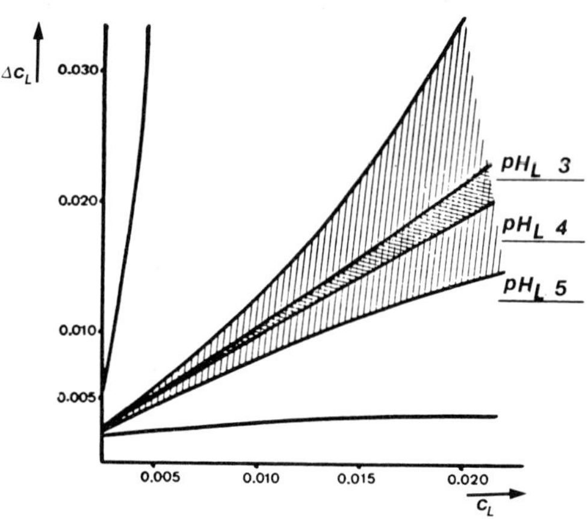



*Fig. 3 Tolerancediagram for the concentration of the leading ion in various operational systems for a divergence less than 1 % in $T_x$. The pH has been varied between 3.0 and 5.0.*

Conclusion

As has been reported elsewhere (7) the diameter of the narrow
bore tube of an analytical isotachophoretic equipment has almost
no influence on the quantitative evaluation of isotachophoretic
analyses, assuming the influence of temperature is small. This
means that not too big differences in inner diameter of the column are allowed, or approximately similar current densities
have to be chosen. The stabilization of the electric driving current determines mainly the divergence in zonelength in a specific operational system, as finally  seen  in the isotachopherogram.
This paper shows that, especially at neutral pH, differnces in
both the concentration of the leading electrolyte and its pH
are allowed, again if the electric driving current is properly
stabilised.


References


1    Everaerts F.M., Mikkers F.E.P, Verheggen Th.P.E.M. (editors),
     Analytical Chemistry Symposia Series, Volume 6, Elsevier

Scientific Publishing Company, Amsterdam, Oxford, New York, 1, (1981).

2   Everaerts F.M., Beckers J.L, Verheggen Th.P.E.M., Isotachophoresis, J. Chromatog. Libr., Volume 6, Elsevier Scientific Publishing Company, Amsterdam, Oxford, New York, (1976).

3   Svendsen P., Schafer-Nielsen C, in Electrophoresis 79, B.Radola (editor), Walter de Gruyter, Berlin-New York, 265, (1980).

4   Mikkers F, Verheggen Th, Ringoir S., Prot.Biol.Fluids, $\underline{27}$, 727, (1979).

5   Mikkers F., Everaerts F., Peek J., J. Chromatog.,$\underline{168}$, 293, (1979).

6   Mikkers F., Everaerts F., Peek J., J. Chromatog.,$\underline{168}$, 317, (1979).

7   Verheggen Th.,Mikkers F., Everaerts F., J. Chromatog., $\underline{132}$, 205, (1977).

8   Everaerts F., Verheggen Th.,Mikkers F., J. Chromatog., $\underline{169}$, 21, (1979).

9   Beckers J.L., Thesis, Eindhoven University of Technology, 5600 MB Eindhoven, The Netherlands, (1973).

10  Boček P., Deml M., Janák J., J. Chromatog., $\underline{91}$, 829, (1974).

11  Oerlemans F., De Bruyn C., Mikkers F., Verheggen Th., Everaerts F., J.Chromatog. Biomed. Appl., in press, (1981).

12  Anhalt E, Holloway C.J., in Analytical Chemistry Symposia Series, Vol.6, Everaerts F.M., Mikkers F.E.P, Verheggen Th. P.E.M. (ed), Elsevier Scientific Publishing Company, Amsterdam, Oxford, New York, 159, (1981).

13  Groot C.K., Graduation Report, Eindhoven University of Technology, 5600 MB Eindhoven, The Netherlands, (1980).

THE ANALYSIS OF HUMAN SERUM PROTEINS BY CAPILLARY
ISOTACHOPHORESIS


Christopher J. Holloway, Wolfgang Heil
Institute of Clinical Biochemistry, Medical School,
D3000 Hannover, Federal Republic of Germany


Eberhard Henkel
Department of Clinical Chemistry II, Medical School,
D3000 Hannover, Federal Republic of Germany


Introduction

Isotachophoresis is most commonly associated with the analysis
of smaller molecules, such as organic acids, nucleotides etc.
However, this electrophoretic method should have a resolution
and separating power comparable with that of isoelectric
focussing, and might thus be considered as a method for the
analysis of proteins, the major advantage over methods in gels
being that the commercial isotachophoresis equipment has
built-in detectors, providing an immediate picture of the
protein patterns. Although some work has been undertaken for
the analysis of serum proteins, the applications have, to date
been quite limited, as demonstrated in the following overview.


Overview of Previous Work in this Field

Clinical Applications of Analytical Capillary Isotachophoresis
have been reviewed by Holloway [1]. Possible applications
include the high-resolution analysis of serum proteins. The
group of Kjellin started relatively early with the analysis of
serum proteins with isotachophoresis [2,3,4,5]. The immediate

754

relevance appeared to be in connection with neurological problems, and particular emphasis was placed on the nature of the
IgG zones. The separation of the various components in serum
could be improved tremendously by the addition of spacers in
the form of carrier ampholytes and amino acids [6,7,8]. The
use of capillary isotachophoresis for the analysis of serum
and cerebrospinal fluid (CSF) in neurological disorders, and
in particular multiple sclerosis, has been thoroughly examined
by Delmotte [9,10,11]. The advantages of the method were well
demonstrated: the total analysis time is short - less than 20
minutes, the sensitivity for detection of serum proteins is
high, the reproducibility and resolution is good. A further
advantage is that no pretreatment of samples is necessary. In
some cases, particular serum proteins are of interest, as for
instance in the work of Hedlund [12,13], where the IgG subclasses were investigated in patients with multiple myeloma.
Proteins can be investigated not only in serum or CSF, but
also in urine, in soluble cell fractions, or even in sweat [14].

Possibly one of the most interesting, and at the same time unexplained effects in isotachophoresis, is that reported by
Gallop and Hambleton [15]. Indirectly, this report inspired
the present work. Rabbits infected with a variety of franci-
sella tularensis showed in the serum isotachopherograms an
additional zone, absent in healthy control animals. This zone
increased with the progress of the disease. The authors noted
that this phenomenon was observable much earlier in the course
of the disease than conventional parameters, such as alkaline
phosphatase or zinc levels in serum. The question which came
to our mind following this publication is whether more infectious diseases result also in alterations in the serum isotachopherogram. This could prove useful in the clinical laboratory for rapid differential diagnosis. In our laboratories,
we have thus been collecting sera from patients with a variety
of infectious and non-infectious disorders, running isotacho-

pherograms, and attempting to find some order in the pattern
obtained. The present work discusses some preliminary methodological details and a few of our data.


Materials and Methods

Analyses were performed on the LKB 2127 Tachophor fitted with
a PTFE capillary of dimensions 23 cm length and 0.5 mm internal
diameter. The capillary block was maintained at 20$^{\circ}$C during
the runs. The leading electrolyte system consisted of 5 mmol/l
MES with 10 mmol/l ammediol as counter ion, with 0.25% HPMC
to reduce electroendosmotic effects. The terminator was 10
mmol/l EACA with 10 mmol/l Ammediol as buffer, corrected to
pH 11 by the addition of saturated barium hydroxide solution.
The electrolytes were filtered before use, and stored in
sealed vessels to prevent pH changes. The electrolytes were
replaced in the instrument approximately hourly. The initial
conditions employed for the separation were constant voltage
(7 kV) until the current dropped to 40 μA. The instrument was
then set to the constant current mode (40 μA) for the remainder of the run. Total analysis time was of the order of 20 min.
The spacers employed were ampholytes (Ampholine[®], LKB, pH
ranges 7-9, 8-9.5 and 9-11) diluted to 1%. In addition,
specific spacers in the form of amino acids were employed,
from stock solutions of 1 mmol/l. Exact details are given in
the relevant figures for the spacers used for particular runs.
Detection was by UV-signal at 280 nm.


Results and Discussion

The leading and terminating buffer systems employed in the
present worklead to considerable problems. The high pH involved is sensitive to changes arising from the absorption of

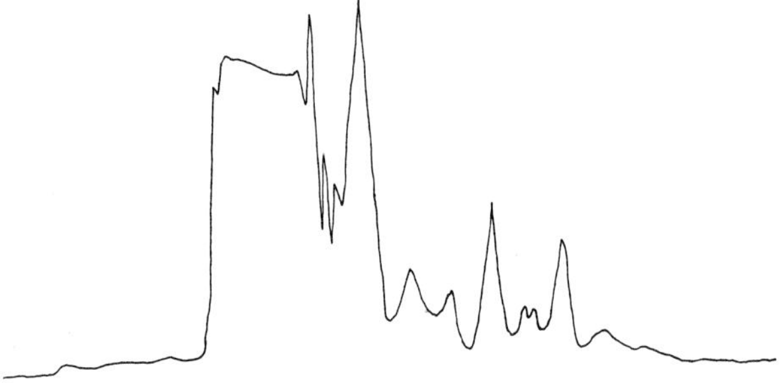

Figure 1: Isotachopherogram of serum from a normal person, spaced only with ampholyte. 5 µl of serum, diluted 1:10 with distilled water were injected into the instrument, together with 1 µl of 1% ampholyte, pH range 9-11. Resolution of the isotachopherogram is obtained predominantly in the lower mobility region (right) corresponding to the IgG fractions.

$CO_2$ from the air. It must be emphasised that reproducible results were only obtained when great care was taken with the preparation of the buffers, and frequent changes of buffer in the instrument vessels. The isotachopherograms obtained without added spacers were similar to those described previously in the literature. Major zones are obtained, corresponding to those in normal electrophoresis (albumin, $\alpha_1$, $\alpha_2$, ß, γ). A variety of combinations of carrier ampholytes can be selected. We found a pH range of 9-11 to provided the most satisfactory resolution, particularly of the low-mobility end of the isotachopherograms. This is illustrated in Figure 1. In order to provide a more defined picture, we opted for the use of specific amino acid spacers. The idea behind this was twofold. Firstly, the amino acids should provide spacing, and hence more detailed information from the isotachopherograms. Secondly, and equally important, is the fact that the amino acid spacers provide a reference for various regions of the pattern. Thus, for example, we can consider a particular zone or group of zones between two particular spacers. Figure 2 shows how

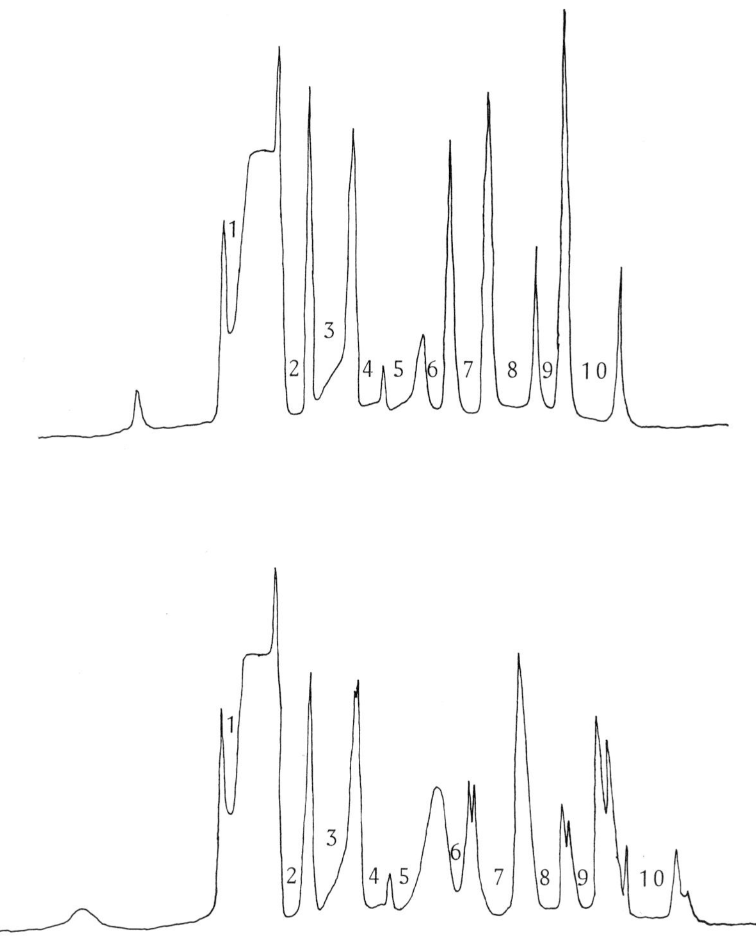

Figure 2: Isotachopherograms of a serum sample with spacers.
A serum has been selected, which contains relatively large
amounts of most protein fractions. In the upper picture, 5 µl
of a 1:10 diluted serum have been injected together with 1 µl
of a spacer solution containing the following amino acids,
each at a concentration of 0.25 mg/ml. The spacing zones are
marked by numbers, corresponding to the following: 1 = glycylglycine, 2= asparagine, 3 = threonine, 4 = glutamine, 5 = methionine, 6 = phenylalanine, 7 = glycine, 8 = valine, 9 = leucine,10 = ß-alanine. The other operating conditions are given
in the materials and methods. In the lower isotachopherogram,
1 µl of 1% ampholyte pH 9-11 has been added.

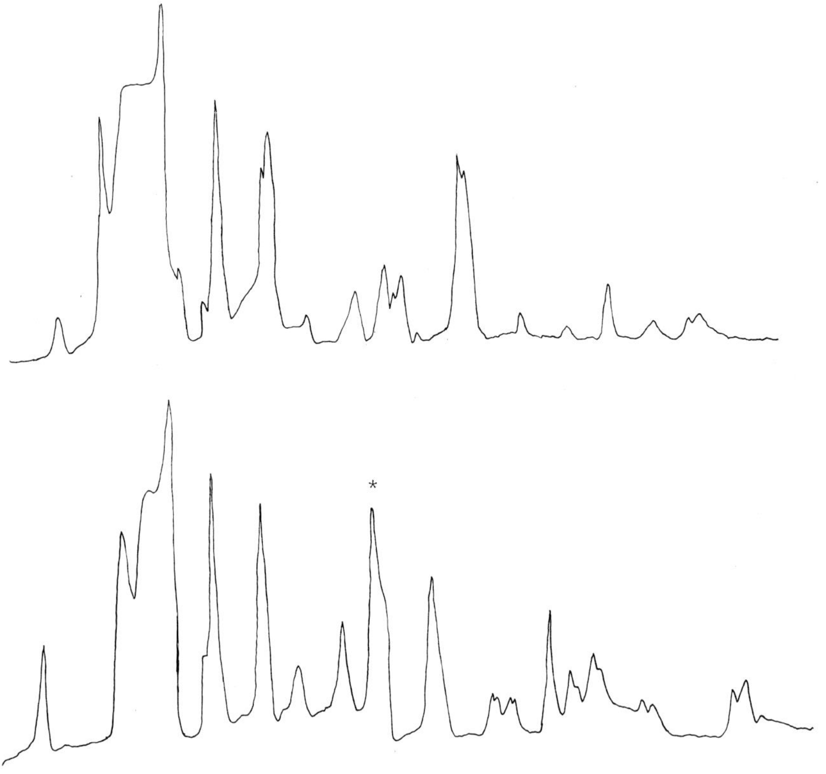

Figure 3: The upper isotachopherogram shows serum from a normal control, analysed under conditions similar to those given
in Figure 2. The lower picture shows the isotachopherogram run
from the serum of a patient with an undefined disease, probably
infectious. There were no abnormalities in the immunoelectrophoresis, but here we see changes in the character of the
protein pattern, particularly the increase in the zone marked
with a *. Since this zone runs between phenylalanine and glycine, spacers 6 and 7 in Figure 2, we call this protein zone
region 6/7.

well resolved the isotachopherogram can be, just using 10
amino acids as spacers. When ampholytes are used in addition,
the isotachopherogram is well resolved into around 20 protein
zones in normal serum. Our long-term aims are to fractionate
the various zones, and identify the individual proteins. At

present, however, we are content to assimilate data, to determine whether trends in the isotachopherograms can be seen,
which correlate with particular disorders.

Very often, we analyse sera which show no abnormalities in the
conventional electrophoresis or immunoelectrophoresis. The
isotachopherogram in Figure 3 (lower picture) is a case in
question. Compared with our normal controls, there are definite abnormalities in the isotachopherogram. This may be the
effect observed by Gallop and Hambledon [15], as mentioned in
the overview, of the appearance of a zone following a particular infectious disease. At present, of course, we are not
able to identify or characterise these proteins, and a hindrance may somtimes be the lack of a definite diagnosis.

In this preliminary work, we have concentrated on the IgG
patterns in various diseases, since the position of these
serum components in the isotachopherogram is well known. Even
at this early stage, we are able to see distinct differences
in pattern according to the type of disease. Figure 4 shows
the isotachopherogram run from the serum of a patient with
cirrhosis of the liver. Apart from the fact that IgG in general is increased, the individual zones seem to split into
several minor components. The exact pattern of this polyclonal
IgG is quite characteristic for liver cirrhosis, and differs
from other cases of increased IgG.

Quite a different case is seen in Figure 5, where the isotachopherograms of a normal control are compared with that of
a patient with a paraproteinaemia in the IgG region. The monoclonal nature of the IgG is quite apparent. Paraproteinaemias
in the IgM region have also been analysed (not shown here),
and such cases serve particularly to idetify the regions of
the isotachopherograms where these proteins are situated. We
have also examined the separation of the four IgG subclasses

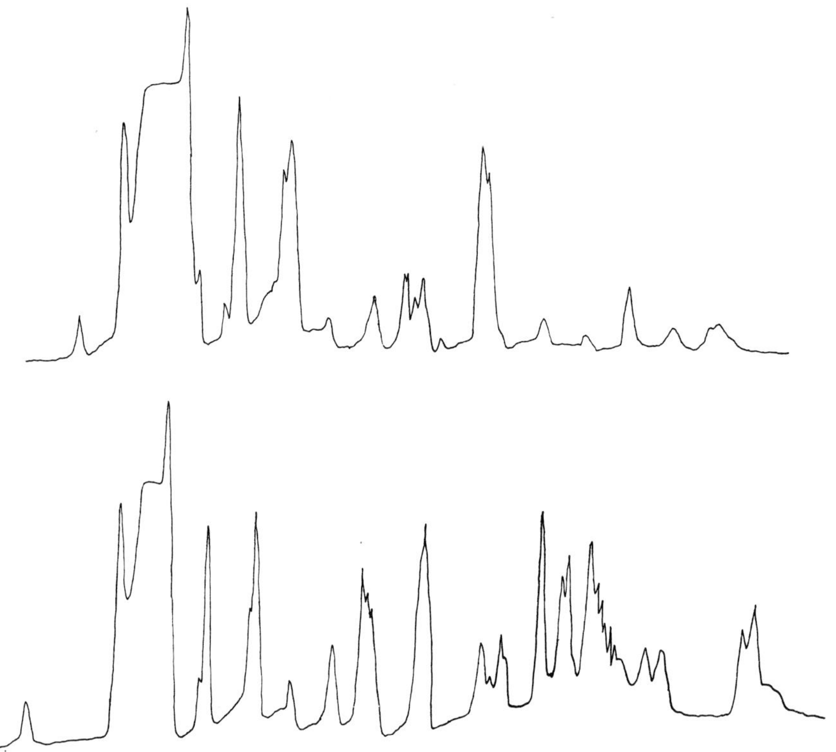

Figure 4: The upper isotachopherogram is run from the serum of
a normal person, under conditions as given in Figures 2 and 3.
The lower picture is typical of the pattern obtained from the
serum of a patient with advanced cirrhosis of the liver. The
region 6/7 is increased, as was the undefined case in Figure 3.
However, the most striking alteration is in the IgG region between the spacers 9 and 10 (leucine and ß-alanine). This
differs quite clearly from cases of generally increased IgG.


by isotachophoresis, as was also done by Hedlund [12,13]. Pure
IgG shows more or less a single zone, as seen in Figure 6.
Carrier ampholytes provide a partial, but not complete reso-
lution of the subclasses. The best separation is achieved if
the amino acids glutamine, leucine and ß-alanine are added.
Hedlund then attributed the subclasses to those peaks shown in

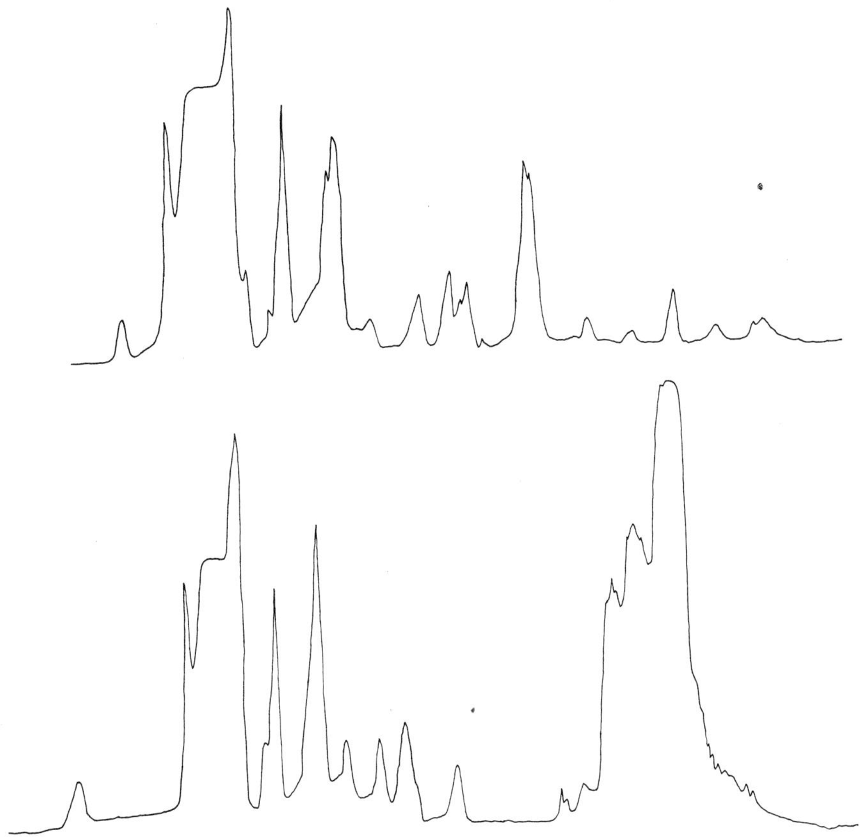

Figure 5: Isotachopherograms run from  the sera of a normal
person (upper picture) and a patient with paraproteinaemia in
the IgG region (lower picture). This monoclonal IgG is seen
at the extreme lower-mobility end of the isotachopherogram,
and can be differentiated quite clearly from non-specific
increases in IgG, or the polyclonal-type of pattern shown in
the previous Figure.

the right-hand picture of Figure 6. However, we disagree that
the peak running in front of glutamine is IgG. The injection
of increasing amounts of IgG leads to a proportionate increase
of the three peaks labelled by Hedlund as IgG 2, 1 and 3.
However, peak 4 does not increase proportionately, and at

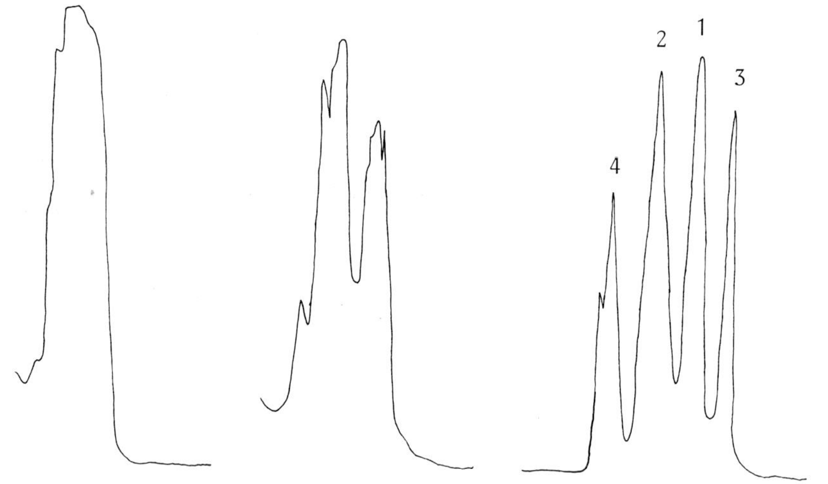

Figure 6: Resolution of IgG into its four subclasses. The isotachopherogram at the far left shows the pattern obtained from
pure human IgG without any spacer. The middle picture represents the pattern obtained after the addition of 1 µl of a 1%
ampholyte spacer, pH range 9-11. In the right-hand picture, the
amino acids glutamine, leucine and ß-alanine. The subclasses
are marked by numbers, whereby part of the subclass 4 peak
derives from impurities in the buffers, as established by the
fact that increasing the injected amount of sample does not
result in a proportionate increase in this zone. As discussed
in the text, we do not agree with this attribution of the four
IgG subclasses, as defined by Hedlund [12,13].


least part of this peak must be due to impurity constituents
of the buffers. Furthermore, examination of the complete serum
isotachopherograms in Figure 2 shows that glutamine runs at
relatively high mobility (spacer 4). Thus, we believe that the
first peak obtained in the isotachopherogram of IgG is not
subclass 4, but an impurity, possible a different immunoglobulin or even albumin. Subclass 4 is such a minor constituent
of IgG, that it is probably overshadowed by one of the other
subclasses. Work is proceeding in our laboratories to further
purify IgG and examine this discrepancy in more detail.

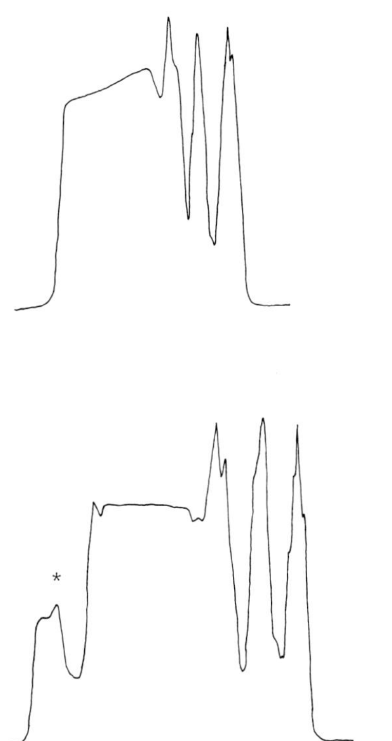

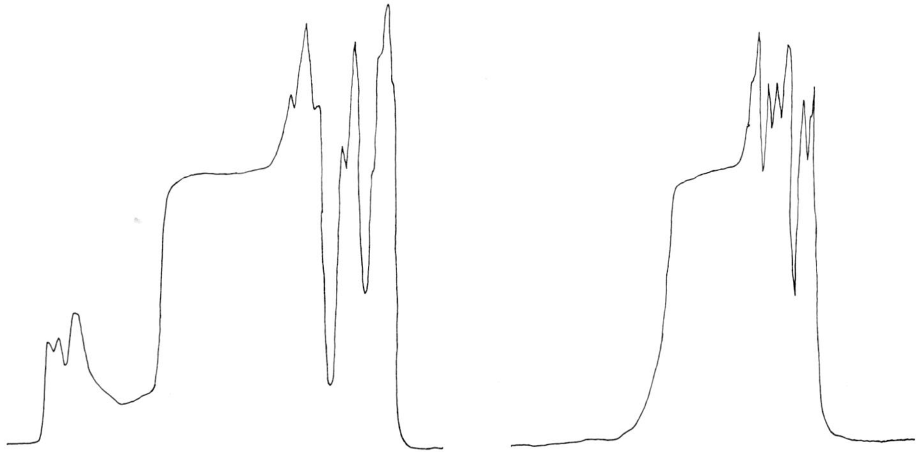

Figure 7: The upper left iso-tachopherogram shows the result obtained with normal CSF. 3 µl of CSF were injected without added spacer. The centre-left picture shows the result obtained with a morphologically defined neurinoma. The peaks running before albumin (*) represent the obvious deviation from the normal pattern. The two lower pictures represent suspected neurinomas. In fact, at autopsy, it proved that only the patient shown in the lower left-hand picture was a neurinoma.

(note: Neurinoma is a benign-type tumour, which originates through proliferation of the Schwann cells. Hence this tumour is sometimes called a Schwannoma. It grows under the skin to a size of approx. 2-3 cm. Its origin is generally around the nervus acusticus, ie. 8th brain nerve).

Protein patterns in CSF have been of particular interest in neurological disorders. Comparisons with the patterns in serum have led to data concerning permeability of the blood/CSF barrier. In the course of our routine investigations of serum proteins, we have also run isotachopherograms of CSF samples from patients with diverse neurological diseases. An interesting result was obtained from the CSF of a patient with a neurinoma (Schwannoma), as shown in Figure 7. Compared with the normal pattern, zones were obtained running at higher mobility than albumin. At the time of running this sample, two further patients were under consideration for possible neurinomas. Isotachopherograms were run of CSF from both patients (shown in the lower part of Figure 7). Only one of the samples was similar to that of the first neurinoma patient, exhibiting at least three additional zones at higher mobility than albumin. The isotachopherogram run from the other sample was more or less normal. It transpired from precise morphological investigation that only one of the two patients, namely that whose CSF gave the isotachopherogram with the additional zones, had a neurinoma. It would be premature from these few data to propose a direct presence of these zones in CSF from patients with neurinomas. Nevertheless, the medical world is continuously on the lookout for possible tumour markers, and perhaps the high resolution offered by isotachophoresis could be exploited in this direction.

The present work has dealt with the isotachophoretic analysis of total proteins in serum or CSF, except for the investigations of IgG as an isolated fraction. Work is proceeding in our laboratories with other isolated proteins, notably the lipoproteins (see contribution in this volume). This research is very much in the preliminary stage, but we believe that isotachophoresis will prove useful in the coming years for the detailed analysis of serum proteins.

Acknowledgments

The authors are grateful to the Gesellschaft der Freunde der Medizinischen Hochschule Hannover and to LKB instrument GmbH Gräfelfing for financial support. The cooperation of many colleagues is gratefully acknowledged. A special note of thanks is due to Prof. I. Trautschold for his encouragement, to Mrs. S. Husmann-Holloway and Dr. E. Borriss for their cooperation with the IgG analyses, and to the Dept. of Neurology for supplying the samples of CSF.

The presentation of this work was made possible by a travel grant from the Deutsche Forschungsgemeinschaft.

References

1.   Holloway, C.J.: in Elektrophorese Forum '80, ed. Radola, B.J., Proceedings TU Munich 225-237 (1980)

2.   Kjellin, K.G., Hallander, L., Moberg, U.: J. Neurol. Sci. 26, 617-622 (1975)

3.   Kjellin, K.G., Siden, A.: Adv. Exp. Med. Biol. 545-559 (1978)

4.   Kjellin, K.G., Hallander, L.: J. Neurol. 221, 225-233 (1979)

5.   Kjellin, K.G., Hallander, L.: J. Neurol. 221, 235-244 (1979)

6.   Catsimpoolas, N., Kenney, J.: Biochim. Biophys. Acta 285 287-292 (1972)

7.   Arlinger, L.: in Progress in Isoelectric Focussing and Isotachophoresis, ed. Righetti, P.G., Elsevier, 331-340 (1975)

8.   Uyttendaele, K., De. Groote, M., Blaton, V., Peeters, H., Alexander, F.: in Protides of the Biological Fluids, Pergamon Press 743-747 (1975)

9.   Delmotte, P.: in Electrofocussing and Isotachophoresis, ed. Radola, B.J., Graesslin, D., de Gruyter 559-564 (1977)

10.  Delmotte, P.: Science Tools 24, 33-41 (1977)

11.  Delmotte, P.: in Electrophoresis '78, ed. Catsimpoolas, N. Elsevier 115-134 (1978)

12.  Hedlund, K.W., Wistar, R., Nichelson, D.: J. Immunol. Methods 25, 43-48 (1979)

13.  Hedlund, K.W., Nichelson, D.E.: J. Chromatogr. 162, 76-80 (1979)

14. Uyttendaele, K., De Groote, M., Blaton, V., Peeters, H.,
    Alexander, F.: J. Chromatogr. $\underline{132}$, 261-266 (1977)
15. Gallop, R., Hambledon, P.: Science Tools $\underline{26}$, 64-66
    (1979)

<u>ISOTACHOPHORETIC ASSESSMENT OF ENZYME IMMUNOGLOBULIN</u>
<u>CONJUGATES USED IN ENZYME IMMUNOASSAY</u>

S. Linpisarn, P.M.S. Clark*, L. J. Kricka and  T.P. Whitehead
Dept. Clinical Chemistry, Wolfson Research Laboratories,
Queen Elizabeth Medical Centre, Birmingham, B15 2TH, U.K.
*author to whom reprint requests should be addressed


<u>Abstract</u>

An isotachophoretic (IT) method for the assessment of enzyme-
immunoglobulin conjugates (horse radish peroxidase –
antiferritin) has been developed.  Two methods of conjugate
preparation, batch to batch variation and equivalent
commercial conjugates were studied by isotachophoresis.
The enzymic and immunological properties of the fractionated
conjugates were assessed using an enzyme immunoassay (EIA) for
serum ferritin.  Fractions could be demonstrated that were
associated with poor and with good analytical performance in
the enzyme immunoassay.  The stability of the conjugates was
also investigated.  Only minor changes were found in the
isotachopherograms or the analytical behaviour of specimens
stored at $-20^{O}C$ or subjected to repeated freezing and
thawing, though changes were found in those conjugates stored
at $4^{O}C$ and room temperature.  This study has demonstrated the
use of analytical isotachophoresis in the monitoring of
production and the assessment of conjugates.  It is
anticipated that this type of study will lead to an
improvement in the quality of enzyme-protein conjugates and
hence to an improvement in the characteristics of assays
utilising such conjugates.

## Introduction

Enzyme-protein conjugates are widely used in the clinical laboratory as reagents in histochemistry and in enzyme immunoassay (1,2).  Ideally a conjugate should possess the full biological activity of its constituent proteins.  However, in practice, conjugation procedures produce a variety of polymeric products in which the biological activity of the constituent proteins has been reduced to various extents, either by the coupling process or by steric factors within the conjugate.  Usually only a partial purification of this mixture is undertaken using either chromatography (gel filtration, ion-exchange or affinity chromatography) or simple molecular filtration (1,3) and the presence of inactive conjugates in this mixture reduces its usefulness as an analytical reagent.  Previous studies of the products of conjugation reactions have utilised low resolution analytical techniques such as gel filtration and thus little is known about the number and properties of the products formed in such reactions.

The objective of this study was therefore to apply a high resolution analytical technique, isotachophoresis, to the study of conjugates.  The conjugation of horse radish peroxidase to anti-human ferritin (immunoglobulin fraction) was chosen for study.  Two methods of conjugation, batch to batch variation, and the stability of the conjugates were studied using isotachophoresis.  In a parallel study the enzymic and immunological properties of the conjugates were assessed in an enzyme immunoassay for serum ferritin.

## Materials and Methods

### Analytical isotachophoresis

This was carried out using an LKB Tachophor 2127 (LKB, Croydon, UK) with a 43 cm capillary at $15^{\circ}$C and with UV (254 nm) and thermal detectors. An initial current of 150 µA was used until a voltage of 20 kV was reached and the current was then reduced to 50 µA. The leading electrolyte was 2(N-morpholino)ethane sulphonic acid (MES: 10 mmol/l; Sigma Chemical Co., Poole, UK) containing hydroxypropyl methyl cellulose (HPMC, 0.5% w/v, LKB) and adjusted to pH 9.2 with 2-amino-2-methyl-1,2-propanediol (ammediol, 1 mmol/l, Sigma). The terminating electrolyte was ε-aminocaproic acid (EACA, 10 mmol/l, Sigma) and adjusted to pH 10.3 with barium hydroxide (BDH, Poole, UK). A spacer solution (10 µl of ampholine pH 3.5-10 LKB, plus 40 µl of amino acid mixture made up to 1 ml with distilled, deionized water: (amino acid mixture comprised glycine, 4 mg, BDH, valine, 4 mg Sigma and β-alanine 3.6 mg Sigma in 1 ml deionized water) was used. Conjugate (5 µl) and spacer solution (2 µl) was injected. All analyses were performed in duplicate. All reagents were prepared using distilled, deionized water. The deionized water was prepared using an Amberlite (monobed resin MB-1, BDH) column. Stock 1% w/v HPMC was dissolved in cold water and was dialyzed against 10 volumes of deionized water for 72 h at $4^{\circ}$C with three changes of water.

### Enzyme immunoassay for ferritin

A solid phase sandwich type enzyme immunoassay using ferritin standards prepared from human spleen was performed according to the method of Linpisarn et al. (4).

## Preparation of conjugates

Rabbit anti-human ferritin (IgG fraction) was obtained from
Dako Immunoglobulin Ltd., Copenhagen F, Denmark.  Horse
radish peroxidase (Hughes and Hughes Co, Romford, Essex, UK)
was purified by gel filtration to give a preparation with a
Reinheit Zahl (A403 nm/A280 nm) of more than 2.8.

An antiferritin horse radish peroxidase conjugate was
prepared by a modified two-stage periodate coupling
procedure (5).  The same reagents were used to prepare
several batches of the conjugate.

The glutaraldehyde coupling method of Boorsma and Kalsbeck(6)
was also used to produce an antiferritin - horse radish
peroxidase conjugate.

Crude conjugates were purified using an Ultrogel ACA 34
column (82 x 1.5 cm, Bio-Rad Laboratories Ltd, Bromley, UK),
equilibrated and eluted with phosphate buffered saline
(0.015 mol/l phosphate buffer pH 7.2 containing 0.15 mmol/l
NaCl).  UV absorbance (280 nm and 403 nm) and enzymic
activity were determined on each fraction(4), and the crude
conjugate and the conjugate fractions were also analysed by
isotachophoresis.

For each of the conjugation experiments the immunological
and enzymic activity of individual fractions and pooled
fractions corresponding to antiferritin - horse radish
peroxidase conjugates were assessed using an enzyme
immunoassay for serum ferritin (4).

## Molar ratio of IgG to HRP

The molar ratio of IgG to HRP in each fraction of purified conjugates prepared by the two methods was calculated from the absorbance measurements at 280 and 403 nm (7).


## Commercial conjugates

The conjugates (anti-ferritin-peroxidase) were obtained from Dako, Mercia Brocades Ltd, Weybridge, UK. (2 batches of 0.01 mol/l in 15 mmol/l $NaN_3$, pH 7.2) and from F. Hoffmann-La Roche & Co Ltd, Basle, Switzerland (in 0.1 mol/l Tris/HCl pH 7.5 with 10 g/l bovine serum albumin). The conjugate solutions from both commercial preparations were dialysed against PBS, overnight at $4^{\circ}C$. Each conjugate was analysed by isotachophoresis and enzyme immunoassay for ferritin as described in the previous sections.


## Stability Study

A freshly prepared conjugate (periodate coupling method) was analysed by isotachophoresis and then a portion of the crude conjugate purified as described previously. The purified conjugate was aliquoted and either stored in plastic vials at $-20^{\circ}C$, $4^{\circ}C$, room temperature or subjected to repeated freezing and thawing. The crude conjugate was similarly stored. The aliquots of both crude and purified conjugate were analysed by isotachophoresis and by enzyme immunoassay after 7, 14 and 21 days of storage.

## Results

## Analysis of Conjugates

The isotachopherograms and EIA standard curves of the conjugates produced by the periodate coupling procedure are shown in Figure 1. The IT results for the unpurified conjugate and the same conjugate after purification by gel filtration are shown in Figure 1 a, b. More than 10 UV absorbing components were demonstrated in the latter, which was a single peak by gel filtration.

Three of the four batches of conjugate produced by the periodate coupling procedure had similar isotachopherograms (cf Batch 1, Fig 1b) and similar EIA standard curves (cf Fig 1b).

However, the isotachopherogram of Batch 4 showed the presence of additional non-UV absorbing components (indicated by arrows in Fig 1c), and the sensitivity of an EIA using this conjugate was inferior to that found with the other three batches.

Individual column fractions corresponding to conjugate were also analysed by IT and the molar ratio of IgG : HRP determined. The molar ratio of the fractions ranged from 1:1.35 to 1:2.79. The IT results from two fractions with different molar ratios (1:1.62 and 1:1.31) are shown in Figure 1d and e. No consistent relationship could be demonstrated between either the IT traces or the results of EIA analysis of the conjugates and the molar ratio of IgG to peroxidase. However, the results of EIA analyses did seem to depend upon the height of peak 3 relative to the first group of proteins (the peaks labelled 1 in Fig 1b). Reduction in the height of peak 3 relative to peak 1 was

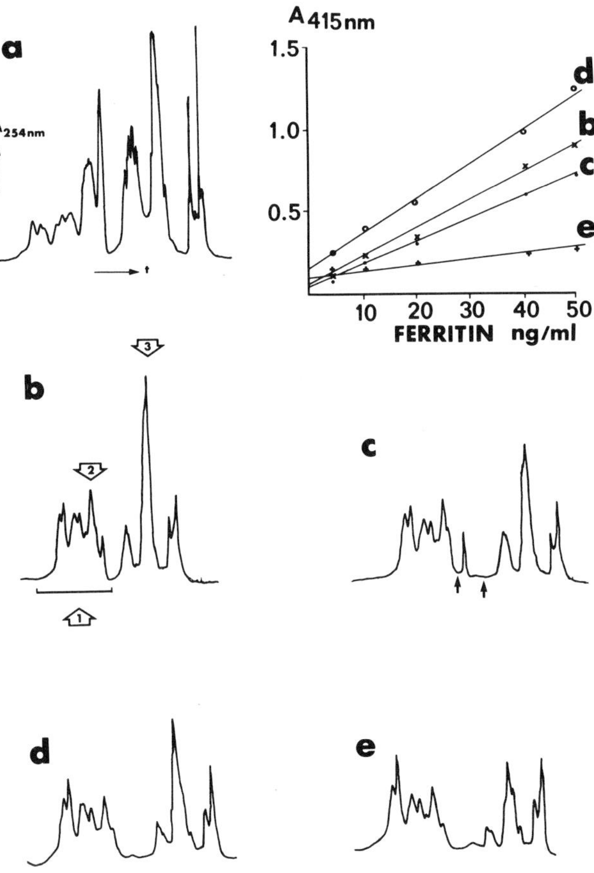


**Figure 1** Isotachopherograms of peroxidase – antiferritin conjugates produced by the periodate coupling procedure and EIA standard curves for ferritin obtained using these conjugates. a; Unpurified conjugate (batch 1) and b; the same conjugate after purification by gel-filtration chromatography. c; Purified conjugate (batch 4). Individual gel column fractions of a purified conjugate with molar ratios of IgG : HRP of 1 : 1.62 (d) and 1 : 1.31 (e).

accompanied by a fall in the sensitivity of the EIA analysis (slope of the standard curve). In addition an increase in the height of peak 2 was associated with a decrease in EIA sensitivity.

The glutaraldehyde preparation (Figure 2a) showed a similar pattern of proteins on IT analysis, though the relative proportions of the peaks did differ (peak 3 was lower and peak 2 higher than in the periodate conjugate cf 1d, e) and the EIA standard curve obtained with this conjugate was inferior to those obtained with the periodate conjugates.

The IT traces of two commercially available conjugates are shown in Figures 2b and c. The IT trace of the Dako products shows a similar protein pattern to the laboratory produced conjugates and although there was an increase in peak 3 relative to the others there was also an increase in peak 2.

The IT of the Roche product differs in the early part of the trace due to the presence of albumin which is added to the conjugate as a stabiliser. Thus it is difficult to compare the IT trace of this conjugate with the traces of the pure conjugates shown in Figures 1 and 2.


Stability Study

Storage of the purified conjugate at, for example, room temperature for seven days greatly reduced its performance in an EIA, as judged by the slope of the standard curve (Figure 3). This was accompanied a decrease in peak 3 and an increase in all the proteins in peak 1 on IT analysis. In addition a new UV absorbing constituent appeared (peak 4)

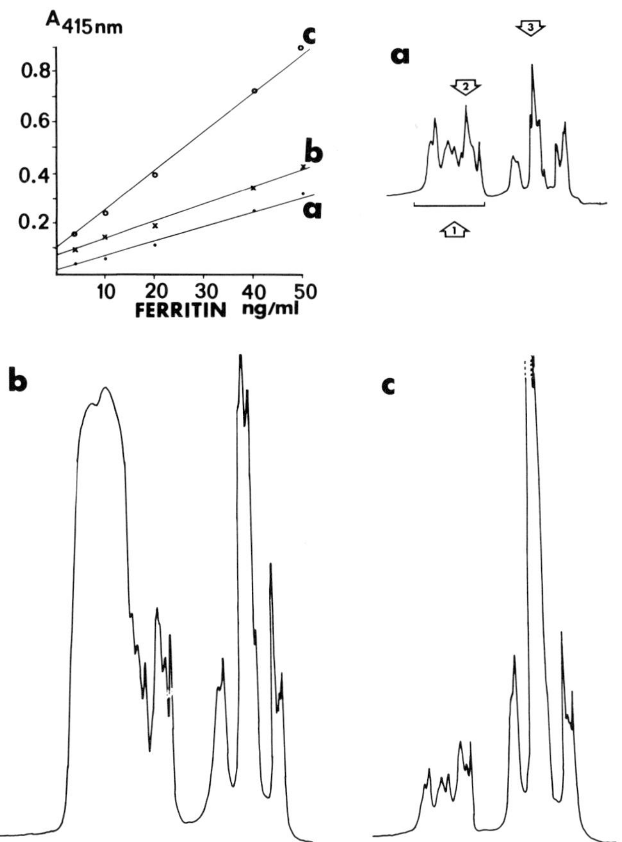



<u>Figure 2</u>  Isotachopherograms of peroxidase – antiferritin conjugates and EIA standard curves for ferritin obtained using these conjugates:
a    Conjugate prepared by glutaraldehyde coupling procedure
b    Conjugate obtained from Roche, and
c    Conjugate purchased from Dako.

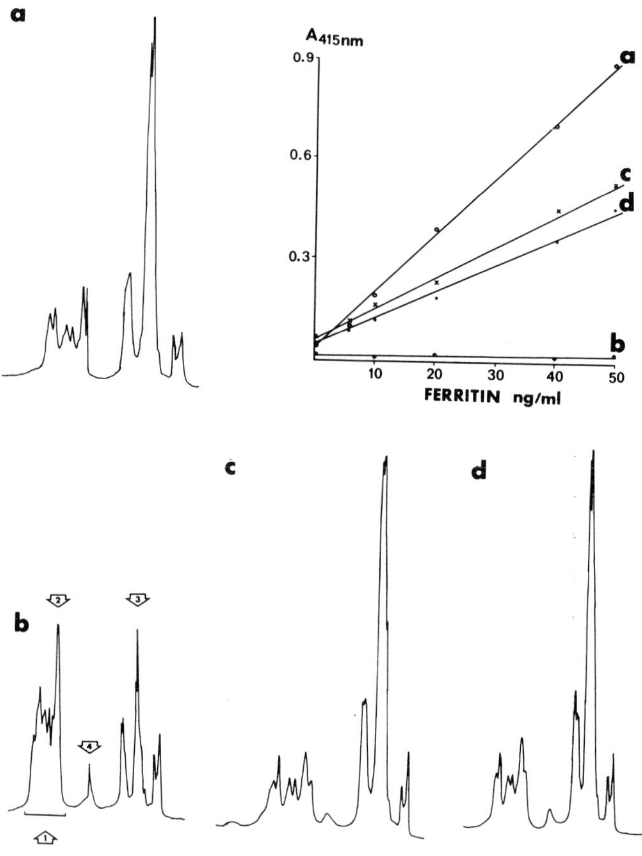



**Figure 3**  Isotachopherograms and EIA standard curves obtained using conjugates stored under different conditions a: Fresh b: conjugate stored at room temperature for 7 days, c; stored frozen at $-20^{\circ}C$ for 21 days, d, repeatedly frozen $(-20^{\circ}C)$ and thawed.

(Figure 3b). The changes in the IT traces of portions of the periodate conjugate which had been either stored frozen or repeatedly frozen and thawed were similar, and were less dramatic than those found in the conjugate stored at room temperature.

<u>Discussion</u>

IT analysis has been used to demonstrate the multiplicity of components present in enzyme-immunoglobulin conjugates produced by two different protein coupling procedures and isolated by gel filtration chromatography. The immunoglobulin fraction and horse radish peroxidase preparation used in this study both proved to be complex mixtures when analysed by IT. Thus it is not surprising that the conjugation procedure should have resulted in so many conjugates with different isotachophoretic mobilities.

The two different protein coupling procedures using the same reagents yielded conjugates with similar protein patterns on IT analysis. Batch to batch variation of the periodate method was also studied and with the exception of one batch of conjugate similar isotachopherograms were obtained. Comparison of the isotachopherograms and performance in EIA of differently prepared conjugates, different batches of conjugate and different gel column fractions has allowed the tentative identification of the protein components which may correspond to the more active (peak 3) and also the less active (peak 2) conjugates. To a lesser extent this is also shown in the results of the analysis of the commercial conjugate. Confirmation of these assignments should be provided by the semi-preparative isotachophoresis studies currently underway in this laboratory.

Individual gel column fractions of the conjugate prepared by the periodate coupling procedure contained up to 10 components and consecutive fractions had very different molar ratios of IgG to peroxidase (Figure 1). Considering that all these fractions came from a single peak of conjugate on gel filtration chromatography it is unlikely that all these conjugates contain different numbers of IgG and peroxidase molecules. Instead they may correspond to a single molecular weight species of conjugate in which either the immunoglobulin and peroxidase are linked _via_ different (amino acid) residues and/or the tertiary or quaternary structure of the conjugates differ because of conformational changes induced by the conditions of the coupling procedure. The calculated molar ratios of IgG to peroxidase in the periodate conjugate fractions would therefore appear to be misleading. This may be due to the introduction of many UV absorbing groups in the initial treatment of the immunoglobulin with the fluorodinitrobenzene (8) blocking reagent and this may invalidate the basis of this type of calculation. Other workers (9) have also noted that the ratio of IgG to peroxidase may not be useful in predicting the quality of a conjugate and recommend that testing by EIA is preferable.

IT has also demonstrated characteristic changes in components of a conjugate following storage under different conditions (Figure 3) and these changes have been related to changes in the performance of the conjugate in an EIA. These studies confirm evidence from other studies with proteins that storage of conjugate at $-20^{\circ}C$ is preferable for long term stability and show that the effects of repeated freezing and thawing for a limited number of times has no deleterious effects.

Improvements in the purity and quality of enzyme immunoglobulin conjugates will inevitably lead to improvements in the characteristics (eg, sensitivity) of assay procedures employing such conjugates. The high resolution of analytical isotachophoresis matches the complexity of the mixture of products produced in coupling reactions and it offers a means of assessing the purity of the reactants (enzyme, immunoglobulin) and monitoring the products of a coupling reaction and their stability. Extension of these studies to include preparative isotachophoresis should facilitate the production of highly purified conjugates with high enzyme and immunological activity.

<u>Acknowledgements</u>

The financial support of the Department of Health and Social Security is gratefully acknowledged.

<u>References</u>

1.    Kennedy, J.H.,Kricka, L.J. and Wilding, P. Clin Chim Acta <u>70</u>, 1-31 (1976).

2.    Wisdom, G.B. Clin Chem <u>22</u>, 1243-1255 (1976).

3.    Saunders, G.C. (1979). <u>In</u> Immunoassays in the Clinical Laboratory (R M Nakamura, W R Dito and E S Tucker III, <u>Eds</u>). Alan R Liss Inc. New York, pp 100-118.

4.    Linpisarn, S., Kricka, L.J., Kennedy, J.H. and Whitehead T.P. Ann Clin Biochem, <u>18</u>, 48-53, (1981).

5.    Nakane, P.K. and Kawaoi, A. J Histochem Cytochem, <u>22</u>, 1084-1091 (1974).

6.    Boorsma, D.M. and Kalsbeck, G.L. J Histochem Cytochem, <u>23</u>, 200-207 (1975).

7.    Yamashita, S., Yamamoto, N. and Yasuda, K. Acta Histochem Cytochem, <u>9</u>, 277-233 (1976).

8.    Weast, R.C. (Ed) (1987) Handbook of Chemistry and
      Physics.  59th Edition, CRC Press, West Palm Beach,
      Florida, p. C-164.

9.    Hagenhaars, A.M., Kumpers, A.J. and Nagel, J. (1980).
      In Imunoenzymatic Assay Techniques (Malvano, R., Ed)
      Martinus Nijhoff Publishers, London, pp 16-27.

ANALYTICAL CAPILLARY ISOTACHOPHORESIS APPLIED TO THE STUDY OF
NUCLEOTIDE-DEPENDENT ENZYMATIC PROCESSES

Sabine Husmann-Holloway
Institute of Microbiology, Medical School, D3000 Hannover,
Federal Republic of Germany


Ehrhard Anhalt, Joachim Lüstorff and Christopher J. Holloway
Institutes of Clinical Biochemistry and Physiological
Chemistry, Medical School, D3000 Hannover, Federal Republic
of Germany


1.    Types of Reaction in which Isotachophoresis can usefully
      be applied

The simplest conceivable enzymatic process is that shown in
scheme *(1)*, in which a single substrate, S, is converted to a
product, P, the reaction being catalysed by a single enzyme,
E:

$$S \xrightarrow{\quad E \quad} P \tag{1}$$

The rate of the enzymatic process can be followed as a function
of the decreasing concentration of substrate, that is:

$$v = -\frac{d[S]}{dt}$$

or as a function of increasing product concentration:

$$V = +\frac{d[P]}{dt}$$

In the absence of any branched or consecutive process linked
to the reaction *(1)*, the two rate expressions should be equal.
Thus, it is irrelevant which process is followed. From the
analytical point-of-view, that parameter which is easiest to

782

monitor, S or P, is chosen. Typically, photometric techniques
are chosen, since spectral properties of one or other of the
reaction components can be exploited. Analytical capillary
isotachophoresis could, of course, also be employed as a
technique for measuring the changes in concentration of S or
P (assuming these species carry a charge), but this would be
highly inefficient compared with photometric and other more
conventional techniques. Much of the early work describing the
application of isotachophoresis to the study of enzyme reactions involved precisely this problem, and the applications put
forward were primarily of academic interest.

In practice, one rarely works with absolutely pure enzymes,
and contaminating activities can interfere with the reaction
of primary interest. One such instance is where the product,
P, is further transformed into another product, R:

$$S \xrightarrow{\ \ E_1\ \ } P \xrightarrow{\ \ E_2\ \ } R \tag{2}$$

If the reaction parameter chosen was the increasing amount of
P, then severe errors will occur due to the useage of this
component for the formation of R. In such cases, S is preferable as a monitoring parameter. Reaction $(2)$ is an example of
a consecutive process. In the case of parallel processes, S
is a less suitable parameter than the product, as seen from
scheme $(3)$:

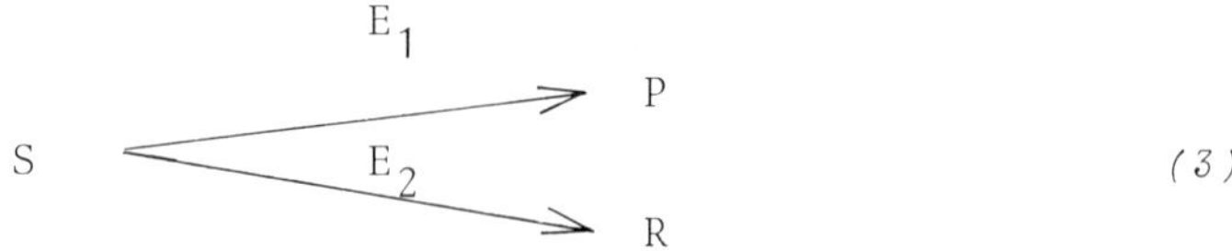


$$\tag{3}$$

Of course, in measuring a single parameter, only one of the
reactions can be assessed. The optimal situation is achieved
when all reaction components can be analysed. If reaction
schemes $(2)$ and $(3)$ are combined into one or more parallel

simultaneous and consecutive reactions, then no single reaction component gives a measure of any one of the processes.
Such can very often be the case in unpurified cell extracts,
although it is surprising how often this point is totally ignored. It is common to find investigations which blandly
assume that the reaction being studied runs in glorious isolation, with no influence from any other components of the
biological material.

We have actually quoted the simplest of cases in schemes $(1)$,
$(2)$ and $(3)$. Most reactions do not start with a single substrate, but are more likely to be as shown in scheme $(4)$:

$$S_1 + S_2 \xrightarrow{E} P_1 + P_2 \tag{4}$$

Needless to say, the arguments put forward for the complications arising from consecutive and parallel reactions are much
amplified in multiple substrate reactions.

In the present work, we shall confine ourselves to a discussion of the possible applications of analytical capillary isotachophoresis to enzymatic processes involving nucleotides as
substrates. Nucleotides have proven to be well suited to analysis by isotachophoresis, as has recently been reviewed by
Holloway and Lüstorff [1].


2.    Hydrolysis of ATP by Nucleotide Pyrophosphatase

2.1   Introduction
When thinking of nucleotides, one thinks first of ATP. The
simplest reaction involving this nucleotide is its cleavage
by hydrolytic processes. ATPases present a wide spectrum of
enzymes, which catalyse the reaction:

$$ATP \longrightarrow ADP + P_i$$

This reaction can be monitored in a number of ways, for example, unreacted ATP can be assayed by bioluminescence, ADP can be coupled to the PK/LDH enzyme system, and monitored photometrically by the reaction of NADH to NAD. Inorganic phosphate, $P_i$, can be analysed colorimetrically. Provided that this reaction occurs in isolation, there is no need for a more involved analytical procedure.

We have considered another hydrolytic process involving ATP which is not as simple as ATPase, namely the nucleotide pyrophosphatase, which acts on pyrophosphate linkages in nucleotides or dinucleotides, but not on inorganic pyrophosphate. Here, two reactions are, at least in principle, possible:

$$ATP \longrightarrow ADP + P_i \longrightarrow AMP + 2P_i$$

$$ATP \longrightarrow AMP + PP_i \tag{6}$$

The first reaction involves initial cleavage of the ß-γ pyrophosphate linkage of ATP, yielding ADP as a primary product, which can, in turn, be hydrolysed to AMP and a further inorganic phosphate. Cleavage of the α-ß linkage as shown in the second reaction yields AMP and inorganic pyrophosphate. The question to be answered was whether only one or both of these reactions is possible. Clearly, only analysis of a single reaction component, ATP or $P_i$ for example, would not give the whole answer.

## 2.2  Materials and Methods

ATP in the concentration range 5-10 mmol/l was incubated with nucleotide pyrophosphatase (Sigma) in a tris/HCl buffer, pH 7.5, at a reaction temperature of $37^{\circ}C$. At various intervals, aliquots of the reaction mixture (1-10 µl) were removed and analysed isotachophoretically. The instrument employed was the LKB Tachophor 2127, fitted with a PTFE capillary of length 63 cm, and thermostatted at $10^{\circ}C$ for the analysis. The leading and terminating electrolytes were chloride and hexanoate, respectively, each at a concentration of 5 mmol/1. The leading system was buffered to pH 3.9 by the addition of ß-alanine, and 0.25% HPMC was used to reduce electroendosmosis. The separations were carried out initially at 100 µA constant current up to 15 kV, where the current was reduced to 50 µA for the rest of the run. The total analysis time was of the order of 30 minutes. Detection of zones was by UV signal at 254 nm. Quantification of zones was done against calibration plots constructed from zone widths after injection of the incubation mixture with known amounts of each sample. Injection of the incubation mixture into the instrument led to immediate quenching of the reaction, and it was established that the enzyme employed contained no contaminating activity such as 5'-nucleotidase or inorganic pyrophosphatase.

2.3 Results and Discussion

The isotachopherogram in Figure 1 shows the separation achieved with a mixture of around 2 nmol each of the possible reaction components given in the schemes (6). Inorganic pyrophosphate is seen as the non-UV-absorbing zone just before ATP, which in turn is separated from ADP by the inorganic phosphate zone. AMP is the UV-absorbing component at lowest mobility. The isotachopherogram in Figure 2 shows the result of an actual incubation of ATP with nucleotide pyrophosphatase whereby the total nucleotide content is of the order of 5 nmol. After 2 hours of reaction, a considerable amount of phosphate, pyrophosphate and AMP has been formed, as shown in the left hand picture of Figure 2. Very little ADP is seen. After more than 3 hours of reaction, as seen in the right hand picture, most of the ATP has been consumed, and has been converted to AMP, $P_i$ and $PP_i$. Qualitatively, we see immediately that both cleavage of the α-ß and of the ß-γ pyrophosphate linkages of ATP occurs. The kinetics of reaction are shown in Figure 3. If we consider the situation at various periods of time, the stoichiometry of the process can be examined. This is summarised in Table I. It can be seen that the amount of AMP + ADP

UV-signal at 254 nm

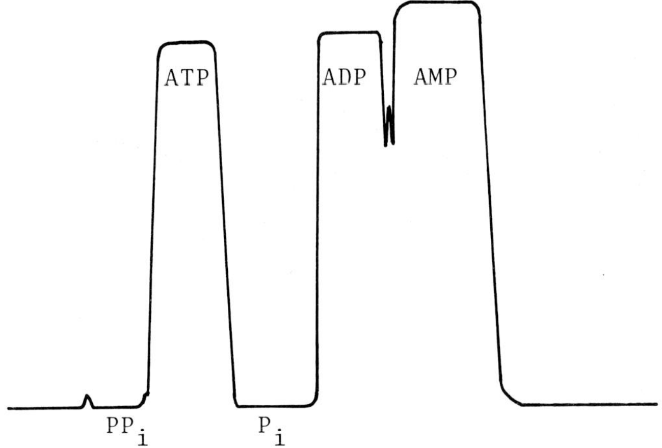


Figure 1: Isotachopherogram of a mixture of inorganic pyro-phosphate (PPi), inorganic phosphate (Pi), and the three nucleotides, ATP, ADP and AMP involved in the reactions in scheme *(6)*. For further details see text.

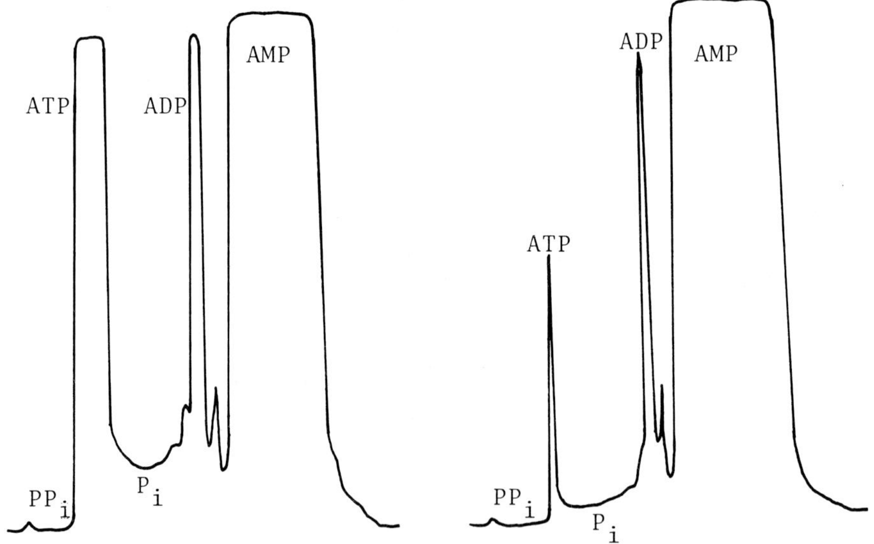


Figure 2: Isotachopherograms run at periods during the hydrol-ysis of ATP by nucleotide pyrophosphatase (left, 2 hours, and right 3 hours). Initial ATP concentration was 0.6 mmol/1. 10 µl of sample were injected, so that the total nucleotide in each isotachopherogram is 6 nmol. Peaks other than those marked derive from impurities in the system. For further details see text.

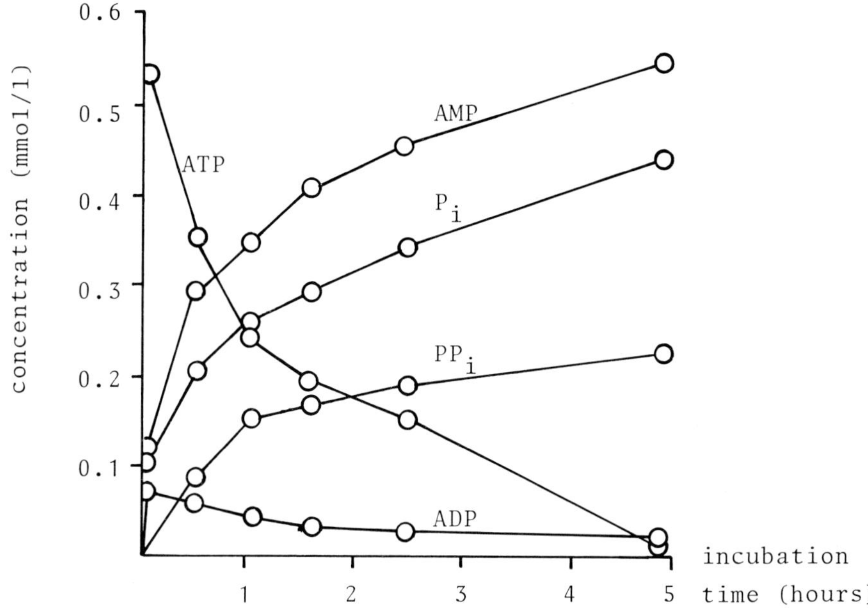



Figure 3: Kinetics of hydrolysis of ATP catalysed by nucleotide pyrophosphatase, derived from isotachopherograms as
shown in Figure 2. Conditions are as given in Figure 2 and in
the text.

Table I: Stoichiometry of the hydrolysis of ATP at the $\alpha$-$\beta$
and $\beta$-$\gamma$ linkages, derived from kinetic experiments such as
that shown in Figure 3. A full explanation is given in the
text.

| hours | ATP | ATP consumed | ADP | AMP | ADP + AMP | $PP_i$ | $P_i$ | PPi + 0.5 Pi |
|---|---|---|---|---|---|---|---|---|
| | (nmol) | (nmol) | (nmol) | (nmol) | (nmol) | (nmol) | (nmol) | (nmol) |
| 0.03 | 5.3 | 0.7 | 0.6 | 0.5 | 1.1 | 0.0 | 1.1 | 0.5 |
| 0.57 | 3.5 | 2.5 | 0.4 | 2.2 | 2.6 | 1.0 | 2.0 | 2.0 |
| 1.12 | 2.4 | 3.6 | 0.4 | 3.1 | 3.5 | 1.6 | 2.5 | 2.9 |
| 1.68 | 1.9 | 4.1 | 0.3 | 3.6 | 3.9 | 1.8 | 2.9 | 3.3 |
| 2.27 | 1.5 | 4.5 | 0.3 | 4.2 | 4.5 | 2.0 | 3.5 | 3.8 |
| 4.90 | 0.2 | 5.8 | 0.3 | 5.0 | 5.3 | 2.8 | 4.4 | 5.0 |

formed corresponds well to the amount of ATP consumed. As expected from scheme *(6)*, the quantity $PP_i + 1/2 \, P_i$ is equal to the amount of AMP formed. It is also interesting to note that just about twice as much $P_i$ is formed as $PP_i$. This is the case is the two reactions shown in scheme *(6)* run to the same degree. It would thus appear that the nucleotide pyrophosphatase recognises and cleaves each pyrophosphate linkage of ATP equally well.


3.     Adenylate Kinase Reaction with Substrate Analogues


3.1    Introduction

The system described in section 2, nucleotide pyrophosphatase could, in principle have been analysed by conventional methods although this would, admittedly have been very tedious indeed. A case where most conventional methods fail is when analogues of normal substrates are employed in enzymatic processes. This is often done in order to establish characteristics of the active site. An example of this is the reversible adenylate kinase reaction, shown in scheme *(7)*:

$$\text{ADP} + \text{ADP} \xrightarrow{\text{ADK}} \text{ATP} + \text{AMP} \qquad (7)$$

In either direction, the reaction can be followed by coupling to NAD(P)/NAD(P)H-dependent reactions, so that the assay is photometric. If, however, not ATP, ADP and AMP are employed, but analogues of these nucleotides, the coupled enzymatic assays fail, since the analogues are not substrates for the coupled enzymes. Clearly, a quantitative chromatographic or electrophoretic method is needed to analyse the reaction components. Isotachophoresis proves to be suitable.

The substrate analogues under consideration here are with an additional phosphate group substituted at the 3'-ribose position. Thus the analogues of ATP, ADP and AMP can be represent-

ed by *p*ATP, *p*ADP and *p*AMP. The 3'-phosphate does not take part
directly in the reaction, but serves merely as a substituent
group, with a resulting altered geometrical and electronic
form of the substrate molecule.

In reaction scheme *(7)*, one of the ADP molecules can be considered as donor of the phosphate group, and the other ADP
molecule as acceptor. The question of interest in this work
was how the analogue *p*ADP reacts in the system.

## 3.2    Materials and Methods

The procedures were essentially as described under section 2.2.
Initial concentrations of ADP or *p*ADP in the range 5-10 mmol/l
were selected, and the enzyme employed was adenylate kinase
from rabbit muscle (Boehringer Mannheim, FRG).

## 3.3    Results and Discussion

The isotachopherograms in Figure 4 show mixtures of ATP, ADP
and AMP or *p*ATP, *p*ADP and *p*AMP with about 2 nmol of each nucleotide. Incubation of ADP with adenylate kinase led to a
disproportionation, with resulting production of ATP and AMP.
With *p*ADP alone, there was no disproportionation. However,
when ADP and *p*ADP were incubated together with the enzyme,
some *p*ATP was formed in addition to the products of disproportionation of ADP (Figure 5). This can only be explained by
the following reaction:

$$\text{ADP} + p\text{ADP} \xrightarrow{\text{ADK}} p\text{ATP} + \text{AMP} \tag{8}$$

It would thus appear that *p*ADP is able to receive a phosphate
group, but not to act as donor. ADP is the specific donor
substrate. Equally, in the reverse reaction, AMP is the specific acceptor, whereas analogues such as *p*ATP can act as
donors. The two ADP sites thus do not seem to be equivalent.
X-ray crystallographic studies have also shown that the ADP
donor (= AMP acceptor) site is deeply buried in the enzyme,

UV signals at 254 nm

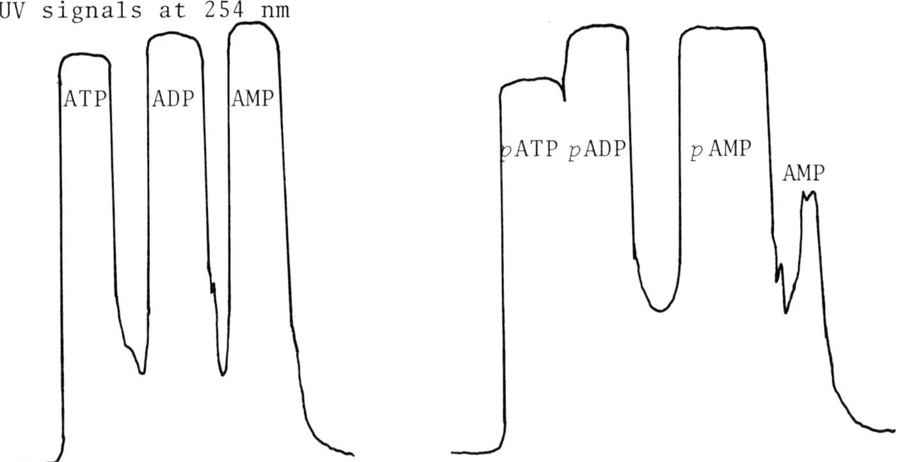


Figure 4: Isotachopherograms showing mixtures of ATP, ADP and AMP (left hand picture) and their 3'-phospho-analogues (right hand picture). Approximately 2 nmol of each nucleotide were injected into the instrument. The analogue mixture is seen to contain a small quantity of AMP as impurity.

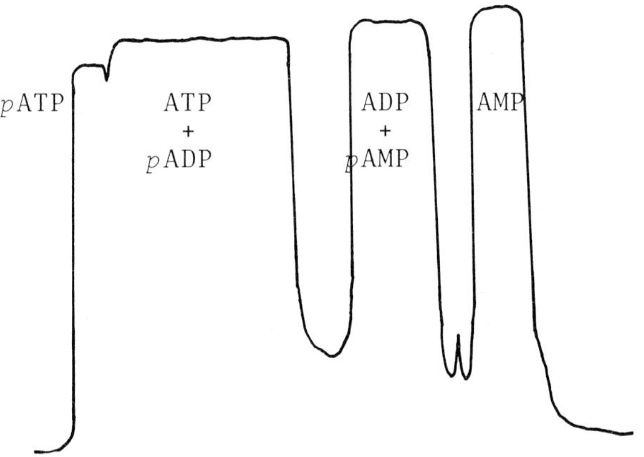


Figure 5: Isotachopherogram showing the result of incubating ADP, pADP and adenylate kinase. The initial concentrations of ADP and pADP were 0.5 and 0.6 mmol/l, respectively. Under these analytical conditions, the triphosphate components form a mixed zone, as do the diphosphates. In fact, the diphosphate zone does not contain any pAMP, since dADP cannot act as phosphate donor in the reaction.

whereas the ADP acceptor (= ATP donor) site is situated at the surface of the enzyme. The ADP donor site must thus be subject to more structural restrictions than the ADP acceptor site, explaining why analogues can function at one site and not at the other.


4.    Glucuronic Acid Pathway

4.1    Introduction
The application of isotachophoresis to the study of the glucuronic acid pathway has been the subject of several publications by our group (summarised in reference [1]), and will only be reviewed briefly here as an example of multiple parallel and consecutive processes.

UDP-glucuronic acid (UDPGA) is the glucuronate donor for the enzyme system UDP-glucuronyltransferase, which is responsible for the majority of detoxication in the mammalian organism. As shown in scheme *(9)*, an acceptor (toxin), R, receives a glucuronic acid residue from the donor molecule, the product, R-GA is generally non-toxic, and can be excreted from the body.

$$R + UDPGA \xrightarrow{\text{UDPGTase}} R\text{-}GA + UDP \qquad (9)$$

In studies with this system, inaccuracies can arise from the fact that the donor molecule UDPGA is hydrolysed by nucleotide pyrophosphatase as follows:

$$UDPGA \xrightarrow{\text{NPPase}} UMP + GA\text{-}1\text{-}P \qquad (10)$$

where GA-1-P is glucuronic acid 1-phosphate. The nucleotide pyrophosphatase further acts on UDP, the primary product of the glucuronidation reaction, so that not this product, but UMP only is seen. As we have demonstrated in earlier work,

792

UDPGA, the other uridine nucleotides, GA-1-P, $P_i$ and glucuronic acid itself can all be analysed in a single run together
with the glucuronidated product R-GA. In the physiological
situation, however, not a single acceptor, but a whole group
of acceptors is biotransformed by glucuronidation simultaneously, a situation which was, up till now, difficult to emu-
late in vitro. This can be represented by the following group
of reactions:

$$R_1 + UDPGA \longrightarrow R_1\text{-}GA + UDP \qquad (11)$$
$$R_2 + UDPGA \longrightarrow R_2\text{-}GA + UDP$$
$$\cdots$$
$$\cdots$$
$$R_n + UDPGA \longrightarrow R_n\text{-}GA + UDP$$

Provided that mixtures of R-glucuronides can be analysed, the
glucuronidation of the various acceptors, $R_1$, $R_2$......$R_n$, can
be followed simultaneously by isotachophoresis.


4.2    Materials and Methods
The enzyme reactions were carried out as described previously.
The operational system was as in section 2.2, except that a
leading electrolyte pH of 3.3 was optimal for the separation
of the ß-glucuronides.


4.3    Results and Discussion
There are any number of acceptors which could have been chosen
for this work. The most common class of acceptor, however,
contains a hydroxyl group as acceptor on an aromatic ring. In
the present work, therefore, we have selected a range of substituted phenols and the isomers of naphthol. Figure 6 shows
the result of glucuronidation of mixtures of various acceptors
as stated in the legends. In all cases, the glucuronides were
well separated from one another. Even in the case of 1- and
2-naphthol, a quantitative resolution of the glucuronides is
possible from zone widths and heights. The mathematical trea-

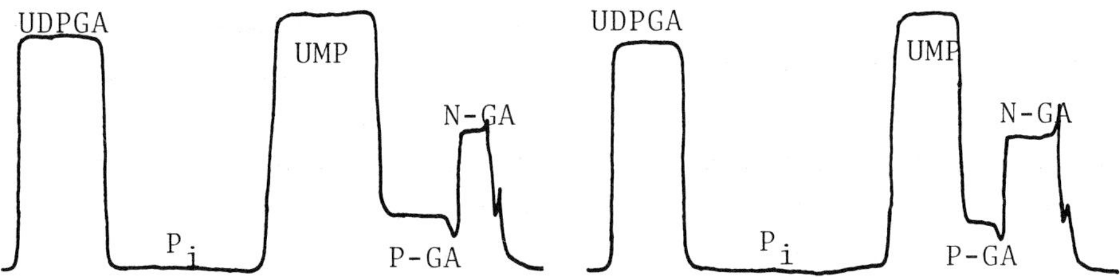


Figure 6: Isotachopherograms showing the result of glucuronidation of phenol and 1-naphthol. The glucuronide products are
represented by P-GA and N-GA, respectively. In the left hand
picture, the initial concentrations of phenol and 1-naphthol
in the incubation mixture were 2 mmol/l and 0.5 mmol/l, respectively. The amounts of the two glucuronides after 24 hours
incubation are comparable. In the right hand picture, the
concentrations of phenol and 1-napthol were 2 mmol/l and
1 mmol/l, respectively. The glucuronidation of phenol is inhibited, demonstrating the higher affinity towards 1-naphthol
as acceptor.

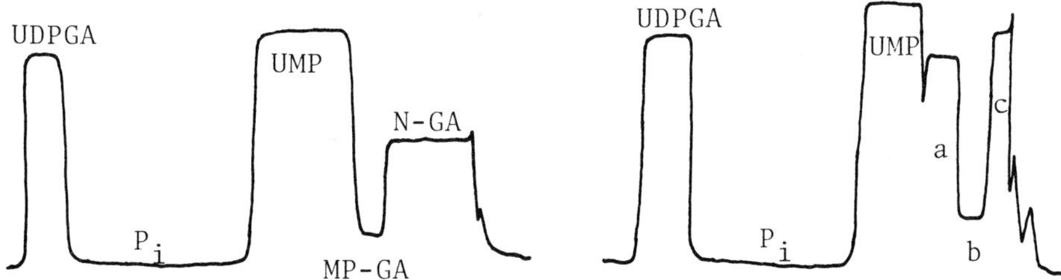


Figure 7: The left hand isotachopherogram shows the result of
the simultaneous glucuronidation of 4-methylphenol and 1-nap-
hthol. Initial acceptor concentrations were each 2 mmol/l.
Although 4-methylphenol alone is glucuronidated more rapidly
than 1-naphthol, the situation is reversed here, showing the
higher affinity of enzyme towards naphthol as acceptor. The
right hand isotachopherogram shows the result of simultaneous
glucuronidation of three acceptors. The marked zones are:
a) 4-nitrophenylglucuronide, b) 4-methylphenylglucuronide and
c) 1-naphthylglucuronide.

The isotachopherograms in these figures represent UV signals
at 254 nm. The highest mobility is at the left hand side of
each picture.

tment of this system has already been given in our earlier
work. The most important consequence of the present work is
that competition studies between various acceptors can be
carried out. Thus, for example, when 1-naphthol and phenol
are offered simultaneously as acceptors, the former is predominantly glucuronidated, demonstrating its higher affinity
with respect to the enzyme. Another interesting example is
seen in the simultaneous glucuronidation of 4-methylphenol
and 2-naphthol. When incubated alone with the enzyme, 4-
methylphenol has a much higher rate of reaction than 2-naph-
thol. However, when both acceptors are offered simultaneously,
naphthol is glucuronidated at the cost of 4-methylphenol.
Again, this can only be explained in terms of higher affinity
of the enzyme towards naphthol as acceptor.

From the point-of-view of the present discussion, a useful
bonus is offered from the isotachophoretic data. Thus, apart
from the mixtures of glucuronides, the other reactants and
products of reaction are observed. Some of these products
derive from the hydrolytic action of UDPGA, an activity which
is hopefully excluded in purification procedures of UDP-
glucuronyltransferase. Thus, during the purification of the
enzyme, the various constituents contaminating the enzyme of
main interest can be assessed at the same time as the main
enzyme of interest is assayed. Further examples of mixtures of
ß-glucuronides are given in Figure 7; the mixtures are explained in the legends.


5.    Conclusions

Analytical capillary isotachophoresis is a useful method for
investigating enzymes forming a complex reaction scheme. A
particularly useful application is when nucleotides and their
derivatives are the reaction constituents. Apart from the
fact that most if not all of the components can be analysed in

a single run, the technique offers the advantage that the untreated sample can be injected directly from the incubation
mixture, since conditions of analysis can be chosen such that
the enzyme proteins do not migrate and disturb the analysis.
Injection into a relatively acid electrolyte system as used
commonly for nucleotide analysis generally causes quenching
of the enzymatic reaction.


Acknowledgments

We would like to thank the Gesellschaft der Freunde der Medizinischen Hochschule Hannover and LKB Instrument GmbH for
their support in these projects. A particular note of thanks
is also due to Professor I. Trautschold for his general
advice and encouragement.
A special acknowledgment is due to Dr. E. Borriss and Prof. J.
Potel for their support and enthusiasm.

Reference

1.  Holloway, C.J., Lüstorff, J.: Electrophoresis $\underline{1}$, 129-136
    (1980)

CONDITIONS FOR THE SEPARATION OF ADENINE NUCLEOTIDES AND THEIR
3'-PHOSPHATE DERIVATIVES IN ANALYTICAL CAPILLARY
ISOTACHOPHORESIS


Joachim Lüstorff and Christopher J. Holloway

Institutes of Clinical Biochemistry and Physiological
Chemistry, Medical School, D3000 Hannover, Federal Republic
of Germany


Introduction

Analytical capillary isotachophoresis has proved to be a suitable technique for the qualitative and quantitative estimation
of nucleotides. This subject has been reviewed recently by
Holloway and Lüstorff [1]. In early work associated with the
analysis of nucleotide mixtures by isotachophoresis, a leading electrolyte system with a pH lower than 4 was generally
selected. Under these conditions, the thermal step heights of
the nucleotides, and hence their relative mobilities, differ
most [2]. Indeed, most of the naturally-occurring nucleotides
can be separated under such conditions. In previous work, we
were interested in the analysis of adenine nucleotides with a
further phosphate group at the 3'-ribose position, in connection with studies of substrate specificity of adenylate kinase.
It was found that with a leading electrolyte pH of the order
of 4, mixed zones were formed between some of the substances,
such that a complete quantitative analysis was not possible
[3,4]. In the continuation of this work, we have attempted to
vary the operating conditions such that a complete resolution
of all the adenine nucleotides is possible.

798

Abbreviations

adenosine 5'-triphosphate (ATP) = *pppA*
adenosine 5'-diphosphate (ADP) = *ppA*
adenosine 5'-monophosphate(AMP) - *pA*
adenosine 3'-phosphate 5'-triphosphate = *pppAp*
adenosine 3'-phosphate 5'-diphosphate = *ppAp*
adenosine 3'-phosphate 5'-monophosphate = *pAp*
$P^1,P^5$ Di(adenosine-5')pentaphosphate = A*ppppp*A


Materials and Methods

The nucleotides were obtained from the usual commercial sources, except for *pppAp* and *ppAp*, which were synthesised by the
method of Michelson [5]. The substances were dissolved in
distilled water as stock solutions to a concentration of
1 mmol/l.

As specific non-UV-absorbing spacers, the following substances
were used at a concentration of 1 mmol/l: succinate, pyruvate,
lactate, acetate, trichloroacetate, glyoxylate, propionate,
butyrate, pentanate, hexanoate, glutamate, aspartate, phosphate and thiophosphate.

The following leading and terminating electrolyte systems were
employed in the present study:

1) pH 3.9
   Leading system: 10 mmol/l Cl⁻, counter ion ß-alanine,
   0.25% hydroxypropylmethylcellulose (HPMC)
   Terminating system: 10 mmol/l hexanoic acid

2) pH 5.3
   Leading system: 10 mmol/l Cl⁻, counter ion pyridine,
   0.25% HPMC
   Terminating system: 10 mmol/l cacodylic acid

3) pH 5.7

Leading system: 10 mmol/l Cl⁻, counter ion histidine, 0.25% HPMC

Terminating system: 10 mmol/l cacodylic acid

4) pH 6.2

Leading system: 10 mmol/l Cl⁻, counter ion histidine, 0.25% HPMC

Terminating system: cacodylic acid

Analyses were performed on the LKB 2127 Tachophor, fitted either with a capillary of length 23 cm or 61 cm. Internal diameter of the capillary was 0.5 mm. The capillary block was maintained in all cases at 20°C. Detection was by UV-signals at 254 or 280 nm, and in some cases by conductimetric detector with differential signal.

Many different running conditions were employed in the present work. Those given here are relevant to the isotachopherograms depicted in the Figures.

pH 3.9: Constant voltage (15 kV) until the current dropped to 80 μA. Switchover to constant current (50 μA) for the remainder of the run. Samples passed the detectors at a potential of about 9 kV. Total analysis time 35 min. Capillary length 61 cm.

pH 5.3: Constant current (40 μA). Samples passed the detectors at a potential around 3 kV. Total analysis time around 40 min. Capillary length 23 cm.

pH 5.7: (a) Constant current (40 μA). Samples passed the detectors at around 3 kV. Analysis time 30 min. Capillary length 23 cm.

(b) Constant current initially 100 μA up to a voltage of 12.5 kV. Current was then reduced to 50 μA for the remainder of the run. Samples passed the detectors at 8-13 kV. Total analysis time 40 min. Capillary length 61 cm.

(c) Constant current (200 µA) up to 10 kV. Current reduced to 100 µA for the rest of the run. Total analysis time around 5 min. Capillary length 23 cm.

pH 6.2: Constant current (40 µA). Samples passed the detector
at around 3 kV. Total analysis time 17-20 min. Capillary length 23 cm.

Various amounts of mixtures of the nucleotides were injected
according to the length of capillary employed, and the operating conditions. A check whether the steady state had been
reached was whether the zone lengths seen in the UV- and
conductivity-signals were identical. In the LKB instrument,
the conductivity detector is placed some way beyond the UV-
detector. The general order-of-magnitude was 1-10 nmol total
nucleotide injected into the instrument.


Results and Discussion


Figure 1 shows the analysis of the three physiological adenine
nucleotides together with their 3'-phosphate derivatives. The
compound ApppppA is a specific inhibitor of adenylate kinase,
and could be employed in kinetic studies of this enzyme. It
was thus interesting to obtain conditions where this nucleotide does not disturb the total analysis. The conditions employed in Figure 1 were those of our previous work [3,4],
namely with a leading electrolyte pH of 3.9. Although 254 nm
is the normally recognised wavelength for nucleotides, 280 nm
could be of interest, since larger differences are seen in the
UV-transmission signals, providing easier identification of
zones. At pH 3.9, mixed zones are formed between nucleotides
with the same number of phosphate groups. ApppppA forms at
least a partially mixed zone with pppAp. In Figure 2, ApppppA
was excluded from the mixture, but varying total nucleotide
quantities were injected, in order to establish which of the
minor peaks derived from impurities in the electrolytes. A

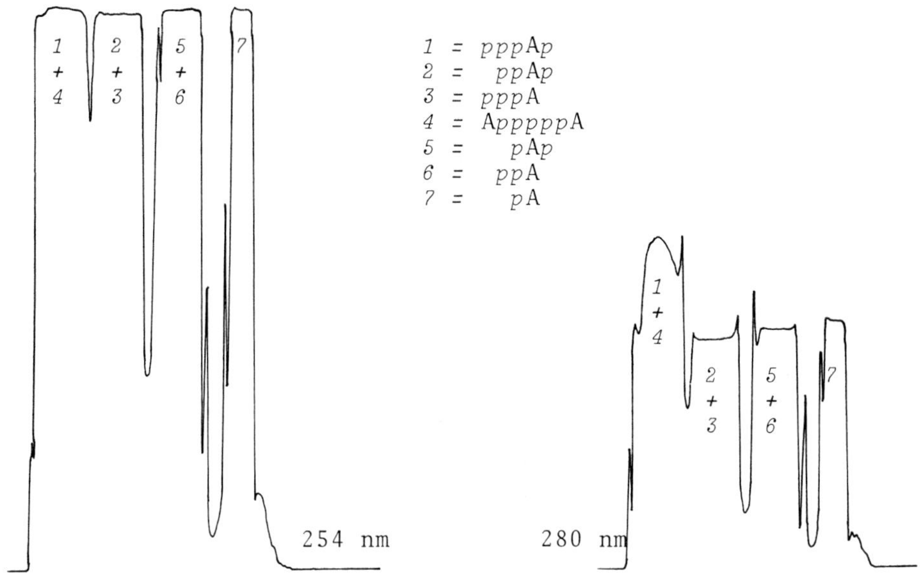



Figure 1: Isotachopherograms of mixtures of nucleotides as indicated with a leading electrolyte pH of 3.9. Detection was by
UV-signals at 254 and 280 nm, as shown. Under these analytical
conditions, mixed zones are formed between nucleotides with
the same number of phosphate groups, ie. between *ppAp* and *pppA*
and between *pAp* and *ppA*. *ApppppA* forms a mixed zone with *pppAp*
whereby this zone is not perfectly rectangular at 280 nm.

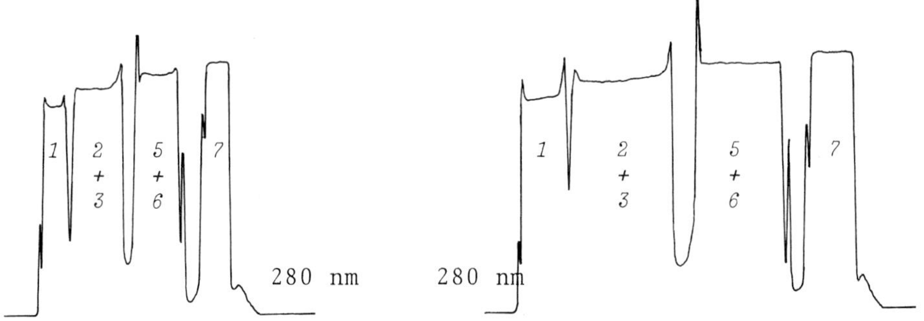



Figure 2: Isotachopherograms of nucleotide mixtures with a
leading electrolyte pH of 3.9. UV detection was at 280 nm. The
mixture is similar to that in Figure 1, except that *ApppppA*
has been excluded. Twice as much sample was injected in the
right-hand picture compared with the left-hand picture to establish which components derive from impurities in the buffer.

802

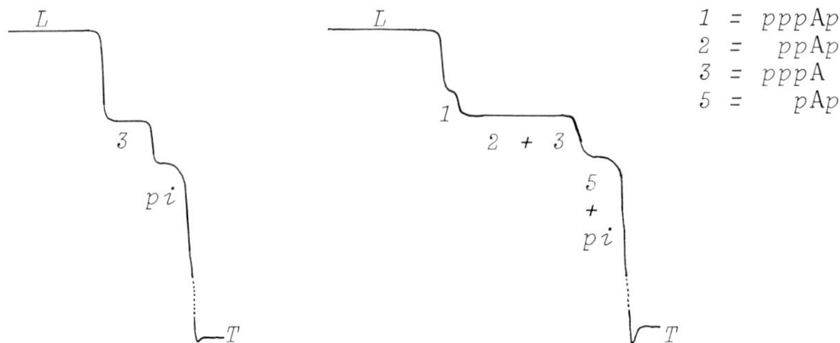


Figure 3: Separations with a leading electrolyte pH of 5.3.
Only conductivity signals are shown, since the counter ion,
pyridine, absorbs in the UV region. The left-hand isotachopherogram shows the separation of *pppA* from inorganic phosphate (*pi*). In the right-hand isotachopherogram, *pppAp*, *ppAp*
and *pAp* are present in addition to *pppA* and *pi*. Only one
additional zone is observed, owing to the fact that mixed
zones are formed between *ppAp* and *pppA* and betwwen *pAp* and *pi*.

characteristic of all measurements carried out at a wavelength
of 280 nm is the "spikes" obtained at the edges of the major
UV-absorbing zones. These apparently derive from the buffers,
since they do not increase in size on injection of more sample.
The influence on the order of separation on increasing pH to
5.3 is demonstrated in Figure 3. Since pyridine was employed
as the common counter ion, no UV-measurements were made. The
conductivity signals show that, in contrast to pH 3.9, *pAp*
and *ppA* are separated one from another, although *ppAp* and
*pppA* still form a mixed zone. A grave disadvantage of these
operating conditions is that *pAp* now forms a mixed zone with
inorganic phosphate.

A relatively minor increase in pH of the leading electrolyte
to 5.7 has a drammatic influence on the separation of the
nucleotides, as demonstrated in Figure 4. At 254 nm, all seven
substances are resolved from one another. Inorganic phosphate
forms a discrete zone between *pAp* and *ppA*. Under these conditions, however, the transmission signals of all the nucleotides

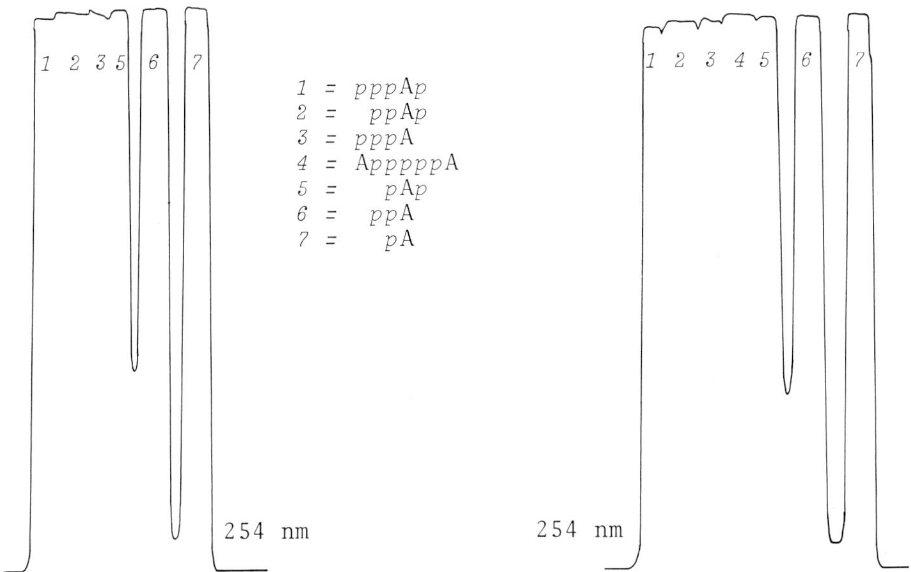



Figure 4: Isotachopherograms showing UV-signals at 254 nm with the mixtures as indicated. The separation was carried out in a capillary of length 61 cm, with a leading electrolyte pH of 5.7. The two pictures only differ in the presence (right-hand) or absence (left-hand) of ApppppA. Since the transmission signals obtained are similar for all the nucleotides at this wavelength, non-UV-absorbing spacers were employed to improve the resolution (see Figure 6).

Figure 5: Isotachopherograms equivalent to those shown in Figure 4, except that detection was at 280 nm. At this pH of 5.7, no mixed zones are formed between any of the nucleotides studied. The presence of the "spikes" provides good definition between the zones.

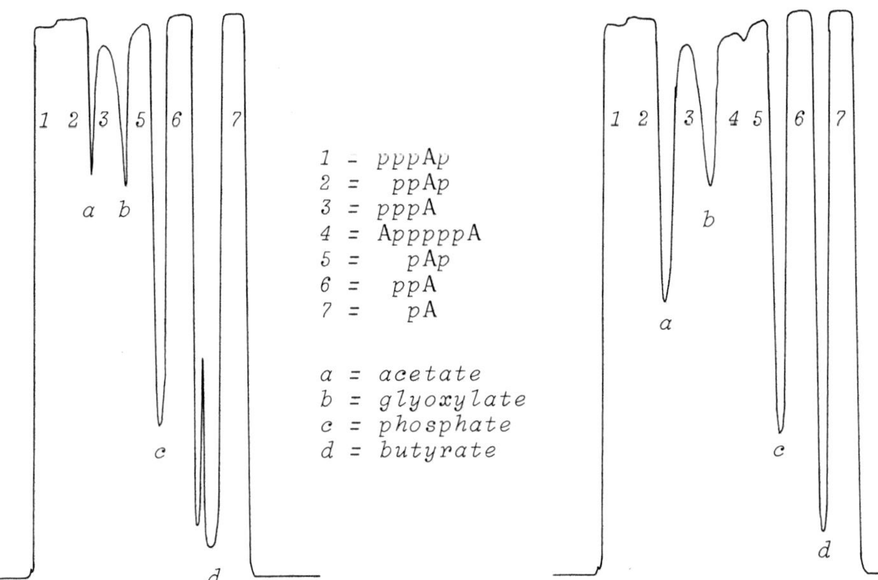



Figure 6: Separation of nucleotides with a leading electrolyte pH of 5.7. The conditions are similar to those in Figure 4, except that non-UV-absorbing spacers have been added to improve the resolution of the picture. The spacers run as indicated by the italic letters. In some cases, these spacers form partial mixed zones with the nucleotides.

are rather similar. Better differentiation is obtained at 280 nm, as shown in Figure 5. Again the "spikes" are observed at the edges of the zones, which do not detract from the quantification of the nucleotides. Improved resolution can be obtained at 254 nm by the addition of non-UV-absorbing spacers to the nucleotide mixture, as depi cted in Figure 6. Of the substances listed in the materials and methods, succinate was found to migrate before pppAp. Pyruvate migrates between the zones 1 and 2 in Figure 6, but tends to mix with the pppAp zone, and was thus not employed as a spacer. Acetate is shown as a suitable spacer between ppAp and pppA. Lactate and glyoxylate ran between zones 3 and 4, whereby the latter provided the better separation. Both phosphate and thiophosphate run as shown in Figure 6 between zones 5 and 6. The carboxylic

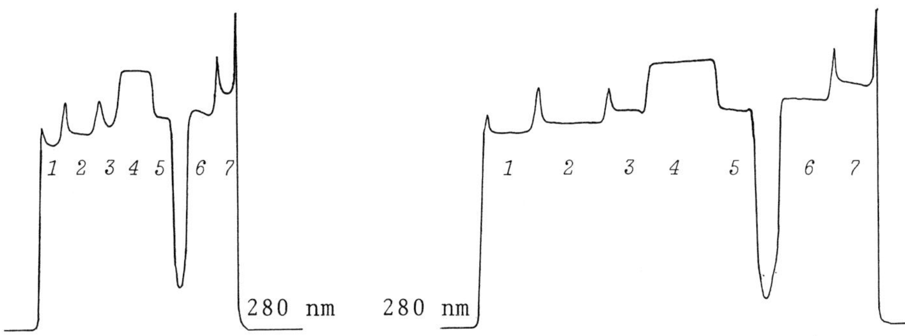

Figure 7: Isotachopherograms run at pH 5.7 (leading electro-lyte) in a 61 cm capillary, with detection at 280 nm. In the right-hand picture, twice as much sample has been injected. The "spikes" are hardly increased in size, so that these do not derive directly from the sample. The marker spikes can be useful to provide exact definition between two UV-absorbing zones of similar transmission. *ApppppA* exhibits higher trans-mission than the other nucleotides, presumably owing to the two adenine residues in the molecule. The separation time in this case was of the order of 30 minutes.

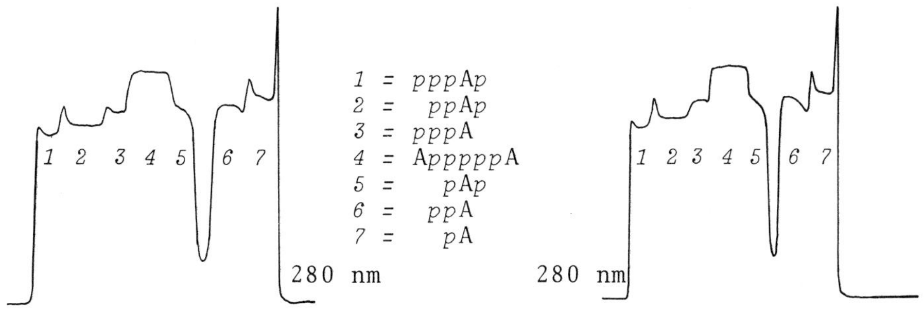

Figure 8: Isotachopherograms run under conditions similar to those in Figure 8, except that a capaillary of length 23 cm was employed. In the left-hand picture, conditions were set such that the analysis time was also of the order of 30 mins. In the right-hand picture, the total analysis time was only of the order of 5 mins. The resolution obtained is inferior to that in the 61 cm capillary. However, in many cases, the high speed separation shown in the right-hand picture may be ade-quate for practical purposes. Analytical conditions are given in the materials and methods. Further discussion is provided in the text.

acids, propionate, butyrate, pentanoate and hexanoate all migrated between $ppA$ and $pA$. There was no particular preference between any of these. Glutamate and aspartate also run between $ppA$ and $pA$.

Conditions of current and voltage and capillary length were varied considerably at pH 5.7, to investigate the possibility of rapid analyses. Although a 61 cm capillary was generally employed, as shown in Figure 7, attempts were made also using a capillary of length 23 cm. Of course, the load capacity of the shorter capillary is more limited. Conditions were first set to give a similar run time of around 30 minutes, as shown in the left-hand picture of Figure 8. The resolution was definitely inferior to that obtained in the 61 cm capillary. However, conditions were also applied under which a complete run time of the order of 5 minutes was necessary. The steady state had probably not been completely achieved, but the separation may be considered adequate for many practical purposes.

Finally, conditions of leading electrolyte were further varied up to pH 6.2. The isotachopherograms in Figure 9 show that these conditions were inferior to those at pH 5.7. At pH 6.2, $pppA$ forms a mixed zone with $pAp$, although the other components of the mixture were separated one from another. Figure 9 also demonstrates how the combination of UV- and conductivity-detectors can be exploited to ascertain whether the steady state has been achieved. Since these two detectors are situated some distance apart in the capillary, identical zone lengths obtained from the two signals help to establish suitable separation conditions. The differential conductivity signal is useful for quantification of zone lengths.

The present work shows that the operating conditions in analytical capillary isotachophoresis are critical for the quality of the separation. The conditions finally chosen will depend

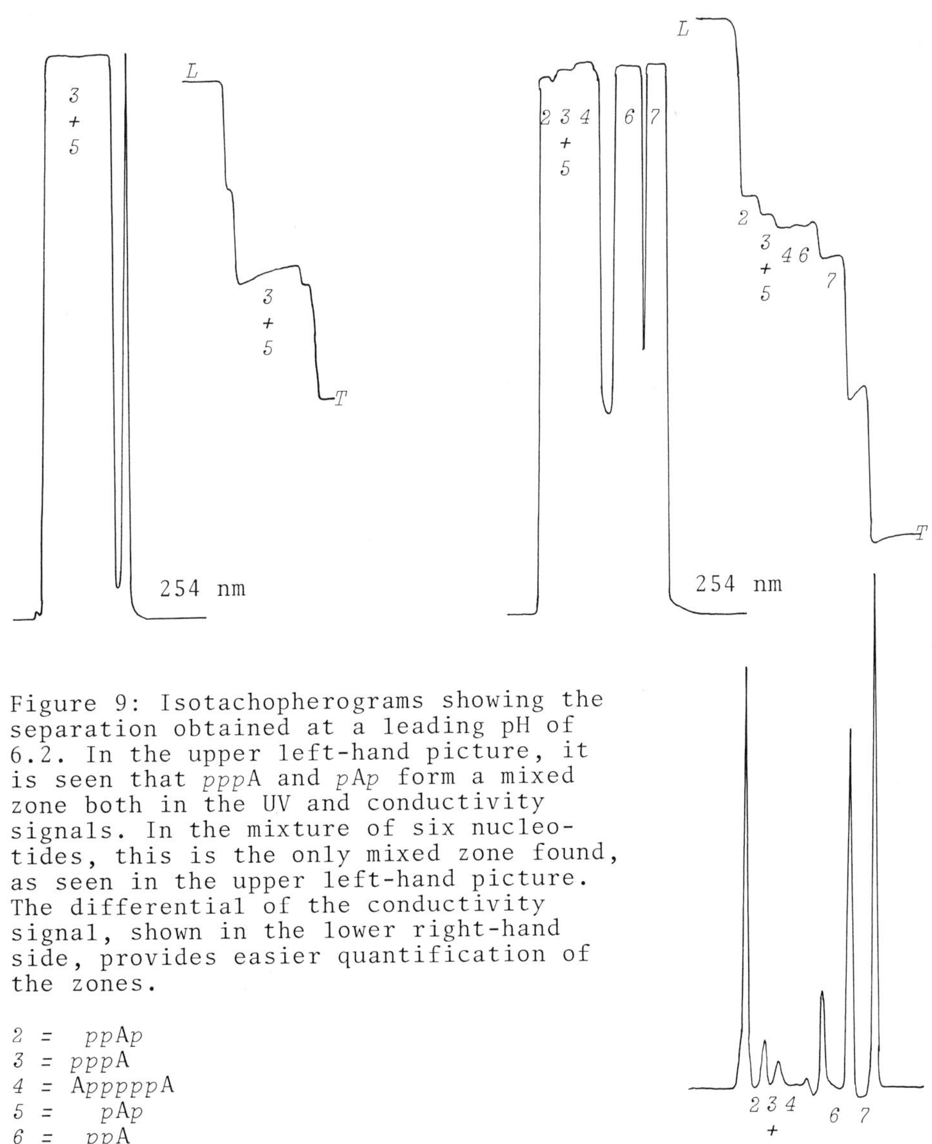



Figure 9: Isotachopherograms showing the separation obtained at a leading pH of 6.2. In the upper left-hand picture, it is seen that *ppp*A and *pAp* form a mixed zone both in the UV and conductivity signals. In the mixture of six nucleotides, this is the only mixed zone found, as seen in the upper left-hand picture. The differential of the conductivity signal, shown in the lower right-hand side, provides easier quantification of the zones.

2  =  *ppAp*
3  =  *pppA*
4  =  *AppppppA*
5  =  *pAp*
6  =  *ppA*
7  =  *pA*

on the nature of the sample, and on possible components of the mixture which should not migrate in the system, such as enzyme proteins etc. The operating conditions can also be varied to avoid migration of one or more of the nucleotides. This can be

useful if some of the nucleotides are present in great excess
over those of primary interest for the analysis. Thus, for
example, hexanoate could be chosen as a terminator at pH 5.7
if it is intended that AMP should not migrate in the system.

The results of the present analysis can be summarised by the
following scheme, whereby nucleotides separated by a diagonal
stroke form mixed zones under those conditions:

| pH 3.9 | pH 5.3 | pH 5.7 | pH 6.2 |
|---|---|---|---|
| $pppAp$ | $pppAp$ | $pppAp$ | $pppAp$ |
| $pppA/ppAp$ | $pppA/ppAp$ | $ppAp$ | $ppAp$ |
|  |  | $pppA$ |  |
| $ppA/\ pAp$ | $pAp$ | $pAp$ | $pppA/pAp$ |
| $pA$ | $ppA$ | $ppA$ | $ppA$ |
|  | $pA$ | $pA$ | $pA$ |

Acknowledgments

Thanks are due to the Gesellschaft der Freunde der Medizinische
Hochschule Hannover, and to LKB Instrument GmbH, Gräfelfing
for their kind support.

The presentation of this work was made possible by a travel
grant from the Deutsche Forschungsgemeinschaft.

References

1.   Holloway, C.J., Lüstorff, J.: Electrophoresis 1, 129-136
     (1980)

2.   Everaerts, F.M., Beckers, J.L., Verheggen, T.P.E.M.:
     Isotachophoresis, Theory, Instrumentation and Application
     J. Chromatogr. Library, Vol. 6, Elsevier, (1976)

3.   Lüstorff, J., Holloway, C.J.: in ITP '80, ed. Everaerts,
     F.M., Elsevier, (1981) in press.

4.   Holloway, C.J., Anhalt, E., Lüstorff, J.: in Elektrophor-
     ese Forum '80, ed. Radola, B.J., Proceedings Munich,
     359-366 (1980)

5.   Michelson, A.M.: Biochim. Biophys. Acta, 114, 460-468
     (1966)

# ISOTACHOPHORETIC ANALYSIS OF HIGH DENSITY LIPOPROTEINS

Marina Bojanovski and Christopher J. Holloway
Institutes of Clinical Biochemistry and Physiological
Chemistry, Medical School, D3000 Hannover, Federal Republic
of Germany

## Introduction

The clinical problem of arterio-sclerosis and resulting
cardiac infarction is integrally connected with amounts and
composition of lipoprotein in the organism. Both from the
clinical and from the biochemical point-of-view, therefore,
there is an increasing requirement for sensitive and reliable
techniques to analyse these proteins. The present work is a
preliminary investigation of the applicability of analytical
capillary isotachophoresis to lipoproteins and their delipidated derivatives, with particular reference to samples of
high density lipoprotein from members of a family in northern
Germany with an extremely rare congenital hyper-alpha hyper-
beta lipoproteinaemia.

## Materials and Methods

Patient material:
The patients tested in this pilot study were two female
members (M.B. and G.G.) of a family with an exceptionally rare
hyper-$\alpha$ hyper-$\beta$ lipoproteinaemia. As a reference, twelve
normal persons in the age range 20-40 years were taken. Blood
was collected from the volunteers, who had not eaten for the
previous 12 hours. The blood was collected in serum monovettes
and allowed to clot, and the serum was separated for the

preparation of the high density lipoprotein (HDL) fractions.

Preparation of HDL fractions:
5 ml of serum were adjusted to a density of 1.073 g/ml by the addition of KBr and centrifuged for 24 hours at 250 000 x g in a Ti 70.1 rotor (Ultracentrifuge Beckman L5-65). The floatation layer, containing principally the low density and very low density lipoprotein fractions, were cut from the cellulose nitrate tubes and rejected from the present study. The bulk liquid was adjusted to a density of 1.21 g/ml with KBr and a further centrifugation was carried out for 24 hours at 250 000 x g. The floatation layer at this density contained the high density lipoprotein (HDL), which was cut from the bulk liquid, diluted to an appropriate volume for the tubes with KBr solution at 1.21 g/ml, and further centrifuged under the above conditions for 24 hours in order to purify the HDL from albumin and other serum proteins. The final floatation layer was cut from the tubes, and divided into two portions (a) for analysis as lipoprotein and (b) for delipidation. Approximately half of the total floatation layer (1-2 ml) was dialysed against 5 litres of an aqueous solution of triton X-100 (1%) overnight to remove KBr from the samples. The HDL suspension was then concentrated to half its original volume by dialysis against PEG 20 000 molecular weight. The protein in the HDL samples was determined by the method of Lowry et al modified according to Wang and Smith [1] for samples which contain the detergent triton. The final protein concentration was of the order of 1.5-2 mg/ml.

Delipidation of the HDL samples:
The remaining portions of the final floatation layers were dialysed against distilled water, and were then lyophilised. The dried samples were delipidated by solvent extraction with a 2:1 mixture of chloroform and methanol. The amount of this solvent mixture employed for each extraction was 1 ml per mg protein isolated. Four extractions were carried out, each for

one hour at 4$^{\circ}$C. Between extractions, the samples were left to stand at -20$^{\circ}$C for optimal stability. The delipidated samples were dried under a nitrogen stream, and were suspended in a 1% aqueous solution of triton X-100 to a final concentration of 1.5-2 mg/ml, as determined by the modified Lowry method.

Isotachophoretic analyses:
The analyses were performed on the LKB Tachophor 2127, fitted with a PTFE capillary of length 63 cm and internal diameter 0.5 mm, thermostatted at 20$^{\circ}$C. A leading electrolyte system of 5 mmol/1 MES, 10 mmol/1 ammediol and 1% triton X-100 was used with a pH around 8.95. The terminating electrolyte was 10 mmol/1 EACA, 10 mmol/1 ammediol, adjusted to pH 11 by the addition of saturated Ba(OH)$_2$ solution. The separation conditions were initially constant voltage (21 kV) until the current had dropped to 50 µA. The instrument was then reset to constant current conditions (50 µA) for the remainder of the run, the samples passing the detectors between 23 and 28 kV. Detection was by thermal signals and UV at 254 or 280 nm. The lower wavelength proved more satisfactory for this application. The total analysis time was of the order of 35 minutes. The amount of lipoprotein or apolipoprotein injected into the instrument was of the order of 5-10 µl (equivalent to about 10 µg protein). Three basic types of analysis were carried out: (1) protein samples without additions, (2) protein samples with ampholytes and (3) protein samples with ampholyte and amino acid markers. The most satisfactory ampholyte proved to be LKB Ampholine$^{®}$ pH range 8-9.5. 1µl of a 1% solution was satisfactory. The amino acids were used to provide reproducible markers between zones. These aminoacids were injected individually to give zone lengths of at least 30 seconds, and the heights of the thermal plateaus relative to the terminator EACA were measured. As described in the results, a mixture of seven aminoacids was selected to provide the spacer markers. The UV-pattern of the proteins from the patients was compared with those from the normal controls.

812

Results

Apart from ampholyte spacers (1% Ampholine® pH range 8-9.5),
amino acids were employed as markers for the isotachopherograms. As described in the methods, sufficient amino acid was
injected to obtain a stable thermal plateau of at least 30
seconds duration. From these data, the reference unit value
(RU-value) was calculated from the following expression:

$$RU = \frac{h_X - h_L}{h_T - h_L} \cdot 100$$

where $h_X$, $h_L$ and $h_T$ are the thermal step heights of the amino
acid, leading electrolyte and terminating electrolyte, respectively. The RU-value is thus expressed as a percentage relative to the terminating signal. The data for the 9 amino acids
employed are summarised in Table I.

Table I: Reference unit values for 9 amino acids employed as
         spacers, calculated as described in the text.

| Amino acid | Abbreviation | RU-Value |
|---|---|---|
| Glycylglycine | Glygly | $3.7 \pm 0.5$ |
| Asparagine | Asn | $11.7 \pm 0.3$ |
| Serine | Ser | $16.8 \pm 0.4$ |
| Threonine | Thr | $17.6 \pm 0.4$ |
| Glutamine | Gln | $22.3 \pm 0.6$ |
| Methionine | Met | $24.5 \pm 0.7$ |
| Glycine | Gly | $30.6 \pm 0.8$ |
| Valine | Val | $42.7 \pm 0.8$ |
| ß-Alanine | Ala | $70.4 \pm 1.0$ |

The ampholyte yielded a linear increase in thermal
signal in the range of RU-value 13-68%

UV-signals at 254 nm

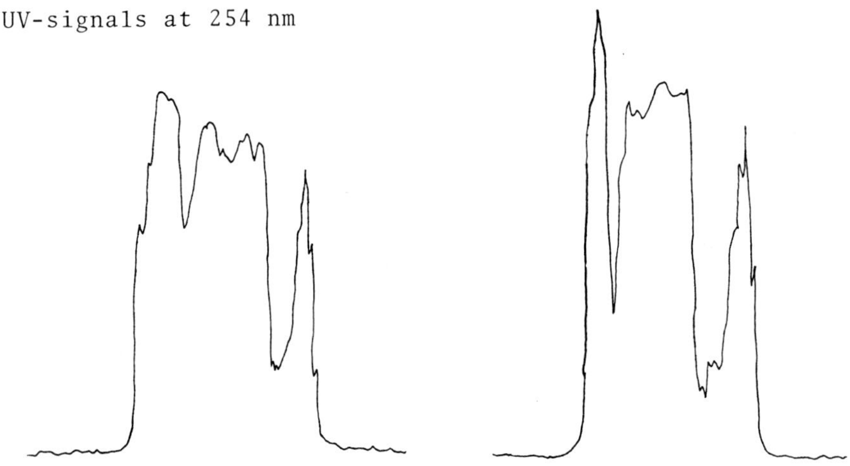

Figure 1: Isotachopherograms of HDL fractions from a normal
person, and a patient with hyper-alpha hyper-beta lipoproteinaemia (left and right pictures, respectively).
The samples employed were the HDL fraction before delipidation.
Around 10 µl of the fraction were injected, equivalent to
17 µg of protein. No ampholyte or amino acid spacers were
added. The left hand side of the isotachopherograms represents
the highest mobility.


Between 15 and 20 µg of HDL protein proved satisfactory for
the isotachophoretic analysis. The HDL fractions from the two
pathological cases of hyper-alpha hyper-beta lipoproteinaemia
and the twelve normal controls were injected individually into
the instrument without any further spacer additives. Examples
of the results are shown in Figure 1. The patterns obtained
between the normal persons and pathological cases differed,
whereas the twelve controls themselves were similar, and the
two patients gave virtually identical isotachopherograms. In
all cases, three main UV-absorbing regions were detected,
obviously made up of several components. In order to resolve
the picture, spacers were employed. Figure 2 shows the influence of various spacers on the isotachopherogram of HDL from
a normal control. Although several ampholytes were tested, a

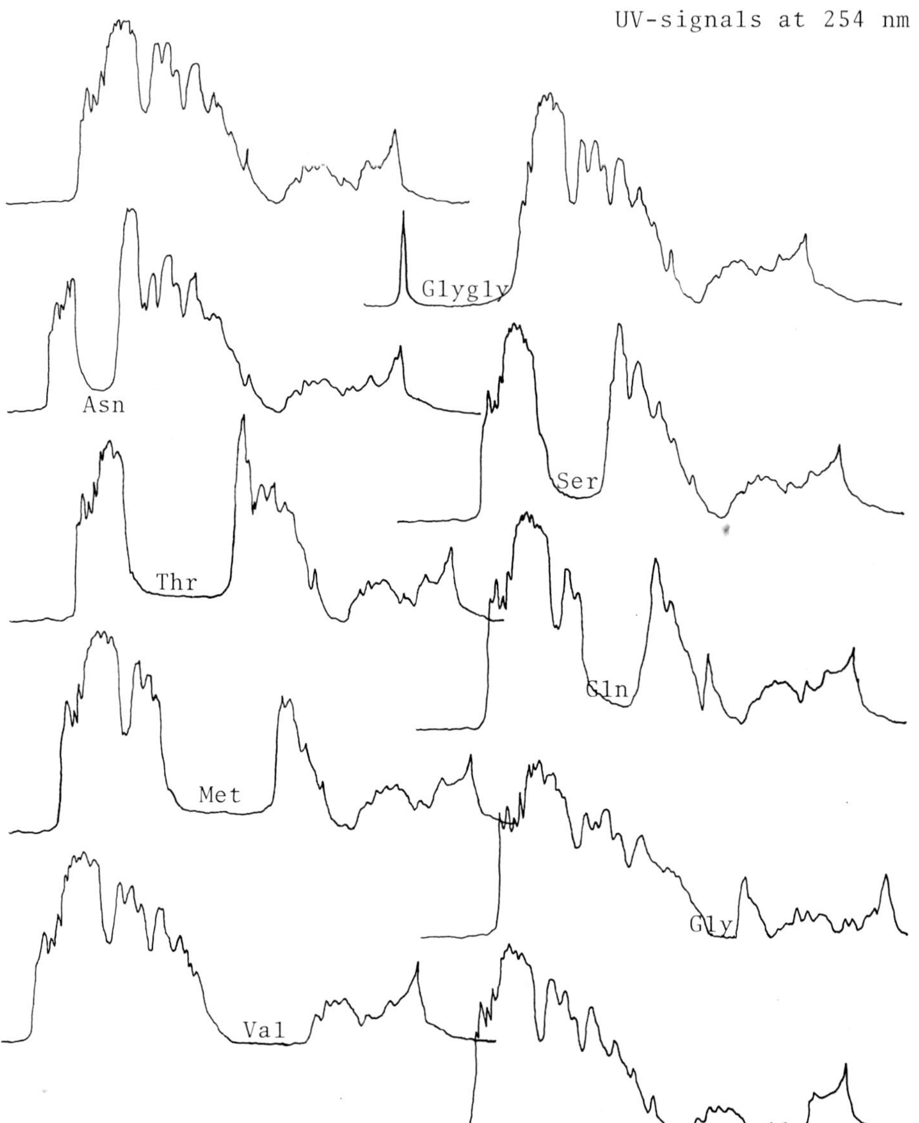


Figure 2: Isotachopherograms of HDL from a normal person with added ampholyte and amino acid spacers.

9µl of the HDL fraction were spaced with 1µl of ampholyte 1% pH range 8-9.5 (upper left picture). The remaining pictures show where amino acids space. In addition to ampholyte, about 10 nmol of each amino acid were injected.

1% Ampholine solution, pH range 8-9.5 gave the best result. Various amino acids and the peptide glycylglycine were also added, to provide specific spacers. Thus, indirectly, reference unit values for various HDL zones could be estimated from the position of those zones relative to the specific amino acid spacers.

Again, the isotachopherograms of the HDL fractions from the twelve control persons with ampholyte and amino acid spacers were similar, but differed from the patterns obtained from the two pathological cases. However, it turned out that the pattern obtained with HDL altered with ageing of the samples. Thus, delipidated apo-HDL samples were employed for the further work. Apo-HDL derived isotachopherograms were very similar to those obtained with the fresh HDL samples, so that we must concluded that a large degree of delipidation is effected by the triton detergent in which the HDL fractions were stored, and in the buffer (leading) for the isotacho-phoretic separations.

The best overall resolution was obtained using apo-HDL samples together with ampholyte and seven of the specific spacers listed in Table I, namely, glycylglycine, asparagine, threonine glutamine, glycine, valine and alanine. Typical isotacho-pherograms are shown in Figure 3 with these spacers. Again, apo-HDL from the control persons gave virtually identical patterns, differing from those obtained with the samples from the two patients. On the one hand, it was apparent that the relative amounts of the various regions separated by the amino acid markers differed between normal and pathological. Perhaps more significant, however, was the fact that the absolute patterns in some regions of the isotachopherograms were quite dissimilar. The zones between the spacers 1 and 2 (glygly and asn) were quite characteristic, whereby the sharp peak of high UV absorbance in the pathological case was found in both patients, and was reproducibly obtained. Furthermore, a UV-

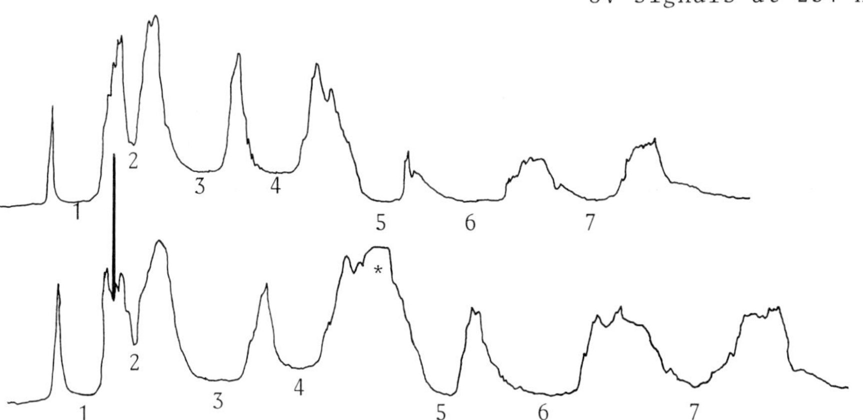

Figure 3: Isotachopherograms of delipidated HDL fractions from a normal person (upper picture) and from a patient with hyper-alpha hyper-beta lipoproteinaemia (lower picture).

9 µl of apo-HDL were injected (around 15 µg protein) together with 1 µl 1% ampholyte, pH range 8-9.5 and 5 nmol each of the following amino acid spacers; 1=glygly, 2=asn, 3=thr, 4=gln, 5=gly, 6=val, 7=ala. Apart from the relative amounts of each protein region, significant differences are seen in the zones between the spacers 1 and 2, and an extra zone is seen in the pathological sample at slightly higher mobility than the spacer 5 (marked with *).

absorbing zone was found in the pathological cases, which is completely absent in the normal controls, occurring at slightly higher mobility than the spacer 5 (gly). Although other differences are seen in the isotachopherograms, the above-mentioned anomalies are the most obvious.


Discussion


The possible clinical applications of isotachophoresis have been discussed by Holloway [2]. Although this technique is attracting increasing interest for the study of plasma proteins researchers in the field of lipoproteins have not, to date, considered isotachophoresis. An earlier report [3] had attempted the analysis of human apo-HDL polypeptides, but the information contained in this work was meagre, the resolution obtained was poor, and the electrolyte systems unsuitable. In that case, ß-alanine was employed as terminator, and we now know from our work, that at least part of the HDL migrates behind this amino acid in this pH range.

The interest in high density lipoproteins has intensified in recent years, since the awareness that this fraction is inversely related to coronary artery disease [4,5,6,7,8]. HDL exhibits a wide heterogeneity in its protein composition. A number of immunochemically defined lipoprotein families have been found in HDL, which have been labelled as LpA, LpB, LpC, LpD, LpE, LpF and LpG, whereby heterogeneity is found even in individual lipoprotein families. The families LpD to LpG are minor components in normal fasting serum. In patients with dyslipoproteinaemias there can be displacement of the proportions of the Lp-families.

Recently, a family was discovered in northern Germany, many of whose members show extremely high levels of cholesterol in combination with hyper-alpha and hyper-beta lipoproteinaemia. A preliminary case report on this family has been presented [9], but to date it has been impossible to state whether this is a completely new syndrome, or simply a combination of the two hyper-lipoproteinaemias. One of the aims of the present work was to examine whether analytical capillary isotachophoresis could provide more information on this point, since conventional methods applied to lipoprotein analysis have not yet provided conclusive data. Initially, the HDL fraction was selected, owing to its better (compared with LDL) solubility.

The results of the present study can be summarised as follows: Analytical capillary isotachophoresis provides a high resolution separation of components of HDL and delipidated HDL. This resolution is improved by the addition of a mobility gradient in the form of carrier ampholytes, and regions of the isotachopherograms can be defined by the addition of specific spacers in the form of amino acids. The pattern obtained is reproducible, both for multiple analyses of single samples, and between different samples from normal control persons. It would seem to be prefereable to work with delipidated HDL, since with HDL itself, changes are observed in the isotachopherograms on ageing of the samples. Significant differences have been observed between the patterns of apo-HDL from normal control persons and members of the family described in this work with the rare congenital hyper-alpha hyper-beta lipoproteinaemia. The differences are not seen throughout the pattern, but are restricted more or less to two distinct regions, whereby at least one zone is detected in the pathological samples which is absent in the normal controls. It is attractive (but premature at this stage) to propose that the pathological HDL contains a protein component which is virtually or even totally absent in normal HDL.

The work described here will be continued, with particular reference to other  members of the northern German family, and other dyslipoproteinaemias.


Acknowledgments

Grateful thanks are due to the Gesselschaft der Freunde der Medizinischen Hochschule Hannover and to LKB Instrument GmbH for support. The pathological samples were kindly supplied by Prof. H. Canzler. We would also like to thank Dr. D. Bojanovski for useful advice and discussions. The cooperation of many colleagues, and particularly Dr. J. Lüstorff is appreciated, as is the expert technical assistance of Mr. B. Büssenschütt, Mrs. R. Goertz and Mrs. I. Haeger.

The presentation of this work was made possible by a travel grant from the Deutsche Forschungsgemeinschaft.

# References


1. Wang, C.S., Smith, R.L.: Anal Biochem 63, 414-417 (1975)

2. Holloway, C.J.: Elektrophorese Forum '80, Proceedings, T.U. Munich, (1980) pp. 225-237

3. Rosseneu, M.Y., Blaton, V., Caster, H., Peeters, H., Kopwillem, A.: Protides of the Biological Fluids, Pergamon Press, 22, 697-700 (1975)

4. Levy, R.I.: Lipids 13, 911-913 (1978)

5. Barboriak, J.J., Anderson, A.J., Rimm, A.A., King, J.F.: Metabolism 28, 735-738 (1979)

6. Pownall, H.J., Morrisett, J.D., Sparrow, J.T., Smith, L.C. Shepherd, J., Jackson, R.L., Gotto, A.M.: Lipids 14, 428-434 (1979)

7. Assmann, G., Schriewer, H., Schulte, H., Oberwittler, W.: Internist 21, 202-212 (1980)

8. Kostner, G.M.: Wiener klin. Wschr. 92 665-672 (1980)

9. Canzler, H., Bojanovski, D.: Abstract, Proceedings of the Arteriosclerosis Meeting, Houston, (1979)

CAPILLARY ISOTACHOPHORESIS AS AN ANALYTICAL MONITOR DURING
THE SYNTHESIS OF THE C-TERMINAL PENTAPEPTIDE OF BOMBININ


Christopher J. Holloway and Klaus Friedel
Institutes of Clinical Biochemistry and Physiological
Chemistry, Medical School, D3000 Hannover, Federal Republic
of Germany


Introduction

Bombinin is a tetracosapeptide component of the defence secretion of the european toad (Bombina variegata) with the primary
structure:
Gly-Ile-Gly-Ala-Leu-Ser-Ala-Lys-Gly-Ala-Leu-Lys-Gly-Leu-Ala-
Lys-Gly-Leu-Ala-Gln-His-Phe-Ala-Asn-NH$_2$
This peptide has surface active properties, which can be explained by its unusual structural characteristics: there is
no C-terminal carboxylic acid in the molecule. The terminal
residue of asparagine is present as an amide. Furthermore, the
C-terminal region of the peptide contains more amino acids
with polar side-chains, whereas the N-terminal direction exhibits more amino acids with non-polar side-chains. As a step
in the synthesis of bombinin, the C-terminal pentapeptide
Gln-His-Phe-Ala-Asn-NH$_2$ has first been prepared, and the
purity of the product and intermediates has been examined by
analytical capillary isotachophoresis.


Materials and Methods

The synthesis of the peptides will be described elsewhere. It
is useful for the present work to know that the protective
group employed during the syntheses (Z-group) was benzyloxy-

carbonyl. The amino acids to be coupled were activated by
coupling to p-nitrophenol as an ester in the case of alanine,
phenylalanine and glutamine. Histidine was coupled by the
carbodiimide method.

Isotachophoretic analyses were performed on the LKB 2127
Tachophor fitted with a PTFE capillary of length 43 cm and
cross section 0.5 mm. Runs were carried out at $10^{o}C$, detection
being by UV-absorption at 254 nm and by thermal signals. The
samples were separated as cations with a leading electrolyte
of 5 mmol/1 KOH corrected to pH 4.9 by the addition of the
common counter ion acetate. 0.25% HPMC was used to reduce electroendosmotic effects. The terminating buffer was 5 mmol/1
ß-alanine corrected to pH 5.1 by the addition of acetic acid.
Runs were carried out at constant current, initially 100 µA
up to a voltage of 15 kV, and then reduced to 50 µA for the
remainder of the analysis. Total analysis time was of the
order of 14 minutes, and the terminator passed the detectors
at a voltage of 10-11 kV. 5-50 nmol of the samples were injected into the instrument for the analyses.


Results


Isotachopherograms of the five peptides involved in the present
study together with the primary structures are shown in Figure
1 as UV-signals at 254 nm. $Asn-NH_2$ is seen as a non-UV-absorbing zone between two spikes arising from impurities in the
buffer systems. The dipeptide $Ala-Asn-NH_2$ is also found as a
single non-UV-absorbing zone. The tripeptide exhibits minor
UV-absorption due to the introduction of the phenylalanine
into the molecule. Both the tetra- and pentapeptides run as
multiple zones, showing considerable impurity, which was not
detectable in thin-layer chromatograms. From the step heights
of the thermal signals (not shown in Figure 1), the reciprocal

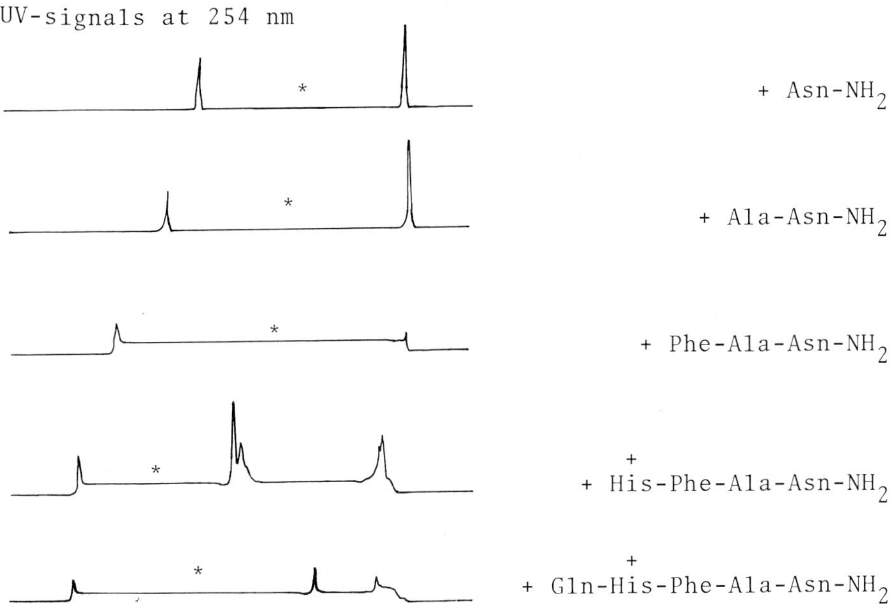



Figure 1: Isotachopherogram of the C-terminal pentapeptide of bombinin and its precursors, together with the structures of the peptides and charge distribution. The zones corresponding to peptides are marked with a *. Other zones are due to impurities in the buffers or samples. Impuruties are first seen in the tetra- and pentapeptides, ie after introduction of histidine into the structure.

reference unit values (RRU values were calculated according to the following expression:

$$ RRU = \frac{h_T - h_L}{h_X - h_L} $$

where $h_T$, $h_L$ and $h_X$ are the step heights of the thermal signal due to terminating electrolyte, leading electrolyte, and sample, respectively. A signal of at least 30 seconds duration was set as the criterion for these calculations. The RRU value obtained for each peptide is listed in Table I.

Table I: Reciprocal reference unit values (RRU), relative to
the terminator, derived from thermal step heights for the
five peptides involved in the present study.

| Peptide | RRU-value |
|---|---|
| $Asn-NH_2$ | $3.25 \pm 0.04$ |
| $Ala-Asn-NH_2$ | $2.08 \pm 0.03$ |
| $Phe-Ala-Asn-NH_2$ | $1.40 \pm 0.02$ |
| $His-Phe-Ala-Asn-NH_2$ | $2.34 \pm 0.03$ |
| $Gln-His-Phe-Ala-Asn-NH_2$ | $2.28 \pm 0.03$ |

Since under the conditions of the operating system (pH ca 5),
the peptides are at least 99.99% in charged form (pK values
of the order 8.8 to 9.7), the decrease in RRU value from the
mono- to the tripeptide must be derived from the increasing
molecular size. The sudden increase in mobility exhibited by
the tetrapeptide over the tripeptide must be due to the additional charge of the histidine residue. At pH 5, at least 90%
of the imidazol group is present in protonated form.

Figure 2 shows the isotachopherograms of some mixtures of the
peptides. The tetra-, tri- and dipeptides are shown each together with the respective precursor. In all cases, a separation is achieved. Moreover, the nature of the isotachopherogram of a mixture of tri- and tetra-peptide demonstrates that
the tripeptide is not the impurity found in the isotachopherogram of the tetrapeptide.


Discussion


Isotachophoresis has been suggested as an analytical tool for
estimation of the purity of peptides [1,2]. In contrast to the
few previous publications pertaining to peptide analysis, the
present work is concerned with an application as a monitor
during the synthesis of peptides for purity control. This is

UV-signals at 254 nm

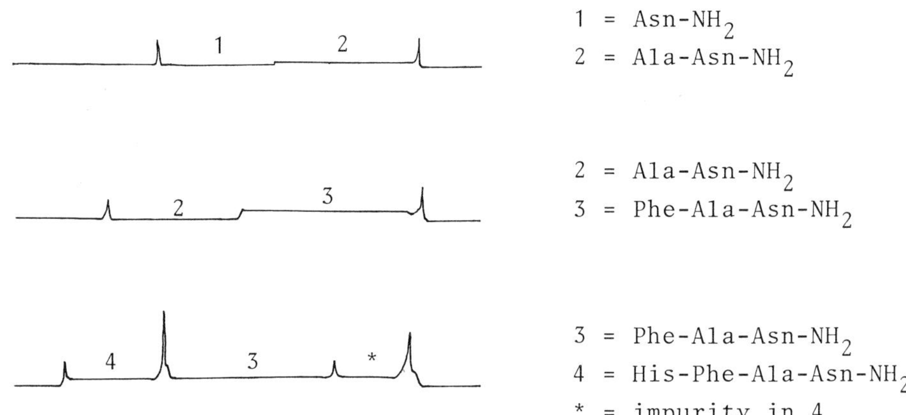



Figure 2: Isotachopherograms showing mixtures of peptides, as indicated. The fact that each peptide is well separated from its precursor demonstrates that isotachophoresis is a valid purity control method in this case. It is particularly interesting to note that the tripeptide is not the impurity detected in the tetrapeptide.

important to the peptide chemist, since each impurity left in a sample when carrying out a further coupling step can lead to undesireable by-products. In the present work, it is shown that the introduction of histidine into the peptide is associated with the production of impurities. In a previous publication [3], we have speculated as to the possible nature of these impurities. The most important consequence for our work at least is the fact that the synthesis-strategy can be modified to attempt to prevent the production of these impurities. Although this work is by no means complete, we can state at this stage that analytical capillary isotachophoresis seems to be useful for the assessment of these peptides, and fills an important gap in this field. Thin layer chromatography lacks the sensitivity required, as was found in practice by the fact that many impurities found in the isotachopherograms were not

detected by thin layer chromatography. The method of isoelectric focussing is also excluded in this case, since the peptides of interest are not zwitterionic in nature.


Acknowledgments

We would like to acknowledge the support of the Gesellschaft
der Freunde der Medizinischen Hochschule Hannover and LKB
Instrument GmbH. Thanks are also due to Professor I.
Trautschold for his advice and encouragement. This work was
carried out with the expert technical assistance of Mrs. I.
Haeger.

The presentation of this work was made possible by a travel
grant from the Deutsche Forschungsgemeinschaft.

References

1.   Kopwillem, A., Chillemi, F., Bosisio-Righetti, A.B.,
     Righetti, P.G.: in Protides of the Biological Fluids,
     ed. Peeters, H. 21, 657-660 (1974)

2.   Kopwillem, A., Moberg, U., Westin-Sjödahl, G., Lundin, R.
     Sievertsson, H.: Aanl. Biochem., 67, 166-181 (1975)

3.   Friedel, K., Holloway, C.J.: Electrophoresis, 2 (1981)
     in press.

AN ALTERNATIVE OUTLOOK ON ELECTROKINETIC CELL SEPARATIONS

Alexander Kolin

Molecular Biology Institute,University of California
Los Angeles,California 90024,USA.

Recent developments in cell biology,molecular genetics,virology
and immunology have increased interest in improvements of cell
separation methods.- Electrophoresis has been used for separation of somatic cells although it is less suitable for such biological materials than for separation of macromolecular ions.
Because the maintenance of a high electric field intensity in
electrophoresis requires an electrolyte of low electric conductivity, the cells must be maintained during the separation process in a medium whose ionic strength may be as low as 1/10 of
that of its normal environment. The artificial environment impairs the viability of cells and leads to cell deterioration
and clumping. Special media must be developed to minimize such
impediments to separation of viable cells ( 1 ).

Ideally one would want to use a method which would maintain
the cells in their natural high-ionic strength medium during
separation. This possibility is offered by electromagnetophoresis ( EMP )( 2,3 ). Although this phenomenon entails migration
of particles in an electric field, it is distinct from electrophoresis. It requires the presence of a transverse magnetic
field and the particles need not be electrically charged.In fact,

unidirectional particle migration can be obtained with an <u>alternating</u> current in a perpendicular alternating magnetic field of equal frequency and phase. Since the migration speed is proportional to the <u>current</u>, rather than electric field strength, it is possible, and in fact advantageous to use a high conductivity suspension medium of high ionic strength. Body fluids would be ideally suited as suspension media.

The EMP force $\vec{F}$ on a spherical particle of radius a is

$$\vec{F} = 2\pi a^3 \left( \frac{\sigma' - \sigma''}{\sigma' + 2\sigma''} \right) \cdot \left[ \vec{B} \times \vec{J} \right] \tag{1}$$

and the migration velocity $\bar{v}$ is

$$\vec{v} = \left[ \vec{B} \times \vec{J} \right] \cdot \left( \frac{\sigma' - \sigma''}{\sigma' + 2\sigma''} \right) \frac{a^2}{3\eta} \, , \tag{2}$$

where $\vec{J}$ is the current density far from the particle, $\vec{B}$ the magnetic field strength, $\sigma''$ the electrical conductivity of the solution, $\sigma'$ the electrical conductivity of the particle and $\eta$ the viscosity of the solution.

Horizontal electromagnetophoresis can be combined with vertical gravitational sedimentation. Both, sedimentation and EMP of a spherical particle obey Stokes' law. The result is an oblique sedimentation at an angle $\theta$ which is given by the ratio of the EMP- and sedimentation- velocities. This ratio turns out to be independent of the particle radius and the fluid viscosity:

$$\tan \theta = \frac{3 \left( \frac{\sigma' - \sigma''}{\sigma' + 2\sigma''} \right) \left| J \times B \right|}{2 \left( \rho' - \rho'' \right) \left| g \right|} \, , \tag{3}$$

where $\rho'$ and $\rho''$ are the densities of the particles and the fluid ,respectively, and g is the magnitude of the gravitational field intensity. This result suggests the possibility of separating such particles in a continuous process independently of their radius and of the viscosity of the medium on the basis of their differences in density and electrical conductivity ( or ion permeability of cell membranes ).

To summarize, The parameters on which EMP particle separation can be based are : overall electrical conductivity, size, shape and cell membrane permeability. In a conductivity gradient, focusing of particles in zones of conductivity matching that of the particles is theoretically possible. With currently available magnetic fields, methods based on this effect should be effective for particles down to the size of cell organelles ( 1 - 2 micra ). Alternating currents in conjunction with a.c. magnetic fields can be used ,offering the advantage of suppressing electrochemical reactions at the electrodes. Cells can be processed in media of high ionic strength, such as blood serum or other body fluids.

References

(1)     Zeiller,K.,Löser,R.,Pascher,G. and Hannig,K.: Hoppe Seylers  Z.Physiol.Chemie 356,1225-1244 (1975).

(2)     Kolin,A. : Science 117, 134-137 (1953).

(3)     " "      :in Medical Physics Vol.3 (O.Glasser,Ed.,Yearbook Publ.Comp. Chicago,1960) pp.268-274.

(4)     Kovalczyk,J.:Thesis,Polytechnic Inst.Gdansk,Poland(1966).

CHARACTERIZATION OF THE SURFACE PHENOTYPE OF ELECTROPHORETICALLY
FRACTIONATED MOUSE LYMPHOCYTES BY FLOW MICROFLUOROMETRY ANALYSIS

Francis Dumont, Robert Habbersett and Aftab Ahmed.
Merck Institute for Therapeutic Research ,Rahway,NJ, USA.

## Introduction

The separation of thymus-dependent (T) from thymus-independent
(B) lymphocytes is one of the most documented application of
preparative cell electrophoresis (reviewed in 1). Yet, little
is known on the possibility of using this technique for separating subpopulations among these two lymphocyte classes. The
studies summarized in the present report aimed primarily at
exploring this possibility. We undertook to define the surface
phenotype of the cells collected in the fractions resulting
from the free-flow electrophoresis of murine spleen and lymph
node cells by taking advantage of two recent methodological
advances which are:(a) the development of monoclonal antibodies
with exquisite specificities for cell surface antigens usable
as markers of lymphocyte subsets, (b) the availability of flow
microfluorometry (FMF) as a rapid mean of quantification on
large numbers of cells of the binding of these antibodies
coupled to fluorescent molecules (2). Fluoresceinated lectins
were additionally used as probes for the peripheral carbohydrate
composition of the fractionated cells (3).

## Materials and Methods

Continuous free-flow electrophoretic fractionation of cell
suspensions was performed in low ionic strength buffer (4)
using a Hannig-type apparatus, Model FF48, Desaga. Cell fractions
were collected in culture medium, enumerated and incubated
under saturating conditions with various monoclonal or conventional antibodies specific for lymphocyte surface markers or
with fluoresceinated lectins. Fluorescein (Fl)- or Biotin (B)
- conjugated antibodies were used. In the latter case, Fl- or
Rhodamine (Rh)-labelled avidin served as a second step reagent.
Single- and dual-parameter FMF analysis of the stained fractions
was done in FACS IV, Becton-Dickinson cell sorter (2). 20,000
or 50,000 viable cells were analysed in each fraction and the
data were stored and processed in a PDP11 computer. The frequency of cells bearing a given marker as well as the magnitude of
cell surface expression of this marker, which is proportional to
the intensity of fluorescence, could thus be determined.

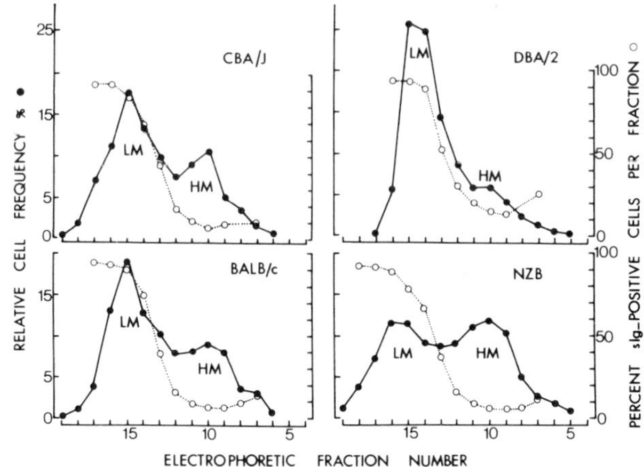



Figure 1: Electropherograms of splenocytes from various mouse
strains.

Results and Discussion

1) FMF analysis of electrophoretically fractionated splenocytes:
In figure 1 are depicted typical cell distribution profiles as
obtained following free-flow electrophoretic fractionation of
splenocytes from various inbred mouse strains. In all cases, two
electrophoretic categories of cells could be distinguished which
were arbitrarily designated as the low-mobility (LM) and high-
mobility (HM) populations. However the proportions of these two
cell types varied from one strain to another. This very probably
reflects the influence of genetic factors located outside the
H-2 region since DBA/2, BALB/c and NZB mice possess the same H-2
haplotype (H-2d). Despite these differences of the electrophoretic patterns, in all four strains examined, the cells stained
with a polyvalent anti-mouse immunoglobulin (Ig) antiserum, a
property of B cells, were considerably enriched in the LM fractions and depleted in the HM fractions。
A more detailed analysis of the surface phenotype of electrophoretically fractionated splenocytes was conducted in CBA/J mice
(figure 2). As expected, the frequency of cells bearing the T
lymphocyte marker Thy1 was low in the LM region but rapidly
rised to 80-85% in the fractions of increasing electrophoretic
mobility (EPM). Also, the distribution curves of the cells positive for each one of three B lymphocyte markers,namely:the Ia
antigen and the heavy chain Ig isotypes mu (IgM) and delta (IgD)
were remarkably parallel to that of Ig+ cells. However, the frequency of Ia+ cells always slightly exceeded that of Ig+ cells
,especially in those fractions intermediate between the LM and

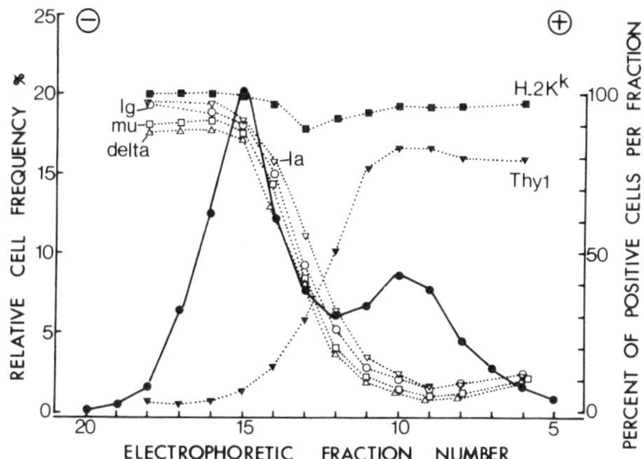


Figure 2: Electropherogram of CBA/J splenocytes (●) and distribution of various B and T cell markers.

HM peaks. The fluorescence intensity profiles obtained with
anti-Ig and anti-Ia antibodies indicated the absence of major
change as a function of EPM in the amount of the corresponding
antigens exposed on the surface of B cells. Two-color fluorescence analysis (5) confirmed these observations and made it
clear that for instance in fraction 13(figure 3A), there was
a small proportion of Ia+ cells possessing low amounts or devoid
of sIg (arrow), while in fraction 16 all cells which were Ia+
were also Ig+. The proportion of double negative cells did not
exceed 5% in this latter fraction.Interestingly, the shape of
the contour map suggested a positive correlation between Ia
and Ig expression.
In the various electrophoretic fractions, the frequencies of
cells bearing the mu or delta isotypes were always slightly
lower than those of Ig+cells (figure 2). In keeping with our
earlier observations (6), the intensities of the fluorescent
stainings for mu or delta were found to vary significantly as
a function of the relative EPM of the fractions. Thus, the mean
fluorescence intensity for mu decreased with the decreasing
EPM of the B cells whereas the mean fluorescence intensity for
delta increased concomitantly. As examplified in figure 3B,
simultaneous staining of the fractions with Fl-anti-mu and with
B-anti-delta antibodies followed by Rh-avidin revealed that most
of the B cells expressed both Ig isotypes on their surface.
However, the cells which stained brightly for mu stained dimly
for delta and vice-versa. In fraction 13, there was a clear-
cut enrichment for the mu-bright delta-dull cells. In contrast,
in the fractions from the cathodic part of the LM peak (e.g.
fraction 16) most of the cells were mu-dull delta-bright. The

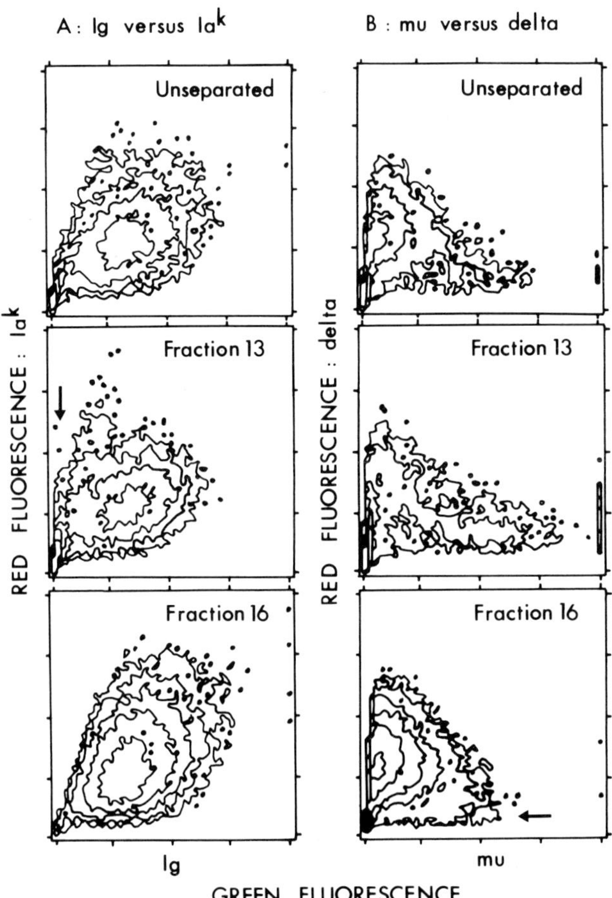



Figure 3: Two-color FMF analysis of CBA/J splenocytes.

contamination of these fractions by double negative cells and
by cells negative for delta but relatively bright for mu
(arrow) was less than 10%. Similar findings were made in BALB/c
and NZB mice.
Therefore, free-flow electrophoresis allows to substancially
enrich for at least two types of splenic B cells which do not
markedly differ by their Ia-Ig phenotype but exhibit distinct
mu(IgM)-delta(IgD) phenotypes: IgM++ IgD+ and IgM+ IgD++.
However, it must be stressed that the IgM++ IgD+ cells recovered in the most anodic LM fractions remain contaminated by up
to 25% of T cells. Thus to obtain these B cells in a more purified form, preparative electrophoresis will have to be

associated with a procedure for removing T cells: e.g. anti-Thy1
and complement cytotoxic treatment. Preliminary experiments have
suggested that these two electrophoretically and phenotypically
distinct B cell subsets also differ in their reactivity to thymus-independent antigens.

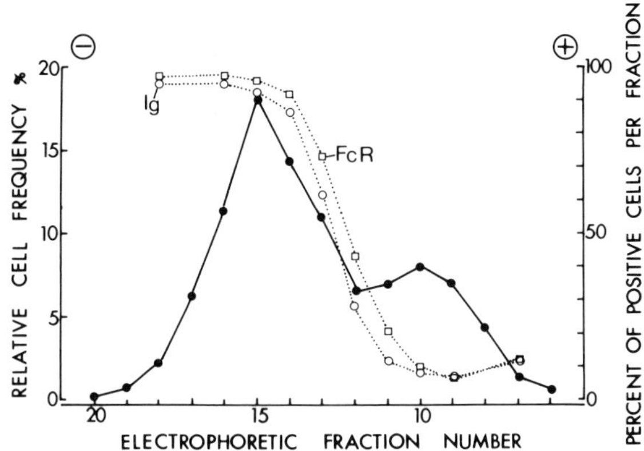


Figure 4: Electropherogram of CBA/J splenocytes (●) and distribution of Ig+ (o) and FcR+ (□) cells.

The expression of the surface receptor for the Fc portion of Ig
(FcR) was also investigated on electrophoretically fractionated
CBA/J splenocytes using a xenogeneic monoclonal antibody ( provided by Dr J. Unkeless). As shown in figure 4, the distribution
of FcR+ cells was similar to that of Ig+ cells with however, like
for Ia, a 10 % excess of FcR+ cells in the intermediate EPM
region. The shape of the FcR fluorescence profile of unseparated
splenocytes denoted some heterogeneity as reflected by a peak
of relatively dull cells and a trail of relatively bright cells.
Although no significant variation in the mean fluorescence intensity as a function of EPM could be seen, the trail of brighter
cells appeared somewhat more pronounced in the anodic part of
the LM peak than in the fractions with lowest EPM (figure 5).
The electrophoretic distribution of another B cell surface
antigen recently discovered by Dr John Kung, was also studied.
This determinant is recognized by a monoclonal antibody (14G8)
and has been shown to be present on a subset of Ig+ cells.
As demonstrated in figure 6, 14G8+ cells were found to display
a distribution pattern clearly different from that of Ig+ cells
and peaked at fraction 14. Moreover, consideration of the ratio
of 14G8+ cells/ Ig+ cells indicated that most of the Ig+ cells
present in the HM region bear the 14G8 antigen as compared to
only 20-30% of the LM Ig+ cells. This suggests that the Ig+ cells
consistently found to contaminate the T-cell-rich fractions represent  a distinct B cell subpopulation.

836

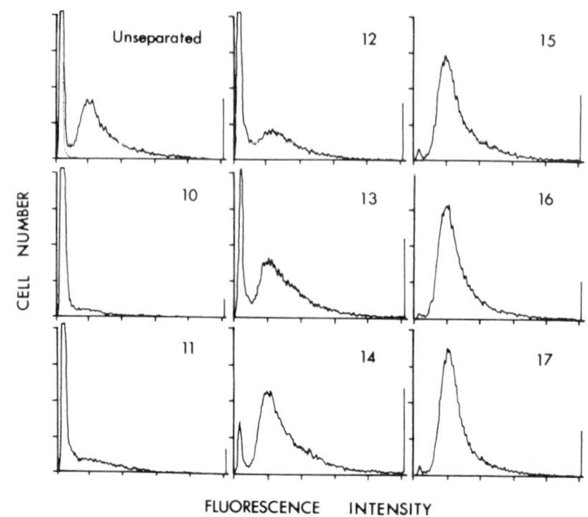


Figure 5: FACS-generated fluorescence profiles of the staining
for FcR of unseparated and electrophoretically
fractionated CBA/J splenocytes.

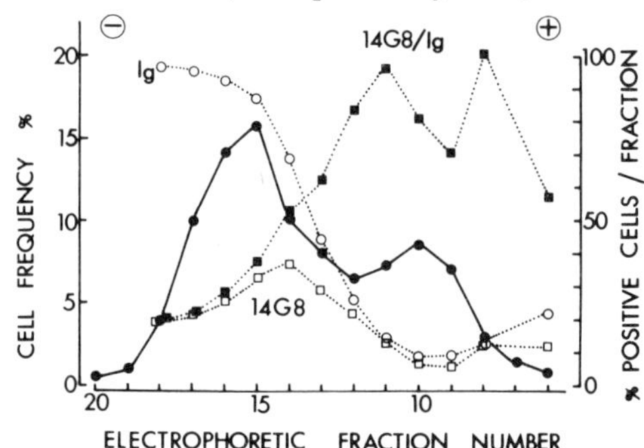


Figure 6: Electropherogram of CBA/J splenocytes (●) and distribution of Ig+ (○) and 14G8+ (□) cells.

It appeared also of special interest to quantify the expression
of H-2 determinants on the various electrophoretic fractions of
splenocytes since these molecules are known to play major functions in cell recognition and lymphocyte interactions. The frequency distribution as a function of EPM of the cells bearing
the H-2K antigen is plotted in figure 2. In the more cathodic
fractions, all the cells were positive for this antigen.In the
LM fractions of increasing EPM (12-13) and to a lesser extent
in the HM fractions, a small percentage of as yet unidentified
H-2K negative cells was detectable. Interestingly, the fluorescence profiles shown in figure 7 demonstrated a broader distribution and a brighter relative mean intensity for the LM

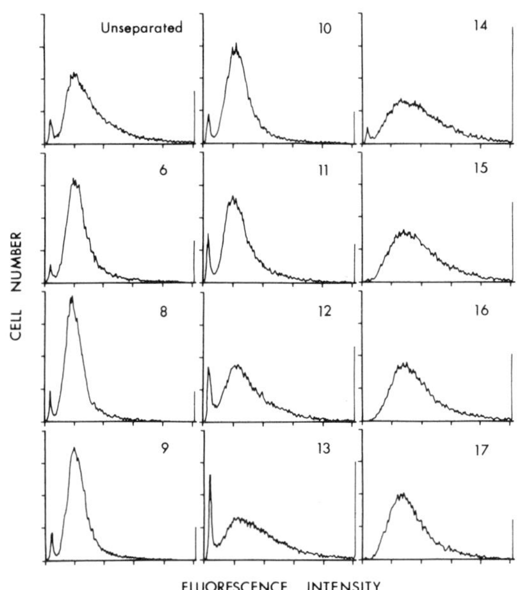



Figure 7: H-2K staining fluorescence profiles of unseparated
and electrophoretically fractionated CBA/J splenocytes

fractions than for the HM fractions. Therefore, B cells are
characterized by a more heterogenous but on average greater
expression of H-2K molecules on their surface than T cells.

2) FMF analysis of electrophoretically fractionated lymph node
cells:
A typical electrophoretic distribution profile of CBA/J lymph
node cells is presented in figure 8. As for splenocytes, a LM
population and a HM population could be clearly distinguished.
However, the relative proportions of these two electrophoretic
cell classes (LM cells: 25%, HM cells: 75%) were markedly different from those recorded in the spleen.
The surface expression of several T cell markers was assessed
on the various fractions harvested. These T cell markers were
present almost exclusively on the HM cells. Concordantly, Ig+
cells were encountered in low frequency in the HM region but
made up the majority of the LM population. The distribution
curves of Lyt1 and Thy1 antigens were closely parallel. However,
in the more anodic fractions the frequency of Lyt1+ cells was
repeatedly found to be higher by 5-10% than the frequency of
Thy1+ cells. This observation converges with a recent report (7)
demonstrating the existence of a small subset of Lyt1+ Thy1-
T cells. The intensity of staining for Lyt1 did not significantly

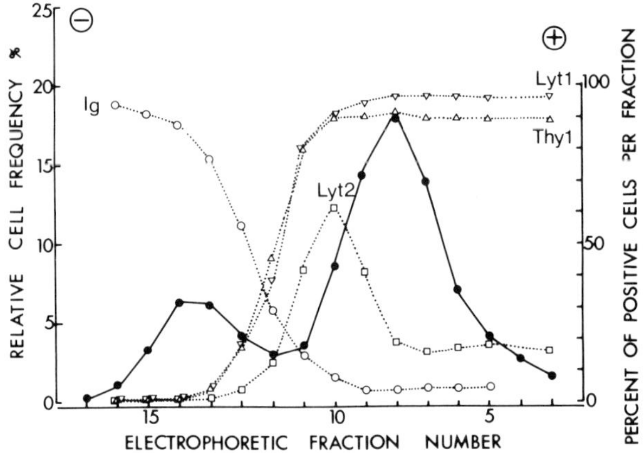



Figure 8: Electropherogram of CBA/J lymph node cells (●) and distribution of various lymphocyte markers.

vary from one fraction to another. However, in the case of Thy1 , a marked shift to brighter fluorescent staining was disclosed in the more cathodic HM fractions (figure 9).

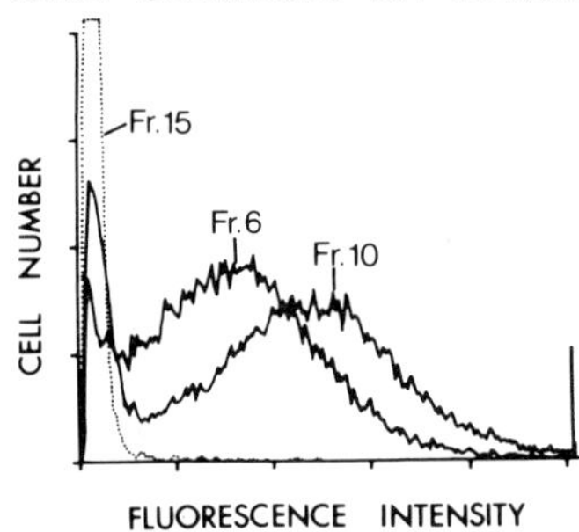



Figure 9: Thy1 fluorescence profiles of three different electrophoretic fractions of CBA/J lymph node cells.

The most interesting finding concerned the distribution of Lyt2+ cells. These cells, which accounted for 20-25% of the unseparated lymph node lymphocytes, were found to make up a distinct peak in the cathodic side of the HM population with a trail extending in the more anodic region. As illustrated by figure 10, two-color fluorescence analysis demonstrated that all the HM cells which are Lyt2+ are also Lyt1+. Noteworthily, the fractions containing the highest frequency of Lyt1+2+ cells were those containing bright Thy1+ cells. The same peculiar electrokinetic behavior of Lyt1+2+ cells as reported here was found in other mouse strains and although not as clearly, with splenocytes. Therefore , free-flow electrophoresis permits the enrichment of two phenotypically distinct peripheral T cell subsets: Lyt1+2+ Thy1++ cells and Lyt1+2- Thy1+ cells. Whether the Lyt1+2+ cells trailing in the anodic HM region represent another subset will

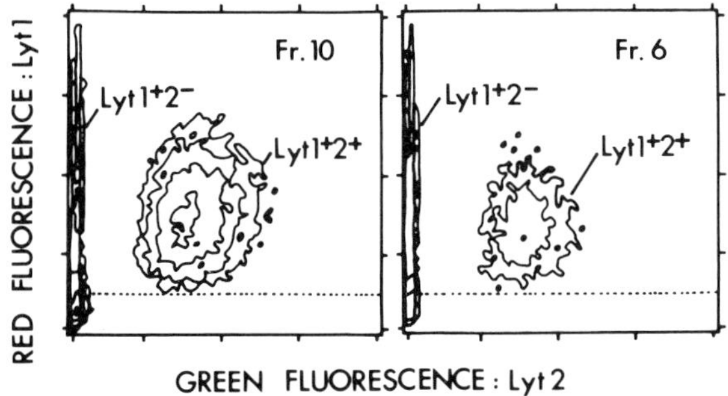


Figure 10: Two-color FMF analysis of two HM fractions of CBA/J
lymph node cells.

have to be determined.
Another approach to characterize the surface phenotype of electrophoretically separated lymphocytes was provided by the study
of the binding of Fl-lectins. In figure 11 are plotted the
frequency distributions of the cells stained with either Maclura
pomifera agglutinin (MPA) which binds to αD-galactosyl residues
or with Limulus polyphemus agglutinin(LPA) which binds to sialic
acid, in the various electrophoretic fractions of CBA/J lymph
node cells. The distribution of MPA+ cells was very similar to
that of Thy1+ cells. This suggests that the MPA binding site
could be used as a marker for murine T lymphocytes. The vast
majority of both LM and HM cells were stained after incubation
with Fl-LPA. LPA negative cells were more abundant in the fractions comprised between the LM and HM peaks. Quite surprisingly
in view of the current belief that the EPM difference between
B and T cells is mainly related to differences in the amount of

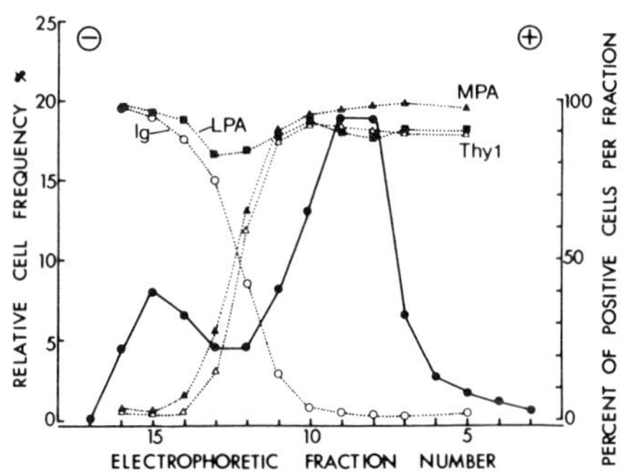


Figure 11: Electropherogram of CBA/J lymph node cells and
distribution of MPA+ and LPA+ cells.

840

sialic acid groups exposed at their periphery ( 8 ), the mean
fluorescence intensity of the LPA staining for LM cells was
consistently found to be higher than for HM cells ( figure 12).
The significance of this seemingly paradoxical finding is
under further investigation.

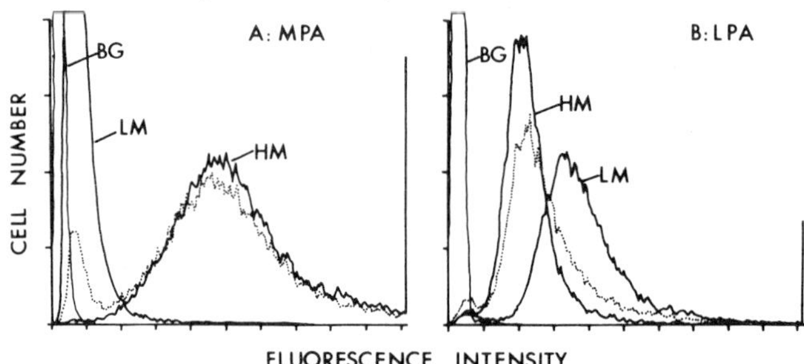


Figure 12: Fluorescence profiles of LM and HM CBA/J lymph
           node cells stained with Fl-MPA (A) or Fl-LPA (B).

References


1. Pretlow,T.G., Pretlow,T.P.: Int. Rev. Cyt. 61, 85-128 (1979).

2. Loken,M.R., Herzenberg,L.A.: Ann.N.Y. Acad.Sci. 254,163 (1975).

3. Sharon,N., Lis,H.: Science 177, 949 (1972).

4. Hannig,K.: Methods in Microbiology 5B,513 (1971).

5. Loken,M.R.,Parks,D.R., Herzenberg,L.A.:J.Hist.Cyt.25,899(1977).

6. Dumont,F.,Ahmed,A.,Habbersett,R.: Clinical applications of
   Cell Electrophoresis, Edit. A.Price, D.Sabolovic,Elsevier(1979)

7. Ledbetter,J.A., Rouse,R.V., Micklem,H.S., Herzenberg,L.A. :
   J. Exp. Med. 152,280 (1980).

8. Nordling,S., Andersson, L.C., Hayry,P. : Science 178,1001
   (1972).

# EVALUATION OF THE ELECTROPHORETIC SEPARABILITY OF TRYPANOSOMA CRUZI PARASITE STAGES


Robert C. Boltz, Jr., Dennis M. Schmatz and P. Keith Murray
Department of Immunology, Merck Institute for Therapeutic Research
Rahway, N.J. 07065, USA


## Introduction

During the life cycle of Trypanosoma cruzi, a series of morphologically distinct stages occur in the insect vector and the final mammalian host (1). These different parasite stages are biochemically and immunologically distinct. In vitro, a range of culture conditions can be used to support the growth of these stages. Isolation of pure parasites from in vitro sources usually results in populations which are morphologically heterogeneous. Various cell separation techniques have been applied to parasitic hemoprotozoa, these have generally been restricted to separating parasites from host cells or cell debris (2-6), and infected from non-infected cells. One of the more successful approaches has been to exploit cell surface charge differences between host and parasite cells to isolate pure populations of parasites by means of anion exchange chromatography (2-5). However, less success has attended efforts to separate individual parasite stages from each other using this technique. To our knowledge the only report of stage purification from heterogeneous mixtures by anion exchange chromatography is a report by Al-Abbassy, Seed and Kreier (7).

Early studies (8-14) using analytical microelectrophoresis indicated surface charge differences both between various stages and species of trypanosomes in various hosts and more recent work by Kreier, Al-Abbassy and Seed (1977) confirmed mobility differences between culture stages of T. cruzi determined by analytical microelectrophoresis (15). In that same year, De Souza et al. (16) characterized the surface charge of the different morphologic forms of T. cruzi by cationic ferritin and microelectrophoresis. Again, significant differences were observed.

842

The authors stated that these differences were of sufficient magnitude to permit separation by preparative electrophoresis.

We are reporting the application of continuous free-flow electrophoresis for the analytical and preparative separation of individual stages of _Trypanosoma_ _cruzi_. Our studies confirm that the various stages of the parasite have a significant surface charge density difference, and that, with the appropriate buffer conditions in our apparatus, successful separations of individual stages can be routinely obtained.


Materials and Methods


1.    Parasites.    _T._ _cruzi_ Y strain, originally obtained from Dr. Felix Pifano, University of Venezuela, now designated MERC 2C (Merck Institute, _Trypanosoma_ _cruzi_ strain 2, from culture) was used in these experiments.    The strain was isolated from a human case of Chagas Disease in Brazil in 1953 by Pereira de Silva and V. Nussenzweig.    Trypomastigote and amastigote stages of the parasite were grown at $37^{\circ}C$ in a tissue culture system using a rat myoblast cell line as host cell as previously described (5).    Axenic culture forms (99% epimastigotes) were grown in Liver Infusion Tryptose media at $27^{\circ}C$ in the presence of 10% $CO_2$ (17).


2. DEAE-Cellulose Purification.    To isolate parasites from myoblast cells and cell debris DEAE-cellulose anion exchange chromatography was conducted as previously described (5).


3. Physicochemical Measurements.    All buffer solutions were made fresh daily. The pH was measured with a Fisher Model 630 pH Meter.    Conductivity was determined with a YSI Model 31 conductivity bridge and the osmolarity with an Osmette Model "S" Osmometer.


4. Continuous Free-Flow Electrophoresis.    Electrophoretic separation was effected in a continuous free-flow electrophoresis apparatus (Desaga FF 48, Desaga GmbH, Heidelberg FRG; Brinkmann Instruments, Inc., Westbury, New York, USA) as described by Hanning _et al._ (18).

In preparing trypanosome suspensions for separation, myoblast cell debris was removed using a phosphate buffer system (PBSG) and DEAE-cellulose chromatography (5). Therefore, the electrophoretic separation buffer of Boltz <u>et al</u>. (19), being similar in composition, was selected for parasite electrophoresis and was adapted to the continuous flow apparatus (see Table 1).

Table 1. Composition of Buffer for Free-Flow Electrophoresis of Trypanosomes

| Component | Chamber Buffer | Electrode Buffer |
|---|---|---|
| KCl | 0.20 g/L | 1.0 g/L |
| $Na_2HPO_4$ | 1.15 g/L | 5.75 g/L |
| $KH_2PO_4$ | 0.20 g/L | 1.0 g/L |
| $MgCl_2 \cdot 6H_2O$ | 0.10 g/L | 0.5 g/L |
| Sucrose | 68.0 g/L | – |
| Dextrose | 10.0 g/L | – |
| pH | 7.4 | 7.4 |
| Conductivity | $1.3 \times 10^{-3}$ MHO/CM @ $22^{\circ}$C | |
| Osmolarity | $318 \times 10^{-3}$ Osmolar | |
| Ionic Strength | I = 0.027 | |

Parasite suspensions were separated at $4^{\circ}$C in a calculated field strength of 110 volts/cm with an approximate spatial resolution of 1.25 mm/fraction.

5. Parasite Fraction Analyses. Fractions of 2 ml were collected for analyses. Visual assessments of parasite numbers, motility and morphology were made in hemocytometer chambers. For detailed microscopical evaluation and for automated cell counting and sizing, parasites were fixed by the addition of an equal volume of 6% glutaraldehyde.

Glutaraldehyde fixed cells were counted using a Coulter ZH Coulter Counter coupled to a Coulter C1000 population accessory and a Coulter X-Y Recorder 4.

844

6. Determination of Infectivity. The infectivity of separated organisms was determined by subcutaneous inoculation into female CF1 mice (Charles River Laboratories). Mice were inoculated within one hour of the separation. Groups of six mice were inoculated with 1,000, 500, 100, 50 and 10 organisms in a volume of 0.2 ml of myoblast culture medium. Thereafter, tail blood wet smears from all mice were examined at regular intervals for the presence of parasites.


Results


1. Electrophoresis Separation Buffer. Recovery of viable organisms may require maintenance of buffer conditions in the physiological range during separation. Thus, the osmolarity and pH were monitored for the phosphate buffer used in these separations (Fig. 1).

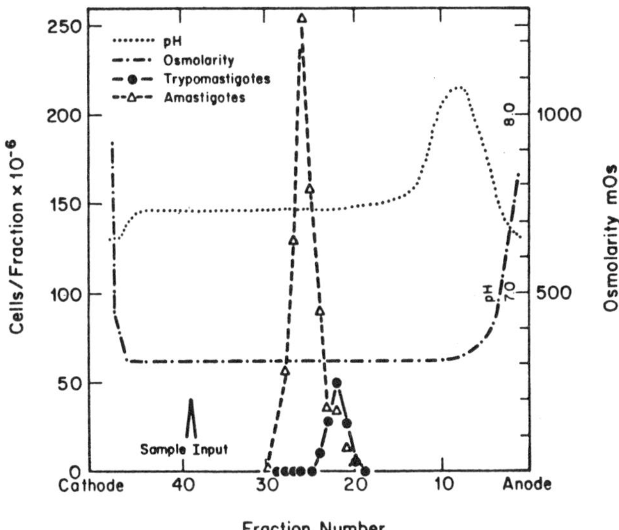


Figure 1. The effect of the electric field on the migration of a heterogeneous population of Trypanosoma cruzi and the physical parameters of the separation buffer.

Under the conditions of the run, a significant increase in buffer pH (to 8.5) was noted occurring around fraction 8. This front was found to migrate ahead of the parasites and the physical properties of the buffer in the parasite zone were within acceptable limits for parasite viability, i.e. pH 7.5 and 310 MOS. Two bands were clearly visible in the separation chamber. Analysis of the collected fractions showed a predominance of trypomastigotes around fraction 22 and amastigotes around fraction 26. There was significant contamination of the trypomastigote peak by amastigotes. The major amastigote peak was essentially pure.

In order to achieve the maximum separation the flow rate was adjusted to .67 mm/sec. This resulted in the leading edge of the trypomastigote peak appearing in the fraction permitting maximum migration but still within physiological limits.

2. Stage Separations. Using the conditions outlined above, good separations were obtained in subsequent runs. Figure 2 shows the average values obtained in three separations. Trypomastigotes were shown to be more electronegative than amastigotes and 80% appeared in fractions 6-10. A predominance of amastigotes (82.4%) occurred in fractions 11-18. It was noted that amastigotes appear to show a greater range of electrophoretic mobility than trypomastigotes. Overall parasite recoveries were good, being invariably above 90% of the starting population.

Figure 3A shows a separation of a sample from a late culture. Cell volume data is shown in Figure 3B as waterfall display. The volume distribution of the starting sample, "S" appears to be a Poisson type distribution of a single population. Fraction 8, however, clearly shows a pronounced shoulder denoting the enrichment of a subpopulation of larger cells. Microscopic examination showed this fraction to be predominantly trypomastigotes. Volume distributions of later fractions (e.g. 11) show that the population of larger cells has disappeared and the smaller cells predominated (amastigotes).

3. Viability and Infectivity. Examination of parasite fractions by phase microscopy showed trypomastigotes to be normal in appearance and vigorously motile. Amastigotes were round, regular and morphologically intact. Infectivity titrations of separated parasites in mice showed separated organisms to be highly infectious

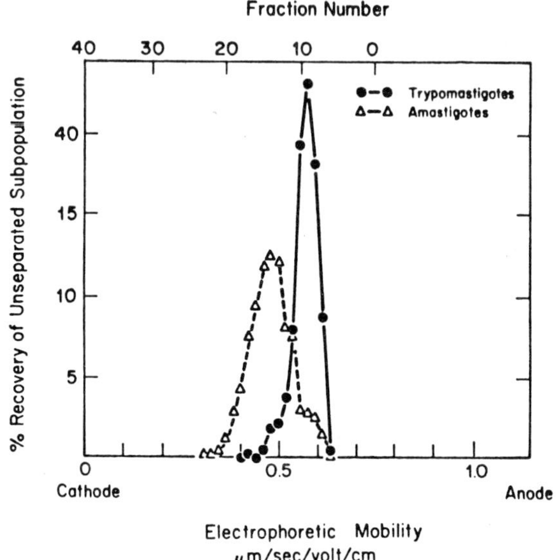



Figure 2. Free-flow electrophoretic distribution of <u>Trypanosoma</u> cruzi subpopulations isolated from myoblast cultures. The profile is an average of 3 experiments and subpopulations were normalized to their respective concentration in the unseparated material.

Table 2. Percentage of Mice Infected with T. cruzi Derived from Myoblast Cultures Before and After Free-Flow Electrophoresis

| | | Trypanosome Population | | |
| No. Parasites Inoculated | No. Mice | Mixed Culture Before Separation | Trypomastigotes* After Separation | Amastigotes* After Separation |
|---|---|---|---|---|
| 480 | (10) | 100 | 100 | 100 |
| 240 | (10) | 100 | 100 | 100 |
| 120 | (10) | 100 | 100 | 100 |
| 60 | (10) | 100 | 100 | 100 |
| 30 | (10) | 100 | 100 | 100 |
| 15 | (10) | 100 | 80 | 100 |

*Trypomastigotes were obtained from electrophoresis fraction 8 and amastigotes from fraction 14.

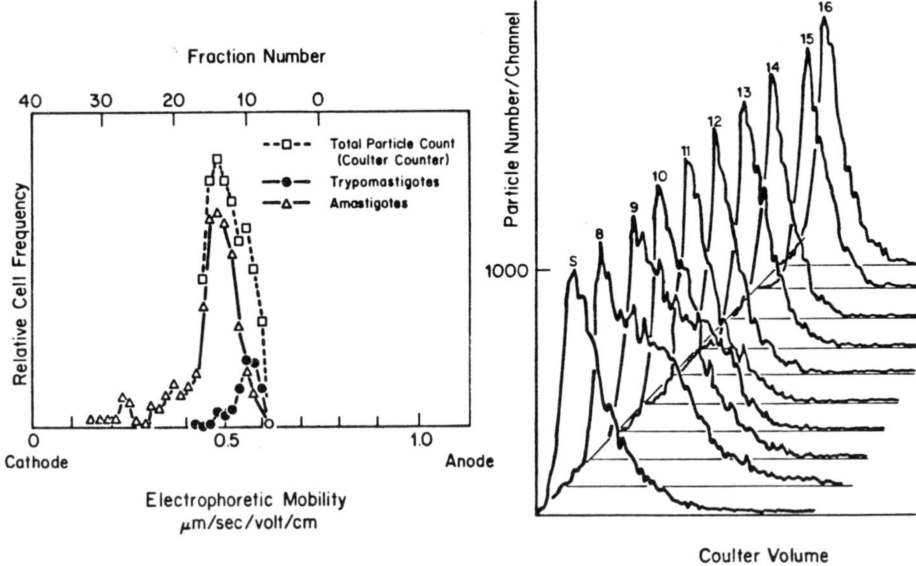



Figure 3A.  Electrophoretic mobility distribution of parasites from a late culture.

Figure 3B.  Coulter volume distributions of separated fractions from the experiment shown in Figure 3A.

(Table 2).  Amastigotes after electrophoresis were equally infectious to the starting mixture and caused 100% infections in mice at a level of 15 parasites/mouse (the lowest dilution tested).  Trypomastigotes were marginally less infective, causing 80% infections at the lowest level tested and 100% at a level of 30 parasites/mouse.

4.    Comparative Electrophoretic Mobility of Epimastigotes.    Axenic cultured organisms comprising 99% epimastigotes were electrophoresed in the continuous free-flow apparatus.   Since it was discovered that this stage had a much higher electrophoretic mobility than either trypomastigotes or amastigotes from myoblast cultures, the curtain buffer flow rate had to be increased to 0.84 mm/sec in order to reduce the residence time in the electric field yet maintain maximum distance separation.   Figure 4 (top) shows that a single peak, centered around fraction 10, was obtained.  In this instance, the recovery was 92%.  In order to determine the

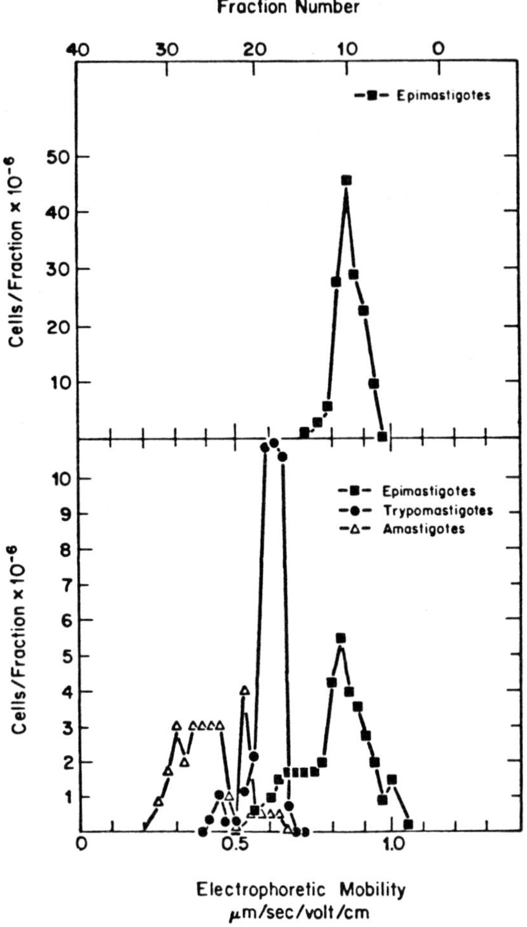



Figure 4. Comparison of the electrophoretic mobility distribution of epimastigotes from axenic culture and trypomastigotes and amastigotes from myoblast cultures.

relative differences in electrophoretic mobility between epimastigotes, trypomastigotes and amastigotes, an artificial population was obtained by mixing myoblast culture stages and axenic stages.

Figure 4 (bottom) illustrates the "analytical" separation which was achieved by electrophoresis of these stages under the flow conditions indicated for epimastigotes. Clear differences in surface charge density can be seen among the stages. Epimastigotes again predominated around fraction 10; trypomastigotes around fraction 18 and amastigotes between fractions 24 and 29. On ascending order of electronegativity amastigotes have least charge, trypomastigotes next and epimastigotes most. In this run, 91% of the epimastigotes appeared in fractions 4 to 16, 92% of trypomastigotes appeared between fractions 16 to 20 and 89% of amastigotes in fractions 21 to 33.

Discussion

Continuous free-flow electrophoresis has been used to analyze the electrophoretic migration distributions of various stages of T. cruzi simultaneously and these studies have confirmed major differences in the surface charge density among stages. The relative mobilities of epimastigotes and trypomastigotes are consistent with those obtained by Kreier et al. (15) on fixed material at high ionic strength (I = .290). The electrophoretic mobility distribution obtained for epimastigotes agrees with that measured by De Souza et al. (16) on T. cruzi, using analytical micro-electrophoresis (same ionic strength I = 0.027), isolated from similar culture conditions. It has been demonstrated that there are sufficient surface charge density differences in amastigotes and trypomastigotes, isolated from myoblast cultures, to permit large scale separation ($> 10^8$ cells/day) without loss in infectivity.

Parasites grown under varied conditions in different media are likely to have qualitative and quantitative biochemical and immunological differences. The continuous free-flow electrophoresis technique described in this paper is able to separate heterogeneous mixtures into pure stages which have been grown under identical conditions and thus allow direct comparisons.

850

## Acknowledgements

One of us (RCB) wishes to thank Dr. K. Hannig for providing facilities in the summer of 1979 during which initial experiments using the buffer (19) in the continuous free-flow electrophoresis apparatus were performed. We also wish to express appreciation to Mr. S. Galuska for providing the epimastigote cultures.

## References

1. Santos-Buch, C.A.: Internat. Rev. of Exp. Path. 19, 61-99 (1979).
2. Alvarenga, N. J., Brener, Z.: J. Parasitol. 65 (5), 814-815 (1979).
3. Danforth, H. D.: The American Society of Parasitologists, Berkeley, California, p. 64 (1980).
4. Lanham, S. M.: Nature 218, 1273-1274 (1968).
5. Schmatz, D. M., Murray, P. K.: J. Parasitol. (in press) (1981).
6. Heidrich, H.-G., Russmann, L., Bayer, B., Jung, A.: Z. Parasitenkd. 58, 151-159 (1979).
7. Al-Abbassy, S. N., Seed, T. M., Kreier, J. P.: J. Parasitol. 58 631-632 (1972).
8. Traube, V. J.: Dtsch. Med. Wochenschr. 31, 1441-1443 (1912).
9. Traube, V. J.: Kolloidchem. Beih. 3, 236-336 (1912).
10. Szent-Gyorgyi, A. V.: Biochem. Zeit. 110, 116-118 (1920).
11. Szent-Gyorgyi, A. V.: Biochem. Zeit. 113, 29-35 (1921).
12. Fischer, F. P., Fischl, V.: Biochem. Z. Band 267, 403-404 (1933).
13. Broom, J. C., Brown, H. C., Hoare, C. A.: Trans. Roy. Soc. Trop. Med. Hyg. XXX (1), 87-100 (1936).
14. Hollingshead, S., Pethica, B. A., Ryley, J. F.: Biochem. J. 89, 123-127 (1963).
15. Kreier, J. P., Al-Abbassy, S. N., Seed, T. M.: Rev. Inst. Med. Trop. Sao Paulo 19 (1), 10-20 (1977).
16. De Souza, W., Arguello, C., Martinez-Palomo, A., Trissl, D., Gonzales-Robles, A., Chiari, E.: J. Protozool. 24, (3), 411-415 (1977).
17. Comargo, E.P.: Rev. Inst. Med. Trop. Sao Paulo 16, 81-87 (1974).
18. Hannig, K., Wirth, H., Meyer, B.-H., Zeiller, K.: Hoppe Seylers Z. Physiol. Chem. 356, 1209-1223 (1976).
19. Boltz, Jr., R. C., Todd, P., Streibel, M. J., Louie, M. K.: Prep. Biochem. 3 (4), 383-401 (1973).

SEPARATION AND CHARACTERIZATION OF HUMAN BONE MARROW CELLS BY DENSITY
GRADIENT ELECTROPHORESIS



Chris D. Platsoucas, Neena Kapoor, Jöern D. Beck, Robert A. Good
and Sudhir Gupta

Memorial Sloan-Kettering Cancer Center, New York, New York   10021



Introduction

Human bone marrow cell suspensions contain pluripotent stem cells,
committed precursor cells of myeloid, erythroid, lymphoid and monocytoid
lineage and mature granulocytes, monocytes and lymphocytes.  Several
methods of cell separation have been employed for fractionation and
characterization of these cells, including isopycnic or equilibrium
density gradient centrifugation (1,2) and sedimentation velocity at unit
gravity (3,4).  Furthermore, centrifugation of human bone marrow cells on
a density cushion of Ficoll-Hypaque (density of 1.077 g/cm$^3$) (5) results
in a cell suspension devoid of reticulocytes or mature erythroid cells,
and containing all the colony forming units <u>in vitro</u> (CFU-c), immature
myeloid and erythroid precursors and mononuclear cells of monocytoid and
lymphoid lineage.

We report here the application of the density gradient electrophoresis
method, for the separation and characterization of human bone marrow cell
suspensions (purified by centrifugation on a Ficoll-Hypaque density
cushion), on the basis of their surface charge.  In the past we applied
the density gradient electrophoresis method for the separation of phenotypically and functionally distinct populations of lymphoid cells in
humans and experimental animals (6-10).

852

## Materials and Methods

Heparinized bone marrow cell suspensions were obtained from healthy adult normal volunteers, by multiple aspirations from the posterior iliac crest. Informed consent was always obtained from the donors according to the declaration of Helsinki. Bone marrow cells suspensions were depleted of mature cells of erythroid lineage by centrifugation on a Ficoll-Hypaque density cushion (5) followed by lysis of remaining erythroid cells by treatment with Tris-buffered ammonium chloride as described elsewhere (11). Density gradient electrophoresis was performed as previously described (12,13). Characterization of the separated cell fractions by cell surface markers and functional tests was carried out by methods described elsewhere (10,11).

## Results and Discussion

A representative separation (of ten experiments) by density gradient electrophoresis of normal human bone marrow cells (purified by centrifugation on Ficoll-Hypaque density cushion) is shown in Figure 1a. The separated cells were pooled in five fractions (I through V), according to their relative position ($R_p$) in the electrophoretic distribution (6,7), in order to permit simultaneous analysis by morphology, cell surface markers and functional tests. Fraction I contained cells of the highest electrophoretic mobility, whereas fraction V contained cells of the lowest mobility. Morphological examination of the cells, after staining with tetrachrome stain, revealed (Figure 1b) that immature cells of myeloid (promyelocytes, myelocytes, metamyelocytes, etc.) and erythroid lineage were significantly enriched in the high-mobility fractions (I and II). In contrast mature cells in general, e.g., small lymphocytes, monocytes and granulocytes (very few) were found to be significantly enriched in the low mobility fractions (IV and V), suggesting an association between surface charge of the cells and the stage of maturation. Large lymphocytes we found to be in the intermediate (III) and low mobility fractions (IV and V). Segmented and band granulocytes were found to be in the intermediate mobility fractions. Recovery after electrophoresis, washing and pooling of the cells was approximately 75%. The viability of the cells after electrophoresis was higher than 95%, as judged by tryptan blue dye exclusion. Recoveries of individual cell types were as follows: immature myeloid (promyelocytes, myelocytes, and metamyelocytes) cells and erythroid precursors (few), 67 ± 7%; segmented and band granulocytes, 79 ± 4%; monocytes, 81 ± 5%; large lymphoid cells, 75 ± 3%; small- and medium-size lymphocytes, 86 ± 3%.

Analysis of the separated cells by surface markers is shown in Figure 1c. These markers are present on various cell types in the bone marrow of lymphoid, myeloid and monocytoid lineage. E-rosette forming cells were found to be significantly enriched in the intermediate- and low-mobility fractions, whereas PNA-positive cells were found only in the low-mobility fractions IV and V (Figure 1d). E-rosette forming cells, bearing Fc receptors for IgM (Tμ cells) were found to be in fractions I through IV. In contrast, E-rosette forming cells bearing Fc receptors for IgG (Tγ cells) were enriched in the low mobility fraction V (Figure 1e).

The proliferative responses of the electrophoretically separated human bone marrow cells to PHA and to allogeneic cells in mixed lymphocyte culture are shown in Figure 1f. Colony-forming units _in vitro_ (CFU-c) were found to be significantly enriched in the high- and intermediate-mobility fractions I, II and III (Figure 1g). Similar results were obtained using two different culture systems, the placental colony stimulating activity system and the human leukocyte feeder layers. Approximately 90% of the CFU-c loaded on the gradient were recovered.

The separation of human bone marrow cells by density gradient electrophoresis was reproducible as shown in Figure 2, where the electrophoretic distribution profiles from three different experiments are superimposed for comparison. Furthermore, the contribution of sedimentation velocity at unit gravity to the electrophoretic separation of the bone marrow cells was investigated. Bone marrow cells were allowed to sediment at unit gravity for 4.5 hours in the 2.5-6.25% Ficoll-inverse 6.35-5.725% sucrose

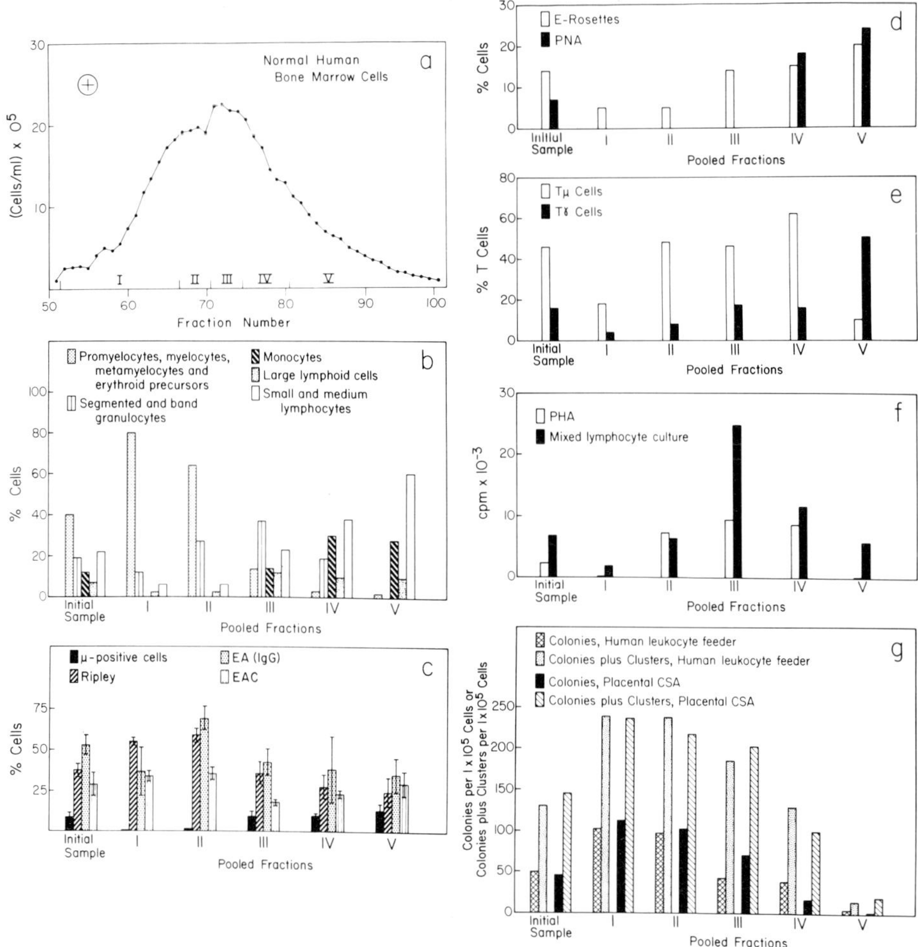

Figure 1. Separation of human bone marrow cells by density gradient electrophoresis: (a) Electrophoretic distribution profile. (b) Characterization of the separated cell fractions by morphology (mean of four experiments). (c) Cell surface markers (mean of five experiments). (d) Peanut agglutinin-positive cells and E-rosette forming cells (mean of four experiments). (e) T cells with Fc receptors for IgM (T$\mu$) and IgG (T$\gamma$). (f) Proliferative responses of the separated cell fractions to PHA and to allogeneic cells in mixed lymphocyte culture. (g) Colony-forming units in vitro, determined using the human leukocyte feeder layer and the placental CSA culture systems.

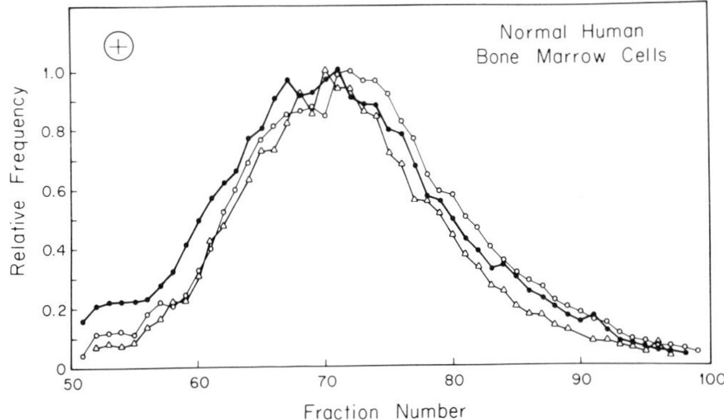

Figure 2.  Superimposed electrophoretic distribution profiles
from three different separations of human bone marrow cells.


density gradient, and the distance of migration and the
distribution were compared with those obtained by density gradient electrophoresis (Figure 3).  When electrophoresed, human bone marrow cells
migrated approximately a distance of 4.5 cm.  In contrast, in the absence
of an electric field (sedimentation velocity at unit gravity), these cells
sedimented a distance of 0.7 cm, because of the relatively low density of
the cells (less than 1.077 g/cm$^3$), and the high density range of the
Ficoll-inverse sucrose density gradient (1.0397-1.048 g/cm$^3$ at 4°C). These
results demonstrate that the contribution of sedimentation velocity at
unit gravity in the separation of human bone marrow cells by density
gradient electrophoresis is small (approximately 15%) and the cells are
separated primarily on the basis of their density of surface charge.


Separation of human bone marrow cells by density gradient electrophoresis,
permitted studies on the induction of differentiation of these cells, by
_In_ _vitro_ treatment with thymopoietin pentapeptide (TP-5) and ubiquitin
(UB).  Because of the low proportions of inducible cells, statistically
significant induction cannot be observed in unfractionated bone marrow
cell suspensions.  _In_ _vitro_ treatment of the separated cells with thymopoietin pentapeptide or ubiquitin (14 hours at 37°C) resulted in significant increase of the proportions of E-rosette forming cells in the

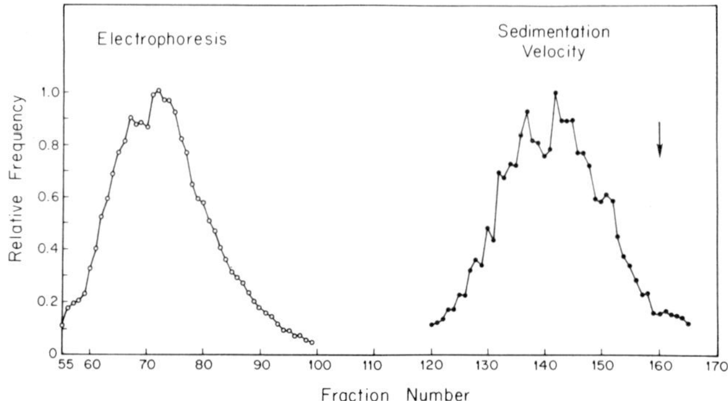

Figure 3. Comparison of density gradient electrophoresis and sedimentation velocity at unit gravity in the (2.5 - 6.25%) Ficoll-inverse (6.35 - 5.725%) sucrose density gradient. The contribution of sedimentation velocity at unit gravity to the electrophoretic separation is small.

intermediate and low mobility fractions IV and V (Figure 4). Tμ cells were significantly increased (five fold) in fraction V, by _in vitro_ treatment with either TP-5 or UB (Figure 4). In contrast this treatment did not

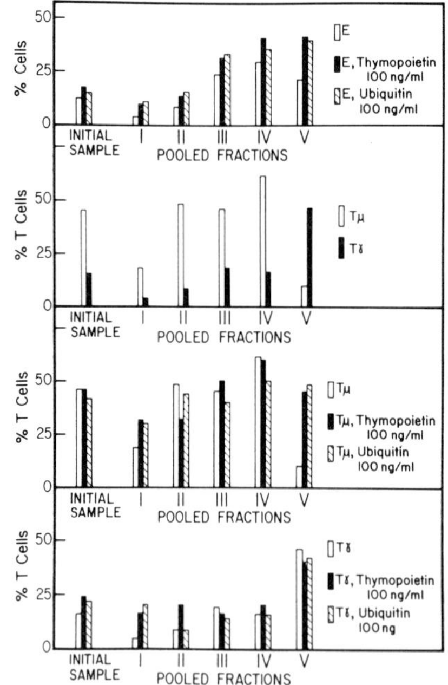

Figure 4. First Panel: Induction of E-rosette forming cells, in electrophoretically separated human bone marrow cell suspensions, by thymopoietin pentapeptide and ubiquitin (14 hours, 37°C).

Second Panel: Separation of bone marrow Tμ and Tγ cells by density gradient electrophoresis. These cells were identified by a double rosetting technique using FITC-labeled sheep erythrocytes and either IgM or IgG antibody-ox erythrocyte complexes (Platsoucas, et al., 1981b).

Third and Fourth Panels: Induction of Tμ but not of Tγ cells, by _in vitro_ treatment with thymopoietin pentapeptide and ubiqunitin.

alter the proportions of Tγ cells.  Proliferative responses to mitogens or allogeneic cells in MLC were significantly augmented (6 fold) by thymopoietin pentapeptide, but not by ubiquitin in the low mobility fraction V (data not shown).  These results demonstrate that T cell precursors in the bone marrow are of low electrophoretic mobility and can be signifiantly enriched by density gradient electrophoresis.  Furthermore, in vitro treatment with ubiquitin, but not with thymopoietin, resulted in increased proportions of IgM-bearing cells in fractions III and IV (Figure 5).  In contrast, both thymopoietin and ubiquitin induced the appearance of cells forming rosettes with mouse erythrocytes (an "early" B cell marker) in the low mobility fractions IV and V (unpublished results).

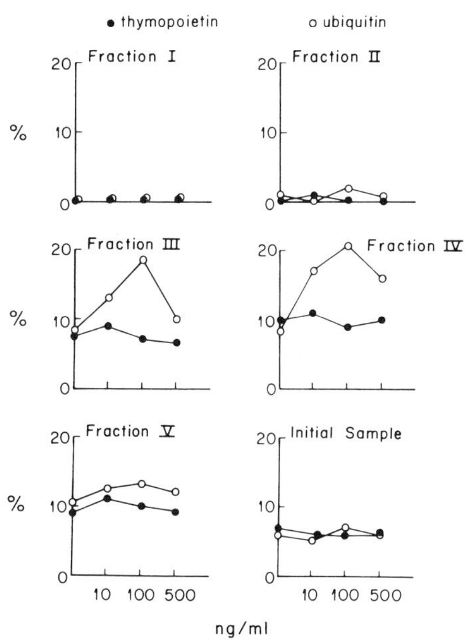


Figure 5.  Induction of IgM-bearing cells, in electrophoretically separated human bone marrow cell suspensions, by ubiquitin.

In conclusion, the density gradient electrophoresis method was applied for the separation and study of human bone marrow cells.  This method will be particularly useful for the characterization and classification of human bone marrow cells in lymphoproliferative disorders and aplastic anemia, as well as for differentiation studies on human bone marrow cells.

Supported in part by grant CH-151 from the American cancer Society, and grants CA-8748 and NS-11457 from the National Institutes of Health.

References

1. Shortman, K.:  J. Exp. Biol. Med. Sci. 46, 375 (1968).
2. Worton, R.G., McCulloch, E.A., Till, J.E.:  J. Cell. Physiol. 74, 171 (1969).
3. Phillips, R.A., Miller, R.G.:  J. Immunol. 105, 1168 (1970).
4. Williams, N., Moore, M.A.S.:  J. Cell. Physiol. 82, 81 (1973).
5. Boyum, A.:  Scand. J. Clin. Invest. 21(Suppl. 97), 77 (1968).
6. Platsoucas, C.D., Griffith, A.L. and Catsimpoolas, N.:  J. Immunol. Methods 13, 145 (1976).
7. Platsoucas, C.D.:  Ph.D. Thesis, Massachusetts Institute of Technology, Cambridge (1978).
8. Platsoucas, C.D. and Catsimpoolas, N.:  J. Immunol. Methods 26, 245 (1979).
9. Platsoucas, C.D., Good, R.A., Gupta, S.:  Proc. Natl. Acad. Sci. USA 76, 1972 (1979).
10. Platsoucas, C.D., Good, R.A., Gupta, S.:  Cell. Immunol. 51, 238 (1980).
11. Platsoucas, C.D., Beck, J.-D., Kapoor, N., Good, R.A. and Gupta, S.: Cell. Immunol. (in press).
12. Boltz, R.C., Todd, P., Streibel, M.J., Louie, M.K.:  Prep. Biochem. 3, 383 (1973).
13. Griffith, A.L., Catsimpoolas, N., Wortis, H.H.:  Life Sci. 16, 1693 (1975).

# FREE-FLOW ELECTROPHORESIS IN MALARIA RESEARCH

Hans-G. Heidrich

Max-Planck-Institut für Biochemie, D-8033 Martinsried, Germany,
and Division of Tropical and Geographic Medicine, University of
New Mexico, Albuquerque, NM 87131, USA

## Introduction

The main goal to be achieved in malaria research is still the
production of a vaccine, in particular against Plasmodium falciparum. There are several candidates for such a vaccine. Sporozoites, the mosquito stage of plasmodia introduced into the host
during the blood meal, should be in first position. Their power
to vaccinate against malaria has been shown in successful clinical vaccination studies (1, 2) in which P. falciparum infected
mosquitoes were allowed to inoculate the immunizing dose of sporozoites into volunteers. Another group of candidates for a
vaccine, already used to immunize simian, rodent, and avian
hosts successfully against P. falciparum, are antigenic preparations containing mature blood stage parasites, in particular
stages such as schizonts or merozoites (3, 4). A third possibility is offered by the production of protective antigenic material as described recently (5) in the form of "wash-off" or
generally soluble components from merozoites or cell cultures.

A crucial prerequisite for an interpretable successful vaccination trial is a vaccine consisting of one of the described constituents but containing no contamination particularly by red
blood cells or red blood cell constituents. In general, the
gentle release of intraerythrocytic mature stages from the red
cells and the separation of the liberated parasites from red
blood cell constituents is a rather difficult task (for evaluation of the difficulties read ref. 6). Similarly the production
of contaminant-free sporozoites in large quantities is hard to
achieve. Nevertheless free-flow electrophoresis in combination

with other techniques has proven to be a valuable tool for producing such preparations and in particular for isolating large quantities of contaminant-free material.

LIBERATION OF INTRAERYTHROCYTIC PARASITES (<u>PLASMODIUM VINCKEI</u>, <u>PLASMODIUM BERGHEI</u>) FROM ERYTHROCYTES AND ELECTROPHORETIC SEPARATION OF FREE PARASITES FROM HOST CELL CONSTITUENTS

Hans-G. Heidrich[a], Lorenz Rüssmann[a], Bettina Bayer[b], and Albrecht Jung[b].

Experimental

Blood with a parasitemia between 30% and 40% was collected from female Nmri-mice in sucrose/citrate solution (made from 75 vol. parts 0.025 M sucrose and 25 vol. parts 130 mM sodium citrate). Red blood cells (RBC) were washed three times in Ringer solution supplemented with 10 mM $Ca^{++}$. 1 ml of a concanavalin A solution (20 mg/ml in Ringer/Ca solution) was added to 1 ml of packed RBCs and after 10 min in ice the aggregated suspension was centrifuged for 10 min at 10,000 g and $4^{o}C$. The resuspended pellet was then passed through a 100 μm and a 20 μm nylon sieve. After 5 min centrifugation at 10,000 g the pellet was washed three times in electrophoresis medium and then resuspended in the same medium to give a suspension of about 7 -10 mg protein/ml.

Free-flow electrophoresis was carried out in a medium composed of 10 mM triethanolamine, 10 mM acetic acid, 0.1 mM $MgSO_4$, and 0.25 M sucrose (pH 7.4 with 2N NaOH). An FFV Apparatus (Bender & Hobein, Munich, Germany) was used at 135 V/cm, 180 mA, $t=5^{o}C$, buffer flow 3 ml/fraction/hr and 5 ml/hr sample input (7). Three distinct bands were seen in the separation chamber.

Enzymes described in the legend to Fig. 1 were determined as described in (8). Total protein was assayed automatically using a ninhydrin method (9). Electron microscopy was performed using standard procedures.

Results and Discussion

The treatment of erythrocytes with concanavalin A resulted in a complete aggregation of cells which was very much enhanced by the relatively high concentration of $Ca^{++}$. When these aggregates were passed through 100 and 20 μm nylon sieves the red cells ruptured into small vesicles thus gently releasing the intraerythrocytic parasites. The resulting mixture of red blood cell ghosts, membrane vesicles, a few remaining intact red blood cells (infected and uninfected), and free parasites in all stages of development was separated electrophoretically (Fig.1). Fractions 5-40 contained cellular material. When analyzed, fractions 5-18 were found to contain free parasites only. This was not obvious from the marker enzyme profile alone (Fig.1), since the particles contained both glutamate dehydrogenase $(NADP^+)$, a marker used for parasites, and glucose-6-phosphate dehydrogenase, the marker used for erythrocyte content. Electron micrographs of these fractions, however, showed only free parasites and no erythrocytes or erythrocyte membranes (Fig.2). Since parasites, in particular immature parasites such as trophozoites and young schizonts, possess food vacuoles which contain erythrocyte content, presence of the marker for the latter in parasite-containing fractions was to be expected. Fractions 21-31, derived from the middle band visible in the electrophoresis chamber, contained all the intact erythrocytes (some infected,

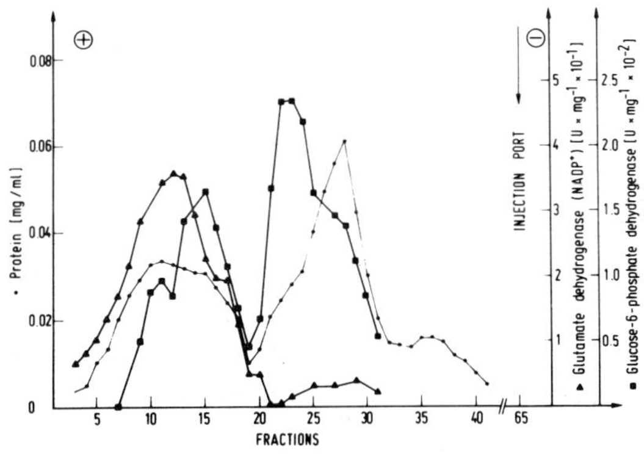



Figure 1: Free-flow electrophoresis isolation of intracellular parasites (P. vinckei). Fractions 5-18 contain free parasites, 20-24 uninfected RBCs, and 25-31 infected RBCs. Protein (●); glutamate dehydrogenase $(NADP^+)$ as parasite marker (▲); glucose-6-phosphate dehydrogenase as RBC marker (■).

most uninfected), erythrocyte ghosts and membrane vesicles, and a few parasites. The components were identified by a combination of enzyme marker analysis (Fig. 1) and electron microscopy (not shown here). Thus a clear separation between free parasites and erythrocytes and their membranes was achieved. Seed and Kreier (10) found previously that free parasites (P. berghei) and rat erythrocytes possess different isoelectric points. This difference appears to result from a different phospholipid composition and different sialic acid content of the plasma membranes (11). These differences in electric surface charges were responsible for the excellent separation described here. Fractions 32-40 also contained free parasites. Their nature could not be clearly elucidated, since the amount of material was too small. Electron micrographs of this fraction, however, suggested that the cells were merozoites.


FREE-FLOW ELECTROPHORESIS SEPARATION OF FREE PARASITES (PLASMODIUM FALCIPARUM) ACCORDING TO STAGES

Hans-G. Heidrich[a,c], John E. K. Mrema[c], Philip Reyes[d], David Vander Jagt[d], and Karl H. Rieckmann[c]


Experimental


P. falciparum parasites at various developmental stages were ob-

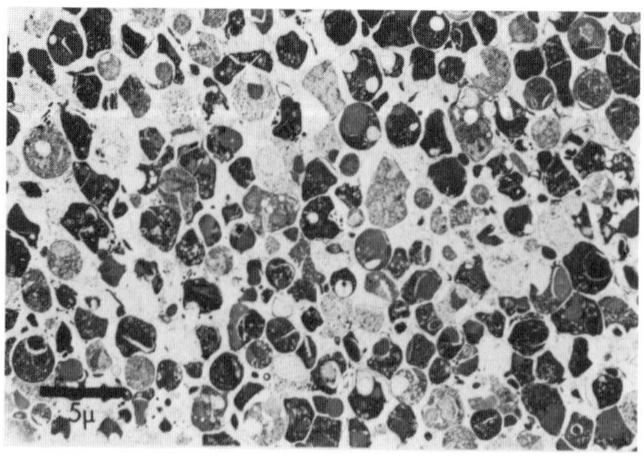

Figure 2: Electron micrograph of free parasites (P. vinckei) from fraction 10 of the electrophoresis run shown in Fig. 1. Bar: 5 μm

tained from continuous cultures grown by the candle jar procedure (12).  Intraerythrocytic mature forms were concentrated with Plasmagel (13, 14).  Liberation of the parasites from erythrocytes was achieved as described in the chapter above, except that the isolation medium was incomplete RPMI 1640 throughout, the lectin phytohemagglutinin instead of concanavalin A (10 mg/ml), and the size of the nylon sieves 100, 50, 20, 10, 5 and 3 μm in sequence.  On average 14 mg of total protein was obtained from 1 ml of original packed cells.

Free-flow electrophoresis was carried out in a medium containing 11 mM triethanolamine, 11 mM acetic acid, 0.5 mM $CaCl_2$, 5mM glucose, and 0.25 M sucrose (pH 7.4 with 1 N KOH).  An FFV Apparatus (Bender & Hobein, Munich, Germany) was used at a buffer flow of 2.85 ml/fraction/hr and a sample input of about 6-7 mg protein/hr (in 3 ml).  Three, or sometimes four, bands were observed in the separation chamber.

Protein was determined according to (15), enzymes as described in (8), and acetylcholine esterase according to (16).  Polyacrylamide gradient gel electrophoresis (PAGE) was carried out in a system similar to Laemmli's (17) with a gradient between 20% and 5% and a stacking gel of 3%, all gels containing 0.1% SDS.  Electron microscopy was performed using routine procedures.


Results and Discussion


The liberation of cultivated P. falicparum parasites from human red blood cells using the lectin aggregation technique was as successful as with P. berghei.  Phytohemagglutinin was used instead of concanavalin A because the isolation medium contained glucose.  Fragmentation of the erythrocytes after passage through the nylon sieves was almost complete.  The free-flow electrophoresis separation was also successful.  Fig 3 shows that the erythrocyte membrane marker acetylcholine esterase was

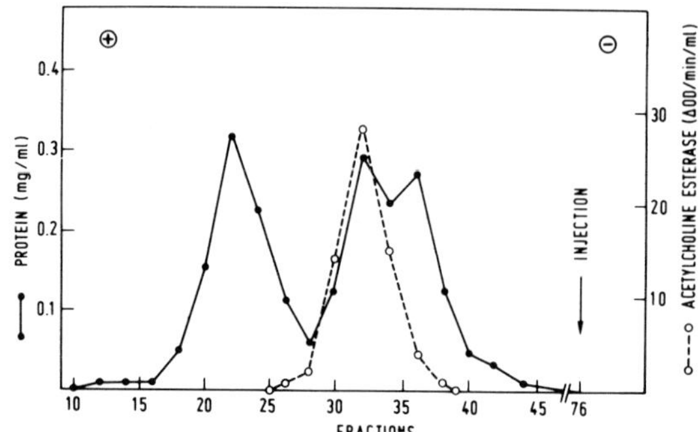

Figure 3: Free-flow electrophoresis isolation of intracellular parasites (P. *falciparum*) after liberation from their host cells. Fractions 12-37 and 33-44 contain free parasites, 28-37 RBCs and their membranes. Protein (●); acetylcholine esterase as RBC membrane marker (o)

only present in fractions 28 - 37. PAGE also proved that red cell contaminants were only present in these fractions: spectrin, an erythrocyte membrane constituent with an apparent molecular weight of about 220,000 - 225,000 daltons, was only seen in these fractions (Fig. 4). All other fractions from the electrophoresis run contained free parasites only.

Morphological analysis of these parasites clearly showed that the different developmental stages had been separated electro-

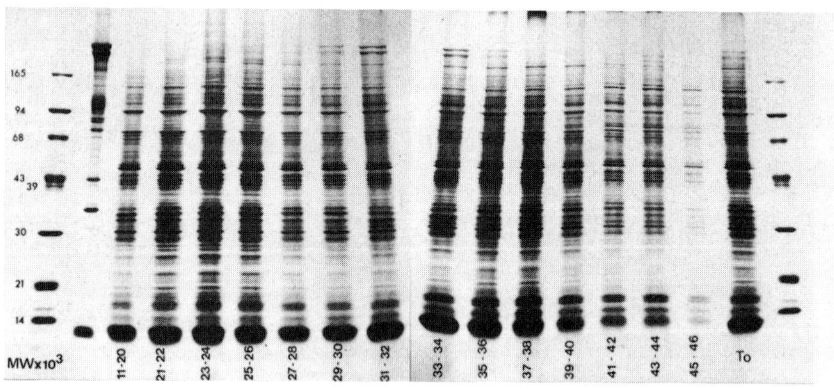

Figure 4: Gradient-SDS-PAGE of the various fractions from Fig.3. Only fractions 28-37 show the spectrin bands (∿200 kdaltons), a marker for RBC membranes. Lane 1: molecular weight standards; lane 2: RBC ghosts (human); lanes 3-16: electrophoresis fractions; 17: starting material prior to electrophoresis; lane 18: standards.

phoretically.  Since asynchronous cultures had been used, all
stages except for ring forms (because of the Plasmagel procedure)
were expected to be present.  Fig. 5 shows electron micrographs
of some fractions.  Young trophozoites were present in fractions
12 - 20 and more mature ones in 20 - 27 (not shown).  Fractions
28 - 37 contained parasitized and non-parasitized erythrocytes
and a large number of ghosts and membrane vesicles.  Schizonts,
not shown here, were present in fractions 35 - 38 and merozoites
in fractions 39 - 44.  The latter possessed a distinct surface
coat and this coat together with the missing parasitophorous
vacuole membrane, cause their low electrophoretic mobility.
This separation was also reflected in the PAGE analysis, although the differences in the gel patterns were mainly quantitative and not qualitative.

The reasons for the different electrophoretical behaviour of
the parasite stages remain to be determined at the molecular
level.  Apparently during development, the membrane surface of
the parasitophorous vacuole is altered and this might account
for the different electrophoretic mobilities of young and ma-

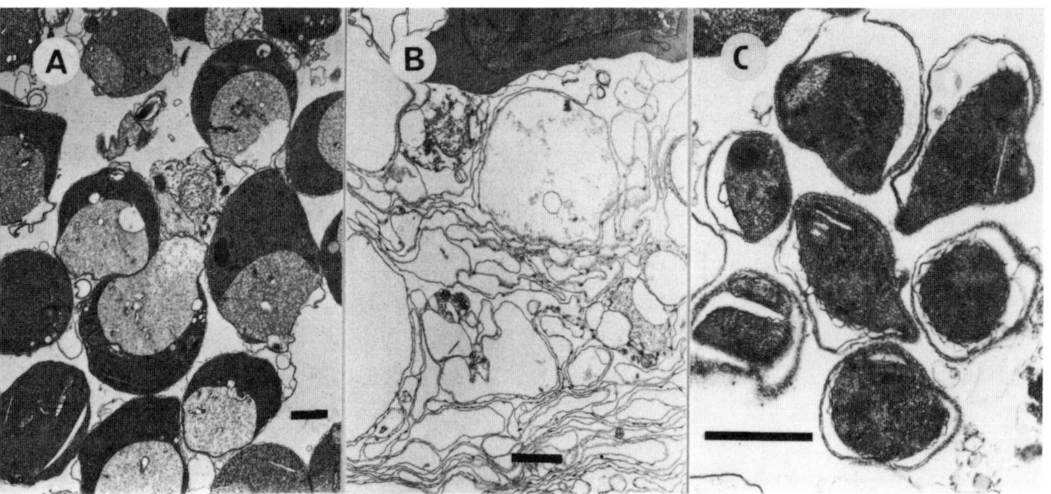

Figure 5: Electron micrographs of some fractions from Fig. 3.
A. Young trophozoites (fractions 12-20). B.  RBCs and RBC ghosts
and vesicles (fraction 28-34). C. Merozoites (fractions 39-44).
Bars: 1 μm.

ture trophozoites and of schizonts. Merozoites possess a slow electrophoretic mobility because of their surface coat. This coat has not yet been characterized chemically due to its lability and the lack of a technique to produce large quantities of intact merozoites.


FREE-FLOW ELECTROPHORESIS ISOLATION OF WHOLE BODY PLASMODIUM BERGHEI SPOROZOITES

Hans-G. Heidrich[a,c], Harry D. Danforth[e], and Richard L. Beaudoin[f]


Experimental

After grinding about 400 female mosquitoes (infected with P. berghei and killed with chloroform) in 5 ml Medium 199, the suspension was layered on top of a two-step (40 and 60% vol/vol) Hypaque gradient (18). Gradients were spun for 12 min at 16,300 g and $4^oC$, and the layer at the interphase was removed and diluted with 1.5 parts of electrophoresis medium. After spinning for 15 min at 16,300 g, the pellet was resuspended in electrophoresis medium to give a suspension of about $6-9 \times 10^6$ sporozoites per ml. Free-flow electrophoresis was carried out in a medium containing 11 mM triethanolamine, 11 mM acetic acid, 0.5 mM $MgCl_2$, 5 mM glucose, and 0.25 M sucrose (pH 7.4 with 1 N KOH). An FFV Apparatus (Bender & Hobein, Munich, Germany) was used at 135 V/cm, 190 mA, $t=5^oC$, buffer flow 2.8 ml/fraction/hr, and 1.5 ml sample injection per hr (i.e. $9-13 \times 10^6$ sporozites). A main and a very faint band were seen in the separation chamber.

Protein was determined and PAGE was performed as described above. Surface iodination was carried out using Iodo-Gen (Pierce) as described in (19). Autoradiograms of the dried gels were performed using routine techniques. Bacterial contamination before

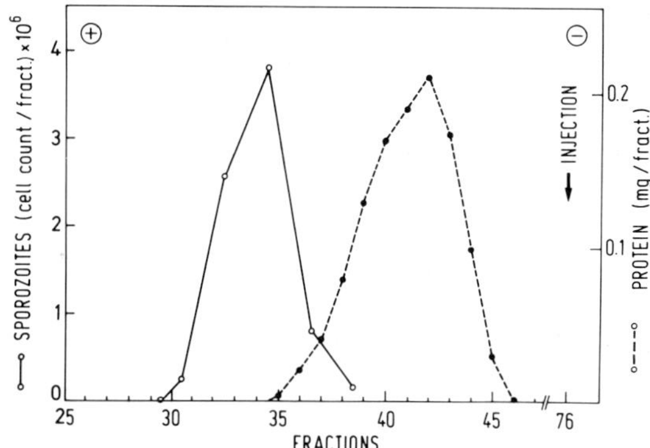



Figure 6: Free-flow electrophoresis isolation of sporozoites (P. berghei). Fractions 31-35 contain sporozoites, 37-45 mosquito debris, mitochondria and unidentified material. Protein ( ● ); sporozoite counts ( ○ ).

and after free-flow electrophoresis was tested by colony counts on agar plates at 24 and 48 hr at a sample dilution of 1:1,000 or 1:10,000. Electron microscopy was performed using routine techniques.


Results and Discussion


Centrifugation of the ground mosquitoes in Hypaque gradients separates the homogenate according to the densities of the particles. Density, however, does not always characterize cells or other bioparticles unequivocally and contaminants may have the same density as the investigated particle. This was observed here. A large amount, in fact almost all the material separated by density gradient centrifugation, was actually contaminating material and was separated from the sporozoites during the electrophoresis run. The main band observed in the separation chamber and collected in fractions 37 - 45 (Fig. 6) contained only mitochondria, mitoplasts and other unidentified membrane vesicles or debris (Fig. 7A). Virtually all the sporozoites and no contaminants were present in fractions 31 - 35 of the electrophoresis run (Fig. 7B). On average 6 - 8 x 10$^6$ sporozoites were isolated from 400 mosquitoes. Their surface appeared not to be altered. The serological status was tested by indirect immunofluorescence using serum from mice immunized

868

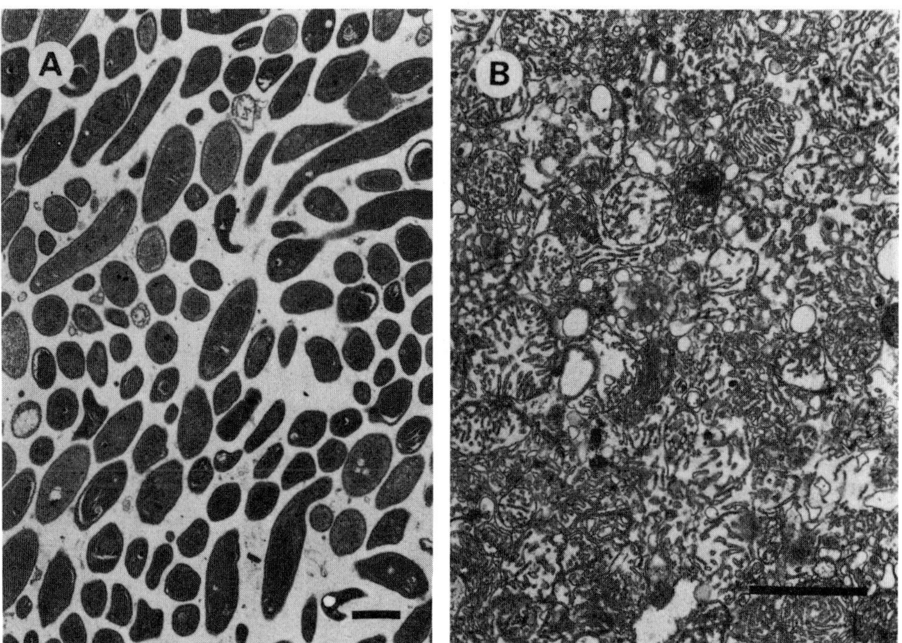

Figure 7: Electron micrographs of fraction pools from Fig. 6.
A. Fractions 31 - 35 contain sporozoites. B. Contaminants are
in fractions 37 - 45. Bars: 1 µm.


with P. berghei sporozoites. A combination of surface iodination, SDS-PAGE, and autoradiography showed a protein with an
apparent molecular weight of about 44,000 daltons to be the predominant constituent of the sporozoite membrane. According to
results of others (20, 21), when sporozoites were incubated
with a monoclonal antibody specific for this Pb 44 antigen,
they were neutralized and failed to infect mice upon inoculation. Besides this 44,000 dalton surface constituent, iodination of the isolated sporozoites labelled three or four other
components whose nature and importance as protective antigens
remain to be investigated. An important step for obtaining
clean sporozoite preparations was the electrophoretic removal
of the bacterial contamination. The contaminating bacteria
showed different electrophoretic mobilities according to the
different insectories from which the mosquitoes originated
(data not shown here).

## Acknowledgements

This work was supported by the World Health Organization (M2/181/155) and the United States Agency for International Development (AID/DPSE-C-0036 and AID/DPSE-C-0068).

## Affiliations

[a]Max-Planck-Institut für Biochemie, D-8033, Martinsried, Germany
[b]University of Tübingen, D-7400 Tübingen, Germany
[c]Division of Tropical Medicine, UNM, Albuquerque, NM 87131, USA
[d]Department of Biochemistry, UNM, Albuquerque, NM 87131, USA
[e]Biomedical Research Institute, Rockville, MD 20852, USA
[f]Naval Medical Research Institute, Bethesda, MD 20014, USA

## References

1.  Nussenzweig, R.: Int. J. Nucl. Med. Biol. 7, 89-96 (1980).

2.  Rieckmann, K. H., Beaudoin, R. L., Cassells, J. S., and Sell, K. W.: Bull. WHO 57 (Suppl. 1), 261-165 (1979).

3.  Rieckmann, K. H., Mrema, J. E. K., and Campbell, G. H.: J. Parasitol. 64, 750-752 (1978).

4.  Siddiqui, W. A., Kan, S., Kramer, K., Case, S., Palmer, K. and Niblack, J. F.: Nature 289, 64-66 (1981).

5.  Grothaus, G. D., and Kreier, J. P.: Infect. Immun. 28, 245-253 (1980).

6.  Hamburger, J. and Kreier, J.P.: Malaria (Immunology and Immunization) 1980. Academic Press, New York, London, San Francisco. p. 1-65.

7.  Heidrich, H.-G., Rüssmann, L., Bayer, B., and Jung, A.: Z. Parasitenkd. (Parasitol. Res.) 58, 151-159 (1979).

8.  Picard-Maureau, A., Hempelmann, E., Krammer, G., Jackisch, R., and Jung, A.: Z. Tropenmed. Parasitol. 26, 406-416 (1975).

9.  Heidrich, H.-G.: Advances in Automated Chemistry. Proc. Technicon, Int. Congr. 1970. p. 347-350.

10. Seed, Th. and Kreier, J. P.: Infect. Immun. $\underline{14}$, 1339-1347 (1976).

11. Rock, R. C., Standefer, J. C., Cook, R. T., Little, W., and Printz, H.: Comp. Biochem. Physiol. $\underline{38}$, 425-437 (1971).

12. Trager, W. and Jensen, J.D.: Science $\underline{193}$, 673-675 (1976).

13. Pasvol, G., Wilson, R. J., Smalley, M. E., and Brown, J.: Ann. Trop. Med. Parasitol. $\underline{72}$, 87-88 (1978).

14. Reese, R. T., Langreth, S. G., and Trager, W.: Bull. WHO $\underline{57}$, 53-61 (1979).

15. Bradford, M.: Anal. Biochem. $\underline{72}$, 248-254 (1976).

16. Steck, T. L. and Kast, J.A.: Meth. Enzymol. $\underline{31}$, 172-180 (1974).

17. Laemmli, U.: Nature (London) $\underline{227}$, 680-685 (1979).

18. Pacheo, N. D., Strome, C. P. A., Mitchell, F., Bawden, M. P., and Beaudoin, R. L.: J. Parasitol. $\underline{65}$, 414-417 (1978).

19. Markwell, M. A. K. and Fox, F.: Biochemistry $\underline{17}$, 4807-4817 (1978).

20. Yoshida, N., Nussenzweig, R., Potocnjak, P., Nussenzweig, V., and Aikawa, M.: Science $\underline{207}$, 71-73 (1980).

21. Potocnjak, P., Yoshida, N., Nussenzweig, R. S., and Nussenzweig, V.: J. Exp. Med. $\underline{151}$, 1504-1513 (1980).

# SEPARATION OF FUNCTIONING MAMMALIAN CELLS BY DENSITY-GRADIENT ELECTROPHORESIS


Paul Todd, W.C. Hymer, Lindsay D. Plank, Gary M. Marks,
M. Elaine Kunze, Vincent Giranda

Althouse Laboratory, The Pennsylvania State University,
University Park, Pennsylvania 16802

J.N. Mehrishi
Department of Medicine, Addenbrookes Hospital, University of
Cambridge, Cambridge, England CB2 2QQ


## Introduction

One of the promises of preparative cell electrophoresis is its ability to purify functioning, living animal cells in useful quantities. One of several such methods is density-gradient electrophoresis. When a low-ionic-strength, isotonic gradient of Ficoll and sucrose is used in a glass column at $4^0$C, the physical factors that determine the upward velocity of cells combine to produce a nearly-constant migration rate. Density-gradient electrophoresis has been applied to suspended cells prepared from freshly dispersed tissues, cultured-cell mono-layers, and naturally suspended cells from body fluids (1, 2, 3, 4). It is the purpose of this report to briefly describe recent work performed at the Pennsylvania State University con-sisting of further physical characterization of density-gradient electrophoresis and the results of a small number of recent applications to living cells. The most popular uses of preparative cell electrophoresis have been in the field of immunology, and this is true of density-gradient electrophor-esis (4, 5). Some rather more unusual applications will be discussed here. All of the research reported here used the

apparatus and methods described by Boltz et al. (6). Columns
without a central cooling finger were used.


Electrophoretic Cell Migration in a Ficoll Gradient

As cells migrate upward through a decreasing Ficoll concentration in the electrophoresis column they encounter changes in
viscosity, density, and conductivity. In addition, Ficoll itself affects electrophoretic mobility (1, 6, 7). The effect
of Ficoll was determined quantitatively by the microscopic
method of cell electrophoresis using the Zeiss cytopherometer
and determination of mobilities from complete velocity para-
bolas corrected for asymmetric electroosmotic backflow (2, 8).
A linear increase of the mobilities of various erythrocytes
with increasing Ficoll concentration was found. From the
remaining known and measured properties of the cells and solutions (9) it is possible to predict the upward migration velocity of cells under the influence of the combined forces of
gravity and the applied electric field. The resulting mathematical relationship, when integrated numerically, produces a
migration plot (distances vs. time) such as that shown in Figure 1. Because migration velocity depends slightly, but
roughly linearly, upon migration distance, an exponential function is expected, but, as Figure 1 indicates, this is a slowly
rising exponential function which is indistinguishable from a
straight line within the accuracy of most distance measurements. The electrophoretic mobility, $\mu$, normally defined as
migration velocity per unit field strength in $\mu$m-cm/V-sec, was
therefore determined in these studies by dividing the distance
migrated by the electrophoresis time and dividing this ratio
by the constant field strength.

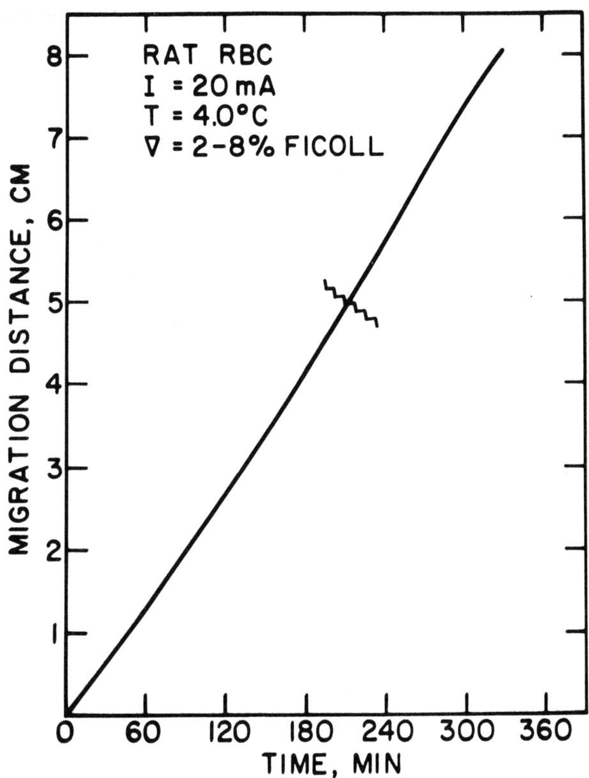



Figure 1. Plot of migration distance vs. time for upward migration of erythrocytes in a Ficoll gradient. The graph was determined by numerical integration of the relationship between cell velocity and the varying properties of the gradient: viscosity, density, conductivity, and Ficoll concentration. Below the break shown near the middle of the curve the numerical integration is approximated satisfactorily by an exponential function resulting from the assumption that cell velocity depends linearly on migration distance.

## Effect of Cell Cycle Phase on Electrophoretic Mobility

Cells from a long-term cultured epithelial line designated "T-1" (10, 11) were cultured in monolayers in plastic flasks and suspended in electrophoresis buffer by detachment with

874

EDTA.  After subjecting the suspended cells to density-gradient
electrophoresis, fractions were collected, and the DNA content
of individual cells was measured by flow cytometry using the
staining method of Trujillo et al. (12) and flow cytometry with
a Bio/Physics Systems, Inc. Cytofluorograf (13).  The cellular
DNA distribution of individual fractions revealed  that low-
mobility fractions were enriched in cells in the G2 phase of
the cell cycle (Figure 2).  However, the mobility distributions
of cells in different phases of the cycle are broad (14), and
the data of Figure 2 reveal only, as found by Forrester (14),
that there is a general trend toward lower mobilities as cells
progress through the cycle.

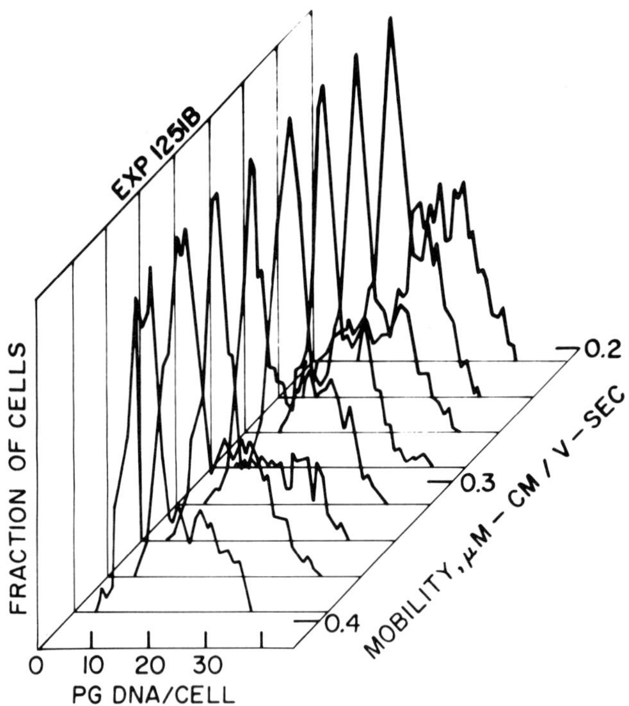



Figure 2.   Cellular DNA distributions of cultured human T-1
            cells in fractions collected after density-gradient
            electrophoresis.  DNA content per cell was cali-
            brated biochemically (15).  Higher mobility frac-
            tions were depleted, while lower mobility fractions
            were enriched, with respect to cells in the G2 phase
            of the cell cycle.

Urokinase-producing Cells in Human Embryonic Kidney Cell
Cultures

Human urokinase, a plasminogen activator produced in the kidney, is being considered as a therapeutic agent for thrombosis
(16).  Electrophoretically purified cultured human embryonic
kidney cells are considered a potentially efficient source of
this material (17), and electrophoretic purification attempts
have been reported (18).  Cells from early-passage culture,
designated "HFK-18", were propagated in monolayers in plastic
flasks and suspended in electrophoresis buffer by detachment
with EDTA.  After subjecting the suspended cells to density-
gradient electrophoresis, fractions were collected and cultured
in complete medium.  Just before these cultures became confluent the complete medium was replaced by a high-glycine, incomplete medium into which urokinase-producing cells secreted their
product.  Urokinase activity was measured in CTA units by a
calibrated fibrin-plate method (19) every 3-4 days after this
medium change.  Figure 3 indicates that most electrophoretic
fractions contain cells that produce urokinase and, in this
particular culture, the highest mobility fractions produced the
highest levels of urokinase activity per cell.


Freshly Dispersed Lymphocytes from the Irradiated Mouse Spleen

Radiation therapy often leads to lymphopenia in treated cancer
patients (20).  It had previously been found that lymphocyte
population changes in thymus lymphocytes of irradiated mice
could be followed by analytical electrophoresis as a function
of time after irradiation and that a clear rise and fall of the
proportion of high-mobility (presumably cortical) cells occurs
(21).  Similarly, spleens of control and irradiated mice were
excised 9 hr after exposure to 5.5 Gy of $^{60}$Co gamma radiation;
lymphocytes were teased away from stromal tissue and purified
by the magnetic removal of macrophages and the removal by

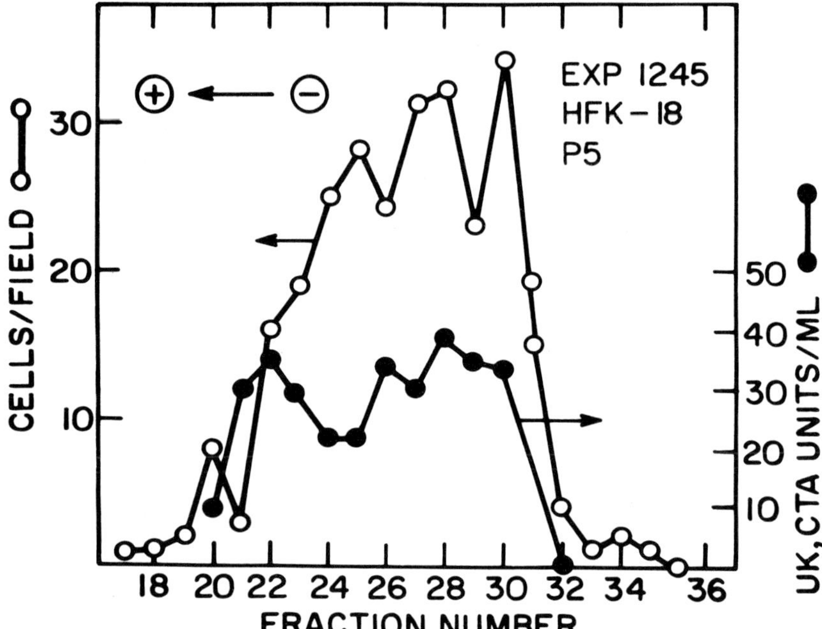



Figure 3.  Density-gradient electrophoretic profile of cultured human embryonic kidney cells "HFK-18" at the fifth passage in vitro.  Cells per field (O) in each fraction was determined by phase-contrast microscopy of cultures in dishes; urokinase activity (●) was determined by a fibrin-plate method (19).  High mobility cells showed a high specific activity.

centrifugation of erythrocytes (22).  The resulting lymphocyte suspensions were subjected to density-gradient electrophoresis, fractions were collected,  and the number of cells per fraction in the 5.0-6.3 μm diameter range was determined by Coulter counter.  The results shown in Figure 4, which is a distribution of cell number vs. electrophoretic fraction, indicates that there was a preferential loss of cells from the low-mobility cell population after irradiation.  The high-mobility fraction has been reported to consist mainly of T-lymphocytes in adult mice (23).  From this result we conclude that non-T-lymphocytes are preferentially reduced in number in the spleens of irradiated mice.

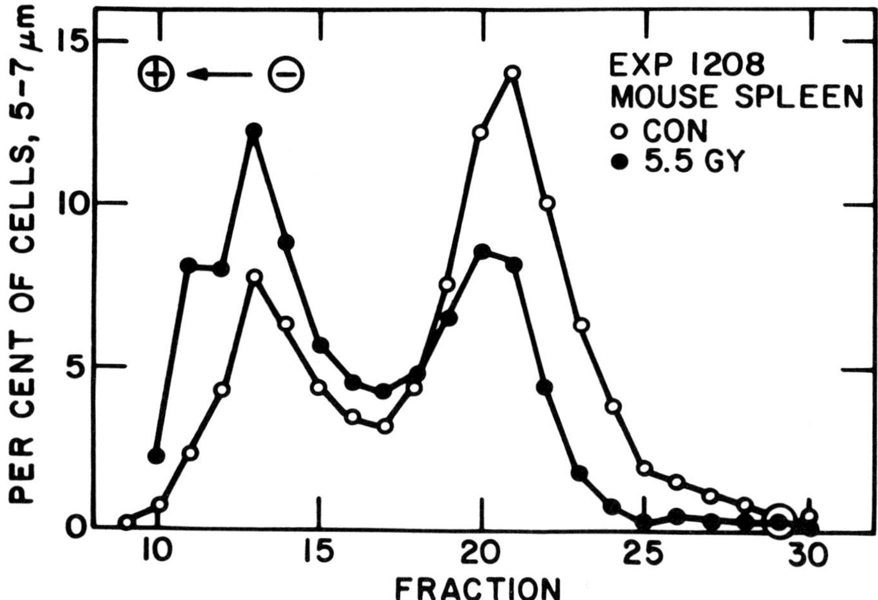


Figure 4.   Density-gradient electrophoretic profile of 5.0-6.3
µm cells in fractions collected from columns to
which were applied splenic lymphocytes from control
(O) and irradiated (●) mice.  The percent cells in
the low-mobility (presumably non-T) population was
preferentially reduced 9 hr after exposure to 5.5
Gy.

Suspended Somatotrophin-producing Cells from the Rat Anterior
Pituitary

Purified populations of live somatotrophin-producing cells have

numerous potential applications, including implantation for

growth enhancement (24), production of mRNAs for gene cloning,

investigation of single-cell responses to releasing agents

(25), the study of other functions of somatotrophs (26), and

hormone production _in_ _vitro_ (27).  Preparative cell electro-

phoresis is one of the methods that is potentially applicable

to this particular cell purification problem.  Somatotrophin-

secreting cells constitute about 30   percent of the cells in

the mammalian adenohypophysis so their electrophoretic mobility

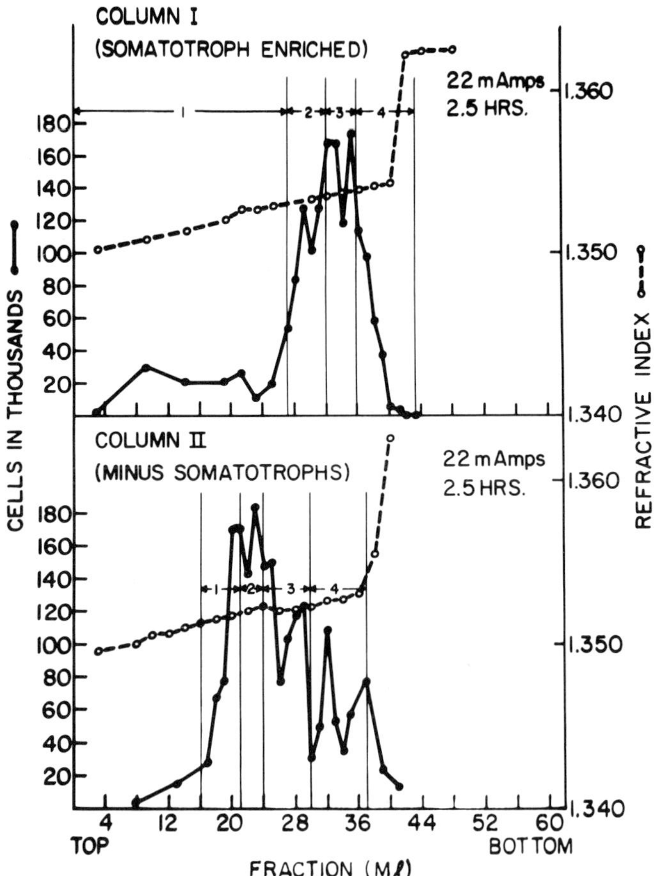



Figure 5. Electrophoretic profiles of suspended rat anterior pituitary cells after enrichment (upper panel) or depletion (lower panel) of somatotrophs by centrifugation. Bands 1, 2, 3, and 4 define fractions that were pooled for radioimmunoassay of hormone per 1000 cells. Band 4 in the upper panel was richest in somatotrophs, and their specific activity was nearly twice that of centrifugally purified somatotrophs.

cannot be determined by analytical cell electrophoresis methods. Heavily granulated rat somatotrophs were separated to ca. 80% purity by a single centrifugation procedure (28) consisting of layering freshly dispersed cells over 28% bovine serum albumin (BSA) (density 1.069 $g/cm^3$) which in turn was layered over a

dense, 34% BSA, layer.  Centrifugation produced two rat pituitary cell bands, band I at the top interface and band II at
the intermediate interface.  Cells in band II were enriched
somatotrophs.  The electrophoretic mobility distribution of
these partially purified cells was determined by density-
gradient electrophoresis.  At the same time electrophoretic
fractions were collected and assayed for growth hormone content
so that specific mobilities could be associated with specific
levels of hormone production per cell.  Two electrophoresis
columns were used to simultaneously separate cells in centrifugal bands I and II.  The separation profiles (Figure 5) reveal
obvious differences in the mobility distributions between bands
I and II.  Radioimmunoassay of pooled fractions of cells from
this separation experiment showed that the specific activity of
cells from band 4 (lowest mobility group) in the upper panel of
Figure 5 was 872 ng of growth hormone per 1000 cells, whereas
pooled fractions represented by bands 1, 2, and 3 had 300 $\pm$
50 ng/1000 cells.  Cells from all fractions from column II
(lower panel of Figure 5) had only 30 ng/1000 cells.  These
data suggest that a small pool of somatotrophs that contain
relatively large quantities of hormone can be separated from
other cell types on the basis of their low electrophoretic
mobility.


Discussion and Summary

The fact that cell migration in Ficoll density-gradient elec-
trophesis is nearly linear allows meaningful separations and
analyses to be performed by density-gradient electrophoresis.
In specific research projects it was found that the electrophoretic mobility of logarithmically growing cells from a long-
term cultured human cell line depends on cell cycle phase.
Preparative electrophoresis makes this experiment possible in
the absence of artificial synchronization of the cells.  Electrophoretic subpopulations of cells from human embryonic kidney

are enriched in cells that produce urokinase, a plasminogen
activator used in thrombolytic therapy. Preparative electrophoresis concentrates larger numbers of these senescing cells
than could be done by the selection of single clones. Electrophoretic separation of freshly dispersed lymphocytes from the
irradiated mouse spleen indicates a preferential loss of low-
mobility ("non-T") cells. Freshly dispersed cells from the rat
anterior pituitary have a low-mobility subpopulation that is
rich in somatotrophs that contain growth hormone after electrophoretic purification. Preparative electrophoresis permits the
post-separation use of functional assays to unequivocally identify classes of cells having known electrophoretic mobilities.
These research examples also indicate the utility of preparative electrophoresis, especially including density-gradient
electrophoresis, in producing populations of functioning cells
for further study or use in the production of cell products.


## Acknowledgments

We thank Mrs. Gertrud Barsch, Mrs. Mary Hershey,   Mr. Michael
Hatfield, and Mr. Charles Goolsby for excellent technical
assistance.  This work was supported by National Aeronautics
and Space Administration Contracts NAS 9-15583, NAS 9-15566,
and NAS 9-15584, and by Public Health Service Grant R01-CA-24090
from the National Cancer Institute.


## References

1.  Boltz, R. C., Jr., Todd, P., Gaines, R. A., Milito, R. P.,
    Docherty, J. J., Thompson, C. J., Notter, M. F. D.,
    Richardson, L. S., Mortel, R.:  J. Histochem. Cytochem.
    24, 16-23 (1976).

2.  Thompson, C. J., Docherty, J. J., Boltz, R. C., Jr.,
    Gaines, R. A., Todd, P.:  J. gen. Virol. 39, 449-461
    (1978).

3.  Hammerstedt, R. H., Keith, A. D., Boltz, R. C., Jr., Todd, P.: Arch. Biochem. Biophys. $\underline{194}$, 565-580 (1979).

4.  Platsoucas, C. D., Griffith, A. L., Catsimpoolas, N.: J. immunol. Meth. $\underline{13}$, 145-152 (1976).

5.  Platsoucas, C. D., Good, R. A., Gupta, S.: Proc. natl. Acad. Sci. U.S. $\underline{76}$, 1972-1976 (1979).

6.  Boltz, R. C., Jr., Todd, P.: In Electrokinetic Separation Methods (Righetti, P. G., van Oss, C. J., Vanderhoff, J. W., eds.) Elsevier/North-Holland, Amsterdam (1979) pp. 229-250.

7.  Brooks, D. E.: J. colloid interface Sci. $\underline{43}$, 714-726 (1973).

8.  Gaines, R. A.: Thesis, The Pennsylvania State University, (1981).

9.  Boltz, R. C., Jr., Todd, P., Streibel, M. J., Louie, M. K.: Prep. Biochem. $\underline{3}$, 383-401 (1973).

10. Barendsen, G. W., Beusker, T. L. J., Vergroesen, A. J., Budke, L.: Radiat. Res. $\underline{13}$, 841-849 (1960).

11. Nelson-Rees, W. A., Flandermeyer, R. R., Daniels, D. W.: Science $\underline{209}$, 720-722 (1980).

12. Trujillo, T. T., Van Dilla, M. A.: Acta Cytologica $\underline{16}$, 26-30 (1972).

13. Leary, J. F., Todd, P., Wood, J. C. S., Jett, J. H.: J. Histochem. Cytochem. $\underline{27}$, 315-320 (1979).

14. Brent, T., Forrester, J. A.: Nature $\underline{215}$, 92-93 (1967).

15. Todd, P.: Radiat. Res. $\underline{61}$, 288-297 (1975).

16. Urokinase Pulmonary Embolism Trial: J. Amer. Med. Assoc. $\underline{214}$, 2163-2172 (1970).

17. Barlow, G. H., Rueter, A., Tribby, I.: In Proteases and Biological Control (Reich, E., Rifkin, D. B., Shaw, E., eds.) Cold Spring Harbor Press, Cold Spring Harbor, NY (1975) pp. 325-331.

18. Allen, R. E., Rhodes, P. H., Snyder, R. S., Barlow, G. H., Bier, M., Biguzzi, P. E., Van Oss, C. J., Knox, R. J., Seaman, G. V. F., Micale, F. J., Vanderhoff, J. W.: Sep. Purif. Meth. $\underline{6}$, 1-28 (1977).

19. Astrup, T., Müllertz, A.: Arch. Biochem. Biophys. $\underline{40}$, 346-351 (1952).

20. Sternswärd, J., Jondal, M., Vanky, F., Wigzell, H., Sealy, R.: Lancet $\underline{1}$, 1352-1356 (1972).

21. Mehrishi, J. N., Hiesche, K. D., Révész, L.: Studia Bio-physica $\underline{73}$, 63-69 (1978).

22. Bøyum, A.: Scand. J. clin. Lab. Invest. $\underline{21}$, suppl. 97, 77 (1968).

882

23. Catsimpoolas, N., Griffith, A. L., Gupta, S., Good, R. A., Platsoucas, C. D.: In Electrophoresis '79 (Radola, B. J., ed.) W. de Gruyter, Berlin/New York (1980) pp. 607-622.

24. Weiss, S. J., Berglund, R., Turpen, C., Hymer, W. C.: Fed. Proc. $\underline{36}$, 363 (1977).

25. Hymer, W. C., Page, R., Kelsey, R. C., Augustine, E. C., Wilfinger, W., Ciolkosz, M.: In Synthesis and Release of Adenohypophyseal Hormones (McKerns, K. W., Jutis'z, M., eds.) Plenum, NY (1979) pp. 126-168.

26. Wallis, M.: Nature $\underline{284}$, 512 (1980).

27. Tashjian, A. H.: Biotechnol. Bioeng. $\underline{11}$, 109 (1969).

28. Hymer, W. C., McShan, W. H.: Cell Biol. $\underline{17}$, 67-86 (1963).

# REVIEW OF THE NASA ELECTROPHORESIS PROGRAM

Robert S. Snyder
NASA/Marshall Space Flight Center
Huntsville, Alabama  35812

## Introduction

The NASA electrophoresis program is part of an investigation
into fluid transport and chemical processes with the principal
objective to determine the influence of gravity on separation
processes, such as electrophoresis.  Over the past several
years, this has involved a variety of tasks including:
1) evaluation of available electrophoresis apparatus and the
capability of this instrumentation to satisfy the needs of the
biomedical community; 2) understanding the limits that flow
disturbances and particle interactions impose on the separation
and 3) development of new designs or operating protocol to
evaluate improvements in resolution or throughput for laboratory or space use.  Working with NASA in these investigations
have been university and industry scientists, most of whom
are presenting their recent results at this meeting.

On Earth, electrophoresis of proteins and macromolecules is
usually carried out in the presence of a gel or other stabilizing medium.  Although the gel offers an advantage of physical sieving, its primary purpose is to prevent convective
mixing of the separating molecules due to temperature gradients
in the buffer.  The gel structure restricts the quantities
that can be separated, complicates the extraction of the
fractions, and also precludes use of the technique for biological cells and particles.

One technique for minimizing some of these restrictions for

cell electrophoresis is the use of density gradients in columns. Vertical columns contain a density gradient to stabilize the fluid against buoyancy-induced convection although sedimentation of cells concentrated in fluid droplets remains a problem. Columns oriented horizontally can be rotated slowly to minimize the effects of sedimentation and thermal convection can be reduced by operating at low applied electric fields. Thin wall capillaries are oriented horizontally during the microscopic measurement of cell mobilities as the analytical analog for cells of gel electrophoresis. Thermal convection is small because the capillaries are immersed in a temperature-controlled water bath and the applied field is low. Sedimentation is perpendicular to the electrophoretic migration and mainly limits the duration of the mobility measurement. Although column electrophoresis has enjoyed only limited success in the laboratory, it was a first choice for space, where the fluid could be uniform and stationary with no disturbance due to buoyancy-induced convection or sedimentation.

The first experiments done on Apollo flights,[1,2] exhibited both electrophoresis and electroosmosis and although the process was easily resolved analytically, a good separation of the polystyrene latex microspheres was not achieved. However, no additional disturbances were noted. A low zeta potential coating (methyl cellulose) was applied to the next set of columns flown during the Apollo-Soyuz Test Project in 1975.[3] Although this experiment developed some new problems, all of the electrophoresis experiments were not affected and electroosmotic flow was negligible during the separation of fixed red blood cells. Human kidney cells were also fractionated and analysis of enzymes produced by the separated cells showed some separation of the kidney cells occurred according to function. Column electrophoresis in space has been shown to extend analytical electrophoresis on Earth by providing high resolution of small quantities of sample.[4] However, even in space, column electrophoresis is still a batch process.

Continuous or free flow electrophoresis was developed to allow
continuous insertion of sample and collection of fractions,
permitting fractionation of preparative quantities of material.
Continuous flow electrophoresis has generated most interest as
a separation technique for biological cells but it has introduced problems of operation and apparatus design, many of
which are gravity related.  The selection of the electrophoresis medium is a compromise between the requirements for
viable cells and optimum fractionation.  Biological cells prefer immersion in buffered electrolytes of physiological ionic
strength.  The cells enter the electrophoresis channel at a
predetermined concentration in a fluid medium whose properties
are comparable to the flowing curtain buffer medium.  The
electric field applied to this buffer curtain and sample insertion stream induces an electric current proportional to its
electrical conductivity.  Cells are typically not very mobile
in any electrolyte and high electric fields are required to
achieve any significant migration.  The combination of high
applied voltage and high electrical conductivity results in
the generation of intense heat in the fluid medium.  Fortu-
nately, the mobility of cells increases as the electrophoresis
buffer ionic strength is decreased and advantage of this
phenomenon can be taken provided cell viability is not affected.
A longer residence time in the electrophoresis chamber can
also give better separations but cell survival is reduced in
low ionic strength, isotonic buffers.  It is also important
that the temperature not be allowed to rise sufficiently to
adversely affect cell viability.  This gives a range of
options for operation with cells and buffer medium.


The apparatus design must consider means to remove the heat
from the buffer that do not inhibit the separation efficiency.
Cooling of the chamber walls is commonly done but this aggra-
vates thermal convection which is proportional to the temperature  gradient in the buffer.  The thermal convection dis-
rupts the rectilinear flow through the chamber and distorts

any separation.  Additionally, the temperature gradient causes a gradient in other fluid properties, such as viscosity and electrical conductivity, which further influences flow and particle migration.  By making the chambers very thick gradients in fluid properties and unwanted flows can be minimized, but with a concomitant loss in its efficiency as a preparative electrophoresis device.  The chamber cannot be too thin or the sample stream through the chamber will not contain enough cells to permit adequate throughput in a reasonable length of time. The cross-section of the sample stream should be small both in the direction of the electric field to increase the resolution of separation and perpendicular to the field to reduce the impact of the fluid cross flow (electroosmosis) near the walls of the chamber.  If the sample stream cross-section is a significant fraction of the chamber thickness, then electro-osmotic cross-flow is necessary to compensate for the parabolic (Poiseuille) down flow through the chamber and yield at least one relatively narrow fraction.  The various interacting phenomena described so far lead to operating procedures and apparatus design that must be a compromise.  Since gravity has a role in most of these interactions, weightlessness can be anticipated to offer some advantages.

The NASA Program

NASA became interested in continuous flow electrophoresis of cells and particles several years ago to investigate aspects of operation and design influenced by gravity.  Although buoyancy-induced fluid flow due to thermal and concentration gradients was shown to be negligible in the micro-gravity environment of space during the Apollo experiments, it was clear that weightlessness did not remove all the problems. Operation and apparatus design features, such as sample insertion and collection, wall coatings and selection of materials compatible with sterilization occupied much of the time necessary for preparation of the space experiments. A Beckman Instruments Continuous Particle Electrophoresis (CPE) system designed by Strickler was obtained to establish the best electrophoretic separation in a free fluid achievable on the ground using a standard reproducible particle population for comparison with the space experiment. Tests with the CPE, confirmed that successful operation of a continuous flow electrophoresis apparatus requires careful consideration of the major parameters and their interaction. Increasing the residence time and/or increasing the applied electric field gave increased displacement of the sample being separated. However, as discussed above, these parameters also influenced directly the Joule heating and extent of electroosmotic distortion of the sample. A series of measurements were done with the CPE and a recently purchased Desaga FF48 Free Flow Electrophoresis system designed by Hannig, to clarify some of these interrelationships and determine the separation efficiency of these devices on Earth.[5]

The CPE and FF48 have design features that made this comparison useful. The electrophoresis chambers are comparable in width and length but the thickness of the buffer curtain is 1.5 mm for the CPE and 0.5 mm for the FF48. The CPE has cooling on only one surface and the FF48 cools both major surfaces. The sample insertion stream is approximately 25% of the curtain thickness for the CPE and nearly 60% for the FF48. Considering these factors, the CPE should be influenced mainly by thermal convection and the FF48 should be limited by wall effects. It was anticipated that the comparison would clarify the separate role of these fluid disturbances. The interior chamber walls of each instrument were coated with bovine serum albumin to give a known, reproducible value for electroosmosis. Curtain and sample flow rates were adjusted to low values consistent with a narrow stream at zero electric field. The field was then applied at increasing values and the test particles, formaldehyde fixed cow and turkey red blood cells were collected.  These cells differ significantly both morphology and mobility and the separation analysis consisted of counting the relative number of cells in each collection tube.

Broadened sample collection was clearly obtained at high field strengths and low curtain flow rates.  Thus, optimum separations for each instrument could be compared with separations predicted by the sample's electrophoretic mobility distribution measured analytically using a Rank Brothers (Bottisham, Cambridge) microscope electrophoresis system.  Neither instrument has given the clear separation anticipated from measured single cell mobilities nor have wall effects or thermal convection been identified as the major disturbance to continuous flow electrophoresis.  Since electroosmosis can be modeled and controlled using existing methods and buoyancy-induced convection cannot, additional laboratory experiments were designed and conducted to analyze the role of buoyancy-induced convection in limiting the performance of continuous flow electrophoresis.

Relatively thick chambers with a buffer curtain thickness of 5.5 mm were constructed to test the hydrodynamics of continuous flow electrophoresis under various conditions of Joule heating.[6] A transparent chamber was built with variable cooling on all surfaces in order to compare operation with predicted behavior under various conditions of buffer curtain flow rates and chamber heating.  Data were collected on temperature and velocity fields and their relation to chamber operating variables, e.g., buffer flow rate, electric field intensity, cooling and orientation.  Tests used neutrally buoyant particles and various flow visualization techniques to obtain the velocity fields, small thermocouple probes to measure fluid temperatures and alternating voltage to induce Joule heating without complicating the experimental analysis with electroosmosis and electrophoretic migration of the velocity markers. A development of mathematical models of continuous flow

electrophoresis by Saville has paralleled these experiments.[7]
The first result  of the thick chamber testing was the realization that the laminar flow through the chamber was already
distorted when the input power was two orders of magnitude
less than predicted.  This lack of agreement between analysis
and experiment has required the more precise experiments
described by Rhodes.[6] Fifteen thermocouple probes now measure
the temperatures at the wall and the center-plane of the
flowing buffer and fluorescent tracer molecules plot the flow
velocity profiles at two locations in the chamber.  The
measured temperature and velocity fields are being used to
determine the lowest order disturbance to the fluid and as
input data to test the analytical model.  Although the experiments to date still do not have the precision of temperature
control and measurement necessary to give complete confidence
in the model, the following conclusions, from experiments at
NASA[8] and analysis by Saville,[9] can be made:

1.  Experiments and analysis agree that flow is extremely
sensitive to lateral (horizontal) temperature gradients.
Gradients less than $0.1C^{\circ}/cm$ are sufficient to distort uniform
base flow in the thick (5.5 mm) chamber.  To provide an estimation of this sensitivity, it has not been possible to control buffer temperatures by merely using coolant chambers on
all surfaces and the next transparent chamber will have a
controlled pathway for coolant fluid to minimize these lateral
gradients.

2.  The chamber is also sensitive to the axial temperature gradient established in the flowing curtain buffer.
Stable axial gradients with the warmer fluid over the denser,
cooler fluid must be maintained to avoid various modes of hydrodynamic instability.

3.  A two-dimensional model of electroosmosis has shown
how this cross flow changes the normally rectilinear flow by
transporting the colder fluid near the wall toward the cathode
and the warmer central fluid toward the opposite electrode.

This could set up a large circulation in the chamber and experiments will be conducted to verify this flow.

4. The computer model has been used to predict the separability of populations with known mobility distributions in the existing continuous flow electrophoresis devices. Although the model accounts for the variation of important fluid parameters with temperature, entrance and edge effects and polarization of membranes, there are still deficiencies in using the model to predict operation in these devices.

Experience with these various configurations of continuous flow electrophoresis devices has suggested two designs that could compensate for the major deficiencies of the existing devices. The "moving wall" electrophoresis device[10] takes advantage of the negligible electroosmosis due to methyl cellulose and chamber walls that move with the velocity of the buffer curtain. Thus, elimination of electroosmosis and Poiseuille flow removes two major sources of fluid disturbances in continuous flow electrophoresis. Methyl cellulose has been applied to belt loops of Mylar that carry the buffer through the chamber and sample particles have been transported through the chamber without distortion. Effort is now underway to separate and collect sample in this device to evaluate its performance in the laboratory. The remaining source of fluid disturbance, buoyancy-induced convection, could be eliminated by operating this system in space. A laboratory prototype system has been built and preliminary results are available.

The second concept, a "Thermally Stabilized Electrophoresis Chamber"[11] is an outgrowth of observations that the rectilinear flow of buffer through the cooled chamber is extremely sensitive to temperature gradients in the horizontal plane and along the major axis of the chamber. Although a stable axial gradient (warmer fluid on top of colder fluid) is desired, it has not been possible to consistently eliminate lateral gradients or prevent destabilizing axial gradients in the present laboratory transparent electrophoresis chamber using prevalent cooling methods. Recognizing that the thermal conductivity of all transparent materials is poor, the use of an efficient heat conductor, copper, has been incorporated in a new design. A copper chamber should equilibrate lateral temperature gradients sufficiently because its thermal conductivity is almost three orders of magnitude higher than glass. In addition, the cooling fluid can be channeled to provide a stabilizing axial gradient. The copper will require coating with a thin (approximately 0.001 inch) film of electrically insulating material to maintain the electric field in the chamber without severely compromising its thermal conductivity. A disadvantage of the copper chamber is the inability to see the sample material traverse the chamber. However, knowledge of the fluid behavior in the chamber gained from extensive modeling and experimentation in addition to mounting glass windows to observe the sample at the inlet and separated bands just before collection should suffice for most applications.

An additional advantage of the material selection is the capability to efficiently sterilize the entire system with high temperature steam. Use of the autoclave and various gaseous and chemical sterilants has not been consistently successful in sterilizing either the CPE or FF48 because of their many polymeric components. Although viable cells can be separated and promptly analyzed, culturing of these cells for subsequent analysis has generally resulted in overgrowth of the cultures with bacterial contaminants in the electrophoresis chamber. Having experienced stress corrosion cracking of plexiglass and polycarbonate materials components due to exposure to ETO/Freon and various chemical sterilants, it is not possible to assure sterility without using glass, metals, such as copper, and other materials with suitable properties.

The copper chamber concept is being tested in two ways. The

CPE has been modified to provide copper plates and improved cooling for both major surfaces.  The CPE will then be operated at higher field strengths and/or lower buffer flow rates to determine the increased efficiency of the new design.  A new electrophoresis facility has also been built that will give the best comparison of a thermally stabilized device and existing separators.

As the behavior of the heated fluid passing through the chamber during electrophoresis becomes understood, the resolution of separation achievable by electrophoresis in a characterized chamber becomes important.  The neutrally buoyant polystyrene latex particles used to map the fluid flow then must be replaced by monodisperse particle populations with narrow and stable electrophoretic mobility distributions that are distinguishable by methods other than electrophoresis.  Although polystyrene latex of different size proved to be good particles to characterize the process during the early tests, the ultimate aim of our electrophoresis work is to separate biological cells.  It has been proposed that the characteristics of biological cells should be retained as much as possible in the standard particles.  The most commonly used "standard" particle in analytical electrophoresis is the red blood cell and they have been adopted for our recent tests.  Different animal species provide red cells of different size, shape and electrophoretic mobility and they can be fixed with aldehydes to give them stability for long periods of time.

It is important that the mobility of the test particles be known with precision.  The microscope electrophoresis device was selected for the measurements principally because of the extensive red cell measurements accomplished with this instrument.  However, repeated use has also emphasized its major limitation, dependency on an observer.  The requirement for more precise mobility measurements on larger numbers of both test particles and viable cells resulted in a survey of available automated methods that could be applied to this technique.[12] The automated analytical electrophoresis apparatus, (AAEA),[13] designed by Bartels was selected over a variety of other approaches as best able to meet the requirements. Basically, the AAEA takes the thin-walled cylindrical electrophoresis chamber and adds phase contrast optics to the microscope. A computer coordinates the tracking of individual cells, whose images are projected onto a vidicon, permits rapid taking of multiple measurements for each cell, and calculates the electrophoretic mobility. The standard error of the cell's mobility is minimized by making multiple measurements. The precision of this instrument has identified an additional limitation of this technique, namely, that fluid and particle trajectories in the cylindrical capillary are not as stable as expected from visual observation. However, this will be evaluated as a later task.

A major design objective of electrophoresis devices has been to increase the amount of sample separated in a given time interval. It has been proposed that the particle concentration that can be separated by electrophoresis in a weightless environment is significantly higher than on Earth because sedimentation and particle interactions compounded by buoyancy induced convection will not occur. The limitations in initial particle concentration imposed by droplet formation and sedimentation on density gradient and continuous flow electrophoresis have been observed. Generally, investigators are hesitant to use high input cell concentrations and operate their devices at the lower levels reported in the literature.

A systematic evaluation of the maximum particle concentration that can be separated in static fluid and continuous flow electrophoresis is underway. Using standardized particles and fluids, a series of experiments have been done to find the maximum packing of particles that can be supported in stationary and flowing liquid systems. Increased stability of

the cell suspension has been achieved by lowering the surface
tension of the supporting solution, thereby decreasing the
van der Waal's attraction.[14]

New methods for doing isoelectric focusing in a free fluid is
also part of the NASA program.  The recycling isoelectric
focusing apparatus, developed by Bier,[15] has the objective of
large scale, continuous flow separation with increased concern
directed toward the fluid dynamics of the entire process.
This apparatus has been designed to separate the major elements
of the process, such as focusing, cooling and measurement of
operational parameters.  Filter elements have been inserted
into the focusing chamber to minimize fluid disturbances
while permitting cross-flow of the ampholytes and macromolecular species.  Although the filters stabilize the flow through
the device, the mechanical problems of the filter, protein
absorption and resultant non-uniform cross flow, and electroosmosis along all surfaces are limitations.  Mathematical
modeling has begun to evaluate the hydrodynamics of this
process with the same rationale used for the comparable
analysis of continuous flow electrophoresis.  This analysis
has clarified the steady state in isoelectric focusing using
electrochemically defined ampholytes and, in particular,
identified the importance of controlling electroosmosis.  These
studies and associated experiments will establish the capabilities of the apparatus and operation with various ampholyte
systems.

Finally, deficiencies in our knowledge of apparatus design,
fluid dynamics and particle interactions do not provide the
total reason why few cell populations have been fractionated
into clearly defined groups by electrophoresis.  Microscope
electrophoresis apparatus measuring individual cell mobilities
has yielded only broad spreads in mobility spectra for many
cell populations that are acknowledged to have diverse
functions.  A method to enhance the mobility difference of a

specific cell subgroup is being investigated by Rembaum and his colleagues at the Jet Propulsion Laboratory and University of Alabama in Birmingham. Hydrophilic polymeric microspheres of various size, composition and surface charge can have specific antibodies chemically bound to their surfaces. The antibody-coated microspheres will then couple only to those cells that normally react to the antibody and a new particle consisting of the cells coated with microspheres is created. Human and turkey fixed red blood cells whose mobilities over-lapped were clearly separated when the human red blood cells were essentially covered with the immunomicrospheres.[16] Re-sults are now being achieved with viable cell populations and new methods developed to couple and decouple the microspheres without modification to the host cell.

Flight Experiment Plans

As laboratory experiments and studies identify specific fea-tures of the separation process that can be evaluated or im-proved in a weightless environment, concepts and designs for a space experiment are formulated and submitted to NASA. Experiment proposals in electrophoresis are now being reviewed and it is anticipated that as the Space Shuttle becomes oper-ational, electrophoresis experiments based upon some of the work discussed in this paper will be accepted for flight. These experiments will have established a clear need for weightlessness and accomplished the supporting laboratory work.

Several years ago, NASA requested industrial participation in space by offering to fly experiments designed and built by industry. The McDonnell-Douglas Astronautics Company (MDAC) agreed to support their own design and construction of a con-tinuous flow electrophoretic separator that should be ready to go into space in late 1982. The major features of their apparatus compared with existing devices are a thicker chamber with considerably longer length and narrower collection ports.

Tests on the ground using internal temperature and flow sensors
have shown improved operation in the inverted mode, i.e.,
sample and buffer inserted at the bottom of the apparatus.
Separations recently done in their laboratory using low applied electric fields and innovative means to control electroosmosis and chamber cooling fluid have yielded greater separation distances but with some band broadening.  MDAC will
investigate the increased particle concentrations that can be
separated in space without density mismatch between sample
and buffer that occurs on Earth.  MDAC has received active
support from one major pharmaceutical company and interest
from several others in the development of their system although details are presently lacking because of the proprietary
nature of this commercial venture.

Summary

The research described in this paper and others sponsored by
NASA have emphasized basic understanding of phenomena and the
development of apparatus or procedures that improve some
specific aspect of electrophoresis.  This effort has an under-
lying basis of evaluating the need for weightlessness but the
results have also clarified or aided electrophoresis in the
laboratory.  New instruments have now been designed and working
models built for testing with standard sample populations.
These instruments will need to show improvements in resolution
or throughput and be supported by analysis to gain continued
support.  Some concepts will then be carried through to construction of the entire system and testing with samples
that are more important to separate.  This program must relate
the expected improvement in separation performance to the
needs of the chemical and biomedical community.  It is anticipated that the combined program of laboratory experiments,
theoretical modeling and space experiments will lead us to
these accomplishments.

References

1.  McKannan, E.C.:  NASA TM X-64611 (1971).

2.  Snyder, R.S., et. al.:  Sep. Purif. Methods, $\underline{2}$, 259 (1974).

3.  Micale, F.J., Vanderhoff, J.W. and Snyder, R.S.: Sep. Purif. Methods, $\underline{5}$, 361 (1976).

4.  Allen, R.E., et. al.:  Sep. Purif. Methods, $\underline{6}$, (1977).

5.  McGuire, J.K., Snyder, R.S.:  Submitted for Publication.

6.  Rhodes, P.H.:  NASA TM-78178 (1979).

7.  Saville, D.A.:  Physicochemical Hydrodynamics, $\underline{2}$, 893 (1978).

8.  Rhodes, P.H., NASA TM in preparation.

9.  Saville, D.A.:  Physicochemical Hydrodynamics, $\underline{4}$, 297 (1980).

10.  Rhodes, P.H.:  Electrophoresis '81 (1981).

11.  Rhodes, P.H., Miller, T.Y., Snyder, R.S.:  Electro-phoresis '81 (1981).

12.  Brooks, D.E.:  University of Oregon Health Sciences Center, Final Report, Contract NAS8-31386 (1976).

13.  Olson, G.B., Bartels, P.H., Bartels, H.G., Brooks, D.E., Seaman, G.V.F.:  Procedings of the Society of Photo-Optical Instrumentation Engineers, $\underline{232}$, 54 (1980)

14.  Omenyi, S.N., Snyder, R.S., van Oss, C.J., Absolom, D.R., Neumann, A.W.:  J. Colloid Interface Sci., in press.

15.  Bier, M., Egen, N.B., Allgyer, T.T., Twitty, G.E., Mosher, R.A.:  Peptides, Structure and Biological Function, Pierce Chemical Co., Rockford, Il., 79 (1978).

# NUMERICAL ANALYSIS OF CONTINUOUS FLOW ELECTROPHORESIS

Percy H. Rhodes and Robert S. Snyder
NASA/Marshall Space Flight Center
Huntsville, Alabama  35812

## Introduction

Continuous flow electrophoresis is a fractionation process
which is performed within a free-flowing film of aqueous electrolyte medium confined in a long rectangular chamber of high
aspect ratio.  The broad faces of the chamber confine the
fluid to flow as a "curtain" between the two parallel plates.
The sample, which is continuously injected into the curtain
as a finely drawn stream, is separated under the influence of
a lateral dc field produced by flanking electrodes.

If true separations based on electrophoretic mobility are to
be realized, fully developed laminar flow must be maintained
in the electrophoresis chamber.  While this base flow is parabolic in the narrow plane (thickness) of the chamber, it is
ideally uniform across nearly the entire width of the plane
of separation, except at the lateral edges where the no-slip
condition on the side walls exists.  However, it has been
noted that small thermal gradients across the width of the
chamber can cause the base flow to become nonuniform and hence
degrade separation.  These very small lateral gradients, which
are sufficient to cause the separating sample streams to
change, are difficult to detect in a typical chamber configuration.  A wide-gap (5.5 mm thick) chamber was therefore
used to intensify these thermal disturbances.  Sufficient
temperature and flow measurements were then taken to verify
an analytical model of the process.  With the mathematical
model, it is possible to characterize small buoyancy-driven

disturbances which are important but not measurable in ground-based electrophoresis chambers.

A mathematical model is now being used to study gravity effects on the electrophoresis process and to predict the optimum separations that can be achieved. Of more importance to the electrophoresis community, the results of parametric studies using the analytical model will be useful in the development of more efficient separation chambers for the laboratory. This paper describes how one such analytical model was developed and experimentally verified and suggests uses for the model in the study and development of continuous flow electrophoresis.

Experimental Apparatus and Procedures

The experimental phase of the work essentially consisted of observing and recording flow and temperature fields in a wide-gap (5.5 mm thickness) electrophoresis-type chamber. These results were necessary to verify the analytical model.

The flow chamber was made of plexiglass with cooling chambers on the front, rear, and sidewalls. Figure 1 shows the flow chamber details. The buffer (electrolyte) entered the chamber at one end through a single entry port and exited at the other end through a single exit port. The coolant entered the cooling chambers at the bottom and exited at the top. No effort was made to route the flow of coolant through the coolant chambers. A degree of thermal control was achieved, however, by manipulating the respective flow rates in the coolant chambers. Figure 2 shows the experimental apparatus and Figure 3 shows a schematic of the flow channel without its accessory apparatus and denotes the chamber orientational and geometrical directions. Note that the x-y plane is the chamber center plane, while the x-z plane is the transverse plane.

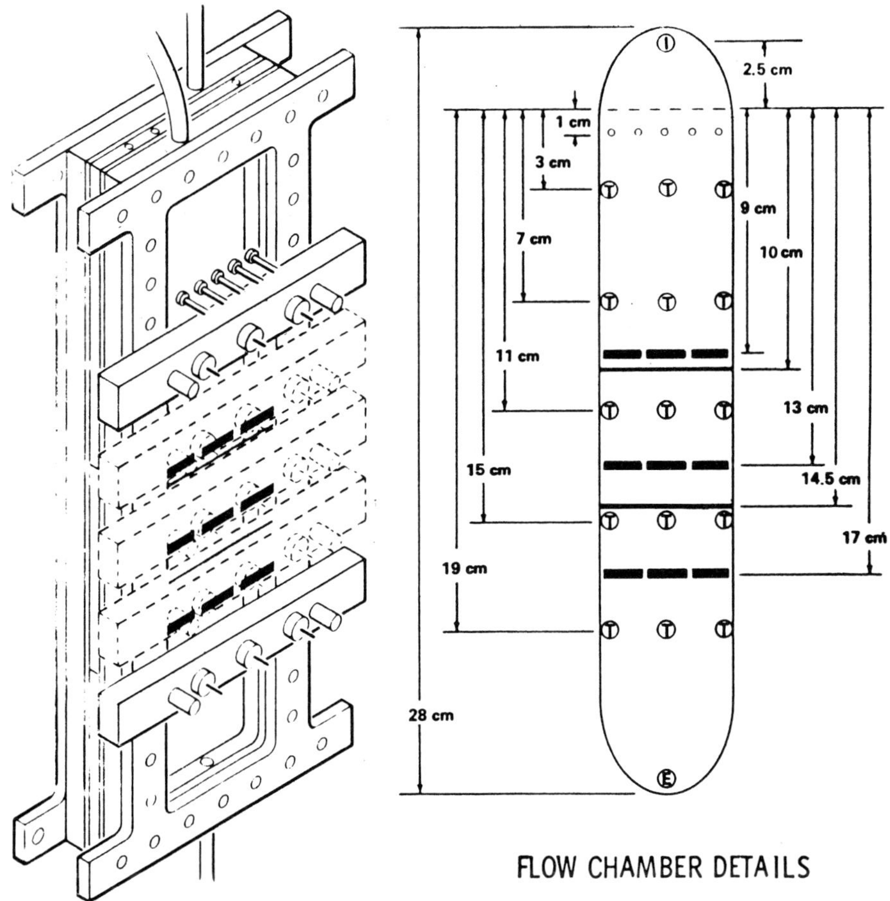



**LEGEND**

| | |
|---|---|
| ○ PSL INJECTION PORTS | COOLING: FRONT, BACK, AND SIDEWALLS |
| ⊤ THERMOCOUPLE PROBES | ELECTRODES: PAIRS MOUNTED |
| ▬ INDIVIDUAL PAIRS OF ELECTRODES | OPPOSITE EACH OTHER IN FRONT |
| ▬ FLOW VISUALIZATION ACTIVATION FIBER | AND REAR FACES (.2 cm X 1.4 cm) |
| ① FLOW INLET | POWER: 400 Hz |
| Ⓔ FLOW OUTLET | TEMPERATURE MEASUREMENT: |
| CHAMBER SPECIFICATIONS | MOVABLE THERMOCOUPLE PROBES |
| TOTAL LENGTH — 28 cm | MOUNTED IN REAR FACE |
| WIDTH — 5.1 cm | FLOW VISUALIZATION: PSL AND |
| THICKNESS — 0.55 cm | FLUORESCENCE OF FLUID MEDIUM. |

Figure 1.   Schematic of Flow Chamber

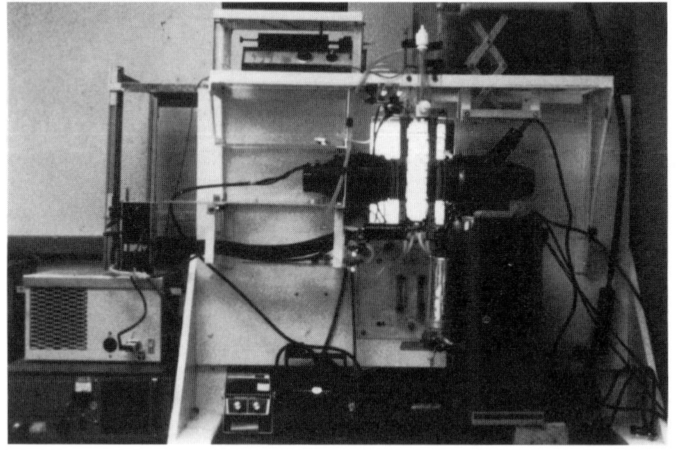

Figure 2.  Experimental Apparatus

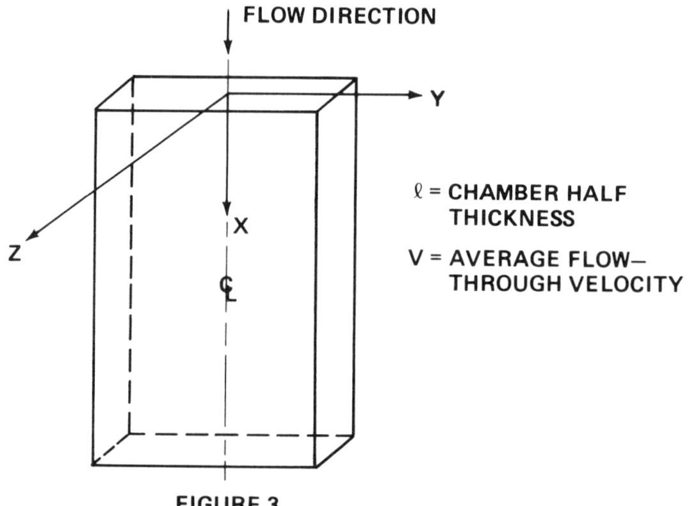


**FIGURE 3**

| Coordinate Axis | Geometrical Description | Orientational Direction |
|---|---|---|
| x | length | axial |
| y | width | lateral |
| z | thickness | transverse |

Figure 3.  Schematic of the Flow Channel

The arrangement of the thermocouple probes is shown in Figure 1. A total of 15 thermocouples were mounted in sets of three on movable bars which were moved into and out of the flow curtain. An additional thermocouple was mounted in the flow inlet to monitor the curtain inlet temperature. The transverse temperature profile and heat transfer through the chamber wall were determined by proper positioning of the temperature probes. The temperature readings were scanned and recorded within 5 seconds using an HP 3495-H scanner.

The velocity profiles at the center plane were determined by a method developed for that purpose. Two small fibers (0.2 mm diameter) were stretched across the width of the flow chamber at the center plane, as shown in Figure 1. Fluorescence of the fluid medium immediately surrounding the fiber was achieved by flowing an antivator fluid through the fiber. Flow of the fluorescent buffer adjacent to the fiber then described the velocity profile in the region of the fiber. Polystyrene latex particles (PSL) were also used on occasion to further describe the flow in the chamber. The PSL was injected through five ports, as shown in Figure 1. Quantitative measurements of velocity were made in the vicinity of the fibers by taking timed sequential photographs.

The data-taking sequence started with a desired flow or velocity field in the chamber. The flow was considered established when temperature changes of less than $0.10^{\circ}C$ were observed over a period equal to the flow residence time of the chamber. However, in some cases, due to the capricious nature of the flows, this standard had to be compromised. After a flow was established, temperature and flow data were taken and printed out using an HP 9835 A computer and associated printer while a photographic sequence was taken of the flow field in the chamber. The photographs and printed data formed a simultaneous record of the flow event.

904

## Analytical Model Development

Symmetrical flow about the center plane of the chamber is
assumed. The properties of the fluid medium are assumed to be
spatially constant and depend only on the average temperature
of the chamber. The fluid may be considered incompressible,
with the buoyancy effect being represented by the Boussinesq
approximation. The transverse velocity component is considered negligible when compared to the lateral (v) and axial
(u) components so that

$$\vec{w} = u\hat{i} + v\hat{j}.$$

The low Reynolds numbers involved justify the dropping of the
inertia terms so that the Navier-Stokes equations assume the
form

$$\rho\frac{\partial \vec{w}}{\partial t} = -\,\text{grad }p + \mu\nabla^2\vec{w} - \rho\vec{g}\beta(T - \bar{T}) \tag{1}$$

$$\text{div }\vec{w} = 0, \tag{2}$$

where $\bar{T}$ is a reference temperature.

The energy equation for incompressible flow is given by

$$\rho c\,\frac{DT}{Dt} = k\nabla^2 T + w_o, \tag{3}$$

where $w_o$ is the energy dissipated in the fluid per unit volume.
The boundary conditions are evaluated at the inside walls of
the chamber.

Investigation of the flow and temperature fields in an electrophoresis-type chamber requires the solution of the preceding
flow and heat transport equations. A full three-dimensional
solution of these equations would be quite time consuming
and unnecessary since for free flow electrophoresis the
sample must necessarily be confined to the vicinity of the
center plane of the chamber. Therefore, these equations are
solved only for flow at the center plane of the chamber.
This is done by estimating the local velocity and temperature
profiles in the transverse plane, from which the preceding

equations may be reduced from three dimensions to two dimensions and solved for flow at the center plane.

To estimate the axial velocity profile in the transverse plane, we consider fully developed flow between two parallel plates of constant temperature. The governing equations above are then reduced to a tractable form which may be solved for the axial velocity.[1,2] Then the local axial velocity is given by:[2]

$$u'(x,y,z,t,T) = u_o(x,y,t)(1-z^2)\left\{1 + \left[k_1 - k_2(5 - z^2)\right]N_3 R_e T(x,y,t)\right\}, \qquad (4)$$

where $k_1$ and $k_2$ are constants ($\frac{4}{15}$ and $\frac{1}{18}$ respectively), the buoyancy term $N_3 = \frac{g\beta\ell}{v^2}$ acts to either blunt or extend the characteristic parabolic profile in the transverse plane, the Reynolds number $R_e = \frac{V\ell}{v}$, the transverse temperature difference $T(x,y,t) = T_c(x,y,t) - T_w(x,y,t)$, $u_o(x,y,t)$ is the temperature independent component of axial velocity at the center plane, and z is a nondimensional transverse coordinate relative to the chamber half thickness $\ell$.

To estimate the transverse temperature profile, we assume a fourth-order polynomial in z and apply boundary and symmetry conditions to obtain

$$T'(x,y,z,t) = T(x,y,t) - \left[2T(x,y,t) + \frac{1}{2}\frac{\partial T}{\partial z}\Big|_{z=1}\right]z^2 \qquad (5)$$
$$+ \left[\frac{1}{2}\frac{\partial T}{\partial z}\Big|_{z=1} + T(x,y,t)\right]z^4 + T_w(x,y,t).$$

The temperature gradient at the wall, $\frac{\partial T}{\partial z}\Big|_{z=1}$, represents the transverse heat transfer at the wall and, like the wall temperature, is a boundary condition. If we use a heat transfer coefficient, h, equation (5) becomes

$$T' = T\left[1 - \left(2 - \frac{h}{2}\right)z^2 + \left(1 - \frac{h}{2}\right)z^4\right] + T_w, \qquad (6)$$

where $\frac{\partial T}{\partial z}\Big|_{z=1} = -hT.$

The heat transfer coefficient was evaluated by substituting temperatures measured at the chamber center line into an abbreviated energy equation (less lateral conduction terms) and solving for h while using local velocities so that

$$h = f(x,y) .$$

So with estimates of the axial velocity in the transverse plane and temperature profiles we can reduce the three-dimensional governing equations to two dimensions. Using the chamber flow thickness, $2\ell$, and the average flow velocity, V, as reference dimensions, equations (1) through (3) become, in extended form,

$$\frac{\partial u'}{\partial t} = -\frac{\partial p}{\partial x} - N_3(T' - \bar{T}) + \frac{1}{R_e}\nabla^2 u' \tag{7}$$

$$\frac{\partial v'}{\partial t} = -\frac{\partial p}{\partial y} + \frac{1}{R_e}\nabla^2 v'$$

$$\frac{\partial T'}{\partial t} = \frac{1}{R_e P_{tn}}\nabla^2 T' - \frac{\partial(u'T')}{\partial x} - \frac{\partial(v'T')}{\partial y} + N_1$$

$$\frac{\partial u'}{\partial x} + \frac{\partial v'}{\partial y} = 0 \quad ,$$

where

$$N_1 = \frac{w_o \ell^2}{k R_e P_{tn}}$$

$P_{tn}$ = Prandtl number

k = thermal conductivity of the buffer.

The parameter $\bar{T}$ is a reference temperature determined so that across any horizontal plane, i.e., perpendicular to the gravity vector, no net change of momentum results from buoyancy effects. The result is that $\bar{T}$ is simply the average temperature in each horizontal layer.

$$\bar{T} = \bar{T}(x) \quad .$$

Now, if we assume parabolic lateral flows in the chamber cross section,

$$v'(x,y,z,t) = v(x,y,t)(1 - z^2) \, ,$$

we can, with equations (4) and (6), integrate over the chamber transverse dimension and hence eliminate the z dependence from the governing equations. Thus equations (7) become two-dimensional expressions in x and y which represent the flow and temperature fields in the chamber center plane. These equations become, after integration:

$$\frac{\partial u}{\partial t} = -\frac{3}{2}C_x\frac{\partial p}{\partial x} + \frac{1}{R_e}\left[\frac{\partial^2 u}{\partial x^2} + \frac{\partial^2 u}{\partial y^2} - 3u\left(1 + \frac{2N}{45}\right)\right]$$

$$- N_3 C_x\left[\frac{4}{3}T + \frac{hT}{10} + \frac{3}{2}(T_w - \overline{T})\right]$$

$$\frac{\partial v}{\partial t} = -\frac{3}{2}\frac{\partial p}{\partial y} + \frac{1}{R_e}\left[\frac{\partial^2 v}{\partial x^2} + \frac{\partial^2 v}{\partial y^2} - 3v\right] \tag{8}$$

$$\frac{\partial T}{\partial t} = \frac{1}{R_e P_{tn}}\left[\frac{\partial^2 T}{\partial x^2} + \frac{\partial^2 T}{\partial y^2} + \frac{15}{8+h}\left(\frac{\partial^2 T_w}{\partial x^2} + \frac{\partial^2 T_w}{\partial y^2}\right) - \frac{15}{8+h}hT\right]$$

$$- \frac{15}{8+h}\left[\frac{8}{15}P\frac{\partial(uT)}{\partial x} + \frac{3}{2}C_o\frac{\partial(uT_w)}{\partial x} + \frac{8}{15}Q_o\frac{\partial(vT)}{\partial y}\right.$$

$$\left. + \frac{2}{3}\frac{\partial(vT_w)}{\partial y}\right] + \frac{15}{8+h}N_1$$

$$C_o\frac{\partial u}{\partial x} + \frac{\partial v}{\partial y} = 0 \, ,$$

where

$$Q_o = \frac{6}{7} + \frac{h}{14}$$

$$u(x,y,t) = C_o U_o$$

$$N = R_e N_3$$

$$C_x = 1 - 90N$$

$$C_o = 1/C_x$$

$$P = \frac{1620 - 8N + h(135 + N)}{21(90 - N)}$$

Equations (8) must be solved numerically for the velocity and temperature fields in the chamber center plane. The method of solution follows that delineated in References 3 and 4. The technique, based on the marker-and-cell method, uses a Eulerian finite-difference formulation with pressure, velocity, and temperature as the primary dependent variables.

After initiation of the program, as shown in the flow chart of Figure 4, the steps involved in completing one calculational cycle are (1) computing estimates for the new temperature and velocities for the entire mesh from the governing equations [equations (8)], which in the finite difference form involve only the previous time values for the contributing pressures, temperatures, and velocities, and (2) adjusting these velocities iteratively to satisfy the continuity equation by making appropriate changes in the pressure field. In the iteration, each cell in the mesh is considered successively and is given a pressure change that drives its instantaneous velocity divergence to zero. Finally, when convergence has been reached, the velocity, temperature, and pressure fields are at the advanced time level and may be used as guesses for the next cycle. The cycle is repeated until convergence to a steady state solution is achieved.

Results

The results are essentially a correlation of analytical and experimental results, in particular, velocity and temperature fields, which will be compared to show the validity of the mathematical model.

Figure 5 shows the experiment data sheet for run #814. The left-hand portion of this sheet shows temperatures measured at the wall and along the center plane of the chamber at locations noted in Figure 1. The right side of the sheet gives the curtain and coolant flow rates, power applied to

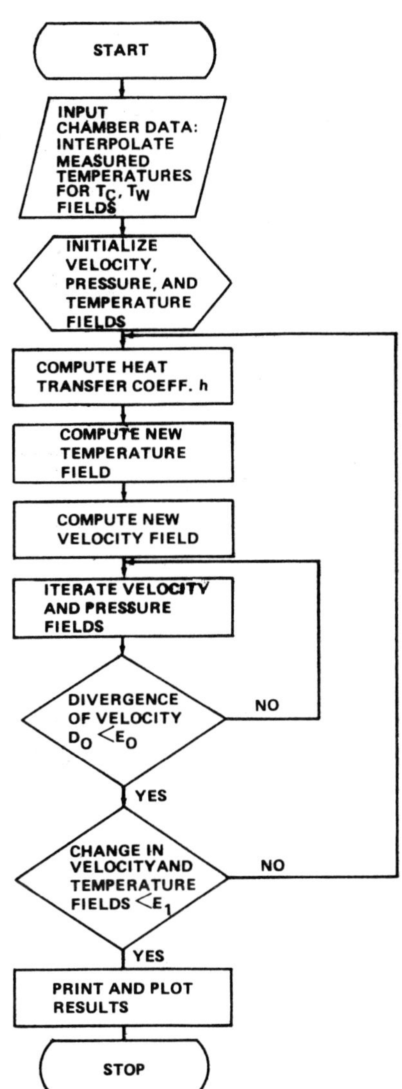


Figure 4.   Program Flow Chart

910

```
****************************************************************************
  RUN NUMBER: 8/4 BASE FLOW                                  Cooler off
****************************************************************************

            THE TIME IS-10:58:00                   THE DATE IS-09/26/80

      CURTAIN INLET TEMPERATURE:    27.34  Deg C    CURTAIN FLOW:    10.0  Ml/min

 6.81     WALL TEMPERATURES Deg C
                                                    COOLANT FLOW
15.23  27.66      27.46      27.76                  RIGHT SIDE-  75  Ml/min
                                                    FRONT FACE-  157  Ml/min
                                                    REAR  FACE-  150  Ml/min
       27.83      27.66      27.90                  LEFT  SIDE-  50  Ml/min

                                                    POWER INPUT
       27.89      27.66      27.93                  VOLTS-  0.0
                                                    MILLIAMPS- 0.0

       27.85      27.71      27.95                  ELECTRODES- 0,00,0,00,0,00,0,00

                                                    PHOTOGRAPHY:Roll # 8 FRAME # 11,12,13,4
       27.86      27.73      27.90
****************************************************************************

            THE TIME IS-10:59:00                   THE DATE IS-09/26/80

      CURTAIN INLET TEMPERATURE:    27.37  Deg C    CURTAIN FLOW:    10.0  Ml/min

      CENTER PLANE TEMPERATURES Deg C
                                                    COOLANT FLOW
       27.68      27.37      27.80                  RIGHT SIDE-  75  Ml/min
                                                    FRONT FACE-  157  Ml/min
                                                    REAR  FACE-  150  Ml/min
       27.88      27.49      27.95                  LEFT  SIDE-  50  Ml/min

                                                    POWER INPUT
       27.88      27.54      27.98                  VOLTS-  0.0
                                                    MILLIAMPS- 0.0

       27.88      27.61      28.00                  ELECTRODES- 0,00,0,00,0,00,0,00

                                                    PHOTOGRAPHY:Roll # 8 FRAME # 11,12,13,14
       27.88      27.66      27.95
****************************************************************************
```

Figure 5.  Experiment Data Sheet

the different electrode combinations and photographic sequence.
Note the adverse axial temperature gradients (cooler at the
top) at both the wall and center plane and the lateral variation in temperatures, i.e., cooler along the chamber center
line than at the side walls.  Note also that no power was
applied to the electrodes during this measurement.  The numbers
at the upper left-hand corner of the page represent times in
seconds between the three photographs (Figure 6) taken to show
the velocity profile.[5]

Note that the velocity is retarded at the chamber side walls
with apparent backflow near the right side wall.  This phenomenon is typical of all of the results we have observed; i.e.,

(a) Flow profile  
    at T = 0

(b) Flow profile  
    at T = 6.81  
    sec

(c) Flow profile  
    at T = 15.23  
    sec

Figure 6.  Velocity Profile Sequence

the velocity is retarded in that region of the chamber cross section where the temperatures are higher than the average for that particular cross section.  This is, of course, to be expected, as can be seen by referring to the buoyancy term of equation (7).  Table 1 is a computer printout of the calculated axial velocity field.  Note that the agreement with the observed velocity profiles of Figure 6 is excellent.  A more obvious comparison may be made by referring to the plot of the computer results shown in Figure 7.  Table 2 is a printout of the lateral velocity Field.

The photographic sequence of Figure 6 allows us to determine a quantitative comparison between the calculated and observed velocity profiles.  The comparison is made for the maximum velocity and proceeds as follows:

1.  A scale factor F is calculated by measuring the distance between two prominent points on the chamber and photograph, respectively -- here we used the electrodes, so

$$F = \frac{4.0 \text{ cm}}{1.70 \text{ cm}} = 2.35 .$$

912

```
PROGRAM COMPLETE USING 1633 CYCLES
RUN NUMBER 814.0000
Dx= 1.9139          Dy= 1.6694
Dt= .0100 E= .0010 E1= .0010
N3= 21.1048 Re= 1.9396
```

THE CHAMBER  AXIAL VELOCITIES ARE

$\longrightarrow y$

|      | 1    | 2    | 3    | 4    | 5    | 6    | 7    | 8    | 9    | 10   | 11    | 12    | 13   |
|------|------|------|------|------|------|------|------|------|------|------|-------|-------|------|
| 1    | -.02 | 1.25 | 1.48 | 1.46 | 1.45 | 1.48 | 1.50 | 1.50 | 1.53 | 1.68 | 1.84  | 1.74  | .22  |
| 2    | .01  | 1.31 | 1.61 | 1.72 | 1.78 | 1.81 | 1.81 | 1.78 | 1.71 | 1.50 | 1.24  | .82   | .02  |
| 3    | .02  | 1.26 | 1.79 | 2.03 | 2.17 | 2.23 | 2.25 | 2.15 | 1.76 | 1.08 | .48   | -.16  | -.00 |
| 4    | .02  | 1.09 | 1.96 | 2.36 | 2.58 | 2.69 | 2.70 | 2.35 | 1.58 | .63  | -.09  | -.83  | .02  |
| 5    | .01  | .86  | 2.08 | 2.66 | 2.97 | 3.12 | 3.05 | 2.40 | 1.38 | .35  | -.48  | -1.34 | .02  |
| 6    | .01  | .63  | 2.14 | 2.91 | 3.31 | 3.47 | 3.30 | 2.42 | 1.26 | .20  | -.77  | -1.75 | .03  |
| 7    | .00  | .43  | 2.14 | 3.09 | 3.57 | 3.74 | 3.51 | 2.45 | 1.22 | .11  | -1.00 | -2.07 | .03  |
| 8    | .00  | .28  | 2.10 | 3.19 | 3.76 | 3.95 | 3.68 | 2.50 | 1.23 | .04  | -1.16 | -2.32 | .02  |
| 9    | .00  | .16  | 2.03 | 3.24 | 3.90 | 4.12 | 3.84 | 2.58 | 1.27 | .00  | -1.29 | -2.48 | .02  |
| 10   | .00  | .03  | 1.98 | 3.29 | 4.01 | 4.26 | 3.99 | 2.68 | 1.31 | -.04 | -1.37 | -2.56 | .02  |
| 11   | -.00 | -.09 | 1.92 | 3.31 | 4.09 | 4.38 | 4.11 | 2.76 | 1.34 | -.07 | -1.41 | -2.61 | .02  |
| 12   | -.00 | -.20 | 1.85 | 3.31 | 4.15 | 4.46 | 4.21 | 2.83 | 1.36 | -.08 | -1.41 | -2.63 | .02  |
| 13   | -.00 | -.28 | 1.78 | 3.29 | 4.17 | 4.51 | 4.28 | 2.89 | 1.39 | -.08 | -1.40 | -2.64 | .01  |
| 14   | -.00 | -.34 | 1.70 | 3.26 | 4.18 | 4.55 | 4.32 | 2.94 | 1.41 | -.06 | -1.37 | -2.63 | .01  |
| 15   | -.00 | -.37 | 1.63 | 3.22 | 4.17 | 4.56 | 4.35 | 2.97 | 1.44 | -.02 | -1.33 | -2.63 | .00  |
| 16   | -.00 | -.39 | 1.58 | 3.18 | 4.15 | 4.56 | 4.35 | 3.00 | 1.46 | .07  | -1.32 | -2.64 | -.00 |
| 17   | -.00 | -.40 | 1.54 | 3.14 | 4.13 | 4.54 | 4.35 | 3.03 | 1.51 | .12  | -1.32 | -2.64 | -.01 |
| 18   | -.00 | -.39 | 1.51 | 3.09 | 4.09 | 4.51 | 4.33 | 3.05 | 1.55 | .12  | -1.30 | -2.62 | -.01 |
| 19   | -.00 | -.37 | 1.49 | 3.04 | 4.05 | 4.48 | 4.30 | 3.06 | 1.58 | .11  | -1.28 | -2.59 | -.01 |
| 20   | -.00 | -.33 | 1.47 | 3.00 | 4.00 | 4.44 | 4.27 | 3.06 | 1.59 | .11  | -1.26 | -2.56 | -.01 |
| 21   | -.00 | -.29 | 1.47 | 2.97 | 3.96 | 4.40 | 4.24 | 3.05 | 1.59 | .12  | -1.24 | -2.53 | -.01 |
| 22   | -.00 | -.24 | 1.48 | 2.93 | 3.91 | 4.36 | 4.21 | 3.03 | 1.59 | .12  | -1.21 | -2.49 | -.01 |
| 23   | -.00 | -.19 | 1.49 | 2.89 | 3.87 | 4.32 | 4.17 | 3.01 | 1.59 | .13  | -1.18 | -2.45 | -.01 |
| 24   | -.00 | -.12 | 1.50 | 2.86 | 3.82 | 4.26 | 4.12 | 2.98 | 1.57 | .14  | -1.14 | -2.39 | -.01 |
| 25   | -.00 | -.05 | 1.50 | 2.82 | 3.76 | 4.21 | 4.08 | 2.95 | 1.55 | .16  | -1.10 | -2.34 | -.02 |
| 26   | .00  | .02  | 1.51 | 2.78 | 3.70 | 4.15 | 4.03 | 2.92 | 1.54 | .18  | -1.05 | -2.27 | -.02 |
| 27   | .00  | .08  | 1.52 | 2.75 | 3.64 | 4.09 | 3.98 | 2.88 | 1.52 | .20  | -.99  | -2.21 | -.02 |
| 28   | .00  | .14  | 1.53 | 2.71 | 3.58 | 4.02 | 3.93 | 2.86 | 1.52 | .23  | -.93  | -2.13 | -.02 |
| 29   | .00  | .19  | 1.54 | 2.68 | 3.51 | 3.95 | 3.87 | 2.83 | 1.52 | .26  | -.87  | -2.05 | -.02 |
| 30   | .00  | .23  | 1.54 | 2.64 | 3.44 | 3.88 | 3.81 | 2.81 | 1.52 | .29  | -.81  | -1.97 | -.02 |
| 31   | .00  | .25  | 1.53 | 2.60 | 3.38 | 3.80 | 3.75 | 2.79 | 1.53 | .33  | -.73  | -1.88 | -.02 |
| 32   | .00  | .27  | 1.52 | 2.56 | 3.32 | 3.73 | 3.69 | 2.77 | 1.54 | .37  | -.66  | -1.78 | -.02 |
| 33   | .00  | .28  | 1.51 | 2.52 | 3.27 | 3.67 | 3.63 | 2.76 | 1.54 | .41  | -.59  | -1.69 | -.02 |
| 34   | .00  | .29  | 1.50 | 2.47 | 3.23 | 3.62 | 3.60 | 2.73 | 1.54 | .44  | -.52  | -1.58 | -.02 |
| 35   | .00  | .30  | 1.48 | 2.42 | 3.17 | 3.57 | 3.58 | 2.70 | 1.53 | .47  | -.46  | -1.47 | -.02 |
| 36   | .00  | .32  | 1.48 | 2.38 | 3.11 | 3.52 | 3.56 | 2.65 | 1.52 | .50  | -.39  | -1.35 | -.02 |
| 37   | .00  | .34  | 1.48 | 2.35 | 3.04 | 3.46 | 3.52 | 2.59 | 1.51 | .53  | -.34  | -1.24 | -.02 |
| 38   | .00  | .37  | 1.48 | 2.34 | 3.00 | 3.40 | 3.45 | 2.54 | 1.50 | .54  | -.29  | -1.14 | -.01 |
| 39   | .00  | .38  | 1.48 | 2.33 | 2.97 | 3.36 | 3.39 | 2.51 | 1.49 | .55  | -.26  | -1.08 | -.01 |
| 40   | -.46 | .38  | 1.48 | 2.33 | 2.97 | 3.36 | 3.39 | 2.51 | 1.49 | .55  | -.26  | -1.08 |      |

x (downward arrow at left, beside rows)

Fiber (arrow at row 27)

TABLE 1.  PRINTOUT OF AXIAL (U) VELOCITY FIELD

THE CHAMBER LATERAL VELOCITIES ARE

$\longrightarrow$ Y

| | 1 | 2 | 3 | 4 | 5 | 6 | 7 | 8 | 9 | 10 | 11 | 12 |
|---|---|---|---|---|---|---|---|---|---|---|---|---|
| | 0.0 | -.03 | .01 | .20 | .51 | .86 | 1.22 | 1.51 | 1.66 | 1.45 | .86 | |
| 2 | 0.0 | -.03 | -.12 | -.34 | -.62 | -.91 | -1.17 | -1.40 | -1.53 | -1.34 | -.80 | 0.00 |
| 3 | 0.0 | .05 | -.10 | -.36 | -.69 | -1.06 | -1.44 | -1.74 | -1.79 | -1.46 | -.84 | 0.00 |
| 4 | 0.0 | .14 | -.01 | -.28 | -.62 | -1.02 | -1.40 | -1.60 | -1.49 | -1.12 | -.62 | 0.00 |
| 5 | 0.0 | .19 | .08 | -.17 | -.50 | -.85 | -1.16 | -1.23 | -1.07 | -.82 | -.46 | 0.00 |
| 6 | 0.0 | .19 | .14 | -.07 | -.34 | -.63 | -.85 | -.87 | -.77 | -.63 | -.37 | 0.00 |
| 7 | 0.0 | .17 | .16 | .02 | -.20 | -.42 | -.59 | -.61 | -.57 | -.48 | -.29 | 0.00 |
| 8 | 0.0 | .13 | .16 | .08 | -.07 | -.24 | -.37 | -.41 | -.41 | -.36 | -.21 | 0.00 |
| 9 | 0.0 | .10 | .16 | .12 | .02 | -.10 | -.21 | -.27 | -.29 | -.26 | -.15 | 0.00 |
| 10 | 0.0 | .11 | .16 | .14 | .07 | -.01 | -.11 | -.17 | -.19 | -.15 | -.08 | 0.00 |
| 11 | 0.0 | .11 | .16 | .16 | .11 | .05 | -.03 | -.08 | -.10 | -.07 | -.04 | 0.00 |
| 12 | 0.0 | .09 | .15 | .16 | .13 | .09 | .03 | -.01 | -.02 | -.02 | -.01 | 0.00 |
| 13 | 0.0 | .07 | .14 | .16 | .14 | .12 | .08 | .04 | .03 | .02 | .00 | 0.00 |
| 14 | 0.0 | .05 | .11 | .14 | .14 | .13 | .11 | .08 | .06 | .04 | .01 | 0.00 |
| 15 | 0.0 | .03 | .09 | .12 | .13 | .13 | .12 | .10 | .08 | .05 | .01 | 0.00 |
| 16 | 0.0 | .02 | .06 | .10 | .12 | .12 | .12 | .11 | .09 | .02 | .00 | 0.00 |
| 17 | 0.0 | .01 | .04 | .08 | .10 | .11 | .11 | .09 | .06 | .01 | .01 | 0.00 |
| 18 | 0.0 | -.01 | .02 | .05 | .07 | .08 | .09 | .07 | .03 | .03 | .02 | 0.00 |
| 19 | 0.0 | -.02 | -.00 | .03 | .05 | .06 | .07 | .06 | .03 | .04 | .03 | 0.00 |
| 20 | 0.0 | -.03 | -.02 | .01 | .03 | .05 | .06 | .06 | .04 | .05 | .03 | 0.00 |
| 21 | 0.0 | -.04 | -.04 | -.01 | .02 | .04 | .06 | .06 | .05 | .05 | .03 | 0.00 |
| 22 | 0.0 | -.04 | -.05 | -.02 | .01 | .03 | .05 | .07 | .06 | .06 | .04 | 0.00 |
| 23 | 0.0 | -.05 | -.06 | -.03 | -.00 | .03 | .06 | .07 | .08 | .07 | .04 | 0.00 |
| 24 | 0.0 | -.06 | -.07 | -.04 | -.01 | .03 | .06 | .08 | .09 | .08 | .05 | 0.00 |
| 25 | 0.0 | -.06 | -.07 | -.05 | -.01 | .03 | .07 | .09 | .11 | .10 | .05 | 0.00 |
| 26 | 0.0 | -.06 | -.07 | -.04 | -.00 | .04 | .08 | .11 | .12 | .10 | .06 | 0.00 |
| 27 | 0.0 | -.06 | -.07 | -.04 | .01 | .06 | .10 | .12 | .13 | .11 | .06 | 0.00 |
| 28 | 0.0 | -.05 | -.06 | -.03 | .02 | .07 | .12 | .14 | .14 | .12 | .07 | 0.00 |
| 29 | 0.0 | -.04 | -.05 | -.02 | .03 | .09 | .13 | .15 | .15 | .13 | .07 | 0.00 |
| 30 | 0.0 | -.03 | -.04 | -.01 | .05 | .10 | .15 | .17 | .16 | .14 | .08 | 0.00 |
| 31 | 0.0 | -.03 | -.02 | .01 | .06 | .11 | .16 | .18 | .17 | .15 | .08 | 0.00 |
| 32 | 0.0 | -.02 | -.01 | .02 | .07 | .12 | .17 | .19 | .18 | .15 | .08 | 0.00 |
| 33 | 0.0 | -.01 | .00 | .04 | .08 | .13 | .17 | .18 | .18 | .15 | .09 | 0.00 |
| 34 | 0.0 | -.01 | .01 | .05 | .09 | .13 | .16 | .18 | .18 | .15 | .09 | 0.00 |
| 35 | 0.0 | -.01 | .00 | .04 | .09 | .13 | .14 | .17 | .18 | .15 | .10 | 0.00 |
| 36 | 0.0 | -.02 | -.01 | .02 | .07 | .11 | .13 | .17 | .18 | .15 | .10 | 0.00 |
| 37 | 0.0 | -.02 | -.02 | .00 | .05 | .09 | .12 | .16 | .17 | .14 | .10 | 0.00 |
| 38 | 0.0 | -.02 | -.02 | -.02 | .02 | .06 | .10 | .14 | .14 | .13 | .09 | 0.00 |
| 39 | 0.0 | -.01 | -.02 | -.01 | .00 | .03 | .07 | .09 | .09 | .08 | .05 | 0.00 |
| 40 | 0.0 | 0.00 | 0.00 | 0.00 | 0.00 | 0.00 | 0.00 | 0.00 | 0.00 | 0.00 | 0.00 | 0.00 |

TABLE 2.  PRINTOUT OF LATERAL (V) VELOCITY FIELD

```
THE CHAMBER TEMPERATURES ARE            ─────────▶ Y
       1      2      3      4      5      6      7      8      9     10     11     12     13
  1  -.00   -.00   -.01   -.02   -.03   -.03   -.04   -.03   -.01    .02    .04    .04    .02
  2  -.02   -.06   -.05   -.04   -.03   -.03   -.02   -.03   -.05   -.08   -.10   -.10   -.02
  3  -.02   -.09   -.07   -.05   -.03   -.02   -.03   -.07   -.09   -.08   -.06   -.02   -.00
  4  -.01   -.08   -.08   -.06   -.04   -.03   -.03   -.05   -.02    .03    .07    .09    .01
  5  -.00   -.06   -.07   -.06   -.04   -.03   -.02   -.02    .02    .07    .08    .09    .02
  6   .00   -.04   -.07   -.07   -.05   -.04   -.02   -.00    .03    .06    .08    .09    .02
  7   .01   -.02   -.06   -.06   -.06   -.04   -.02   -.00    .03    .05    .07    .08    .03
  8   .01   -.01   -.06   -.08   -.07   -.06   -.04   -.01    .01    .04    .06    .08    .04
  9   .02   -.00   -.05   -.07   -.07   -.06   -.04   -.02   -.00    .02    .05    .07    .04
 10   .03    .00   -.06   -.08   -.09   -.09   -.07   -.05   -.02    .02    .06    .08    .04
 11   .04    .01   -.06   -.10   -.11   -.11   -.09   -.07   -.03    .01    .05    .07    .04
 12   .04    .01   -.06   -.10   -.12   -.13   -.11   -.08   -.04    .00    .04    .07    .05
 13   .05    .02   -.06   -.11   -.14   -.14   -.13   -.10   -.05   -.01    .03    .07    .05
 14   .05    .02   -.06   -.11   -.14   -.15   -.13   -.10   -.06   -.01    .02    .06    .05
 15   .05    .02   -.07   -.12   -.15   -.16   -.15   -.12   -.07   -.02    .01    .05    .05
 16   .05    .01   -.06   -.11   -.14   -.15   -.14   -.11   -.07   -.03    .00    .04    .05
 17   .05    .01   -.07   -.12   -.15   -.16   -.15   -.13   -.09   -.06   -.01    .03    .05
 18   .04    .00   -.06   -.12   -.15   -.16   -.15   -.12   -.09   -.05   -.01    .03    .05
 19   .04    .00   -.06   -.11   -.14   -.15   -.14   -.12   -.08   -.04   -.00    .03    .05
 20   .04   -.00   -.06   -.10   -.13   -.14   -.13   -.11   -.08   -.04   -.00    .03    .05
 21   .03   -.00   -.05   -.10   -.12   -.13   -.12   -.11   -.07   -.04   -.00    .03    .05
 22   .03   -.00   -.05   -.09   -.12   -.13   -.12   -.11   -.07   -.03   -.00    .03    .05
X23   .03   -.00   -.05   -.09   -.11   -.12   -.11   -.10   -.07   -.03   -.00    .03    .05
 24   .03   -.00   -.05   -.08   -.11   -.12   -.11   -.10   -.07   -.03   -.00    .02    .05
 25   .03   -.00   -.04   -.08   -.10   -.11   -.10   -.09   -.06   -.03   -.00    .02    .05
 26   .03   -.00   -.04   -.07   -.10   -.11   -.10   -.09   -.06   -.03   -.01    .02    .05
 27   .03   -.00   -.04   -.07   -.09   -.11   -.10   -.09   -.06   -.03   -.01    .01    .05
 28   .03   -.00   -.03   -.07   -.09   -.10   -.10   -.09   -.06   -.04   -.02    .01    .05
 29   .03    .00   -.03   -.06   -.09   -.10   -.10   -.09   -.07   -.04   -.02    .00    .05
 30   .03    .00   -.03   -.06   -.09   -.10   -.10   -.09   -.07   -.04   -.03   -.00    .05
 31   .03    .00   -.03   -.06   -.08   -.10   -.10   -.09   -.07   -.04   -.03   -.01    .05
 32   .03    .00   -.03   -.06   -.08   -.09   -.10   -.09   -.07   -.05   -.03   -.01    .05
 33   .03    .00   -.03   -.06   -.08   -.09   -.09   -.09   -.07   -.05   -.04   -.01    .05
 34   .03    .00   -.03   -.06   -.08   -.09   -.09   -.09   -.07   -.05   -.03   -.01    .05
 35   .02   -.00   -.03   -.05   -.08   -.09   -.09   -.09   -.06   -.04   -.03   -.01    .05
 36   .02   -.00   -.03   -.05   -.08   -.09   -.09   -.09   -.06   -.04   -.03   -.01    .05
 37   .02   -.00   -.03   -.05   -.07   -.09   -.09   -.08   -.05   -.03   -.02   -.01    .05
 38   .01   -.00   -.03   -.05   -.07   -.08   -.09   -.07   -.05   -.03   -.02   -.00    .05
 39   .01   -.00   -.03   -.04   -.06   -.07   -.07   -.06   -.04   -.02   -.01   -.00    .05
 40  -.00   -.00   -.03   -.04   -.06   -.07   -.07   -.06   -.04   -.02   -.01   -.00   -.00


THE MEASURED CHAMBER TEMPERATURES ARE
       1                                    7                                   13
  1  -.03                                 -.03                                 -.03

  9   .02                                 -.09                                  .04

 17   .05                                 -.17                                  .05

 25   .03                                 -.12                                  .05

 33   .03                                 -.10                                  .05

 40   .00                                 -.07                                  .05
```

TABLE 3.   PRINTOUT OF CALCULATED AND MEASURED TEMPERATURES

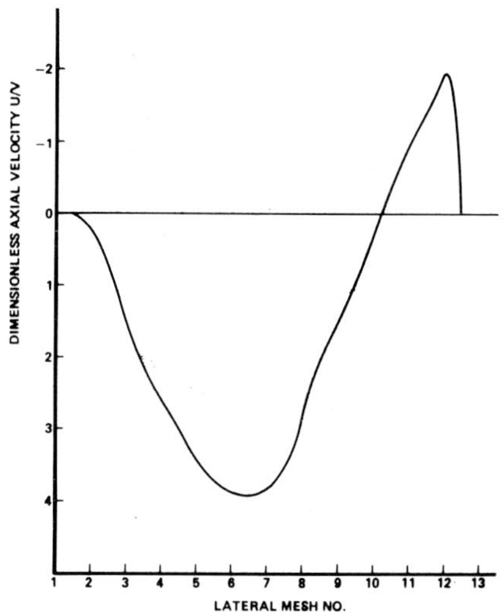


Figure 7.    Velocity Plot of Computer
             Results at Axial Position No. 31

2.  The maximum displacements taken from the photographs
    of Figure 6 are, respectively:

$$x_1 = .40 \text{ cm} \times 2.35 = 0.94 \text{ cm}$$
$$x_2 = 1.05 \text{ cm} \times 2.35 = 2.47 \text{ cm}$$
$$x_3 = 1.80 \text{ cm} \times 2.35 = 4.23 \text{ cm}$$

3.  From Figure 5, the times at which the photographs
    were taken are given by:

$$T_1 = 0$$
$$T_2 = 6.81 \text{ sec}$$
$$T_3 = 15.23 \text{ sec.}$$

4.  Therefore, the velocities are given by:

$$\bar{V}_1 = \frac{x_2 - x_1}{T_2 - T_1} = 0.22 \text{ cm/sec}$$
$$\bar{V}_1 = \frac{x_3 - x_2}{T_3 - T_2} = 0.21 \text{ cm/sec}$$

916

5. The dimensionless velocities are then given by:

$$V_1 = \frac{\overline{V}_1}{V_o} = 3.67$$

$$V_2 = \frac{\overline{V}_2}{V_o} = 3.50 \ ,$$

where $V_o = 0.06$ cm/sec is the average velocity in the chamber

Although the fiber is located approximately at axial position #28, the average position during each velocity measurement is 31 and 32, respectively; there the calculated velocities are:

$$V_1 = 3.80$$
$$V_2 = 3.73$$

The discrepancy between the measured and calculated velocities is 3.4% and 6.2% respectively, which is good considering the assumptions made in calculations and the crude method of velocity measurement.

Table 3 shows a comparison between calculated and measured temperature differences (T). There is good agreement between the calculated and measured temperature fields; however, some slight discrepancy occurs at the top of the chamber. This is probably due to inexact entrance conditions used in the model which uses a rectangular grid and therefore does not match the curved entrance region of the actual flow chamber.

Conclusions

Despite the assumptions made in developing the two-dimensional flow equations, good agreement is seen to exist between the analytical model and the experiment. Many other comparisons have been made, all of which show similar agreement.

It has been observed that very small lateral gradients on the order of $0.05^{\circ}$C/cm can cause perturbations to the uniform

base flow in the chamber.  It is true that we used a 0.55 cm
thick chamber; however, since buoyancy-induced disturbances
depend linearly on chamber thickness, we would expect temperature gradients on the order of $0.20^{\circ}$C/cm to be critical in a
0.15 cm chamber.

With the analytical model, it is possible to characterize
small disturbances which are important but not measurable in
ground-based electrophoresis chambers.  The model will be used
to study gravity effects on the electrophoresis process and to
predict the optimum separations that can be achieved.  The
results of parametric studies using the analytical model will
be useful in the development of more efficient separation
chambers for use in the laboratory.

References

1.  Ostrach, S.:  ESA Publication 114 (1976).

2.  Saville, D. A.:  Final Report NAS-8-31349, Princeton
    University (1978).

3.  Hirt, C. W. et al.:  Report LA-5852, Los Alamos Scientific
    Laboratory (1975).

4.  Stephani, L. M., Butler, T. D.:  Report LA-6014,
    Los Alamos Scientific Laboratory (1975).

5.  Rhodes, P. H., in preparation.

HIGH-RESOLUTION CONTINUOUS-FLOW ELECTROPHORESIS IN THE REDUCED
GRAVITY ENVIRONMENT


Percy H. Rhodes
NASA/Marshall Space Flight Center
Huntsville, Alabama 35812, USA


Introduction

Operating flowing or contained fluid electrophoretic separation devices in a reduced gravity environment will eliminate buoyancy-induced disturbances to the process, as the Apollo 16 (1) and ASTP (2) static electrophoresis experiments have clearly shown. There are, however, other significant fluid problems that have historically limited continuous-flow electrophoresis on Earth which are not related to gravity. The no-slip condition on the chamber wall causes a parabolic profile to develop in the direction of buffer flow (with zero velocity at the chamber wall and maximum velocity at the chamber center plane), thus causing sample residence time to vary locally in the thickness of the chamber. In addition, a lateral flow across the width of the chamber (perpendicular to buffer flow), called electroosmosis, exists when charged walls are present. As a result of laminar flow, a particle traveling through the separation chamber at or near the center plane will be deflected less than an electrophoretically similar particle moving through at some distance from the center plane. Therefore, an initially regular pattern of injected sample will be distorted into a convex shape when viewed in the direction of sample migration. On the other hand, electroosmotic flow increases the deflection of particles at or near the center plane so that concave patterns are formed when viewed in the direction of sample migration. These phenomena combine to produce crescent-shaped distortions, the curvature of which is determined by the flow that predominates --laminar flow or electroosmosis. Therefore, unless exact compensation exists, a bending of the injected sample band will result -- an artifact which we will call flow distortion. The condition for exact compensation for these flow effects is the equality of sample and wall zeta potential. Since the wall usually has only one zeta potential, only one fraction can be in focus. This interaction of electroosmotic and laminar flow effects was first reported by Strickler and Sacks in 1970 (3) and, as they state, provides a "signature" of separation performance.

Disregarding these nongravity-related problems, most concepts for separation devices designed for operation in space are simply adaptations of existing instruments which include these deficiencies and are not designed specifically for re- duced gravity operation. The only exception to this has been the work of Bier (4) and Strickler (5) who developed the Spider cell and Deflected Laminar Electrophoresis chamber con- cepts. Their proposed devices, however, were intended to im- prove only throughput and not resolution with respect to ground-based instruments. It is our contention that an order of magnitude improvement in resolution should be sought over ground-based operation to justify the various complexities of a space experiment.

Therefore, new concepts should be developed which will opti- mize performance in such a manner as to make space processing of biologicals feasible. This paper proposes one such concept.


Impact of Flow Disturbances

To estimate the magnitude of the aforementioned flow disturb-

ances, we make use of a very simple mathematical model which
has been developed (6). The model assumes that two independent orthogonal flows exist in the separation chamber, one
being fully developed laminar flow longitudinally in the chamber gap and the second being electroosmotic flow in the chamber
cross section. The electrical conductivity and fluid viscosity are considered to be linear functions of temperature.
With the respective velocity fields known, the highest and
lowest mobility particles entering a typical rectangular
collection port can be found and this result translated into
resolution. Therefore, the estimate for resolution is the
best possible considering that all other spurious effects,
including diffusion and any gravity-related effect, have been
neglected. The resolution expression, derived in Reference 6,
gives the mobility difference that a device is capable of resolving in terms of an "ideal" separation performance for
electrophoretic separation, plus a term that represents the
degradation of resolution by zeta potential mismatch and thermal distortion due to the change of fluid viscosity with
temperature.

To appreciate the importance of zeta potential matching of
the sample and wall, numerical values can be applied to give
the following hypothetical separation. The mean mobility of
a sample distribution at $0^{\circ}$C is 0.5 µm cm/V sec. The signif-
cant parameters for the chamber are: the ratio of the injected
sample band thickness to chamber thickness is 0.5, the width
of a typical collection port is 0.05 cm, the chamber residence
time is 200 seconds and applied electric field is 50 V/cm.
Note that here we have chosen a 0.5 cm thick chamber since
current concepts for space application consist of simply increasing the gap thickness of the separation chamber. A plot
of resolution versus zeta potential matching is given in
Figure 1.

922

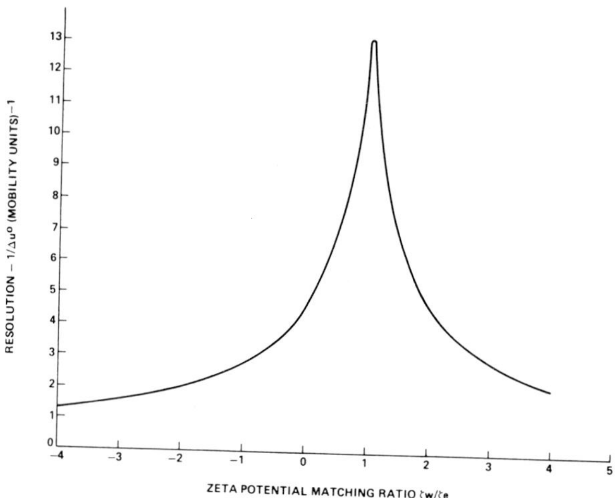



Figure 1.

As Figure 1 shows, operating at $\xi_w/\xi_e = 1$, as one must, can be
quite a precarious position; i.e., small changes in zeta potential matching can lead to large degradations in resolution.


Proposed Concepts

Most methods proposed to date for solving the flow distortion
problem have merely sought to compensate for the disturbance
rather than remove it.  Operating in such a manner is rather
like a balancing act and is not conducive to reliable operation.  In addition, only one component of the sample will be
in "focus" at any one time.

The thick chamber for use in zero gravity is a method which
does not involve compensation.  The increase in resolution
would be achieved by keeping the sample away form the chamber
walls via the increased chamber thickness and, hence, reduce

sample band distortion.  However, the resolution decreases
with increasing chamber thickness for constant power dissipation in the chamber because the increase in thickness will
also dictate a lower applied voltage for constant mid-plane
temperature.  Thus, merely increasing the chamber thickness
without considering the impact on other operational parameters
does not increase the resolution of separation.

Another method to eliminate flow disturbances was advanced by
Kolin and Ellerbroek (7).  This method, which might be thought
of as the ultimate in compensation, uses a cross flow to neutralize electroosmosis and relies on thermal convection to
blunt the parabolic flow-through profile sufficiently so that
the center-plane region of the chamber will be distortion free.
The power levels necessary to cause the required deformation
of the parabolic flow-through profile are many times the power
level limit that will disrupt electrophoresis in conventional
separation chambers.  Also, the exact and uniform counter flow
along the length of the chamber necessary to counter electroosmosis would be very difficult to achieve.  While this scheme
is possible theoretically, it is practically impossible to
implement successfully.

Probably the most ingenious idea yet employed to compensate
for chamber flow distortions was developed by Strickler and
Sacks in 1972 (8).  The concept consisted of coating longitudinal sections of the inner chamber wall with materials having
different zeta potentials and sectioning the respective electrodes so that the electrical field could be independently
applied to each section.  By controlling the electrical field
strength in each section, flow distortions (crescent for-
mations) created in a previous section could be compensated
in a subsequent section simply by turning a control knob to
change the applied voltage.  The "focusing" process, however,
had to be controlled by visual observation through a cross-
section illuminator which revealed the crescent-shaped band

cross sections.  Although this concept is theoretically sound and workable, it has not found great acceptance in the field. It is possible that the system requires a constant operator interface to maintain precise "focusing."  As Figure 1 shows, small changes in zeta potential matching can cause large changes in resolution.

It, therefore, appears that new methods and design concepts must be sought to overcome the problem of sample stream distortion, leading to a new device which would offer unique capabilities for operation in reduced gravity and offer a significant improvement in resolution over similar ground-based machines.  Our attempt to satisfy this requirement has led us to the concept of a moving wall.  This concept is not new but, to our knowledge, has never been successfully implemented. The moving walls entrain the fluid to flow as a rigid body, hence eleminating Poiseuille flow.  All of the sample throughout the chamber thickness is thus exposed to the imposed electric field for the same period of time, while electroosmosis has been eliminated through the use of film-forming latexes.  The zeta potential of the latex has been altered by prior treatment of the particle surface.  Methylcellulose absorbs strongly on these latexes, yielding a zeta potential near zero.  Since both sources for the disturbances have been eliminated, no compensation is required.  The system separates like a static device while providing throughput like a conventional continuous-flow system.  In addition, no limitation is placed on the usable fraction of the chamber thickness.

As early as 1972, McCreight and Griffin of General Electric (9) demonstrated a concept which has been regarded in some quarters as being equivalent to the moving wall concept.  Their process would be carried out in a conventional-type free-flow chamber in the following manner:

1. Buffer would enter the chamber while entraining a stream of sample until the sample stream extended the entire chamber length.

2. The buffer flow would then be stopped and the electric field turned on long enough to achieve the desired degree of separation.

3. After the separation, the electric field would be turned off and the fluid laterally displaced, parallel (or antiparallel) to the field direction, in order to rectify any distortion resulting from the separation process or electroosmosis.

4. The buffer flow-through would then be resumed and the separated bands displaced into collection ports.

Although such a scheme is mechanically simpler than the moving wall idea, it suffers from a host of conceptual deficiencies. These are:

1. The scheme is a discontinuous, batch flow process which by design can never attain steady state -- a condition crucial to reliable separation performance. Experiments in the Beckman Continuous Particle Electrophoresis apparatus showed that the sample stream was disturbed by stopping and starting of buffer flow through the chamber.

2. Even the slightest variation in electrical field strength along the length of the chamber would cause a distortion which could not be subsequently compensated for by the lateral flow. These variations in electrical field strength do not pose a problem in a continuous process where the particles of sample essentially see the field in a Lagrangian way as they pass through the chamber.

3. The compensating laminar flow in the chamber would need to be perpendicular to the direction of buffer flow, which would necessitate entrance and exit ports along the entire length of the chamber. Also, very precise metering of buffer cross flow would be required. Indeed, whether a flow could be made uniform over the entire length of a 10- to 60-cm chamber is extremely doubtful and, to say the least, is as challenging as installing moving walls.

4. The efficiency of the batch process would be lower
than the moving wall continuous process because of chamber
thickness utilization limitations. Nothing prevents samples
from approaching the wall, where their flow-through velocity
will approach zero -- a circumstance which could require long
sample exchange times.

Concept Evaluation

The first step in developing the moving wall concept was simply to build a laboratory model which would show entrainment
of a sample stream and, in particular, whether any disturbance
could be observed that could be attributed to the moving wall.
The second step would, naturally, be to produce a separation
in the device, while the third step would be to develop a
collection system and collect separated fractions.

Apparatus

The uniqueness of the device is essentially in the chamber
walls, which are two continuous Mylar belts of 0.003 in.
thickness. The belts are carried on rollers and are supported
by rigid chamber subwalls. Figure 2 shows a photograph and
sketch of one such wall assembly. Note that the roller contains teeth which engage in perforations in the belt to assure
that no belt slippage occurs. The wall assemblies each contain an internal cooling chamber to dissipate Joule heating.
The outer wall substratum, which contacts the Mylar film, is
perforated with small holes connected to a partial vacuum to
assure adherence of the Mylar to its flat support. Figure 3
shows a schematic of the system. The chamber sidewalls fit
tightly against belts to prevent leakage from the sides of
the separation chamber. A slotted 20-gauge hypodermic needle
serves to inject the sample into the total thickness of the

chamber. The entire assembly is housed in a water-tight enclosure, with the belts being driven through shafts which extend through the compartment walls. Note that belt synchronization is assured by two gears outside the compartment which are driven by a Sage syringe pump motor. Figure 4 shows the complete system.

## Test Results

We have completed the first step in our evaluation of the scheme and have found no detrimental effects of moving the chamber walls with regard to an injected sample stream. In fact, we have found that the moving belts entrain the sample and buffer in a most uniform way, and sample band broadening was not observed unless leakage was allowed to occur at the side walls. Figure 5 shows a sample stream being carried through the chamber; note the uniformity and straightness of the band. Figure 6 shows an end view of the chamber; note

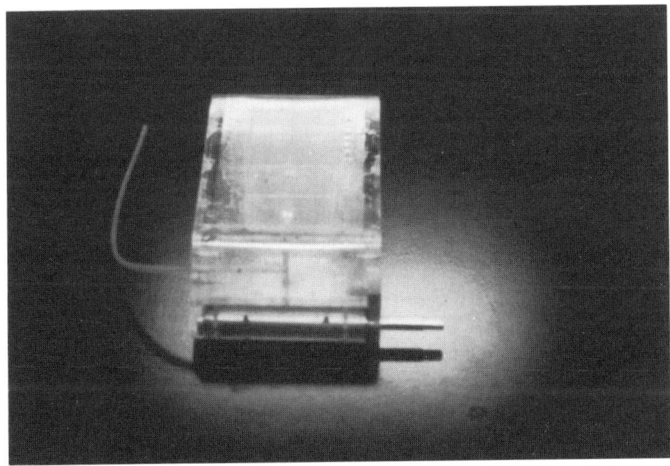

Figure 2.

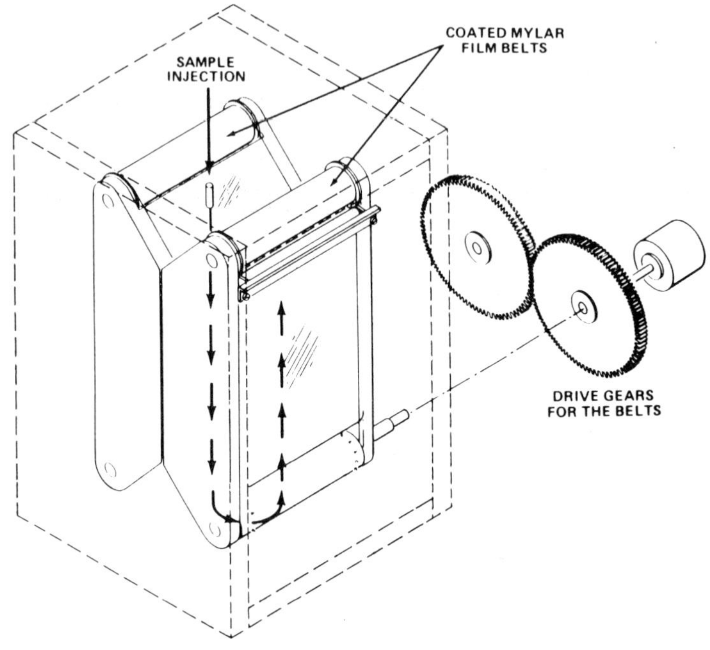


Figure 3.   SCHEMATIC OF MOVING WALL TEST SYSTEM

Figure 4.

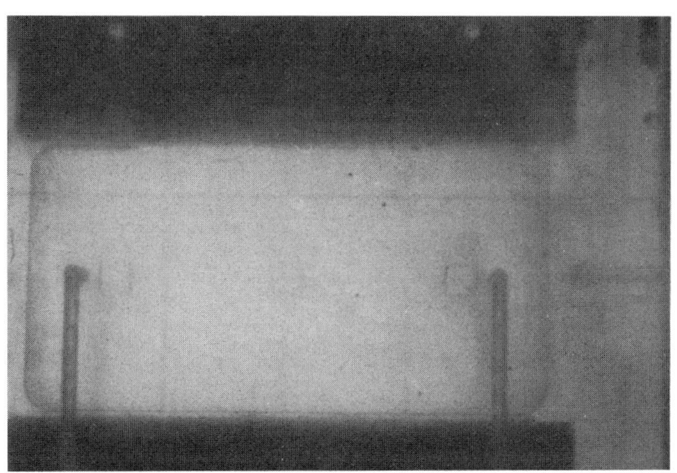

Figure 5.

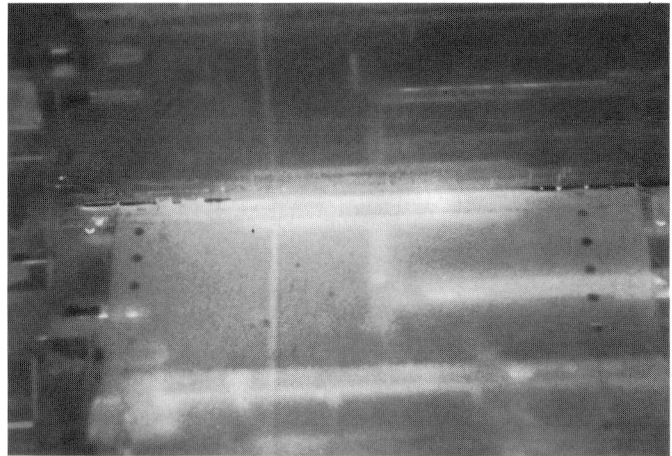

Figure 6.

that the sample completely fills the chamber thickness and
note also the "wrap-around" stream just below it.  The "wrap-
around" effect will, of course, be eliminated with the addition of a collection system.  Figure 7 shows a broadened
stream resulting from leakage through a side wall.  This shows
that leakage of the buffer through the side wall past the
moving belts could be a design consideration which needs careful consideration.


Future Testing


The second step in testing -- that of obtaining a separation--
could be very difficult to achieve on Earth.  Since the belts
entrain the buffer to move as a rigid body, there is no fluid
velocity with respect to the walls (after steady-state conditions have been reached).  If we consider that the buoyancy

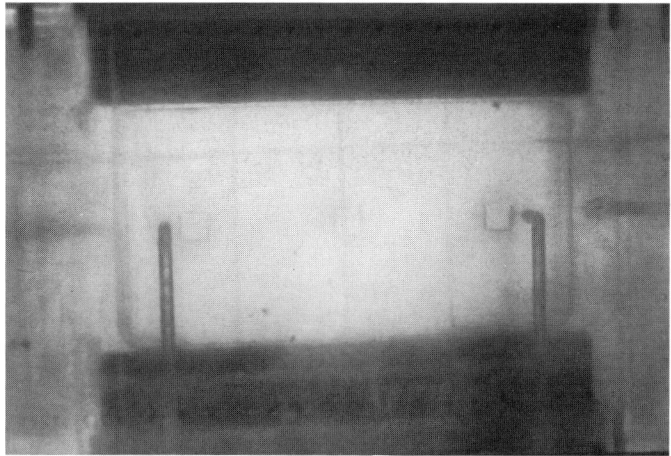

Figure 7.

effect can be represented by the dimensionless parameter

$$\frac{g\beta\ell}{V^2} \ (T - \bar{T}) \ ,$$

where

    g = gravitational acceleration
    $\beta$ = coefficient of thermal expansion
    $\ell$ = chamber half thickness
    V = average flow velocity in the chamber
    T = local temperature
    $\bar{T}$ = a reference temperature

then we can see that as the flow velocity V approaches zero, the buoyancy disturbance can become quite large. Operating the chamber in a horizontal orientation might provide the thermal stability needed to achieve a separation in Earth gravity; however, it is obvious that the concept could only reach its full potential in weightlessness.

932

Conclusion

In designing equipment for space application, reliability of
operation should be the overriding consideration.  Therefore,
any method which controls a disturbance by compensation
should be avoided, if possible, even if this means that the
machine design must be more complex.  A device is only built
once but may be operated many times; therefore, operational
ease and reliance should take precedence over design simplicity.  Whether the moving wall will be worthy of space
application remains to be seen, but it is safe to say that
this is the type of concept which is needed for space application -- a concept which eliminates major operational difficulties and offers unique capabilities for operation in
space.


References


1.  Snyder, R.S., et. al., Sep. Purif Methods $\underline{2}$, 259,
    (1973).

2.  Allen, R.E., et. al. NASA TM X-58173, (1976).

3.  Strickler, A., Sacks, T. Preparative Biochem. $\underline{3}$, 269,
    (1973).

4.  Bier, M., et. al. J. Colloid and Interface Sci. $\underline{55}$,
    196, (1976).

5.  Strickler, A., AIAA Paper No. 77-233, (1977).

6.  Rhodes, P.H., NASA TM-78158, (1977).

7.  Kolin, A., Ellerbroek, B.L., Sep. Purif. Methods $\underline{8}$,
    1, (1979).

8.  Strickler, A., Sacks, T., Annals of N.Y. Acad. Sci.
    $\underline{209}$ (1973).

9.  Griffin, R., General Electric, Final Report, Contract
    NAS8-28365, (1977).

APPLICATION OF THE AUTOMATED ELECTROPHORESIS MICROSCOPE SYS-
TEM TO VARIOUS BIOLOGICAL TESTS


George B. Olson, Martha McFadden and Peter H. Bartels
Department of Microbiology, University of Arizona
Tucson, Arizona  85721


Introduction

The automated electrophoresis microscope system (AEMS) is a
computer operated instrument designed for the rapid acquisition and analysis of the electrophoretic mobilities of cells.
The system, its operational and operational software, its
operation and precision, have been described previously (1-4).
The purpose of this paper is to describe the application of
the AEMS to various problems in cell biology and the clinical
laboratory.  These include:  (i) detection and quantitation
of virus-altered cells in a cell population, (ii) detection
and quantitation of cells altered by conjugated monospecific
antibody, and (iii) application of the macrophage electrophoretic migration test (MEM).


Materials and Methods

<u>Protocol for detecting virus-altered cells</u>.  Blood was col-
lected from human subjects in EDTA (1.5 mg Na$_2$ EDTA per ml of
blood) on day of use, diluted 1 part blood plus two parts
cell balanced salt solution (CBSS) and separated on Ficoll-
Hypaque gradients.  The erythrocytes (RBC) were removed,
washed and suspended to 0.5% in CBSS.  One-tenth ml quantities
of various influenza virus preparations were mixed with 0.5
ml portions of RBC and incubated at 4°C for 45 min, then in-

cubated at 37°C overnight. Cells were washed and a portion
was tested to determine if the cells could be agglutinated
with the same virus preparations. Virus-treated and non-
treated RBC were washed in 0.15 M NaCl·HCO$_3$ buffer pH 7.4 and
suspended in various portions to a final concentration of
4 x 10$^6$ cells/ml. Cell electrophoresis was done at 40V, 25°C
in a cylinder chamber with platinum electrodes at a depth of
500 μ in the chamber.

Virus preparations included HON1, H1N1 and H3N2 influenza
viruses.

<u>Protocol for detecting cells tagged with monospecific anti-
body</u>. Antisera, with specificity directed against Thy-1 antigens of murine T cells and immunoglobulins of human and murine
B cells were treated with saturated ammonium sulfate to obtain
the gamma globulins of the sera. The gamma globulins were
conjugated with materials such as DNP, poly-L-aspartic acid
and poly-L-glutamic acid to change the isoelectric points of
the globulins without altering the immunologic specificity of
the globulins. For example, 15 mg of gamma globulin, 1 mg
poly-L-aspartic acid and 5 mg EDAC were mixed at a pH of 5.5
for 1 hr at 37°C. The reaction was stopped by the addition
of 15 mg glycine.

Peripheral blood lymphocyte (PBL) preparations, murine thymocytes and laboratory-maintained cultures of human and murine
cells, prepared by standard methods to a concentration of
1 x 10$^7$ cells in 1 ml of CBSS was mixed with 0.2 ml conjugated
antiserum. In a direct labelling test, the cells were incubated at 4°C for 30 min with the conjugated primary antiserum.
In an indirect labelling test, the cells were incubated at 4°C
for 30 min with the primary antiserum, washed three times in
CBSS and then incubated at 4°C for 30 min with a conjugated
secondary antibody preparation. In both tests, after the reaction of cells with a conjugated antiserum, the cells were

washed three times in CBSS and suspended to a concentration of
$4 \times 10^6$ cells/ml in 0.15 M NaCl·$HCO_3$ pH 7.4.  Cell electrophoresis was done as described above.

Protocol for the MEM test.  In various experiments the MEM
test was done using (i) previously prepared migration inhibition factors (MIF) in one and two-stage operations with guinea
pig macrophages, murine macrophages and treated RBC (5-7);
(ii) freshly-obtained PBL and macrophages in a one-stage test
(8), and (iii) freshly-obtained PBL in a two-stage test with
treated RBC (8,9).  Basically; (1) macrophages were obtained
by collecting peritoneal exudate cells from guinea pigs and
mice three days after the animals had been injected with min-
eral oil.  Cells were washed free of oil droplets and the
macrophages were incubated with either PBL or MIF preparations
in Eagles' medium essential medium (MEM) at pH 7.2 for 90 min
at 23°C (equivalent to stage one).  (2) MIF preparations used
in other studies were evaluated.  The MIF preparations had
been obtained from sensitized guinea pig cells by incubating
$2.5 \times 10^7$ peritoneal exudate cells per ml medium for 24 hr at
37°C in the presence of the sensitizing antigen.  The supernatants were then removed, lyophilized and stored at -20°C
until used.  These preparations are assumed to be free of the
MIF inhibitor factor (11).  (3) RBC pretreated with tannic
acid, glutaraldehyde or formalin according to normal procedures (9,10) were tested for their effectiveness as indicator
cells.  Formalin-treated SRBC ($SRBC_f$) were chosen as the indicator cells for all two-stage MEM tests.  The second stage
was done by mixing MIF preparations or supernatants from the
first stage with $5 \times 10^7$ $SRBC_f$ in MEM pH 7.2 for 90 min at
23°C.  Cell electrophoresis was done as described above.


Results and Discussion

Detection and quantitation of virus-altered cells.  Figure 1

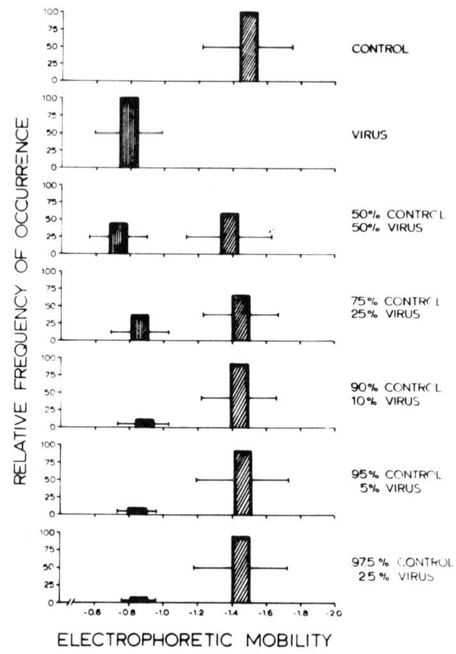



Fig. 1  Electrophoretic mobilities of cell populations containing normal human erythrocytes, influenza virus-altered erythrocytes and mixtures of normal and virus-altered cells.  Cell electrophoresis done at 40V, at a depth of 500 µm in a cylinder chamber with platinum electrodes in NaCl·HCO$_3$ buffer pH 7.4. Statistic in figure represents standard deviation of the mean.

shows the electrophoretic mobility distributions of virus-treated and untreated human RBC.  Treatment of cells with influenza virus removes the sialic acid residues from the plasma membrane.  This reduces the absolute electrophoretic mobility from 1.489 ± .052 µm/sec$^{-1}$/V$^{-1}$/cm to 0.792 ± .052 (500 µ depth in the chamber).  The error boundaries given here and in the following data are based upon the standard error of the mean as derived from the mean square error term of an analysis of variance (12).  The presence of virus-treated RBC mixed with normal RBC at various proportions can be detected by cell electrophoresis.  Figure 1 also shows the capability of the AEMS to detect the cells, analyze the electrophoretic mobilities and create subsets in the data to correctly show the existence of as few as 2.5% virus-altered cells in a cell population. In each case, the electrophoretic mobilities of the normal and virus-altered cell subsets are not changed significantly, and the number of cells found in each cell population does not differ from what can be expected in the mixing of such cells by normal laboratory procedures.

In a second series of experiments, human erythrocytes were mixed with dilutions of different strains of Influenza viruses (HON1, H1N1 and H3N2). The concentrations of the viruses were expressed only by hemagglutination titers which ranged from 640 to 2560 for the undiluted preparations. Therefore, the goal of this experiment was to determine if the electrophoretic mobilities of RBC treated with different virus preparations would reflect a trend which might yield information as to the kinetics of the virus-receptor reaction. Analyses of data show all virus preparations reduce the electrophoretic mobilities of RBC and the degree of reduction in three of four cases corresponds to the virus titer (Fig. 2). As the virus is diluted, the electrophoretic mobility of the RBC returns to normal. The kinetics of the reaction appear to be similar in all cases. An analysis of variance shows that the mobilities of the virus-treated cells differ from untreated cells, at the 95% confidence limits, at a $10^{-4}$ dilution for one virus preparation, $10^{-3}$ dilution for a second virus preparation and $10^{-2}$ dilutions for the remaining two virus preparations. These data show a direct relationship to the hemagglutination titers for the virus preparations.

<u>Detection and quantitation of cells altered by conjugated monospecific antibody</u>. Figure 3 presents data from a study in which a portion of murine thymocytes was tagged with rabbit anti Thy-1 globulin conjugated to poly-L-aspartic acid. The mean electrophoretic mobility of 150 untagged cells was 1.308 ± 0.045 whereas the mean mobility of 150 cells tagged with the negatively-charged antibody was 1.610 ± 0.045. The 23.7%

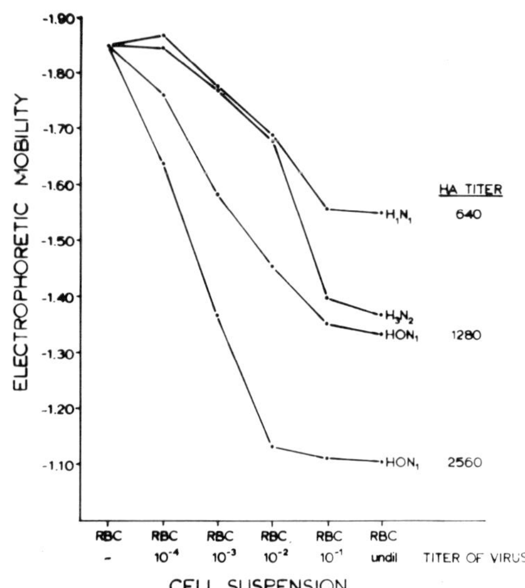


Fig. 2   Electrophoretic mobilities of human erythrocytes altered with different types and dilutions of Influenza viruses.
Cell electrophoresis done at 40V, at a depth of 500 µm in a
cylinder chamber with platinum electrodes in NaCl·HCO₃ buffer
pH 7.4.

change in electrophoretic mobility is significant at a P <0.01
level.  The change in mobilities within the cell population is
reflected in the histograms for electrophoretic mobilities.
The relative number of cells in two intervals in the higher
electrophoretic mobility range increases from 20% to 60%.

A second approach to ascertaining the alteration of cell mobility by the use of a conjugated antibody is seen in Figure 4.
The mobility of human B cells (Daudi cells) maintained in cell
culture was calculated:  (1) without any modification of the
cell membrane, (ii) after treatment in an indirect labelling
technique with unconjugated primary antibody and unconjugated
secondary antibody and (iii) after treatment in an indirect
labelling technique with unconjugated primary antibody and secondary antibody conjugated with poly-L-aspartic acid.  Mean

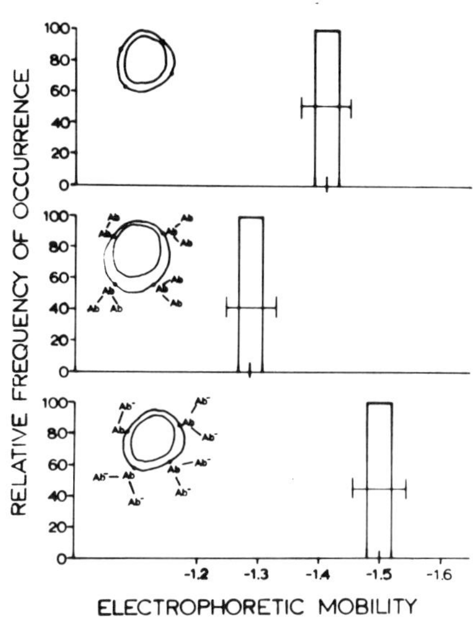



Fig. 4  Electrophoretic mobilities of human B cells (Daudi) and Daudi cells labelled by an indirect labeling technique with antibody designed to alter the mobility of tagged cells.


Table I  Electrophoretic mobilities of various cells before and after reaction with monospecific antibody.

| Type Cell | No Ab | Mobilities With: Indirect Test | Direct Test | % Effect |
|---|---|---|---|---|
| Human PBL | -1.145 | -1.240 | | 9 > |
| Daudi Cells | -1.409 | -1.501 | | 7 > |
| Daudi Cells | -2.018 | -2.144 | | 6 > |
| Mouse Thymus | -2.143 | -2.227 | | 4 > |
| Mouse Thymus | -1.873 | | -2.305 | 23 > |
| Mouse Thymus | -1.208 | | -1.502 | 25 > |


and (ii) less efficiently conjugated secondary antibody.


<u>Application of the MEM test</u>.  Preliminary work on the MEM test has been concerned with (i) comparing the effectiveness of

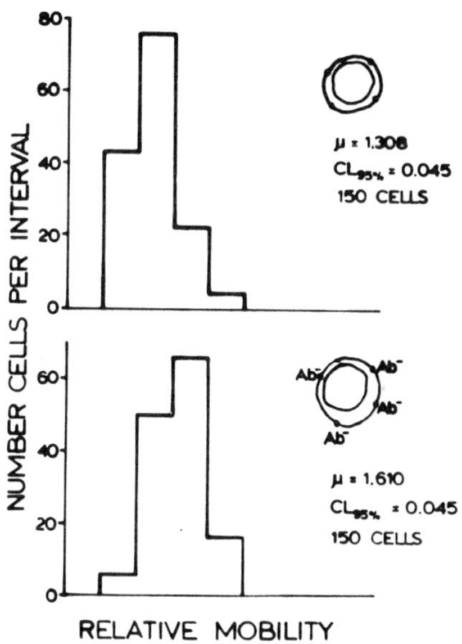



Fig. 3  Relative mobilities of murine thymocytes labelled in a
direct labeling technique with monospecific antibody designed
to alter the mobility of tagged cells.

electrophoretic mobilities for these cells are 1.409 ± 0.045,
1.28 ± 0.045 and 1.497 ± 0.045, respectively.  The changes in
mobility reflect the changes in relative surface charges on
the membranes caused in the first test by the accumulation of
antibody with an isoelectric point of 6 - 6.5 and in the second test by the accumulation of an increased negative charge
associated with the poly-L-aspartic conjugated antibody.

Table I compares results obtained in several experiments employing the indirect and direct techniques for labelling.  A
larger increase in electrophoretic mobility was noted when
cells were tagged with poly-L-aspartic acid conjugated to monospecific primary antibody (direct method) than with a poly-L-
aspartic acid secondary antibody (indirect method).  The reason for this result is not known, but possible explanations include (i) a loss of affinity in the secondary antibody reaction

Table III compares the effects of equal concentrations of MIF and guinea pig albumin preparations when added to human $RBC_f$. MIF preparations cause a 7 - 12% decrease in electrophoretic mobility whereas purified albumin preparations induce a 9 - 12% increase in electrophoretic mobilities.

Table III   Comparison of electrophoretic mobilities of human erythrocytes after incubation with MIF and other factors.

| | | Concentration of Additive | | | | |
| Type Cell | None | 0.5 mg | 1.0 mg | 2.0 mg | % Effect |
|---|---|---|---|---|---|
| human $RBC_f$ + MIF | -1.672 | -1.651 | | <u>-1.552</u> | 7 < |
| human $RBC_f$ + MIF | -1.703 | | | -1.507 | 12 < |
| human $RBC_f$ + albumin | -1.542 | <u>-1.726</u> | -1.602 | -1.627 | 12 > |
| human $RBC_f$ + albumin | -1.627 | <u>-1.779</u> | -1.692 | -1.645 | 9 > |

Based upon these results, $SRBC_f$ were selected for use as the indicator cells in analyzing the responsiveness of PBL obtained from guinea pigs sensitized against Brucella antigens.  PBL were obtained three weeks after sensitization and (i) mixed with antigen and $SRBC_f$ in a one-stage immediate assay, (ii) mixed with antigen to cause the release of the factor into the supernatant which was then immediately assayed in a two-stage test, and (iii) mixed with antigen to obtain a supernatant which was then stored before assay in a two-stage test.  All results indicate a decrease in the electrophoretic mobility of $SRBC_f$ with the largest decreases of 17 - 29% observed in the immediate two-stage test (Table IV).  The two-stage test of stored supernatant appeared to yield the more consistent results.

The final test of the AEMS to be reported in this paper is to show its ability to detect changes which occur in lymphocytes of guinea pigs infected with Herpes virus type 1.  Six animals were given interdermal injections of 0.1 ml of Herpes virus on

various indicator cells when mixed in the two-stage test with
known MIF preparations, (ii) comparing the electrophoretic mobilities of human $RBC_f$ when mixed in the two-stage test with
known concentrations of MIF and other additives, and (iii) comparing the electrophoretic mobilities of $SRBC_f$ in the two-
stage test when mixed with sensitized lymphocytes and antigens
or with supernatants removed from the cell-antigen preparations.

Table II  Comparison of the electrophoretic mobilities of different indicator cells for the MEM test.  Cells are incubated
with different concentrations of MIF preparations.

| | | Concentration of MIF | | | |
| Indicator Cell | None | 0.5 mg | 1.0 mg | 2.0 mg | % Effect |
| --- | --- | --- | --- | --- | --- |
| guinea pig macrophage | -1.167 | -1.215 | -1.178 | -1.189 | |
| guinea pig macrophage | -1.167 | | -1.226 | -1.208 | |
| guinea pig macrophage | -2.094 | | -2.099 | -2.112 | |
| mouse macrophage | -1.355 | <u>-1.184</u> | | | 12 < |
| mouse macrophage | -1.353 | -1.265 | -1.275 | <u>-1.264</u> | 7 < |
| mouse macrophage | -1.353 | <u>-1.273</u> | | | 6 < |
| mouse macrophage | -1.491 | -1.512 | | | |
| human $RBC_f$ (30) | -1.756 | <u>-1.687</u> | | -1.877 | 4 < |
| (90) | -1.669 | | | <u>-1.581</u> | 5 < |
| (150) | -1.672 | -1.651 | | <u>-1.552</u> | 7 < |
| (270) | -1.703 | | | <u>-1.507</u> | 12 < |

Table II presents the electrophoretic mobilities obtained with
guinea pig macrophages, murine macrophages and human $RBC_f$ when
mixed with various concentrations of stored MIF preparations.
MIF caused up to a 12% decrease in the electrophoretic mobilities of murine macrophages and human erythrocytes.  No effect
was observed with guinea pig macrophages at any of the concentrations of MIF tested.  The failure to detect a response with
guinea pig macrophages may reflect a subclinical health or nutritional problem associated with the colony of guinea pigs (13).

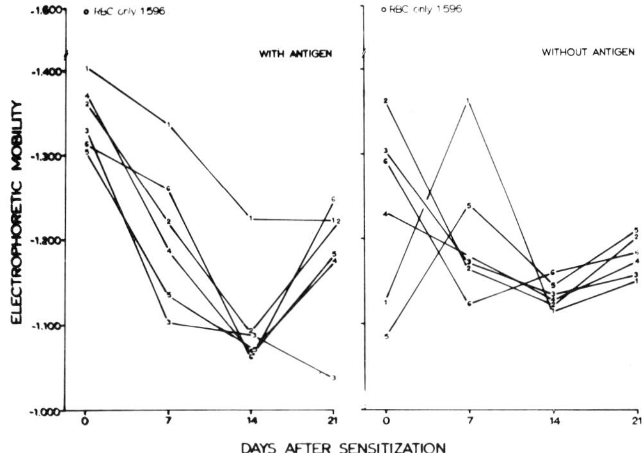



Fig. 5  Two-stage MEM test to detect the reactivity of lymphocytes taken from guinea pigs at various times after infection with Herpes virus type 1.

phoretic mobility of $SRBC_f$.  This decrease is less variable when the lymphocytes are cultured with antigen than without antigen, and (ii) in addition to the decrease of electrophoretic mobilities observed with nonsensitized cells (day 0), addition of specific antigen to sensitized lymphocytes causes a further decrease in electrophoretic mobility.  This second event, which constitutes the accepted MEM test, showed that four of the six animals on day 7 contained a substance which depressed the electrophoretic mobility of $SRBC_f$ on an average of 7.3%.  Data was analyzed by a two-way factoral analysis of variance using an inmixed Model III with antigen as a fixed factor and animals as a random factor (Table V).


Conclusions


The AEMS has been applied to a wide variety of biological problems to measure normal and altered electrophoretic mobilities of cells.  The capability of the AEMS to acquire data for

Table IV  Comparison of electrophoretic mobilities in the MEM test. $SRBC_f$ are treated with lymphocytes and supernatants obtained from guinea pigs sensitized against Brucella antigens.

| $SRBC_f$ Only | Immediate Assay | | | | Supernatant Stored | |
| | One-Stage | | Two-Stage | | Two-Stage | |
| Mobility | Mobility | % Effect | Mobility | % Effect | Mobility | % Effect |
|---|---|---|---|---|---|---|
| -1.471 | -1.324 | 10 < | -1.182 | 20 < | | |
| -1.471 | -1.309 | 11 < | -1.284 | 13 < | | |
| -2.601 | -2.396 | 8 < | -1.861 | 29 < | | |
| -1.479 | -1.290 | 12 < | -1.231 | 17 < | | |
| -1.479 | -1.342 | 9 < | | | | |
| -1.707 | | | | | -1.471 | 13 < |
| -1.707 | | | | | -1.454 | 15 < |
| -1.707 | | | | | -1.502 | 12 < |
| -1.707 | | | | | -1.503 | 12 < |
| -1.707 | | | | | -1.398 | 18 < |
| -1.707 | | | | | -1.421 | 17 < |

day 0.  Peripheral blood lymphocytes were collected on days 0, 7, 14 and 21.  Separated lymphocytes were cultured with and without formalin-killed Herpes type 1 viral antigens, according to the standard one-stage procedure.  Supernatants were reacted later with $SRBC_f$ in the two-stage test.  Electrophoretic mobility readings for 400 cells per test, with and without antigen, for each animal at each time period were made.  A total of 20,800 electrophoretic mobilities were recorded.

Guinea pigs injected with the virus showed the typical skin lesions of indurated erythema by day 2 - 3, necrosis from day 4 - 9 and rejection of the necrotic lesions by day 12.  Lymphocytes obtained from these animals and employed in the MEM test show the following results (Fig. 5):  (i) lymphocytes taken from the animals prior to injection with the virus seem to release substances to the supernatant which decrease the electro-

large sample sizes makes it feasible to include all appropriate controls.  This will allow one to make statistically valid determinations and approach problems which heretofore have not been tractable.

Table V  Analysis of variance of MEM data from day 7.

| Source of Variation | Sum of Squares | df | Mean Squares | f Value | Alpha |
|---|---|---|---|---|---|
| Total | 405.368035 | 4799 | | | |
| Treat-ment | 34.7674909 | 11 | | | |
| Antigen-MIF(A) | 4.25064655 | 1 | 4.25064655 | $^1_5 8.422$ | >0.025 to <.05 |
| Animals (B) | 27.9934133 | 5 | 5.59868266 | $^5_{4788} 72.33$ | <0.001 |
| AB | 2.5234107 | 5 | 0.504686213 | $^5_{4788} 6.520$ | <0.001 |
| Error Term | 370.600544 | 4788 | .07740195 | | |

Analysis done on electrophoretic mobility measurements at 500 μm depth in chamber.  Analysis done using an unmixed Model III with factor A (antigen-MIF) as fixed and factor B (animals) as random factor.  The 95% confidence limits of the electrophoretic mobility was based upon the $\sqrt{\text{mean square}}$ of AB  term divided by $\sqrt{\text{number of observations}}$.  95% CL = $X_n$ ± 0.069.


References


1.  Bartels, P. H. and Bier, M.: NASA contract #NAS8-31386, NASA George C. Marshall Space Flight Center, Huntsville, Alabama, April (1975).

2.  Olson, G. B., Bartels, P. H., Bartels, H. G., Brooks, D. E. and Seaman, G. V. F.: SPIE. <u>232</u>, 54-61 (1980).

3.  Bartels, P. H., Olson, G. B., Brooks, D. E. and Seaman, G. V. F.: Accepted Cell Biophysics (1981).

4.  Bartels, P. H. and Olson, G. B.  This journal (1981).

5.  Field, E. J. and Caspary, E. A.: J. Clin. Path. <u>24</u>, 179-184 (1971).

6.  Caspary, E. A. and Field, E. J.: Brit. Med. J. $\underline{2}$, 613-618 (1971).

7.  Lubran, M. M.: J. Clin. & Lab. Sci. $\underline{4}$, 121-124 (1974).

8.  Pritchard, J. A. V., Moore, J. L., Sutherland, W. H. and Joslin, C. A. F.: Brit. Med. J. $\underline{1}$, 823 (1972).

9.  Shenton, B. K. and Smith, B. M.: INSERM Symp. 11, 323-333 (1980).

10. Seaman, G. V. F.: The Red Blood Cell, Vol. II.  Academic Press, Inc., N. Y.  (1975).

11. Cohen, S. and Yoshida, T.: J. Immunol. $\underline{119}$, 719-728 (1977).

12. Sokal, R. R. and Rohlf, O. J.: Biometry.  W. H. Freeman and Company, San Francisco (1969).

13. Field, E. J.: INSERM Symp. 11, 293-302 (1980).

OPERATIONAL PARAMETERS FOR CONTINUOUS FLOW ELECTROPHORESIS
OF CELLS

Janice K. McGuire and Robert S. Snyder
NASA/Marshall Space Flight Center
Huntsville, Alabama  35812


Introduction

Large quantities of pure cell populations are of major interest
to the biomedical community.  Preparative electrophoresis has
the potential for supplying large quantities of cells purified
on the basis of their surface charge, a characteristic usually
related to their function.  NASA became interested in improving
the electrophoresis technique by taking it into the microgravity environment of space.  Weightlessness did not remove
all the problems of electrophoresis although buoyancy induced
fluid flows due to thermal and concentration gradients were
negligible.  The basic adjustable parameters had to be investigated for understanding and optimization.

A Beckman continuous particle electrophoresis apparatus (CPE)
was obtained for comparison with space experiments and to
determine the optimum conditions for ground-based electrophoresis.  Previous work by these authors (1) has determined
the optimum field and curtain flow rate for our standard cell
populations in phosphate buffer.

Two other readily adjustable parameters are discussed in this
presentation, the concentration and flow rate of the sample.
Additionally, a postulation by Ostrach (2) that an upflow
electrophoresis chamber would be more efficient than a downflow chamber is investigated.  The upflow chamber's increased
efficiency was suggested to be due to reduced convection with
the buffer warmed by Joule heat during its transit through the
chamber.

The study of sample concentration was combined with the CPE operation in upflow and downflow.  Various concentrations of equal parts of fixed cow and turkey red blood cells were separated.  Sample flow rate was varied in the upflow mode in an effort to improve the resolution using the optimum concentration level where separation began to break down in an effort to push this upper limit to the same level as the downflow.

Materials and Methods

Experiments were run in CPE apparatus.  For the upflow experiment, the chamber was inverted and the stainless steel collecting tubes were replaced with teflon tubing.  Figure 1 shows the CPE in the downflow mode and Figure 2 shows in the upflow. A phosphate buffer, used for all experiments, contained $NaH_2PO_4$ $7H_2O$, 1.76 mM, $KH_2PO_4$, 0.367 mM; and $Na_2EDTA.2H_2O$, 0.336 mM in deionized water.  Sample material was vena puncture red blood cells (RBCs) fixed and stored in formaldehyde.

The CPE chamber was coated with 2% bovine serum albumin, crystalline A grade, Calbio Chem-Behring Corp., La Jolla, CA. Sample was injected with a syringe and pump, Model 335, Sage Instrument, Inc., Cambridge, MA.  A magnetic stirrer, situated above the syringe which contained the sample and a small magnetic stirring bar, kept the cells in uniform suspension during sample injection.  Cooling water, curtain buffer, and electrode rinse buffer were kept in a $4^O$C water bath.  RBCs were counted in a bright line hemacytometer, Model # 1492, American Optical, Buffalo, NY.

Washed RBCs were diluted to the desired concentration in curtain buffer.  Flow rates were adjusted, sample injection started, and the chamber allowed to cool and equilibrate for 10 minutes.  Sample injection rate was 1.5 ml/hr; curtain flow rate was 25 ml/min.  Curtain fractions were then collected for base measurements with no electric field.  The field of 65 V/cm

was applied and conditions again allowed to equilibrate for
10 minutes before fractions with field were collected.  Fraction collection time was 10 minutes.  After each experiment,
sample flow rate and electric field were terminated and the
apparatus was flushed.

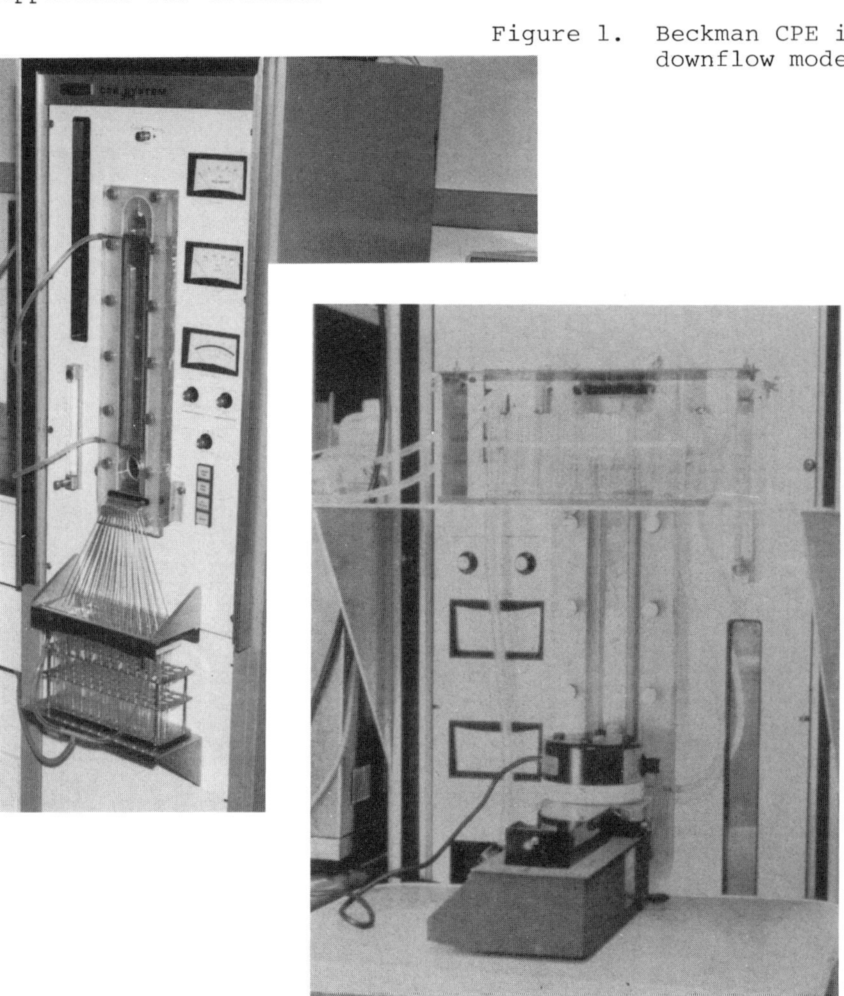

Figure 1.  Beckman CPE in the
downflow mode

Figure 2.  Beckman CPE in
the upflow mode

950

Results

Downflow runs were straightforward except for the appearance
of clumps of RBCs at concentrations of 1 and 2 x $10^9$/ml.
This phenomena was seen frequently in runs with no field and
less frequently in runs with field.  The same clumping was
seen in upflow at a concentration of 8 x $10^8$ RBCs/ml and above.
Figure 3 shows clumps in a sample stream of 1 x $10^9$ down with
no field.

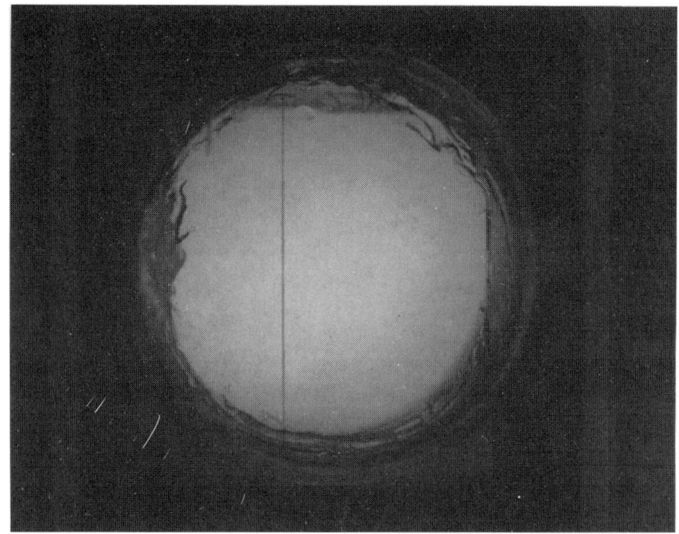

Figure 3.  Clumps in sample stream of 1 x $10^9$ RBC/ml
with no field

Inversion of the chamber for upflow studies brought on additional problems.  Beginning at concentrations of $8 \times 10^8$ RBCs/ml, the sample fountained from the top of the input port and accumulated in the area directly beneath it.  Figure 4 shows this phenomena.  Sedimented cells were subsequently picked up by the curtain buffer flow and carried on through the electrophoresis process resulting in a normal range of cell recovery rates.

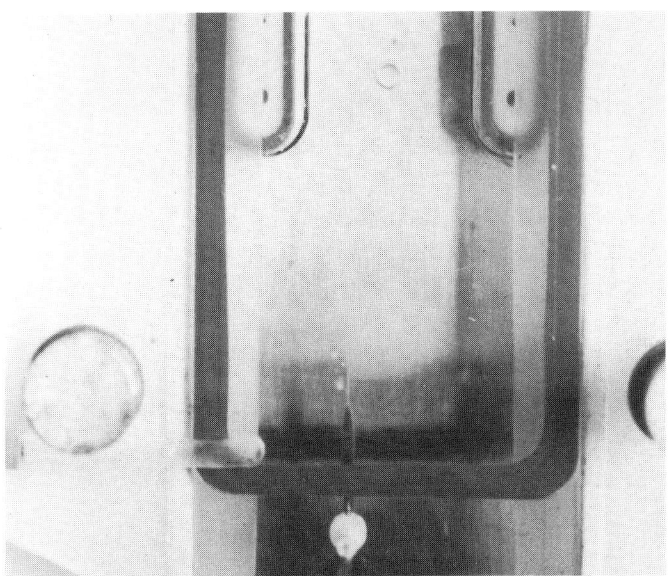

Figure 4.  Sample sedimenting from the input port at a concentration of $8 \times 10^8$ RBC/ml in upflow

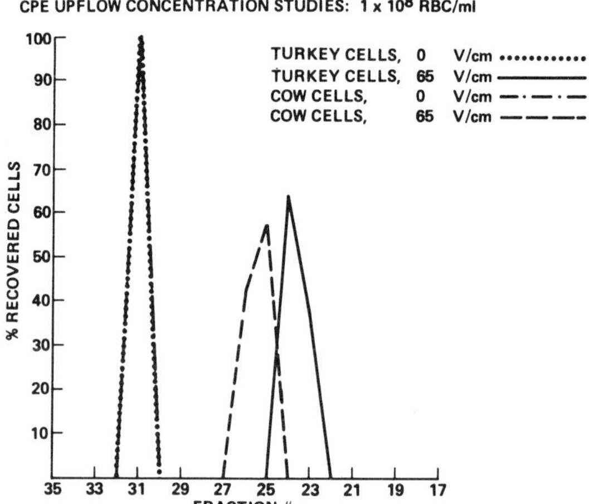


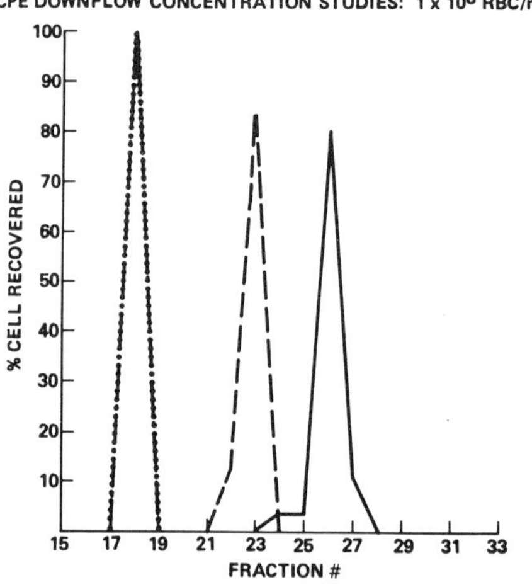


Figure 5.   Separation at $1 \times 10^8$ RBCs/ml

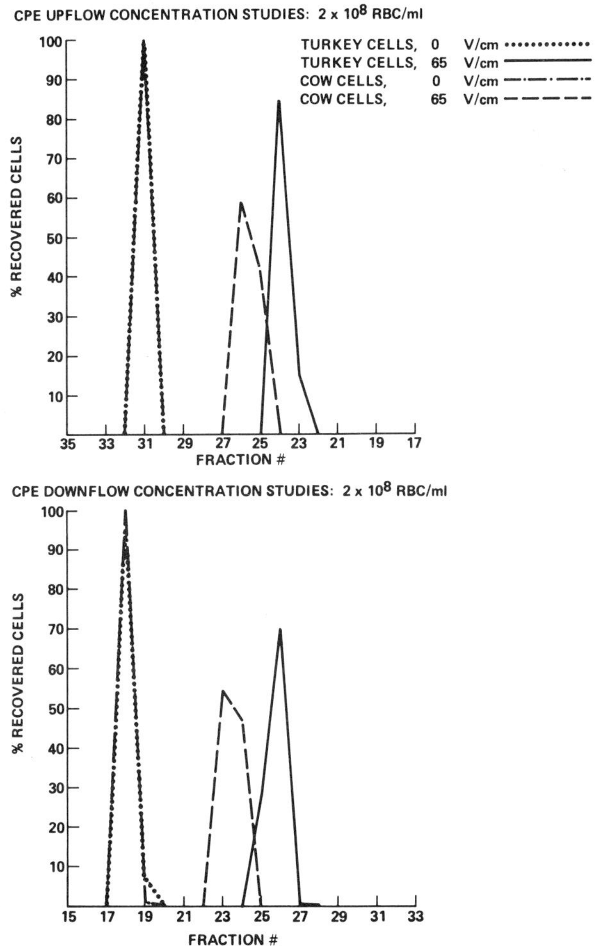



Figure 6.   Separation at $2 \times 10^8$ RBCs/ml

954

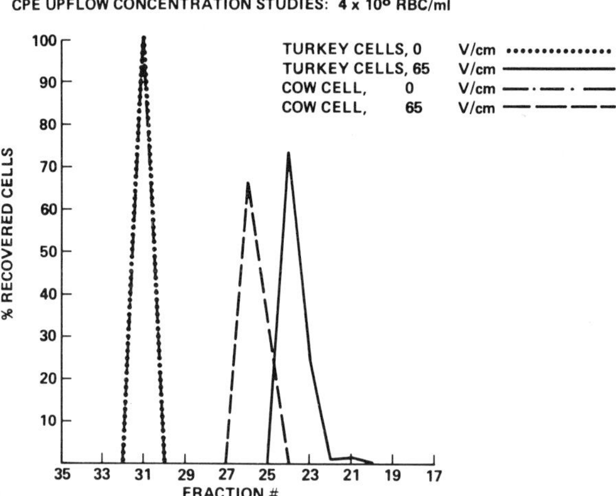


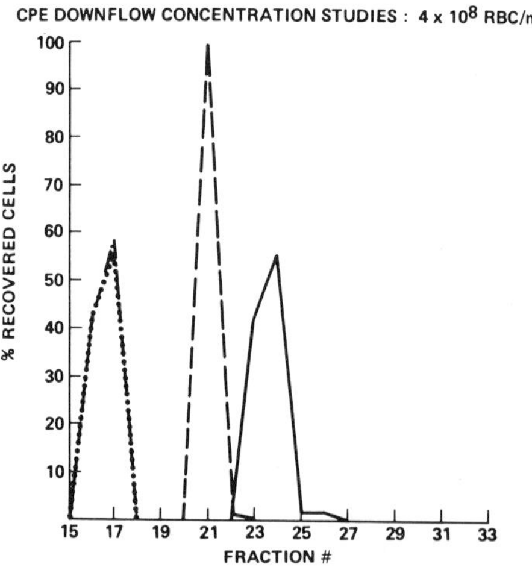


Figure 7.  Separation at 4 x 10$^8$ RBCs/ml

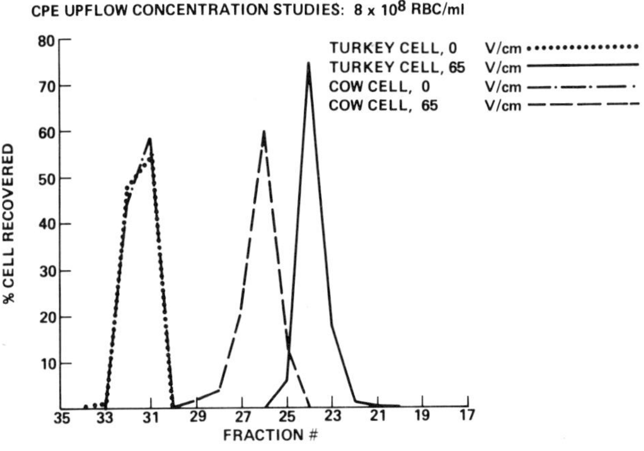



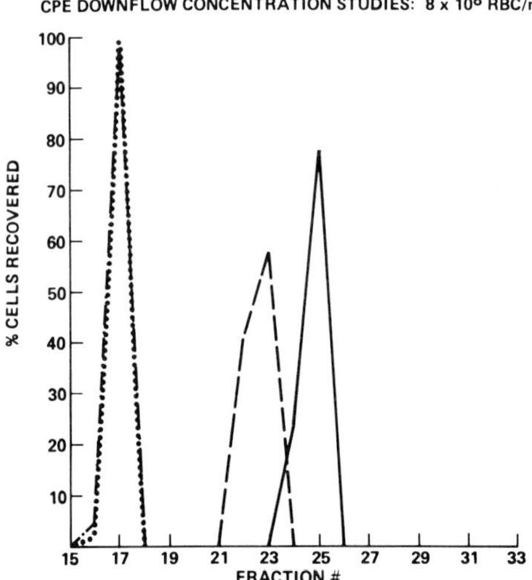



Figure 8.  Separation at $8 \times 10^8$ RBCs/ml

Another problem which appeared in the upflow configuration was
a wavering of the sample stream.  A wave of synchronous
falling drops from close groups of collection tubes resulted
in a differential pressure drop across the chamber that
pulled the sample stream to one side.  As the wave passed the
stream would return to its normal position.  A repetition of
the cycle produced a wavering of the stream.  This wavering
occurred over the last 70 mm of the sample stream and over
3 mm of the width of the chamber.  Disruption of the wavering
pattern was accomplished by constricting selective tubes,
coupled with rapid variation of the curtain flow rate thus
increasing the back pressure reflected into the chamber.

Separations at concentrations of 1, 2, 4, and 8 x $10^8$ RBC/ml
showed little difference in the upflow or downflow configuration.  Turkey cells generally moved 8 fractions and cow cells
5 or 6.  Graphs of representative separations are presented
in Figures 5 thru 8.  Photos of separation at a sample concentration of 2 x $10^8$ RBCs/ml is presented in Figure 9.

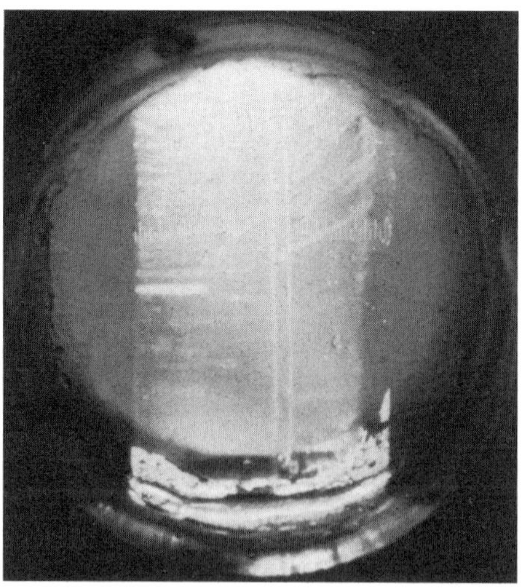

Figure 9.  Separation at 2 x $10^8$ RBCs/ml

Broadening of bands is very apparent at concentrations above
$8 \times 10^8$. A separation is obtained through $2 \times 10^9$ RBCs/ml in
downflow but breaks down in upflow with considerable overlap
and broadening of bands. Figure 10 shows this breakdown point
graphically. Photos of the upflow and downflow separations
at $2 \times 10^9$ RBCs/ml are presented in Figures 11 and 12.

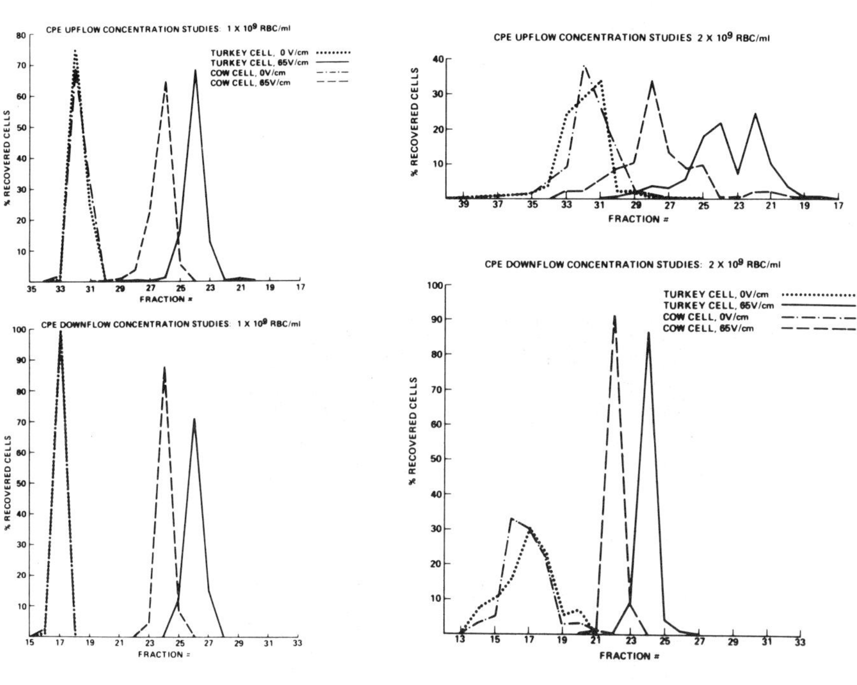


Figure 10.  Separations at 1 and 2 x $10^9$ RBCs/ml

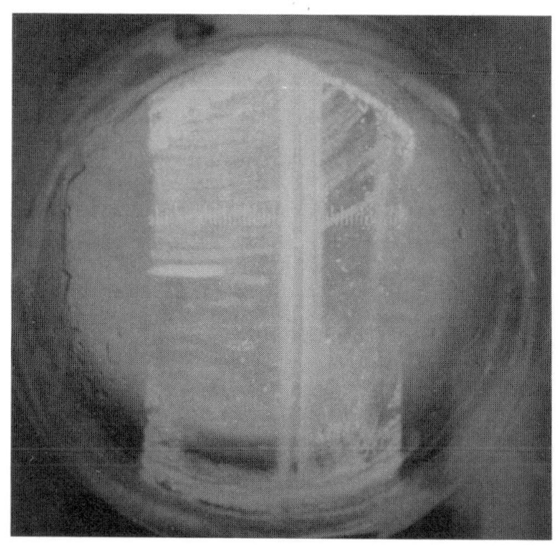

Figure 11.  Upflow separation at 2 x $10^9$ RBCs/ml

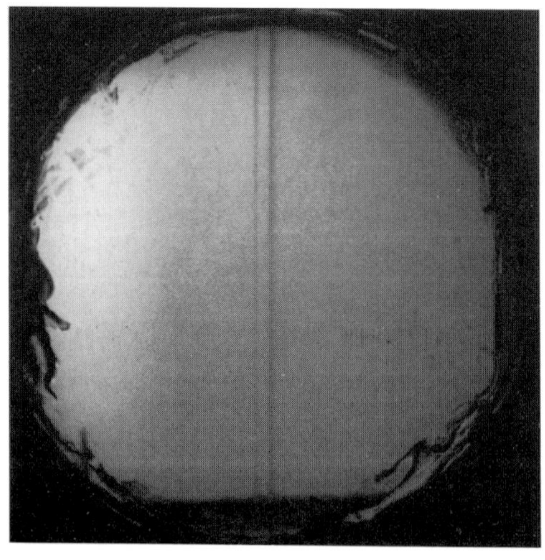

Figure 12.  Downflow separation at 2 x $10^9$ RBCs/ml

In an effort to improve the upflow separation to the level of the downflow, sample flow rates were tested using the upper limit concentration of $1 \times 10^9$ RBCs/ml, and also $2 \times 10^8$ RBCs/ml. $1 \times 10^9$ RBCs was run at 0.5, 1.0, 1.5 and 2.0 ml/hr sample flow rate and at 20 and 25 ml/min curtain flow rate with no improvement in the separation. $2 \times 10^8$ RBCs/ml was run at 0.5, 1.0 and 1.5 ml/hr sample flow rates and 20 and 25 ml/min curtain flow rates. Figure 13 shows some improvement at 1.0 ml/hr versus the 1.5 ml/hr used in previous experiments.

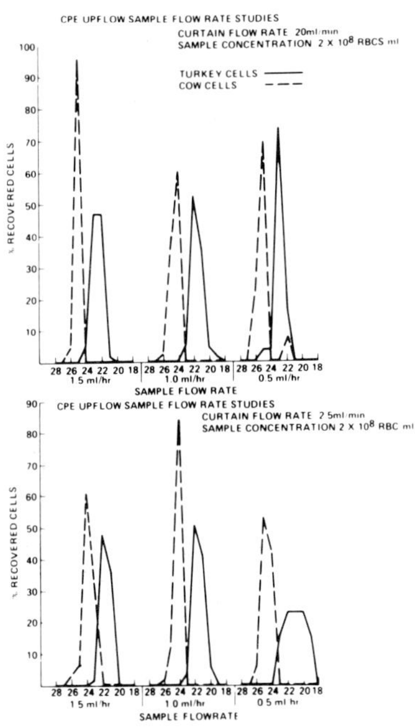


Figure 13. Sample flow rate studies at $2 \times 10^8$ RBCs/ml concentration

## Conclusions

The maximum concentration of cells that can be separated in
the Beckman CPE in our optimum conditions in downflow is
$2 \times 10^9$ RBCs/ml.  In the upflow mode it is $1 \times 10^9$ RBCs/ml.
Sedimentation at the sample input is a problem in upflow at
high concentrations.  Upflow resolution can be improved by
decreasing sample flow rate to 1 ml/hr at $2 \times 10^8$ RBCs/ml but
no improvement is seen at $1 \times 10^9$ RBCs/ml over a range of
flow rates.

## References


1.  McGuire, J. K., Snyder, R. S.:  in press.

2.  Ostrach, S.:  J. Chromat. __140__, 187 (1977).

REDUCTION OF DROPLET FORMATION AND SEDIMENTATION OF FIXED
ERYTHROCYTES IN STATIONARY AND FLOWING SYSTEMS

S. N. Omenyi*, and R. S. Snyder
NASA/Marshall Space Flight Center
Huntsville, Alabama  35812


C. J. van Oss

Department of Microbiology
State University of New York at Buffalo
Buffalo, New York  14214


D. R. Absolom, and A. W. Neumann
Department of Mechanical Engineering
University of Toronto
Toronto, Canada  M5S 1A4


Introduction

A major design objective of electrophoresis devices has been
to increase the amount of sample separated in a given time
interval.  It has been proposed that the particle concentration that can be separated by electrophoresis in a weightless
environment will be significantly higher because sedimentation and particle interactions compounded by buoyancy-induced
convection will not occur.  On earth, the application of a
homogeneous sample zone into a liquid column stabilized by a
density gradient can be accompanied by hydrodynamic disturbances termed droplet sedimentation or streaming.[1-6]  The
occurance of this phenomena results in the lower layers being
contaminated by particles trapped in falling droplets.  Excessive mixing of layers therefore occurs and separation
resolution is degraded.

*NRC Resident Research Associate

The factors contributing to the above phenomena are not well understood. The effects of particle concentration, the composition and density of supporting gradients and the diffusion coefficients of solutes have been identified and studied.[2,4,5,6] Yet the actual load supportable is not predicted by the available theory. In a systemmatic study, we have identified additional factors affecting the stability of suspension layers., viz., the effects of particle surface properties which have not been considered heretofore. Two methods are described for the reduction of droplet formation and sedimentation at the interface between the suspension layer and the supporting liquid as observed in zone electrophoresis. The first method involves the use of additives which lower the surface tension of the suspending solution and which are not toxic to the particle. Using the maximum sample concentration supportable on a liquid cushion as a criteria of stability, we showed that by choice of appropriate amounts of dimethyl sulfoxide (DMSO), suspension concentrations which are about a factor or two larger can be suspended. The second method involves the use of a flowing liquid cushion. The importance of surface properties of particles was further emphasized by the choice of correct combinations of liquid cushion and particle suspension flow rates. Increased suspension layer stability was also realized. Using erythrocytes from various sources and liquids, such as $D_2O$ and Ficoll, maximum particle concentrations for stable layers were measured sequentially using DMSO and in a flowing suspension liquid.[8]


Experimental Technique


The technique employed to determine maximum sample concentration for a stable sample zone was as follows: A known concentration of a selected sample suspended in 0.15M NaCl was layered carefully on the liquid cushion (2% (w/v) Ficoll

solution or DMSO) using a pipette.  At the end of the layering process (taken as the zero time), the timing was commenced
while the sample zone - liquid cushion interface was observed
microscopically at a magnification of 40X (Figure 1).  The
time at which droplets of cells started to form at the interface was taken as the droplet formation time, designated by
$T_1$.  This experiment was repeated for different concentrations
of the given sample suspension, layered on a fresh supply of
the same test liquid, and the data plotted as particle concentrations in particles/ml against droplet formation times.
The cell concentrations below which droplet formation was
less probable for an extended period of time was taken as the
critical concentration for droplet formation.

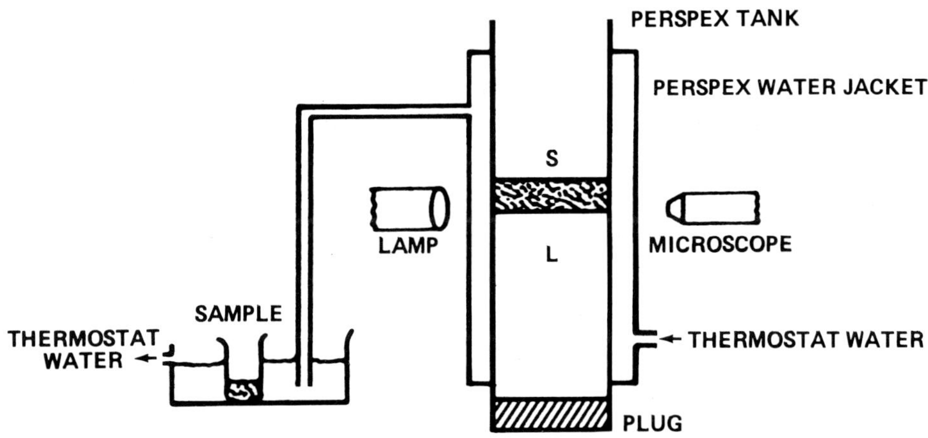


Figure 1.

In the absence of DMSO, the data showed that it was possible
to determine the maximum concentration of particles that
could remain stable on a given liquid cushion (Figure 2).
For dilute suspensions, droplet formation did not occur for

long periods of time.  For concentrated suspensions, droplets occurred within less than a minute after being layered in the suspension fluid.  Since our major objective was to determine the maximum suspension concentration that could be handled, the effect of using surface tension lowering agents was considered and the above experiment was repeated using DMSO of varying concentrations up to 15%.

Figure 3 shows the relation between maximum particle concentration and concentration of DMSO for human erythrocytes on $D_2O$.  Between 11 and 14% DMSO, critical suspension concentrations that are about a factor of two larger than observed previously were possible.  This figure also shows that very large concentrations of DMSO were not advantageous.  We can therefore see that by the choice of appropriate surface tension lowering agents that stability is possible.  Graphs similar to figure 3 were obtained using other erythrocytes and conditions at the maximum cell concentration.  The surface tension of the cells were derived and shown to be close to 65 ergs/$cm^2$.  By modifying the surface tension of the liquids to values equal to those of the cells, maximum cell concentrations were obtainable.

Apart from the use of additives the liquid cushion could be made to flow and the critical suspension concentration determined as a function of the flow conditions.  The technique used to study flowing sample streams stabilized against convection was similar to that developed by Mel and called the stable-flow free boundary technique.  The experimental device, shown in Figure 4, consisted of a glass migration chamber with inner dimensions of 0.7 x 7.0 x 21.3 cm.  The inside diameter of each of the seven teflon inlet tubes was 1.5 mm while the inner diameter of the sample inlet glass capillary, sealed in tygon tubing, was 0.5 mm.  The diameter of each of the nine outlet teflon tubes was 1 mm ID.  With reservoirs

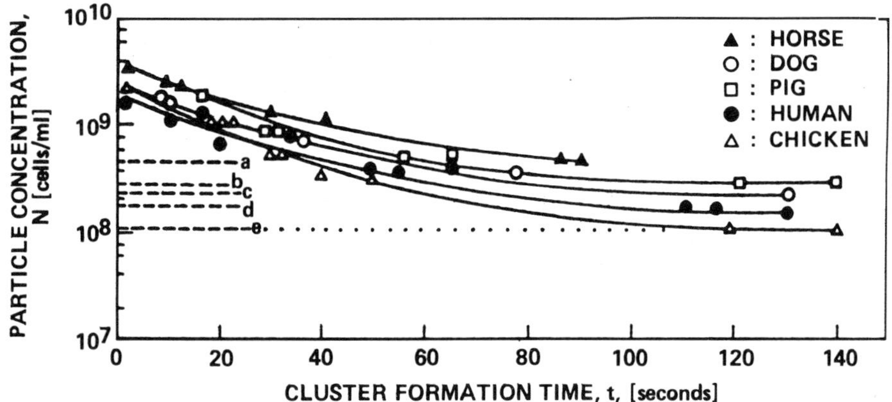

**THE VARIATION OF CLUSTER FORMATION TIMES WITH PARTICLE CONCEN-TRATIONS, FOR 2% FICOLL SOLUTION.  THE BROKEN LINES REPRESENT THE CRITICAL PARTICLE CONCENTRATIONS FOR CLUSTER FORMATION, AND GIVEN AS: a) HORSE, $4.5 \times 10^8$; b) DOG, $2.6 \times 10^8$; c) PIG, $2.7 \times 10^8$; d) HUMAN, $1.8 \times 10^8$; AND e) CHICKEN, $1.1 \times 10^8$ CELLS/ml.**

Figure 2

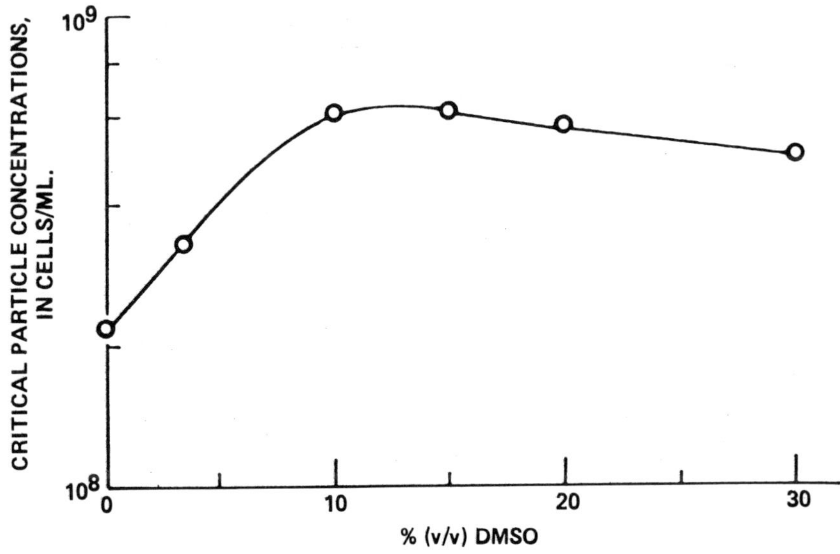

Figure 3

placed about 45 cm above the migration chamber, liquid was
fed into the migration chamber by gravity. 2% (w/v) Ficoll
solution was first admitted into the migration chamber followed by 0% Ficoll solution until a sharp interface appeared
between the two liquids and was in line with the sample inlet
tube. The liquid flow rate was determined by collecting a
known volume of the liquid in a given time. Then a sample
suspension, the concentration of which had previously been
determined, was injected into the flow chamber at the interface between the two liquids, using a syringe pump.

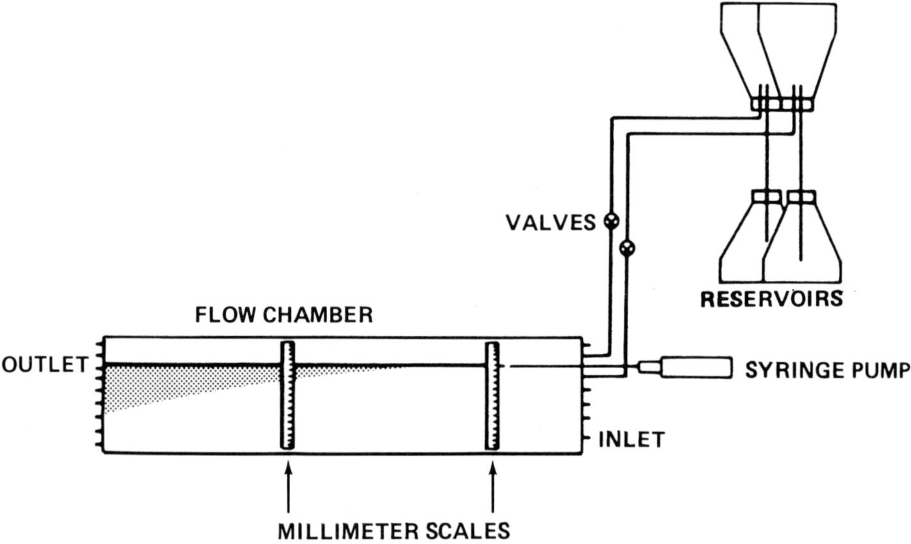


Figure 4.

The conditions for onset of instability for different samples
or liquid cushion flow rates were measured. As can be seen
in figures 5 and 6, it is clear that by increasing the liquid
cushion flow rate at a given value the supportable suspension
concentration was increased for a given particle material.
Also, increased suspension concentrations led to increased

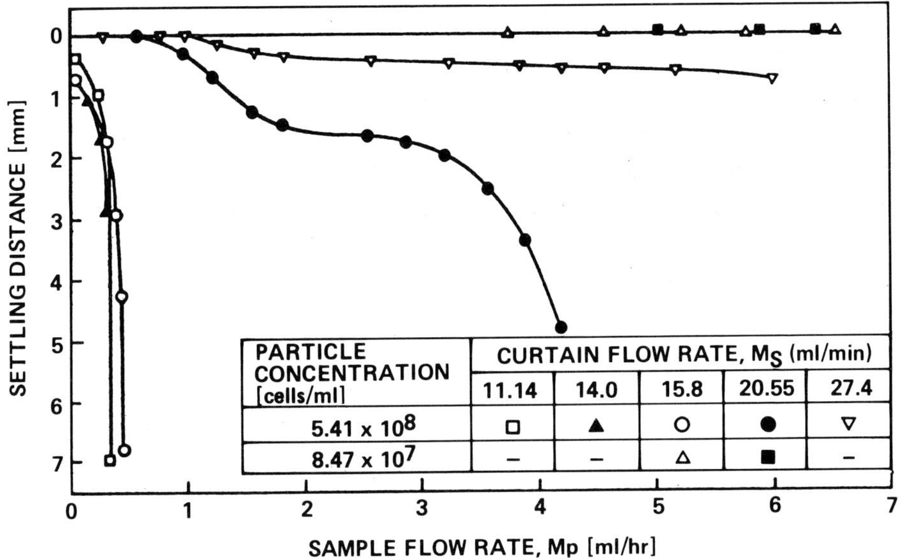



| PARTICLE CONCENTRATION [cells/ml] | CURTAIN FLOW RATE, $M_S$ (ml/min) | | | | |
|---|---|---|---|---|---|
| | 11.14 | 14.0 | 15.8 | 20.55 | 27.4 |
| $5.41 \times 10^8$ | □ | ▲ | ○ | ● | ▽ |
| $8.47 \times 10^7$ | – | – | △ | ■ | – |

**DEPENDENCE OF SEDIMENTATION OF FIXED TURKEY RED BLOOD CELLS ON CURTAIN FLOW RATE AND SAMPLE FLOW RATE IN 2% FICOLL SOLUTION**

Figure 5.

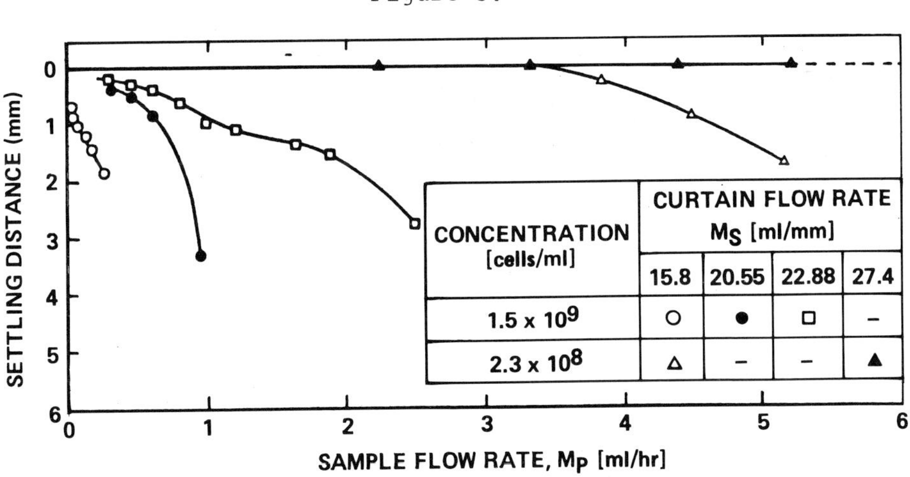



| CONCENTRATION [cells/ml] | CURTAIN FLOW RATE $M_S$ [ml/mm] | | | |
|---|---|---|---|---|
| | 15.8 | 20.55 | 22.88 | 27.4 |
| $1.5 \times 10^9$ | ○ | ● | □ | – |
| $2.3 \times 10^8$ | △ | – | – | ▲ |

**DEPENDENCE OF SEDIMENTATION OF FIXED CHICKEN RED BLOOD CELLS ON CURTAIN FLOW RATE AND SAMPLE FLOW RATE, IN 2% FICOLL SOLUTION**

Figure 6.

instability.  For these considerations, the point at which
the settling distance ceases to be zero was taken as the condition for the onset of instability.

Figures 7 and 8 confirmed the results of figure 2 that the
maximum supportable suspension concentration depends on the
cell type.  Maximum instability was observed for large cells
(turkey) whereas the smallest (horse, pig) showed the best
stability.  The importance of particle size and type together
with the flow condition are therefore important in considering the conditions for maximizing suspension stability.
Independent experiments have shown that concentrations higher
by a factor of 2 or 3 are achieved by the use of a flowing
cushion than for a stationary case.[8]


Conclusion

The dependence of droplet formation on particle concentration
was confirmed.  It was also shown that it was possible to
measure the maximum particle concentration supportable by a
liquid cushion by a simple layering technique.  We showed
for the first time that it was possible to reduce (though
not completely eliminate) the droplet formation by the use
of DMSO which acted as a surface tension lowering agent and
which will not be toxic to viable cells planned for future
use.  Suspended concentrations more than a factor of 2 larger
were obtained.  In the absence of surface tension lowering
additives, we also showed that it was possible to use the
flowing suspended liquid to reduce droplet formation and increase the suspendable particle concentration.

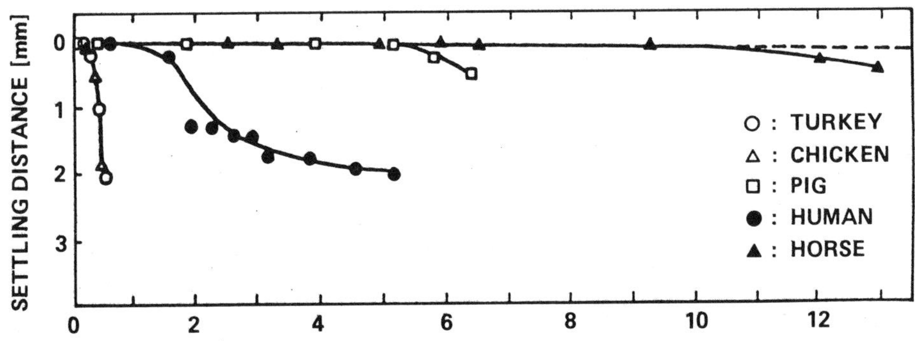



STABILITY OF SAMPLE STREAMS AT CONSTANT CURTAIN FLOW RATE OF $M_S$ = 8.16 ml/min FOR DIFFERENT PARTICULATE MATERIALS AT A CONSTANT PARTICLE CONCENTRATION OF 3.45 x $10^8$ CELLS/ml, ON 2% FICOLL SOLUTION

Figure 7.

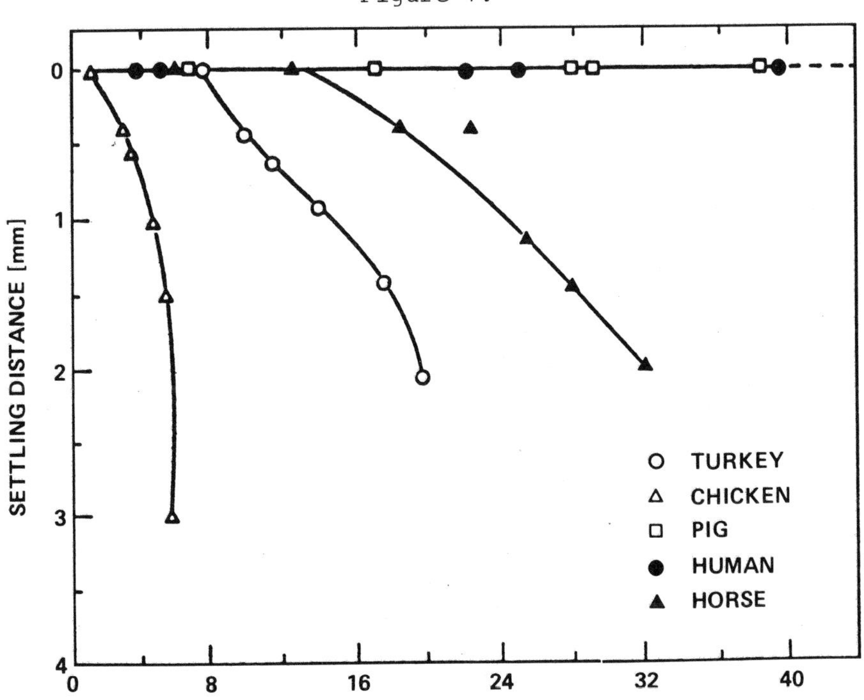



STABILITY OF SAMPLE STREAMS AT CONSTANT CURTAIN FLOW RATE OF $M_S$ = 23.08 ml/min FOR DIFFERENT PARTICULATE MATERIALS AT A CONSTANT PARTICLE CONCENTRATION OF 3.45 x $10^8$ cells/ml, ON 2% FICOLL SOLUTION

Figure 8.

References

1.    Brakke, M.K., Arch. Biochem. Biophys., 55, 175 (1955).

2.    Nason, P., Schumaker, V., Halsall, B., and Schevedes, J., Biopolymers, 7, 241 (1969).

3.    Meuwissen, J.A., and Heirwegh, K.P., Biochem. Biophys. Res. Commun., 41, 675 (1970).

4.    Peterson, E.A., and Evans, W.H. Nature, 214, 824 (1967).

5.    Plesset, M.S., and Winet, H., Nature, 248, 441 (1974).

6.    Halsall, H.B., and Schumaker, V.N., Biochem. Biophys. Res. Commun., 43, 601 (1971).

7.    Omenyi, S.N., Snyder, R.S., van Oss, C.J., Absolom, D.R., and Neumann, A.W., J. Coll. Interface Sci., in press.

8.    Omenyi, S.N., Snyder, R.S. and Rhodes, P.H., submitted for publication.

DESIGN CONSIDERATIONS  OF A THERMALLY STABILIZED CONTINUOUS
FLOW ELECTROPHORESIS CHAMBER


Percy H. Rhodes, Teresa Y. Miller and Robert S. Snyder
NASA/Marshall Space Flight Center
Huntsville, Alabama  35812


## Introduction

The concept of the "thermally stabilized electrophoresis
chamber" is an outgrowth of observations that the rectilinear
base flow through the cooled chamber is extremely sensitive
to any lateral temperature gradients which exist on the chamber inner walls.  The uniform velocity front was seen to be
affected by lateral gradients as small as $0.04^\circ$C/cm in our
0.55 cm thick test chamber and the plastic walls of a conventional continuous flow electrophoresis chamber support significantly higher gradients.  Therefore, recognizing that the
thermal conductivity of all transparent materials is poor, the
use of an efficient heat conductor, such as copper, has been
incorporated in a new design.  The copper chamber walls should
equilibrate lateral temperature gradients since the thermal
conductivity of copper is almost three orders of magnitude
better than glass.  The cooling fluid can also be channeled
through passages in the copper to provide a small axial thermal gradient from the bottom to the top of the chamber.  The
copper requires coating with a film of electrically insulating
material to maintain the electric field in the chamber, but
the coating should be as thin as possible so the thermal conductivity of the copper will not be degraded.  A disadvantage
of the copper chamber is the inability to see the sample
material traverse the chamber.  However, incorporation of
glass windows to observe the sample inlet and separated bands
just before collection should suffice for most applications.
A further advantage of the material selection is the capability

to efficiently sterilize the entire system with high temperature steam. Use of the autoclave and various gaseous and chemical sterilants has not been consistently successful in sterilizing either the CPE or FF48 for use with viable cells because each instrument contains critical polymeric components.

Materials and Methods

A plexiglass flow visualization chamber was used to investigate the effects of lateral temperature gradients on chamber performance (Figure 1). The 28 cm x 5 cm x 0.55 cm chamber is equipped with front, rear and sidewall cooling chambers. Curtain electrolyte enters the chamber through a port at the top and exits through a port at the bottom. The temperature of the curtain electrolyte is monitored by a thermocouple probe mounted at the entry port. Coolant fluid enters each cooling chamber through a bottom port and exits at the top and some degree of thermal control was obtained by altering the flow rate of coolant through the various cooling chambers. Fifteen thermocouples mounted in sets of three on movable bars (Figure 2) were used to monitor temperature changes in the curtain buffer, predominately at the center plane and the walls of the chamber. The temperatures were scanned and recorded within five seconds using an HP 3495-H scanner.

Velocity profiles at the chamber's center plane were determined using a recently developed fluorescent flow visualization technique. A 0.2 mm diameter fiber stretched across the width of the chamber at the center plane served as a method for allowing ionic contact between an activator solution and the (fluorescently treated) curtain buffer. The fluorescence of the curtain buffer was achieved after activation by using two ultraviolet lights positioned on either side of the chamber. The flow profile of the fluorescing fluid medium in the center-plane of the chamber accurately defined the velocity profile from established temperature gradients.

Figure 1.

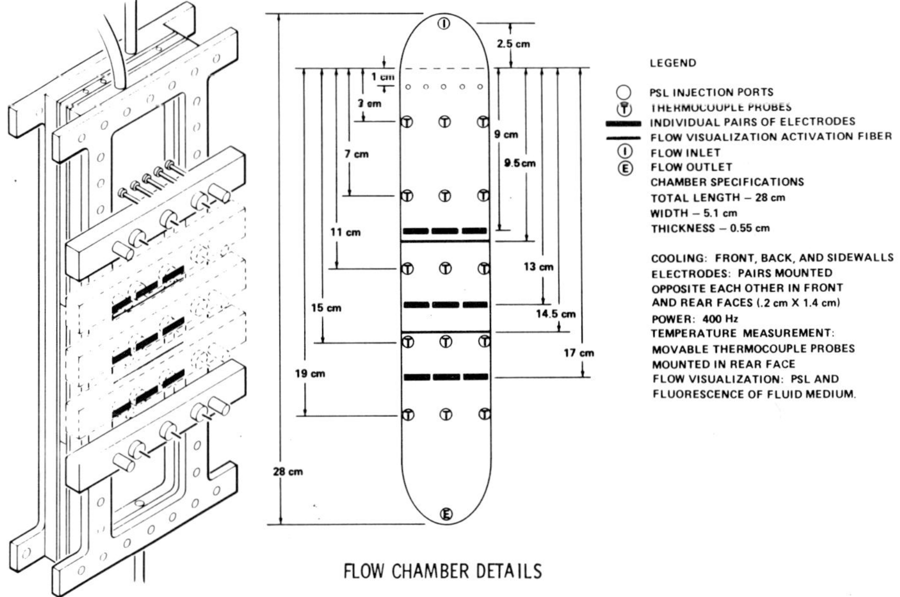



FLOW CHAMBER DETAILS

## Results and Discussion

The following examples show correlations between chamber wall temperatures and the resulting velocity profiles in the chamber. The profiles were initiated at the small fibers stretched across the width of the chamber in the locations shown in Figure 2. The temperature data shown in Figures 3, 4, and 5 correspond to the positions of the thermocouple array.

Figure 3 shows uniform base flow which is typical of what one would expect in a free flow electrophoresis chamber. The flow is observed at the center plane at the lower fiber and is essentially uniform across the width of the chamber except at the side walls. Note, however, that the velocity profile is not exactly symmetrical, i.e., a slightly greater retardation of flow on the right can be seen. The temperature gradient

```
CURTAIN INLET TEMPERATURE:   26.98  Deg C      CURTAIN FLOW:    10.0  Ml/min

        WALL TEMPERATURES Deg C
                                                       COOLANT FLOW
     26.90        26.93        27.00            RIGHT SIDE-    75  Ml/min
                                                FRONT FACE-   157  Ml/min
                                                REAR  FACE-   150  Ml/min
     27.10        27.10        27.15            LEFT  SIDE-    50  Ml/min

                                                POWER INPUT
     27.12        27.10        27.15            VOLTS-   0.0
                                                MILLIAMPS- 0.0

     27.12        27.07        27.17          ELECTRODES- 0,00,0,00,0,00,0,00

                                              PHOTOGRAPHY:Roll # 8 FRAME # 3,4,5
     27.12        27.07        27.12
*****************************************************************************
        THE TIME IS-10:46:00                     THE DATE IS-09/26/80
                                                   Cooler   off
CURTAIN INLET TEMPERATURE:   26.95  Deg C      CURTAIN FLOW:    10.0  Ml/min

     CENTER PLANE TEMPERATURES Deg C
                                                       COOLANT FLOW
     26.93        26.88        27.00            RIGHT SIDE-    75  Ml/min
                                                FRONT FACE-   157  Ml/min
                                                REAR  FACE-   150  Ml/min
     27.12        27.05        27.17            LEFT  SIDE-    50  Ml/min

                                                POWER INPUT
     27.12        27.05        27.17            VOLTS-   0.0
                                                MILLIAMPS- 0.0

     27.12        27.07        27.17          ELECTRODES- 0,00,0,00,0,00,0,00

                                              PHOTOGRAPHY:Roll # 8 FRAME # 3,4,5
     27.12        27.07        27.15
*****************************************************************************
```

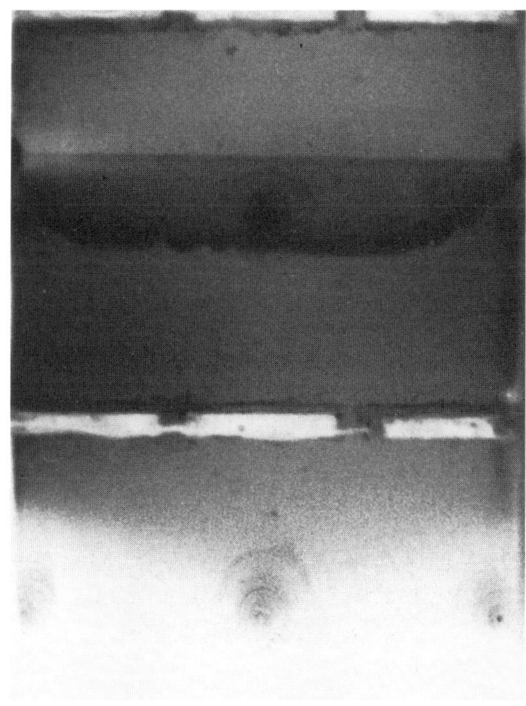

Figure 3.

976

for this region can be calculated from the fourth row of temperatures and is,

$$\frac{27.17 - 27.07}{2.55} = 0.039^{O}C/cm$$

For this experiment a lateral temperature gradient as small as $0.04^{O}C/cm$ has caused visual distortion of the symmetric base flow.

Figure 4 shows two velocity profiles in the chamber.  Note that the symmetrical base flow of the upper profile corresponds to the rather symmetrical temperature distribution of the upper portion of the chamber while the lower velocity profile exhibits a higher flow rate on the left side of the chamber corresponding to the small laterial gradients in that region.  This gradient, approximately $0.03^{O}C/cm$, is sufficient to cause appreciable distortion to the uniform base flow even though there is still rectilinear flow in the chamber.  Even though this temperature gradient is smaller than the previous case, it nevertheless causes a larger pertubation.  This can be explained by noting that the latter case extends across the total chamber width while the gradient associated with the previous case is localized.

Figure 5 depicts a case where rectilinear flow has been disrupted by lateral gradients as shown by the onset of reverse flow on the right side of the chamber.  The lower temperatures along the chamber center line are responsible for the higher flow rate in this region, while the slightly higher temperatures on the right side of the chamber retard the flow sufficiently so that back flow results.  Lateral gradients of the order of $0.10^{O}C/cm$ are therefore seen to be sufficient to cause this flow pertubation which would preclude reliable electrophoretic separation.

```
CURTAIN INLET TEMPERATURE:    22.60  Deg C    |    CURTAIN FLOW:      10.0  Ml/min

       WALL TEMPERATURES Deg C
                                                          COOLANT FLOW
   22.39        22.38        22.38              RIGHT SIDE-   75  Ml/min
                                                FRONT FACE-  157  Ml/min
                                                REAR  FACE-  150  Ml/min
   22.25        22.18        22.28              LEFT  SIDE-   50  Ml/min

                                                POWER INPUT
   22.25        22.10        22.08              VOLTS-  0
                                                MILLIAMPS- 0

   22.25        22.18        22.30              ELECTRODES-

                                                PHOTOGRAPHY:Roll # 18 FRAME # 4,5,6,78
   22.05        22.10        22.18
****************************************************************************************
              THE TIME IS-10:08:00                        THE DATE IS-10/09/80
CURTAIN INLET TEMPERATURE:    22.60  Deg C    |    CURTAIN FLOW:      10.0  Ml/min

   CENTER PLANE TEMPERATURES Deg C
                                                          COOLANT FLOW
   22.20        22.38        22.23              RIGHT SIDE-   75  Ml/min
                                                FRONT FACE-  157  Ml/min
                                                REAR  FACE-  150  Ml/min
   22.08        22.18        22.03              LEFT  SIDE-   50  Ml/min

                                                POWER INPUT
   22.10        22.08        22.00              VOLTS-  0
                                                MILLIAMPS- 0

   22.03        22.03        22.03              ELECTRODES-

                                                PHOTOGRAPHY:Roll # 18 FRAME # 4,5,6,78
   21.95        21.98        21.93
****************************************************************************************
```

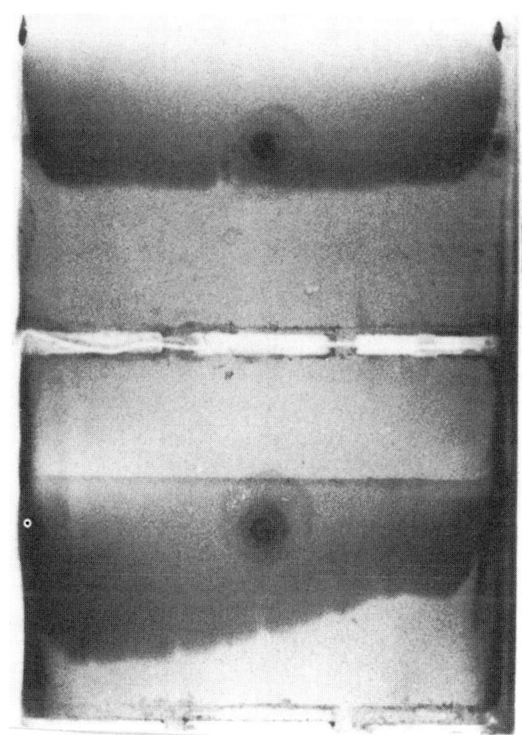

Figure 4.

```
CURTAIN INLET TEMPERATURE:   27.34  Deg C  | CURTAIN FLOW:    10.0  Ml/min

     WALL TEMPERATURES Deg C               |
                                           |       COOLANT FLOW
     27.66       27.46       27.76         |   RIGHT SIDE-   75  Ml/min
                                           |   FRONT FACE-  157  Ml/min
                                           |   REAR  FACE-  150  Ml/min
     27.83       27.66       27.90         |   LEFT  SIDE-   50  Ml/min
                                           |
                                           |       POWER INPUT
     27.85       27.66       27.93         |   VOLTS-  0.0
                                           |   MILLIAMPS- 0.0
                                           |
     27.85       27.71       27.95         | ELECTRODES- 0,00,0,00,0,00,0,00
                                           |
                                           | PHOTOGRAPHY:Roll # 8 FRAME # 11,12,13,4
     27.88       27.73       27.90         |
*****************************************************************************

          THE TIME IS-10:59:00                     THE DATE IS-09/26/80

CURTAIN INLET TEMPERATURE:   27.37  Deg C  | CURTAIN FLOW:    10.0  Ml/min

     CENTER PLANE TEMPERATURES Deg C       |
                                           |       COOLANT FLOW
     27.68       27.37       27.80         |   RIGHT SIDE-   75  Ml/min
                                           |   FRONT FACE-  157  Ml/min
                                           |   REAR  FACE-  150  Ml/min
     27.88       27.49       27.95         |   LEFT  SIDE-   50  Ml/min
                                           |
                                           |       POWER INPUT
     27.88       27.54       27.98         |   VOLTS-  0.0
                                           |   MILLIAMPS- 0.0
                                           |
     27.88       27.61       28.00         | ELECTRODES- 0,00,0,00,0,00,0,00
                                           |
                                           | PHOTOGRAPHY:Roll # 8 FRAME # 11,12,13,14
     27.88       27.66       27.95         |
*****************************************************************************
```

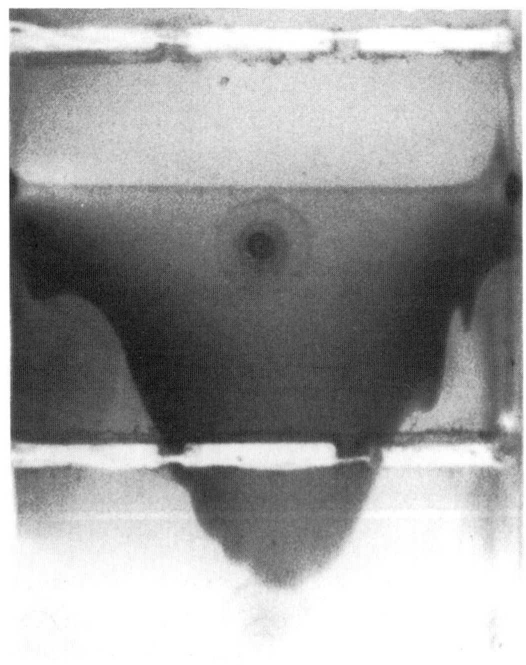

Figure 5.

These examples show that small lateral thermal gradients can
disrupt the uniform base flow in an electrophoresis type chamber and may cause performance degradation even though rectilinear flow still exists.  In particular, the flow pertubations
are seen to depend upon the temperature distributions on the
chamber wall.  Therefore, the more uniform the wall temperatures, the more uniform we can expect the velocity profiles
to be.  This conclusion has led us to propose that the flow
chamber walls be made of a high conductivity material.

To further verify thermal effects on increased resolution
and throughput, a chamber has been constructed which incor-
porates the Beckman CPE collection system and the Beckman CPE
front plate modified by replacing the 39 x 6.5 cm glass insert
with a copper insert of the same dimensions.

A plexiglass back plate with comparable copper insert has
been designed to replace the standard all plexiglass CPE back
plate (Figure 6).  Two vertical channels in the back plate
contain the electrode chambers which communicate with the
separation ·chamber through two 28 x 0.2 cm slots in the copper
insert.  Two 2.5 mm diameter spectra/pore semi-micro cellulose membranes separate the electrode chambers from the curtain buffer flow.

A major problem encountered in the development of the metal
walled electrophoresis chamber was the application of a thin,
buffer resistant coating which would provide electrical insulation without impairing the thermal conductivity of the
copper walls.  The performance of several coatings recommended
for copper were evaluated according to the above criteria.
Vinylidene Fluoride Power Resin ("Kynar", Penwalt Corp.) in
Dimethyl Formamide; "Formvar" (Monsanto); a "Butvar", (Mon-
snato), "Epon 1007" (Shell Oil Co.) and a "Methylon 75103"
(General Electric) formulation for· metal can coatings; R-4-
3117 silicone conformal coating (Dow Corning); and Class H

Figure 6.

insulating varnish (P. D. George Company) were preceded where appropriate by an Iridite #7P (Allied Research Products) pre-coat and then applied to copper test strips by dipping. The Kynar film was crosslinked using a 10 mega rad dose of gamma radiation from a $Co^{60}$ source. A fluidized bed technique was employed in the application of a 71000 epoxy resin coating (Armstrong).  A 3 mil teflon tape, ("Temp-R-Tape", DuPont) was applied directly to clean copper strips.  All coated test strips were exposed to 0.15M NaCl solutions for varying intervals not exceeding 30 days, and examined periodically for evidence of corrosion.  The more resistant coatings, R-4-3117 silicon, 71000 epoxy and teflon Temp-R-Tape were subsequently evaluated in the modified CPE chamber.  Both the R-4-3117 silicon and 71000 epoxy, applied at film thicknesses of approximately .001 inch, failed to completely insulate the copper wall inserts.  Adquate insulation was provided by .003 inch teflon tape, but the configuration of the electrode slots did not prove compatible with application of the tape. Two alternative methods for alievating the insulation problem are being evaluated.  Insulating Delrin inserts are being epoxied into the electrode slots, so that teflon tape need only be applied to the flat copper surfaces.  Another method for eliminating the slot coating problem is to redesign the chamber configuration, positioning the electrodes along the side walls thus leaving only the flat front and back walls for insulation.

THE SHEEP ERYTHROCYTES ELECTROPHORETIC MOBILITY : EFFECT OF
SUPERNATANTS OF LYMPHOCYTES STIMULATED WITH VARIOUS IMMUNO-
POTENTIATORS.


Nobuya Hashimoto, Masaki Ageshio, Shunichi Nose, Shoichi
Horita, Toshiko Kobayashi and Masakazu Abe.
Department of Internal Medicine,
Jikei University School of Medicine,
Minato-ku, Tokyo.


Introduction

The sensitized lymphocytes stimulated with antigen or non
specific mitogen result in production of the substances with
bilogical activities, lymphokines.  One of them, called
macrophage electrophoretic slowing factor, has the property
of reduction of electrophoretic mobility of normal guinea pig
macrophages.  The macrophage electrophoretic mobility test
was first described as a method for the detection of malignant diseases by Field and Caspary (1).  However, there are
considerable difficulties in the operation of the macrophage
electrophoretic mobility test due to the continual requirement of freshly prepared guinea pig macrophages and selection
of the macrophages used for measurement.
A new modification was introduced by Porzsolt and Ax (2),
and Jenssen and Shenton (3).  The guinea pig macrophages as
indicator cells were replaced by tanned sheep erythrocytes.
This test system is simple and reliable.  Using this test
system, we have confirmed that electrophoretic mobility of
tanned sheep erythrocytes was reduced by supernatants from
sensitized lymphocytes incubated with PPD, and not changed
in incubation with PPD and lymphocytes from tuberculin test
negative subjects (4).

The aim of the present study is to investigate effect of the
supernatants of human lymphocytes stimulated with the various
immunopotentiators on electrophoretic mobility of sheep
erythrocytes.


Materials and Methods


Lymphocytes were isolated from human peripheral blood and
devided into two test sample.  To one test sample, mitogens
or immunopotentiators were added and cultured at 37ºC.  Then
the supernatant of this incubation mixture was incubated with
the indicator cells, i.e. the tanned sheep erythrocytes.  As
a control, supernatant from another test sample to which
immunopotentiator was not added was prepared. A lymphocyte
suspension was adjusted in $10^6$ cells in 0.1 ml of electrophoretic buffer.
As indicator cells, sheep erythrocytes, tanned and stabilized
with sulphosalicylic acid (ETS), were obtained as a lyophilized preparation from Behringwerke, West Germany (Courtesy
of the Japan Hoechst Co., Tokyo).  ETS suspension was adjusted to $10^7$ cells in 0.1 ml of electrophoretic buffer.
Measurements of erythrocyte electrophoretic mobility were
performed in an analytical cell microelectrophoretic apparatus, manufactured by the Sugiura Laboratory Co., Tokyo.  All
measurements were carried out at a potential difference of 40V
and current of 200 $\mu$A and at a temperature of 25ºC.
The time for each of 20 to 30 cells from each sample to cross
one square of the microscope grid was determined, each cell
being timed in both directions of the field (1 grid square =
34.3 $\mu$), and recorded by an electric digital stop-watch with
an automatic printer (reading to 0.01 sec).  Any measurements
in which the difference in timing for the two directions of
the field was greater than 10% were discarded.  The cells in

focus in the stationary layer were measured.  The mean
actual migration time was measured.  The percentage slowing
in mobility was expressed by the formula :

$$100 \left( \frac{T_2 - T_1}{T_1} \right) = \% \text{ slowing}$$

$T_1$ : mean time of the sample
$T_2$ : mean time of the control

The electrophoretic buffer used was the triethanolamine solution devised by Hannig (5).

As immunopotentiators, the following stimulating agents were
used.  PHA (phytohaemagglutinin, HA16, Wellcome) was used
at 30 $\mu$g / 15 $\mu$l.  ConA (concanavalin A, EY Lab.,) was used
at 5 $\mu$g / 0.1 ml.  OK-432 is a streptococcal preparation
(Picibanil), produced by incubating cultures of the low-
virulent Su strain of Streptococcus pyogenes A3 of human
origin.  PSK is protein-bound polysaccharide preparation extracted from Coriolus versicolor of Basidiomycetes (Krestin).
Levamisole is a derivative from imidazothiazol, well-known as
antihelmenthic drug.  PSK and Levamisole were used at a
concentration of 1 $\mu$g, separately.  Poly A:U is a synthetic
double-strand polyribonucleotide polyadenylate polyuridylate,
able to stimulate antibody production and cell-mediated immune response. Poly I:C is a double-strand polyribonucleotide
polyinosinate-polycytidylate, able to stimulate production of
interferon.  These were used in each at 2 ug concentrations.
Sodium periodate was recently reported an inducer of lymphokines (6).  Methyl $B_{12}$ was described as an immunomodulator
for mitogenic activity of lymphocytes (7), which was used at
10 $\mu$g.  Lentinan is a polysaccharide extracted from Lentinus
edodes, having antitumor effect, which was used at 10 $\mu$g.


Results and Discussion


The effects of treatment of sheep erythrocytes with tannic

Table 1.  Mean electrophoretic mobilities of sheep
          erythrocytes

| Cells | Mobilities ($\mu$/sec/v/cm) |
|---|---|
| Fresh sheep erythrocytes | $-1.54\pm0.14$  (100) |
| Tanned sheep erythrocytes | $-1.55\pm0.10$  (300) |
| ETS | $-1.57\pm0.12$  (510) |
| ETS+PHA | $-1.58\pm0.14$  (100) |
| ETS+ConA | $-1.57\pm0.18$  (100) |

ETS : tannic acid treated sulphosalicylic acid stabilized

sheep erythrocytes (supplied by Behringwerke)


acid and/or sulphosalicylic acid are summarized in Table 1.
Freshly drawn sheep erythrocytes, tanned sheep erythrocytes
prepared by the method of Stavitsky (8) and ETS showed almost
same electrophoretic mobilities.  Furthermore, even ETS to
which PHA and ConA were added for 45 min, 37ºC, did not have
their surface changed.  In subsequent work we have utilized
only ETS.
To investigate the effect of supernatants from lymphocytes
stimulated with immunopotentiator on electrophoretic mobility
of sheep erythrocytes, OK-432 was first used.  Various concentrations of OK-432 were added to the normal lymphocyte
suspension.  In 1/10 KE of concentration slowing of electrophoretic mobility of erythrocytes showed high percentages
(1.83 ± 4.19%) (Fig. 1).
In PSK mean percentage slowing of erythrocyte electrophoretic
mobility was 0.94 ± 6.70% in 12 normal subjects.  The effect
of PSK on electrophoretic mobility of erythrocytes is not
significant.
Percentage slowing of electrophoretic mobility of sheep erythrocytes in Levamisole-treated lymphocytes from normal subjects was 0.83 ± 2.21%.  It is similar to the result of PSK.
In Vit. B$_{12}$, percentage slowing of erythrocyte electrophoretic

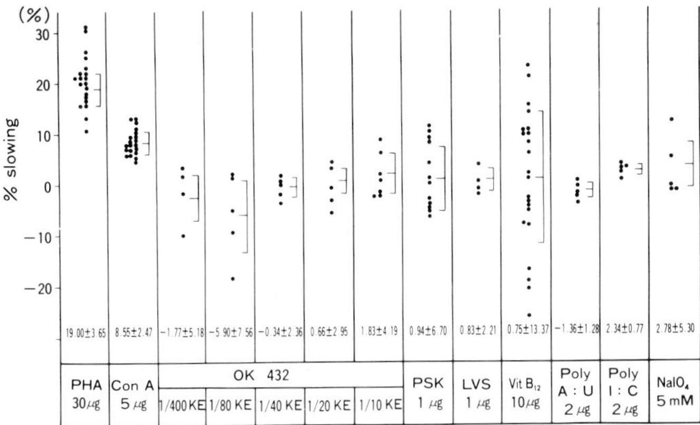

Fig. 1.  % slowing of EMT in the lymphocytes treated with
mitogens and immunopotentiators

mobility showed wide distribution, -26.02 to 23.43%.  Mean
percentage slowing of 22 normal subjects was 0.75 ± 13.37%.
It does not mean a significance as effect of stimulation with
Methyl $B_{12}$.
The effects of supernatants of normal lymphocytes stimulated
with Poly A:U, Poly I:C and Sodium periodate on erythrocyte
electrophoretic mobility were the results that mean percentage slowing was -1.36 ± 1.28% in Poly A:U, 2.34 ± 0.77% in
Poly I:C and 2.78 ± 5.30% in Sodium periodate, respectively.
Among these immunopotentiators the supernatant of lymphocytes
stimulated with Sodium periodate has  a  slightly increased
percentage slowing.
The results obtained for lymphocytes from normal subjects
stimulated with these immunopotentiating drugs are summarized
in Fig. 1.  As we have previously decribed, the percentage
slowing  of sheep erythrocytes by supernatants from normal
lymphocytes stimulated with PHA and ConA was  obviously increased (19.00 ± 3.65% in PHA and 8.55 ± 2.47% in ConA) (9).

988

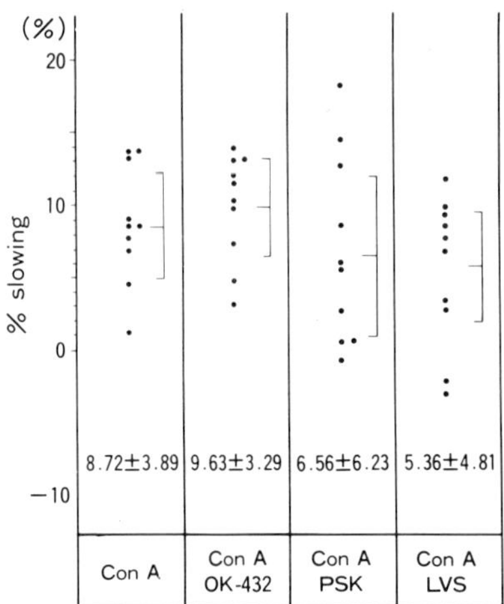


Fig. 2. % slowing of EMT in the normal lymphocytes stimulated with ConA and immunopotentiators


In contrast to this, other immunopotentiators did not
strongly affect erythrocyte electrophoretic mobility added
to the supernatant of lymphocytes stimulated with them.
Furthermore, to investigate immunopotentiating action of
OK-432, PSK and Levamisole, these drugs were added to the
lymphocytes pretreated with ConA.  Fig. 2 shows the result.
In group of ConA and OK-432, mean percentage slowing was
slightly higher than  the ConA control group, but not significantly. In both group of ConA · PSK and ConA · Levamisole,
percentage slowing  were slightly decreased, compared  with
controls.
Subsequently, the   erythrocyte electrophoretic mobility test
was performed for supernatant of the cancer patients lymphocytes stimulated with the immunopotentiators.  The patients
with well-defined malignant diseases were admitted in our
University Hospital.  The result is shown in Fig. 3.

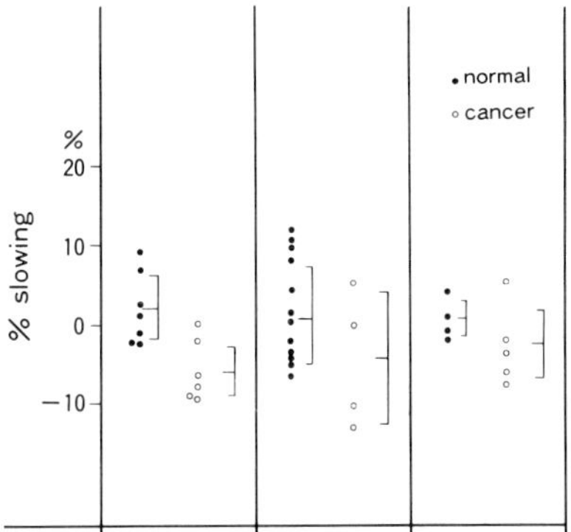

| normal | 1.83±4.19 | 0.94±6.70 | 0.83±2.21 |
| cancer | −6.07±3.67 | −4.16±8.27 | −2.27±4.39 |
| Immuno-potentiators | OK-432 | PSK | LVS |

Fig. 3.  % slowing of EMT in the lymphocytes from normal
subjects and cancer patients stimulated with OK-432,
PSK and LVS

Electrophoretic mobility of sheep erythrocytes exposed to
the cultures of cancer lymphocytes stimulated with OK-432
was faster than that of non-treated lymphocytes from cancer
patients.  Mean percentage slowing was -6.07 ± 3.67%.  The
result on  stimulating the cancer lymphocytes with PSK was
similar.  Mean percentage slowing was -4.16 ± 8.27%.  Also
using Levamisole, the percentage slowing of electrophoretic
mobility test in cancer patients was reduced.  Mean percentage slowing was -2.27 ± 4.39%.  In any group of the immunopotentiators, percentage slowing showed low levels in cancer
lymphocytes stimulated with the immunopotentiators, comparing
with the normal lymphocytes stimulated with the same immunopotentiators.
Effect of Lentinan was more obviously, different from other
immunopotentiators.  Table 2 shows the result for percentage
slowing of sheep erythrocyte electrophoretic mobility by the

Table 2.  % slowing of EMT in the normal lymphocytes
stimulated with Lentinan

| No. | subjects | mean electrophoretic time | | % slowing |
| --- | --- | --- | --- | --- |
| | | non-treated | Lentinan | |
| 1 | 38 M | 3.65 | 3.96 | 8.49 |
| 2 | 32 M | 3.53 | 3.77 | 6.80 |
| 3 | 34 M | 4.02 | 4.43 | 10.19 |
| 4 | 31 M | 4.41 | 4.94 | 12.02 |
| 5 | 33 M | 3.27 | 3.32 | 1.53 |
| 6 | 31 M | 3.27 | 3.78 | 15.59 |
| 7 | 25 M | 3.27 | 3.49 | 6.74 |
| 8 | 30 M | 3.27 | 3.90 | 19.26 |
| 9 | 23 M | 3.19 | 3.39 | 6.27 |
| 10 | 28 M | 4.14 | 4.44 | 7.25 |
| 11 | 40 M | 3.60 | 3.62 | 0.55 |
| 12 | 23 M | 3.30 | 3.55 | 7.58 |
| 13 | 21 M | 4.21 | 4.40 | 4.51 |
| 14 | 40 M | 4.30 | 4.42 | 2.79 |
| 15 | 45 M | 4.45 | 4.66 | 4.72 |
| 16 | 38 M | 4.10 | 4.19 | 2.20 |
| 17 | 39 M | 3.38 | 3.65 | 8.00 |
| 18 | 35 M | 3.84 | 4.18 | 8.85 |
| mean ± SD | | | | 7.41 ± 4.64 |

supernatants of lymphocytes from 18 normal subjects stimulated with Lentinan.  Mean percentage slowing was 7.41 ± 4.64%
and it is similar to the result obtained by effect of ConA.
Then we stimulated the lymphocytes from collagen disease
patients with Lentinan.  The patients were 4 systemic lupus
erythematosis, 8 rheumatoid arthritis, 3 progressive systemic
sclerosis, 1 polymyositis and 3 Sjögren's syndrome.  Using a
mixture of the patient lymphocytes and Lentinan, the electrophoretic mobility test was carried out.  Table 3 shows the
results. The percentage slowing of electrophoretic mobility
of sheep erythrocytes did not increase and was 0.81 ± 4.01%.
It is equivalent to the value obtained from supernatant of
the lymphocytes from collagen disease without Lentinan.  Subsequently, the lymphocytes from patients with malignant diseases were used.  Their clinical diagnosis and the result are
shown in table 4.  The results were similar to the affects of

Table 3. % slowing of EMT in the lymphocytes from collagen disease patients stimulated with Lentinan

| patients | diagnosis | activity | mean electrophoretic time | | % slowing |
|---|---|---|---|---|---|
| | | | non-treated | Lentinan | |
| S.A 17F | SLE | remitted | 4.63 | 4.63 | 0 |
| K.H 34F | SLE | remitted | 3.42 | 3.36 | −1.75 |
| F.U 30F | SLE | remitted | 3.94 | 3.88 | −1.48 |
| Y.F 31F | SLE | active | 3.95 | 4.28 | 7.12 |
| Y.M 28F | RA | remitted | 4.20 | 4.22 | 0.48 |
| Y.T 64F | RA | remitted | 4.36 | 4.23 | −2.98 |
| T.K 40F | RA | remitted | 3.10 | 3.20 | 3.23 |
| S.O 56F | RA | active | 4.18 | 4.09 | −2.15 |
| M.G 59F | RA | remitted | 3.37 | 3.63 | 7.72 |
| T.S 47M | RA | remitted | 3.20 | 3.21 | 0.31 |
| Y.S 36F | RA | active | 3.23 | 3.19 | −1.23 |
| S.O 54F | RA | active | 4.08 | 4.10 | 0.55 |
| S.U 32M | PM | active | 3.80 | 4.21 | 10.78 |
| K.O 62F | PSS | active | 4.27 | 4.22 | −1.17 |
| M.S 46F | PSS | active | 4.15 | 4.03 | −2.89 |
| M.K 21F | PSS | remitted | 3.89 | 4.03 | 3.70 |
| C.S 54F | Sjögren | remitted | 4.30 | 4.24 | −1.40 |
| M.O 40F | Sjögren | active | 4.42 | 4.23 | −4.30 |
| mean ± SD | | | 0.81 ± 4.01 | | |

Table 4. % slowing of EMT in the lymphocytes from malignancy patients stimulated with Lentinan

| patients | diagnosis | treatment | mean electrophoretic time | | % slowing |
|---|---|---|---|---|---|
| | | | non-treated | Lentinan | |
| H.S 43F | gastric cancer | chemotherapy | 4.08 | 4.07 | −0.25 |
| T.Su 54M | gastric cancer | chemotherapy | 4.32 | 4.28 | −0.92 |
| J.O 30M | pulm. cancer | chemotherapy | 3.27 | 3.24 | −0.92 |
| T.Sh 57M | pulm. cancer | radiation | 4.43 | 4.17 | −5.87 |
| S.T 66F | breast cancer | chemotherapy | 3.21 | 3.25 | 1.25 |
| T.T 60F | breast cancer | chemotherapy | 4.47 | 4.41 | −1.34 |
| Y.A 61M | pancreas cancer | chemotherapy | 3.36 | 3.20 | −4.76 |
| M.M 58M | colon cancer | chemotherapy | 3.90 | 3.95 | 1.28 |
| W.A 42F | clear cell cancer | chemotherapy | 3.25 | 3.26 | 0.31 |
| To.S 53F | pharyng. cancer | radiation | 3.19 | 3.21 | 0.62 |
| K.T 15F | pharyng. cancer | chemotherapy | 3.29 | 3.36 | 2.13 |
| S.M 51M | Hodgkin | chemotherapy | 4.10 | 4.28 | 4.39 |
| K.F 20M | non Hodgkin | chemotherapy | 3.45 | 3.29 | −4.64 |
| K.K 56F | non Hodgkin | chemotherapy | 3.37 | 3.27 | −2.97 |
| S.H 57M | myeloma | chemotherapy | 3.97 | 4.01 | 1.01 |
| H.M 58F | myeloma | chemotherapy | 4.09 | 3.94 | −3.67 |
| T.Y 56M | myeloma | chemotherapy | 4.10 | 3.99 | −2.68 |
| M.O 33F | myeloma | chemotherapy | 4.88 | 4.64 | −4.92 |
| H.K 38M | myeloma | chemotherapy | 3.28 | 3.18 | −3.05 |
| mean ± SD | | | −1.32 ± 2.72 | | |

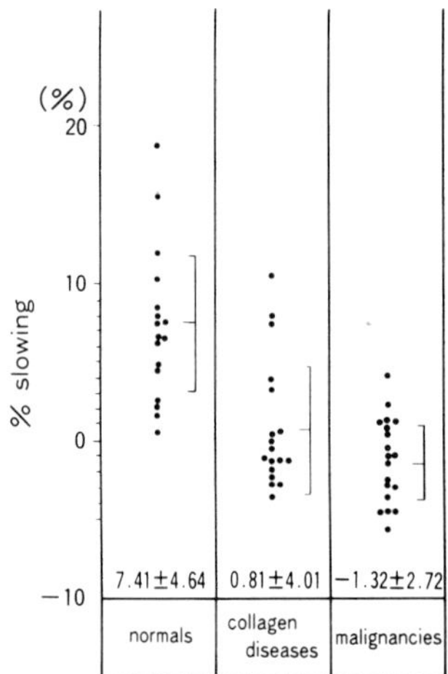

Fig. 4. % slowing of EMT in the lymphocytes from normal
subjects, collagen disease and malignancy patients
stimulated with Lentinan


Lentinan on lymphocytes from collagen diseases. The percentage slowing was less reduced, -1.32 ± 2.72%. The results for
affects of Lentinan are summarized in Fig. 4. When Lentinan
was used as an immunopotentiator on lymphocytes the decreased
percentage slowing was obvious in the collagen disease and
the malignancy.
In this aspect, we suggest that lymphokines which affect
electrophoretic mobility of sheep erythrocytes, may be unable
to be released from the lymphocytes from the patients with
collagen diseases and malignancies even by stimulating with
Lentinan, or that another lymphokine which is different from
electrophoretic slowing factor may be produced from the
lymphocytes from these patients by stimulation with Lentinan.


Conclusion

The present study showed affects of supernatant of lymphocytes stimulated with various immunopotentiators on electrophoretic mobility of sheep erythrocytes. In supernatants of the normal lymphocytes stimulated with PHA and ConA, percentage slowing of electrophoretic mobility of sheep erythrocytes was high. In supernatants of the normal lymphocytes stimulated with OK-432, PSK, Levamisole, Vit. $B_{12}$, Poly A:U, Poly I:C and Sodium periodate, percentage slowing was not changed. Among various immunopotentiators, affects of Lentinan on electrophoretic mobility was obvious. Although percentage slowing has been increased in supernatant of normal lymphocytes stimulated with Lentinan, it was rather reduced in mixture of Lentinan and the lymphocytes from collagen diseases and malignancies.

References


(1)  Field, E. J. and Caspary, E. A., : Lancet ii, 1337, (1970)

(2)  Porzsolt, F. et al., : Behring Inst. Mitt., 57, 128, (1975)

(3)  Jenssen, H. I. & Shenton, B. K., : Acta biol. med. germ. 34, 29, (1975)

(4)  Hashimoto, N., Physico-Chemical Biology, : 23, 209, (1979)

(5)  Hannig, K., : Methods in microbiology ed. by Norris & Ribbons, 5B, 513, (1979)

(6)  Bressler, J. P., et al., Cell, Immunol., : 54, 274, (1980)

(7)  Takimoto, G., et al., : Igaku-no-Ayumi, 112, 588, (1980)

(8)  Stavitsky, A., : J. Immunol., 72, 360, (1954)

(9)  Hashimoto, N., Physico-Chemical Biology : 24, 209, (1980)

# ELECTRICAL INSULATION OF ELECTROPHORETIC FLOWS AND A 'RAIN-BOX'

J.O.N.Hinckley

Electrophoresis Consultant, POB 7104, Beaumont, Texas 77706,
USA.

## Introduction

In electrophoresis it is sometimes desirable to have continuous or intermittent
buffer flow, while the electric field is applied. Such flows may be into or
out of the apparatus, between parts of it, or combinations of these. Examples
are sample introduction and removal in continuous transverse flow methods,
introduction and removal of the transverse flow of buffer itself, removal of
product in gel and free solution or density gradient batch methods, transloc-
ation of gradients, counterflow to balance one or more zones or displacement
electrophoretic (transphoretic, isotachophoretic (1)) compartments, removal
and replenishment of electrode vessel buffer, anolyte and catholyte interchange, and recycling of buffer coolant if contacting the electrophoretic
system at two point of different potential or grounding it. These buffer flows
are electrical conduction paths and are therefore also electroosmotically active
unless preventive measures are taken. It is frequently desirable therefore, and
essential in some applications, to divest such buffer flows of their current
carrying property, and so to eliminate or substantially reduce shunting of the
separation column through ground or through the apparatus, incidentally improving safety. This may be achieved in a number of ways, which are discussed below, and an example of one such, a 'rain-box', which was constructed and used, is described.

## Methods of Flow Insulation

First, continuity of a flowing buffer line can be interrupted by co-moving
insulators. This class includes injection and removal of aliquots of fluid
insulator, which may be a gas or non-polar immiscible liquid. The effectiveness of gas insulation is familiar from its use to separate aliquots being
transported through the tubes of some autoanalysers, although the purpose
there is not electrical. Unless the flow of buffer is itself used as a source
of work, say against gravity, such systems would require a driving mechanism. The same applies to trains of solid insulating spheres in a pipe, or
chains of dynamic seals and the like.

A different way of moving the solid insulator is to use a peristaltic pump,
where the closed part of a suitably roller-flattened elastic tube has very
high electrical resistance. Some arrangements of this kind could even be
driven by the buffer flow at high pressure differentials, directly or otherwise.
A close relative, mechanically, is non-pumping intermittent interruption of
the flow at several points, such that at least one of the on-off valves is

closed at any time. Unless the intervalve pockets are very flexible, and
there are flexible ampullae fore and aft of the valve train, a separate depuls-
ing device may be required. Not only peristaltic pumps, but others with
insulating parts may be used, such as piston pumps, vane and impellor pumps,
and to some extent the screw and gear pumps, progressing cavity pumps
bridging this type and the peristaltic methods discussed.

A particular variant deserves mention, which might be called an 'isometer'(2).
Limited volume unidirectional flow, and unlimited reciprocating flow, may be
accommodated with a semisolid insulator with negligible volume elasticity as
follows. Take a rigid vessel with two apertures, one of which leads to a
blind flexible bag inside the vessel and whose potential volume is probably
at least that of the vessel ( or to one side of a vessel's internal floppy
dividing septum), the other aperture being to the complementary trans-septal
space. If there are no gas inclusions, and if hydrostatic pockets are suitably
eliminated, flow into one aperture and out of the other is equal and simul -
taneous, but the flexible insulator prevents electrical continuity. A homologue
of this hydraulic quasi-capacitor is a back-to-back syringe pair. As with
syringe systems, equality of output and input can be achieved with combinations of several inputs and outputs, allowing segregation of flows. The
attraction of flexible bag isometers is simplicity, for limited volume runs.
It can also save tank space in microgravity continuous electrophoresis planned
for industrial use in the Shuttle (3), as previously proposed for similar use
in continuous radial transphoretic separation (2). If buffer can be reused or
renewed by dyalisis or continuous ionexchange, or using servo-systems, then
a stream-switching valve can cooperate with unidirectional pumped flow to
give well nigh continuous operation with electrical insulation. But for true
continuous recycling, an insulating pump may be a simpler and better choice
if pulsing can be eliminated or tolerated easily. Isometers allow compliant
(and usually non-insulating) pumps to be used for impelling the flow, with
pulse avoidance, while the isulation burden is shouldered by the non-pumping isometer.

A second class of methods is related to the first, if we generalize the intermediate insulator and the insulating buffer-retaining duct  as constituting a
single phase through which a dispersed phase of buffer is caused to move.
This phase is probably fluid, rather than merely flexible or elastic. We again
have the same choice of driving force - either the insulating fluid may be
used as the source of movement, causing the buffer to fragment into droplets,
from tips or ducts by a wind or venturi or local suction, to be deposited in
a recipient buffer pool by coalescence or centrifugal force, say, or, conver se-
ly, the buffer itself may be the energy source. Thus the buffer could be
pumped conventionally or by a gravity head, or it could incorporate a pump
as a centrifugal rotating 'shower-head' or sprinkler, or it could use electrostatic attraction between donor and recipient buffer bodies to transport
droplets, as most high voltage electrophoreticists will have observed at one
time or another - the transported current is negligible compared to most
ionically conducted aqeous currents. For microgravity applications, the centrifugal head would substitute gravitational fall, illustrating a geherality: in
space the absence of gravity liberates the use of radial symmetries, which,

mutatis mutandis, are also necessary for the centrifugal generation of artificial 'gravity'.

If we imagine not one stream of droplets, but optionally many others in parallel, caused to descend perhaps by gravity, where the fluid may be air, we
may justify calling such a device a 'rain-box', and extend this nickname to
that general class of devices.

Principles of a Rain-box

At its simplest, the rain-box principle is a droplet-forming disperser or nozzle,
with trav el of the droplet from the donor buffer to the open recipient buffer
pool through an insulating fluid. The design and performance are affected by
several parameters. High interfacial energy between the two types of medium
promotes rapid droplet formation and coalescence with the recipient pool,
but puts an extra burden on the pumping method. So will the fineness of the
buffer dispersion, which is promoted by fine nozzles or jets ( if any) and
results in fast droplet formation. Speed of droplet formation reduces the
minimum residence time of buffer in the rainbox, and therefore its dimensions.
Dispersion is slowed by increased viscosity of the buffer or insulating fluid,
or both, and may therefore benefit from an increase in the number of nozzles
or operation at higher te mperature. There is a trade-off on droplet fineness,
especially if gravity or initial droplet momentum is the moving force for
droplets, for then the surface area to weight ratio increases and there is
increased slowing of droplets in a viscous insulating fluid. For gravity
rain-boxes, buffer-insulator density difference reduction also slows travel
of droplets. Interestingly, for very dense and low viscosity insulators, there
may be an advantage in running upside down, with droplets rising, to exploit
a large density difference for normal low solute percentage buffers. Hold-up
of droplets as a cloud at the recipient pool end ,when fine droplets are used,
and where external pumping (droplet receives initial momentum) predominates
over gravity or centrifugation as a method of droplet transport. For high voltage
insulation, it is advantageous, for a given dielectric breakdown strength, and
given volume resistivity of the insulator and surface resistivity of the box
walls, to increase distances between droplets and between pools, even if
each nozzle must less frequently generate a droplet. If ionization or tracking
breakdown does occur, whether due to excess voltage or other causes, then
the rain-box will become a lightning-box, and noisy current will flow. If the
insulator is a liquid there may be some current leakage, in addition to the
electrostatic leak, by to-and-fro migration of charge-carrying particles and
electroconvection - a phenomenon familiar to those who deal in capacitor
oils, and liquid insulators of high voltage heads. There are then many
trade-offs for a particular application where an ideal and rigorous design is
sought.

Liquid insulators have greater viscosity than gas insulators, but have an
advantage, especially for free solution systems which prefer fixed volumes,
of lower compressibility. A general loss of efficiency owing to reduced
interfacial energy, lower density difference, and increased viscosity, may be
expected as the price of isometric use of an essentially inelastic fluid.

We may anticipate a related problem, the removal of insulator from buffer repatriated to the recipient pool . This acquisition by buffer of included insulator may occur by emulsion or solution, and dissolved insulator may redeposit undesirably on adsorptive or colder parts of the apparatus, especially interfering with electrode performance and affecting wall zeta-potentials - one of the time-honored reasons why greases, such as silicone grease, should be meticuluously segregated from electrophoretic buffers. Likewise dissolved gases from gaseous insulators may outgas on warmer or lower pressure parts, with attendant undesired volume or resistance changes. It would seem then that liquid rainboxes should be run colder ,and gaseous ones warmer or at lower pressure, than critical parts of the apparatus, to reduce such effects. Segregation of air bubbles from the continuing post-rainbox buffer flow can be achieved by using a hydrophilic filter - a stratagem adopted for removing electrolytic gas from electrode vessel effluents for recycling, in the ingenious phase-separators designed for NASA's Apollo-Soyuz Test Project  static zone electrophoresis experiment (ASTP MA-011)(4), and previous flights.

I n the absence of terrestrial gravity, or in the effective microgravity of free-fall in orbiting satellites, where electrophoresis has been done with a view to preparative bioprocessing manufacture, we must use pumps to give droplets initial momentum, or use electrostatic transport (which is not dificult if tranport is between two different voltages anyway, and that difference is high). There are two types of centrifugal shower-head pumps. First, the head alone can turn or reciprocate, scattering droplets outward, but only imparting initial momentum. In the second type, the whole body rotates, including the insulating medium and droplets in radial flight, so the droplets continue to experience centrifugation in flight. Alternatively a whole terrestrial symmetry rainbox may be swung around an axis, conceptually as an element of a radial device. Radial symmetry in space has another appeal: the donor and recipient walls, in the absence of gravity or centrifugation, and using say other means of droplet propulsion, can be used to hydrophilically coalesce, and hydrophobically segregate the buffer pools, a concept which draws upon the spectacular exploitation of interfacial energy to manipulate liquid and multiple fluid phases by workers at Martin Marietta Corporation (5).

Such symmetry is acceptable terrestrially if droplet transport requires neither gravity to impell it, nor is interfered with by gravity(assuming say a negligible buffer-insulator density difference, or centrifugal forces in excess). This last condition is commonly a limitation, so if gravity imposes a polarity of sym-metry and movement, we may as well also use it to best effect. But this does not preclude apparatus capable of multiple orientation, and the spherical sym-etry devices are in many cases very suitable for such terretsrial use. One such was proposed, and would be usable for electrophoretic modular apparatus soon to be commercially available (6): a shower-head droplet generator on a tube-stalk on one corner of a box, at a largely equal angle to each of the three walls, which are also the planes parallel to any of the orientations in which the electrophorator can function, would produce a 'rain' in any orient-ation of use. For that matter, a gravity-polarised design can be made a bi-directional flow rain-box, if the collecting floors of two opposed rain-boxes are fused, and the floor removed for that matter.

There are doubtless many further devices, some perhaps already described in related technologies, others to be devised, which could achieve similar aims. This discussion has left out of account all kinds of bucket fountains, archimedean screws, and other gadgets which await the ingenuity of a latter day Hero of Alexandria, and are not necessarily the worse for a debt to the limited means of archaic  or amateur engineering.

## Rain-box Design Considerations

The described rain-box was designed for use in a part of a recycling flow circuit where constant volume was not mandatory. This allowed the easy first choice of atmospheric air as the insulating fluid. The choice of flow rate, some 200 ml a minute was dictated by the intended application. It was the flow required for each electrode vessel of an Endless Belt Electrophoresis continuous free solution preparator, similar in design to that designed and pioneered by Kolin and Luner (7), and using phosphate buffer. Maxima were some 1800 volts and 230 mA, a bit over 400 watts. Whereas the original concept (7) used a single Mariotte bottle of some 60 liters above head level to feed both electrode vessels, which in turn ended in adjustable level driplines, and included a number of other accessories and actual and putative devices (now described in later papers of Kolin), this writer was unable to obtain a mandate to abandon gravitationally driven flows and therefore resorted to the interim expedient of using overflowing constant heads for the source and sink of each electrode vessel. In order to be able to reuse buffer of approximately constant conductivity, without making up 60 liters a day, this head system's effluents were returned to the source head of the electrode vessel of opposite sign, to gain pH stability. Such a figure-of-eight circuit was doubtless the incorrectly designed intent of the otherwise ingenious flow system of the ASTP MA-011 (4), discussed further elsewhere (6). The figure-of-eight flow system used compliant centrifugal pumps to regain gravitational energy of height at source head, and used rain-boxes of several designs ( one combined with a sink constant head) to confer insulation in the vessel-to-vessel cross flows. They should therefore withstand the full available voltage and more, and still pass 200 ml a minute. This would allow interchange of electrode liquors without an electrical shunt, and the volume of buffer, including that required by the continuous filters and filter-charcoal dye-removers, was very much less than 60 liters. The whole provisional experimental laboratory system was exhibited at FASEB in 1976 with its heads and rainboxes, and in operation, shortly before the project was terminated ( for non-technical reasons). Slow streak drift on that occasion was typical of operation in surroundings which are not carefully stabilized (to $\frac{1}{2}°C$, according to Luner (8)), and with buffer not equilibrated thereto for perhaps days, by reason of the differing flow resistance of tubing with some $2\frac{1}{2}\%/°C$ viscosity variation of water (8) – the major objection to compliant and gravitational pumps, in a system like this where streak drift is a sensitive amplifier of small differences of electrode vessel pressures, since flow-resistive membranes are avoided for classical reasons in the apparatus as then devised by Kolin and Luner. Thus streak-drift is no fault of the rain-boxes.

The chosen 200 ml per minute flow compelled multiple drip lines in parallel rather than a single stream of droplets, and presumably the overlap of drip formation, detachment, and fusion with the recipient buffer pool, is such as to smoothe out most of any pressure, volume, or electrostatic noise. Tube length and bore, dependent on application, was found empirically to be that which avoided unbroken flow for a few inches of insulating air, for the anticipated head of a few feet at most. Spacing of tubes was intended to avoid coalescence of drops into a stream from any two or more tubes. The tube mouths were raised slightly above the retaining board to aid in application of adhesive, cleaning, freedom from sediment obstruction. A baffle was introduced at the box influx lest jet flows caused central tubes to start streaming. Likewise a slant could be used in the insulating phase compartment to reduce splashing, if the box were correspondingly deepened. Following the success of this design, there seems little reason why the tube-populated area of the ceiling board could not be scaled up or down, for like heads and viscosities, according to the particular flow need. A cylinder could simply replace the present arbitrarily closed box. Tube siting near the wall was avoided to discourage wall trickling. A hydrophilic mesh bubble remover in the recipient pool could probably be used with benefit, but was not in fact used. It seemed desirable to make the box of clear colorless plastic for ease of observation and assembly checking, but glass or heavy acrylic might be a more robust choice and easier to clean and maintain – the illustrated device was a quick answer to an immediate problem and makes no pretension to engineering quality. Metal was avoided because of corrosion, electrocorrosion, gas generation, and contamination of buffer, especially in presence of electric fields. Glass fiberboard was considered to confer a desired rigidity against the pressure head, and was in any case conveniently provided with holes (electrical connector board). Influx and efflux tubes were wide like the system plumbing, to avoid extra flow resistance, and were attached by solvent irrigation of the contacting tube flange – part of a hollow disposible syringe plunger. A spiracle was drilled high in a sidewall, and an air vent tube was attached to it and raised above the head, to avoid flooding spillage of electrically live buffer. The vent could be used with a filter against microorganisms, and serves to avoid pressure and volume effects on an otherwise enclosed volume of air, which might affect rain-box or other functions.

Materials and Equipment

1) Box: 'Q-Tips Photoholder', Chesebrough-Ponds Inc., Greenwich, Conn., 06830, obtainable in USA at most pharmacies.
2) Microtubing for droplet formation: ID 0.023",OD 0.038", Intramedic Polyethylene, Clay Adams Division of Becton Dickinson & Co., Parsippany, N.J. 07054.
3) Silicone rubber sealant: white filled two-part RTV-11, GEC, Waterford, N,Y. 12188.
4) Silicone glue: "Aquarium Sealer" one -part clear acetic-liberating RTV, Dow-Corning Corp., Midland, Michigan 48640.
5) Baffle and supports: Rohm and Haas Grade A cast clear acrylic 1/8"

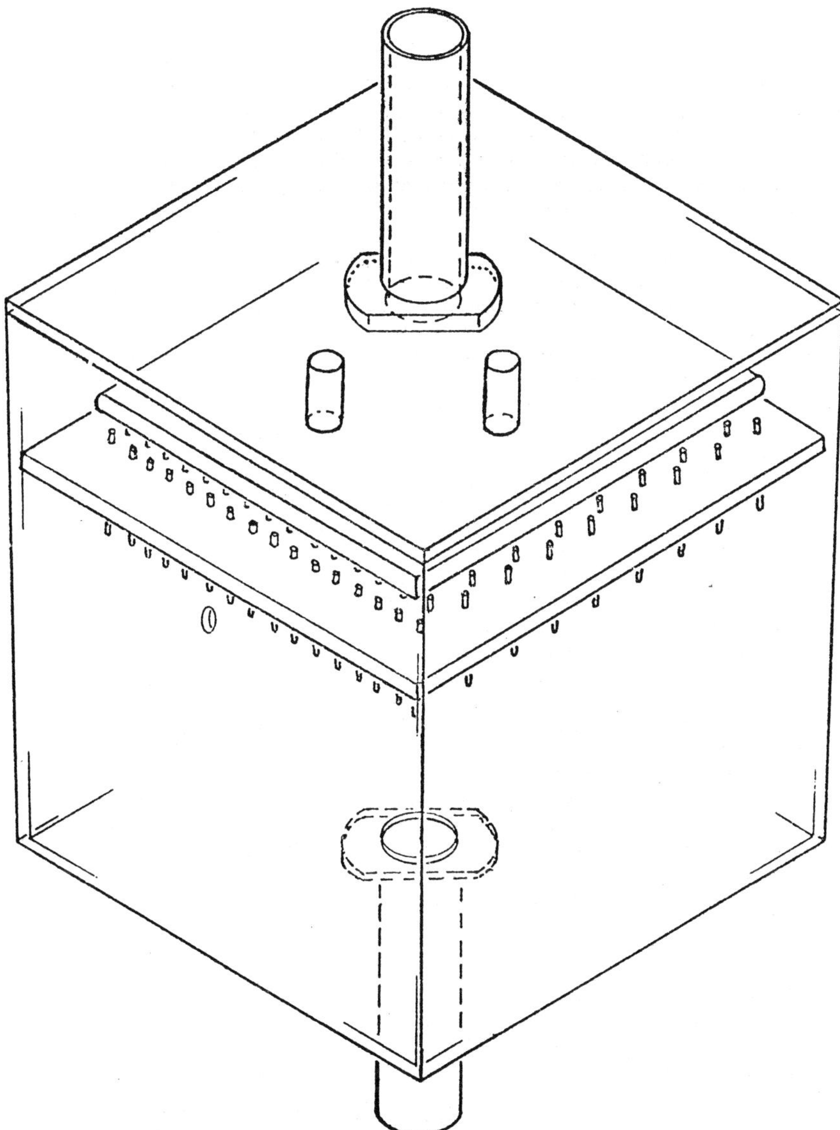

Figure 1. 'Rain-box' electrophoretic buffer flow insulator, using air insulation.

1002

sheet and $\frac{1}{4}$" rod, glued by irrigation with chloroform.
6) Glass fiberboard: Green-tint multihole "Vectorboard"169P84-062, Vector, Sylmar, California.
7) Disposible syringe parts: Gillette Scimitar 10 ml syringe, Gillette Ltd., U.K..
8) Power supply: current-regulated custom-built 0-30 kV, 1 mA max, (similar to original LKB-supplied PSU for LKB Tachophor), Brandenburg Ltd., Thornton Heath, Surrey, U,K.
9) Meter: Model 245 Digital Multimeter, Data Precision Corp., Wakefield, Mass. 01880
10) Mariotte bottle of buffer, suitable plumbing, electrodes at T-junctions on inflow and outflow pipes of rain-box, suitable electrical insulation of apparatus from ground and shorting, chain of three 3.7 megohm high voltage resistors between high tension output and low tension return of power supply , clamp for flooding rain-box from below, plastic basin for rainbox efflux.

Testing and Results

The following tests were done at the time, and being rather less than exhaustive, do not claim to do more than indicate the possibilities of high voltage insulation by such rain-boxes. Thus the experiment here reported does not test at more than 1000 volts, although the functioning device at over 1500 volts was also used in the dark and without noise or discharge, for what that is worth. It is felt, however that the device and principle have been shown to work sufficiently to be worth reporting in the context of a discussion of the whole class of devices; and oportunity will be sought to retets this type of rain-box.

Using the Mariotte bottle, a flow of approximately constant head buffer was allowed to flow at a representative flow rate ( in the order of more than 200 ml per minute). The lower efflux electrode was connected to low tension return of the power supply through the battery powered digital ammeter, which has a resolution of one microampere. The power supply also has a micro-ammeter, but the voltmeter was out of commission. The influx electrode was connected to a tap above the lowest of the three identical resistors in the chain. Efflux was reduced by clamping , causing the floor buffer level to rise until the air insulator was excluded.

In normal function, before flooding, there was no registered current through the rain-box (i.e.maximum one microampere), but 280 microamperes in the resistor chain implying a total chain voltage of 3,108 volts or so, that is 1036 volts across each resistor, and therefore across the rain-box. It follows that at 1kV the rain box has a flowing resistance of at least 1000 megohms, which is hardly cause for complaint. When the rain-box was flooded, only at the last minute when the level almost touched the microtubes, was there current in the rain-box circuit: the full 280 microamperes, implying a flooded rainbox resistance of some 13000 ohms, and most of this must be of the tubing.

## Conclusion

It is hoped that in a survey of insulation methods, the ease and usefulness of flow insulation will have become apparent. One such device, the 'rain-box' has been described in enough detail to encourage its general use, if only as a safety device - there must be many effluents which short circuit or ground in the waste basin or plumbing, many overflows at high potential or capable of carrying even lethal currents. Both the latter could be made safer with rain-boxes. No doubt improved versions with latches and disconnects and made of thicker material would be easier to use and clean, and more durable.

## References


1. Hinckley, J.O.N., Electrophoresis, Cap.5 of *Internat. Rev. Sci.*, Phys. Chem. Ser.2., Vol.13, Analytical Chemistry Part 2, Ed. T.S.West, Butterworths (London, Boston) 1976, 141-186.

2. Hinckley, J.O.N., *AIAA Paper* #74-664, Amer.Inst.Aeronaut.Astronaut., New York, 1974.

3. Marx,W.R., Richman, D., Hawkins, V.H., Lanham, J.W., Weiss, R.A., *Feasibility Study of Commercial Space Manufacturing*, 1977, McDonnell Douglas Astronautics Co., St.Louis, Missouri.
4. Allen, R.E., et al.,*Separ'n. Purif. Methods*, $\underline{6}$(1), 1-59 (1977).

5. Fester, D.A., Eberhardt, R.N,,and Tegart, J.R., *Bioprocessing in Space* (NASA TM X-58191) Proceedings of 1976 Colloquium, Ed. D.R.Morrison, NASA-JSC, Houston 1977, p.37-46.

6. Hinckley, J.O.N., in *Electrophoresis 78*, Ed. N.Catsimpoolas, Elsevier North Holland Inc., 1978, p.167-194.

7. Kolin, A., and Luner, S.J., *Anal. Biochem.*, $\underline{30}$, 111-131 (1969)

8. Luner, S.J., Doctoral Thesis, UCLA, 1969., p.58.


## Acknowledgement


The writer wishes to ackowledge the capable technical assistance of M.R.Barbay, who also suggested use of bubble segregators (4) in the rain-box. The construction of the latter and its testing in 1976, based on antecedent concepts, was done at Helena Laboratories, Beaumont, Texas, under an arrangement with Medical Testing Systems (MTS) of California, abandoned in 1976.

| | |
|---|---|
| Yeast proteins | 266 |
| Zone Immunoelectrophoresis<br>  Assay (ZIA) | 89, 103 |

# New versatility in polyacrylamide gel electrophoresis

Basic PAGE technique has now expanded to include many successful modifications of the methodology. Performance of all of these PAGE techniques and coverage of all stages in the procedures demands a highly versatile equipment system. Look to Pharmacia for a system of unrivalled versatility and reliability.

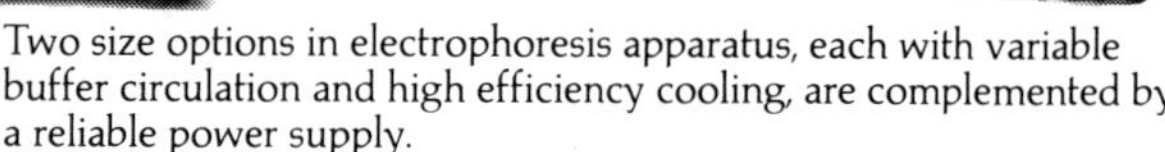

A 72 page book gives complete methodologies and "recipes" for the many electrophoretic techniques which can be performed using the equipment shown here.

Two size options in electrophoresis apparatus, each with variable buffer circulation and high efficiency cooling, are complemented by a reliable power supply.

High and Low Molecular Weight Calibration Kits are available for convenience in molecular weight estimations.

Complete your work by de-staining and drying with reliable, matched equipment.

Cast your own gels or choose from a range of ready made gradient gels for simple, reproducible high resolution work.

# Improvements in isoelectric focusing

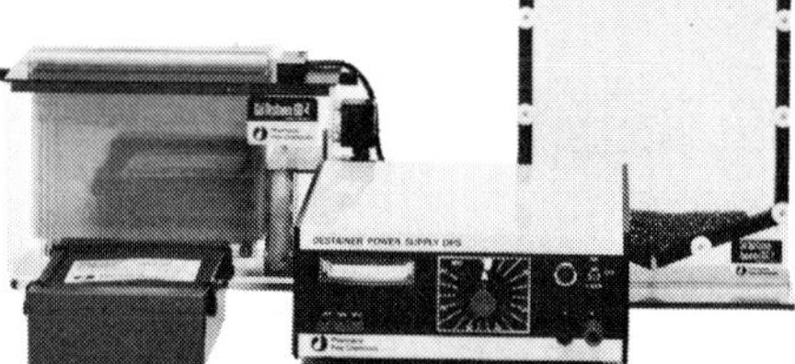

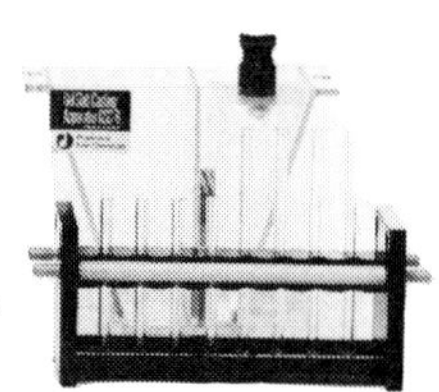

New concepts of high voltage focusing and volthour integration added to significant improvements in carrier ampholytes and agarose support media have made Pharmacia the new leader in IEF technology. Pharmacia's System has been designed with one aim—to extend the scope of this high resolution technique.

The 3000 volt power supply ECPS 3000/150 gives you more speed and resolution with your IEF work. Digital display of volts, current or power gives you more control and ease of operation.

The Volthour Integrator VH-1 monitors and displays elapsed volthours—the clearest parameter of an IEF experiment. Now record, report and repeat your work with high precision.

The Pharmacia pI Calibration Kits allow easy and accurate determination of pH gradients.

The Flat Bed Electrophoresis apparatus FBE 3000 combines the latest design features, including outstanding cooling capacity, for IEF, preparative IEF and immunoelectrophoresis with minimum investment.

Pharmalyte® carrier ampholytes, Agarose IEF and Sephadex® IEF are specially formulated to give you improved results at analytical or preparative scale.